Applied Mathematical Sciences
Volume 68

Springer
New York
Berlin
Heidelberg
Barcelona
Budapest
Hong Kong
London
Milan
Paris
Santa Clara
Singapore
Tokyo

Applied Mathematical Sciences

1. *John:* Partial Differential Equations, 4th ed.
2. *Sirovich:* Techniques of Asymptotic Analysis.
3. *Hale:* Theory of Functional Differential Equations, 2nd ed.
4. *Percus:* Combinatorial Methods.
5. *von Mises/Friedrichs:* Fluid Dynamics.
6. *Freiberger/Grenander:* A Short Course in Computational Probability and Statistics.
7. *Pipkin:* Lectures on Viscoelasticity Theory.
8. *Giacoglia:* Perturbation Methods in Non-linear Systems.
9. *Friedrichs:* Spectral Theory of Operators in Hilbert Space.
10. *Stroud:* Numerical Quadrature and Solution of Ordinary Differential Equations.
11. *Wolovich:* Linear Multivariable Systems.
12. *Berkovitz:* Optimal Control Theory.
13. *Bluman/Cole:* Similarity Methods for Differential Equations.
14. *Yoshizawa:* Stability Theory and the Existence of Periodic Solution and Almost Periodic Solutions.
15. *Braun:* Differential Equations and Their Applications, 3rd ed.
16. *Lefschetz:* Applications of Algebraic Topology.
17. *Collatz/Wetterling:* Optimization Problems.
18. *Grenander:* Pattern Synthesis: Lectures in Pattern Theory, Vol. I.
19. *Marsden/McCracken:* Hopf Bifurcation and Its Applications.
20. *Driver:* Ordinary and Delay Differential Equations.
21. *Courant/Friedrichs:* Supersonic Flow and Shock Waves.
22. *Rouche/Habets/Laloy:* Stability Theory by Liapunov's Direct Method.
23. *Lamperti:* Stochastic Processes: A Survey of the Mathematical Theory.
24. *Grenander:* Pattern Analysis: Lectures in Pattern Theory, Vol. II.
25. *Davies:* Integral Transforms and Their Applications, 2nd ed.
26. *Kushner/Clark:* Stochastic Approximation Methods for Constrained and Unconstrained Systems.
27. *de Boor:* A Practical Guide to Splines.
28. *Keilson:* Markov Chain Models—Rarity and Exponentiality.
29. *de Veubeke:* A Course in Elasticity.
30. *Shiatycki:* Geometric Quantization and Quantum Mechanics.
31. *Reid:* Sturmian Theory for Ordinary Differential Equations.
32. *Meis/Markowitz:* Numerical Solution of Partial Differential Equations.
33. *Grenander:* Regular Structures: Lectures in Pattern Theory, Vol. III.
34. *Kevorkian/Cole:* Perturbation Methods in Applied Mathematics.
35. *Carr:* Applications of Centre Manifold Theory.
36. *Bengtsson/Ghil/Källén:* Dynamic Meteorology: Data Assimilation Methods.
37. *Saperstone:* Semidynamical Systems in Infinite Dimensional Spaces.
38. *Lichtenberg/Lieberman:* Regular and Chaotic Dynamics, 2nd ed.
39. *Piccini/Stampacchia/Vidossich:* Ordinary Differential Equations in $\mathbf{R}^n$.
40. *Naylor/Sell:* Linear Operator Theory in Engineering and Science.
41. *Sparrow:* The Lorenz Equations: Bifurcations, Chaos, and Strange Attractors.
42. *Guckenheimer/Holmes:* Nonlinear Oscillations, Dynamical Systems and Bifurcations of Vector Fields.
43. *Ockendon/Taylor:* Inviscid Fluid Flows.
44. *Pazy:* Semigroups of Linear Operators and Applications to Partial Differential Equations.
45. *Glashoff/Gustafson:* Linear Operations and Approximation: An Introduction to the Theoretical Analysis and Numerical Treatment of Semi-Infinite Programs.
46. *Wilcox:* Scattering Theory for Diffraction Gratings.
47. *Hale et al:* An Introduction to Infinite Dimensional Dynamical Systems—Geometric Theory.
48. *Murray:* Asymptotic Analysis.
49. *Ladyzhenskaya:* The Boundary-Value Problems of Mathematical Physics.
50. *Wilcox:* Sound Propagation in Stratified Fluids.
51. *Golubitsky/Schaeffer:* Bifurcation and Groups in Bifurcation Theory, Vol. I.
52. *Chipot:* Variational Inequalities and Flow in Porous Media.
53. *Majda:* Compressible Fluid Flow and System of Conservation Laws in Several Space Variables.
54. *Wasow:* Linear Turning Point Theory.
55. *Yosida:* Operational Calculus: A Theory of Hyperfunctions.
56. *Chang/Howes:* Nonlinear Singular Perturbation Phenomena: Theory and Applications.
57. *Reinhardt:* Analysis of Approximation Methods for Differential and Integral Equations.
58. *Dwoyer/Hussaini/Voigt (eds):* Theoretical Approaches to Turbulence.
59. *Sanders/Verhulst:* Averaging Methods in Nonlinear Dynamical Systems.
60. *Ghil/Childress:* Topics in Geophysical Dynamics: Atmospheric Dynamics, Dynamo Theory and Climate Dynamics.

(continued following index)

Roger Temam

Infinite-Dimensional Dynamical Systems in Mechanics and Physics

Second Edition

With 13 Illustrations

Springer

Roger Temam
Laboratoire d'Analyse Numérique
Université Paris Sud
Orsay 91405
France

The Institute for Scientific Computing
and Applied Mathematics
Indiana University
Bloomington, IN 47405 USA

Editors

J.E. Marsden
Control and Dynamical Systems, 116-81
California Institute of Technology
Pasadena, CA 91125
USA

L. Sirovich
Division of Applied Mathematics
Brown University
Providence, RI 02912
USA

Mathematics Subject Classification (1991): Primary: 35B99, 35K60, 35Q30, 35Q53, 58F25, 76F99
Secondary: 76D05, 78A05, 78A40, 80A30, 92D15

Library of Congress Cataloging-in-Publication Data
Temam, Roger.
Infinite-dimensional dynamical systems in mechanics and physics / Roger Temam.—2nd ed.
p. cm.—(Applied mathematical sciences; 68)
Includes bibliographical references and index.
ISBN 0-387-94866-X (hardcover: acid-free paper)
1. Differentiable dynamical systems. 2. Boundary value problems. 3. Nonlinear theories. I. Title. II. Series: Applied mathematical sciences (Springer-Verlag New York Inc.); v. 68.
QA1.A647 vol. 68 1997
[QA614.8]
510 s—dc20
[515.3′52] 96-33318

Printed on acid-free paper.

Production managed by Robert Wexler; manufacturing supervised by Jeffrey Taub.
Typeset by Asco Trade Typesetting Ltd., Hong Kong.
Printed and bound by Maple-Vail Book Manufacturing Group, York, PA.
Printed in the United States of America.

9 8 7 6 5 4 3 2 1

ISBN 0-387-94866-X Springer-Verlag New York Berlin Heidelberg SPIN 10549226

TO CLAUDETTE

Preface to the Second Edition

Since publication of the first edition of this book in 1988, the study of dynamical systems of infinite dimension has been a very active area in pure and applied mathematics; new results include the study of the existence of attractors for a large number of systems in mathematical physics and mechanics; lower and upper estimates on the dimension of the attractors; approximation of attractors; inertial manifolds and their approximation. The study of multilevel numerical methods stemming from dynamical systems theory has also developed as a subject on its own. Finally, intermediate concepts between attractors and inertial manifolds have also been introduced, in particular the concept of inertial sets.

Whereas we attempted, in the first edition, to cover the subject in an exhaustive way, this goal rapidly appeared to be unthinkable for the second edition. Hence, beside a number of minor alterations and improvements, this second edition includes a limited number of new topics which have been selected in a rather arbitrary way. The additions include a number of sections and subsections concerning the existence of attractors for specific systems for which *the semigroup is not compact* (Sections III.3.2, IV.6, 7, 8, VI.4.2), and two chapters on inertial manifolds and the approximation of attractors and inertial manifolds. Sections III.3.2 and VI.4.2. concern the existence and dimension of attractors for boundary-driven flows, the dimension being reduced from an exponential function of the Reynolds number to a polynomial function. Sections IV.6, 7, 8 address the existence of an attractor in the absence of compactness. Sections IV.6 and 7 are related to reversible wave equations, namely, the weakly damped nonlinear Schrödinger equation and the Korteweg–de Vries equation. In Section IV.8 we show how to prove the existence of the attractor in an unbounded case (the physical domain is unbounded which produces a lack of compactness). The techniques used in these three sections depart in a significant way from the techniques used in the rest of the book.

The new Chapters IX and X are related to inertial manifolds and approximations. In Chapter IX we give, with a different proof, a new result of the existence of inertial manifolds generalizing that of Chapter VIII. This new result applies to the non-self-adjoint case and allows us to relate the concept of inertial manifolds to the concept of slow manifolds encountered in meteorology and oceanography. Finally, Chapter X addresses the approximation of the attractor by smooth finite-dimensional manifolds and at an exponential order (when inertial manifolds are not known to exist); it also produces convergent sequences of simple ("explicit") finite-dimensional manifolds approximating an exact inertial manifold when it exists.

This second edition has benefited from comments from a number of people; unable to mention all of them, I would like to mention, in particular, C. Foias, M. Jolly, O. Manley, S. Wang, X. Wang, and A. Debussche, O. Goubet, A. Miranville, I. Moise, and R. Rosa. The latter have also directly contributed to the writing of the new additions (A.D., Chapters IX and X; O.G., Section IV.6; A.M., Sections IV.3.2 and VI.4.2; I.M., Section IV.7; R.R., Sections IV.7 and 8), and I address my special thanks to them. Finally, I would like to thank Teresa Bunge and Danièle Le Meur who handled the typing kindly and efficiently.

Roger Temam
Université Paris Sud

Preface to the First Edition

The study of nonlinear dynamics is a fascinating question which is at the very heart of the understanding of many important problems of the natural sciences. Two of the oldest and most notable classes of problems in nonlinear dynamics are the problems of celestial mechanics, especially the study of the motion of bodies in the solar system, and the problems of turbulence in fluids. Both phenomena have attracted the interest of scientists for a long time; they are easy to observe, and lead to the formation and development of complicated patterns that we would like to understand. The first class of problems are of finite dimensions, the latter problems have infinite dimensions, the dimensions here being the number of parameters which is necessary to describe the configuration of the system at a given instant of time. Besides these problems, whose observation is accessible to the layman as well as to the scientist, there is now a broad range of nonlinear turbulent phenomena (of either finite or infinite dimensions) which have emerged from recent developments in science and technology, such as chemical dynamics, plasma physics and lasers, nonlinear optics, combustion, mathematical economy, robotics, ...

In contrast to linear systems, the evolution of nonlinear systems obeys complicated laws that, in general, cannot be arrived at by pure intuition or by elementary calculations. Given a dynamical system starting from a particular initial state, it is not easy to predict if the system will evolve towards rest or towards a simple stationary state, or if it will go through a sequence of bifurcations leading to periodic states or to quasi-periodic states or even to fully chaotic states. The mathematical problem here is the study of the long-term behavior of the system ($t \to \infty$) corresponding to the practical problem of determining which "permanent" state will be observed after a short transient period in, say, a wind tunnel or an electrical circuit. In an attempt to predict the long-term behavior of dynamical systems we encounter several difficulties related to chaos, bifurcation, and sensitivity to initial data:

Chaotic (turbulent) behavior can appear, as well as simple well-ordered states.

We do not know a priori towards which state a given system may evolve, and we do not know when significant changes of the state may occur.

In many cases physical phenomena are stable and small variations in the initial circumstances produce only small variations in the final state. But quoting Maxwell's *Mechanics*, "there are other cases in which small initial variations may produce a very great change in the final state of the system, as when the displacement of the 'points' causes a railway train to run into another instead of keeping its proper course".

Because of its complexity and sensitivity to certain variations, the evolution of a nonlinear system cannot be predicted by mere computations, be it analytical or numerical. They do not offer a satisfactory solution, even if they produce a feasible one; nonlinear phenomena are global and there is a need for a more geometrical view of the phenomena which could provide the proper guidelines for the computations. The limits of the computational methods have been pointed out by Poincaré in his classic work on differential equations; he showed the need to marry analytic and geometric methods, and although he was concerned with asymptotic analytic methods, this also applies to numerical methods since the difficulty is inherent in the problem.

In recent years there has been considerable work on dynamical systems theory. Probably this is due to the favorable convergence of several factors:

The need for the understanding of new phenomena appearing in new areas of science and technology.

The increase in computing power, producing more insight into the behavior of dynamical systems and into the description of chaotic behavior.

New ideas and new mathematical tools such as the work of S. Smale on attractors; the mechanism proposed by D. Ruelle and F. Takens for the explanation of turbulence; the popularization of fractal sets due to B. Mandelbrot and others after him. In related areas we can mention the Kolmogorov–Arnold–Moser theory relating chaos and nonintegrability of Hamiltonian systems; the period-doubling mechanism for mappings of the interval $[0, 1]$; and the associated number discovered by M. Feigenbaum.

Following the ideas of S. Smale, D. Ruelle, and F. Takens, the chaotic behavior of a dissipative dynamical system can be explained by the existence of a complicated attractor to which the trajectories converge as $t \to \infty$; this set can be a fractal, like a Cantor set or the product of a Cantor set and an interval. This attractor is the natural mathematical object describing the observed nonstationary flow, and its complicated structure is the cause (or one of the causes) of the perceived chaos (or the apparent chaos). An understanding of these sets is of course necessary for a better understanding of the flow that they describe, and for the discovery of the laws and structures of the flow underlying the small-scale chaos. Already finite-dimensional systems (i.e., those whose state is described by a finite number of parameters) lead to complicated attractors as shown by the classic example of the Lorenz attractor in space dimension 3.

The study of turbulence in finite-dimensional systems suggests that the level of complexity of the phenomena increases with the level of complexity of the system. Thus we may wonder what is the level of complexity of infinite-dimensional dynamical systems such as those arising in continuum mechanics or continuum physics; this is one of the questions addressed in this book: we will see that, fortunately, the number of degrees of freedom of such systems is finite although it may be high. Thus in infinite dimensions the complexity of motion is due at the same time to the large (but finite) number of degrees of freedom and to the possible chaotic behavior of some of them.

The material treated in this book is limited to infinite-dimensional dissipative dynamical systems; very few finite-dimensional systems are presented, and then only to serve as reference examples or model problems; most of the systems considered are derived from evolutionary partial differential equations associated with boundary-value problems. Of course, in the study of such systems we are not only faced with the difficulties of nonlinear dynamics, but also with the difficulties related to evolutionary partial differential equations. Unlike the case of ordinary differential equations, no general theorem of existence and uniqueness of solutions exists for such problems and each partial differential equation necessitates a particular study. As a rule, a proper treatment of such problems necessitates the use of several different function spaces. It is one of our aims in this book to combine the problems and methods of both theories, dynamical systems and evolutionary partial differential equations, in an attempt to fill the gap between them and to make all the aspects of the theory accessible to the nonspecialist. We did not try to develop an abstract setting, but rather we deliberately chose to study specific equations and to remain close to their physical context.

For each equation, the following questions are addressed:

Existence and uniqueness of the solution and continuous dependence on the initial data. These preliminary results are not new, but they are part of the definition of the dynamical system and most of the necessary tools are also needed in other parts of the study. We have thus included these results for the sake of completeness.

Existence of absorbing sets. These are sets which all the orbits corresponding to the different initial data eventually enter. The existence of such sets is a step in the proof of the existence of an attractor. It is also an evidence (or a consequence) of the dissipative nature of the equation. In finite dimensions J.E. Billoti and J.P. La Salle [1] propose it as a definition of dissipativity; unfortunately, some difficulties specific to infinite dimensions make it, in that case, a less natural definition of dissipativity.

Existence of a compact attractor. It is shown that the equation possesses an attractor towards which all the orbits converge. Several adjectives are attributed to this attractor: we call it the *global* or *universal* attractor since it describes all the possible dynamics that a given system can produce; we also call it the *maximal* attractor since it is maximal (for the inclusion relation) among all bounded attractors. This set $\mathscr{A}$ attracts all bounded sets and it is an invariant set for the flow: by this we mean that any point of $\mathscr{A}$ belongs to a complete orbit lying in $\mathscr{A}$; this is particularly unusual in infinite dimensions where the backward initial-value problem is usually not well posed. The three names, global, universal, and maximal attractor, have the same meaning; we did not want to favor one of them and, at this point, leave the choice to the reader.

Finite dimensionality and estimate of the dimension of the attractor. Another aspect of dissipativity is that the attractor has a finite dimension, so that the observed permanent regime depends on a finite number of degrees of freedom. This was first proved for a delay evolution equation by J. Mallet-Paret [1] and in the case of the Navier–Stokes equations by C. Foias and R. Temam [1]. Here, besides showing the finite dimensionality of the attractor, we actually estimate an upper bound of its dimension in terms of the physical data and, in some specific cases, there are indications that these bounds are physically relevant. For instance, in the three-dimensional turbulent fluid flows we recover exactly the estimate of the number of degrees of freedom predicted by the Kolmogorov theory of turbulence (see P. Constantin, C. Foias, O. Manley, and R. Temam [1]).

* * *

Let us now describe the content of this book in its chronological order. The first chapter is a general introduction which further develops the present Preface, the motivations for this work, and a description of the main results. It also contains a "User's Guide" which is intended for more physics-oriented readers who are interested in the questions discussed in this book, but not in all of the mathematical aspects. In Chapter I we introduce the general results and concepts on invariant sets and attractors. This chapter contains all the basic definitions and properties and, in particular, a general criterion of existence of a global attractor which we use repeatedly in the sequel. This chapter contains, as an illustration, simple examples drawn from ordinary differential equations, in particular, the well-known Lorenz model. Another example treated there is that of fractal interpolation producing an interesting application in infinite dimensions which does not require a complicated functional framework. Chapter II is a technical chapter containing some elements of functional analysis: function spaces, linear operators, linear evolution equations of first and second order in time. This chapter is conceived as a reference chapter; it is not suggested that it be read in its entirety before reading the subsequent chapters, but rather that it be read "locally" as needed.

Chapters III and IV contain a systematic study of several infinite-dimensional dynamical systems arising in chemistry, mechanics, and physics. In Chapter III we consider the reaction–diffusion equations; pattern formation equations; fluid mechanics equations which include the Navier–Stokes equations in space dimension 2; and some other equations: magneto-hydrodynamics, thermohydraulics, fluid driven by its boundary, and geophysical flows (flows on a manifold). Chapter IV deals with damped nonlinear wave equations. For each equation we provide the appropriate functional setting and the results of existence and uniqueness of solutions; we prove the existence of absorbing sets for various norms, as needed, and we show the existence of a maximal attractor. This is the mathematical object describing all the possible dynamics (behaviors) of the system. Chapters III and IV also contain some technical results (like the injectivity of the semigroup which is equivalent to backward uniqueness), or some aspects that we mention without developing them thoroughly: e.g., regularity results and stability of attractors with respect to perturbations.

In Chapter V we introduce some new tools and results which we then apply in Chapter VI to all the equations considered earlier. The central theme of Chapter V is that of Lyapunov exponents and Lyapunov numbers. The chapter starts with a technical section on linear and multilinear algebras which recapitulates some known results and gives some extensions. We then introduce the Lyapunov exponents and show their relation to the distortion of volumes generated by the semigroup on the attractor. The chapter also contains general results concerning the Hausdorff and fractal dimensions of attractors which were proved in P. Constantin, C. Foias, and R. Temam [1]. These results, when properly applied, show the finite dimensionality of the attractor, and they allow a sharp explicit estimate of the dimension in terms of some quantities directly related to the physical problems. For attractors which are expected to be complicated (fractal) sets, the dimension is one of the few mathematical pieces of information on the geometry of such sets. On the physical and numerical sides, this dimension gives one an idea of the number of degrees of freedom of the system, and therefore of the number of parameters and the size of the computations needed in numerical simulations.

As indicated above, Chapter VI contains a systematic application of these results to the equations of Chapters III and IV and to the Lorenz equation which serves as a simple model. We derive various estimates for the Lyapunov exponents, the evolution of the volume element in the phase space, and, most important for the dimension of the attractor, the bound on the dimension being expressed as explicitly as possible in terms of physical entities.

Chapter VII contains some miscellaneous topics: the extension of the results to non-well-posed problems, and some extensions regarding unstable manifolds leading to a detailed description of the global attractor of a semigroup possessing a Lyapunov function; and on the other hand, to a briefly sketched method for deriving lower bounds on dimensions of attractors.

The last chapter, Chapter VIII, is devoted to inertial manifolds, new mathematical objects which have been recently introduced in relation with the study of the long-time behavior of dynamical systems. These are finite-dimensional Lipschitz manifolds, which attract exponentially all the orbits; they are positively invariant for the flow and contain of course the global attractor. This question is the object of much current investigation, and we thought it was desirable to include here some typical recently proved results.

The book ends with an Appendix providing some collective Sobolev inequalities which are used in Chapter VI for refined estimates of the traces of certain linear operators (the linearized operator corresponding to the first variation equation).

Most of the topics developed here are new or have appeared recently. Relevant questions which are not developed here include those of partly dissipative systems, and of nonautonomous systems. These aspects are currently being investigated and will appear in articles elsewhere (see M. Marion [2], J.M. Ghidaglia and R. Temam [6]). Also the theory of attractors for stochastic differential equations (see A. Bensoussan and R. Temam [1], [2], P. Malliavin [1]) has not yet been approached.

In conclusion, I would like to thank all those who helped in the realization of this book through encouragement, advice, or scientific exchanges: Peter Constantin, Jean-Michel Ghidaglia, Jack Hale, Martine Marion, Basil Nicolaenko, David Ruelle, Jean-Claude Saut, Bruno Scheurer, George Sell, and I.M. Vishik. More particularly, I would like to thank Ciprian Foias for a continued and friendly collaboration on which part of this book is based, and Oscar Manley who undertook the considerable task of improving the English language.

Madame Le Meur typed the manuscript struggling through hundreds of pages while being introduced to the word processor. I would like to thank her for her kind and patient cooperation.

Roger Temam
Université Paris Sud

Contents

CHAPTER II
Elements of Functional Analysis 43

CHAPTER III
Attractors of the Dissipative Evolution Equation of the First Order in Time: Reaction–Diffusion Equations. Fluid Mechanics and Pattern Formation Equations 82

CHAPTER V
Lyapunov Exponents and Dimension of Attractors 335

CHAPTER VI
Explicit Bounds on the Number of Degrees of Freedom and the Dimension of Attractors of Some Physical Systems 380

General Introduction. The User's Guide

Introduction

In this General Introduction we intend to focus on the general motivations and general ideas underlying this work, separating them from the mathematical technicalities, and thus developing further the presentation in the Preface.

The increase of interest in turbulence and chaos is due, in part, to the emergence of new mathematical ideas and concepts, as well as, in some cases, new insights into these concepts (e.g., attractors, fractal sets, Feigenbaum cascades ...). It is also due, in part, to the increase in computing power which brings us closer to obtaining solutions of complex realistic problems. Also, turbulence continues to be an important factor in several scientific and technological areas, especially in sophisticated technologies.

Beside the desire to understand better nonlinear dynamics and chaos as theoretical problems of mathematical physics, there are two major issues arising in connection with practical applications:

On the theoretical side, we would like to understand the mechanisms causing chaos.

On the computational side, we would like to know how to compute chaotic situations, and what to compute when chaos has appeared.

These are the questions that we will address in a general manner in this General Introduction (Sections 1 to 4). This Introduction ends with a section (Section 5) which serves as a user's guide for the more physics-oriented reader, who does not want to enter into all the mathematical aspects and who wants to reach rapidly the most physically relevant results.[1]

[1] A more general overview of chaos and nonlinear dynamics can be found in J. Gleick [1].

1. Mechanism and Description of Chaos. The Finite-Dimensional Case

We consider the solution $u = u(t)$ of a differential equation

$$\frac{du(t)}{dt} = F(u(t)), \tag{1.1}$$

with initial data

$$u(0) = u_0, \tag{1.2}$$

and we are interested in the behavior as $t \to \infty$ of $u(t)$. The variable $u = u(t)$ belongs to a linear space H called the *phase space*, and F is a mapping of H into itself. It is understood that the knowledge of $u = u(t)$ permits a complete description of the state of the system at a given time.

Two cases are worth considering:

The finite-dimensional case, where

$$u = u(t) \in H = \mathbb{R}^N.$$

The infinite-dimensional case, where

$$u = u(t) \in H = a\ Hilbert\ space.$$

Although the finite- and infinite-dimensional cases have several common points as is shown below, there are nevertheless some significant differences between the two cases.

First, we address finite-dimensional systems. Many systems of type (1.1) are worth considering in finite dimensions. In two dimensions we consider, for example, a pendulum in a vertical plane (see Figure 1.1, and Section I.2.1). The state is described, at each time t, by $u(t) = \{\theta(t), d\theta(t)/dt\} \in \mathbb{R}^2$, where $\theta(t)$ is the angle of the pendulum with, say, the downward vertical. More generally, with N coupled pendula moving in the same vertical plane, we obtain a system

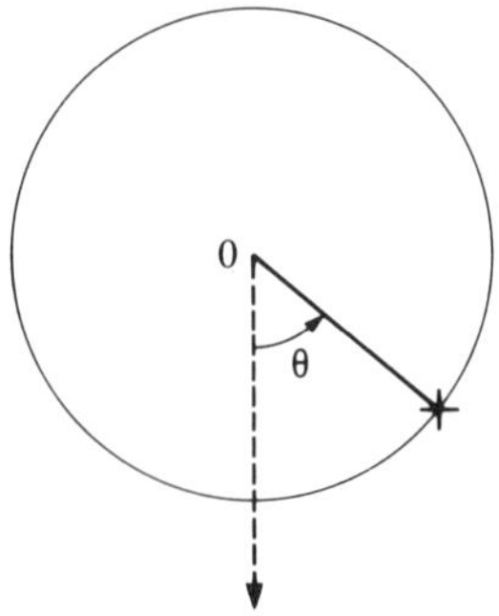

Figure 1.1. Pendulum in a vertical plane.

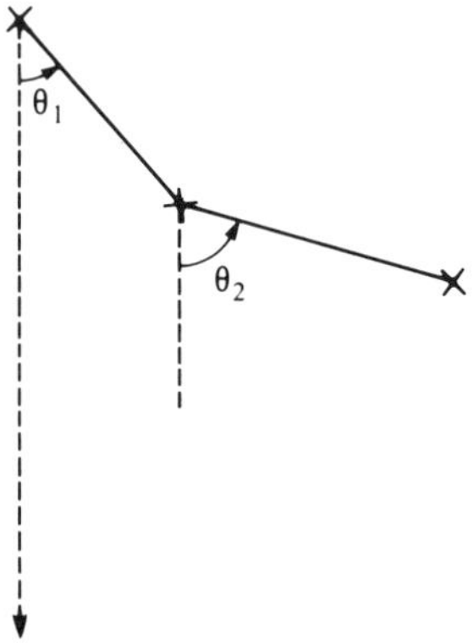

Figure 1.2. Coupled pendula.

in dimension $2N$, where

$$u(t) = \left\{\theta_i(t), \frac{d\theta_i(t)}{dt}, i = 1, \ldots, N\right\},$$

$\theta_i(t)$ being the angle of the ith pendulum with the downward vertical (Figure 1.2). Such systems can be considered as simplified models of the Josephson junction (see, for instance, J. Chandra [1] and A. Newell [1]).

In robotics, certain mechanical components, e.g., articulated arms, lead naturally to nonlinear systems of the form (1.1), $u(t) \in \mathbb{R}^N$ where N is the number of degrees of freedom of the robot mechanism. Some methods of image processing using attractors also lead to finite-dimensional dynamical systems (see, e.g., M.F. Barnsley [2]).

Similarly, in system theory (see P. Faurre and M. Robin [1], J.L. Lions [3]), the control of systems with a nonlinear feedback leads to a nonlinear dynamical system of the form (1.1): if $u = u(t)$ is the control, $y = y(t)$ the state, and $u = \Phi(y)$ the feedback law, then we have

$$\frac{dy}{dt} = f(y, u) = f(y, \Phi(y)).$$

Hence a system (1.1) for the state y (Figure 1.3). Finally, a large class of

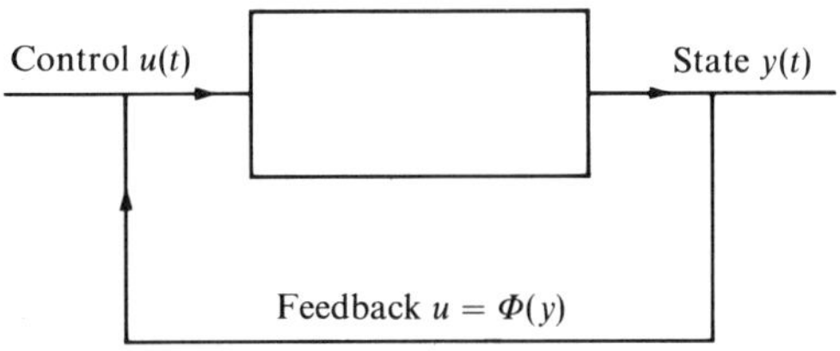

Figure 1.3. Nonlinear dynamics and control.

finite-dimensional dynamical systems is generated by the Galerkin (or other) approximations of infinite-dimensional systems; one of the most celebrated systems of this form is the Lorenz system which is a three-mode Galerkin approximation of the thermohydraulics equations (see Sections I.2.3 and III.3.5).

From the mathematical point of view, in considering a system such as (1.1), (1.2), we are interested in the behavior of $u(t)$ as $t \to \infty$. Equivalently, from the physical point of view, we are interested in predicting/describing the state of the system after "a small transient period". In the general situation, F depends on some physical parameter λ

$$F(u) = F_\lambda(u).$$

The phenomena that are observed depend on λ; the role of λ and the step of the transitions in the states vary considerably from one system to another, but in a very schematic generic way we can describe what happens as follows:

Step 1. For λ small, say $\lambda < \lambda_1$, there exists a unique stationary solution, i.e., a unique solution $u = u_1^s$ to equation

$$F_\lambda(u) = 0. \tag{1.3}$$

This stationary solution is stable and attracts all the orbits, i.e.,

$$u(t) \to u_1^s \qquad \text{as} \quad t \to \infty,$$

for any solution u of (1.1), (1.2), i.e., for any u_0.

Step 2. For larger values of λ, say $\lambda_1 < \lambda < \lambda_2$, other solutions to (1.3) appear, i.e., other stationary solutions $u_2^s, u_3^s, \ldots,$ appear, while u_1^s looses its stability. We say that a bifurcation of stationary solutions has occurred at $\lambda = \lambda_1$. Typically, for $t \to \infty$, $u(t)$ will converge to one of the stationary solutions

$$u(t) \to u_2^s \qquad \text{as} \quad t \to \infty,$$

the limit solution depending on u_0; each stationary solution possesses a basin of attraction and attracts all the solutions of (1.1), (1.2) which start from within the basin of attraction.

Step 3. When λ is increased further, $\lambda_2 < \lambda < \lambda_3$, then a Hopf bifurcation can occur; in this case, the flow never becomes stationary. Instead, we have

$$u(t) - \varphi(t) \to 0 \qquad \text{as} \quad t \to \infty,$$

where φ is a *time-periodic* solution of (1.1), of period $T > 0$:

$$\frac{d\varphi}{dt}(t) = F(\varphi(t)), \qquad \forall t \in \mathbb{R}, \tag{1.4}$$

$$\varphi(t + T) = \varphi(t). \tag{1.5}$$

The value λ_2 of λ where the Hopf bifurcation has occurred is a priori unknown, as is the solution φ of (1.4) and even *the period T.* In every instance detailed analysis is necessary to determine these quantities.

From that point on, when λ varies, we can either observe a Feigenbaum cascade of period doubling or the system can go directly to Step 4. When observed, the Feigenbaum cascade corresponds to a sequence of functions $\varphi = \varphi_j$, periodic with period $2T, 2^2T, \ldots, 2^jT$.

Step 4. For larger λ, $\lambda_3 < \lambda < \lambda_4$, invariant tori can appear, i.e., for $t \to \infty$,

$$u(t) - \varphi(t) \to 0, \tag{1.6}$$

where φ is now a quasi-periodic solution of (1.4) of the form

$$\varphi(t) = g(\omega_1 t, \ldots, \omega_n t), \tag{1.7}$$

where g is periodic with period 2π in each variable and the frequencies $\omega_i = 1/T_i$ are rationally independent numbers.

Of course at this stage the flow looks very chaotic, but Fourier analysis will easily show that the system behavior is governed by modes with discrete frequencies ω_i, which is not characteristic of chaotic behavior.

Step 5. Finally, for $\lambda > \lambda_4$, we reach the last stage, that of chaos where $u(t)$ looks completely random for all time, and a Fourier analysis leads to a wide-band continuous spectrum. A mathematical analysis which will be carried out many times in the course of this book shows that

$$u(t) \to X \qquad \text{as} \quad t \to \infty \tag{1.8}$$

in the sense that

$$\text{distance of } u(t) \text{ to } X \to 0 \qquad \text{as} \quad t \to \infty.$$

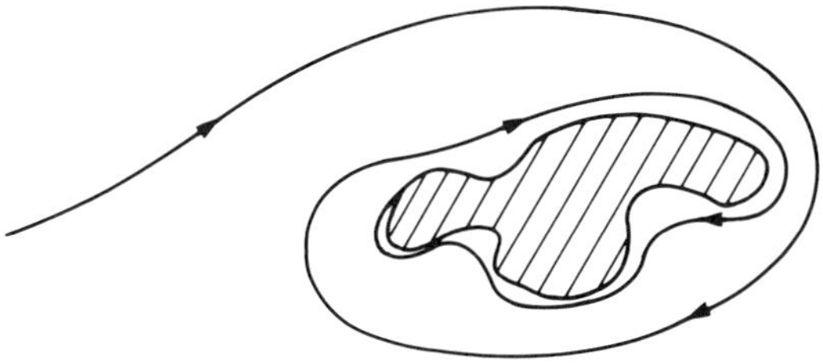

Figure 1.4. Convergence to an invariant set.

Here X is an invariant subset of the phase space H, i.e., it enjoys the remarkable property of being invariant for the semigroup $\{S(t)\}_{t \geq 0}$ associated to (1.1), (1.2):

$$S(t)X = X, \qquad \forall t > 0, \tag{1.9}$$

where $S(t)$ is the mapping

$$S(t)\colon u_0 \in H \to u(t) \in H,$$

u being the solution of (1.1), (1.2).[1] Typically the set X is complicated; it can

[1] As we recall below some invariant sets are trivial (stationary solutions, orbit of time-periodic solutions, ...). However, general invariant sets are very particular objects.

be a fractal set, such as a Cantor set or the product of a Cantor set with an interval. When t is large $u(t)$ wanders on X (or near X), and the complicated convolutions of X give rise to the complicated form of the flow. In a sense, this explains the chaotic appearance of the flow.

This is turbulence from the dynamical systems point of view. It pictures turbulence by chaotic behavior with respect to time (S. Smale, D. Ruelle, and F. Takens).

We emphasize that the sequence of transient states to turbulence, as illustrated by Steps 1 to 5, is very schematic; the situation can be more complicated, the sequence of steps can be different and several parameters can produce competing effects.

2. Mechanism and Description of Chaos. The Infinite-Dimensional Case

In infinite dimensions the phase space H is a Hilbert space; most often H is a function space on some domain $\Omega \subset \mathbb{R}^d$, $d = 1, 2, 3$ (or more), and Ω is the domain in which the given physical phenomena occur, e.g., fluid or solid motion, wave, population growth, ... In such cases $u(t) \in H$ is the symbolic notation for a function of the space variable $x = (x_1, \ldots, x_d)$:

$$\{x \in \Omega \to u(x, t)\}.$$

The governing equation for the phenomena has the same form as (1.1):

$$\frac{du(t)}{dt} = F(u(t)), \tag{2.1}$$

with an initial condition

$$u(0) = u_0. \tag{2.2}$$

In general, (1.8) is the functional form of a dissipative nonlinear partial differential equation with the associated appropriate boundary conditions; and (2.1), (2.2) is the functional setting of the corresponding initial- and boundary-value problem. In Chapters III and IV we show how the boundary-value problems associated with dissipative phenomena in mechanics, physics, chemistry, ..., can be put in the form (2.1), (2.2).

Again, we are interested here in *the behavior of the solutions of* (2.1), (2.2) as $t \to \infty$, and the analogy of (2.1), (2.2) with (1.1), (1.2) allows us to extend easily the previous discussion to the present context.

As in the finite-dimensional case it is convenient and physically relevant to assume that $F(u) = F_\lambda(u)$ depends on a parameter λ: this parameter could be, for instance, a Reynolds number, a Grashof number, ... The same transition to turbulence as in Steps 1 to 5 above can arise: to illustrate this, let us recall two classic experiments.

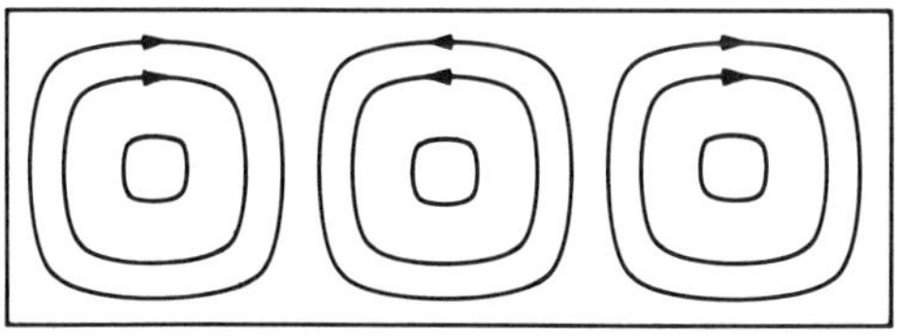

Figure 2.1. Bénard problem.

Example 1 (Bénard Problem). This is the motion of a horizontal layer of fluid heated from below. For the present discussion we can assume the layer to be infinite in the horizontal direction, or finite with appropriate boundary conditions on the vertical calls (see Section III.3.5.). Here the function $u = u(x, t)$ represents the pair $\{v, \theta\}$, where $v(x, t)$ is the velocity of the particle of fluid at point x at time t, and $\theta = \theta(x, t)$ is the temperature, and Ω is the region filled by the fluid. We refer the reader to Section III.3.5 for the representation of the Bénard problem in the form (2.1), (2.2). With other experimental quantities left unchanged, the parameter λ could be the ratio $(T_0 - T_1)/T_0$.[1] If λ is small, $\lambda < \lambda_1$, the fluid remains at rest and the equilibrium solution $u = u_1^s$ corresponds to a pure conduction solution with a linear distribution of the temperature along the vertical axis. For $\lambda_1 < \lambda < \lambda_2$, the pure conduction solution loses its stability and the fluid starts to move, reaching another steady state u_2^s corresponding to the formation of the classical rolls (see Figure 2.1). The time-periodic solution described in Step 3 corresponds to the case in which the boundaries of the rolls start to oscillate in a time-periodic manner.[2] Then, in Step 4 for $\lambda_3 < \lambda < \lambda_4$, the boundaries of the rolls oscillate in a less regular manner, and spectral analysis shows that indeed two or more independent periods have appeared. Finally, in Step 5, corresponding to large values of λ, $\lambda > \lambda_4$, the rolls have disappeared and the flow seems totally unstructured. Actually, the results that we prove in this work tend to indicate that the flow remains structured even for large values of λ, but the flow then resulting from the superposition of a large number of structures looks unstructured.

Example 2 (Flow Past a Sphere). Another classic experiment, where Steps 1 to 5 are indeed observable, is the flow past a sphere.

A sphere is embedded in an incompressible fluid with velocity $\mathbf{U}_\infty$ at infinity. A natural parameter λ is then the Reynolds number

$$\lambda = \text{Re} = \frac{R|\mathbf{U}_\infty|}{\nu},$$

[1] A better choice is the Grashof number; but at this point we refrain from introducing this number.

[2] Period doubling phenomena corresponding to Feigenbaum cascades may occur during this phase, although the observation of this phenomena is recent and delicate.

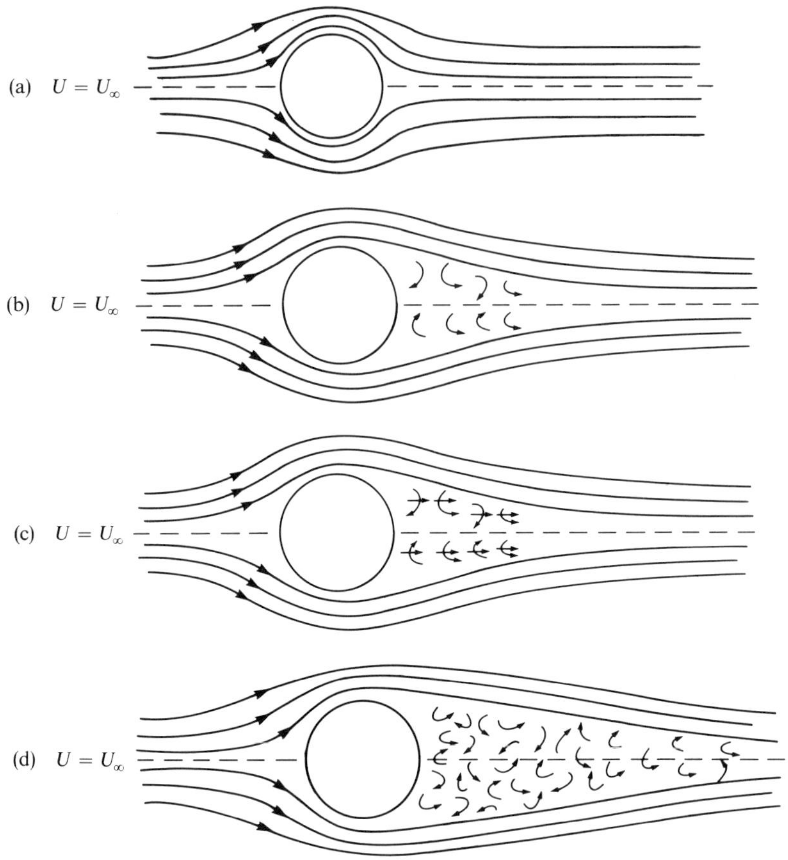

Figure 2.2. Flow past a sphere. (a) Laminar flow (small Reynolds number). (b) Appearance of the von Kármán vortices in the wake behind the sphere (stationary flow). (c) Time-periodic flow: the vortices behind the sphere are moving to the right in an (apparently) time-periodic manner. (d) Fully turbulent flow in the wake behind the sphere at large Reynolds numbers.

where v is the kinematic viscosity and R is the radius of the sphere. In a practical experiment the apparatus (v and R) remains unchanged, and we increase λ by increasing the speed at infinity $U_\infty = |\mathbf{U}_\infty|$. Steps 1 and 2 correspond to cases where, after a short initial transient period, the observed flow is steady. It is fully laminar in Step 1, followed by the appearance of steady von Kármán vortices in Step 2 (see Figure 2.2(a) and (b)). In Step 3, after a Hopf bifurcation has occurred, the flow never becomes stationary again, in fact it is time periodic. Here the von Kármán vortices move to the right and vanish, while reappearing on the left in a seemingly time-periodic manner. In Step 4, when U_∞ is still large, the displacement of the von Kármán vortices to the right is less regular and corresponds to a seemingly quasi-periodic flow. Finally, in Step 5 for large values of U_∞, a fully turbulent completely unstructured flow appears in the wake of the sphere. The flow is nonstationary for all time and can be described only, from the mathematical point of view, by the attractor X that it defines.

Origin of Chaos

As in finite dimensions in general

$$u(t) \to X \qquad \text{as} \quad t \to \infty,$$

where $X \subset H$ is a compact attractor which can be a fractal set. Thus *chaos in time* is attributed to the complicated structure of X, the trajectory $u(t)$ following the set X in the function space.

However, in infinite dimensions, two radically new aspects arise of which as yet little is known:

(i) *First is the existence of chaos in space.*
The example of flow past a sphere shows that turbulence (chaos) can develop in some parts of Ω and not in others. There is at the moment no theory, similar to the attractor approach, to explain the development of turbulence in specific parts of Ω. Other examples include turbulent jets and plumes.

(ii) *Second is the possible existence in space of singularities on some small sets.*
For example, in fluid mechanics (incompressible fluids, thermohydraulics, magnetohydrodynamics), it is not yet known whether or not in three-dimensional space the curl of the velocity, curl $\mathbf{v}$ can become infinite in some parts of the domain Ω occupied by the flow. In fact, J. Leray in 1932 proposed the possible appearance of singularities as an explanation for turbulence; and the introduction of weak solutions to the Navier–Stokes equations in his pioneering work (J. Leray [1], [2], [3]) was intended to handle such situations.

More recently, B. Mandelbrot [1] suggested that the singular set

$$\Sigma(t) = \{x \in \Omega, |\text{curl } \mathbf{v}(x, t)| = +\infty\}$$

is a fractal set. V. Scheffer [1] gave several estimates on the Hausdorff dimension of the singular set; finally, L. Caffarelli, R. Kohn, and L. Nirenberg [1] improved Scheffer's result and gave the best available estimate of the dimension of $\Sigma(t)$, namely:

The one-dimensional Hausdorff measure of $\Sigma(t)$ is 0.

Although less extensively studied than the Navier–Stokes equations, the Kuramoto–Sivashinsky equation in space dimension $d \geq 2$ (see Section III.4.1) and some nonlinear wave equations in $\mathbb{R}^3$ (see Section IV.3) can also produce similar difficulties (possible appearance of singularities in space), and their mathematical nature is not fully understood: for instance, one puzzling aspect is that the existence and uniqueness of solutions of (1.8), (1.9) is not guaranteed for all initial data.

To avoid mathematical difficulties, we limit ourselves mostly to cases where the initial-value problems associated with the equations are well posed. In Sections VII.1 and VII.2, we develop, as an example, an extension of our results to the three-dimensional Navier–Stokes equations; many other equations could be treated along the same lines.

3. The Global Attractor. Reduction to Finite Dimension

The presentation that we adopt in this book applies to both the finite- and infinite-dimensional cases ($\dim H < \infty$, $\dim H = \infty$). As indicated in the Preface the emphasis will be on infinite systems, but when it is useful we will illustrate our general results with finite-dimensional examples.

Hence we consider an equation of the form (2.1) (or (1.1)) and we assume that the initial-value problem (2.1), (2.2) (or (1.1), (1.2)) is well posed for all $t \geq 0$. This allows us to define the semigroup $\{S(t)\}_{t \geq 0}$, i.e., the family of operators:

$$S(t)\colon u_0 \in H \to u(t) \in H. \tag{3.1}$$

The long-time behavior of a solution of (2.1), (2.2) is always described by a functional invariant set $X \subset H$, to which the orbit $u(t)$ converges as $t \to \infty$:

$$S(t)X = X, \qquad \forall t \in \mathbb{R}, \tag{3.2}$$

$$\text{distance}\,(u(t), X) \to 0 \qquad \text{as} \quad t \to \infty. \tag{3.3}$$

In the trivial cases where $u(t)$ converges to a stationary solution u^s as $t \to \infty$, the set X reduced to $\{u^s\}$ enjoys the property (3.2), (3.3). Its dimension is 0. When $u(\cdot)$ converges to a time-periodic solution $\varphi(\cdot)$ of (2.1), then the appropriate set X is the orbit of $\varphi(\cdot)$:

$$X = \{\varphi(t), t \in \mathbb{R}\}. \tag{3.4}$$

Due to the periodicity property of φ, X is in fact a closed curve, and it clearly enjoys the properties (3.2), (3.3).

When a quasi-periodic flow occurs as mentioned in Section 1 (see (1.6), (1.7)), the set X is still the orbit of φ as in (3.4), but now this orbit lies on an n-dimensional torus (n as in (1.7)).

In the general case, the existence of a set X enjoying properties (3.2), (3.3) has to be proved; this existence result is well known and we recall it at the beginning of Chapter I, $X = \omega(u_0)$ being called the ω-limit set of u_0. In all cases it is the mathematical object describing the long-time behavior of the solution u of (2.1), (2.2).

Of course, the complexity of the flow described by X, i.e., produced by u_0 depends in part on the dimension of X and in part on the geometry of X. The higher the dimension is, the more complicated the flow. Similarly, complicated (unsmooth) geometries of X produce complicated flows. Here we are able to "measure" the complexity of a flow through the dimension of X, while the study of the complexity of the flow due to the geometry of X is an untouched question.

One of the results proved in this book is that, for dissipative systems, such sets $X = \omega(u_0)$ exist and they are all part of a "large" compact invariant set $\mathscr{A}$ which attracts *all* the orbits:

$$S(t)\mathscr{A} = \mathscr{A}, \qquad \forall t \in \mathbb{R}.$$

This set $\mathscr{A}$ is called the global (or maximal, or universal) attractor of the system. Dissipativity is characterized here by the existence of an absorbing set, i.e., a bounded set $\mathscr{B}_0$ which all solutions u of (2.1), (2.2), enter, whatever u_0 is; of course, this is a drastic contrast with conservative (Hamiltonian) systems where orbits may fill the whole phase space or regions of it.

The global attractor $\mathscr{A}$ is itself compact and in some sense very "thin". Another aspect of our results shows that not only X but $\mathscr{A}$ itself has finite dimensions, even when H has infinite dimensions; the dimension used here is either the Hausdorff dimension of $\mathscr{A}$ or its fractal dimension, also called the capacity of $\mathscr{A}$. We recall that if the Hausdorff dimension of $\mathscr{A}$ is $\leq N$, then many projectors of dimension $2N + 1$ are injective on $\mathscr{A}$ (in fact, almost all of them, see R. Mañé [1]). Hence the flow on $\mathscr{A}$ can be parametrized by at most $2N + 1$ parameters belonging to a linear space. This proves that all observable flows are finite dimensional and establishes the reduction to finite dimension. Furthermore, in several cases, there are reasons to believe that our estimates of the number of degrees of freedom are physically relevant (at this level of generality, i.e., when we intend to describe all possible flows).

Equations of mathematical physics for which this program is carried out include the following:

the reaction–diffusion equations from chemical dynamics and population growth;

the incompressible Navier–Stokes equations in space dimension 2;

other equations of fluid mechanics: flow on a manifold (e.g., geophysical flows), magnetohydrodynamics, thermohydraulics;

the Kuramoto–Sivashinsky equation;

the Cahn–Hilliard equation;

the sine–Gordon equation;

a nonlinear wave equation of quantum mechanics and other nonlinear wave equations;

the Ginzburg–Landau equation (a nonlinear Schrödinger equation).

4. Remarks on the Computational Aspect

When they are applicable, our results show that the flows can be described by a finite number of parameters even in infinite dimensions, certainly an advantage as far as computations are concerned. In particular, beside yielding information about the geometry of the attractors, the estimates of the dimension of the attractors indicate the actual number of parameters that should be retained for an accurate approximation of a flow.

Nevertheless, despite its reduction to finite dimensions, the computation of large-scale turbulent systems remains a problem of considerable difficulty and its resolution is still in its infancy. Two principal difficulties are as follows:

(i) In many cases the dimension of the attrator $\mathscr{A}$ and thus the number of required parameters is too large for existing computational resources. For instance, in fluid mechanics the number of degrees of freedom (dim $\mathscr{A}$) is of the order of 10^9 in aeronautical and wind-tunnel experiments, and up to order 10^{20} in geophysical flows (meteorology). Compared with the number of words available in the central memory of the most powerful computers at the present time, we see that those computers are not yet sufficient for general fluid mechanics problems. However, they are not too far from being able to handle some significant problems in fluid mechanics (wind tunnel, aeronautics ...), but further reduction of dimension through simplified models remains necessary for more complex flows (meteorology, plasmas, ...).

(ii) Assuming that the computer power is available, there remains the problem of choosing the relevant parameters. In performing numerical computations, in spectral approximations, a natural choice of the parameters is a set of Fourier modes (and the like) or, in finite differences and multigrid methods, the nodal values. We want to emphasize the fact that it is not sufficient to take a number of parameters equal to the number of degrees of freedom (= dimension of $\mathscr{A}$) to describe the flow accurately. The fact that dim $\mathscr{A} = N$ implies that there exist N parameters describing properly the corresponding flow, but it is not at all certain that these are the first N Fourier modes or N regularly spaced nodal values; in fact, there are indications that these are not the most appropriate parameters. It will probably take a much more subtle choice of parameters to reach a proper reduction to finite dimension, i.e., to establish that the number of necessary parameters equals the dimension of the attractor.

Hence the choice of a set of parameters is an open question; the reader is referred to C. Foias, O. Manley, R. Temam, and Y. Treve [1] and to C. Foias and R. Temam [6] where the related concepts of determining modes and determining points is investigated.

Concerning the number of parameters to be retained in numerical computations, we recall that the long-time behavior of Galerkin approximations of a given system can vary drastically with the number of modes, and there is also a need to determine the number of modes necessary for a given computation. Some very preliminary results in this direction are derived in P. Constantin, C. Foias, and R. Temam [3].

Although the question of numerical computations is not tackled in this book, we thought it important to keep it in mind; in part, this is the motivation of this section. Of course, all the preceding remarks apply to the pesent status of computations and computing power and these may change drastically in the coming years.

5. The User's Guide

This section is intended as a user's guide for the more physics-oriented reader who does not want to follow all the mathematical details, but rather wants to reach rapidly the most physically relevant results. To serve this purpose it should be consulted as the reader progresses through the book and not read independently.

The first step in the study of the dynamics of a given physical nonlinear system is to write it in the form of an infinite-dimensional abstract equation such as (2.1). This is now a standard step. Then it is necessary to establish the existence and uniqueness of solutions to (2.1), (2.2), in order to define the semigroup $S(t)$, $t \geq 0$. There are no results which apply generally, but many standard and well-known methods are available and indeed here they are used in Chapters III and IV. All the available methods are based on the derivation of *a priori estimates* for the solutions of (2.1), (2.2). Usually they are obtained by taking the scalar product in H of (2.1) with u or with some function of u. Equations obtained in this way express some conservation laws: conservation of energy or of some other quantity.

After this preliminary step, the next step consists of proving the existence of an attractor, in general, the maximal one. Here this is done repeatedly in Chapters III and IV, by application of a general theorem stated and proved in Chapter I, Theorem 1.1. One of the necessary hypotheses which must be checked is the existence of an absorbing set. Its existence, if any, follows from the same conservation equalities and a priori estimates as that used to establish existence. Furthermore, a compactness property (or at least an asymptotic compactness property for t large) of the semigroup $S(t)$ is necessary (see assumptions (1.12) and (1.13) in Chapter I).

The third class of results proved here is the estimate of the Lyapunov exponents, the exponential decay of the volume element, and the estimate of the dimension of the attractor. They are all derived from an estimate of the trace of a linear operator associated with the first variation equation of (2.1).

Let $u(\cdot)$ be a solution of (2.1) lying in the global attractor $\mathscr{A}$ (or, more generally, in an invariant set X). The corresponding first variation equation for (2.1) is the linearization of that equation around $u(\cdot)$, i.e.,

$$\frac{dU(t)}{dt} = F'(u(t))U(t), \tag{5.1}$$

where F' is the Fréchet differential of F which is assumed to exist. We may supplement this with an initial condition

$$U(0) = \xi. \tag{5.2}$$

If (5.1), (5.2) possesses a solution we so define a linear operator

$$L(t; u_0)\colon \xi \to U(t), \tag{5.3}$$

which for $t > 0$ is the Fréchet differential at u_0 of the semigroup operator

$$S(t)\colon u_0 \to u(t).$$

Then we consider the trace Tr $F'(u(t))$ of the linear operator $F'(u(t))$ and, for every integer m, the number

$$\tilde{q}_m = \limsup_{t\to\infty}\ \operatorname*{Sup}_{u_0 \in X} \left\{ \frac{1}{t} \int_0^t \operatorname*{Sup}_{Q_m} \operatorname{Tr} F'(u(\tau)) \circ Q_m \, d\tau \right\}, \tag{5.4}$$

where Q_m is an arbitrary m-dimensional projector in H (dim $Q_m = m$) and X is the functional invariant set, or attractor, under consideration.[1] Methods suitable for estimating the trace of $F'(u(\tau)) \circ Q_m$ (and thus $\tilde{q}_m$) are developed in Chapter VI. Note that if φ_j, $j = 1, \ldots, m$, is a basis of $Q_m H$, orthonormal in H, then

$$\operatorname{Tr} F'(u(\tau)) \circ Q_m = \sum_{j=1}^{m} (F'(u(\tau))\varphi_j, \varphi_j)_H, \tag{5.5}$$

and we actually have to estimate the right-hand side of (5.5).

In the most favorable cases at the end of the computations we obtain a majorization of $\tilde{q}_m$ of the form

$$\tilde{q}_m \leq -\kappa_1 m^\alpha + \kappa_2. \tag{5.6}$$

It then follows from the general results of Chapter V (see Theorem V.3.3 and also Lemma VI.2.2) that the Hausdorff dimension of X is, at most, equal to $(\kappa_2/\kappa_1)^{1/\alpha}$ and its fractal dimension is less than or equal to $1 + (2\kappa_2/\kappa_1)^{1/\alpha}$. Also, the volume element of dimension $m \geq (\kappa_2/\kappa_1)^{1/\alpha}$ in the function space decays exponentially. Finally, (5.6) also yields a majorization for all the Lyapunov exponents.

[1] Slightly different, more natural, numbers called q_m are considered in Chapters V and VI; $q_m \leq \tilde{q}_m$, but the methods of estimating q_m and $\tilde{q}_m$ produce exactly the same type of bounds for these numbers.

CHAPTER I

General Results and Concepts on Invariant Sets and Attractors

Introduction

This chapter is devoted to the presentation of general results and concepts on invariant sets and attractors. Although the presentation is sufficiently general to include ordinary differential equations, the main objective is, of course, to consider infinite-dimensional dynamical systems. The general results, which will be applied in the following chapters to infinite-dimensional systems of physical interest, are already illustrated in this chapter with simpler examples which do not necessitate any knowledge of partial differential equations, i.e., they are limited to some ordinary differential equations and to some iterative processes.

In Section 1 we recall the basic assumptions and definitions: we introduce a semigroup of operators $S(t)$ which can be either continuous ($t \in \mathbb{R}_+$) or discrete ($t = n \in \mathbb{N}$ and $S(n) = S^n$, $S = S(1)$) and describe the first necessary assumptions on the semigroup. The definition of fixed or equilibrium points is recalled, as is the stable and unstable manifolds of such a point (Section 1.1). Then we introduce the important concepts of functional invariant sets (Section 1.2) and absorbing sets and attractors (Section 1.3). A few essential examples of functional invariant sets are given (including the ω-limit sets), and in Section 1.3 we give a general result which will be used constantly and which ensures the existence of an attractor which is maximal (for the inclusion relation). We conclude Section 1 with a remark on the stability of the concept of attractors; we consider some perturbations of the semigroup and show a convergence result for the associated attractors.

Section 2 illustrates the concepts with some simple examples involving ordinary differential equations. Section 2.1 describes the motion of a pendulum with or without friction. Section 2.2 is devoted to a system due to Gh. Minea which enjoys several common features with the Navier–Stokes equations. Finally, we recall the now classic Lorenz model and show how the results of Section 1 can be applied.

Section 3 deals with the concept of fractal interpolation. This is a method of constructing fractal sets which are attractors of some appropriate mappings in infinite dimension. Although these examples do not correspond to the main preoccupations of this volume, they provide examples of attractors in infinite dimensions which are simple to describe and moreover are interesting in themselves.

1. Semigroups, Invariant Sets, and Attractors

1.1. Semigroups of Operators

We will consider dynamical systems whose state is described by an element $u = u(t)$ of a metric space H. In most cases, and in particular for dynamical systems associated to partial or ordinary differential equations, the parameter t (= the time or the timelike variable) varies continuously in $\mathbb{R}$ or in some interval of $\mathbb{R}$; in some cases, t will take only discreet values, $t \in \mathbb{Z}$ or some subset of $\mathbb{Z}$. Usually the space H will be a Hilbert or a Banach space but, for the moment, there is no need to introduce that assumption.

The evolution of the dynamical system is described by a family of operators $S(t)$, $t \geq 0$, that map H into itself and enjoy the usual semigroup properties:

$$\begin{cases} S(t+s) = S(t) \cdot S(s), & \forall s, t \geq 0, \\ \quad S(0) = I & \text{(Identity in } H). \end{cases} \tag{1.1}$$

If φ is the state of the dynamical system at time s then $S(t)\varphi$ is the state of the system at time $t + s$, and

$$u(t) = S(t)u(0), \tag{1.2}$$

$$u(t+s) = S(t)u(s) = S(s)u(t), \qquad s, t \geq 0. \tag{1.3}$$

In the cases which interest us particularly, the semigroup $S(t)$ will be determined by the solution of an ordinary or a partial differential equation. In the case of ordinary differential equations the general theorems of existence of solutions provide the definition of the operators $S(t)$. In the infinite-dimensional case there are no theorems of existence and uniqueness of solutions which are sufficiently general, and therefore the proof of existence of the operators $S(t)$ and the derivation of their properties may be a necessary preliminary step in the study of a given dynamical system.

The basic properties of the operators $S(t)$ which are needed are given in

Section 1.2 but, we assume at least that,

$$S(t) \text{ is a continuous (nonlinear) operator from } H \text{ into itself, } \forall t \geq 0. \tag{1.4}$$

The operators $S(t)$ may or may not be one-to-one; the injectivity property of $S(t)$ is equivalent to the *backward uniqueness* property for the dynamical system. When $S(t)$, $t > 0$, is one-to-one we denote by $S(-t)$ its inverse which maps $S(t)H$ onto H; we then obtain a family of operators $S(t)$, $t \in \mathbb{R}$, which enjoy the property (1.1) on their domains of definition, $\forall s, t \in \mathbb{R}$. As recalled from the Preface of this book it is a general fact, in infinite dimensions, that even if the operators $S(t)$, $t > 0$, are defined everywhere, the operators $S(t)$, $t < 0$, are not usually defined everywhere in H.

For $u_0 \in H$, the *orbit* or *trajectory* starting at u_0 is the set $\bigcup_{t \geq 0} S(t)u_0$.

Similarly, *when it exists*, an orbit or trajectory ending at u_0 is a set of points

$$\bigcup_{t \leq 0} \{u(t)\},$$

where u is a mapping from $]-\infty, 0]$ into H such that $u(0) = u_0$ and $u(t + s) = S(t)u(s)$, $\forall s, t, s \leq 0, s + t \leq 0, t \geq 0$ (or equivalently $u(t) \in S(-t)^{-1}u_0$, $\forall t \geq 0$). The orbits starting or finishing at u_0 are also called the positive or negative orbits through u_0. A complete orbit containing u_0 is the union of the positive and negative orbits through u_0. For $u_0 \in H$ or for $\mathscr{A} \subset H$, we define the ω-limit set of u_0 (or $\mathscr{A}$), as

$$\omega(u_0) = \bigcap_{s \geq 0} \overline{\bigcup_{t \geq s} S(t)u_0},$$

or

$$\omega(\mathscr{A}) = \bigcap_{s \geq 0} \overline{\bigcup_{t \geq s} S(t)\mathscr{A}},$$

where the closures are taken in H. Analogously, when it exists, the α-limit set of $u_0 \in H$ or $\mathscr{A} \subset H$ is defined as

$$\alpha(u_0) = \bigcap_{s \leq 0} \overline{\bigcup_{t \leq s} S(-t)^{-1}u_0},$$

or

$$\alpha(\mathscr{A}) = \bigcap_{s \leq 0} \overline{\bigcup_{t \leq s} S(-t)^{-1}\mathscr{A}}.$$

It is easy to see that $\varphi \in \omega(\mathscr{A})$ if and only if there exists a sequence of elements $\varphi_n \in \mathscr{A}$ and a sequence $t_n \to +\infty$ such that

$$S(t_n)\varphi_n \to \varphi \quad \text{as} \quad n \to \infty. \tag{1.5}$$

Similarly, $\varphi \in \alpha(\mathscr{A})$ if and only if there exists a sequence ψ_n converging to φ

in H and a sequence $t_n \to +\infty$, such that

$$\varphi_n = S(t_n)\psi_n \in \mathscr{A}, \qquad \forall n.$$

A *fixed point*, or a *stationary point*, or an *equilibrium point* is a point $u_0 \in H$ such that

$$S(t)u_0 = u_0, \qquad \forall t \geq 0.$$

The orbit and the ω and α-limit sets of such a point are of course equal to $\{u_0\}$. If u_0 is a stationary point of the semigroup we define the stable and the unstable manifolds of u_0. The *stable manifold* of u_0, $\mathscr{M}_-(u_0)$, is the (possibly empty) set of points u_* which belong to a complete orbit $\{u(t), t \in \mathbb{R}\}$, $u_* = u(t_0)$ and such that

$$u(t) = S(t - t_0)u_* \to u_0 \qquad \text{as} \quad t \to \infty.$$

The *unstable manifold* of u_0, $\mathscr{M}_+(u_0)$, is the (possibly empty) set of points $u_* \in H$ which belong to a complete orbit $\{u(t), t \in \mathbb{R}\}$ and such that

$$u(t) \to u_0 \qquad \text{as} \quad t \to -\infty.$$

A stationary point u_0 is *stable* if $\mathscr{M}_+(u_0) = \varnothing$ and *unstable* otherwise. A classification of the different types of stable or unstable points for a semigroup (nodes, sinks, saddle points, ...) is given, for instance, in M.W. Hirsch and S. Smale [1].

The Discrete Case

In the discrete case, we consider a mapping S from H into itself and for $n \in \mathbb{N}$, we write

$$S(n) = S^n.$$

If S is injective we can define S^{-1} and for $n \in \mathbb{N}$, $S(-n) = S^{-n}$. The orbit of $u_0 \in H$, the ω-limit set of $u_0 \in H$, or of $\mathscr{A} \subset H$ are defined exactly as above as are the α-limit sets when they exist, with the sole restriction that $t \in \mathbb{Z}$.

Unless otherwise specified the presentation will be unified for the discrete and continuous cases, the general definitions and results will apply for both cases with either $t \in \mathbb{R}$ or $t \in \mathbb{Z}$.

1.2. Functional Invariant Sets

We say that a set $X \subset H$ is positively invariant for the semigroup $S(t)$ if

$$S(t)X \subset X, \qquad \forall t > 0.$$

It is said to be negatively invariant if

$$S(t)X \supset X, \qquad \forall t > 0.$$

When the set is both positively and negatively invariant, we call it an invariant set or a functional invariant set.

Definition 1.1. A set $X \subset H$ is a functional invariant set for the semigroup $S(t)$ if

$$S(t)X = X, \qquad \forall t \geq 0. \tag{1.6}$$

When the operators $S(t)$ are *one-to-one* (backward uniqueness), relation (1.6) implies that $S(-t)$ is defined on X, $\forall t > 0$ and

$$S(t)X = X, \qquad \forall t \in \mathbb{R}. \tag{1.7}$$

A trivial example of an invariant set is a set X consisting of a fixed point, $X = \{u_0\}$, or of any union of fixed points. When it exists, a time-periodic orbit is also a simple example of an invariant set, i.e., if for some $u_0 \in H$ and $T > 0$, $S(T)u_0 = u_0$, then $S(t)u_0$ exists for all $t \in \mathbb{R}$ and

$$X = \{S(t)u_0, t \in \mathbb{R}\} \tag{1.8}$$

is invariant. More generally, any orbit existing for all $t \in \mathbb{R}$ defines an invariant set X, such as (1.8), where u_0 is any point of this orbit; in particular, if the function $t \in \mathbb{R} \to S(t)u_0$ is quasi-periodic (i.e., of the form $g(\omega_1 t, \ldots, \omega_n t)$ where g is periodic with period 2π in each variable and the frequencies ω_j are rationally independent), we obtain invariant tori.

We now describe less obvious invariant sets. First, if u_0 is a stationary point, the stable manifold $\mathcal{M}_-(u_0)$ and the unstable manifold $\mathcal{M}_+(u_0)$, defined above, if not empty, are the union of trajectories defined for all time; thus they are invariant sets. Of particular interest for the understanding of the dynamics (see, for instance, J. Guckenheimer and P. Holmes [1]) are the *heteroclinic* orbits which go from the unstable manifold of a stationary point u_* to the stable manifold of another stationary point $u_{**} \neq u_*$; when $u_{**} = u_*$ such a curve is called a *homoclinic* orbit. The points belonging to a heteroclinic (or a homoclinic) orbit are called *heteroclinic* (or *homoclinic*) *points.* Note that if the operators $S(t)$ are one-to-one two stable (or two unstable) manifolds of distinct fixed points u_*, u_{**} cannot intersect, nor can $\mathcal{M}_-(u_*)$ (or $\mathcal{M}_+(u_*)$) intersect itself. However, the stable (or unstable) manifold of u_* can meet the unstable (or stable) manifold of u_{**} or of u_*, in which case they are the same and they consist of the heteroclinic or homoclinic curves.

Remark 1.1. If X is an invariant set and $u_0 \in X$, then assuming the injectivity of the operators $S(t)$, $S(t)u_0$ exists for all $t \in \mathbb{R}$. It is easy to see that $X_{u_0} = \{S(t)u_0, t \in \mathbb{R}\}$ is itself an invariant set and that two such sets never intersect unless they are identical. The sets X_{u_0}, $u_0 \in X$, then constitute a partition of X into invariant sets. Furthermore, the sets X_{u_0} are *minimal* invariant sets; indeed a proper subset of X_{u_0} cannot be itself invariant since an invariant set containing u_0 must contain $S(t)u_0$, $\forall t \in \mathbb{R}$.

Some other invariant sets that will be of particular interest to us are provided by the following lemma:

Lemma 1.1. *Assume that for some subset* $\mathscr{A} \subset H$, $\mathscr{A} \neq \varnothing$, *and for some* $t_0 > 0$, *the set* $\bigcup_{t \geq t_0} S(t)\mathscr{A}$ *is relatively compact in* H. *Then* $\omega(\mathscr{A})$ *is nonempty, compact, and invariant. Similarly, if the sets* $S(t)^{-1}\mathscr{A}$, $t \geq 0$, *are nonempty and for some* $t_0 > 0$, $\bigcup_{t \geq t_0} S(t)^{-1}\mathscr{A}$ *is relatively compact, then* $\alpha(\mathscr{A})$ *is nonempty, compact, and invariant.*

Proof. Since $\mathscr{A}$ is nonempty, the sets $\bigcup_{t \geq s} S(t)\mathscr{A}$ are nonempty for every $s \geq 0$. Hence the sets $\overline{\bigcup_{t \geq s} S(t)\mathscr{A}}$ are nonempty compact sets which decrease as s increases; their intersection, equal to $\omega(\mathscr{A})$, is then a nonempty compact set. By the characterization (1.5) of $\omega(\mathscr{A})$, it is easy to see that $S(t)\omega(\mathscr{A}) = \omega(\mathscr{A})$, $\forall t > 0$. If $\psi \in S(t)\omega(\mathscr{A})$, then $\psi = S(t)\varphi$, $\varphi \in \omega(\mathscr{A})$, and using (1.1), (1.4) and the sequences φ_n, t_n, provided by (1.5) we find:

$$S(t)S(t_n)\varphi_n = S(t + t_n)\varphi_n \to S(t)\varphi = \psi,$$

which shows that $\psi \in \omega(\mathscr{A})$. Conversely, if $\varphi \in \omega(\mathscr{A})$, we consider again the sequences φ_n, t_n, given by (1.5) and observe that for $t_n \geq t$, the sequence $S(t_n - t)\varphi_n$ is relatively compact in H. Thus there exists a subsequence $t_{n_i} \to \infty$ and $\psi \in H$ such that

$$S(t_{n_i} - t)\varphi_{n_i} \to \psi \qquad \text{as} \quad t \to \infty.$$

It follows from (1.5) that $\psi \in \omega(\mathscr{A})$, and by (1.1) and the continuity (1.4) of the operators $S(\tau)$,

$$S(t_{n_i})\varphi_{n_i} = S(t)S(t_{n_i} - t)\varphi_{n_i} \to S(t)\psi = \varphi \qquad \text{as} \quad n_i \to \infty;$$

thus $\varphi \in S(t)\omega(\mathscr{A})$.

The proof is totally similar for $\alpha(\mathscr{A})$. □

Remark 1.2. This lemma will often be used especially for ω-limit sets. It will provide examples of invariant sets whenever we can show that $\bigcup_{t \geq 0} S(t)\mathscr{A}$ is relatively compact. This set can consist of a single stationary solution u_* if all the trajectories starting from $\mathscr{A}$ converge to u_* as $t \to \infty$; or it can consist of the orbit of a time-periodic solution, or a time-quasi-periodic solution, or even a more complex set. To prove the main assumption that $\bigcup_{t \geq 0} S(t)\mathscr{A}$ is relatively compact, we usually show that this set is bounded if H is finite dimensional or, in infinite dimensions, we show that it is bounded in a space W compactly imbedded in H.

1.3. Absorbing Sets and Attractors

Definition 1.2. An attractor is a set $\mathscr{A} \subset H$ that enjoys the following properties:

(i) $\mathscr{A}$ is an invariant set ($S(t)\mathscr{A} = \mathscr{A}$, $\forall t \geq 0$).

(ii) $\mathscr{A}$ possesses an open neighborhood $\mathscr{U}$ such that, for every u_0 in $\mathscr{U}$, $S(t)u_0$ converges to $\mathscr{A}$ as $t \to \infty$:

$$\operatorname{dist}(S(t)u_0, \mathscr{A}) \to 0 \quad \text{as} \quad t \to \infty.$$

The distance in (ii) is understood to be the distance of a point to a set

$$d(x, \mathscr{A}) = \inf_{y \in \mathscr{A}} d(x, y),$$

$d(x, y)$ denoting the distance of x to y in H. If $\mathscr{A}$ is an attractor, the largest open set $\mathscr{U}$ that satisfies (ii) is called *the basin of attraction* of $\mathscr{A}$. Alternatively, we express (1.9) by saying that $\mathscr{A}$ attracts the points of $\mathscr{U}$. We will say that $\mathscr{A}$ uniformly attracts a set $\mathscr{B} \subset \mathscr{U}$ if

$$d(S(t)\mathscr{B}, \mathscr{A}) \to 0 \quad \text{as} \quad t \to \infty, \tag{1.9}$$

where $d(\mathscr{B}_0, \mathscr{B}_1)$ is now the semidistance of two sets $\mathscr{B}_0, \mathscr{B}_1$:[1]

$$d(\mathscr{B}_0, \mathscr{B}_1) = \operatorname*{Sup}_{x \in \mathscr{B}_0} \operatorname*{Inf}_{y \in \mathscr{B}_1} d(x, y). \tag{1.10}$$

The convergence in (1.9) is equivalent to the following: for every $\varepsilon > 0$, there exists t_ε such that for $t \geq t_\varepsilon$, $S(t)\mathscr{B}$ is included in $\mathscr{U}_\varepsilon$, the ε-neighborhood of $\mathscr{A}$ ($\mathscr{U}_\varepsilon$ = the union of all open balls of radius ε centered in $\mathscr{A}$). When no confusion can occur we simply say that $\mathscr{A}$ attracts $\mathscr{B}$. For instance, we say that $\mathscr{A}$ attracts the bounded sets (or compact sets) of $\mathscr{U}$ if $\mathscr{A}$ uniformly attracts each bounded set (or each compact set) of $\mathscr{U}$. An attractor $\mathscr{A}$ may or may not possess such a property.

In infinite dimensions, where we need to work with different topologies, we can consider sets $\mathscr{A}$ that are attractors in a space W, $W \subset H$: this means that $\mathscr{A} \subset W$, $S(t)\mathscr{A} = \mathscr{A}$, and (ii) is valid with the topology of W, i.e., $\mathscr{U}$ is open in W and the distance in (1.9) is that of W (W a metric space).[2]

A key concept in this study is that of global or universal attractors of a semigroup.

Definition 1.3. We say that $\mathscr{A} \subset H$ is a global (or universal) attractor for the semigroup $\{S(t)\}_{t \geq 0}$ if $\mathscr{A}$ is a compact attractor that attracts the bounded sets of H (and its basin of attraction is then all of H).

[1] We recall that the Hausdorff distance defined on the set of nonempty compact subsets of a metric space is defined by

$$\delta(\mathscr{B}_0, \mathscr{B}_1) = \max(d(\mathscr{B}_0, \mathscr{B}_1), d(\mathscr{B}_1, \mathscr{B}_0));$$

d is not a distance as $d(\mathscr{B}_0, \mathscr{B}_1) = 0$ implies only $\mathscr{B}_0 \subset \overline{\mathscr{B}}_1$; see, for instance, H. Federer [1].

[2] A stronger definition of an attractor is the following (G. Sell): $\mathscr{A}$ is an attractor if $\mathscr{A}$ is the ω-limit set of one of its open neighbors $\mathscr{U}_0$. The basin of attraction of $\mathscr{A}$ is the union of the open sets $\mathscr{U}_0$ such that $\mathscr{U}_0 \supset \mathscr{A}$ and $\omega(\mathscr{U}_0) = \mathscr{A}$.

However, this improvement in the definition is not essential in this book where the emphasis will be on the concept of global attractors defined hereafter.

It is easy to see that such a set is necessarily unique. Also such a set is maximal for the inclusion relation among the bounded attractors and among the bounded functional invariant sets. For this reason we will also call it the maximal attractor.

In order to establish the existence of attractors, a useful concept is the related concept of absorbing sets.

Definition 1.4. Let $\mathscr{B}$ be a subset of H and $\mathscr{U}$ an open set containing $\mathscr{B}$. We say that $\mathscr{B}$ is *absorbing* in $\mathscr{U}$ if the orbit of any bounded set of $\mathscr{U}$ enters into $\mathscr{B}$ after a certain time (which may depend on the set):

$$\begin{cases} \forall \mathscr{B}_0 \subset \mathscr{U}, \qquad \mathscr{B}_0 \text{ bounded,} \\ \exists t_1(\mathscr{B}_0) \text{ such that } S(t)\mathscr{B}_0 \subset \mathscr{B}, \qquad \forall t \geq t_1(\mathscr{B}_0). \end{cases} \tag{1.11}$$

We also say that $\mathscr{B}$ absorbs the bounded sets of $\mathscr{U}$.

The existence of a global attractor $\mathscr{A}$ for a semigroup $S(t)$ implies that of an absorbing set. Indeed, for $\varepsilon > 0$, let $\mathscr{V}_\varepsilon$ denote the ε-neighborhood of $\mathscr{A}$ (i.e., the union of open balls of radius ε centered on $\mathscr{A}$). Then, for any bounded set $\mathscr{B}_0$, $d(S(t)\mathscr{B}_0, \mathscr{A}) \to 0$ as $t \to \infty$; hence $d(S(t)\mathscr{B}_0, \mathscr{A}) \leq \varepsilon/2$ for $t \geq t(\varepsilon)$ and $S(t)\mathscr{B}_0 \subset \mathscr{V}_\varepsilon$ for such t's. This shows that $\mathscr{V}_\varepsilon$ is an absorbing set.

Conversely, we will show below that a semigroup which possesses an absorbing set and enjoys some other properties possesses an attractor.

Remark 1.3.

(i) We can extend the definition of an absorbing set and consider a set $\mathscr{B}$ which absorbs the points of $\mathscr{U}$ or the compact sets of $\mathscr{U}$ (i.e., (1.11) is satisfied with $\mathscr{B}_0 = \{u_0\}$, $u_0 \in \mathscr{U}$, or $\mathscr{B}_0 =$ a compact set, $\mathscr{B}_0 \subset \mathscr{U}$). These extensions will not be necessary for the applications we have in mind and we will use Definition 1.3 for the absorbing set.

(ii) The existence of an absorbing set is related to a *dissipativity* property for the dynamical system, and in the case of ordinary differential equations the expression "the semigroup $S(t)$ dissipates the bounded sets (or the compact sets or the points)" is also used to state that there exists a set $\mathscr{B}$ which absorbs the bounded sets or the compact sets or the points (see J.E. Billoti and J.P. La Salle [1], and J. Hale [1]). However, in infinite dimensions, for some systems which are physically considered as dissipative (e.g., the Navier–Stokes equations in dimension 3, see Chapter VII, Section 1, 2), the existence of an absorbing set is not known and thus it is not clear if this property is as good a definition of dissipativity as in finite dimensions.

If we consider a space $W \subset H$, then we can replace (in Definition 1.3) the bounded and open sets of H by that of W and we obtain the concept of sets which are absorbing in W.

We now show how to prove the existence of an attractor when the existence of an absorbing set is known. Further assumptions on the semigroup $S(t)$ are necessary at this point and we will make one of the two following assumptions (1.12), (1.13):

The operators $S(t)$ are *uniformly compact* for t large. By this we mean that for every bounded set $\mathscr{B}$ there exists t_0 which may depend on $\mathscr{B}$ such that

$$\bigcup_{t \geq t_0} S(t)\mathscr{B}$$

is relatively compact in H.[1] (1.12)

Alternatively, if H is a Banach space, we may assume that $S(t)$ is the perturbation of an operator satisfying (1.12) by a (nonnecessarily linear) operator which converges to 0 as $t \to \infty$. We formulate this assumption more precisely:

H is a Banach space and for every t, $S(t) = S_1(t) + S_2(t)$ where the operators $S_1(\cdot)$ are uniformly compact for t large (i.e., satisfy (1.12)) and $S_2(t)$ is a continuous mapping from H into itself such that the following holds:

For every bounded set $C \subset H$,

$$r_c(t) = \operatorname*{Sup}_{\varphi \in C} |S_2(t)\varphi|_H \to 0 \qquad \text{as} \quad t \to \infty. \tag{1.13}$$

Of course, if H is a Banach space, any family of operators satisfying (1.12) also satisfies (1.13) with $S_2 = 0$.

Theorem 1.1. *We assume that H is a metric space and that the operators $S(t)$ are given and satisfy* (1.1), (1.4) *and either* (1.12) *or* (1.13). *We also assume that there exists an open set $\mathscr{U}$ and a bounded set $\mathscr{B}$ of $\mathscr{U}$ such that $\mathscr{B}$ is absorbing in $\mathscr{U}$.*

Then the ω-limit set of $\mathscr{B}$, $\mathscr{A} = \omega(\mathscr{B})$, is a compact attractor which attracts the bounded sets of $\mathscr{U}$. It is the maximal bounded attractor in $\mathscr{U}$ (for the inclusion relation).

Furthermore, if H is a Banach space, if U is convex,[2] *and the mapping $t \to S(t)u_0$ is continuous from $\mathbb{R}_+$ into H, for every u_0 in H; then $\mathscr{A}$ is connected too.*

[1] In Theorem 1.1, the assumption (1.12) can be replaced by the weaker hypothesis:

$$\text{For some } t_1 > 0,\ S(t_1) \text{ is compact.} \tag{1.12'}$$

Indeed, in the course of the proof of Theorem 1.1, we need to know that $\bigcup_{t \geq t_*} S(t)\mathscr{B}$ is relatively compact in H when $\mathscr{B}$ is an absorbing set. Define t_0 by $S(t)\mathscr{B} \subset \mathscr{B}$, $\forall t \geq t_0$. Then, for $t \geq t_* = t_0 + t_1$, $S(t)\mathscr{B} = S(t_1)S(t - t_1)\mathscr{B}$ is included in $S(t_1)\mathscr{B}$ (as $S(t - t_1)\mathscr{B} \subset \mathscr{B}$), and thus $\bigcup_{t \geq t_*} S(t)\mathscr{B}$ is included in $S(t_1)\mathscr{B}$.

For all of the examples in Chapter III, the proof of (1.12) and (1.12′) are essentially the same. Also note that in (1.13) we assume that the operators $S_1(t)$ satisfy (1.12) and not (1.12′).

[2] With a suitable modification of the proof we may only assume that $\mathscr{U}$ is connected.

PROOF. We first prove the theorem in the simpler case where we assume (1.12).

Since the set $\bigcup_{t \geq t_0} S(t)\mathscr{B}$ is relatively compact, Lemma 1.1 applies and shows that $\omega(\mathscr{B})$ is a nonempty compact invariant set. We then prove that $\mathscr{A} = \omega(\mathscr{B})$ is an attractor in $\mathscr{U}$ and that it attracts the bounded sets of $\mathscr{U}$. We argue by contradiction and assume that for some bounded set $\mathscr{B}_0$ of $\mathscr{U}$, $\operatorname{dist}(S(t)\mathscr{B}_0, \mathscr{A})$ does not tend to 0 as $t \to \infty$; thus there exists $\delta > 0$ and a sequence $t_n \to \infty$ such that

$$\operatorname{dist}(S(t_n)\mathscr{B}_0, \mathscr{A}) \geq \delta > 0, \qquad \forall n.$$

For each n there exists $b_n \in \mathscr{B}_0$ satisfying

$$\operatorname{dist}(S(t_n)b_n, \mathscr{A}) \geq \frac{\delta}{2} > 0. \tag{1.14}$$

Since $\mathscr{B}$ is absorbing, $S(t_n)\mathscr{B}_0$, and hence $S(t_n)b_n$, belong to $\mathscr{B}$ for n sufficiently large (i.e., such that $t_n \geq t_1(\mathscr{B}_0)$). The sequence $S(t_n)b_n$ is relatively compact and possesses at least one cluster point β,

$$\beta = \lim_{n_i \to \infty} S(t_{n_i})b_{n_i} = \lim_{n_i \to \infty} S(t_{n_i} - t_1)S(t_1)b_{n_i}.$$

Since $S(t_1)b_n \in \mathscr{B}$, β belongs to $\mathscr{A} = \omega(\mathscr{B})$ and this contradicts (1.14).

The attractor $\mathscr{A}$ is maximal: if $\mathscr{A}' \supset \mathscr{A}$ is a larger bounded attractor then $\mathscr{A}' \subset \mathscr{B}$ since $S(t)\mathscr{A}' = \mathscr{A}'$ is included in $\mathscr{B}$ for t sufficiently large ($\mathscr{B}$ absorbing); consequently, $\omega(\mathscr{A}') = \mathscr{A}' \subset \omega(\mathscr{B}) = \mathscr{A}$.

Finally, the connectedness of $\mathscr{A}$ follows from Lemma 1.3 below, and the theorem is proved in this case. □

In order to prove the theorem under assumption (1.13) we first prove a slightly modified version of Lemma 1.1.

Lemma 1.2. *If the semigroup* $\{S(t)\}_{t \geq 0}$ *satisfies* (1.1), (1.4) *and* (1.12) *or* (1.13), *then for any bounded set* $\mathscr{B}_0$ *of* H, $\omega(\mathscr{B}_0)$ *is nonempty, compact, and invariant.*

PROOF. If we assume (1.12) then, as observed already, $\bigcup_{t \geq t_0} S(t)\mathscr{B}_0$ is relatively compact and Lemma 1.1 applies directly. When (1.13) is assumed, we make the following remark which will be used repeatedly:

> If φ_n is bounded and $t_n \to \infty$, then $S_2(t_n)\varphi_n \to 0$ and $S_1(t_n)\varphi_n$ is convergent if and only if $S(t_n)\varphi_n$ converges (in which case, the limits are equal). (1.15)

The norm of $S_2(t_n)\varphi_n$ is indeed bounded by $r_c(t)$ where C is the sequence φ_n, $n \in \mathbb{N}$. Thus $S_2(t_n)\varphi_n$ converges to 0, and

$$S(t_n)\varphi_n = S_1(t_n)\varphi_n + S_2(t_n)\varphi_n$$

converges if and only if $S_1(t_n)\varphi_n$ converges.

Using (1.15) and the characterization (1.5) of an ω-limit set we show that the ω-limit set of $\mathscr{B}_0$ for $S(t)$, $\omega(\mathscr{B}_0)$ is equal to the set

$$\omega_1(\mathscr{B}_0) = \bigcap_{s \geq 0} \overline{\bigcup_{t \geq s} S_1(t)\mathscr{B}_0}. \tag{1.16}$$

The definition of $\omega_1(\mathscr{B}_0)$ resembles that of an ω-limit set although S_1 is not a semigroup. A property similar to (1.5) is however valid: $\varphi \in \omega_1(\mathscr{B}_0)$ if and only if there exists a sequence $\varphi_n \in \mathscr{B}_0$ and a sequence $t_n \to \infty$, such that

$$S_1(t_n)\varphi_n \to \varphi \qquad \text{as} \quad n \to \infty.$$

Let us show that $\omega(\mathscr{B}_0) = \omega_1(\mathscr{B}_0)$. Assume that $\varphi \in \omega(\mathscr{B}_0)$; then by (1.5) there exist $\varphi_n \in \mathscr{B}_0$ and a sequence $t_n \to \infty$, such that

$$S(t_n)\varphi_n \to \varphi \qquad \text{as} \quad n \to \infty.$$

Due to (1.15), $S_1(t_n)\varphi_n$ also converges to φ as $n \to \infty$ and $\varphi \in \omega_1(\mathscr{B}_0)$. The inclusion $\omega_1(\mathscr{B}_0) \subset \omega(\mathscr{B}_0)$ is proved in a similar manner, and finally

$$\omega(\mathscr{B}_0) = \omega_1(\mathscr{B}_0).$$

As in the proof of Lemma 1.1 we observe that $\omega_1(\mathscr{B}_0)$ is nonempty and compact since the sets $\overline{\bigcup_{t \geq s} S_1(t)\mathscr{B}_0}$ are nonempty, closed, and decreasing and, by assumption, $\overline{\bigcup_{t \geq t_0} S_1(t)\mathscr{B}_0}$ is compact. Hence $\omega(\mathscr{B}_0)$ is nonempty and compact and there remains to show that $\omega(\mathscr{B}_0)$ is invariant for S.

The inclusion $S(t)\omega(\mathscr{B}_0) \subset \omega(\mathscr{B}_0)$, $\forall t > 0$, is proved exactly as in Lemma 1.1. Let $\psi \in S(t)\omega(\mathscr{B}_0)$, $\psi = S(t)\varphi$, $\varphi \in \omega(\mathscr{B}_0)$. Considering the sequences φ_n, t_n provided by (1.5), we obtain

$$S(t)S(t_n)\varphi_n = S(t + t_n)\varphi_n \to S(t)\varphi = \psi,$$

and thus $\psi \in \omega(\mathscr{B}_0)$. The opposite inclusion $\omega(\mathscr{B}_0) \subset S(t)\mathscr{B}_0$, $\forall t > 0$, necessitates a slightly different argument. Let $\varphi \in \omega(\mathscr{B}_0)$; by (1.5) there exists a sequence $\varphi_n \in \mathscr{B}_0$ and $t_n \to \infty$ such that

$$S(t_n)\varphi_n \to \varphi \qquad \text{as} \quad n \to \infty.$$

For $t_n \geq t$, the sequence $S(t_n - t)\varphi_n$ is of the form

$$S(t_n - t)\varphi_n = S_1(t_n - t)\varphi_n + S_2(t_n - t)\varphi_n .$$

The sequence $S_1(t_n - t)\varphi_n$ is relatively compact in H and contains a converging subsequence

$$S_1(t_{n_i} - t)\varphi_{n_i} \to \psi \qquad \text{as} \quad n_i \to \infty.$$

We infer from (1.15) that $S_2(t_{n_i} - t)\varphi_{n_i}$ converges to 0 and thus

$$S(t_{n_i} - t)\varphi_{n_i} \to \psi \qquad \text{as} \quad n_i \to \infty.$$

This implies that $\psi \in \omega(\mathscr{B}_0)$ and

$$\varphi = \lim_{n_i \to \infty} S(t)S(t_{n_i} - t)\varphi_{n_i} = S(t)\psi$$

belongs to $S(t)\omega(\mathscr{B}_0)$. □

Proof of Theorem 1.1 (*continued*). Lemma 1.2 shows that $\omega(\mathscr{B}) = \mathscr{A}$ is nonempty compact invariant. The fact that $\mathscr{A}$ attracts the bounded sets is proved by contradiction as above: the only difference arises when we prove that $S(t_n)b_n$ is relatively compact; this does not follow immediately from (1.12) but we observe instead that $S_1(t_n)b_n$ is relatively compact and (1.15) implies then that $S(t_n)b_n$ is relatively compact. We prove exactly as above that $\mathscr{A}$ is maximal. □

The proof of Theorem 1.1 will be complete after we show Lemma 1.3.

Lemma 1.3. *If $\mathscr{U}$ is an open convex set and $K \subset \mathscr{U}$ is a compact invariant set which attracts compact sets, then K is connected.*

Proof. The closed convex hull of K, $\overline{\text{conv } K} = \mathscr{B}$ is compact, connected and included in $\mathscr{U}$ and thus K attracts $\mathscr{B}$. If K were not connected we could find two open sets $\mathscr{U}_1$, $\mathscr{U}_2$ with $\mathscr{U}_1 \cap K \neq \emptyset$, $\mathscr{U}_2 \cap K \neq \emptyset$, $K \subset \mathscr{U}_1 \cup \mathscr{U}_2$ and $\mathscr{U}_1 \cap \mathscr{U}_2 = \emptyset$. Since $K \subset \mathscr{B}$, $K = S(t)K \subset S(t)\mathscr{B}$, but $\mathscr{B}$ is connected and since $S(t)$ is continuous, $S(t)\mathscr{B}$ is connected too. Thus $\mathscr{U}_i \cap S(t)\mathscr{B} \neq \emptyset$, $i = 1, 2$, and $\mathscr{U}_1 \cup \mathscr{U}_2$ does not cover $S(t)\mathscr{B}$. Hence, for every $t > 0$, there exists $x_t \in S(t)\mathscr{B}$, $x_t \notin \mathscr{U}_1 \cup \mathscr{U}_2$. Now consider the sequence x_n, $n \in \mathbb{N}$ $(t = n)$. This sequence is relatively compact; this is obvious under assumption (1.12), while under assumption (1.13) we write $x_n = S(n)y_n$, $y_n \in \mathscr{B}$, i.e.,

$$x_n = S_1(n)y_n + S_2(n)y_n.$$

The sequence $S_1(n)y_n$ is relatively compact and (1.15) implies that x_n is relatively compact too. Thus K attracts $\{x_n\}$ and the sequence x_n contains a subsequence (still denoted x_n) which converges to a point $x \in K$; necessarily, $x \notin \mathscr{U}_1 \cup \mathscr{U}_2$ and the contradiction follows. □

Remark 1.4. A slight generalization of Theorem 1.1 can be obtained by replacing hypothesis (1.13) by the weaker hypothesis

> The semi-group $\{S(t)\}_{t \geq 0}$ is *asymptotically compact*
> i.e., for every bounded sequence $\{x_k\}$ in H and every sequence (1.17)
> $t_k \to \infty$ $\{S(t_k)x_k\}_k$ is relatively compact in H.

This hypothesis appears in F. Abergel [2] (where it is called asymptotic precompactness), in O.A. Ladyzhenskaya [8] and in G. Sell and Y. You [3].

It is easy to see that (1.13) implies (1.17) but these conditions are in fact equivalent (see Remark 1.5 below). In most cases in the sequel of this book, we use (1.13). However in a few cases we will actually use (1.17) directly. The proof of Theorem 1.1 remains valid if we just replace (1.12) or (1.13) by (1.17). The proof that $\mathscr{A} = \omega(\mathscr{B})$ is the same as in the case of hypothesis (1.12) (including the proof of a modified version of Lemma 1.1), since all the sequences $\{t_n\}_n$ considered there are such that $t_n \to \infty$. The only change is for proving that $\mathscr{A} = \omega(\mathscr{B})$ is non-empty; this is true here because $\omega(\varphi) \neq \emptyset$, $\forall \varphi \in \mathscr{B}$ thanks to (1.17). We then show similarly that $\mathscr{A} = \omega(\mathscr{B})$ attracts any bounded set $\mathscr{B}_0 \subset H$, and Lemma 1.3 applies as well, showing that $\mathscr{A}$ is connected if $\mathscr{U}$ is convex.

Remark 1.5 (O. Goubet and I. Moise): The following remark supplements Remark 1.4 and it has also a unifying interest: it unifies different results concerning the existence of a global attractor. When H is a uniformly convex Banach space and if we assume the existence of a bounded absorbing set $\mathscr{B}$ as in Theorem 1.1, then the following three conditions are in fact equivalent:

(i) Assumption (1.13) (decomposition $S = S_1 + S_2$)
(ii) Assumption (1.17) (asymptotic compactness)
(iii) There exists a compact set $K \subset H$ such that $d(S(t)\mathscr{B}, K) \to 0$ as $t \to \infty$.

As we said hypothesis (ii) is used in O.A. Ladyzhenskaya [8] and in G. Sell and Y. You [3]; hypothesis (iii) is used by J. Hale [1] for the existence of attractor and hypothesis (i) is used extensively in this book.[1]

For the equivalence of (i), (ii) and (iii), the only nonobvious result is for showing that (ii) implies (i). Let us prove this result. Assuming (ii), Theorem 1.1 and Remark 1.4 imply that $\mathscr{A} = \omega(\mathscr{B})$ is the global attractor for the semigroup, and it is compact. The closed convex hull of $\mathscr{A}$, denoted K is also compact. We consider then the projector Π_K: $H \to K$ from H onto K (a closed convex nonempty set).

For every $\varphi \in H$, we set

$$\begin{cases} S_1(t)\varphi = \Pi_K(S(t)\varphi) \in K \ (compact) \\ S_2(t)\varphi = S(t)\varphi - \Pi_K(S(t)\varphi). \end{cases}$$

All the other properties in (1.13) being easy, we only have to prove that, for every bounded set $C \subset H$, $\operatorname{Sup}_{\varphi \in C} |S_2(t)\varphi|_H \to 0$ as $t \to \infty$. We argue by contradiction and assume the contrary: then there exists a bounded sequence $\{\varphi_j\}$ of H and $t_j \to \infty$ such that $|S_2(t_j)\varphi_j|_H \geq \delta > 0$, for some $\delta > 0$. Since $\mathscr{B}$ is absorbing and using (ii) we see that there exists a subsequence (still denoted t_j) such that $S(t_j)\varphi_j \to \varphi$ as $j \to \infty$. We have $\varphi \in \omega(\mathscr{B}) = \mathscr{A} \subset K$ and since Π_K is continuous, $S_1(t_j)\varphi_j = \Pi_K(S(t_j)\varphi_j) \to \Pi_K\varphi = \varphi$; therefore $S_2(t_j)\varphi_j \to 0$ and we obtain the contradiction. □

[1] In fact, J. Hale [1] uses a weaker version of property (iii) where the compact set may depend on $\mathscr{B}$, $K = K(\mathscr{B})$, and $\mathscr{B}$ is any bounded, positively invariant set. The corresponding property for any bounded set $\mathscr{B}$ not necessarily positively invariant is fully equivalent to asymptotic compactness independently of the existence of an absorbing set. Note also that, assuming (ii) or (iii) or either of the weaker forms of (iii) just described, the existence of a compact attractor is shown provided we assume furthermore that there exists a bounded absorbing set which absorbs the points, i.e., (1.11) holds when $\mathscr{B}_0$ is simply reduced to one point (see Remark 1.3(i)).

1.4. A Remark on the Stability of the Attractors

In this section we consider a family of perturbations $S_\eta(t)$ of the semigroup $S(t)$; these perturbations may be due to various causes: variations of some significant parameter, variations or errors in the data (like coefficients of the differential operators or of the driving forces); they can also come from approximation procedures, for instance, a finite-dimensional approximation necessitated by numerical computations. While the general invariant sets may be totally unstable with respect to perturbations,[1] the attractors possess some stability properties and we are going to describe such a stability result.

The semigroup $\{S(t)\}_{t\geq 0}$ is given satisfying the conditions (1.1), (1.4) above, and we assume that it possesses an attractor $\mathscr{A}$ which attracts an open neighborhood $\mathscr{U}$. The perturbed semigroups depend on a parameter η, $0 < \eta \leq \eta_0$, and are defined as follows. We consider a family of closed subspaces H_η of H which depend on the parameter η, $0 < \eta < \eta_0$, such that

$$\bigcup_{0<\eta\leq\eta_0} H_\eta \text{ is dense in } H.$$

For each $\eta > 0$, we consider a semigroup of operators $\{S_\eta(t)\}_{t\geq 0}$, where $S_\eta(t)$ maps H_η into itself and these operators satisfy the conditions similar to (1.1) and (1.4).

It is assumed that at the limit $\eta \to 0$, the operators $S_\eta(t)$ approximate S uniformly on the product of $\mathscr{U}$ by the bounded sets of $\mathbb{R}_+$:

$$\begin{cases} \text{For every compact interval } I \subset]0, +\infty[, \\ \delta_\eta(I) = \sup\limits_{u_0 \in \mathscr{U}\cap H_\eta} \sup\limits_{t\in I} d(S_\eta(t)u_0, S(t)u_0) \to 0 \quad \text{as} \quad \eta \to 0. \end{cases} \tag{1.18}$$

We also assume that

$$\begin{cases} \text{For every } \eta,\ S_\eta \text{ possesses an attractor } \mathscr{A}_\eta \text{ which attracts} \\ \mathscr{U}' \cap H_\eta, \text{ where } \mathscr{U}' \text{ is an open neighborhood of } \mathscr{A}_\eta \cup \mathscr{A}, \\ \text{independent of } \eta, \end{cases} \tag{1.19}$$

and we have

Theorem 1.2. *Under the above hypotheses, when $\eta \to 0$, $\mathscr{A}_\eta$ converges to $\mathscr{A}$ in the sense of the semidistance d (see* (1.10)*):*

$$d(\mathscr{A}_\eta, \mathscr{A}) \to 0 \quad \textit{as} \quad \eta \to 0. \tag{1.20}$$

PROOF. It is sufficient to show that for every $\varepsilon > 0$, there exists $\eta(\varepsilon)$ and $\tau(\varepsilon)$ such that, for $0 < \eta \leq \eta(\varepsilon)$ and $t \geq \tau(\varepsilon)$,

$$S_\eta(t)(\mathscr{U} \cap \mathscr{U}' \cap H_\eta) \subset \mathscr{V}_\varepsilon(\mathscr{A}), \tag{1.21}$$

[1] This is the case in particular for homoclinic and heteroclinic curves; see, for instance, the examples in Section 2.

where $\mathscr{V}_\varepsilon(\mathscr{A})$ denotes the ε-neighborhood of $\mathscr{A}$. Indeed, since $\mathscr{A}_\eta$ is the ω-limit set of $\mathscr{U}' \cap H_\eta$ (for S_η), (1.21) implies that $\mathscr{A}_\eta \subset \mathscr{V}_\varepsilon(\mathscr{A})$ and thus

$$d(\mathscr{A}_\eta, \mathscr{A}) \leq \varepsilon \qquad \text{if} \quad \eta \leq \eta(\varepsilon). \tag{1.22}$$

We now prove (1.21); there exists $r_0 > 0$ such that $\mathscr{U}'' = \mathscr{U} \cap \mathscr{U}' \supset \mathscr{V}_{r_0}(\mathscr{A})$. Assuming that $\varepsilon < r_0$, and since $\mathscr{A}$ attracts $\mathscr{U}$, we can find τ_0 such that, for $t \geq \tau_0 = \tau_0(\varepsilon)$,

$$S(t)\mathscr{U}'' \subset V_{\varepsilon/2}(\mathscr{A}).$$

We then apply (1.18) with $I = [\tau_0, 2\tau_0]$ and we find $\eta_0 = \eta_0(\varepsilon)$ such that

$$d(S_\eta(t)u_0, S(t)u_0) \leq \varepsilon/2$$

for every $t \in [\tau_0, 2\tau_0]$, $0 < \eta \leq \eta_0(\varepsilon)$, and $u_0 \in \mathscr{U}'' \cap H_\eta$. Thus we have proved (1.21) for every $0 < \eta \leq \eta_0(\varepsilon)$ and every $t \in [\tau_0, 2\tau_0]$. In order to establish (1.21) for every $t \geq 2\tau_0$ (and $\eta \in]0, \eta_0]$) we proceed by induction. We assume that (1.21) holds for $t \in [\tau_0, n\tau_0]$ and we prove it for $t \in [n\tau_0, (n+1)\tau_0]$. For such a t we write $t = (n-1)\tau_0 + \tau$, $\tau \in [\tau_0, 2\tau_0]$, and if $u_0 \in \mathscr{U}'' \cap H_\eta$ and $\eta \in]0, \eta_0]$:

$$S_\eta(t)u_0 = S_\eta(\tau)S_\eta((n-1)\tau_0)u_0. \tag{1.23}$$

By the induction assumption, $S_\eta((n-1)\tau_0)u_0$ belongs to $\mathscr{V}_\varepsilon(\mathscr{A}) \subset \mathscr{V}_{r_0}(\mathscr{A}) \subset \mathscr{U}''$ and then (1.21) (which is valid on the interval $[\tau_0, 2\tau_0]$) and (1.23) give $S_\eta(\tau)u_0 \in \mathscr{V}_\varepsilon(\mathscr{A})$. The result follows. □

2. Examples in Ordinary Differential Equations

In this section we consider classical examples for ordinary differential equations; we apply the results of Section 1 and we describe the corresponding invariant sets and attractors.

2.1. The Pendulum

We start with the very simple example of the pendulum, or equivalently, the problem of a point moving without friction on a vertical circle. This example is not typical of the evolutionary problem that we will consider since there is no friction and no dissipation of energy but it is still worth mentioning it.

If θ is the angle with the downward vertical, then the equation of motion is

$$\theta'' + \beta \sin\theta = 0, \tag{2.1}$$

$\beta = g/r$, g the gravity, r the radius of the circle, and $\theta' = d\theta/dt$, $\theta'' = d^2\theta/dt^2, \ldots$. This gives (after multiplication by θ' and integration):

$$\tfrac{1}{2}\theta'^2 - \beta\cos\theta = c_0 \qquad (= \tfrac{1}{2}\theta_0'^2 - \beta\cos\theta_0), \tag{2.2}$$

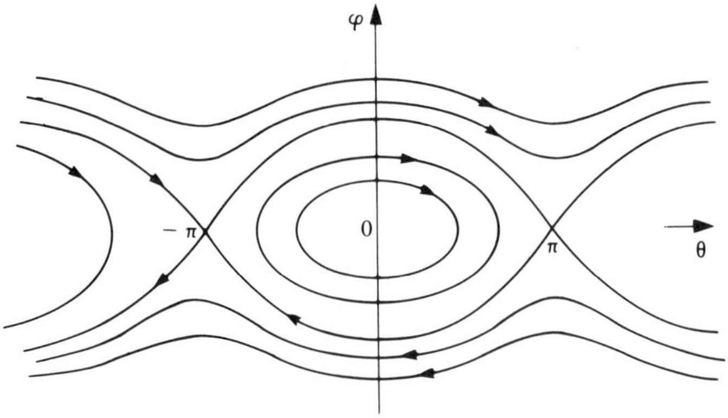

Figure 2.1. Trajectories in the phase space.

$\theta'_0 = \theta'(0)$, $\theta_0 = \theta(0)$. Setting $\varphi = \theta'$, we rewrite (2.1) as a first-order system

$$\begin{cases} \theta' = \varphi, \\ \varphi' = -\beta \sin \theta. \end{cases} \tag{2.3}$$

The equilibrium points are given by (2.1), $\sin \theta = 0$, $\theta = k\pi$, k integer, corresponding to the lowest and highest points A, B, on the circle. By studying the equation (2.2) in the (θ, φ) plane (the phase space), we see that if $|c_0| < \beta$, the motion is periodic around A (and the maximum value of $\theta(t)$ is θ_*, $\cos \theta_* = |c_0|/\beta$); if $|c_0| > \beta$, $\theta(t)$ increases (or decreases indefinitely). Finally, if $\beta = |c_0|$, the initial kinetic energy of the point is exactly the energy necessary to attain the highest point B and this point is reached in an infinite time. In this case, the trajectory is part of the homoclinic curve which connects the stable manifold of B with its unstable manifold.[1]

Except when $\beta = |c_0|$, the ω-limit set of a point (θ_0, φ_0) is precisely the corresponding orbit; when $\beta = |c_0|$, the ω-limit set is the point $(\theta = \pi, \varphi = 0)$. The invariant sets are the two equilibria points, the two homoclinic curves and the trajectories corresponding to the cases $|c_0| \neq \beta$. In this nondissipative case there is no absorbing set and no attractor.

The Pendulum with Friction

If we consider that the point is moving on the circle with a friction proportional to the velocity, then (2.1) is replaced by

$$\theta'' + \alpha\theta' + \beta \sin \theta = 0 \tag{2.4}$$

[1] We may identify $-\pi$ and π and consider that θ takes its values on the one-dimensional torus. We may also consider that θ takes its values in $\mathbb{R}$ in which case the curve joining the points $k\pi$, to $(k \pm 2)\pi$ (k odd) are heteroclinic curves.

(α, $\beta > 0$). After multiplication by θ' we obtain

$$\frac{d}{dt}\left(\frac{\theta'^2}{2} + \beta(1 - \cos\theta)\right) = -\alpha\theta'^2, \tag{2.5}$$

which shows that no periodic orbit exists. Setting $\varphi = \theta'$ we write the system

$$\begin{cases} \theta' = \varphi, \\ \varphi' = -\beta\sin\theta - \alpha\varphi. \end{cases} \tag{2.6}$$

The stationary points are again the points A and B ($\theta = 0$ and π). As usual, the stability of these points is studied by determining the eigenvalues at these points of the Jacobian matrix of the right-hand side of (2.6), i.e.,

$$\begin{pmatrix} 0 & 1 \\ -\beta\cos\theta & -\alpha \end{pmatrix}. \tag{2.7}$$

At the point B ($\theta = \pi$) we find two real eigenvalues of opposite signs. We thus have a one-dimensional stable and a one-dimensional unstable manifold; according to the usual terminology B is a saddle point. At the point A ($\theta = 0$) we find two real negative eigenvalues if $\alpha^2 - 4\beta \geq 0$ and two conjugate imaginary eigenvalues with a negative real part if $\alpha^2 - 4\beta < 0$; thus A is a node or a sink.

The phase portrait in the φ, θ plane follows from these considerations and also from a comparison with the case $\alpha = 0$ (Figure 2.1), the role of the term $-\alpha\varphi$ being to bring the curves closer to the axis $\varphi = 0$. Figure 2.2 gives a partial picture of the phase portrait. The stable manifolds of the points $(2k + 1)\pi$ are invariant sets (which are bounded or unbounded depending on whether we consider that $\theta \in$ one-dimensional torus $\mathbb{T}$ or $\theta \in \mathbb{R}$). The stable manifold of the points π and $-\pi$ intersects the axis $\theta = 0$, respectively, at

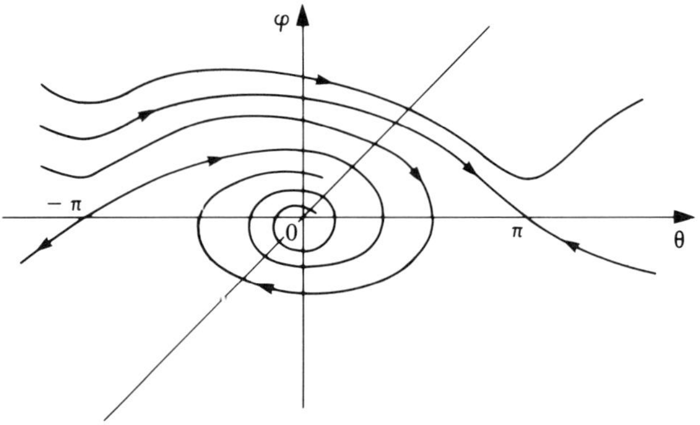

Figure 2.2. Partial picture of the trajectories in the phase space ($\alpha > 0$).

the points $\bar{\varphi}_1, -\bar{\varphi}_1$. The attractor (compact if $\theta \in \mathbb{T}$) consists of the points A, B and of the heteroclinic curves which connect them.

Remark 2.1. We can also assume that a constant torque is applied and this leads to the equation

$$\theta'' + \alpha\theta' + \beta \sin \theta = \delta. \tag{2.8}$$

The study is much more complicated in this case; see some indications and references in I. Guckenheimer and P. Holmes [1] and in J. Chandra [1]. As will be seen later, equations of this type also arise in physics; in particular, in connection with the Josephson junction (see Chapter IV).

2.2. The Minea System

This is a very simple system of ordinary differential equations in $\mathbb{R}^3$ introduced by Gh. Minea [1]. This system possesses some similarities with the Navier–Stokes equations (a quadratic nonlinearity, orthogonality property of the nonlinear term, see below and in Chapter III) and it gives indications about these equations.

The system reads

$$\begin{cases} u_1' + u_1 + \delta(u_2^2 + u_3^2) = 1, \\ u_2' + u_2 - \delta u_1 u_2 = 0, \\ u_3' + u_3 - \delta u_1 u_3 = 0, \end{cases} \tag{2.9}$$

where $\delta > 0$ plays the role of the Reynolds number in fluid mechanics. The right-hand side (which is that in the reference quoted above) can be replaced by any other vector of $\mathbb{R}^3$; it is also easy to imagine a similar system in dimension $n > 3$. Another interesting case is that where the linear terms u_1, u_2, u_3 in the left-hand side of (2.9) are replaced by $\lambda_1 u_1, \lambda_2 u_2, \lambda_3 u_3, \lambda_i > 0$; in this case, the analysis is however slightly more complicated than hereafter.

We first prove that $u(t) = (u_1(t), u_2(t), u_3(t))$ remains bounded as $t \to \infty$, and that there exists an absorbing set. By multiplying the jth equation by u_j and summing for $j = 1, 2, 3$, we find that the cubic terms disappear (this is the orthogonality property of the nonlinear term mentioned above) and there remains

$$\begin{gathered} \frac{1}{2}\frac{d}{dt}|u|^2 + |u|^2 = u_1 \le |u| \le \frac{1}{2}|u|^2 + \frac{1}{2}, \\ \frac{d}{dt}|u|^2 + |u|^2 \le 1, \\ |u(t)|^2 \le |u(0)|^2 \exp(-t) + 1 - \exp(-t). \end{gathered} \tag{2.10}$$

Hence

$$\limsup_{t\to+\infty} |u(t)| \le 1, \tag{2.11}$$

and there exists an *absorbing set*, for instance, any ball $B(0, \rho)$ centered at 0 of radius $\rho > 1$: if $\mathscr{B}_0$ is a bounded set of $H = \mathbb{R}^3$ included in a ball $B(0, R)$, then $S(t)\mathscr{B}_0 \subset B(0, \rho)$ for $t \ge t(\mathscr{B}_0)$,

$$t(\mathscr{B}_0) = \log \frac{R^2}{\rho^2 - 1}. \tag{2.12}$$

We also note that the balls $B(0, \rho)$ for $\rho \ge 1$ are positively invariant, i.e.,

$$S(t)B(0, \rho) \subset B(0, \rho), \qquad \forall t \ge 0.$$

The existence of a *maximal bounded attractor* follows from Theorem 1.1. Assumption (1.4) is easy, while (1.12) is a consequence of (2.11).

The study of the dynamics can be reduced to the two-dimensional case by observing that if $u_2(0) = 0$ (or $u_3(0) = 0$), then $u_2(t) = 0$, $\forall t > 0$ (or $u_3(t) = 0$, $\forall t > 0$). More generally, we can introduce the cylindrical coordinates r, θ, $u_2 = r \cos\theta$, $u_3 = r \sin\theta$ and deduce from the last two equations (2.10) that

$$\theta' = 0, \qquad \theta(t) = \theta(0), \qquad \forall t, \tag{2.13}$$

and

$$r' + r - \delta u_1 r = 0, \tag{2.14}$$

which is the same equation as that governing u_2 and u_3.

The fixed points are the point A ($u_1 = 1, u_2 = u_3 = 0$) and a circle of points C

$$u_1 = \frac{1}{\delta}, \qquad u_2^2 + u_3^2 = \frac{\delta - 1}{\delta^2},$$

which appear as soon as $\delta > 1$. As usual, the stability of the equilibrium points is determined by the study, at these points, of the eigenvalues of the Jacobian matrix of the mapping

$$\begin{pmatrix} u_1 \\ u_2 \\ u_3 \end{pmatrix} \mapsto \begin{pmatrix} -u_1 - \delta(u_2^2 + u_3^2) \\ -u_2 + \delta u_1 u_2 \\ -u_3 + \delta u_1 u_3 \end{pmatrix}.$$

We obtain the matrix

$$\begin{pmatrix} -1 & -2\delta u_2 & -2\delta u_3 \\ \delta u_2 & -1 + \delta u_1 & 0 \\ \delta u_3 & 0 & -1 + \delta u_1 \end{pmatrix}.$$

At A the eigenvalues are -1 and $-1 + \delta$ twice; A is stable for $\delta < 1$ and for

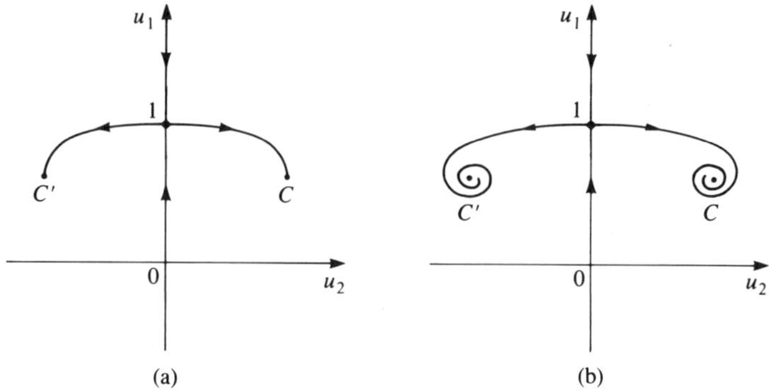

Figure 2.3. Trajectories in the (u_1, u_2) phase space: (a) for $1 < \delta < 9/8$; (b) for $\delta > 9/8$.

$\delta > 1$, A possesses a one-dimensional stable manifold—which is the axis $u_2 = u_3 = 0$—and a two-dimensional unstable manifold. At the points C there is one eigenvalue $\lambda = 0$—which due to (2.13) is not surprising—and two eigenvalues which are negative for $1 < \lambda < 9/8$ and are conjugate imaginary numbers for $\lambda > 9/8$ with a negative real part.

In Figure 2.3 the phase portrait is drawn in the (u_1, u_2) plane which is sufficient by (2.13). For $\delta < 1$, the attractor is reduced to the point A which attracts all the orbits. For $\delta > 1$, all the orbits converge to the point C or C' except if $u_2(0) = 0$, in which case the orbit converges to A; for $\delta > 9/8$ the orbits spiral around C. The maximal attractors consist of A, the points C (in the different planes $\theta =$ const.), and the heteroclinic curves which go from A to the points C.

2.3. The Lorenz Model

The third example in ordinary differential equations is the system proposed by E.N. Lorenz [1] as an evidence (or an indication) of the limits of predictability in weather prediction. This system is a three-mode Galerkin approximation (one in velocity and two in temperature) of the Boussinesq equations for fluid convection in a two-dimensional layer heated from below. The equations are

$$\begin{cases} x' = -\sigma x + \sigma y, \\ y' = rx - y - xz, \\ z' = -bz + xy, \end{cases} \tag{2.15}$$

where σ, r, b are three positive numbers representing, in this order, the Prandtl and Rayleigh numbers and the aspect ratio. After the change of variables

$x \mapsto x$, $y \mapsto y$, $z \mapsto z - r - \sigma$, the system becomes

$$\begin{cases} x' + \sigma x - \sigma y = 0, \\ y' + \sigma x + y + xz = 0, \\ z' + bz - xy = -b(r + \sigma). \end{cases} \tag{2.16}$$

We set $H = \mathbb{R}^3$, $u = (x, y, z)$, and considering the form (2.16) of the equation we prove that $u(t)$ remains bounded as $t \to \infty$ and that there exists an absorbing set; we write

$$\frac{1}{2}\frac{d}{dt}|u|^2 + \sigma x^2 + y^2 + bz^2 = b(r + \sigma)z$$

$$\leq (b - 1)z^2 + \frac{b^2}{4(b - 1)}(r + \sigma)^2,$$

$$\frac{d}{dt}|u|^2 + 2l|u|^2 \leq \frac{b^2}{b - 1}(r - \sigma)^2, \qquad l = \min(1, \sigma).$$

$$|u(t)|^2 \leq |u(0)|^2 \exp(-2lt) + \frac{b^2}{4l(b - 1)}(r + \sigma)^2(1 - \exp(-2lt)). \tag{2.17}$$

It follows from (2.17) that

$$\limsup_{t \to +\infty} |u(t)| \leq \rho_0, \qquad \rho_0 = \frac{b(r + \sigma)}{2\sqrt{l(b - 1)}}, \tag{2.18}$$

and that there is an *absorbing set*, for instance, any ball $B(0, \rho)$ centered at 0 of radius $\rho > \rho_0$: if $\mathscr{B}_0$ is a bounded set of $H = \mathbb{R}^3$, included in a ball $B(0, R)$, then $S(t)\mathscr{B}_0 \subset B(0, \rho)$ for $t \geq t(\mathscr{B}_0)$,

$$t(\mathscr{B}_0) = \frac{1}{2l} \log \frac{R^2}{\rho^2 - \rho_0^2}.$$

We also note that the balls $B(0, \rho)$, for $\rho > \rho_0$ and even $\rho = \rho_0$ are positively invariant, i.e., that $S(t)B(0, \rho) \subset B(0, \rho)$, $\forall t \geq 0$.

The existence of a *maximal bounded attractor* follows from Theorem 1.1. Assumption (1.4) is easy and (1.12) is a consequence of (2.17).

The structure of the maximal attractor is not well understood in general, and the following indications on its structure are based on numerical evidences and on the study of the stable and unstable manifolds of the equilibrium points. It is convenient to return to the variables x, y, z in (2.15). For $r \leq 1$, there is only one stationary point, the origin (0, 0, 0). For $r > 1$, there are three stationary points, the origin and the points C, C',

$$x = y = \pm\sqrt{b(r + 1)}, \qquad z = r - 1.$$

As usual the stability of the fixed points is made by determining the eigenvalues at these points of the Jacobian matrix of the right-hand side of (2.15), i.e.,

$$\begin{bmatrix} -\sigma & \sigma & 0 \\ r - z & -1 & -x \\ y & x & -b \end{bmatrix}. \tag{2.19}$$

For $r < 1$, the eigenvalues are all negative, the origin is a stable fixed point, and all the orbits converge to 0 as $t \to \infty$; the maximal attractor is reduced to that point. For $r > 1$, one eigenvalue is positive, the origin is unstable with a one-dimensional unstable manifold and a two-dimensional stable manifold. For $1 < r < r_h$,

$$r_h = \sigma(\sigma + b + 3)/(\sigma - b - 1)^1$$

the nontrivial stationary points C, C' are stable and every orbit $u(t)$ converges to one of these points. The maximal attractor consists of C, C' and the unstable manifold of the origin, which connects 0 to C and C'. There is an intermediate value r_1, $1 < r_1 < r_h$, such that for $1 < r < r_1$ the eigenvalues of the matrix (2.19) at C and C' are negative, while for $r_1 < r < r_h$ two are conjugate imaginary numbers with a negative real part; hence for $r_1 < r < r_h$, the unstable manifold of 0 circles around C and C'.

Finally, for $r > r_h$, all the eigenvalues of (2.19) at C, C' are negative and the *three stationary points are unstable*, and the numerical computations indicate the existence of a complicated (strange) attractor. Near C points come in along a line (the stable manifold of C) and go out along a two-dimensional surface (the unstable manifold of C): they spiral out because the eigenvalues with positive real part have a nonvanishing imaginary part. The situation is similar near C'. During the dynamics the point $u(t)$ goes to C along its stable manifold, then spiral out and go to C' along its stable manifold, etc.... For more details, see J. Guckenheimer and P. Holmes [3]; see also C. Foias and R. Temam [2].

3. Fractal Interpolation and Attractors

In this section we develop the concept of fractal interpolation. It introduces interesting and nontrivial attractors in situations which are different from those we will usually encounter in this book: H is not a Banach space, not even a vector space and t is discrete ($S(t) = S^n$, $t = n \in \mathbb{N}$). These cases arise in the theory of fractal interpolation where the purpose is to interpolate points

[1] We assume $\sigma > b + 1$ as in E.N. Lorenz [1]. In this reference, the numerical experiments are made with $\sigma = 10$, $r = 28$, $b = \frac{8}{3}$.

(x_i, y_i) by a continuous function which is not necessarily differentiable so that its graph could be a set of noninteger dimension (see M.F. Barnsley [1]).

3.1. The General Framework

Let K be a compact metric space endowed with the distance function $d(x, y)$, $x, y \in K$. We denote by H the set of all nonempty closed (compacts) subsets of K. We know that H is a compact metric space for the Hausdorff distance

$$\delta(\mathscr{A}, \mathscr{B}) = \operatorname{Max}\left\{\operatorname*{Sup}_{x\in\mathscr{A}}\operatorname*{Inf}_{y\in\mathscr{B}} d(x, y), \operatorname*{Sup}_{x\in\mathscr{B}}\operatorname*{Inf}_{y\in\mathscr{A}} d(x, y)\right\}, \tag{3.1}$$

which is in fact defined whenever $\mathscr{A}$ and $\mathscr{B}$ are subsets of K (compare with (1.10) and see H. Federer [1]).

We are given a set of continuous functions w_n from K into itself, $1 \le n \le N$, N given, and we associate with these functions the mapping S from H into itself defined by

$$S(\mathscr{A}) = \bigcup_{n=1}^{N} w_n(\mathscr{A}), \qquad \forall \mathscr{A} \in H. \tag{3.2}$$

As usual, an invariant set for S is a subset A of H such that

$$SA = A. \tag{3.3}$$

We can also consider attractors in the sense of Definition 1.2 and, actually, Theorem 1.1 applies and provides *the existence of a maximal attractor*: indeed assumptions (1.1) and (1.12) are obviously satisfied (H is compact) and we only have to check (1.4), i.e., the continuity of S in H; this follows from the relation

$$\delta(S\mathscr{B}_0, S\mathscr{B}_1) \le \eta(\delta(\mathscr{B}_0, \mathscr{B}_1)), \qquad \forall \mathscr{B}_0, \mathscr{B}_1 \in H, \tag{3.4}$$

where, for $r > 0$, $\eta(r) = \operatorname{Sup}_{j=1,\ldots,N} \eta_j(r)$ and

$$\eta_j(r) = \operatorname*{Sup}_{x, y \in K, d(x,y) \le r} d(w_j(x), w_j(y)). \tag{3.5}$$

The proof of (3.4) is elementary and we notice that η (like the η_j) is an increasing function and $\eta(r) \to 0$ as $r \to 0$ by the continuity assumption.

We note that if all the functions w_n are contractions, i.e., there exists q, $0 \le q < 1$, such that

$$d(w_n(x), w_n(y)) \le q \cdot d(x, y), \qquad \forall x, y \in K, \quad 1 \le n \le N,$$

then $\eta_n(r) = \eta(r) = q \cdot r$ and (3.4) becomes

$$\delta(S\mathscr{B}_0, S\mathscr{B}_1) \le q \cdot \delta(\mathscr{B}_0, \mathscr{B}_1), \qquad \forall \mathscr{B}_0, \mathscr{B}_1 \in H. \tag{3.6}$$

Hence S is a strict contraction in H and it is a general fact that, for a strict contraction, any bounded invariant set (in particular, the maximal attractor) is reduced to one point, namely, the unique fixed point of the mapping. Indeed, if X is compact and $SX = X$, there exists $x, y \in X$ such that

$$\begin{aligned}\text{diameter of } X &= \text{diameter of } SX \\ &= \delta(S(x), S(y)) \\ &\leq q\delta(x, y) \\ &\leq q \text{ diameter of } X.\end{aligned}$$

Necessarily, the diameter of X vanishes, X is reduced to one point which is obviously the fixed point of S.

Remark 3.1. Following M.F. Barnsley [1], we call a set $\{K; w_1, \dots, w_N\}$ an *iterative system*, and when the corresponding mapping S is contractive we will say that the iterative system is *hyperbolic.*

3.2. The Interpolation Process

We consider the set of data points $(x_i, y_i) \in \mathbb{R}$, $i = 0, \dots, N$, and we set $I = [x_0, x_N]$. As indicated above we are concerned with continuous functions $f: I \mapsto \mathbb{R}$ which interpolate the data according to

$$f(x_i) = y_i, \qquad i = 0, \dots, N. \tag{3.7}$$

We choose two finite numbers a, b, such that

$$-\infty < a < y_i < b < +\infty, \qquad \forall i = 0, \dots, N,$$

and we write $K = I \times [a, b]$ which is endowed with the distance

$$d((u, v), (u', v')) = \max\{|u - u'|, |v - v'|\}, \qquad \forall (u, v), (u', v') \in K.$$

Set $I_n = [x_{n-1}, x_n]$ and let $L_n: I \mapsto I_n$ be contractive homeomorphisms $(1 \leq n \leq N)$ such that

$$L_n(x_0) = x_{n-1}, \qquad L_n(x_N) = x_n, \qquad 1 \leq n \leq N, \tag{3.8}$$

$$|L_n(u) - L_n(v)| \leq l|u - v|, \qquad \forall u, v \in I, \tag{3.9}$$

for some l, $0 \leq l < 1$. Furthermore, we consider continuous mappings $F_n: K \mapsto [a, b]$ $(1 \leq n \leq N)$ such that

$$F_n(x_0, y_0) = y_{n-1}, \qquad F_n(x_N, y_N) = y_n \tag{3.10}$$

and for some q, $0 \leq q < 1$,

$$|F_n(u, v) - F_n(u, v')| \leq q|v - v'|, \qquad \forall u \in I, \quad \forall v, v' \in [a, b]. \tag{3.11}$$

We then consider the functions $w_n: K \mapsto K$ for $n = 1, \ldots, N$, defined by

$$w_n(x, y) = (L_n(x), F_n(x, y)), \tag{3.12}$$

and the corresponding mapping $S: H \mapsto H$ defined as in (3.2). The existence of a maximal attractor A for S follows as in Section 3.1, but the assumptions above do not imply that S is a contraction and it is not obvious that A reduces to a single point (= an element of H). Neverhteless, this is true and we can prove more.

Theorem 3.1. *Under the above assumptions and, in particular,* (3.2) *and* (3.8) *to* (3.12), *the maximal attractor A of S is reduced to the graph of a continuous function f from I into $[a, b]$ which satisfies* (3.7).

The theorem will be proved in Section 3.3 and we now conclude this section with some examples and comments.

EXAMPLES. The first example is that in which the L_n and F_n are affine functions

$$L_n(x) = x_{n-1} + \frac{(x_n - x_{n-1})}{(x_n - x_0)}(x - x_0) = a_n x + h_n, \tag{3.13}$$

$$F_n(x, y) = \alpha_n y + b_n x + k_n, \tag{3.14}$$

where α_n, b_n, k_n are chosen to ensure (3.10). More precisely, we choose $\alpha_n \in]-1, +1[$ and then determine recursively the b_n and k_n by the formula

$$b_n = (y_n - y_{n-1} - \alpha_n(y_N - y_0))/(x_N - x_0),$$

$$k_n = y_{n-1} - \alpha_n y_0 - b_n x_0, \qquad n = 1, \ldots, N.$$

The other example described here is that in which the L_n are still affine and defined by (3.13), while the F_n are written as

$$F_n(x, y) = \alpha_n y + q_n(x), \tag{3.15}$$

where again $\alpha_n \in]-1, +1[$, while q_n is a real continuous function on $[x_0, x_N]$ such that $F_n(x_0, y_0) = y_{n-1}$, $F_n(x_N, y_N) = y_n$; with $q_n(x) = b_n x + k_n$ we recover (3.14).

It is easy to see that assumptions (3.8)–(3.11) are satisfied in both examples. The reader is referred to M.F. Barnsley [1] for a numerical algorithm for the construction of the interpolation function and for related computer graphics.

Remark 3.2. The reader is also referred to M.F. Barnsley [1] for several other properties of these interpolation processes. Let us mention in particular a property for the moments: for the interpolation processes given in examples

(3.13)–(3.15), all the moments of f

$$\int_{x_0}^{x_N} x^n f(x)\,dx$$

and more generally the integrals

$$f_{n,m} = \int_{x_0}^{x_N} x^n (f(x))^m\,dx$$

can be evaluated explicitly in terms of integrals of the q_k's.

3.3. Proof of Theorem 3.1

(i) We can associate with A ($\subset H$) a subset of K, namely,

$$\mathscr{A} = \bigcup_{\mathscr{B} \in A} \mathscr{B} \subset K.$$

We observe that this set $\mathscr{A}$ is closed and is therefore itself an element of H: indeed if x_j is a sequence in $\mathscr{A}$, then $x_j \in \mathscr{B}_j$ for some $\mathscr{B}_j \in A$, $\forall j$. The sequence x_j is relatively compact in K and the sequence $\mathscr{B}_j$ relatively compact in H (K, H compacts), and there exists a subsequence j_k, and $x_* \in K$, $\mathscr{B}_* \in H$ such that

$$x_{j_k} \to x_* \quad \text{in } K,$$
$$\mathscr{B}_{j_k} \to \mathscr{B}_* \quad \text{in } H \qquad \text{as} \quad j_k \to \infty.$$

This implies $\{x_{j_k}\} \to \{x_*\}$ in H, and $\delta(\{x_{j_k}\}, \mathscr{B}_{j_k})$ $(= 0) \to \delta(\{x_*\}, \mathscr{B}_*)$; thus, since $\mathscr{B}_*$ is closed, $x_* \in \mathscr{B}_*$. On the other hand, A is closed in H, therefore $\mathscr{B}_* \in A$ so that $x_* \in \mathscr{A} = \bigcup_{\mathscr{B} \in A} \mathscr{B}$ and this set is closed. Clearly, (3.3) implies

$$S\mathscr{A} = \mathscr{A}. \tag{3.16}$$

The proof is now divided into several steps.

(ii) First, we show that the projection of A (or $\mathscr{A}$) onto I is the whole interval I,

$$P_x \mathscr{A} = I, \tag{3.17}$$

where P_x: $K \mapsto I$ is the projector $(a, b) \mapsto a$, $\forall (a, b) \in K$.

Let $\hat{I} = P_x \mathscr{A}$ and observe that since $A = SA = \bigcup_{n=1}^N w_n(A)$, $\hat{I} = P_x A$,

$$\hat{I} = P_x \bigcup_{n=1}^N w_n(A) = \bigcup_{n=1}^N L_n(\hat{I}). \tag{3.18}$$

We consider the general framework of Section 3.1 with $\{K; w_1, \dots, w_N\}$, replaced by $\{I; L_1, \dots, L_N\}$; we can define an operator Σ similar to S which maps the set H_I of nonempty compact subsets of I into itself:

$$\Sigma(G) = \bigcup_{n=1}^{N} L_n(G), \qquad \forall G \in H_I.$$

Due to (3.4) and (3.9), Σ is a strict contraction in H_I, hence with the terminology of Section 3.1 we have defined a hyperbolic iterative system. Now (3.18) expresses that $\Sigma\hat{I} = \hat{I}$, $\hat{I}$ is a fixed point of Σ and we have seen in Secton 3.1 that Σ possesses a unique attractor which is its unique fixed point; since $\Sigma I = I$, we conclude that $\hat{I} = I$.

(iii) We now prove that $\mathscr{A}$ is a graph above I, i.e., that

$$\Pi_\alpha \cap \mathscr{A} \text{ is reduced to one point}, \qquad \forall \alpha \in I, \tag{3.19}$$

where Π_α is the line $x = \alpha$ in $\mathbb{R}^2$.

We start with $\alpha = \alpha_0, \ldots, \alpha_N$, and first $\alpha = \alpha_0$. If $(x, y) \in E_0 = \Pi_{x_0} \cap \mathscr{A}$ then $x = x_0$ and since $A = SA$, $(x, y) \in S(x', y')$ for some $(x', y') \in \mathscr{A}$.

$$\begin{cases} (x, y) = w_n(x', y'), \qquad x = L_n(x'), \qquad y' = F_n(x', y') \\ \text{for some } (x', y') \in A \text{ and some } n \in \{1, \ldots, N\}. \end{cases}$$

Necessarily, $n = 1$, $x' = x_0$, and $(x', y') \in E_0$, which amounts to saying that $S^{-1}E_0 \subset E_0$, and therefore with (3.16) the restriction of $S \cap E_0$ to E_0, which reduces to w_1, satisfies $w_1(E_0) = E_0$. By (3.11) w_1 is a strict contraction, its only invariant set is its unique fixed point, and since $w_1(x_0, y_0) = (x_0, y_0)$ (see (3.10)) we conclude that $E_0 = \{x_0, y_0\}$.

We prove in exactly the same manner that $E_N = \Pi_{x_N} \cap \mathscr{A}$ is reduced to the point (x_N, y_N), and we then have to consider $E_n = \Pi_{x_n} \cap \mathscr{A}$, $n = 1, \ldots, N - 1$. Any point $(x, y) \in E_n$ is of the form $w_j(x', y')$ for some j and some $(x', y') \in \mathscr{A}$; necessarily, $j = n$ and $x' = x_N$ or $j = n + 1$ and $x' = x_0$. If $x' = x_N$, then $(x', y') \in E_N$ and $y' = y_N$, whereas if $x' = x_0$, then $(x', y') \in E_0$ and $y' = y_{N-1}$. Thus $y = F_n(x_N, y_N)$ or $F_{N+1}(x_0, y_0)$ and (due to (3.10)) these two values are the same and are equal to y_n. Finally,

$$\Pi_{x_n} \cap \mathscr{A} = (x_n, y_n), \qquad n = 0, \ldots, N. \tag{3.20}$$

We now establish (3.19). Consider

$$\lambda = \text{Max}\{|s - t|, (x, s) \in \mathscr{A}, (x, t) \in \mathscr{A} \text{ for some } x \in I\}.$$

It was observed at the beginning that $\mathscr{A}$ is compact and therefore this supremum is achieved at some pair of points $(\hat{x}, s)$, $(\hat{x}, t)$. The result follows from (3.20) if $\hat{x}$ is one of the points x_n; hence, we can assume that $\hat{x}$ belongs to the interval $]x_{n-1}, x_n[$ for some n. Then

$$(\hat{x}, s) = w_n(L_n^{-1}(\hat{x}), u),$$

$$(\hat{x}, t) = w_n(L_n^{-1}(\hat{x}), v),$$

where $(L_n^{-1}(\hat{x}), u)$ and $(L_n^{-1}(\hat{x}), v)$ belong to $\mathscr{A}$. Hence

$$\begin{aligned}\lambda = |s - t| &= |F_n(L_n^{-1}(\hat{x}), u) - F_n(L_n^{-1}(\hat{x}), v)| \\ &\leq \text{(by (3.9))} \\ &\leq q|u - v| \leq q\lambda.\end{aligned}$$

This implies $\lambda = 0$ and $\mathscr{A}$ is the graph of a function $f: I \mapsto [a, b]$ which obeys (3.7).

There remains to show that f is continuous. We observed in (i) that its graph $\mathscr{A}$ is closed and this guarantees that f is continuous.

The proof is complete. □

CHAPTER II
Elements of Functional Analysis

Introduction

In this chapter we present some general tools and basic results of functional analysis which will be used frequently in the sequel; most results are recalled without proofs. This chapter is not conceived as an introduction to the next chapters; *it is not necessary, and it is not suggested that this chapter be read completely* before reading the subsequent ones. It should be viewed as a technical reference to be read "locally" as needed.

Section 1 contains the description of various functions spaces including Sobolev spaces. In Section 2 we recall a few results on linear operators and we present concrete examples of linear operators, namely, those associated to a few classical elliptic boundary value problems. Sections 3 and 4 contain basic results on the existence, uniqueness, and regularity of solutions of linear evolution equations of the first order in time (Section 3) and of the second order in time (Section 4).

1. Function Spaces

1.1. Definition of the Spaces. Notations

We denote by Ω an open set of $\mathbb{R}^n$ with boundary Γ. In general, some regularity property of Ω will be assumed. We will assume that either

$$\Omega \text{ is Lipschitz,} \tag{1.1}$$

i.e., Γ is locally the graph of a Lipschitz function, or

$$\Omega \text{ is of class } \mathscr{C}^r, \qquad r \geq 1 \text{ to be specified,} \tag{1.2}$$

i.e., Γ is a manifold of dimension $n-1$ of class $\mathscr{C}^r$. In both cases we assume that Ω is locally on one side of Γ.

The generic point of $\mathbb{R}^n$ is denoted by $x = \{x_1, \dots, x_n\}$, $y = \{y_1, \dots, y_n\}, \dots$, the Lebesgue measure on $\mathbb{R}^n$ is denoted by $dx = dx_1 \dots dx_n$, and $d\Gamma$ represents the surface measure on Γ. For $T > 0$ given, we write $Q = \Omega \times (0, T)$, $\Sigma = \Gamma \times (0, T)$, $d\Sigma = d\Gamma\, dt$.

We denote by $\mathscr{C}(\Omega)$ (resp. $\mathscr{C}^k(\Omega)$, $k \in \mathbb{N}$ or $k = \infty$) the space of real continuous functions on Ω (resp. the space of k times continuously differentiable functions on Ω); $\mathscr{C}(\bar{\Omega})$ (resp. $\mathscr{C}^k(\bar{\Omega})$) represents the space of real continuous functions on $\bar{\Omega}$ (resp. the space of k times continuously differentiable functions on Ω). The spaces of real $\mathscr{C}^\infty$ functions on Ω with a compact support in Ω is denoted $\mathscr{C}_0^\infty(\Omega)$ or $\mathscr{D}(\Omega)$ as in the theory of distributions of L. Schwartz [1]; $\mathscr{D}'(\Omega)$ is the space of distributions on Ω.

For $1 \leq p < \infty$, $L^p(\Omega)$ is the space of (classes of) real functions on Ω which are L^p for the Lebesgue measure dx. It is a Banach space for the norm

$$\|u\|_{L^p(\Omega)} = \left(\int_\Omega |u(x)|^p \, dx \right)^{1/p}.$$

For $p = \infty$, $L^\infty(\Omega)$ is the space of (classes of) real functions on Ω which are measurable and essentially bounded; it is also a Banach space for the norm

$$\|u\|_{L^\infty(\Omega)} = \sup_{x \in \Omega} \operatorname{ess} |u(x)|.$$

When $p = 2$, $L^2(\Omega)$ is a Hilbert space for the scalar product

$$(u, v) = \int_\Omega u(x) \cdot v(x) \, dx, \tag{1.3}$$

the corresponding norm being denoted $|u|$,

$$|u| = \|u\|_{L^2(\Omega)} = \{(u, u)\}^{1/2}. \tag{1.4}$$

We will use the following notation for the partial differential derivatives of a function:

$$\begin{cases} D_i u = \dfrac{\partial u}{\partial x_i}, \qquad 1 \leq i \leq n, \\ D^\alpha u = D_1^{\alpha_1} \dots D_n^{\alpha_n} u = \dfrac{\partial^{\alpha_1 + \cdots + \alpha_n} u}{\partial x_1^{\alpha_1} \dots \partial x_n^{\alpha_n}}, \\ \alpha = \{\alpha_1, \dots, \alpha_n\} \in \mathbb{N}^n, \qquad [\alpha] = \alpha_1 + \cdots + \alpha_n. \end{cases} \tag{1.5}$$

We introduce the Sobolev spaces, which will be considered in more detail, in Subsection 1.1.2. For $m \in \mathbb{N}$, $1 \leq p \leq \infty$, $W^{m,p}(\Omega)$ is the space of functions u in $L^p(\Omega)$ whose distribution derivatives of order $\leq m$ are in $L^p(\Omega)$. This is a Banach space for the norm

$$\|u\|_{W^{m,p}(\Omega)} = \sum_{[\alpha]\leq m} \|D^\alpha u\|_{L^p(\Omega)}.$$

When $p = 2$ we write $W^{m,2}(\Omega) = H^m(\Omega)$ and this is a Hilbert space for the scalar product

$$((u, v))_{H^m(\Omega)} = \sum_{[\alpha]\leq m} (D^\alpha u, D^\alpha v).$$

Of course, similar spaces can be defined on any open set other than Ω, in particular, on $\mathcal{Q} = \Omega \times]0, T[\subset \mathbb{R}^{n+1}$.

If X is a Banach space, $1 \leq p < \infty$ and $-\infty \leq a < b \leq +\infty$, then $L^p(a, b; X)$ is the space of (classes of) L^p functions from (a, b) into X which is Banach for the norm

$$\|f\|_{L^p(a,b;X)} = \left(\int_a^b \|f(t)\|_X^p \, dt\right)^{1/p}.$$

For $p = \infty$, $L^\infty(a, b; X)$ is the space of (classes of) measurable functions from (a, b) into X which are essentially bounded; the space is Banach for the norm

$$\|f\|_{L^\infty(a,b;X)} = \sup_{t\in(a,b)} \text{ess}\, \|f(t)\|_X.$$

Similarly, when $-\infty < a < b < +\infty$, we denote by $\mathscr{C}([a, b]; X)$ the space of continuous functions from $[a, b]$ into X, and by $\mathscr{C}^k([a, b]; X)$, $k \in \mathbb{N}$, the space of k times continuously differentiable functions from $[a, b]$ into X. They are Banach spaces for the norms

$$\|f\|_{\mathscr{C}([a,b];X)} = \operatorname*{Sup}_{t\in[a,b]} \|f(t)\|_X,$$

$$\|f\|_{\mathscr{C}^k([a,b];X)} = \sum_{j=0}^{k} \left\|\frac{d^j f}{dt^j}\right\|_{\mathscr{C}([a,b];X)}.$$

1.2. Properties of Sobolev Spaces

The spaces $H^m(\Omega)$ and $W^{m,p}(\Omega)$ have already been defined; they contain $\mathscr{C}^\infty(\bar{\Omega})$ and even $\mathscr{C}^m(\bar{\Omega})$. The closure of $\mathscr{C}_0^\infty(\Omega)$ in $H^m(\Omega)$ (resp. $W^{m,p}(\Omega)$) is denoted by $H_0^m(\Omega)$ (resp. $W_0^{m,p}(\Omega)$). In particular, we will frequently use the spaces $H^1(\Omega)$, $H_0^1(\Omega)$

$$H^1(\Omega) = \{u \in L^2(\Omega), D_i u \in L^2(\Omega), 1 \leq i \leq n\},$$

$$H_0^1(\Omega) = \text{the closure of } C_0^\infty(\Omega) \text{ in } H^1(\Omega).$$

They are both Hilbert spaces for the scalar product

$$((u, v))_{H^1(\Omega)} = (u, v) + \sum_{i=1}^{n} (D_i u, D_i v).$$

When Ω is bounded, or at least bounded in one direction,[1] we have the Poincaré inequality

$$|u| \le c_0(\Omega)\left\{\sum_{i=1}^{n} |D_i u|^2\right\}^{1/2}, \qquad \forall u \in H_0^1(\Omega), \tag{1.6}$$

which implies that $H_0^1(\Omega)$ is a Hilbert space for the scalar product

$$((u, v)) = \sum_{i=1}^{n} (D_i u, D_i v) \tag{1.7}$$

and that the corresponding norm

$$\|u\| = \{((u, u))\}^{1/2}$$

is equivalent to the norm induced by $H^1(\Omega)$.

We now recall the important properties of Sobolev spaces; they are embodied in the density, embedding, compactness, and trace theorems. For the proofs the reader is referred to R.S. Adams [1], R. Dautray and J.L. Lions [1], J.L. Lions and E. Magenes [1], V.J. Maz'ja [1].

Density Theorems

If Ω is an open set of $\mathbb{R}^n$ of class $\mathscr{C}^m$, $m \ge 1$, $1 \le p < \infty$, then $\mathscr{C}^m(\bar{\Omega})$ is dense in $W^{m,p}(\Omega)$. (1.8)

This density result is also valid under weaker regularity assumptions on Ω; in particular, it is valid whenever

There exists a continuous linear prolongation operator $\Pi \in \mathscr{L}(W^{m,p}(\Omega), W^{m,p}(\mathbb{R}^n))$, $(\Pi u)(x) = u(x)$ for almost every $x \in \Omega$. (1.9)

This implies, when Ω is sufficiently regular, that

$$\begin{cases} W^{m,p}(\Omega) \text{ is dense in } W^{m-1,p}(\Omega), \\ H^m(\Omega) \text{ is dense in } H^{m-1}(\Omega). \end{cases}$$

Embedding Theorems

Let p be real, ≥ 1 and, for the moment only, let $m \ge 1$ be an integer. Assume that $\Omega \subset \mathbb{R}^n$ is an open set, not necessarily bounded, which is sufficiently regular, for instance, of class $\mathscr{C}^{m+1}$. Then

If $1/p - m/n > 0$, $W^{m,p}(\Omega) \subset L^q(\Omega)$, $1/q = 1/p - m/n$, *and the embedding is continuous*

$$\|u\|_{L^q(\Omega)} \le c(m, n, p, \Omega)\|u\|_{W^{m,p}(\Omega)}; \tag{1.10}$$

If $1/p - m/n = 0$, *the functions in* $W^{m,p}(\Omega)$ *are in* $L^q_{\text{loc}}(\Omega)$, *i.e.,* L^q *on any*

[1] That is, Ω is included in the set limited by two hyperplanes orthogonal to that direction.

bounded subdomain of Ω, $\forall q$, $1 \le q < \infty$,

$$\|u|_{\mathcal{O}}\|_{L^q(\mathcal{O})} \le c(m, n, \mathcal{O}, \Omega)\|u\|_{W^{m,p}(\Omega)}, \qquad \forall \mathcal{O} \text{ bounded} \subset \Omega; \tag{1.11}$$

If $1/p - m/n < 0$, then we write $m - n/p = k + \alpha$, $k =$ the integer part of $m - n/p$, $0 \le \alpha < 1$, *and the functions in* $W^{m,p}(\Omega)$ *belong to* $\mathscr{C}^{k,\alpha}(\mathcal{O})$, $\forall \mathcal{O} \subset \Omega$ *bounded, and*

$$|D^k u(x) - D^k u(y)| \le c(m, n, p, \mathcal{O}, \Omega)\|u\|_{W^{m,p}(\Omega)}|x - y|^\alpha, \quad \forall x, y \in \mathcal{O} \subset \Omega, \text{ bounded.} \tag{1.12}$$

The space $\mathscr{C}^{k,\alpha}(\mathcal{O})$ is the space of function in $\mathscr{C}^k(\bar{\mathcal{O}})$ whose derivatives of order k, $D^j u$, $[j] = k$, are Hölder continuous with exponent α

$$\operatorname*{Sup}_{\substack{x, y \in \mathcal{O} \\ x \ne y}} \frac{|D^j u(x) - D^j u(y)|}{|x - y|^\alpha} < \infty, \qquad \forall j, \quad [j] = k. \tag{1.13}$$

This is a Banach space when endowed with the norm

$$\|u\|_{\mathscr{C}^{k,\alpha}(\Omega)} = \|u\|_{\mathscr{C}^k(\bar{\mathcal{O}})} + \operatorname*{Sup}_{\substack{[j]=k \\ x, y \in \mathcal{O} \\ x \ne y}} \frac{|D^j u(x) - D^j u(y)|}{|x - y|^\alpha}. \tag{1.14}$$

When Ω is bounded the embedding of $W^{m,p}(\Omega)$ into $\mathscr{C}^{k,\alpha}(\Omega)$ is continuous according to (1.12).

The embedding properties above are valid when $\Omega = \mathbb{R}^n$. For $\Omega \ne \mathbb{R}^n$, they are valid for less regular domains than $\mathscr{C}^{m+1}$ domains; they are valid whenever (1.9) is satisfied.

When u belongs to $W_0^{m,p}(\Omega)$ the function $\tilde{u}$ (which is equal to u in Ω and to 0 in $\complement\Omega$) belongs to $W^{m,p}(\mathbb{R}^n)$, and hence the embedding properties above (and (1.10)–(1.12)) are valid for u in $W_0^{m,p}(\Omega)$ without any regularity assumption for Ω.

Compactness Theorems

Let Ω be any *bounded* set of class $\mathscr{C}^1$ (or satisfying (1.9), $m = 1$). Then

The embedding $W^{1,p}(\Omega) \subset L^{q_1}(\Omega)$ is compact for any q_1, $1 \le q_1 < \infty$, if $p \ge n$, and for any q_1, $1 \le q_1 < q$, $(1/q = 1/p - 1/n)$ if $1 \le p < n$. (1.15)

If $p > n$, the embedding $W^{1,p}(\Omega) \subset \mathscr{C}^{0,\alpha_1}(\Omega)$ is compact, $\forall \alpha_1 < \alpha = 1 - n/p$. (1.16)

With the same values of q_1 and α_1, the embeddings

$$\mathring{W}^{1,p}(\Omega) \subset L^{q_1}(\Omega) \qquad \text{if } p \le n,$$

$$\mathring{W}^{1,p}(\Omega) \subset \mathscr{C}^{0,\alpha_1}(\Omega) \quad \text{if } p \le n,$$

are compact for any bounded open set Ω. (1.17)

Trace Theorems

If Ω is smooth (for instance, of class $\mathscr{C}^{m+1}$) and a function u belongs to a Sobolev space $W^{m,p}(\Omega)$, then we can define the trace of u on Γ which coincides with the value of u on Γ when u is smooth (say $u \in \mathscr{C}^1(\bar{\Omega})$). More generally, if $\mathbf{v} = \{v_1, \ldots, v_n\}$ is the unit outward normal on Γ, we can also define the traces on Γ of certain normal derivatives $\partial^j u/\partial \mathbf{v}^j, j = 1, \ldots, m-1$. For the sake of simplicity, we restrict ourselves to the case $p = 2$; the trace theorems concerning the spaces $W^{m,p}(\Omega)$ will be recalled when needed.

For $m = 1$:

Let Ω be a bounded set of class $\mathscr{C}^1$. There exists a linear continuous operator $\gamma_0 \in \mathscr{L}(H^1(\Omega), L^2(\Gamma))$ such that

$$\gamma_0 u = u|_\Gamma, \qquad \forall u \in \mathscr{C}^1(\bar{\Omega}). \tag{1.18}$$

The space $L^2(\Gamma)$ is the space of (classes of) real functions which are L^2 on Γ for the measure $d\Gamma$. We also have

$$H_0^1(\Omega) = \{u \in H^1(\Omega), \gamma_0 u = 0\} = \text{the kernel of } \gamma_0. \tag{1.19}$$

The space $\gamma_0(H^1(\Omega))$ is not the whole space $L^2(\Gamma)$; it is denoted by $H^{1/2}(\Gamma)$ and this space can be endowed, for instance, with the quotient norm

$$\|\varphi\|_{H^{1/2}(\Gamma)} = \operatorname*{Inf}_{\gamma_0 u = \varphi} \|u\|_{H^1(\Omega)}, \tag{1.20}$$

which makes it a Hilbert space. Its dual is denoted by $H^{-1/2}(\Gamma)$.

The following are similar results for $m \geq 2$:

Let Ω be a bounded set of class $\mathscr{C}^{m+1}$. There exist linear continuous operators $\gamma_0, \ldots, \gamma_{m-1} \in \mathscr{L}(H^m(\Omega), L^2(\Gamma))$, such that

$$\gamma_j u = \left.\frac{\partial^j u}{\partial \mathbf{v}^j}\right|_\Gamma, \qquad \forall u \in \mathscr{C}^{m+1}(\bar{\Omega}), \quad j = 0, \ldots, m-1. \tag{1.21}$$

$$\begin{aligned} H_0^m(\Omega) &= \{u \in H^m(\Omega), \gamma_j u = 0, j = 0, \ldots, m-1\} \\ &= \text{the kernel of } \gamma_0 \times \cdots \times \gamma_{m-1}. \end{aligned} \tag{1.22}$$

The space $\gamma_j(H^m(\Omega))$, which is not $L^2(\Gamma)$, is denoted by $H^{m-j-1/2}(\Gamma)$ and can be endowed with a quotient norm similar to (1.20). Its dual is denoted by $H^{-m+j+1/2}(\Gamma)$. Alternative definitions and norms for the spaces $H^{r/2}(\Gamma)$, $r \in \mathbb{N}$, are indicated below; they show in particular that the definitions above are consistent.

1.3. Other Sobolev Spaces

We now indicate some properties of other Sobolev spaces.

Noninteger m

Let X and Y be two Hilbert spaces, $X \subset Y$, X dense in Y, the injection being continuous. The interpolation theory, which is briefly recalled in Section 2,[1] provides a family of Hilbert spaces denoted by $[X, Y]_\theta$, $0 \leq \theta \leq 1$, such that $[X, Y]_0 = X$, $[X, Y]_1 = Y$ and

$$X \subset [X, Y]_\theta \subset Y, \tag{1.23}$$

the injections in (1.23) being continuous, and each space dense in the succeeding one. The norm on $[X, Y]_\theta$ is such that

$$\|u\|_{[X,Y]_\theta} \leq c(\theta)\|u\|_X^{1-\theta}\|u\|_Y^\theta, \qquad \forall u \in X, \quad \forall \theta \in [0, 1]. \tag{1.24}$$

By interpolation between $H^m(\Omega)$ and $H^{m+1}(\Omega)$ we can thus define for $\alpha \in]0, 1[$

$$H^{m+\alpha}(\Omega) = [H^{m+1}(\Omega), H^m(\Omega)]_{1-\alpha}. \tag{1.25}$$

We can also define $H^{m+\alpha}(\Omega)$, $0 < \alpha < 1$, $m \in \mathbb{N}$, by appropriate interpolation between $H^{m+1}(\Omega)$ and $L^2(\Omega)$ (or $H^{m-1}(\Omega), \ldots$), and the definitions are equivalent (i.e., the corresponding spaces are isomorphic).

The density, embedding, compactness, and trace theorems quoted earlier extend without modifications to the spaces $H^s(\Omega) = W^{s,2}(\Omega)$, $s \in \mathbb{R}_+$. The compactness theorems (1.15)–(1.17) can be completed as follows:

If Ω is a *bounded* set of class $\mathscr{C}^1$ (or satisfying (1.9) with $m = 1$), the embedding of $H^{s_1}(\Omega)$ into $H^{s_2}(\Omega)$ is compact, $\forall s_1, s_2$, $0 \leq s_2 < s_1$. (1.26)

We can also define spaces which are intermediate between Sobolev spaces $W^{m,p}(\Omega)$, $W^{m+1,p}(\Omega)$, $m \in \mathbb{N}$, and thus obtain the family of spaces $W^{s,p}(\Omega)$, $s \in \mathbb{R}_+$, $p > 1$. However, there are different methods which produce different (nonequivalent) definitions of these spaces; see J.L. Lions [1], J.L. Lions and J. Peetre [1], and the bibliographical indications in the comments of J.L. Lions and E. Magenes [1] (Ch. I, §17). These results will be recalled only when needed.

Spaces $H^s(\Gamma)$

By using charts on Γ and partitions of unity subordinated to the covering of Γ by charts, we can define spaces $W^{m,p}(\Gamma)$, $H^m(\Gamma)$, $m \in \mathbb{N}$, or even $W^{s,p}(\Gamma)$, $H^s(\Gamma)$, $s \in \mathbb{R}$, $p > 1$. For $s = r/2$, $r \in \mathbb{N}$, the definitions are consistent with the definitions above.

[1] Most of Section 2 can be read independently of Section 1.

Space Periodic Functions

Sometimes we consider spaces of real functions which are defined on $\mathbb{R}^n$, and are periodic with period $L_j > 0$ in each direction $x_1, \dots, x_n$,

$$u(x + L_j e_j) = u(x), \qquad \forall x, j = 1, \dots, n, \tag{1.27}$$

where $e_1, \dots, e_n$, is the canonical basis of $\mathbb{R}^n$. In this case, we denote the period by Ω,

$$\Omega =]0, L_1[\times \cdots \times]0, L_n[,$$

and we denote by $H^m_{\text{per}}(\Omega)$ (or also $W^{m,p}_{\text{per}}(\Omega)$) the space of restrictions to Ω of periodic functions (in the sense of (1.27)) which are in $H^m(\mathcal{O})$ (or $W^{m,p}(\mathcal{O})$) on every bounded open set $\mathcal{O}$. Using the trace theorems, we can show that $H^1_{\text{per}}(\Omega)$ is the space of u in $H^1(\Omega)$ such that the traces $\gamma_0 u$ on the corresponding faces of Ω are equal;[1] this is a Hilbert subspace of $H^1(\Omega)$. Similarly, $H^m_{\text{per}}(\Omega)$ is the space of u in $H^m(\Omega)$ such that the traces $\gamma_j u$ on the corresponding faces of Ω are equal—if j is even or opposite, if j is odd (this is due to the opposite orientation of ν on the corresponding faces), $j = 1, \dots, m - 1$.

To study the spaces $H^m_{\text{per}}(\Omega)$ we can use the Fourier series expansion

$$u(x) = \sum_{k \in \mathbb{Z}^n} u_k \exp\left(2i\pi k \cdot \frac{x}{L}\right), \tag{1.28}$$

with $\bar{u}_k = u_{-k}$ (so that u is real) and

$$\frac{x}{L} = \left\{\frac{x_1}{L_1}, \dots, \frac{x_n}{L_n}\right\}, \qquad k \cdot \frac{x}{L} = k_1 \frac{x_1}{L_1} + \cdots + k_n \frac{x_n}{L_n}. \tag{1.29}$$

Then u is in $L^2(\Omega)$ if and only if

$$\|u\|^2_{L^2(\Omega)} = |\Omega| \sum_{k \in \mathbb{Z}^n} |u_k|^2 < \infty, \qquad |\Omega| = L_1 \dots L_n,$$

and u is in $H^s_{\text{per}}(\Omega)$, $s \in \mathbb{R}_+$, if and only if

$$\sum_{k \in \mathbb{Z}^n} (1 + |k|^2)^s |u_k|^2 < \infty. \tag{1.30}$$

Furthermore, the square root of expression (1.30) induces on $H^s_{\text{per}}(\Omega)$ a norm equivalent to that of $H^s(\Omega)$.

We denote by $\dot{L}^2(\Omega)$ and $\dot{H}^m(\Omega)$ the space of functions u in $L^2(\Omega)$ or $H^m(\Omega)$ such that

$$\int_\Omega u(x)\, dx = 0.$$

Then $\dot{H}^m_{\text{per}}(\Omega)$ or $\dot{H}^s_{\text{per}}(\Omega)$, $m \in \mathbb{N}$, $s \in \mathbb{R}_+$, is the space of functions u in $L^2(\Omega)$

[1] $\Gamma_j = \Gamma \cap \{x_j = 0\}$ and $\Gamma_{j+n} = \Gamma \cap \{x_j = L_j\}$ are called corresponding faces of Ω (or Γ). Two corresponding points of Γ are two points with the same coordinates, except the jth ones which are equal to 0 and L_j ($j = 1, \dots, n$).

satisfying (1.28), (1.30), and

$$u_0 = \frac{1}{|\Omega|} \int_\Omega u(x)\, dx = 0. \tag{1.31}$$

We have on $\dot{H}^1_{\text{per}}(\Omega)$ a Poincaré inequality similar to (1.6)

$$|u| \le c_0'(\Omega) \|u\|, \qquad \forall u \in \dot{H}^1_{\text{per}}(\Omega), \tag{1.32}$$

and this shows that $\dot{H}^1_{\text{per}}(\Omega)$ is Hilbertian for the scalar product $((\cdot, \cdot))$ defined in (1.7), and $\|u\| = \{((u, u))\}^{1/2}$ is a norm on this space equivalent to that induced by $H^1(\Omega)$.

1.4. Further Properties of Sobolev Spaces

We now describe some further properties of Sobolev spaces which will be used on certain occasions.

First is the following result due to J. Deny and J.L. Lions [1]:

If Ω is bounded and Lipschitz and if a distribution u on Ω has all its first derivatives in $L^2(\Omega)$, then u belongs to $L^2(\Omega)$ (i.e., $u \in H^1(\Omega)$), and there exists a constant c depending on Ω such that

$$|u|_{L^2(\Omega)/\mathbb{R}} = \operatorname*{Inf}_{k \in \mathbb{R}} |u + k|_{L^2(\Omega)} \le c(\Omega) |\text{grad } u|_{L^2(\Omega)^n}. \tag{1.33}$$

By reiteration of this result we see that

If Ω is bounded and Lipschitz and if a distribution u on Ω has all its derivatives of order m in $L^2(\Omega)$, then u belongs to $H^m(\Omega)$, and there exists a constant c depending only on Ω such that

$$\begin{aligned} |u|_{H^{m-1}(\Omega)/\mathcal{P}_{m-1}} &= \operatorname*{Inf}_{q \in \mathcal{P}_{m-1}} |u + q|_{H^{m-1}(\Omega)} \\ &\le c(\Omega) \sum_{[\alpha]=m} |D^\alpha u|_{L^2(\Omega)}, \end{aligned}$$

where $\mathcal{P}_{m-1}$ is the space of polynomials of degree $\le m - 1$. (1.34)

Generalized Poincaré Inequalities

Based on (1.33), other forms of the Poincaré inequality which generalize (1.6), (1.32) can be obtained, in $H^1(\Omega)$, and in $H^m(\Omega)$, $m > 1$. For instance, for $H^1(\Omega)$,

Let Ω be a bounded and Lipschitz set in $\mathbb{R}^n$, and let p be a continuous seminorm on $H^1(\Omega)$ which is a norm on the constants ($p(a) = 0$, $a \in \mathbb{R} \Rightarrow a = 0$). Then there exists a constant c depending only on Ω such that

$$|u|_{L^2(\Omega)} \le c(\Omega)\{|\text{grad } u|_{L^2(\Omega)^n} + p(u)\}, \qquad \forall u \in H^1(\Omega). \tag{1.35}$$

Equivalently, (1.35) expresses the fact that

$$|\text{grad } u|_{L^2(\Omega)^n} + p(u) \tag{1.36}$$

is a norm on $H^1(\Omega)$ which is equivalent to the usual norm. We can take, for instance, the following expressions for p:

$$\begin{cases} p(u) = \left| \int_\Omega u(x)\, dx \right|, \\ p(u) = |u|_{L^2(\omega)}, \quad \omega \subset \Omega, \quad \text{meas } \omega > 0, \\ p(u) = \left\{ \int_\Gamma |\gamma_0 u|^2\, dt \right\}^{1/2}, \quad \text{or} \\ p(u) = \left\{ \int_{\Gamma_0} |\gamma_0 u|^2\, d\Gamma \right\}^{1/2}, \quad \Gamma_0 \subset \Gamma, \quad \text{meas } \Gamma_0 > 0, \quad \text{etc.} \ldots \end{cases} \tag{1.37}$$

Similarly, for $H^m(\Omega)$, $m > 1$:

Let Ω be a bounded and Lipschitz set in $\mathbb{R}^n$, and let p be a continuous seminorm on $H^m(\Omega)$ which is a norm on $\mathscr{P}_{m-1}$ ($p(\varphi) = 0$, $\varphi \in \mathscr{P}_{m-1} \Rightarrow \varphi = 0$). Then there exists a constant c depending only on Ω such that

$$|u|_{H^{m-1}(\Omega)} \le c \left\{ \sum_{[\alpha]=m} |D^\alpha u|_{L^2(\Omega)} + p(u) \right\}, \qquad \forall u \in H^m(\Omega). \tag{1.38}$$

Equivalently, (1.38) expresses the fact that

$$\sum_{[\alpha]=m} |D^\alpha u|_{L^2(\Omega)} + p(u) \tag{1.39}$$

is a norm on $H^m(\Omega)$ which is equivalent to the usual norm. Several norms or seminorms p similar to those in (1.37) can be thought of.

Agmon's Inequalities

If $\Omega \subset \mathbb{R}^n$ and $s = n/2$, then we are in the limit case for Sobolev embeddings (see Sections 1.1 and 1.3) and $H^s(\Omega) \subset L^p(\Omega)$, $\forall p < \infty$, but $H^s(\Omega)$ is not included in $L^\infty(\Omega)$. The following inequalities, which would follow from $H^s(\Omega) \subset L^\infty(\Omega)$ (if this were true), and interpolation inequalities are however valid and due to S. Agmon [1]:

Assuming that $\Omega \subset \mathbb{R}^n$ is of class $\mathscr{C}^n$, there exists a constant c depending only on Ω such that

$$|u|_{L^\infty(\Omega)} \le \begin{cases} c\,|u|^{1/2}_{H^{(n/2)-1}(\Omega)}\,|u|^{1/2}_{H^{(n/2)+1}(\Omega)}, & \forall u \in H^{(n/2)+1}(\Omega) \quad \text{if } n \text{ is even}, \\ c\,|u|^{1/2}_{H^{(n-1)/2}(\Omega)}\,|u|^{1/2}_{H^{(n+1)/2}(\Omega)}, & \forall u \in H^{(n+1)/2}(\Omega) \quad \text{if } n \text{ is odd}. \end{cases} \tag{1.40}$$

Other inequalities of this type (involving other spaces H^m) are also valid.

Multiplicative Algebra

We have

If $\Omega \subset \mathbb{R}^n$ is of class $\mathscr{C}^m$, and $m > n/2$, then $H^m(\Omega)$ is a multiplicative algebra. There exists a constant c depending only on Ω such that if $u, v \in H^m(\Omega)$, then $u \cdot v \in H^m(\Omega)$ and

$$|u \cdot v|_{H^m(\Omega)} \leq c(\Omega) |u|_{H^m(\Omega)} |v|_{H^m(\Omega)}. \tag{1.41}$$

Truncations

We conclude this section by recalling the definition of the truncation operators and their properties.

If $s \in \mathbb{R}$, then we write classically

$$s_+ = \max(s, 0), \qquad s_- = \max(-s, 0),$$

$$s = s_+ - s_-, \qquad |s| = s_+ + s_-.$$

If u is a real function on Ω, we define the functions u_+ and u_-,

$$u_+(x) = (u(x))_+, \qquad u_-(x) = (u(x))_-, \qquad x \in \Omega.$$

If $u \in L^2(\Omega)$ (or $L^p(\Omega)$, $1 \leq p \leq \infty$), it is clear that u_+, u_- are in $L^2(\Omega)$ (or $L^p(\Omega)$, $1 \leq p \leq \infty$) with

$$\|u_+\|_{L^p(\Omega)} \leq \|u\|_{L^p(\Omega)}. \tag{1.42}$$

If $u \in H^1(\Omega)$ (or $W^{1,p}(\Omega)$, $1 \leq p \leq \infty$), then u_+, u_- are also in $H^1(\Omega)$ (or $W^{1,p}(\Omega)$) and, furthermore, according to G. Stampacchia [1] for a.e. $x \in \Omega$:

$$\frac{\partial u_+}{\partial x_i}(x) = \begin{cases} \dfrac{\partial u}{\partial x_i}(x) & \text{if } u(x) > 0, \\ 0 & \text{if } u(x) \leq 0. \end{cases}$$

Hence

$$\left\| \frac{\partial u_+}{\partial x_i} \right\|_{L^p(\Omega)} \leq \left\| \frac{\partial u}{\partial x_i} \right\|_{L^p(\Omega)}, \tag{1.43}$$

$$\|u_+\|_{W^{1,p}(\Omega)} \leq \|u\|_{W^{1,p}(\Omega)}, \tag{1.44}$$

and similar results hold, of course, for u_-.

2. Linear Operators

We recall a few facts about linear operators associated with a bilinear form, and give a few examples of such operators associated with classical elliptic boundary-value problems.

2.1. Bilinear Forms and Linear Operators

Let V be a Hilbert space (scalar product $((u, v))$, norm $\|u\|$) with dual V' (norm $\|\cdot\|_*$).[1] We are given a bilinear continuous form a on V. We can associate with a a linear continuous operator A from V into V', $A \in \mathscr{L}(V, V')$, as follows: for every $u \in V$, the application

$$v \mapsto a(u, v)$$

from V into $\mathbb{R}$ is linear continuous, and thus defines an element $\xi = \xi_u \in V'$. We then denote by A the mapping

$$u \mapsto \xi_u$$

from V into V', and it follows from the properties of a that A is linear continuous, and if (by the continuity of a)

$$|a(u, v)| \leq M \|u\| \|v\|, \qquad \forall u, v \in V, \tag{2.1}$$

then

$$\|A\|_{\mathscr{L}(V,V')} \leq M. \tag{2.2}$$

Conversely, to a given linear continuous operator $A \in \mathscr{L}(V, V')$ we can associate a bilinear continuous form a on V by setting

$$a(u, v) = \langle Au, v \rangle, \qquad \forall u, v \in V, \tag{2.3}$$

where $\langle \cdot, \cdot \rangle$ is the scalar product between V and V'.

If $a(u, v) = ((u, v))$ is the scalar product of V, then $A = \Lambda$ is the canonical isomorphism of V onto V'. In the general case, we will assume that a is coercive, i.e.,

$$\exists \alpha > 0, \qquad a(u, u) \geq \alpha \|u\|^2, \qquad \forall u \in V. \tag{2.4}$$

We then have the Lax–Milgram theorem:

Theorem 2.1. *If a is a bilinear continuous coercive form on V then A is an isomorphism from V onto V'.*

Proof. In order to show that A is onto, we must solve

$$Au = f \tag{2.5}$$

for f given in V', and u unknown in V. But for any $\rho > 0$, (2.5) is equivalent to the equation in V

$$u = u - \rho \Lambda^{-1}(Au - f). \tag{2.6}$$

[1] That is, for every $l \in V'$, $\|l\|_* = \operatorname*{Sup}_{\substack{u \in V \\ u \neq 0}} \dfrac{|\langle l, u \rangle|}{\|u\|}$

The mapping $u \mapsto Tu = u - \rho\Lambda^{-1}(Au - f)$ is affine continuous from V into itself. For every u_1, u_2 in V we have

$$\begin{aligned}\|Tu_1 - Tu_2\|^2 &= \|u_1 - u_2 - \rho\Lambda^{-1}A(u_1 - u_2)\|^2 \\ &= \|u_1 - u_2\|^2 + \rho^2\|\Lambda^{-1}A(u_1 - u_2)\|^2 \\ &\quad - 2\rho((u_1 - u_2, \Lambda^{-1}A(u_1 - u_2))) \\ &= \|u_1 - u_2\|^2 + \rho^2\|A(u_1 - u_2)\|_*^2 - 2\rho a(u_1 - u_2, u_1 - u_2) \\ &\leq \text{(with (1.34), (1.38))} \\ &\leq (1 + \rho^2 M^2 - 2\rho\alpha)\|u_1 - u_2\|^2. \end{aligned} \tag{2.7}$$

Thus if $\rho < 2\alpha/M^2$, T is a strict contraction in V and has a unique fixed point u which is the solution of (2.6), i.e., (2.5). This defines A^{-1}; the continuity of A^{-1} then follows from the relations ($u = A^{-1}f$):

$$\alpha\|u\|^2 \leq a(u, u) = \langle Au, u\rangle = \langle f, u\rangle \leq \|f\|_*\|u\|. \tag{2.8}$$

Hence

$$\|A^{-1}\|_{\mathscr{L}(V',V)} \leq \frac{1}{\alpha}, \tag{2.9}$$

and the theorem is proved. □

In the applications to partial differential equations, another space H is provided such that, $V \subset H$, V dense in H, the injection being continuous. The scalar product and the norm in H are denoted (u, v), $|u|$. By duality, if H' is the dual of H, then the adjoint i^* of the identity is injective, $i^*(H')$ is dense in V', and we can identify H' to a dense subspace of V'. In general, but not always, it is also convenient to identify H to its dual H', thanks to the Riesz representation theorem and we obtain

$$V \subset H \equiv H' \subset V', \tag{2.10}$$

where each space is dense in the following, the injections being continuous. We observe that

$$\langle f, v\rangle = (f, v), \qquad \forall f \in H, \quad \forall v \in V. \tag{2.11}$$

The bilinear form a being continuous and coercive, we define the domain of A in H as

$$D(A) = \{u \in V, Au \in H\}.$$

This allows us to consider A as a linear unbounded operator in H with domain $D(A)$. Because of (2.3), (2.4), the operator A is *strictly positive*

$$\langle Au, u\rangle = a(u, u) \geq \alpha\|u\|^2 > 0, \qquad \forall u \neq 0. \tag{2.12}$$

The space $D(A)$ can be endowed with the graph norm $\{|Au|^2 + |u|^2\}^{1/2}$

or more simply with the norm $|Au|$ which is equivalent: indeed, we infer from (2.8):

$$\begin{aligned} \|u\| &\le \frac{1}{\alpha}\|Au\|_* \le \frac{c_1}{\alpha}|Au|, \\ |u| &\le \frac{c_1^2}{\alpha}|Au|, \end{aligned} \tag{2.13}$$

where c_1 is the norm of the injection of V into H which is also that of the injection of H into V'

$$|u| \le c_1 \|u\|, \qquad \forall u \in V. \tag{2.14}$$

It is easy to see that $D(A)$ is a Hilbert space for either norm, and that A is also an isomorphism of $D(A)$ onto H.

Spectral Properties

When the form a is symmetric the operator A is self-adjoint (from V into V' and as an unbounded operator in H)

$$\langle Au, v\rangle = \langle Av, u\rangle = a(u, v), \qquad \forall u, v \in V. \tag{2.15}$$

Its inverse A^{-1} is also self-adjoint (in H).

Considering A as a strictly positive self-adjoint unbounded operator in H we can utilize the spectral theory (see, for instance, F. Riesz and B.S. Nagy [1], K. Yosida [1]) which allows us, for instance, to define the powers A^s of A, for $s \in \mathbb{R}$ (see below). We restrict ourselves to the case where

$$\text{The injection of } V \text{ in } H \text{ is compact.} \tag{2.16}$$

In this case, A^{-1} can be considered as a self-adjoint compact operator in H, and we can use the elementary spectral theory of self-adjoint compact operators in a Hilbert space (see, for instance, R. Courant and D. Hilbert [1]). We infer that there exists a complete orthonormal family of H, $\{w_j\}_{j \in \mathbb{N}}$ made of eigenvectors of A

$$A^{-1}w_j = \mu_j w_j, \qquad \forall j \in \mathbb{N},$$

where the sequence μ_j is decreasing and tends to 0. It is clear that $w_j \in D(A)$, $\forall j$, and setting $\lambda_j = \mu_j^{-1}$ we obtain

$$\begin{cases} Aw_j = \lambda_j w_j, \qquad j = 1, \dots, \\ 0 < \lambda_1 \le \lambda_2, \dots, \qquad \lambda_j \to \infty \quad \text{as} \quad j \to \infty. \end{cases} \tag{2.17}$$

The family w_j is orthonormal in H, and orthogonal for a in V:

$$\begin{cases} (w_j, w_k) = \delta_{jk} = \text{the Kronecker symbol}, \\ a(w_j, w_k) = \langle Aw_j, w_k\rangle = \lambda_j \delta_{jk}, \qquad \forall j, k. \end{cases} \tag{2.18}$$

In particular, if $a(u, v) = ((u, v))$ is the scalar product of V

$$\begin{cases} ((w_j, w_k)) = \lambda_j \delta_{jk}, \\ ((w_j, w_k))_* = ((A^{-1} w_j, w_k)) = \dfrac{1}{\lambda_j} \delta_{jk}, \qquad \forall j, k. \end{cases} \tag{2.19}$$

Powers of A

As indicated above, when a is symmetric, A is a closed positive self-adjoint unbounded operator in H and the spectral theory of these operators allows us to define the powers A^s of A for $s \in \mathbb{R}$ (see F. Riez and B.S. Nagy [1], K. Yosida [1]). For every $s > 0$, A^s is an unbounded self-adjoint operator in H with a dense domain $D(A^s) \subset H$. The operator A^s is strictly positive and injective. The space $D(A^s)$ is endowed with the scalar product and the norm

$$\begin{cases} (u, v)_{D(A^s)} = (A^s u, A^s v), \\ |u|_{D(A^s)} = \{(u, u)_{D(A^s)}\}^{1/2}, \end{cases} \tag{2.20}$$

which makes it a Hilbert space and A^s is an isomorphism from $D(A^s)$ onto H. For $s = 1$, we recover $D(A)$ and for $s = \frac{1}{2}$, $D(A^{1/2}) = V$.

We define $D(A^{-s})$ as the dual of $D(A^s)$ ($s > 0$) and A^s can be extended as an isomorphism from H onto $D(A^{-s})$. Alternatively, $D(A^{-s})$ can be endowed with the scalar product and the norm in (2.20) where s is replaced by $-s$. We obtain, finally, an increasing family of spaces $D(A^s)$, $s \in \mathbb{R}$,

$$D(A^{s_1}) \subset D(A^{s_2}), \qquad \forall s_1, s_2 \in \mathbb{R}, \quad s_1 \geq s_2. \tag{2.21}$$

Each space is dense in the following one, the injection is continuous, and $A^{s_1 - s_2}$ is an isomorphism of $D(A^{s_1})$ into $D(A^{s_2})$, $\forall s_1, s_2 \in \mathbb{R}$, $s_1 > s_2$.

In the self-adjoint compact case these spaces and operators are very easy to characterize by utilization of the spectral basis of A.

In this case, for positive real s, we have

$$D(A^s) = \left\{ u \in H, \sum_{j=1}^{\infty} \lambda_j^{2s} (u, w_j)^2 < \infty \right\},$$

and for negative s, $D(A^s)$ is the completion of H for the norm

$$\left\{ \sum_{j=1}^{\infty} \lambda_j^{2s} (u, w_j)^2 \right\}^{1/2}.$$

For $s \in \mathbb{R}$, the scalar product and the norm of $D(A^s)$ in (2.20) can be written alternatively as

$$(u, v)_{D(A^s)} = \sum_{j=1}^{\infty} \lambda_j^{2s} (u, w_j)(v, w_j),$$

$$|u|_{D(A^s)} = \left\{ \sum_{j=1}^{\infty} \lambda_j^{2s} (u, w_j)^2 \right\}^{1/2},$$

and for $u \in D(A^s)$ we can write:

$$A^s u = \sum_{j=1}^{\infty} \lambda_j^s (u, w_j) w_j. \tag{2.22}$$

Finally, in the compact case, the embedding of $D(A^s)$ into $D(A^{s-\varepsilon})$ is compact, $\forall s \in \mathbb{R}$, $\forall \varepsilon > 0$.

Notions of Interpolation

Let X and Y be two Banach spaces, $X \subset Y$, X dense in Y, the injections being continuous. Then we can define intermediate Banach spaces between X and Y; and there are several different methods of doing it, see, for instance, J.L. Lions and J. Peetre [1], J.L. Lions and E. Magenes [1].

The simplest case is that in which X and Y are Hilbert spaces and the injection of X into Y is compact. In this case, the framework described above provides the definition of the intermediate spaces which are called the interpolation spaces. We set

$$V = X, \qquad H = Y,$$

$$a(u, v) = ((u, v))_X = \text{the scalar product in } X,$$

then, as noticed above $D(A^{1/2}) = X$, $D(A^0) = Y$, and the spaces $D(A^s)$, $s \in [0, \frac{1}{2}]$, are intermediate spaces between X and Y. We write

$$D(A^{(1-\theta)/2}) = [X, Y]_\theta, \qquad \forall \theta \in [0, 1], \tag{2.23}$$

and we endow this space with the Hilbert scalar product

$$((u, v))_{[X,Y]_\theta} = ((A^{(1-\theta)/2} u, A^{(1-\theta)/2} v))_Y.$$

All the interesting properties of the interpolation spaces follow in this case from the properties above of $D(A^s)$ and A^s. In particular, we easily infer from expression (1.53) of the norm of $D(A^s)$, the interpolation inequality

$$\|u\|_{[X,Y]_\theta} \leq c(\theta) \|u\|_X^{1-\theta} \|u\|_Y^\theta, \qquad \forall u \in X, \quad \forall \theta \in [0, 1], \tag{2.24}$$

which is even valid in this case with $c(\theta) = 1$.

Another important property of interpolation spaces is that for two given pairs of spaces X_0, Y_0 and X_1, Y_1 as above, if L is a linear continuous operator from X_0 into X_1 and from Y_0 into Y_1, then L is also linear continuous from $[X_0, Y_0]_\theta$ into $[X_1, Y_1]_\theta$, $\forall \theta \in [0, 1]$.

2.2. "Concrete" Examples of Linear Operators

We now describe a few classical (and typical) examples of linear operators associated with elliptic eigenvalue problems. We start with one-dimensional problems.

Example 2.1 (The Dirichlet Problem in $(0, L)$). Let $\Omega =]0, L[\subset \mathbb{R}$; we consider the Dirichlet problem on $]0, L[$:

$$-\frac{d^2u}{dx^2} + \lambda u = f \quad \text{in }]0, L[, \tag{2.25}$$

$$u(0) = u(L) = 0. \tag{2.26}$$

We briefly recall the weak formulation of this problem. If u is a smooth solution of (2.25), (2.26), say $u \in \mathscr{C}^2([0, L])$, and if v is a sufficiently regular test function (say, also, $v \in \mathscr{C}^2([0, L])$) satisfying $v(0) = v(L) = 0$, then we multiply (2.25) by v, integrate over $(0, L)$, and integrate by parts. We find that

$$a(u, v) = (f, v), \tag{2.27}$$

where

$$a(u, v) = \int_0^L \left(\frac{du}{dx}\frac{dv}{dx} + \lambda uv\right) dx. \tag{2.28}$$

Conversely, if u is smooth, $u(0) = u(L) = 0$, and (2.27) is valid for every such test function v, then u is the solution of (2.25), (2.26). Now we observe that such a function u is obviously in $H_0^1(]0, L[)$, and that by continuity ($\mathscr{C}^2([0, L]) \cap H_0^1(]0, L[)$ is dense in $H_0^1(]0, L[)$), (2.27) will be valid for every v in $H_0^1(]0, L[)$. This leads to the weak form of the problem:

For f given in $H = L^2(]0, L[)$, find $u \in V = H_0^1(]0, L[)$ such that

$$a(u, v) = (f, v), \qquad \forall v \in V. \tag{2.29}$$

Let us interpret (2.29). For u in V and f in H (or even f in V' = the dual of $H_0^1(]0, L[)$ denoted $H^{-1}(]0, L[)$) saying that (2.29) holds amounts to saying that $Au = f$ in V'. If f is in H (and not only in V'), then u is in $D(A)$. Writing (2.28) with a test function u in $\mathscr{D}(]0, L[)$, we see that u satisfies (2.25) in the distribution sense in $]0, L[$. Thus

$$u'' \left(= \frac{d^2u}{dx^2}\right) = \lambda u - f \in L^2(]0, L[),$$

and u is in $H^2(]0, L[)$. Also, because of the embedding theorem (1.12), $H^1(]0, L[) \subset \mathscr{C}([0, L])$ and by (1.19),

$$H_0^1(]0, L[) = \{u \in H^1(]0, L[), u(0) = u(L) = 0\}.$$

No more information is contained in (2.29) and we conclude that the domain of A in H is

$$D(A) = H^2(]0, L[) \cap H_0^1(]0, L[),$$

and $Au = -u'' + \lambda u$ for u in $D(A)$.

For $\lambda > 0$, all the assumptions of Theorem 1.1 are satisfied and this theorem implies that (2.29) possesses a unique solution; A is strictly positive, A is an isomorphism from V onto V' and from $D(A)$ onto H.

A Poincaré inequality similar to (1.6) is valid

$$|u| \leq c_0 |u'|, \qquad \forall u \in H_0^1(]0, L[), \tag{2.30}$$

and shows that the coercivity property (2.2) of a (and the positivity of A) also holds for $\lambda = 0$ and even for $\lambda \leq 0$, $-c_0^2 < \lambda < 0$.

The eigenvalues and eigenvectors of A can easily be determined. Let us assume for the sake of simplicity that $\lambda = 0$.[1] Then the eigenvalues and eigenvectors (eigenfunctions) are

$$\lambda_k = \Pi^2 \frac{k^2}{L^2}, \qquad w_k = \sqrt{\frac{2}{L}} \sin \Pi \frac{kx}{L}, \qquad k \geq 1. \tag{2.31}$$

The best (smallest) constant c_0 in (2.30) is given by

$$\operatorname*{Inf}_{u \in H_0^1(]0, L[)} \frac{|u'|^2}{|u|^2} = \lambda_1 = \frac{\Pi^2}{L^2} \quad \left(\geq \frac{1}{c_0^2} \right).$$

EXAMPLE 2.2 (The Periodic Boundary Condition on $(0, L)$). For $\Omega =]0, L[\subset \mathbb{R}$, we consider the same equation (2.25) with periodic boundary conditions

$$-\frac{d^2 u}{dx^2} + \lambda u = f \quad \text{in }]0, L[, \tag{2.32}$$

$$u(0) = u(L), \qquad \frac{du}{dx}(0) = \frac{du}{dx}(L). \tag{2.33}$$

Let u be a smooth solution of (2.32), (2.33) and let v be a smooth test function (say, $v \in \mathscr{C}^2([0, L])$) such that $v(0) = v(L)$. We multiply (2.32) by v, integrate over $]0, L[$, and integrate by parts and we find that

$$a(u, v) = (f, v), \tag{2.34}$$

with $a(u, v)$ as in (2.28). Conversely, it is elementary that if u is smooth, $u(0) = u(L)$, and if (2.34) is valid for every smooth test function v such that $v(0) = v(L)$, then u is a classical solution of (2.32), (2.33).

This leads to the weak formulation of the problem. Observe, as above, that $H^1(]0, L[) \subset \mathscr{C}([0, L])$ (by 1.12)) and let

$$V = \{v \in H^1(]0, L[), v(0) = v(L)\} = H^1_{\text{per}}(]0, L[).$$

The weak form of (2.32), (2.33) is:

For f given in $H = L^2(]0, L[)$, find $u \in V$ such that

$$a(u, v) = (f, v), \qquad \forall v \in V. \tag{2.35}$$

[1] For $\lambda \neq 0$, we just have to add λ to all the eigenvalues.

If $\lambda > 0$, all the assumptions of Theorem 1.1 are satisfied and we conclude from this theorem that for f given in H there exists a unique u in V satisfying (2.35).[1] By writing (2.35) with test functions v in $\mathscr{D}(]0, L[)$, we see that (2.32) is satisfied in the distribution sense and thus

$$u'' = \lambda u - f \in L^2(]0, L[), \qquad u \in H^2(]0, L[).$$

Then again writing (2.35) with more general test functions v, we find that (2.33) is also satisfied (note that $u'(0)$ and $u'(L)$ make sense since $H^2(]0, L[) \subset \mathscr{C}^1([0, L])$).

The domain of A in H is the set of u in V such that $Au = f$ belongs to $L^2(]0, L[)$; thus by (2.32)

$$D(A) = H^2_{\text{per}}(]0, L[).$$

For u in $D(A)$, $Au = -u'' + \lambda u$.

The eigenvalues and eigenvectors of A are well known (and easily computed):

$$w_0 = \frac{1}{\sqrt{L}}, \qquad \lambda_0 = \lambda \quad \text{and for} \quad k \geq 1,$$

$$w_{2k} = \sqrt{\frac{2}{L}} \cos 2\Pi \frac{kx}{L}, \qquad w_{2k+1} = \sqrt{\frac{2}{L}} \sin 2\Pi \frac{kx}{L}, \tag{2.36}$$

$$\lambda_{2k} = \lambda_{2k+1} = \frac{4\Pi^2 k^2}{L^2} + \lambda.$$

The problem (2.32), (2.33) cannot be solved with $\lambda = 0$ unless we demand that

$$\int_0^L f(x)\, dx = 0,$$

i.e., $f \in \dot{L}^2(]0, L[)$, and we seek u in $\dot{H}^1_{\text{per}}(]0, L[)$. In this case, we set

$$H = \dot{L}^2(]0, L[), \qquad V = \dot{H}^1_{\text{per}}(]0, L[),$$

and the same results as above are valid with $\lambda = 0$. In particular, the eigenvalues of A are the same as in (2.36), except for λ_0, and we find

$$\operatorname*{Inf}_{u \in \dot{H}^1_{\text{per}}(]0, L[)} \frac{|u'|^2}{|u|^2} = \lambda_1 = \frac{4\Pi^2}{L^2}. \tag{2.37}$$

We have a Poincaré inequality similar to (1.32) with the best constant

$$c_0' = \frac{1}{\sqrt{\lambda_1'}} = \frac{L}{2\Pi}.$$

[1] It suffices that $f \in V'$ ($\supset H$) but the space V' is now a complicated space and we do not consider such f's.

This inequality guarantees the coercivity of a on $\dot{H}^1_{\text{per}}(]0, L[)$ for $\lambda = 0$ and even for $\lambda < 0$, $-L/2\Pi < \lambda < 0$; the coercivity property allows us to apply Theorem 1.1 and proves the positivity of A.

EXAMPLE 2.3 (The Neumann Problem on $(0, L)$). We consider the Neumann problem on $]0, L[$:

$$-\frac{d^2u}{dx^2} + \lambda u = f \quad \text{in }]0, L[, \tag{2.38}$$

$$\frac{du}{dx}(0) = \frac{du}{dx}(L) = 0. \tag{2.39}$$

Let u be a smooth solution of (2.38), (2.39) and let v be a smooth test function (say, $v \in \mathscr{C}^2([0, L])$). We multiply (2.38) by v, integrate over $]0, L[$, integrate by parts, and we find that

$$a(u, v) = (f, v) \tag{2.40}$$

with $a(u, v)$ as in (2.28). Conversely, if u is a smooth function satisfying (2.40) for every smooth v, then u is a classical solution of (2.38), (2.39).

The weak form of (2.38), (2.39) is

For f given in $H = L^2(]0, L[)$, find u in $V = H^1(]0, L[)$ which satisfies

$$a(u, v) = (f, v), \qquad \forall v \in V. \tag{2.41}$$

For u in V and f in H, saying that (2.41) holds for every v in V, amounts to saying that $Au = f$ in V' and thus $u \in D(A)$. Alternatively, we can interpret this relation as follows: we write (2.41) with $v \in \mathscr{D}(]0, L[)$ and we find that (2.38) is satisfied in the distribution sense in Ω. It also follows from (2.38) that

$$u'' = \lambda u - f \in L^2(]0, L[), \qquad \text{i.e.,} \quad u \in H^2(]0, L[).$$

Thus $u \in \mathscr{C}^1([0, L])$ (by (1.12)), and writing (2.41) with an arbitrary test function v we conclude that u satisfies (2.39) as well, and there is no further information in the relations (2.41). We infer from this that the domain of A in H is

$$D(A) = \{u \in H^2(]0, L[), u'(0) = u'(L) = 0\},$$

and

$$Au = -u'' + \lambda u, \qquad \forall u \in D(A).$$

For $\lambda > 0$, the form a is bilinear continuous coercive on V and Theorem 1.1 provides the existence and uniqueness of solution of (2.41).

The eigenvalues and eigenvectors of A are easy to compute. We have

$$\begin{aligned} \lambda_0 &= \lambda, & w_0 &= \frac{1}{\sqrt{L}}, \\ \lambda_k &= \frac{\Pi^2 k^2}{L^2} + \lambda, & w_k &= \sqrt{\frac{2}{L}} \cos \Pi \frac{kx}{L}, \qquad k \geq 1. \end{aligned} \tag{2.42}$$

For $\lambda = 0$, we cannot solve (2.38), (2.39) unless f satisfies

$$\int_0^L f(x)\,dx = 0, \tag{2.43}$$

i.e., $f \in \dot{L}^2(]0, L[)$, and we seek u in $\dot{H}^1(]0, L[)$. In this case, we set

$$\dot{H} = \dot{L}^2(]0, L[) = \{f \in L^2(]0, L[), \text{(2.43) is satisfied}\},$$
$$\dot{V} = \dot{H}^1(]0, L[) = H^1(]0, L[) \cap \dot{H},$$

and the same results as above are valid for $\lambda = 0$. In particular, the eigenvalues of A are the same as in (2.42), λ_0, w_0 being removed. We then have a Poincaré-type inequality since

$$\operatorname*{Inf}_{u \in V} \frac{|u'|^2}{|u|^2} = \lambda_1 = \frac{\Pi^2}{L^2}. \tag{2.44}$$

This inequality implies that a is coercive on $\dot{H}^1(]0, L[)$ for $\lambda = 0$, or even for $-2\Pi/L < \lambda \leq 0$.

Example 2.4 (The Dirichlet Problem in a Bounded Domain of $\mathbb{R}^n$). We denote by Ω a bounded domain of class $\mathscr{C}^2$ of $\mathbb{R}^n$, with boundary Γ. We consider the homogeneous Dirichlet problem in Ω associated with the Laplace operator ($\Delta = \sum_{i=1}^n D_i^2 = \sum_{i=1}^n \partial^2/\partial x_i^2$):

$$-\Delta u + \lambda u = f \quad \text{in } \Omega, \tag{2.45}$$

$$u = 0 \quad \text{on } \Gamma. \tag{2.46}$$

If u is a smooth solution of (2.45), (2.46) and v is a smooth test function (say $v \in \mathscr{C}^2(\bar{\Omega})$) which vanishes on Γ, we multiply (2.45) by v and integrate over Ω. By Green's theorem

$$-\int_\Omega \Delta u \cdot v\,dx = -\int_\Gamma \frac{\partial u}{\partial \mathbf{v}} v\,d\Gamma + \int_\Omega \operatorname{grad} u \cdot \operatorname{grad} v\,dx, \tag{2.47}$$

and we obtain

$$a(u, v) = (f, v) \tag{2.48}$$

with

$$a(u, v) = \int_\Omega (\operatorname{grad} u \cdot \operatorname{grad} v + \lambda uv)\,dx. \tag{2.49}$$

Conversely, if u is a smooth function which vanishes on Γ and satisfies (2.48) for every smooth test function v vanishing on Γ, then u is the solution of (2.45), (2.46). This observation leads to the weak formulation of the Dirichlet problem (2.45), (2.46).

For f given in $H = L^2(\Omega)$, find $u \in V = H_0^1(\Omega)$ such that

$$a(u, v) = (f, v), \qquad \forall v \in V. \tag{2.50}$$

We can interpret (2.50) as in dimension 1 although some of the results are now more delicate. If u belongs to V and f belongs to H (or even to V' = the dual of $H_0^1(\Omega)$ denoted $H^{-1}(\Omega)$), then relation (2.50) is equivalent to

$$Au = f \quad \text{in } V'.$$

If $f \in H$ (and not only $f \in V'$), then by definition, $u \in D(A)$. Writing (2.50) with test functions v in $\mathscr{D}(\Omega)$, we find that (2.45) is satisfied in the distribution sense in Ω. Furthermore,

$$\Delta u = \lambda u - f \in L^2(\Omega).$$

There is no further information in (2.50), and hence

$$D(A) = \{u \in H_0^1(\Omega), \Delta u \in L^2(\Omega)\}$$

and for $u \in D(A)$, $Au = -\Delta u + \lambda u$. This characterization of $D(A)$ is valid, in fact, for an arbitrary domain Ω, which need not be regular or bounded. For a bounded regular ($\mathscr{C}^2$, for instance) set Ω, it follows from the regularity theory of elliptic boundary-value problems (see, S. Agmon, A. Douglis, and L. Nirenberg [1], J.L. Lions and E. Magenes [1]), that

$$D(A) = H^2(\Omega) \cap H_0^1(\Omega).$$

For $\lambda > 0$, all the assumptions of Theorem 1.1 are easily verified and we conclude that A is an isomorphism from V onto V' and from $D(A)$ onto H. By using the Poincaré inequality (1.6) we see that this is also true for $\lambda = 0$ and even for $\lambda \leq 0$, $-c_0(\Omega)^2 < \lambda \leq 0$.

Since Ω is bounded, the injection of $H_0^1(\Omega)$ in $L^2(\Omega)$ is compact (see the compactness theorem (1.17) which reduces in this case to Rellich's theorem). There exists an orthonormal basis of $L^2(\Omega)$ which consists of eigenvectors of A; the corresponding eigenvalues are denoted λ_k for $\lambda = 0$, and are equal to $\lambda_k + \lambda$ for other values of λ:

$$-\Delta w_k = \lambda_k w_k, \qquad w_k \in H_0^1(\Omega), \qquad 0 < \lambda_1 \leq \lambda_2 \ldots. \tag{2.51}$$

In general, it is not easy to determine explicitly the eigenfunctions and eigenvalues, except for very particular domains Ω like rectangles or balls. Note that λ_1 provides the best constant for the Poincaré inequality

$$\operatorname*{Inf}_{u \in H_0^1(\Omega)} \frac{\|u\|^2}{|u|^2} = \lambda_1 \qquad \left(\geq \frac{1}{c_0(\Omega)^2}\right). \tag{2.52}$$

Example 2.5 (The Neumann Problem in a Bounded Domain of $\mathbb{R}^n$). We again denote a bounded domain of $\mathbb{R}^n$ with a smooth boundary Γ of class $\mathscr{C}^2$ by Ω, and we consider the Neumann problem in Ω

$$-\Delta u + \lambda u = f \quad \text{in } \Omega, \tag{2.53}$$

$$\frac{\partial u}{\partial \mathbf{v}} = 0 \quad \text{on } \Gamma. \tag{2.54}$$

If u is a smooth solution of (2.53), (2.54) and v is an arbitrary smooth test function (say $u, v \in \mathscr{C}^2(\bar{\Omega})$), we multiply (2.53) by v and integrate over Ω. Due to (2.47) and (2.54) we obtain

$$a(u, v) = (f, v) \tag{2.55}$$

with $a(u, v)$ as in (2.49). Conversely, if u is a smooth function satisfying (2.55) for every v in $\mathscr{C}^2(\bar{\Omega})$, then u is a classical solution of (2.53), (2.54). The weak formulation of this Neumann problem is then

For f given in $H = L^2(\Omega)$, find u in $V = H^1(\Omega)$ satisfying

$$a(u, v) = (f, v), \qquad \forall v \in V. \tag{2.56}$$

If u belongs to V, f belongs to H, and (2.56) is satisfied, then we have $Au = f$ in V' which implies that $u \in D(A)$. We can characterize (2.56) by interpreting properly (2.56). First, we write this relation with test functions v in $\mathscr{D}(\Omega)$ and we conclude that (2.53) is satisfied in the distribution sense in Ω and therefore

$$\Delta u = \lambda u - f \in L^2(\Omega).$$

Thanks to a trace theorem which will not be explicitly described at this point (see, for instance, R. Temam [2]), if u belongs to $H^1(\Omega)$ and $\Delta u \in L^2(\Omega)$, then the trace $\gamma_1 u = \partial u/\partial \mathbf{v}|_\Gamma$ can be defined and belongs to $H^{-1/2}(\Gamma)$ (= the dual of $H^{1/2}(\Gamma)$). Furthermore, a generalized form of Green's theorem (2.47) is valid for every v in $H^1(\Omega)$:

$$-(\Delta u, v) = -\langle \gamma_1 u, \gamma_0 v \rangle + (\operatorname{grad} u, \operatorname{grad} v). \tag{2.57}$$

Returning to (2.56), and using (2.53) and (2.57), we conclude that

$$\langle \gamma_1 u, \gamma_0 v \rangle = 0, \qquad \forall v \in H^1(\Omega).$$

Since $\gamma_1 u \in (H^{1/2}(\Gamma))'$ and $\gamma_0 H^1(\Omega) = H^{1/2}(\Gamma)$ (see Section 1.1.2), this implies

$$\gamma_1 u = 0, \tag{2.58}$$

which is the weak form of (2.54).

We have now fully characterized the domain of A in $L^2(\Omega)$:

$$D(A) = \{u \in H^2(\Omega), \Delta u \in L^2(\Omega), \gamma_1 u = 0\}, \tag{2.59}$$

and, for u in $D(A)$, $Au = -\Delta u + \lambda u$. A supplementary information on $D(A)$ can be obtained by utilization of the regularity theory of elliptic boundary-value problems (see S. Agmon, A. Douglis, and L. Nirenberg [1], J.L. Lions and E. Magenes [1]), namely,

$$D(A) = \{u \in H^2(\Omega), \gamma_1 u = 0\}. \tag{2.60}$$

For $\lambda > 0$, the assumptions of Theorem 1.1 are easily verified and this shows that A is an isomorphism from V onto V' and from $D(A)$ onto H.

Since Ω is bounded the injection of $H^1(\Omega)$ in $L^2(\Omega)$ is compact and hence

A^{-1} is self-adjoint compact. There exists an orthonormal basis of $L^2(\Omega)$ consisting of eigenvectors of A; for convenience, the eigenvalues are written $\lambda_k + \lambda$, $k \geq 0$,

$$\begin{cases} \lambda_0 = 0, \qquad w_0 = 1, \\ -\Delta w_k = \lambda_k w_k, \qquad w_k \in D(A), \qquad k \geq 1, \\ 0 = \lambda_0 < \lambda_1 \leq \lambda_2 \ldots. \end{cases} \tag{2.61}$$

The problem (2.53), (2.54) cannot be solved when $\lambda = 0$ unless f satisfies a necessary condition obtained by integration of (2.53) over Ω:

$$\int_\Omega f(x)\, dx = 0. \tag{2.62}$$

In this case, let

$$H = \dot{L}^2(\Omega) = \{f \in L^2(\Omega),\, f \text{ satisfies (2.62)}\},$$
$$V = \dot{H}^1(\Omega) = H \cap H^1(\Omega).$$

Then the form $a(u, v)$ is coercive on V for $\lambda = 0$, or even $-\sqrt{\lambda_1} < \lambda \leq 0$, λ_1 as in (2.61), or equivalently,

$$\lambda_1 = \operatorname*{Inf}_{u \in \dot{H}^1(\Omega)} \frac{\|u\|^2}{|u|^2}. \tag{2.63}$$

In this case, results similar to those above hold for $\lambda = 0$.

Example 2.6 (The Periodic Boundary Conditions in $\mathbb{R}^n$). Let Ω denote the set $\prod_{j=1}^n]0, L_j[$ in $\mathbb{R}^n$, $L_j > 0$. We denote by Γ_j and Γ_{j+n} the following faces of the boundary Γ on Ω:

$$\Gamma_j = \Gamma \cap \{x_j = 0\}, \qquad \Gamma_{j+n} = \Gamma \cap \{x_j = L_j\}, \qquad j = 1, \ldots, n.$$

We consider the boundary-value problem

$$-\Delta u + \lambda u = f \quad \text{in } \Omega, \tag{2.64}$$

$$u|_{\Gamma_j} = u|_{\Gamma_{j+n}}, \qquad j = 1, \ldots, n, \tag{2.65}$$

$$\left(-\frac{\partial u}{\partial \nu}\bigg|_{\Gamma_j} =\right) \frac{\partial u}{\partial x_j}\bigg|_{\Gamma_j} = \frac{\partial u}{\partial x_j}\bigg|_{\Gamma_{j+n}} \left(= \frac{\partial u}{\partial \nu}\bigg|_{\Gamma_{j+n}}\right). \tag{2.66}$$

When (2.65), (2.66) are satisfied we say that u is Ω periodic. Let u be a classical solution of (2.64)–(2.66), and let v be a smooth test function ($v \in \mathscr{C}^2(\bar{\Omega})$) satisfying conditions similar to (2.65). We multiply (2.64) by v, integrate over Ω, and use Green's formula (2.47). We obtain

$$a(u, v) = (f, v), \tag{2.67}$$

where a is the same as in (2.49). Conversely, if u is a smooth function (say $u \in \mathscr{C}^2(\Omega)$) which satisfies (2.65) and (2.67) for every smooth function v satisfying (2.65), then we can show that u is a classical solution of (2.64)–(2.66). The weak formulation of this problem is then

For f given in $H = L^2(\Omega)$, find u in $V = H^1_{\text{per}}(\Omega)$, such that

$$a(u, v) = (f, v), \qquad \forall v \in V. \tag{2.68}$$

If $u \in V$ and $f \in H$ satisfy (2.68), then $Au = f$ in V' and $u \in D(A)$. In order to interpret (2.68) we first write this relation with $v \in \mathscr{D}(\Omega)$ and we see that (2.64) is satisfied in the distribution sense in Ω. In particular,

$$\Delta u = \lambda u - f \in L^2(\Omega).$$

Hence $\gamma_1 u$ makes sense and we can utilize the generalized Green's formula (2.57). With (2.64), (2.67), and (2.57) we find that

$$\gamma_1 u|_{\Gamma_j} = -\gamma_1 u|_{\Gamma_{j+n}}, \qquad j = 1, \ldots, n, \tag{2.69}$$

which is the weak form of (2.66). We conclude that

$$D(A) = \{u \in H^1_{\text{per}}(\Omega),\ \Delta u \in L^2(\Omega),\ (2.69) \text{ is satisfied}\}.$$

We can improve this result by using the regularity theory of elliptic boundary-value problems, or more simply in this case, with an elementary calculation on Fourier series. This gives

$$D(A) = H^2_{\text{per}}(\Omega).$$

Theorem 1.1 applies; A is an isomorphism from V onto V' and from $D(A)$ onto H.

The eigenvalues and eigenfunctions are well known in this case; for convenience, we write them in the form $\lambda_k + \lambda$, w_k, $\tilde{w}_k$, with $k \in \mathbb{N}^n$:

$$\begin{cases} \lambda_0 = 0, \qquad w_0 = \dfrac{1}{\sqrt{|\Omega|}}, \qquad |\Omega| = L_1 \ldots L_n, \\ \lambda_k = 4\Pi^2\left(\dfrac{k_1^2}{L_1^2} + \cdots + \dfrac{k_n^2}{L_n^2}\right), \\ w_k = \sqrt{\dfrac{2}{|\Omega|}} \cos 2\Pi \dfrac{kx}{L}, \qquad \tilde{w}_k = \sqrt{\dfrac{2}{|\Omega|}} \sin 2\Pi \dfrac{kx}{L}, \end{cases} \tag{2.70}$$

where we have written $kx/L = k_1 x_1/L_1 + \cdots + k_n x_n/L_n$.

For $\lambda = 0$, the problem (2.64)–(2.66) can be solved only if (2.62) is satisfied. We then set

$$H = \dot{L}^2(\Omega),$$
$$V = \dot{H}^1_{\text{per}}(\Omega).$$

The form a is, for $\lambda = 0$, coercive on this space V and results similar to those above are valid under these conditions when $\lambda = 0$. We can write the optimal constant in the Poincaré inequality (1.32)

$$\operatorname*{Inf}_{u \in \dot{H}^1_{\text{per}}(\Omega)} \frac{\|u\|^2}{|u|^2} = 4\Pi^2 \min_{1 \leq j \leq n} \left(\frac{1}{L_j^2}\right) \left(\geq \frac{1}{c_0'(\Omega)^2}\right).$$

3. Linear Evolution Equations of the First Order in Time

Our objective in this section is to recall the general framework and the basic results of existence of solutions for linear evolution equations of the first order in time of the type

$$\frac{du}{dt} + Au = f.$$

We consider an operator A (independent of time) similar to those described in Section 2 and f is allowed to depend on time. In Section 3.1 we indicate in what sense such an equation is understood. In Section 3.2 we state a result of existence, uniqueness, and regularity of solutions. Further results are given in Section 3.3.

3.1. Hypotheses

The framework is that described in Section 2. We are given two Hilbert spaces V and H, $V \subset H$, V dense in H, the injection being continuous. The scalar product and the norm on V and H are, respectively, denoted by $((\cdot, \cdot))$, $\|\cdot\|$, $(\cdot, \cdot)$, $|\cdot|$. We can identify H with its dual H', and H' with a dense subspace of the dual V' of V (norm $\|\cdot\|_*$), so that

$$V \subset H \subset V', \tag{3.1}$$

where the injections are continuous and each space is dense in the following one.

We are also given a biliner continuous form $a(u, v)$ on V which is coercive,

$$\exists \alpha > 0, \qquad a(u, u) \geq \alpha \|u\|^2, \qquad \forall u \in V. \tag{3.2}$$

With this form we associate, as in Section 2, the linear operator A which is an isomorphism from V onto V' and from $D(A) \subset H$ (its domain in H) onto H.

We consider an interval of time $[0, T]$, $T > 0$, and we are given $u_0 \in H$ (for instance), and a function f from $[0, T]$ in H satisfying, for example,

$$f \in L^2(0, T; H). \tag{3.3}$$

We seek a function u from $[0, T]$ in V which is the solution to the initial-value problem

$$\frac{du}{dt} + Au = f \quad \text{on } (0, T), \tag{3.4}$$

$$u(0) = u_0. \tag{3.5}$$

Since we will not be considering very regular solutions of such a problem, we must first clarify in what sense relations like (3.4) and (3.5) are understood.

Strictly speaking, an equation like (3.4) (and the similar ones arising for nonlinear problems) is satisfied in the sense of distributions with values in V'. For the theory of vector-valued distributions, see L. Schwartz [2]; hereafter, we will need only the following simple lemma:

Lemma 3.1. *Let X be a given Banach space with dual X' and let u and g be two functions belonging to $L^1(a, b; X)$. Then the following three conditions are equivalent:*

(i) *u is almost everywhere equal to a primitive function of g, i.e., there exists $\xi \in X$ such that*

$$u(t) = \xi + \int_a^t g(s)\,ds, \qquad \textit{for a.e.} \quad t \in [a, b]. \tag{3.6}$$

(ii) *For every test function $\varphi \in \mathscr{D}(]a, b[)$,*

$$\int_a^b u(t)\varphi'(t)\,dt = -\int_a^b g(t)\varphi(t)\,dt \qquad \left(\varphi' = \frac{d\varphi}{dt}\right). \tag{3.7}$$

(iii) *For each $\eta \in X'$,*

$$\frac{d}{dt}\langle u, \eta\rangle = \langle g, \eta\rangle \tag{3.8}$$

in the scalar distribution sense on $]a, b[$.

If (i)–(iii) *are satisfied we say that g is the (X-valued) distribution derivative of u, and u is almost everywhere equal to a continuous function from $[a, b]$ into X.*

This lemma is borrowed from R. Temam [2] (Ch. III, Lemma 1.1) to which the reader is referred for the proof.

This lemma gives a sense to (3.4) if, as will be the case below, u is in $L^2(0, T; V)$. Indeed, A being an isomorphism from V into V', $Au \in L^2(0, T; V')$; by (3.2), $f \in L^2(0, T; V')$ and $u \in L^2(0, T; V')$. Thus $u' = f - Au$ in the distribution sense in V'. In such a case, Lemma 3.1 also implies that u is (a.e. equal to) a function continuous from $[0, T]$ into V' and (3.5) makes sense too.

3.2. A Result of Existence and Uniqueness

Theorem 3.1. *The assumptions are those in Section* 3.1. *For* u_0 *given in* H *and* f *given in* $L^2(0, T; V')$, *there exists a unique solution* u *of* (3.4), (3.5) *such that*

$$u \in L^2(0, T; V) \cap \mathscr{C}([0, T]; H), \tag{3.9}$$

$$u' \in L^2(0, T; V'). \tag{3.10}$$

Proof. This result is proved, for instance, in J.L. Lions and E. Magenes [1], or R. Temam [2] (Ch. III, §1). We give only a sketch of the proof which emphasizes some points needed in the sequel. □

The existence is proved by the Faedo–Galerkin method. We assume, for simplicity, that V is a separable space and we consider a sequence of linearly independent elements of V, $w_1, \ldots, w_m, \ldots$, which is complete in V. For each m we define an approximate solution u_m of (3.4), (3.5) as follows:

$$u_m(t) = \sum_{i=1}^{m} g_{im}(t) w_i, \tag{3.11}$$

$$\frac{d}{dt}(u_m, w_j) + a(u_m, w_j) = \langle f, w_j \rangle, \qquad j = 1, \ldots, m, \tag{3.12}$$

$$u_m(0) = u_{0m}, \tag{3.13}$$

where u_{0m} is, for example, the orthogonal projection in H of u_0 on the space spanned by $w_1, \ldots, w_m$.

Equations (3.12), (3.13) are equivalent to an initial-value problem for a linear finite- (m-)dimensional ordinary differential equation for the g_{im}. The existence and uniqueness is obvious. The function u_m is in $\mathscr{C}([0, T]; V)$ with u'_m in $L^2(0, T; V)$.

Then we multiply (3.12) by g_{jm}, add these relations for $j = 1, \ldots, m$, which gives

$$(u'_m, u_m) + a(u_m, u_m) = \langle f, u_m \rangle. \tag{3.14}$$

We infer from (3.13) an a priori estimate for u_m

$$u_m \text{ belongs to a bounded set of } L^2(0, T; V) \cap L^\infty(0, T; H). \tag{3.15}$$

By weak compactness, we find a subsequence still denoted by u_m and a u in $L^2(0, T; V) \cap L^\infty(0, T; H)$ such that

$$u_m \to u \quad \text{in } L^2(0, T; V) \text{ weakly}, \tag{3.16}$$

$$u_m \to u \quad \text{in } L^\infty(0, T; H) \text{ weak-star}. \tag{3.17}$$

We then pass to the limit in (3.12), (3.13) and we find that

$$\frac{d}{dt}(u, v) + a(u, v) = \langle f, v \rangle, \qquad \forall v \in V,$$

in the distribution sense in $(0, T)$. According to Lemma 3.1, this is equivalent to (3.4). Lemma 3.1 shows that $u' \in L^2(0, T; V')$; thus u is almost everywhere equal to a continuous function from $[0, T]$ into V', and (3.5) follows by a passage to the limit in (3.13).

There remains to show that u is unique and that $u \in \mathscr{C}([0, T]; H)$. This follows from a result of J.L. Lions and E. Magenes [1]; a direct proof is given in R. Temam [2] (Lemma 1.2, Ch. III):

Lemma 3.2. *Let V, H, V' be three Hilbert spaces, each space included and dense in the following one as in* (3.1), *V' being the dual of V. If a function u belongs to $L^2(0, T; V)$ and its derivative u' belongs to $L^2(0, T; V')$, then u is almost everywhere equal to a function continuous from $[0, T]$ into H and we have the following equality which holds in the scalar distribution sense on* $(0, T)$:

$$\frac{d}{dt}|u|^2 = 2\langle u', u\rangle. \tag{3.18}$$

Note that equality (3.18) is meaningful since the functions

$$t \mapsto |u(t)|^2, \qquad t \mapsto \langle u'(t), u(t)\rangle$$

are both integrable (i.e., L^1) on $(0, T)$.

The proof of the uniqueness follows easily from this lemma. Let u and v be two solutions of (3.4), (3.5) satisfying (3.9), (3.10) and let $w = u - v$. Then w belongs to the same spaces as u and v, and

$$w' + Aw = 0, \qquad w(0) = 0. \tag{3.19}$$

Taking the scalar product of the first equality (3.19) with $w(t)$ we find

$$\langle w'(t), w(t)\rangle + a(w(t), w(t)) = 0.$$

Then using (3.18) and (3.2) we see that

$$\frac{d}{dt}|w(t)|^2 + 2a(w(t), w(t)) = 0, \tag{3.20}$$

$$\frac{d}{dt}|w(t)|^2 \leq 0,$$

$$|w(t)|^2 \leq |w(0)|^2 = 0, \qquad t \in [0, T],$$

and hence $u(t) = v(t)$ for each t.

3.3. Regularity Results

We indicate two regularity results for u which will be used frequently in the nonlinear case.

Theorem 3.2. *The hypotheses are those of Theorem* 3.1. *We assume, furthermore, that*

$$f, f' \in L^2(0, T; H), \tag{3.21}$$

$$u_0 \in D(A). \tag{3.22}$$

Then the solution u of (3.4), (3.5), *given by Theorem* 3.1, *also satisfies*

$$u \in \mathscr{C}([0, T]; D(A)), \tag{3.23}$$

$$u' \in L^2(0, T; V) \cap \mathscr{C}([0, T]; H), \tag{3.24}$$

$$u'' \in L^2(0, T; V'). \tag{3.25}$$

PROOF. The principle of the proof consists of observing that $u' = v$ satisfies a similar equation obtained by the differentiation of (3.4)

$$v' + Av = f', \tag{3.26}$$

$$v(0) = u'(0) = f(0) - Au_0. \tag{3.27}$$

Due to (3.21) and Lemma 3.1, f is in $\mathscr{C}([0, T]; H)$ and $f(0)$ is well defined; thus with (3.22), $v(0) = u'(0) \in H$. The estimates necessary for the proof of (3.24)–(3.27) are derived from the equations obtained by time differentiation of the Galerkin approximation u_m of u, in (3.21), (3.22). Then, for (3.23), we observe that

$$Au(t) = f(t) - u'(t),$$

and since $f - u' \in \mathscr{C}([0, T]; H)$ and A is an isomorphism from $D(A)$ on H, u is in $\mathscr{C}([0, T]; D(A))$. □

The last result is

Theorem 3.3. *The assumptions are those of Theorem* 3.1. *We assume, furthermore, that*

$$a \text{ is symmetric}, \ a(u, v) = a(v, u), \qquad \forall u, v \in V, \tag{3.28}$$

$$\text{the injection of } V \text{ in } H \text{ is compact}, \tag{3.29}$$

$$u_0 \in V, \qquad f \in L^2(0, T; H). \tag{3.30}$$

Then the solution u of (3.4), (3.5) *given by Theorem* 3.1 *also satisfies*

$$u \in L^2(0, T; D(A)) \cap \mathscr{C}([0, T]; V), \tag{3.31}$$

$$u' \in L^2(0, T; H). \tag{3.32}$$

PROOF. For the proof we proceed as for Theorem 3.1, but with a particular choice of the functions w_j in the Faedo–Galerkin method. Because of (3.28), (3.29), there exists an orthonormal basis of H consisting of the eigenvectors of A

$$Aw_j = \lambda_j w_j, \qquad \forall j \tag{3.33}$$

(see (2.17)), and we implement the Faedo–Galerkin method with these functions. We obtain further a priori estimates on u_m as follows: for each j, we multiply (3.22) by $\lambda_j g_{jm}$ and add these relations for $j = 1, \ldots, m$. By (2.17) and (2.18)

$$\begin{cases} (u_m', \lambda_j w_j) = ((u_m', w_j)), \\ a(u_m, \lambda_j w_j) = (Au_m, \lambda_j w_j) = (Au_m, Aw_j), \\ (f, \lambda_j w_j) = (f, Aw_j), \end{cases} \tag{3.34}$$

and therefore we obtain

$$a(u_m', u_m) + |Au_m|^2 = (f, Au_m),$$

$$\begin{aligned} \frac{1}{2}\frac{d}{dt}a(u_m, u_m) + |Au_m|^2 &= (f, Au_m) \\ &\leq |f|\,|Au_m| \\ &\leq \tfrac{1}{2}|f|^2 + \tfrac{1}{2}|Au_m|^2, \end{aligned} \tag{3.35}$$

$$\frac{d}{dt}a(u_m, u_m) + |Au_m|^2 \leq |f|^2. \tag{3.36}$$

This implies

$$\begin{aligned} a(u_m(t), u_m(t)) + \int_0^t |Au_m|^2\, ds &\leq a(u_m(0), u_m(0)) + \int_0^t |f|^2\, ds \\ &\leq a(u_0, u_0) + \int_0^T |f|^2\, ds, \quad \forall t \in [0, T], \end{aligned} \tag{3.37}$$

and

$$u_m \text{ remains in a bounded set of } L^2(0, T; D(A)) \cap L^\infty(0, T; V). \tag{3.38}$$

At the limit we find that u belongs to $L^2(0, T; D(A)) \cap L^\infty(0, T; V)$; then (3.4) and the properties of A imply (3.22).

It remains to show that u is continuous from $[0, T]$ in V. First, we deduce from Lemma 3.3 below that u is weakly continuous from $[0, T]$ into V, i.e.,

$$t \mapsto ((u(t), v)) \text{ is continuous}, \qquad \forall v \in V.$$

Similarly,

$$t \to a(u(t), v) \text{ is continuous}, \qquad \forall v \in V.$$

Then we observe that an equality similar to (3.35) is valid for u (its proof relying on Lemma 3.2 suitably applied):

$$\frac{d}{dt}a(u, u) + 2|Au|^2 = 2(f, Au) \quad \text{on } (0, T). \tag{3.39}$$

Together with the previous results this relation shows that the function

$$t \mapsto a(u(t), u(t))$$

is continuous on $[0, T]$. The continuity of u from $[0, T]$ to V, for the topology of the norm, follows (note that $\{a(\varphi, \varphi)\}^{1/2}$ is a norm on V equivalent to $\|\varphi\|$). □

Lemma 3.3. *Let X and Y be two Banach spaces such that*

$$X \subset Y \tag{3.40}$$

with a continuous injection.

If a function φ belongs to $L^\infty(0, T; X)$ and is weakly continuous with values in Y, then φ is weakly continuous with values in X.

This result is proved in W.A. Strauss [1] (see also R. Temam [2], (Lemma 1.4, Chap. III)).

3.4. Time-Dependent Operators

In view of the applications in Chapter VI it is useful to extend the above results to the case where the operator A (or equivalently the form a) depends on t.

The spaces are like those in Section 3.1 and (3.1) is verified; instead of (3.2) we are given a family $a(t; u, v)$ of bilinear continuous forms on V, $t \in [0, T]$, such that

For every $u, v \in V$, $t \to a(t; u, v)$ is a measurable function; (3.41)

There exists $M = M_T < \infty$, such that $|a(t; u, v)| \leq M_T \|u\| \, \|v\|$, $\forall u, v \in V$, a.e. $t \in [0, T]$; (3.42)

There exists $\alpha > 0$ such that $a(t; u, u) \geq \alpha \|u\|^2$, $\forall u \in V$, a.e. $t \in [0, T]$. (3.43)

To each of these forms we associate the linear operator $A(t)$ which is an isomorphism form V onto V'; it is also an isomorphism from $D(A(t)) \subset V$ onto H.

Given f in $L^2(0, T; H)$ as in (3.3) and u_0 in H, we seek a function u from $[0, T]$ into V, which is the solution to the initial-value problem

$$\frac{du(t)}{dt} + A(t)u(t) = f(t) \quad \text{on } (0, T), \tag{3.44}$$

$$u(0) = u_0. \tag{3.45}$$

The analog of Theorem 3.1 holds:

Theorem 3.4. *The hypotheses are those given above and in particular* (3.1), (3.41)–(3.43). *For u_0 in H and f given in $L^2(0, T; V')$ (or $L^2(0, T; H)$), there exists a unique solution u of* (3.44), (3.45) *such that*

$$u \in L^2(0, T; V) \cap \mathscr{C}([0, T]; H), \tag{3.46}$$

$$u' \in L^2(0, T; V'). \tag{3.47}$$

As Theorem 3.1, this result is classical and its proof differs very slightly from that of Theorem 3.1; we omit the details.

In a similar manner, we can also extend Theorems 3.2 and 3.3. To extend Theorem 3.2 we formulate assumptions (3.41)–(3.43) above and, furthermore,

$$\text{For every } u, v \in V, \text{ the function } t \to a(t; u, v) \text{ is absolutely continuous with derivative } a'(t; u, v); \tag{3.48}$$

For a.e. $t \in [0, T]$, $a'(t; u, v)$ is a bilinear continuous form on V and there exists $M_1 = M_{1T} < \infty$ such that

$$|a'(t; u, v)| \leq M_{1T}\|u\|\,\|v\|, \qquad \forall u, v \in V, \qquad \text{a.e.} \quad t \in [0, T]. \tag{3.49}$$

Due to (3.48), the form $a(t; u, v)$ and the operator $A(t)$ are now well defined for every $t \in [0, T]$. In particular, $D(A(t)) \subset V$ is well defined, $\forall t \in [0, T]$.

We then have

Theorem 3.5. *The hypotheses are those of Theorem* 3.4 *and, furthermore,* (3.48), (3.49). *We also assume that*

$$f, f' \in L^2(0, T; H), \tag{3.50}$$

$$u_0 \in D(A(0)). \tag{3.51}$$

Then the solution u of (3.44), (3.45) *satisfies, besides* (3.46), (3.47),

$$u(t) \in D(A(t)), \forall t \in [0, T] \text{ and } t \to A(t)u(t) \text{ is continuous from } [0, T] \text{ into } H. \tag{3.52}$$

$$u' \in L^2(0, T; V) \cap \mathscr{C}([0, T]; H), \tag{3.53}$$

$$u'' \in L^2(0, T; V'). \tag{3.54}$$

Remark 3.1. In the most interesting examples we have

$$D(A(t)) = D(A) \text{ is independent of } t. \tag{3.55}$$

In this case, (3.42) reduces to

$$u \in \mathscr{C}([0, T]; D(A)).$$

Finally, we can extend Theorem 3.3 as follows:

Theorem 3.6. *The hypotheses are those of Theorem* 3.4. *We also assume that* (3.48), (3.49), (3.55) *are satisfied and*

$$a(t; u, v) = a(t; v, u), \qquad \forall u, v \in V, \quad \forall t \in [0, T], \tag{3.56}$$

$$u_0 \in V, \qquad f \in L^2(0, T; H). \tag{3.57}$$

Then the solution u of (3.44), (3.45) *also satisfies*

$$u \in \mathscr{C}([0, T]; V) \cap L^2(0, T; D(A)). \tag{3.58}$$

Like Theorem 3.4, Theorems 3.5 and 3.6 are easy extensions of the corresponding results of the time-independent case.

4. Linear Evolution Equations of the Second Order in Time

In this section we recall some results on the existence of solutions and some properties of linear evolution equations of the second order in time

$$\frac{d^2u}{dt^2} + \alpha\frac{du}{dt} + Au = f.$$

The operator A is similar to those described in Section 2; it is independent of time and self-adjoint, f depends on time and $\alpha \in \mathbb{R}$. The hypotheses and the main results are given in Section 4.1. Section 4.2 contains some complements.

4.1. The Evolution Problem

The framework is again as described in Section 2. We are given two Hilbert spaces V and H, $V \subset H$, V dense in H, the injection being continuous. The scalar product and the norm in V and H are, respectively, denoted by $((\cdot, \cdot))$, $\|\cdot\|$, $(\cdot, \cdot)$, $|\cdot|$. We identify H with its dual H' and H' with a dense subspace of the dual V' of V (norm $\|\cdot\|_*$); hence

$$V \subset H \subset V',$$

where the injections are continuous and each space is dense in the following one.

We are also given a bilinear continuous form $a(u, v)$ on V which is symmetric and coercive,

$$\exists \alpha_0 > 0, \qquad a(u, v) \geq \alpha_0 \|u\|^2, \qquad \forall u \in V. \tag{4.1}$$

With this form we associate (as in Section 2) the linear operator A which is now self-adjoint. As indicated in Section 2, we can define the powers A^s of A for $s \in \mathbb{R}$ which operate on the spaces $D(A^s)$. We set here

$$V_{2s} = D(A^s), \qquad s \in \mathbb{R}, \tag{4.2}$$

which is a Hilbert space for the scalar product and the norm (2.30)

$$\begin{aligned} (u, v)_{2s} &= (A^s u, A^s v), \qquad \forall u, v \in D(A^s), \\ |u|_{2s} &= \{(u, u)_{2s}\}^{1/2}. \end{aligned} \tag{4.3}$$

Given $T > 0, 0 < T < \infty$, and f, u_0, u_1 satisfying

$$f \in L^2(0, T; H), \qquad u_0 \in V_1, u_1 \in H, \tag{4.4}$$

we want to solve the initial-value problem ($\alpha \in \mathbb{R}$)

$$\frac{d^2u}{dt^2} + \alpha\frac{du}{dt} + Au = f \quad \text{on } (0, T), \tag{4.5}$$

$$u(0) = u_0, \qquad \frac{du}{dt}(0) = u_1. \tag{4.6}$$

We have

Theorem 4.1. *The hypotheses are those above, in particular, f, u_0, u_1 are given satisfying* (4.4). *Then there exists a unique solution u of* (4.5), (4.6) *such that*

$$u \in \mathscr{C}([0, T]; V), \qquad u' = \frac{du}{dt} \in \mathscr{C}([0, T]; H). \tag{4.7}$$

PROOF. We give the principle of the proof: we assume, for simplicity, that V is a separable space and consider a sequence of linearly independent elements of V, $w_1, \ldots, w_m, \ldots$, which is total in V. Using the Faedo–Galerkin method we define for each m an approximate solution u_m of (4.5), (4.6)

$$u_m(t) = \sum_{i=1}^{m} g_{im}(t) w_i, \tag{4.8}$$

$$\frac{d^2}{dt^2}(u_m, w_j) + \alpha \frac{d}{dt}(u_m, w_j) + a(u_m, w_j) = (f, w_j), \qquad j = 1, \ldots, m, \tag{4.9}$$

$$u_m(0) = u_{0m}, \qquad u'_m(0) = u_{1m}, \tag{4.10}$$

where u_{0m} (resp. u_{1m}) is the projection in V (resp. H) of u_0 (resp. u_1) onto the space spanned by $w_1, \ldots, w_m$.

Equations (4.8)–(4.10) are equivalent to a linear initial-value problem for an ordinary (finite-dimensional) differential equation. They possess a unique solution defined for all time and in particular on $[0, T]$; the functions u_m, u'_m are in $\mathscr{C}([0, T]; V)$, u''_m is in $L^2(0, T; V)$.

A priori estimates are obtained by multiplying (4.9) by g'_{jm} and summing these relations for $j = 1, \ldots, m$. We obtain

$$(u''_m, u'_m) + \alpha |u'_m|^2 + a(u_m, u'_m) = (f, u'_m), \tag{4.11}$$

$$\frac{d}{dt}\{|u'_m|^2 + a(u_m, u_m)\} + 2\alpha |u'_m|^2 = 2(f, u'_m) \leq |f|^2 + |u'_m|^2. \tag{4.12}$$

It follows, with the use of the Gronwall lemma, that

$$\begin{aligned} &u_m \text{ belongs to a bounded set of } L^\infty(0, T; V), \\ &u'_m \text{ belongs to a bounded set } L^\infty(0, T; H). \end{aligned} \tag{4.13}$$

Thus there exists a subsequence, still denoted u_m, and u, such that

$$u \in L^\infty(0, T; V), \qquad u' \in L^\infty(0, T; H), \tag{4.14}$$

$$\begin{cases} u_m \to u & \text{in } L^\infty(0, T; V) \text{ weak-star}, \\ u'_m \to u' & \text{in } L^\infty(0, T; H) \text{ weak-star} \qquad \text{as} \quad m \to \infty. \end{cases} \tag{4.15}$$

Then passing to the limit in (4.8)–(4.10) we see that u is a solution of (4.5), (4.6) which satisfies (4.14).

To conclude the proof of existence there remains to show the continuity properties (4.7).

It follows from Lemma 3.3 and (4.14) that u is weakly continuous from

$[0, T]$ in V. Similarly, we infer from (4.5) that

$$u'' = f - \alpha u' - Au$$

and $u'' \in L^2(0, T; V')$, since $f \in L^2(0, T; H)$, $u' \in L^\infty(0, T; H)$, $u \in L^\infty(0, T; V)$ which implies $Au \in L^\infty(0, T; V')$. Lemma 3.1 then shows that u is continuous from $[0, T]$ in V', Lemma 3.3 and (4.14) imply that u' is weakly continuous from $[0, T]$ in H.

We deduce from Lemma 4.1 below that u satisfies an equation similar to (4.12), namely

$$\frac{d}{dt}\{|u'|^2 + a(u, u)\} + 2\alpha|u'|^2 = 2(f, u').$$

This shows that the function

$$t \mapsto |u'(t)|^2 + a(u(t), u(t))$$

is continuous on $[0, T]$. In conjuction with the above properties of weak continuity, we conclude that $u \in \mathscr{C}([0, T]; V)$ and $u' \in \mathscr{C}([0, T]; H)$.

For the proof of uniqueness, let u and v be two solutions of (4.5), (4.6) satisfying (4.7) and let $w = u - v$. The function w is in the same spaces (4.7) and satisfies

$$w'' + \alpha w' + Aw = 0,$$

$$w(0) = 0, \qquad w'(0) = 0.$$

Thanks to Lemma 4.1 below, we see that

$$\frac{d}{dt}\{|w'|^2 + a(w, w)\} + 2\alpha|w'|^2 = 0.$$

If $\alpha \geq 0$, we obtain

$$|w'(t)|^2 + a(w(t), w(t)) \leq 0, \qquad \forall t, \tag{4.16}$$

and $w(t) = 0$ for $t \in [0, T]$. If $\alpha < 0$, we write

$$\frac{d}{dt}\{|w'|^2 + a(w, w)\} \leq -2\alpha\{|w'|^2 + a(w, w)\}.$$

Using the Gronwall lemma, we again obtain (4.16).

The proof of Theorem 4.1 will be complete after we prove

Lemma 4.1. *We assume that w is such that*

$$w \in L^2(0, T; V), \qquad w' \in L^2(0, T; H), \tag{4.17}$$

and

$$w'' + Aw \in L^2(0, T; H). \tag{4.18}$$

Then, after modification on a set of measure zero, u is continuous from $[0, T]$ into V, u' is continuous from $[0, T]$ into H and, in the sense of distributions *on* $]0, T[$,

$$(w'' + Aw, w') = \frac{1}{2}\frac{d}{dt}\{|w'|^2 + a(w, w)\}. \tag{4.19}$$

Proof. We first prove (4.19). This relation is obvious if w is sufficiently regular, say $w \in \mathscr{C}^2([0, T]; V)$. Under the present assumptions we observe that (4.19) is a local property, and its suffices to prove it for the functions $\tilde{w}$ from $\mathbb{R}$ into V equal to θw on $(0, T)$ and to 0 on $\mathbb{R}\backslash[0, T]$; here θ is a truncation function, $\mathscr{C}^\infty$ from $\mathbb{R}$ into $[0, 1]$, equal to 0 on $\mathbb{R}\backslash[0, T]$ and to 1 on some subinterval of $]0, T[$. It is clear that such a function $\tilde{w}$ belongs to $L^2(\mathbb{R}; V)$, $\tilde{w}' \in L^2(\mathbb{R}, H)$, and $\tilde{w}'' + A\tilde{w} \in L^2(\mathbb{R}; H)$. By regularization of $\tilde{w}$, we obtain a function $\tilde{w}_\varepsilon = \rho_\varepsilon * \tilde{w}$ (ρ_ε is a $\mathscr{C}^\infty$ mollifier) which is $\mathscr{C}^\infty$ from $\mathbb{R}$ to V. The relation (4.19) for $\tilde{w}_\varepsilon$ is obviously satisfied on the whole line $\mathbb{R}$

$$(\tilde{w}_\varepsilon'' + A\tilde{w}_\varepsilon', \tilde{w}_\varepsilon') = \frac{1}{2}\frac{d}{dt}\{|\tilde{w}_\varepsilon'|^2 + a(\tilde{w}_\varepsilon, \tilde{w}_\varepsilon)\}. \tag{4.20}$$

We pass to the limit $\varepsilon \to 0$ in (4.20) and find the same relation for $\tilde{w}$; finally, by restriction to $(0, T)$ we obtain (4.19) for θw.

By assumption, the left-hand side of (4.19) is L^1 on $(0, T)$. Due to (4.1) and the symmetry of a, $\{a(w, w)\}^{1/2}$ is a norm on V equivalent to the initial one and we infer from (4.19) that w is in $L^\infty(0, T; V)$; also, $w'' \in -Aw + L^2(0, T; H)$ belongs to $L^2(0, T; V')$. Lemma 3.1 implies that w is weakly continuous from $[0, T]$ into H, w' is weakly continuous from $[0, T]$ into V'; then thanks to Lemma 3.3, w is weakly continuous from $[0, T]$ into V and w' is weakly continuous from $[0, T]$ into H. Again using (4.19), we see that the real function

$$t \to |w'(t)|^2 + a(w(t), w(t))$$

is continuous on $[0, T]$. It is then easy to conclude that w is strongly continuous from $[0, T]$ into V, and w' is strongly continuous from $[0, T]$ into H.

The proof is complete. □

Remark 4.1. An important difference with the evolution equations of first order, considered in Section 3, is that (4.5), (4.6) can be solved backward in time as well, and hence on the interval $[-T, T]$. For the solution of (4.5), (4.6) on $[-T, 0]$, we simply observe that if we perform the change of variable $t/-t$, this has only, for effect, to change α into $-\alpha$ in (4.5), and Theorem 4.1 applies.

4.2. Another Result

We can generalize Theorem 4.1 as follows:

Theorem 4.2. *Let $s \in \mathbb{R}$. Assume that V, H, a are given as above and that*

$$f \in L^2(0, T; V_s), \qquad u_0 \in V_{s+1}, \quad u_1 \in V_s. \tag{4.21}$$

Then there exists a unique function u satisfying (4.5), (4.6) *and such that*

$$u \in \mathscr{C}([0, T]; V_{s+1}), \qquad u' \in \mathscr{C}([0, T]; V_s). \tag{4.22}$$

Furthermore, the following equalities (*of energy type*) *hold*:

$$\frac{1}{2}\frac{d}{dt}(|u|_{s+1}^2 + |u'|_s^2) + \alpha|u'|_s^2 = (f, u')_s. \tag{4.23}$$

PROOF. It suffices to apply Theorem 4.1 replacing the spaces V, H, V' $(= V_{-1})$ by V_{s+1}, V_s, V_{s-1} (in this case, it is V_s which is identified with its dual while V_{s-1} is identified with the dual of V_{s+1}; note that, as indicated in Section 2, A is an isomorphism from V_{s+1} onto V_{s-1}). The existence and uniqueness of u satisfying (4.22) follows.

For (4.23), we apply Lemma 4.1, more precisely (4.19) when $s = 0$, and we proceed as above for the other values of s. For $s = 0$, (4.19) gives

$$\begin{aligned}(f, u') &= (u'' + \alpha u' + Au, u')\\ &= \frac{1}{2}\frac{d}{dt}\{|u'|^2 + a(u, u)\} + \alpha|u'|^2\\ &= \frac{1}{2}\frac{d}{dt}\{|u'|^2 + |u'|_1^2\} + \alpha|u'|^2.\end{aligned}$$

Hence (4.23). □

4.3. Time-Dependent Operators

In view of the applications in Chapter VI it is useful to extend the above results to the case where the operator A depends on t.

The equation that we consider reads

$$\frac{d^2u(t)}{dt^2} + \alpha\frac{du(t)}{dt} + Au(t) + A_1(t)u(t) = f(t), \tag{4.24}$$

and the initial conditions are still

$$u(0) = u_0, \qquad \frac{du}{dt}(0) = u_1. \tag{4.25}$$

The difference between (4.24) and (4.5) lies only in the adjunction of the term $A_1(t)u(t)$. The operator A is the same as in (4.5), the hypotheses on f, u_0, u_1 are still (4.4). Concerning the operators $A_1(t)$, we assume that for a.e. $t \in [0, T]$, $A_1(t) \in \mathscr{L}(V, H)$ and

There exists $M = M_T < \infty$ such that $|A_1(t)|_{\mathscr{L}(V,H)} \leq M_T$, a.e. $t \in [0, T]$. (4.26)

For the initial-value problem (4.24), (4.25), we have the following analog of Theorem 4.1 which is proved in essentially the same manner.

Theorem 4.3. *The hypotheses are those above, in particular* (4.26); f, u_0, u_1 *are given satisfying* (4.4). *Then there exists a unique solution* u *of* (4.24), (4.25) *such that*

$$u \in \mathscr{C}([0, T]; V), \qquad u' = \frac{du}{dt} \in \mathscr{C}([0, T]; H). \tag{4.27}$$

Remark 4.2. As before we can change t to $-t$ and solve problem (4.24), (4.25) on the interval $[-T, 0]$ if f is given in $L^2(-T, 0; H)$, or on the interval $[-T, T]$ if f is given in $L^2(-T, T; H)$.

CHAPTER III

Attractors of the Dissipative Evolution Equation of the First Order in Time: Reaction–Diffusion Equations. Fluid Mechanics and Pattern Formation Equations

Introduction

Our aim in this chapter is to describe some nonlinear evolution equations of the first order in time, which arise in mechanics and physics. In each case, we present briefly the physical model and the governing equations; then we present the mathematical setting of the equations which leads to the introduction of the corresponding semigroup $\{S(t)\}_{t\geq 0}$. Once the semigroup is defined, we address the following questions:

The nonlinear stability of the problem. This amounts to proving that the solutions of the evolution problem remain bounded as $t \to \infty$.

The existence of absorbing sets. This is a necessary mathematical step for the utilization of the results of Chapter I. From the physical point of view, it is also an "evidence" of the dissipativity of the problem (see Remark I.1.3).

The existence of a global attractor. Once the necessary properties of the semigroup are established we may apply the general results of Chapter I and particularly Theorem 1.1. That theorem produces the existence of an attractor which is maximal (for the inclusion) among the bounded attractors and among the bounded functional invariant sets; it fully describes the long-term behavior of the solutions of the equations. We have indicated in the General Introduction of this book that, for infinite-dimensional dynamical systems, in particular those derived from nonlinear partial differential equations, there is no general result for the existence of solution. Therefore the existence and uniqueness of a solution (i.e., the definition of the semigroup $S(t)$) is a part of the study of a given dynamical system, and for the sake of completeness we briefly present the corresponding results of existence and uniqueness. In fact, with all the tools already available, that requires few additional developments.

The chapter is organized as follows. In Section 1 we consider reaction–diffusion equations which include chemical reaction equations, neural network equations, ... We start with a simple equation with a polynomial nonlinearity (Section 1.1) and then we consider equations which leave a region invariant (and which sometimes are well posed only in such a region).

Section 2 is devoted to the Navier–Stokes equations of viscous incompressible fluids in space dimension 2. The three-dimensional case leading to further mathematical difficulties is considered in Chapter VII. In Section 3, we then consider several equations in fluid mechanics. Section 3.1 is devoted to the study of an abstract equation which contains (and slightly generalizes) the abstract equation associated with the two-dimensional Navier–Stokes equations. The equation of Section 3.1 includes the Navier–Stokes equations with a nonhomogeneous Dirichlet boundary condition (considered in Section 3.2), the magnetohydrodynamics equations (Section 3.3), the equations of the flow on a manifold (Section 3.4), and finally, with some appropriate adaptations to the proofs, the thermohydraulic equations (Section 3.5).

Section 4 contains the study of some pattern formation equations, namely, the Kuramoto–Sivashinsky equation (Section 4.1) and the Cahn–Hilliard equation (Section 4.2). Finally, in Section 5, we treat some semilinear equations with a monotone leading term.

We conclude this chapter with the proof (in Section 6) of a backward uniqueness result which, when applicable, implies the *injectivity* of the operators $S(t)$. Section 6.1 contains an abstract result and Section 6.2 contains the applications.

1. Reaction–Diffusion Equations

In this section we study the attractors associated with reaction–diffusion equations. In Section 1.1 we consider a single equation with a polynomial nonlinearity, and in Section 1.2 we consider more general systems which include the nerve equations (Hodgkin–Huxley equations and Fitz–Hugh–Nagumo equations), superfluid equations, some equations arising in combustion, and equations related to the Belousov–Zhabotinsky reactions in chemical dynamics.

Section 1.1.1 contains the description of the equations (of Section 1.1); Section 1.1.2 gives the results on absorbing sets and the existence of the maximal attractor. Section 1.1.3 contains a technical result which will be used constantly, the uniform Gronwall lemma; Section 1.1.4 contains a sketch of the proof of existence and uniqueness of solutions for the problem under consideration.

In Section 1.2, we first describe the equations (Section 1.2.1); then in Section 1.2.2 we establish the existence of absorbing sets and of a maximal attractor in the case of Dirichlet boundary conditions. Section 1.2.3 contains the examples mentioned above drawn from mathematical biology, physics, and chemistry. Finally, Section 1.2.4 gives an extension of the results of Section 1.2.2 to other boundary conditions, *Neumann* or space-periodic boundary conditions.

1.1. Equations with a Polynomial Nonlinearity

1.1.1. Description of the Equation and the Semigroup

Let Ω denote an open bounded set of $\mathbb{R}^n$ with boundary Γ and let g be a polynomial of odd degree with a positive leading coefficient

$$g(s) = \sum_{j=0}^{2p-1} b_j s^j, \qquad b_{2p-1} > 0. \tag{1.1}$$

We are interested in the following boundary-value problem involving a scalar function $u = u(x, t)$:

$$\frac{\partial u}{\partial t} - d\Delta u + g(u) = 0 \quad \text{in} \quad \Omega \times \mathbb{R}_+, \tag{1.2}$$

$$u = 0 \quad \text{on } \Gamma, \tag{1.3}$$

where $d > 0$ is given. If we consider an initial-value problem, then we supplement (1.2), (1.3) with an initial condition

$$u(x, 0) = u_0(x), \qquad x \in \Omega, \tag{1.4}$$

where u_0 is given. For the mathematical setting of this problem we write $H = L^2(\Omega)$, $V = H_0^1(\Omega)$, and we note the following theorem which follows from the general results of existence and uniqueness of solutions for parabolic equations (see, for instance, A. Friedman [1], J.L. Lions [2]).

Theorem 1.1. *For u_0 given in H, there exists a unique solution u of* (1.2)–(1.4) *which satisfies*

$$u \in L^2(0, T; H_0^1(\Omega)) \cap L^{2p}(0, T; L^{2p}(\Omega)), \qquad \forall T > 0, \tag{1.5}$$

$$u \in \mathscr{C}(\mathbb{R}_+; H). \tag{1.6}$$

The mapping $u_0 \to u(t)$ is continuous in H. If, furthermore, $u_0 \in H_0^1(\Omega)$, then u belongs to

$$\mathscr{C}([0, T); V) \cap L^2(0, T; H^2(\Omega)), \qquad \forall T > 0.$$

The proof of this theorem will be sketched in Section 1.1.4, and some complements to this result will also be given. Theorem 1.1 is however sufficient to define the semigroup $S(t)$; we set

$$S(t)\colon u_0 \in H \mapsto u(t) \in H. \tag{1.7}$$

The basic properties I.(1.1) and I.(1.4) are satisfied and in view of an application of Theorem I.1.1, we will now prove the uniform compactness property of the operators $S(t)$ and the existence of an absorbing set.

1.1.2. Absorbing Sets and Attractors

Usually, proving the existence of absorbing sets amounts to proving a priori estimates; for the equation under consideration, we will prove the existence of absorbing sets in H and V.

Using the Young inequality we infer from (1.1) the existence of a constant $c_1' > 0$ such that

$$\left| \sum_{j=0}^{j=2p-2} b_j s^{j+1} \right| \leq \tfrac{1}{2} b_{2p-1} s^{2p} + c_1', \qquad \forall s,$$

and hence

$$\tfrac{1}{2} b_{2p-1} s^{2p} - c_1' \leq g(s)s \leq \tfrac{3}{2} b_{2p-1} s^{2p} + c_1', \qquad \forall s \in \mathbb{R}. \tag{1.8}$$

We multiply (1.2) by $u = u(x, t)$ and integrate over Ω. Using (1.3) and the Green formula we find

$$\frac{1}{2}\frac{d}{dt}|u|^2 + d\|u\|^2 + \int_\Omega g(u)u\,dx = 0. \tag{1.9}$$

We have written $|\cdot|$ and $\|\cdot\|$ for the norms in $L^2(\Omega)$ and $H_0^1(\Omega)$,

$$|u| = \left(\int_\Omega u^2\,dx\right)^{1/2}, \qquad \|u\| = \left(\sum_{i=1}^{n} \left|\frac{\partial u}{\partial x_i}\right|^2\right)^{1/2}.$$

Thanks to (1.8) we obtain

$$\frac{d}{dt}|u|^2 + 2d\|u\|^2 + \int_\Omega b_{2p-1}\, u^{2p}\,dx \leq 2c_1'|\Omega|, \tag{1.10}$$

$|\Omega| =$ the measure (volume) of Ω. Due to the Poincaré inequality (see II.(1.6)), there exists a constant $c_0 = c_0(\Omega)$, such that

$$|u| \leq c_0\|u\|, \qquad \forall u \in H_0^1(\Omega),$$

and, setting $c_2' = 2c_1'|\Omega|$, we infer from (1.10) that

$$\frac{d}{dt}|u|^2 + \frac{2d}{c_0^2}|u|^2 \leq c_2'.$$

Using the classical Gronwall lemma[1] we see that

$$|u(t)|^2 \leq |u_0|^2 \exp\left(-\frac{2d}{c_0^2}t\right) + \frac{c_2' c_0^2}{2d}\left(1 - \exp\left(-\frac{2d}{c_0^2}t\right)\right). \tag{1.11}$$

[1] In Section 1.1.4, we recall the classical Gronwall lemma and derive an extension of it, the uniform Gronwall lemma.

Thus

$$\limsup_{t\to+\infty} |u(t)| \le \rho_0, \qquad \rho_0^2 = \frac{c_2' c_0^2}{2d}. \tag{1.12}$$

There exists an absorbing set $\mathscr{B}_0$ in H, namely, any ball of H centered at 0 of radius $\rho_0' > \rho_0$. If $\mathscr{B}$ is a bounded set of H, included in a ball $B(0, R)$ of H, centered at 0 of radius R, then $S(t)\mathscr{B} \subset B(0, \rho_0')$ for $t \ge t_0(\mathscr{B}; \rho_0')$

$$t_0 = \frac{c_0^2}{2d} \log \frac{R^2}{(\rho_0')^2 - \rho_0^2}. \tag{1.13}$$

We also infer from (1.10), after integration in t, that

$$\begin{aligned} &2d \int_t^{t+r} \|u\|^2\, ds + \int_t^{t+r} \int_\Omega b_{2p-1} u^{2p}\, dx\, ds \\ &\le rc_2' + |u(t)|^2, \qquad \forall r > 0. \end{aligned} \tag{1.14}$$

With (1.12) we conclude that

$$\begin{aligned} &\limsup_{t\to\infty} \left\{ 2d \int_t^{t+r} \|u\|^2\, ds + \int_t^{t+r} \int_\Omega b_{2p-1} u^{2p}\, dx\, ds \right\} \\ &\le rc_2' + \rho_0^2, \qquad \forall r > 0, \end{aligned} \tag{1.15}$$

and if $u_0 \in \mathscr{B} \subset B(0, R)$ and $t \ge t_0(\mathscr{B}, \rho_0')$, then

$$2d \int_t^{t+r} \|u\|^2\, ds + \int_t^{t+r} \int_\Omega b_{2p-1} u^{2p}\, dx\, ds \le rc_2' + (\rho_0')^2. \tag{1.16}$$

Remark 1.1. Relation (1.9) was derived under the implicit assumption that u is sufficiently regular. However, using the methods of Section II.3.2, this relation can be fully proved for the weak solution given by Theorem 1.1 (see Section 1.1.4.); the same remark applies to relation (1.17) below.

Absorbing Set in $H_0^1(\Omega)$

We now prove the existence of an absorbing set in $H_0^1(\Omega)$ and the uniform compactness of the $S(t)$. For that purpose we need another energy-type equality, similar to (1.9): it is obtained by multiplying (1.2) by $-\Delta u$ and integrating over Ω.

We have, using (1.3) and the Green formula,

$$-\int_\Omega \Delta u \frac{\partial u}{\partial t}\, dx = \sum_{i=1}^n \int_\Omega \frac{\partial u}{\partial x_i} \frac{\partial^2 u}{\partial t \partial x_i}\, dx = \frac{1}{2} \frac{d}{dt} \|u\|^2,$$

$$-\int_\Omega \Delta u g(u)\, dx = -\int_\Omega g(u) \Delta u\, dx = \sum_{i=1}^n \int_\Omega g'(u) \left(\frac{\partial u}{\partial x_i} \right)^2 dx.$$

Hence

$$\frac{1}{2} \frac{d}{dt} \|u\|^2 + d|\Delta u|^2 + \int_\Omega g'(u)(\operatorname{grad} u)^2\, dx = 0. \tag{1.17}$$

As in (1.8), we can prove with repeated applications of the Young inequality that there exists $c_3' > 0$ such that

$$\frac{2p-1}{2} b_{2p-1} s^{2p-2} - c_3' \le g'(s) = \sum_{j=1}^{2p-1} j b_j s^{j-1}$$
$$\le \tfrac{3}{2}(2p-1) b_{2p-1} s^{2p-2} + c_3',$$
$$\forall s \in \mathbb{R}. \qquad (1.18)$$

We also infer from general results on the Dirichlet problem in Ω (see Chapter II) that $|\Delta u|$ is, on $H_0^1(\Omega) \cap H^2(\Omega)$, a norm equivalent to that induced by $H^2(\Omega)$; therefore, there exists a constant $c_1 = c_1(\Omega)$ depending on Ω such that

$$\|u\| \le c_1 |\Delta u|, \qquad \forall u \in H_0^1(\Omega). \qquad (1.19)$$

Setting $c_4' = \frac{1}{2}(2p-1) b_{2p-1} > 0$ we then deduce from (1.18)

$$\frac{1}{2}\frac{d}{dt}\|u\|^2 + \left(\frac{d}{c_1^2} - c_3'\right)\|u\|^2 + c_4' \int_\Omega u^{2p-2} (\operatorname{grad} u)^2 \, dx \le 0. \qquad (1.20)$$

In particular,

$$\frac{d}{dt}\|u\|^2 \le 2c_3' \|u\|^2. \qquad (1.21)$$

If u_0 is in $V\ (= H_0^1(\Omega))$, then the usual Gronwall lemma shows that

$$\|u(t)\|^2 \le \|u_0\|^2 \exp(2c_3' t), \qquad \forall t > 0. \qquad (1.22)$$

A bound valid for all $t \in \mathbb{R}_+$ is obtained by application of the uniform Gronwall lemma (see Section 1.1.3); for an arbitrary fixed $r > 0$, we find

$$\|u(t+r)\|^2 \le \frac{\kappa}{r} \exp(2c_3' r), \qquad t \ge t_*, \qquad (1.23)$$

provided

$$\int_t^{t+r} \|u(s)\|^2 \, ds \le \kappa, \qquad \forall t \ge t_*. \qquad (1.24)$$

An explicit value of κ can be derived from (1.10), and the computation above, when $t_* = 0$. Hence (1.23) provides a uniform bound for $\|u(t)\|$, $t \ge r$, while (1.21) provides a uniform bound for $\|u(t)\|$ for $0 \le t \le r$. For our purpose, it is simpler and sufficient to set $t_* = t_0$ (as in (1.13)), in which case, the value of κ is given by (1.16),

$$\kappa = \frac{1}{2d}(r c_2' + \rho_0'^2). \qquad (1.25)$$

It follows that the ball of V centered at 0 of radius ρ_1 is absorbing in V, where

$$\rho_1^2 = \frac{\kappa}{r} \exp(2c_3' r), \qquad \kappa \text{ as in (1.25)},$$

and if u_0 belongs to the ball $B(0, R)$ of H centered at 0 of radius r, then $u(t)$

enters this absorbing set denoted $\mathscr{B}_1$ at a time $t \leq t_0 + r$, and remains in it for $t \geq t_0 + r$. At the same time, this result provides the uniform compactness of $S(t)$: any bounded set $\mathscr{B}$ of H is included in such a ball $B(0, R)$, and for $u_0 \in \mathscr{B}$ and $t \geq t_0 + r$, t_0, r as above, $u(t)$ belongs to $\mathscr{B}_1$ which is bounded in $H_0^1(\Omega)$ and relatively compact in $L^2(\Omega)$.

In conclusion, the assumptions I.(1.1), I.(1.4), I.(1.12) are satisfied and we have proved the existence of an absorbing set $\mathscr{B}_0$ in H. Theorem I.1.1. applies with $\mathscr{U} = H$ and we have

Theorem 1.2. *Let Ω denote an open bounded set of $\mathbb{R}^n$ and let g denote a polynom satisfying* (1.1). *The semigroup $S(t)$ associated with the boundary-value problem* (1.2)–(1.4) *possesses a maximal attractor $\mathscr{A}$ which is bounded in $H_0^1(\Omega)$, compact and connected in $L^2(\Omega)$. Its basin of attraction is the whole space $L^2(\Omega)$, $\mathscr{A}$ attracts the bounded sets of $L^2(\Omega)$.*

Remark 1.2. Assume that in (1.20), $d > c_1^2 c_3'$, c_1 given by (1.19) and c_3' given by (1.18). Then $\|u(t)\|^2$ decreases exponentially to 0 as $t \to \infty$,

$$\|u(t)\|^2 \leq \|u(0)\|^2 \exp\left(-2\left(\frac{d}{c_1^2} - c_3'\right)\right).$$

Thus all the trajectories converge to 0. This is of course the trivial case as far as the dynamics is concerned. The condition $d > c_1^2 c_3'$ which leads to this situation means that the nondissipative part of the nonlinear term ("measured" by c_3') is small compared to the dissipativity coefficient d.

The same conclusion is also valid if $g'(s) \geq 0$, $\forall s$ (with the same proof). This is true, for instance, if the equation is linear $p = 1$ and $b_0 = 0$ ($g(s) = b_1 s$, $b_1 > 0$); more generally it is also true for any monotone increasing function g.

Note that the fact that all trajectories converge to 0 does not necessarily imply that the universal attractor reduces to the point $\{0\}$. It can indeed contain, for instance, one (or many) heteroclinic curve(s).

Remark 1.3. Let us mention here an interesting variant of the proof of absorbing sets providing an absorbing ball for which the *entering time is independent of the initial data, however large the norm is of this initial data* (B. Nicolaenko). This property, which is due to the particular form (the rapid growth) of the nonlinear term, will not reappear in general in the subsequent examples:

We begin with the absorbing ball in $L^2(\Omega)$. Thanks to the Hölder inequality

$$\int_\Omega u^{2p}\, dx \geq |\Omega|^{-p/p'} \left(\int_\Omega u^2\, dx\right)^p$$

($1/p' = 1 - 1/p$). Therefore we infer from (1.10) that

$$y' + \gamma y^p \leq \delta,$$

where $y = |u|^2$, $\gamma = b_{2p-1}|\Omega|^{-p/p'}$, $\delta = 2c_1'|\Omega|$. A variant of the Gronwall lemma proved in Section 5 (Lemma 5.1) shows that this differential inequality implies

$$y(t) \leq \left(\frac{\delta}{\gamma}\right)^{1/p} + \frac{1}{(\gamma(p-1)t)^{1/(p-1)}}, \qquad \forall t > 0.$$

Let ρ_2 be any number strictly larger than $(\delta/\gamma)^{1/p}$ and

$$T_0 = \frac{1}{\gamma(p-1)}\left(\rho_2^2 - \left(\frac{\delta}{\gamma}\right)^{1/p}\right)^{-(p-1)}.$$

The above relations show that for any set $\mathscr{B}$ of $L^2(\Omega)$, bounded or not, $S(t)\mathscr{B}$ is included in the ball $\mathscr{B}_2$ of $L^2(\Omega)$ centered at 0 of radius ρ_2, if $t \geq T_0$, T_0 as above. Then we replace $\mathscr{B}_0'$ by $\mathscr{B}_2$ in the above proof of the existence of an absorbing set in $H_0^1(\Omega)$ and we obtain a similar result in $H_0^1(\Omega)$.

Remark 1.4. The example given above is a model problem for which we did not seek the most general assumptions. For instance, with very slight modifications in the proof, we can assume that the coefficients of g depend on x and that g contains a zeroth-order term

$$g(x, s) = \sum_{j=0}^{2p-1} b_j(x)s^j,$$

provided $b_j \in L^\infty(\Omega)$, $j = 0, \ldots, 2p-1$, and

$$b_{2p-1}(x) \geq \beta > 0, \qquad \text{a.e.} \quad x \in \Omega.$$

More general functions $g(x, s)$ and other boundary conditions can also be considered; see a related situation in Section 1.2.5.

Remark 1.5. Let us briefly indicate the few modifications of the proofs which are necessary if we leave g as in (1.1), but replace the boundary condition (1.3) by

$$\frac{\partial u}{\partial \mathbf{v}} = 0 \quad \text{on } \Gamma, \tag{1.3$'$}$$

$\mathbf{v}$ being the unit outward normal on Γ, or,

$$\Omega = (0, L)^n \text{ and } u \text{ is } \Omega\text{-periodic.} \tag{1.3$''$}$$

For the functional setting, we let $H = L^2(\Omega)$ but replace $V = H_0^1(\Omega)$ by $V = H^1(\Omega)$ for (1.3′) or $V = H_{\text{per}}^1(\Omega)$ for (1.3″). Theorem 1.1 holds without any modification in its formulation or in its proof. In the study of the existence of an L^2-absorbing set, after (1.10), since the Poincaré inequality is not available, we write

$$\begin{aligned}|u|^2 = \int_\Omega u^2\,dx &\le \text{(by the Hölder's inequality)}\\ &\le \left(\int_\Omega u^{2p}\,dx\right)^{1/p} |\Omega|^{1/p'} \qquad (1/p' = 1 - 1/p)\\ &\le \text{(by the Young inequality)}\\ &\le \tfrac{1}{2} b_{2p-1} \int_\Omega u^{2p}\,dx + c b_{2p-1}^{-p'/p} |\Omega|.\end{aligned}$$

Thus (1.10) implies

$$\frac{d}{dt}|u|^2 + 2d\|u\|^2 + |u|^2 + \tfrac{1}{2} b_{2p-1} \int_\Omega u^{2p}\,dx \le c_2' + c b_{2p-1}^{-p'/p} |\Omega| = c_*. \qquad (1.10')$$

We obtain (1.11)–(1.13) with $c_0^2 = 2d$, and (1.14)–(1.16) with slight modifications to the value of the constants.

For the absorbing set in V, slight modifications to the proof are necessary, since (1.19) is not valid; they are left as an exercise. Finally, we arrive to the same conclusions as in Theorem 1.2.

1.1.3. The Uniform Gronwall Lemma

We present the uniform Gronwall lemma that was implicitly used for the first time by C. Foias and G. Prodi [1] in the context of the Navier–Stokes equations.

Let g, h, y be three locally integrable functions on $]t_0, +\infty[$ that satisfy

$$\frac{dy}{dt} \le gy + h \qquad \text{for} \quad t \ge t_0, \tag{1.26}$$

the function dy/dt being also locally integrable. For the usual Gronwall lemma we multiply (1.26) by

$$\exp\left(-\int_{t_0}^t g(\tau)\,d\tau\right),$$

and observe that the resulting inequality reads

$$\frac{d}{dt}\left(y(t)\exp\left(-\int_{t_0}^t g(\tau)\,d\tau\right)\right) \le h(t)\exp\left(-\int_{t_0}^t g(\tau)\,d\tau\right).$$

Hence, by integration between t_0 and t,

$$\begin{aligned}y(t) \le{}& y(t_0)\exp\left(\int_{t_0}^t g(\tau)\,d\tau\right)\\ &+ \int_{t_0}^t h(s)\exp\left(-\int_t^s g(\tau)\,d\tau\right) ds, \qquad t \ge t_0.\end{aligned} \tag{1.27}$$

This is the usual Gronwall inequality which is useful for bounded values of t. When $t \to \infty$, this relation is not sufficient for our purposes since it allows an exponential growth of y; for instance, for $y \geq 0$, $h = 0$, $g = 1$, we find

$$y(t) \leq y(t_0) \exp(t - t_0).$$

We now present an alternative form of this inequality that provides (under slightly stronger assumptions) a bound valid uniformly for $t \geq t_0$.

Lemma 1.1 (The Uniform Gronwall Lemma). *Let g, h, y, be three positive locally integrable functions on $]t_0, +\infty[$ such that y' is locally integrable on $]t_0, +\infty[$, and which satisfy*

$$\frac{dy}{dt} \leq gy + h \qquad \text{for} \quad t \geq t_0, \tag{1.28}$$

$$\int_t^{t+r} g(s)\,ds \leq a_1, \qquad \int_t^{t+r} h(s)\,ds \leq a_2, \qquad \int_t^{t+r} y(s)\,ds \leq a_3 \quad \text{for} \quad t \geq t_0, \tag{1.29}$$

where r, a_1, a_2, a_3, are positive constants. Then

$$y(t + r) \leq \left(\frac{a_3}{r} + a_2\right) \exp(a_1), \qquad \forall t \geq t_0. \tag{1.30}$$

Proof. Assume that $t_0 \leq t \leq s \leq t + r$. We write (1.28) with t replaced by s, multiply by

$$\exp\left(-\int_t^s g(\tau)\,d\tau\right),$$

and obtain the relation

$$\frac{d}{ds}\left(y(s) \exp\left(-\int_t^s g(\tau)\,d\tau\right)\right) \leq h(s) \exp\left(-\int_t^s g(\tau)\,d\tau\right) \leq h(s).$$

Then by integration between t_1 and $t + r$

$$\begin{aligned} y(t + r) &\leq y(t_1) \exp\left(\int_{t_1}^{t+r} g(\tau)\,d\tau\right) + \left(\int_{t_1}^{t+r} h(s)\,ds\right) \exp\left(\int_t^{t+r} g(\tau)\,d\tau\right) \\ &\leq (y(t_1) + a_2) \exp(a_1). \end{aligned}$$

Integration of this last inequality, with respect to t_1 between t and $t + r$, gives precisely (1.30).

1.1.4. *Proof of Theorem 1.1*

We sketch the proof of existence and uniqueness of solutions for (1.2)–(1.4). We proceed as in Chapter II, Section 3. Let $V = H_0^1(\Omega)$, $H = L^2(\Omega)$, and let A be the linear operator associated to the bilinear form

$$a(u, v) = d \int_\Omega \operatorname{grad} u \cdot \operatorname{grad} v\,dx.$$

We consider the orthonormal basis of H consisting of the eigenvectors of A,

$$Aw_j = \lambda_j w_j, \qquad \forall j,$$

and we implement the Faedo–Galerkin method with these functions. For each integer m we look for an approximate solution u_m of (1.2)–(1.4) of the form

$$u_m(t) = \sum_{i=1}^{m} g_{im}(t) w_i, \tag{1.31}$$

satisfying

$$\left(\frac{du_m}{dt}, w_j\right) + a(u_m, w_j) + (g(u_m), w_j) = 0, \qquad j = 1, \ldots, m, \tag{1.32}$$

$$u_m(0) = u_{0m}, \tag{1.33}$$

where u_{0m} is the orthogonal projection in H of u_0 onto the space spanned by $w_1, \ldots, w_m$. The existence of u_m on some interval $[0, T_m[$, follows from standard results of existence of solutions of ordinary differential equations: that $T_m = +\infty$ is a consequence of these results and of the following a priori estimates.

The first energy-type equality is obtained by multiplying (1.32) by g_{jm} and summing those relations for $j = 1, \ldots, m$. We obtain (1.9) precisely with u replaced by u_m. Then with the same computations as those following (1.9) we conclude that, for $T > 0$ arbitrary,

u_m is bounded independently of m in

$$L^\infty(0, T; H), \qquad L^2(0, T; V) \quad \text{and} \quad L^{2p}(0, T; L^{2p}(\Omega)). \tag{1.34}$$

By weak compactness we find a subsequence still denoted as u_m and u in $L^\infty(0, T; H) \cap L^2(0, T; V) \cap L^{2p}(0, T; L^{2p}(\Omega))$ such that

$$u_m \to u \quad \text{in } L^2(0, T; V) \text{ and } L^{2p}(0, T; L^{2p}(\Omega)) \text{ weakly}, \tag{1.35}$$

$$u_m \to u \quad \text{in } L^\infty(0, T; H) \text{ weak-star}. \tag{1.36}$$

We pass to the limit in (1.32), (1.33) and find that

$$\frac{d}{dt}(u, v) + a(u, v) + (g(u), v) = 0, \qquad \forall v \in V \cap L^{2p}(\Omega).$$

Compared with the linear case, the main difficulty here is showing that $g(u_m)$ converges to $g(u)$ in some weak sense. This follows from a classical compactness argument which will not be presented here; the reader is referred, for instance, to J.L. Lions [2].

Thus u satisfies

$$\frac{du}{dt} + Au + g(u) = 0. \tag{1.37}$$

It shows that $u' = -Au - g(u)$ is in

$$L^2(0, T; V') + L^q(0, T; L^q(\Omega)),$$

q the conjugate exponent of $2p$, $(1/q + 1/2p = 1)$. This space is in duality with

$$L^2(0, T; V) \cap L^{2p}(0, T; L^{2p}(\Omega)),$$

and with a slightly modified version of Lemma II.3.2 we see that u is in $\mathscr{C}([0, T], H)$. Hence $u(0)$ makes sense and, of course, (1.4) follows from (1.33) by passage to the limit $m \to \infty$.

The uniqueness in Theorem 1.1 and the continuous dependence on u_0 are proved by standard methods. Further regularity results for the solution are proved by deriving an energy-type equation similar to (1.17). This relation is obtained on multiplying the relations (1.32) by $\lambda_j g_{jm}$, and using the relations II.(3.24) followed by summing over $j = 1, \ldots, m$. We obtain (1.17) precisely with u replaced by u_m. Repeating the computations which follow (1.17) we see that

$$u_m \text{ is bounded independently of } m \text{ in } L^\infty(\eta, T; V) \text{ and } L^2(\eta, T; H^2(\Omega)), \tag{1.38}$$

with $\eta = 0$ if $u_0 \in H_0^1(\Omega)$, $\eta > 0$, arbitrary ($\eta < T$) if $u_0 \in L^2(\Omega)$. In the limit $m \to \infty$, we find that

$$u \in \mathscr{C}(]0, T]; V) \cap L^2(\eta, T; H^2(\Omega)), \qquad \forall \eta > 0 \qquad \text{if} \quad u_0 \in L^2(\Omega), \tag{1.39}$$

and

$$u \in \mathscr{C}([0, T]; V) \cap L^2(0, T; H^2(\Omega)) \qquad \text{if} \quad u_0 \in V = H_0^1(\Omega). \tag{1.40}$$

The continuity of u (with values in V) and the relation (1.17) are proved by approximation with a slightly modified version of Lemma II.3.2.

Remark 1.6. The injectivity property of $S(t)$, which amounts to a backward uniqueness property for this problem, is true and will be proved in Section 6.

1.2. Equations with an Invariant Region

1.2.1. The Equations and the Semigroup

We again denote by Ω an open bounded set of $\mathbb{R}^n$ with boundary Γ. We consider a boundary-value problem involving a vector function $u = (u_1, \ldots, u_m)$ from $\Omega \times \mathbb{R}_+$ into $\mathbb{R}^m$; u satisfies an equation

$$\frac{\partial u}{\partial t} - D\Delta u + g(u, x) = 0 \qquad \text{in} \quad \Omega \times \mathbb{R}_+, \tag{1.41}$$

where D is a positive diagonal matrix of diffusion coefficients

$$D = \begin{pmatrix} d_1 & 0 & \ldots & 0 \\ 0 & d_2 & \ldots & 0 \\ \ldots & \ldots & \ldots & \ldots \\ 0 & & \ldots & d_m \end{pmatrix}, \qquad d_i > 0,$$

and $g = (g_1, \ldots, g_m)$ is continuous on $\mathbb{R}^m \times \bar{\Omega}$. Equation (1.41) is supplemented with boundary conditions of either Dirichlet or Neumann type or, if $\Omega = \prod_{i=1}^{n}]0, L_i[$, of periodicity type; we can also consider a combination of such conditions where different components of u are subjected to either of these conditions. For the moment we restrict ourselves to a Dirichlet boundary condition

$$u_i = 0 \quad \text{on } \Gamma, \qquad \forall i. \tag{1.42}$$

In the case of an initial-value problem, u also satisfies the initial condition

$$u(x, 0) = u_0(x), \qquad x \in \Omega, \tag{1.43}$$

where u_0 is given.

Three assumptions are made concerning this boundary-value problem. First,

> There exists a closed convex region $\mathscr{D} \subset \mathbb{R}^m$ which is positively invariant. (1.44)

This means that if $u_0(x) \in \mathscr{D}$ for every (or almost every) x in Ω, then $u(x, t) \in \mathscr{D}$ for all $t > 0$ for which the solution of (1.41)–(1.43) exists. The reader is referred to J. Smoller [1] for the derivation of sufficient or necessary and sufficient conditions of geometric and algebraic character on D, $\mathscr{D}$, and g, which guarantee that $\mathscr{D}$ is a positively invariant region for (1.41). The second assumption which has to be proved in all specific examples is that the initial-value problem is well posed if u_0 is in $L^2(\Omega; \mathscr{D})$, i.e., $u_0 \in L^2(\Omega)^m$ and $u_0(x) \in \mathscr{D}$ for a.e. $x \in \Omega$. More precisely

> For every $u_0 \in L^2(\Omega; \mathscr{D})$, the problem (1.41)–(1.43) possesses a unique solution u for all time, $u(t) \in L^2(\Omega; \mathscr{D})$, $\forall t$, $u \in L^2(0, T; H_0^1(\Omega)^m)$, $\forall T > 0$, and the mapping $S(t)\colon u_0 \mapsto u(t)$ is continuous in $L^2(\Omega)^m$, $\forall t \geq 0$.
> Furthermore, if $u_0 \in H_0^1(\Omega)^m \cap L^2(\Omega; \mathscr{D})$, then $u \in \mathscr{C}([0, T]; H_0^1(\Omega)^m) \cap L^2(0, T; H^2(\Omega)^m)$, $\forall T > 0$. (1.45)

Finally, we assume that

> g is continuous and bounded on $\mathscr{D} \times \bar{\Omega}$ (and we set $c_2 = \sup_{u \in \mathscr{D}, x \in \bar{\Omega}} |g(u, x)|$). (1.46)

1.2.2. Absorbing Sets and Attractors

We consider the space $L^2(\Omega)^m$ that is endowed with the usual norm $|\cdot|$, and we denote by V the space $H_0^1(\Omega)^m$ endowed with the norm

$$\|u\| = \left\{ \sum_{i=1}^{m} \int_{\Omega} |\operatorname{grad} u_i|^2 \, dx \right\}^{1/2}. \tag{1.47}$$

We multiply (1.41) by $u = u(x, t)$ and integrate over Ω. By the Green formula

$$-\int_\Omega D\Delta u \cdot u \, dx = \sum_{i=1}^m d_i \int_\Omega |\text{grad } u_i|^2 \, dx. \tag{1.48}$$

The right-hand side of (1.48) is denoted $a(u, u)$, $a(u, v)$ representing the symmetric bilinear form on V associated to this quadratic form.

Hence we obtain

$$\frac{1}{2}\frac{d}{dt}|u|^2 + a(u, u) + \int_\Omega g(u)u \, dx = 0. \tag{1.49}$$

We infer from the Poincaré inequality the existence of a constant $c_3 > 0$ such that[1]

$$|u| \le c_3 \|u\|, \qquad \forall u \in V, \tag{1.50}$$

and since a is coercive on V

$$\begin{aligned} &a(u, u) \ge d_0 \|u\|^2 \ge c_4' |u|^2, \qquad \forall u \in V, \\ &d_0 = \min\{d_i, i = 1, \dots, m\}, \qquad c_4' = d_0/c_3^2. \end{aligned} \tag{1.51}$$

Then we deduce from (1.49) and (1.50)

$$\begin{aligned} \frac{1}{2}\frac{d}{dt}|u|^2 + c_4'|u|^2 &\le c_2 \int_\Omega |u(x)| \, dx \\ &\le \text{(with the Schwarz inequality)} \\ &\le c_2 |\Omega|^{1/2} |u| \\ &\le \frac{c_4'}{2}|u|^2 + \frac{c_2^2}{2c_4'}|\Omega|, \\ \frac{d}{dt}|u|^2 + c_4'|u|^2 &\le \frac{c_2^2}{c_4'}|\Omega|. \end{aligned} \tag{1.52}$$

By integration of (1.52) we find

$$|u(t)|^2 \le |u_0|^2 \exp(-c_4' t) + \frac{c_2^2}{c_4'^2}|\Omega|(1 - \exp(-c_4' t)) \tag{1.53}$$

$$\limsup_{t \to \infty} |u(t)|^2 \le \frac{c_2^2}{c_4'^2}|\Omega|. \tag{1.54}$$

Setting $\rho_0 = (c_2/c_4')|\Omega|^{1/2}$, we see that any ball of H, centered at 0 of radius $\rho_0' \ge \rho_0$, is positively invariant for the semigroup $S(t)$ and if $\rho_0' > \rho_0$, this ball is also absorbing in H. We choose ρ_0' and denote by $\mathscr{B}_0$ the ball $B(0, \rho_0')$ of H. If $\mathscr{B}$ is any bounded set of H included in $B(0, R)$ for some $R > 0$, then we easily

[1] $c_3 = \sqrt{m c_0}$, c_0 as in II.(1.6).

deduce from (1.53) that $S(t)\mathscr{B} \subset \mathscr{B}_0$ for $t \geq t_0 = t_0(\mathscr{B}, \rho_0')$,

$$t_0 = \frac{1}{c_4'} \log \frac{R^2}{\rho_0'^2 - \rho_0^2}. \tag{1.55}$$

Then we rewrite (1.49) as follows

$$\begin{aligned}
\frac{d}{dt}|u|^2 + 2a(u, u) &= -2 \int_\Omega g(u)u \, dx \\
&\leq 2c_2 \int_\Omega |u(x)| \, dx \\
&\leq 2c_2 |\Omega|^{1/2} |u| \\
&\leq (\text{by } 1.51)) \\
&\leq 2c_2 \left\{\frac{a(u, u)}{c_4'}\right\}^{1/2} |\Omega|^{1/2} \\
&\leq a(u, u) + \frac{c_2^2}{c_4'} |\Omega|,
\end{aligned}$$

$$\frac{d}{dt}|u|^2 + a(u, u) \leq \frac{c_2^2}{c_4'} |\Omega|. \tag{1.56}$$

For $r > 0$ fixed, we integrate this relation between t and $t + r$ and obtain

$$\int_t^{t+r} a(u, u) \, ds \leq r \frac{c_2^2}{c_4'} |\Omega| + |u(t)|^2. \tag{1.57}$$

With (1.54) we conclude that

$$\limsup_{t \to \infty} \int_t^{t+r} a(u, u) \, ds \leq \frac{c_2^2}{c_4'} |\Omega| \left(r + \frac{1}{c_4'}\right), \tag{1.58}$$

and if $u_0 \in \mathscr{B} \subset B(0, R)$ and $t \geq t_0(\mathscr{B}, \rho_0')$, then

$$\int_t^{t+r} a(u, u) \, ds \leq r c_4' \rho_0^2 + \rho_0'^2. \tag{1.59}$$

Absorbing Set in $H_0^1(\Omega)$

We now prove the existence of an absorbing set in $H_0^1(\Omega)$. For that purpose, we multiply equation (1.41) by $-\Delta u$ and integrate over Ω.

As in Section 1.1, using the Green formula, we see that

$$-\int_\Omega \Delta u \cdot \frac{\partial u}{\partial t} \, dx = \sum_{i=1}^m \sum_{j=1}^n \int_\Omega \frac{\partial u_i}{\partial x_j} \frac{\partial^2 u_i}{\partial x_j \partial t} \, dx = \frac{1}{2} \frac{d}{dt} \|u\|^2.$$

We then obtain

$$\frac{1}{2} \frac{d}{dt} \|u\|^2 + (D\Delta u, \Delta u) = \int_\Omega g(u) \Delta u \, dx. \tag{1.60}$$

Using the constants c_2 and d_0, introduced in (1.48) and (1.51), we infer from (1.60)

$$\frac{d}{dt}\|u\|^2 + 2d_0|\Delta u|^2 \le c_2 \int_\Omega |\Delta u(x)|\, dx$$
$$\le c_2|\Omega|^{1/2}|\Delta u|$$
$$\le d_0|\Delta u|^2 + \frac{c_2^2}{d_0}|\Omega|,$$

$$\frac{d}{dt}\|u\|^2 + d_0|\Delta u|^2 \le \frac{c_2^2}{d_0}|\Omega|. \tag{1.61}$$

Using (1.19), we have

$$\|u\| \le c_3|\Delta u|, \qquad \forall u \in V, \quad c_3 = \sqrt{m c_1}, \tag{1.62}$$

and (1.61) implies

$$\frac{d}{dt}\|u\|^2 + c_5'\|u\|^2 \le c_6', \tag{1.63}$$

$c_5' = d_0/c_3^2$, $c_6' = c_2^2|\Omega|/d_0$. By integration, (1.63) implies

$$\|u(t)\|^2 \le \|u_0\|^2 \exp(-c_5' t) + \frac{c_6'}{c_5'}(1 - \exp(-c_5' t)) \tag{1.64}$$

and

$$\limsup_{t\to\infty} \|u(t)\|^2 \le \frac{c_6'}{c_5'}. \tag{1.65}$$

The ball of V centered at 0 of radius $\rho_1 = (c_6'/c_5')^{1/2}$, $B_V(0, \rho_1)$ is positively invariant and the same is true of any ball $B_V(0, \rho_1')$, $\rho_1' \ge \rho_1$. It is also easy to deduce from (1.64) that any ball $B_V(0, \rho_1')$, $\rho_1' > \rho_1$, is absorbing in V.

To conclude we now show that the operators $S(t)$ are uniformly compact. We multiply (1.63) by t and obtain

$$\frac{d}{dt}(t\|u\|^2) + c_5' t\|u\|^2 \le c_6' t + \|u\|^2. \tag{1.66}$$

By integration between 0 and r ($r > 0$)

$$\|u(r)\|^2 \le c_6' r + \frac{1}{r}\int_0^r \|u\|^2\, ds$$
$$\le \text{(with (1.57), (1.53), (1.51))}$$
$$\le c_6' r + \frac{c_2^2|\Omega|}{d_0 c_4'} + \frac{1}{r d_0}|u_0|^2 + \frac{c_2^2|\Omega|}{c_4'^2 r d_0}.$$

If u_0 is in a ball $B_H(0, R)$ of H, then $u(r) = S(r)u_0$ is in a ball $B_V(0, R_1)$ of V with

$$R_1^2 = c_6' r + \frac{c_2^2|\Omega|}{d_0 c_4'} + \frac{R^2}{r d_0} + \frac{c_2^2|\Omega|}{r c_4' d_0}.$$

After a certain time $t_1 = t_1(\mathscr{B}, \rho_1')$, which is easily computed from (1.64), we find that $u(t)$ belongs to the absorbing set $\mathscr{B}_1 = B_V(0, \rho_1')$. This shows that

$$S(t)\mathscr{B} \subset B_V(0, \rho_1'), \qquad \forall t \geq t_1,$$

and I.(1.12) is proved.

All the assumptions of Theorem I.1.1. are now satisfied with $H = L^2(\Omega; \mathscr{D})$ and $\mathscr{U} = H$, and we can state a result similar to Theorem 1.2.

Theorem 1.3. *We assume that Ω is an open bounded set of $\mathbb{R}^n$ and that the hypotheses* (1.44)–(1.46) *are satisfied.*

Then the semigroup $S(t)$ associated to the system (1.41)–(1.43) *possesses a maximal attractor $\mathscr{A}$ which is bounded in $H_0^1(\Omega)^m$, compact and connected in $L^2(\Omega; \mathscr{D})$; $\mathscr{A}$ attracts the bounded sets of $L^2(\Omega; \mathscr{D})$.*

Remark 1.7. If g is independent of x, $\mathscr{C}^1$ and bounded on $\mathscr{D}$, and $g(0) = 0$, a remark similar to Remark 1.2 can be made: if the nonlinear term g is in some sense small compared to the smallest dissipativity coefficient d_0 (see (1.51)) then all the trajectories converge to 0 as $t \to +\infty$, i.e., the dynamics are trivial. Indeed, integrating by parts the integral in the right-hand side of (1.60), we find that this integral is equal to

$$-\sum_{i=1}^{m}\sum_{j=1}^{n}\int_\Omega \frac{\partial}{\partial x_j} g_i(u) \frac{\partial u_i}{\partial x_j}\,dx = -\sum_{i,k=1}^{m}\sum_{j=1}^{n}\int_\Omega \frac{\partial g_i}{\partial u_k}\frac{\partial u_k}{\partial x_j}\frac{\partial u_i}{\partial x_j}\,dx$$
$$\leq \beta\|u\|^2,$$

where

$$\beta = \operatorname*{Sup}_{u\in\mathscr{D}}\left\{\sum_{i=1}^{m}\left(\frac{\partial g_i}{\partial u_k}(u)\right)^2\right\}^{1/2}.$$

Hence with (1.62)

$$\frac{d}{dt}\|u\|^2 + \left(\frac{d_0}{c_3^2} - \beta\right)\|u\|^2 \leq 0, \tag{1.67}$$

and if $d_0 > \beta c_3^2$, $u(t) \to 0$ as $t \to \infty$.

Remark 1.8. We could consider examples combining the properties (and difficulties!) of equations (1.2) and (1.41): we could consider a system (1.41)–(1.43) which satisfies (1.44), (1.45) and, instead of (1.46), consider the case where g is unbounded on $\mathscr{D}$ ($\mathscr{D}$ is itself unbounded). The boundedness assumption on g is then replaced by growth assumptions on g that are somehow similar to (1.8), (1.18). However, we will not develop these generalizations here.

1.2.3. Examples

We describe some examples of reaction–diffusion systems satisfying the hypotheses (1.44)–(1.46).

Example 1.1 (The Hodgkin–Huxley Equations). This is the system proposed in 1952 by A.L. Hodgkin and A.F. Huxley [1] to describe the nerve impulse transmission. Here $n = 1$, $\Omega = (0, L)$, and the system is of the form (1.41) with $m = 4$, $u = (u_1, \ldots, u_4)$,

$$\begin{cases} \dfrac{\partial u_1}{\partial t} = d_1 \dfrac{\partial^2 u_1}{\partial x^2} - g_1(u_1, u_2, u_3, u_4), \\[2ex] \dfrac{\partial u_2}{\partial t} = d_2 \dfrac{\partial^2 u_2}{\partial x^2} + k_1(u_1)(h_1(u_1) - u_2), \\[2ex] \dfrac{\partial u_3}{\partial t} = d_3 \dfrac{\partial^2 u_3}{\partial x^2} + k_2(u_1)(h_2(u_1) - u_3), \\[2ex] \dfrac{\partial u_4}{\partial t} = d_4 \dfrac{\partial^2 u_4}{\partial x^2} + k_3(u_1)(h_3(u_1) - u_4). \end{cases} \tag{1.68}$$

We have

$$g_1(u) = -\gamma_1 u_2^3 u_3(\delta_1 - u_1) - \gamma_2 u_4^4(\delta_2 - u_1) - \gamma_3(\delta_3 - u_1),$$
$$\delta_1 > \delta_3 > 0 > \delta_2.$$

Furthermore, $k_i > 0$, $1 > h_i > 0$, $i = 1, 2, 3$. In this model, u_1 represents the electrical potential in the nerve, while u_2, u_3, u_4 represent chemical concentrations and are thus nonnegative. In the original model, $d_2 = d_3 = d_4 = 0$; we follow here the modification in J. Smoller [1] and assume $d_i > 0$, $\forall i$ (see D.S. Jones and B.D. Sleeman [1] for more details concerning these equations).

It is proved in J. Smoller [1] (see Chapter 14), that any rectangle

$$\mathscr{D} = \{(u_1, u_2, u_3, u_4), \alpha_0 \le u_1 \le \alpha_1, 0 \le u_i \le \alpha_i, i = 2, 3, 4\} \tag{1.69}$$

is an invariant region for (1.68), provided $\alpha_1 \ge \delta_1$ (> 0), $\alpha_0 \le \delta_2$ (< 0), and $\alpha_i \ge 1$, $i = 2, 3, 4$; hence (1.44). Assumption (1.46) is obviously satisfied and (1.45) is proved with a slight modification to the results in J. Smoller [1]. The result proved there assumes that u_0 is continuous on $\bar{\Omega}$ (vanishes on $\partial\Omega$ and takes its values in $\mathscr{D}$) and provides $u(\cdot, t)$ continuous in $\bar{\Omega}$. By approximation and passage to the limit it is easy to obtain (1.45), when $u_0 \in L^2(\Omega; \mathscr{D})$. Theorem 1.3 applies and proves the existence of an attractor $\mathscr{A}$ which is supposed to describe the long-term behavior of the impulse transmission in the nerve (the boundary condition is the Dirichlet boundary condition (1.42)).

Example 1.2 (The Fitz-Hugh–Nagumo equations). These equations, introduced in 1961–62 by R. Fitz-Hugh [1] and J. Nagumo *et al.* [1], are also intended to describe the signal transmission across axons; they are slightly

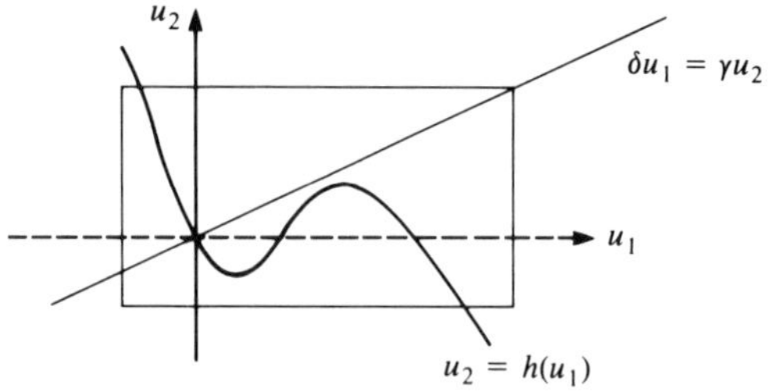

Figure 1.1

simpler than (1.68) and read

$$\begin{cases} \dfrac{\partial u_1}{\partial t} = d_1 \dfrac{\partial^2 u_1}{\partial x^2} + h(u_1) - u_2, \\ \dfrac{\partial u_2}{\partial t} = d_2 \dfrac{\partial^2 u_2}{\partial x^2} + \delta u_1 - \gamma u_2. \end{cases} \tag{1.70}$$

Here $n = 1, \Omega = (0, L), m = 2, u = \{u_1, u_2\}, d_i, \gamma, \delta$ are positive constants and $h(u_1) = -u_1(u_1 - \beta)(u_1 - 1)$ where $0 < \beta < \frac{1}{2}$. The constants d_i can usually be ≥ 0 but we impose $d_i > 0$. As in (1.68), u_1 represents the electrical potential in the axon but u_2 has a more complicated interpretation.

It is proved in J. Smoller [1] that any rectangle

$$\mathscr{D} = \{(u_1, u_2), -\alpha_1 \leq u_1 \leq \alpha_2, -\alpha_3 \leq u_2 \leq \alpha_4\} \tag{1.71}$$

is a positively invariant region provided $\alpha_i \geq 0$, $\forall i$, $\alpha_1 > 0$, $\alpha_2 > 1$, and the edges $u_2 = \alpha_4$ and $u_1 = -\alpha_3$ of $\mathscr{D}$ are, respectively, included in the half-planes $\delta u_1 - \gamma u_2 < 0$ and $\delta u_1 - \gamma u_2 > 0$ and are, respectively, above and below the maximum and minimum of $h(u_1)$ (see Figure 1.1). Assumption (1.44) follows; (1.46) is obvious and (1.45) is a slight extension of the existence result in J. Smoller [1], similar to the extension necessary for the Hodgkin–Huxley equations.

Theorem 1.3 applies and gives the existence of a maximal attractor $\mathscr{A}$. Note that Remark 1.2 is valid in this case, and shows that the dynamics are trivial when d_1 and d_2 are sufficiently large (for γ, δ fixed). In this case, the nerve impulse goes to 0 as $t \to \infty$, but this may not be always the case if d_1 and d_2 are small.

Example 1.3. These equations arise in the study of superconductivity of liquids. We have $m = n \geq 1$, Ω is a bounded domain of $\mathbb{R}^n$, and $u = (u_1, \ldots, u_n)$ is a solution of

$$\frac{\partial u}{\partial t} - D\Delta u = (1 - |u|^2)u, \tag{1.72}$$

D as in (1.41).

It is proved in J. Smoller [1] that any rectangle of $\mathbb{R}^n$

$$\mathscr{D} = \{(u_1, \dots, u_n), \alpha_i \le u_i \le \alpha_1', \forall i\}, \tag{1.73}$$

which contains the disc $|u| \le 1$ in its interior, is invariant if $d_i \neq d_j$ for some indices i, j. If $d_1 = \cdots = d_n$, then any ball

$$\mathscr{D} = \left\{(u_1, \dots, u_n), \sum_{j=1}^{n} u_i^2 \le \delta\right\}, \qquad \delta > 1 \tag{1.74}$$

is invariant. Thus (1.44) is satisfied, (1.46) is obvious, and (1.45) is proved as above. Theorem 1.3 applies.

Example 1.4 (A Problem in Combustion). This example (borrowed from combustion theory) leads to an invariant region $\mathscr{D}$ which is not bounded. We consider an open bounded set Ω of $\mathbb{R}^n$, $n \ge 1$, and for simplicity we restrict ourselves to $m = 2$ although more complicated systems can be treated: the function $u = \{u_1, u_2\}$ is a solution of

$$\begin{cases} \dfrac{\partial u_1}{\partial t} - d_1 \Delta u_1 - (u_2)^p h(u_1) = 0, \\[2ex] \dfrac{\partial u_2}{\partial t} - d_2 \Delta u_2 - (u_2)^p h(u_1) = 0, \end{cases} \tag{1.75}$$

where $h(s) = |s|^\gamma \exp(-\alpha/s)$, and p, α, γ are positive constants, $\gamma < 1$. In (1.75), u_1 represents a temperature and $u_1 \ge 0$ while u_2 (representing a concentration) satisfies $0 \le u_2 \le 1$; we supplement (1.75) with the Dirichlet boundary conditions (1.42).

Condition (1.44) is satisfied by

$$\mathscr{D} = \{(u_1, u_2), u_1 \ge 0, 0 \le u_2 \le 1\}. \tag{1.76}$$

This is proved by using the truncation operators (see Section II.1.4):

If $\varphi \in H^1(\Omega)$, then $\varphi^+ = \max(\varphi, 0)$ is in $H^1(\Omega)$ and for a.e. $x \in \Omega$

$$\frac{\partial \varphi^+}{\partial x_i}(x) = \frac{\partial \varphi}{\partial x_i}(x) \quad \text{if } \varphi(x) > 0, \qquad = 0 \quad \text{if } \varphi(x) \le 0. \tag{1.77}$$

Assuming that u is a sufficiently regular solution of (1.75), we multiply the first equation (1.75) by $u_1^- = (-u_1)^+$. Integrating over Ω and using the Green formula we find

$$\begin{aligned} \int_\Omega u_1^- \frac{\partial u_1}{\partial t} dx &= \int_\Omega u_1^- \left(\frac{\partial u_1^+}{\partial t} - \frac{\partial u_1^-}{\partial t}\right) dx \\ &= \int_\Omega u_1^- \frac{u_1^-}{\partial t} dx \\ &= -\frac{1}{2}\frac{d}{dt}|u_1^-|^2, \end{aligned}$$

$$-\int_\Omega \Delta u_1 u_1^- \, dx = (\text{since } u_1^-(t) \in H_0^1(\Omega), \forall t) \tag{1.78}$$

$$= \int_\Omega (\text{grad } u_1^+ - \text{grad } u_1^-) \cdot \text{grad } u_1^- \, dx$$

$$= -\int_\Omega (\text{grad } u_1^-)^2 \, dx,$$

$$\frac{1}{2}\frac{d}{dt}|u_1^-|^2 + |\text{grad } u_1^-|^2 = -u_1^-|u_2|^p h(u_1) \leq 0.$$

It follows that

$$|u_1^-(t)| \leq |u_1^-(0)|. \tag{1.79}$$

By assumption, $u_1(x, 0) \geq 0$ and $u_1^-(x, 0) = 0$ for a.e. $x \in \Omega$, so that $u_1(x, t) \geq 0$, a.e. $x \in \Omega$. We prove, in a similar manner, that $0 \leq u_2(x, t) \leq 1$ for every $t \geq 0$ and a.e. $x \in \Omega$ if $u_2(x, 0)$ satisfies the same condition: this is shown by successively multiplying the second equation in (1.75) by u_2^- and $(u_2 - 1)^+$ and deriving inequalities similar to (1.79).

The function $u \mapsto |u_2|^p h(u_1)$ is obviously continuous and bounded on $\mathscr{D}$; this property (weaker than (1.46)) is however sufficient for the proof of Theorem 1.3. The existence property (1.45) has been proved by several authors (see, for instance, B. Larrouturou [1]) and relies on the methods used for Theorem 1.1. This shows that Theorem 1.3 is applicable.

Example 1.5. The equations of this last example serve as a model for the Belousov–Zhabotinsky reactions in chemical dynamics (see, for instance, L. Howard and N. Kopell [1], S. Hastings and J. Murray [1]).

Here $m = 3$ and $u = (u_1, u_2, u_3)$ satisfies

$$\begin{cases} \dfrac{\partial u_1}{\partial t} - d_1 \Delta u_1 - \alpha(u_2 - u_1 u_2 + u_1 - \beta u_1^2) = 0, \\[2ex] \dfrac{\partial u_2}{\partial t} - d_2 \Delta u_2 - \dfrac{1}{\alpha}(\gamma u_3 - u_2 - u_1 u_2) = 0, \\[2ex] \dfrac{\partial u_3}{\partial t} - d_3 \Delta u_3 - \delta(u_1 - u_3) = 0, \end{cases} \tag{1.80}$$

where $\alpha, \beta, \gamma, \delta$ are positive constants; u_1, u_2, u_3 denote chemical concentrations.

It is shown in J. Smoller [1] that

$$\mathscr{D} = \{(u_1, u_2, u_3), 0 \leq u_1 \leq a, 0 \leq u_2 \leq b, 0 \leq u_3 \leq c\} \tag{1.81}$$

is invariant, provided

$$a > \max(1, \beta^{-1}), \qquad c > a, \quad b > \gamma c.$$

This shows (1.44); (1.46) is obvious and (1.45) is proved by the same methods as in Examples 1.1 to 1.3. Theorem 1.3 applies and shows the existence of an attractor $\mathscr{A}$ which is maximal in $L^2(\Omega; \mathscr{D})$.

1.2.4. Other Boundary Conditions

As indicated at the beginning of Section 1.2.3, other boundary conditions can be associated with (1.41); in particular, the Neumann boundary condition or, if $\Omega = \prod_{i=1}^{n}]0, L_i[$, the space periodicity boundary condition; we can also consider different boundary conditions of this type for the different components of u.

For the sake of simplicity we will assume that

$$\text{The invariant region } \mathscr{D} \text{ is bounded in } \mathbb{R}^m, \tag{1.82}$$

and we will consider the pure Neumann boundary conditions

$$\frac{\partial u_i}{\partial \mathbf{v}} = 0 \quad \text{on } \Gamma, \qquad \forall i. \tag{1.83}$$

It will be easy to adapt the following modifications to other boundary conditions.

The natural working spaces for (1.83) are then $L^2(\Omega)^m$ and $H^1(\Omega)^m$, and we formulate, in this case, assumptions (1.44), (1.45), (1.46) without any modification except that $H_0^1(\Omega)^m$ is replaced by $H^1(\Omega)^m$ in (1.45). The semigroup operators $S(t)$ are then just defined as the operators

$$S(t)\colon u_0 \in H \to u(t) \in H,$$

that operate in the space $H = L^2(\Omega; \mathscr{D})$ and whose existence is asserted by (1.45).

Due to the simplifying assumption (1.82), the existence of an absorbing set in $H = L^2(\Omega, \mathscr{D})$ is obvious. Then we set

$$c_4 = \operatorname*{Sup}_{\xi \in \mathscr{D}} |\xi|, \tag{1.84}$$

we observe that (1.49) is still valid and, using (1.46), (1.84), we write

$$\left| \int_\Omega g(u)u \, dx \right| \le c_2 c_4 |\Omega|^{1/2}; \tag{1.85}$$

therefore we obtain, instead of (1.56),

$$\frac{d}{dt}|u|^2 + 2a(u, u) \le 2c_2 c_4 |\Omega|^{1/2}. \tag{1.86}$$

Then for $r > 0$ fixed, we have

$$\int_t^{t+r} a(u, u) \, ds \le \tfrac{1}{2} c_4^2 |\Omega| + 2r c_2 c_4 |\Omega|^{1/2}, \tag{1.87}$$

and this inequality (valid for $t \ge 0$) replaces (1.57)–(1.59). Relations (1.60) to

(1.63) remain valid and the consequences of (1.63) hold true with only the slight modifications in the constants due to the replacement of (1.59) by (1.87); concerning (1.62), let us observe that this inequality is still true with a constant c_3 which may be different.

We then deduce the existence of an absorbing set in V for the semigroup $S(t)$ and the fact that the operators $S(t)$ are uniformly compact. We then have the analog of Theorem 1.3.

Theorem 1.4. *We assume that Ω is an open bounded set of $\mathbb{R}^n$, and that the hypotheses* (1.44)–(1.46) *and* (1.82) *are satisfied.*

Then the semigroup $S(t)$ associated with the system (1.41), (1.43), (1.83) *possesses a maximal attractor $\mathscr{A}$ which is bounded in $H^1(\Omega)^m$, compact and connected in $L^2(\Omega; \mathscr{D})$; $\mathscr{A}$ attracts the bounded sets of $L^2(\Omega; \mathscr{D})$.*

Remark 1.9. The presentation in this section follows M. Marion [1]. For other works concerning the attractors for reaction–diffusion equations see A.V. Babin and M.I. Vishik [2], D. Ruelle and N. Kopell [1] and the references in these three articles; for the partly dissipative case (i.e., when some $d_i = 0$), see M. Marion [2].

2. Navier–Stokes Equations ($n = 2$)

Our aim in this section is to study the attractors associated with the Navier–Stokes equations of fluid mechanics. We restrict ourselves to the flow in a bounded domain of $\mathbb{R}^2$. The three-dimensional case is more complicated and will be studied in Chapter VII. Other equations of fluid mechanics are also considered in Section 3.

In Section 2.1 we recall the equations and their mathematical setting; in Section 2.2 we show the existence of absorbing sets and the existence of a maximal attractor. Finally, Section 2.3 contains a sketch of the proof of existence and uniqueness of solution of the equations.

Before entering into the mathematical developments let us explain the physical significance of the attractor and of the results proved here. We consider a viscous fluid driven by stationary volume forces f. It is conjectured on the basis of experimental evidences in fluid mechanics that if, for a given geometry and a given viscosity, the forces are "small", then the flow converges as $t \to \infty$ to a stationary solution. However if, in some appropriate sense, the forces are very strong, then the flow will remain turbulent and time-dependent for all time. If u_0 are the initial data for the velocity field, then the ω-limit set of u_0, $\omega(u_0)$ (associated with the semigroup $S(t)$ defined more precisely below) is the mathematical object describing the long-term behavior of the flow with initial data u_0, i.e., the permanent regime. The maximal attractor $\mathscr{A}$ that we introduce here contains, in particular, all the ω-limit sets corresponding to all the initial data. Therefore it is able to describe all the flows that can be observed corresponding to all the initial data. For this reason it was called by C. Foias and R. Temam (see [2], [3]), the *universal attractor* for the Navier–Stokes equations.

2.1. The Equations and Their Mathematical Setting

We consider the Navier–Stokes equations of viscous incompressible fluids. Let Ω denote an open bounded set of $\mathbb{R}^2$ with boundary Γ. The Navier–Stokes equations in Ω govern the flow of a fluid which fills an infinite cylinder of cross section Ω and moves parallel to the plane of Ω.

The unknowns are $u = (u_1, u_2)$ and p; u is the velocity vector, $u(x, t)$ representing the velocity of the particle at x at time t, and $p(x, t)$ is the pressure at x at time t. The equations are

$$\rho_0\left(\frac{\partial u}{\partial t} + (u \cdot \nabla)u\right) - \nu\Delta u + \nabla p = f, \tag{2.1}$$

$$\operatorname{div} u = 0, \tag{2.2}$$

where $\rho_0 > 0$ is the density of the fluid that is constant, $\nu > 0$ is the kinematic viscosity, and f represents volume forces that are applied to the fluid. Equation (2.1) is the momentum conservation equation and (2.2) is the mass conservation equation (incompressibility condition).

Usually we set $\rho_0 = 1$. Alternatively, equation (2.1) (with $\rho_0 = 1$) can be considered as the *nondimensionalized form* of the Navier–Stokes equations in which case u, p, f are nondimensionalized quantities and ν is replaced by Re^{-1}, Re representing the Reynolds number

$$\mathrm{Re} = \frac{UL}{\nu}, \tag{2.3}$$

where U and L are the typical velocity and length used for the nondimensionalization.

Equations (2.1), (2.2) are supplemented with a boundary condition. Two cases will be considered:

The nonslip boundary condition. The boundary Γ is solid and at rest; thus

$$u = 0 \quad \text{on } \Gamma.^1 \tag{2.4}$$

The space-periodic case. Here $\Omega = (0, L_1) \times (0, L_2)$ and

$$u, p \text{ and the first derivatives of } u \text{ are } \Omega\text{-periodic.}^2 \tag{2.5}$$

[1] If Γ is solid but not at rest, then the nonslip boundary condition is $u = \varphi$ on Γ where $\varphi = \varphi(x, t)$ is the given velocity of Γ. This case will be considered in Section 3.

[2] That is, u and p take the same values at corresponding points of Γ (see Chapter II).

Furthermore, we assume in this case that the average flow vanishes

$$\int_\Omega u\,dx = 0. \tag{2.6}$$

When an initial-value problem is considered we supplement these equations with

$$u(x, 0) = u_0(x), \qquad x \in \Omega. \tag{2.7}$$

For the mathematical setting of this problem we consider a Hilbert space H which is a closed subspace of $L^2(\Omega)^n$ ($n = 2$ here). In the nonslip case,

$$H = \{u \in L^2(\Omega)^n, \text{div } u = 0, u \cdot \nu = 0 \text{ on } \Gamma\}, \tag{2.8}$$

and in the periodic case

$$H = \{u \in \dot{L}^2(\Omega)^n, \text{div } u = 0, u_{i|\Gamma_i} = -u_{i|\Gamma_{i+n}}, i = 1, \dots, n\}.^1 \tag{2.9}$$

We refer the reader to R. Temam [2] for more details on these spaces and, in particular, a trace theorem showing that the trace of $u \cdot \nu$ on Γ exists and belongs to $H^{-1/2}(\Gamma)$ when $u \in L^2(\Omega)^n$ and div $u \in L^2(\Omega)$. The space H is endowed with the scalar product and the norm of $L^2(\Omega)^n$ denoted by $(\cdot, \cdot)$ and $|\cdot|$.

Another useful space is V, a closed subspace of $H^1(\Omega)^n$,

$$V = \{u \in H_0^1(\Omega)^n, \text{div } u = 0\} \tag{2.10}$$

in the nonslip case and, in the space-periodic case,

$$V = \{u \in \dot{H}_{\text{per}}^1(\Omega)^n, \text{div } u = 0\}, \tag{2.11}$$

where $\dot{H}_{\text{per}}^1(\Omega)$ is defined in II.(1.31). In both cases, V is endowed with the scalar product

$$((u, v)) = \sum_{i,j=1}^{n} \left(\frac{\partial u_i}{\partial x_j}, \frac{\partial v_i}{\partial x_j}\right)$$

and the norm $\|u\| = \{((u, u))\}^{1/2}$.

We denote by A the linear unbounded operator in H which is associated with V, H and the scalar product $((u, v))$ as indicated in Section 2.1, Chapter II:

$$(Au, v) = ((u, v)), \qquad \forall u, v \in V.$$

The domain of A in H is denoted by $D(A)$; A is a self-adjoint positive operator in H. Also, A is an isomorphism from $D(A)$ onto H. The space $D(A)$ can be fully characterized by using the regularity theory of linear elliptic systems (see, for instance, R. Temam [2], [3])

$$D(A) = H^2(\Omega)^n \cap V$$

[1] Γ_i and Γ_{i+n} are the faces $x_i = 0$ and $x_i = L_i$ of Γ. The condition $u_{i|\Gamma_i} = -u_{i|\Gamma_{i+n}}$ expresses the periodicity of $u \cdot \nu$; $\dot{L}^2(\Omega)^n$ is the space of u's in $L^2(\Omega)^n$ satisfying (2.6) (see II.(1.31)).

and

$$D(A) = H^2_{\text{per}}(\Omega)^n \cap V$$

in the nonslip and periodic cases; furthermore, $|Au|$ is on $D(A)$ a norm equivalent to that induced by $H^2(\Omega)^n$.

Let V' be the dual of V; then H can be identified to a subspace of V' (see II.(2.10)) and we have

$$D(A) \subset V \subset H \subset V', \tag{2.12}$$

where the inclusions are continuous and each space is dense in the following one.

In the space-periodic case we have $Au = -\Delta u$, $\forall u \in D(A)$, while in the nonslip case we have

$$Au = -P\Delta u, \qquad \forall u \in D(A),$$

where P is the orthogonal projector in $L^2(\Omega)^n$ on the space H. We can also say that $Au = f$, $u \in D(A)$, $f \in H$, is equivalent to saying that there exists p ($\in H^1(\Omega)$) such that

$$\begin{cases} -\Delta u + \operatorname{grad} p = f & \text{in } \Omega, \\ \operatorname{div} u = 0 & \text{in } \Omega, \\ u = 0 & \text{on } \Omega. \end{cases}$$

The operator A^{-1} is continuous from H into $D(A)$ and since the embedding of $H^1(\Omega)$ in $L^2(\Omega)$ is compact, the embedding of V in H is compact. Thus A^{-1} is a self-adjoint continuous compact operator in H, and by the classical spectral theorems there exists a sequence λ_j

$$0 < \lambda_1 \leq \lambda_2, \ldots, \qquad \lambda_j \to \infty,$$

and a family of elements w_j of $D(A)$ which is orthonormal in H and such that

$$Aw_j = \lambda_j w_j, \qquad \forall j. \tag{2.13}$$

The weak form of the Navier–Stokes equations due to J. Leray [1], [2], [3] involves only u. It is obtained by multiplying (2.1) by a test function v in V and integrating over Ω. Using the Green formula (2.2) and the boundary conditions, we find that the term involving p disappears and there remains

$$\frac{d}{dt}(u, v) + \nu((u, v)) + b(u, u, v) = (f, v), \qquad \forall v \in V, \tag{2.14}$$

where

$$b(u, v, w) = \sum_{i,j=1}^{n} \int_\Omega u_i \frac{\partial v_j}{\partial x_i} w_j \, dx, \tag{2.15}$$

whenever the integrals make sense. Actually, the form b is trilinear continuous on $H^1(\Omega)^n$ ($n = 2$) and in particular on V. We have the following inequalities giving various continuity properties of b:

$$
|b(u, v, w)|
\leq c_1 \times \begin{cases}
|u|^{1/2} \|u\|^{1/2} \|v\|^{1/2} |Av|^{1/2} |w|, & \forall u \in V, \quad v \in D(A), \quad w \in H, \\
|u|^{1/2} |Au|^{1/2} \|v\| \, |w|, & \forall u \in D(A), \quad v \in V, \quad w \in H, \\
|u| \, \|v\| \, |w|^{1/2} |Aw|^{1/2}, & \forall u \in H, \quad v \in V, \quad w \in D(A), \\
|u|^{1/2} \|u\|^{1/2} \|v\| \, |w|^{1/2} \|w\|^{1/2}, & \forall u, v, w \in V,
\end{cases} \tag{2.16}
$$

where $c_1 > 0$ is an appropriate constant.

An alternative form of (2.14) can be given using the operator A and the bilinear operator B from $V \times V$ into V' defined by

$$
(B(u, v), w) = b(u, v, w), \qquad \forall u, v, w \in V. \tag{2.17}
$$

We also set

$$
B(u) = B(u, u), \qquad \forall u \in V',
$$

and we easily see that (2.14) is equivalent to the equation

$$
\frac{du}{dt} + \nu A u + B(u) = f, \tag{2.18}
$$

while (2.7) can be rewritten

$$
u(0) = u_0. \tag{2.19}
$$

We assume that f is independent of t so that the dynamical system associated with (2.18) is autonomous

$$
f(t) = f \in H, \qquad \forall t. \tag{2.20}
$$

Existence and uniqueness results for (2.18), (2.19) are well known (see, for instance, R. Temam [2], [3] and see also Section 2.4). The following theorem collects several classical results.

Theorem 2.1. *Under the above assumptions, for f and u_0 given in H, there exists a unique solution u of* (2.17), (2.18) *satisfying*

$$
u \in \mathscr{C}([0, T]; H) \cap L^2(0, T; V), \qquad \forall T > 0.
$$

Furthermore, u is analytic in t with values in $D(A)$ for $t > 0$, and the mapping

$$
u_0 \mapsto u(t)
$$

is continuous from H into $D(A)$, $\forall t > 0$.

Finally, if $u_0 \in V$, then

$$
u \in \mathscr{C}([0, T]; V) \cap L^2(0, T; D(A)), \qquad \forall T > 0.
$$

Some indications for the proof of Theorem 2.1 will be given in Section 2.3. This theorem allows us to define the operators

$$S(t)\colon u_0 \mapsto u(t).$$

These operators enjoy the semigroup properties I.(1.1) and they are continuous from H into itself and even from H into $D(A)$.

2.2. Absorbing Sets and Attractors

We first prove the existence of an absorbing set in H.

A first energy-type equality is obtained by taking the scalar product of (2.18) with u. Using the orthogonality property (see R. Temam [1])

$$b(u, v, v) = 0, \qquad \forall u \in V, \quad \forall v \in H^1(\Omega)^n,[1] \tag{2.21}$$

we see that $(B(u), u) = 0$ and there remains

$$\frac{1}{2}\frac{d}{dt}|u|^2 + \nu\|u\|^2 = (f, u) \leq |f|\,|u|. \tag{2.22}$$

We know that

$$|u| \leq \lambda_1^{-1/2}\|u\|, \qquad \forall u \in V,$$

where λ_1 is the first eigenvalue of A. Hence, we can majorize the right-hand side of (2.22) by

$$\lambda_1^{-1/2}|f|\,\|u\| \leq \frac{\nu}{2}\|u\|^2 + \frac{1}{2\nu\lambda_1}|f|^2,$$

and we obtain

$$\frac{d}{dt}|u|^2 + \nu\|u\|^2 \leq \frac{1}{\nu\lambda_1}|f|^2, \tag{2.23}$$

$$\frac{d}{dt}|u|^2 + \nu\lambda_1|u|^2 \leq \frac{1}{\nu\lambda_1}|f|^2. \tag{2.24}$$

Using the classical Gronwall lemma, we obtain

$$|u(t)|^2 \leq |u_0|^2 \exp(-\nu\lambda_1 t) + \frac{1}{\nu^2\lambda_1^2}|f|^2(1 - \exp(-\nu\lambda_1 t)). \tag{2.25}$$

Thus

$$\limsup_{t\to\infty} |u(t)| \leq \rho_0, \qquad \rho_0 = \frac{1}{\nu\lambda_1}|f|. \tag{2.26}$$

We infer from (2.25) that the balls $B_H(0, \rho)$ of H with $\rho \geq \rho_0$ are positively invariants for the semigroup $S(t)$, and these balls are absorbing for any $\rho > \rho_0$.

[1] More generally, $b(u, v, v) = 0$, $\forall u, v \in H^1(\Omega)^n$, such that $\operatorname{div} u = 0$ in Ω, and $u \cdot \mathbf{v}$ or $v = 0$ on Γ.

We choose $\rho_0' > \rho_0$ and denote by $\mathscr{B}_0$ the ball $B_H(0, \rho_0')$. Any set $\mathscr{B}$ bounded in H is included in a ball $B(0, R)$ of H. It is easy to deduce from (2.25) that $S(t)\mathscr{B} \subset \mathscr{B}_0$ for $t \geq t_0(\mathscr{B}, \rho_0')$, where

$$t_0 = \frac{1}{\nu\lambda_1} \log \frac{R^2}{\rho_0'^2 - \rho_0^2}. \tag{2.27}$$

We then infer from (2.23), after integration in t, that

$$\nu \int_t^{t+r} \|u\|^2 \, ds \leq \frac{r}{\nu\lambda_1} |f|^2 + |u(t)|^2, \qquad \forall r > 0. \tag{2.28}$$

With the use of (2.26) we conclude that

$$\limsup_{t\to\infty} \int_t^{t+r} \|u\|^2 \, ds \leq \frac{r}{\nu^2\lambda_1} |f|^2 + \frac{|f|^2}{\nu^3\lambda_1^2}, \tag{2.29}$$

and if $u_0 \in \mathscr{B} \subset B_H(0, R)$ and $t \geq t_0(\mathscr{B}, \rho_0')$, then

$$\int_t^{t+r} \|u\|^2 \, ds \leq \frac{r}{\nu^2\lambda_1} |f|^2 + \frac{1}{\nu} \rho_0'^2. \tag{2.30}$$

Absorbing Set in V

We continue and show the existence of an absorbing set in V. For that purpose we obtain another energy-type equation by taking the scalar product of (2.18) with Au. Since

$$(Au, u') = ((u, u')) = \frac{1}{2}\frac{d}{dt} \|u\|^2,$$

we find

$$\frac{1}{2}\frac{d}{dt} \|u\|^2 + \nu |Au|^2 + (B(u), Au) = (f, Au). \tag{2.31}$$

We write

$$(f, Au) \leq |f| \, |Au| \leq \frac{\nu}{4} |Au|^2 + \frac{1}{\nu} |f|^2,$$

and using the second inequality (2.16)

$$\begin{aligned} |(B(u), Au)| &\leq c_1 |u|^{1/2} \|u\| \, |Au|^{3/2} \\ &\leq \text{(with the Young inequality)}^1 \\ &\leq \frac{\nu}{4} |Au|^2 + \frac{c_1'}{\nu^3} |u|^2 \|u\|^4. \end{aligned}$$

[1] By the Young inequality

$$ab \leq \frac{\varepsilon}{p} a^p + \frac{1}{p' \varepsilon^{p'/p}} b^{p'}, \qquad \forall a, b, \varepsilon > 0, \quad \forall p, \quad 1 < p < \infty, \quad p' = p/(p-1).$$

Here $p = \frac{4}{3}$ and $\varepsilon/p = \nu/4$, i.e., $\varepsilon = \nu/3$.

Hence

$$\frac{d}{dt}\|u\|^2 + \nu|Au|^2 \le \frac{2}{\nu}|f|^2 + \frac{2c_1'}{\nu^3}|u|^2\|u\|^4, \tag{2.32}$$

and since

$$\|\varphi\| \le \lambda_1^{-1/2}|A\varphi|, \qquad \forall \varphi \in D(A), \tag{2.33}$$

we also have

$$\frac{d}{dt}\|u\|^2 + \nu\lambda_1\|u\|^2 \le \frac{2}{\nu}|f|^2 + \frac{2c_1'}{\nu^3}|u|^2\|u\|^4. \tag{2.34}$$

An a priori estimate of u in $L^\infty(0, T; V)$, $\forall T > 0$, follows easily from (2.34) by application of the classical Gronwall lemma, using the previous estimates on u. We are more interested in an estimate valid for large t. Assuming that u_0 belongs to a bounded set $\mathscr{B}$ of H and that $t \ge t_0(\mathscr{B}, \rho_0')$, t_0 as in (2.27), we apply the uniform Gronwall lemma (Lemma 1.1) to (2.34) with g, h, y replaced by

$$\frac{2c_1'}{\nu^3}|u|^2\|u\|^2, \qquad \frac{2}{\nu}|f|^2, \qquad \|u\|^2,$$

Thanks to (2.25), (2.30) we estimate the quantities a_1, a_2, a_3 in Lemma 1.1 by

$$\begin{cases} a_1 = \dfrac{2c_1'}{\nu^3}\rho_0'^2 a_3, \\[2ex] a_2 = \dfrac{2r}{\nu}|f|^2, \qquad a_3 = \dfrac{r}{\nu^2\lambda_1}|f|^2 + \dfrac{1}{\nu}\rho_0'^2, \end{cases} \tag{2.35}$$

and we obtain

$$\|u(t)\|^2 \le \left(\frac{a_3}{r} + a_2\right)\exp(a_1) \qquad \text{for} \quad t \ge t_0 + r, \tag{2.36}$$

t_0 as in (2.27).

Let us fix $r > 0$ and denote by ρ_1^2 the right-hand side of (2.36). We then conclude that the ball $B_V(0, \rho_1)$ of V, denoted by $\mathscr{B}_1$, is an absorbing set in V for the semigroup $S(t)$. Furthermore, if $\mathscr{B}$ is any bounded set of H, then $S(t)\mathscr{B} \subset \mathscr{B}_1$ for $t \ge t_0(\mathscr{B}, \rho_0') + r$. This shows the existence of an absorbing set in V, namely $\mathscr{B}_1$, and also that the operators $S(t)$ are uniformly compact, i.e., I.(1.12) is satisfied.

Maximal Attractor

All the assumptions of Theorem I.1.1 are satisfied and we deduce from this theorem the existence of a maximal attractor for the Navier–Stokes equations.

Theorem 2.2. *The dynamical system associated with the two-dimensional Navier–Stokes equations, supplemented by the boundary conditions* (2.4) *or* (2.5), (2.6), *possesses an attractor $\mathscr{A}$ that is compact, connected, and maximal in H. $\mathscr{A}$ attracts the bounded sets of H and $\mathscr{A}$ is also maximal among the functional invariant sets bounded in H.*

Remark 2.1 (Physical Significance of $\mathscr{A}$). As indicated in the Introduction to Section 2, although f is independent of time, the permanent flow which is observed for given initial data can be a time-dependent one, if $|f|$ is large. Strictly speaking, this is only a conjecture based on physical evidence and on our present understanding of turbulence. Under this hypothesis, the mathematical object that describes the permanent flow corresponding to an initial data u_0, is the ω-limit set of u_0, $\omega(u_0)$, that may or may not be an attractor. The attractor $\mathscr{A}$ contains, in particular, all the ω-limit sets corresponding to all the initial data $u_0 \in H$, and therefore it provides a "universal" description of all the possible flows corresponding to all the initial data (for f fixed), see C. Foias and R. Temam [2].

Remark 2.2. The existence of an absorbing set in V for the two-dimensional Navier–Stokes equations was first proved (with different terminology) by C. Foias and G. Prodi [1]. The existence of the attractors for these equations was first proved by C. Foias and R. Temam [1].

Remark 2.3. The intensity of the force f is measured by a nondimensional number proportional to the L^2-norm $|f|$ of the forces

$$G = \frac{|f|}{\nu^2 \lambda_1}. \tag{2.37}$$

This number was introduced in C. Foias, O. Manley, R. Temam, and Y. Treve [1] and called the (generalized) Grashof number. Some authors prefer to consider a nondimensional number proportional to the inverse of the kinematic viscosity, i.e., the number

$$\mathrm{Re} = \frac{|f|^{1/2}}{\nu \lambda_1^{1/2}}, \tag{2.38}$$

which is then called the Reynolds number. However, this terminology is not fully justified since there is no typical velocity in a turbulent flow that could play the role of U in (2.3).

The reference velocity that is used in (2.38) if $|f|^{1/2}$, while $\lambda_1^{-1/2}$ plays the role of a typical length. We could, of course, replace $\lambda_1^{-1/2}$ by any other typical length, the diameter of Ω, or $|\Omega|^{1/2}$, where $|\Omega|$ is the measure of Ω (i.e., the area of Ω).

Remark 2.4. We can treat, in exactly the same manner, the case where the Navier–Stokes equations are associated with another classical boundary condition (free boundary)

$$u \cdot \mathbf{v} = 0 \quad \text{and} \quad \operatorname{curl} u \times \mathbf{v} = 0 \quad \text{on } \Gamma. \tag{2.39}$$

The first boundary condition in (2.39) is the nonpenetration boundary condition, while the second one is also known in the form

$$(\sigma \cdot \mathbf{v})_\tau = 0 \quad \text{on } \Gamma,$$

where $\sigma = \sigma(u)$ is the (stress) tensor with components

$$\sigma_{ij}(u) = 2\nu\varepsilon_{ij}(u) - p\delta_{ij}, \qquad \varepsilon_{ij}(u) = \frac{1}{2}\left(\frac{\partial u_i}{\partial x_j} + \frac{\partial u_j}{\partial x_i}\right), \qquad \forall i, j,$$

and $(\sigma \cdot \mathbf{v})_\tau$ is the tangential component on Γ of $\sigma \cdot \mathbf{v}$.

For the functional setting, the space H is the same as in (2.8) while

$$V = \{v \in H^1(\Omega)^n, \operatorname{div} v = 0, v \cdot \mathbf{v} = 0 \text{ on } \Gamma\},$$

$$D(A) = \{v \in H^2(\Omega)^n, \operatorname{div} v = 0, v \cdot \mathbf{v} \text{ and } \operatorname{curl} v \times \mathbf{v} = 0 \text{ on } \Gamma\}.$$

The same results hold and are proved in the same manner. One of the differences in the proof is related to the utilization of the Poincaré inequality in the form $|u| \le \lambda_1^{-1/2} \|u\|$ (see the inequality after (2.22)). We replace this by an inequality

$$|u| \le c\|u\|, \qquad \forall u \in V,$$

which follows readily from II.(1.35) applied with $H^1(\Omega)$ replaced by $H^1(\Omega)^n$ ($n = 2$), and

$$p(u) = \int_\Gamma |u \cdot v| \, d\Gamma.$$

In order to be able to apply II.(1.35) we must show that, if u is a constant, $u = \alpha \in \mathbb{R}^n$ and $\alpha \cdot \mathbf{v} = 0$ a.e. on Γ, then $\alpha = 0$; this is simply due to the fact that if $\mathbf{v}$ is orthogonal to a vector α almost everywhere on Γ, then Γ is included in the union of lines orthogonal to Γ and a set of (one-dimensional) measure 0: this is impossible for a bounded regular region. We also observe that (2.33) remains valid but with λ_1 replaced by λ_1'' (= the first eigenvalue of A in the present case).

The conclusions of Theorems 2.1 and 2.2 are also valid in this case.

2.3. Proof of Theorem 2.1

We conclude Section 2 by giving a sketch of the proof of the results in Theorem 2.1 and some bibliographical references.

The existence of a solution of (2.17), (2.18), that belongs to $L^\infty(0, T; H) \cap L^2(0, T; V)$, $\forall T > 0$, is first obtained by the Faedo–Galerkin method. We implement this approximation procedure with the functions w_j representing the eigenvalues of A (see (2.13)). For each m we look for an approximate solution u_m of the form

$$u_m(t) = \sum_{i=1}^{m} g_{im}(t) w_i$$

satisfying

$$\left(\frac{du_m}{dt}, w_j\right) + \nu a(u_m, w_j) + b(u_m, u_m, w_j) = (f, w_j), \qquad j = 1, \ldots, m, \quad (2.40)$$

$$u_m(0) = P_m u_0, \quad (2.41)$$

where P_m is the projector in H (or V) on the space spanned by $w_1, \ldots, w_m$. Since A and P_m commute, the relation (2.40) is also equivalent to

$$\frac{du_m}{dt} + \nu A u_m + P_m B(u_m) = P_m f. \quad (2.42)$$

The existence and uniqueness of u_m on some interval $[0, T_m[$ is elementary and then $T_m = +\infty$, because of the a priori estimates that we obtain for u_m. An energy equality is obtained by multiplying (2.40) by g_{jm} and summing these relations for $j = 1, \ldots, m$. We obtain (2.22) exactly with u replaced by u_m and we deduce from this relation that

$$u_m \text{ remains bounded in } L^\infty(0, T; H) \cap L^2(0, T; V), \qquad \forall T > 0. \quad (2.43)$$

Due to (2.21) and the last inequality (2.16)

$$\|B(\varphi)\|_{V'} \le c_1 |\varphi| \, \|\varphi\|, \qquad \forall \varphi \in V. \quad (2.44)$$

Therefore $B(u_m)$ and $P_m B(u_m)$ remain bounded in $L^2(0, T; V')$ and by (2.42)

$$\frac{du_m}{dt} \text{ remains bounded in } L^2(0, T; V'). \quad (2.45)$$

By weak compactness it follows from (2.43) that there exists $u \in L^\infty(0, T; H) \cap L^2(0, T; V)$, $\forall T > 0$, and a subsequence still denoted m, such that

$$u_m \to u \text{ in } L^2(0, T; V) \text{ weakly and in } L^\infty(0, T; H) \text{ weak-star}$$

$$\frac{du_m}{dt} \to \frac{du}{dt} \quad \text{in } L^2(0, T; V') \text{ weakly.} \quad (2.46)$$

Due to (2.45) and a classical compactness theorem (see, for instance, R. Temam [2] (Ch. III, Sec. 2)), we also have

$$u_m \to u \quad \text{in } L^2(0, T; H) \text{ strongly.} \quad (2.47)$$

This is sufficient to pass to the limit in (2.40)–(2.42) and we find (2.17), (2.18) at the limit. For (2.18) we simply observe that (2.46) implies that

$$u_m(t) \to u(t)$$

weakly in V' or even in H, $\forall t \in [0, T]$ (see Lemmas II.3.1 and II.3.2).

By (2.17) (or (2.46)), $u' = du/dt$ belongs to $L^2(0, T; V')$ and, by Lemma II.3.2, u is in $\mathscr{C}([0, T]; H)$. The uniqueness and continuous dependence of $u(t)$ on u_0 (in H) follow by standard methods using Lemma II.3.2.

The fact that $u \in L^\infty(0, T; V) \cap L^2(0, T; D(A))$, $\forall T > 0$, is proved by deriving further a priori estimates on u_m. They are obtained by multiplying (2.40) by $\lambda_j g_{jm}$ and summing these relations for $j = 1, \ldots, m$. Using (2.13) we find a relation that is exactly (2.31) with u replaced by u_m. We deduce from this relation that

$$u_m \text{ remains bounded in } L^\infty(0, T; V) \cap L^2(0, T; D(A)), \qquad \forall T > 0. \quad (2.48)$$

At the limit we then find that u is in $L^\infty(0, T; V) \cap L^2(0, T; D(A))$. The fact that u is in $\mathscr{C}([0, T]; V)$ then follows from an appropriate application of Lemma II.3.2.

Finally, the fact that u is analytic in t with values in $D(A)$ results from totally different methods, for which the reader is referred to C. Foias and R. Temam [1] or R. Temam [3]. However, this property was given for the sake of completeness and is never used here in an essential manner.

3. Other Equations in Fluid Mechanics

In this section we consider more general equations in fluid mechanics. Section 3.1 contains the generalization of the results of Section 2 to an abstract equation. Although this abstract equation just looks as a slight extension of (2.18), it allows many applications to fluid mechanics equations which are developed in Section 3.2–3.5. Section 3.2 is devoted to the Navier–Stokes equations with nonhomogeneous boundary conditions (fluid driven by its boundary). Section 3.3 is concerned with the equations of magnetohydrodynamics. In Section 3.4, we consider flows on a manifold that can model geophysical flows. Finally, in Section 3.5, we consider the equations of thermohydraulics.

3.1. Abstract Equation. General Results

We are given two Hilbert spaces V and H with $V \subset H$, V dense in H, the injection of V in H being *compact*. The scalar product and the norm in V and H are denoted by $((\cdot, \cdot))$, $\|\cdot\|$, $(\cdot, \cdot)$, $|\cdot|$. We can identify H with a subspace of V' (= the dual of V), and then $V \subset H \subset V'$ where the injections are continuous and each space is dense in the following one.

We consider a symmetric bilinear continuous form on V, $a(u, v)$, which is coercive

$$a(u, u) \geq \alpha \|u\|^2, \qquad \forall u \in V. \quad (3.1)$$

We associate with a the linear operator A that is an isomorphism from V onto V', and can alternatively be considered as a linear unbounded self-adjoint operator in H with domain

$$D(A) = \{u \in V, Au \in H\}.$$

We have

$$D(A) \subset V \subset H \subset V', \tag{3.2}$$

where the injections are continuous and each space is dense in the following one.

Those assumptions are the general assumptions of Section II.2.1 and the reader is referred to that section for more details. More specific assumptions are as follows (see also J.M. Ghidaglia [1]).

We are given a linear continuous operator R from V into V' which maps $D(A)$ into H and such that there exist $\theta_1, \theta_2 \in [0, 1[$ and two positive constants c_1, c_2 such that

$$|Ru| \leq c_1 \|u\|^{1-\theta_1} |Au|^{\theta_1}, \qquad \forall u \in D(A), \tag{3.3}$$

$$|(Ru, u)| \leq c_2 \|u\|^{1+\theta_2} |u|^{1-\theta_2}, \qquad \forall u \in V. \tag{3.4}$$

We also assume that $A + R$ is coercive on V

$$a(u, u) + (Ru, u) \geq \alpha' \|u\|^2, \qquad \forall u \in V \quad (\alpha' > 0). \tag{3.5}$$

Finally, we are given a bilinear continuous operator B mapping $V \times V$ into V' and $D(A) \times D(A)$ into H such that

$$(B(u, v), v) = 0, \qquad \forall u, v \in V, \tag{3.6}$$

$$|(B(u, v), w| \leq c_3 |u|^{\theta_3} \|u\|^{1-\theta_3} \|v\| \, \|w\|^{\theta_3} |w|^{1-\theta_3}, \qquad \forall u, v, w \in V, \tag{3.7}$$

$$|B(u, v)| + |B(v, u)| \leq c_4 \|u\| \, \|v\|^{1-\theta_4} |Av|^{\theta_4}, \qquad \forall u \in V, \ \forall v \in D(A), \tag{3.8}$$

$$|B(u, v)| \leq c_5 |u|^{\theta_5} \|u\|^{1-\theta_5} \|v\|^{1-\theta_5} |Av|^{\theta_5}, \qquad \forall u \in V, \ \forall v \in D(A), \tag{3.9}$$

where c_3, c_4, c_5, are appropriate constants and $\theta_i \in [0, 1[$, $i = 3, 4, 5$.

For simplicity, we set $B(u) = B(u, u)$ and, for f given in H, we consider the nonlinear evolution equation in H

$$\frac{du}{dt} + Au + B(u) + Ru = f, \tag{3.10}$$

$$u(0) = u_0. \tag{3.11}$$

Here we have a result of existence and uniqueness of solutions which is similar to Theorem 2.1.

Theorem 3.1. *Under the above assumptions and in particular* (3.3), (3.4), *and* (3.6)–(3.9),[1] *for f and u_0 given in H, there exists a unique solution u of* (3.10), (3.11),

$$u \in \mathscr{C}([0, T]; H) \cap L^2(0, T; V), \qquad \forall T > 0. \tag{3.12}$$

[1] Assumption (3.5) is needed in Theorem 3.2 but not in Theorem 3.1.

Furthermore, u is analytic in t with values in $D(A)$ for $t > 0$, and the mapping

$$u_0 \mapsto u(t)$$

is continuous from H into $D(A)$, $\forall t > 0$.

Finally, if $u_0 \in V$, then

$$u \in \mathscr{C}([0, T]; V) \cap L^2(0, T; D(A)), \qquad \forall T > 0.$$

The steps of the proof are the same as in Section 2.3; the details can be found in J.M. Ghidaglia [1].

Theorem 3.1 allows us to define the operators

$$S(t)\colon u_0 \mapsto u(t), \qquad t \geq 0,$$

from H into itself. These operators enjoy the semigroup property I.(1.1) and the continuity property I.(1.4).

Absorbing Sets

We now proceed with the proof of existence of absorbing sets in H and V. The first necessary estimates are obtained by taking the scalar product of (3.10) with u. Due to (3.5), we find

$$\frac{1}{2}\frac{d}{dt}|u|^2 + a(u, u) + (Ru, u) = (f, u). \tag{3.13}$$

There exists a constant c_6 such that

$$\begin{cases} |u| \leq c_6 \|u\|, & \forall u \in V, \\ \|u\| \leq c_6 |Au|, & \forall u \in D(A). \end{cases} \tag{3.14}$$

We use (3.5) and majorize the right-hand side of (3.13) by

$$\begin{aligned} |f|\,|u| &\leq c_6 |f|\,\|u\| \\ &\leq \frac{\alpha'}{2}\|u\|^2 + \frac{c_6}{2\alpha'}|f|^2. \end{aligned}$$

We obtain

$$\frac{d}{dt}|u|^2 + \alpha'\|u\|^2 \leq \frac{c_6}{\alpha'}|f|^2. \tag{3.15}$$

This inequality is similar to (2.23) and the consequences that we deduce from it are the same as in Section 2. In particular, we conclude the existence of a bounded set $\mathscr{B}_0$ of H which absorbs the bounded sets of H.

In order to obtain the existence of an absorbing set in V, we take the scalar product of (3.10) with Au in H. Since

$$\left(Au, \frac{du}{dt}\right) = a(u, \frac{du}{dt}) = \frac{1}{2}\frac{d}{dt}a(u, u),$$

we obtain

$$\frac{1}{2}\frac{d}{dt}a(u, u) + |Au|^2 = (f, Au) - (B(u), Au) - (Ru, Au).$$

We write

$$(f, Au) \le |f||Au| \le \tfrac{1}{6}|Au|^2 + |f|^2,$$

and, using (3.3), (3.9),

$$\begin{aligned}
|(B(u), Au)| &\le c_5|u|^{\theta_5}\|u\|^{2-2\theta_5}|Au|^{1+\theta_5} \\
&\le \tfrac{1}{6}|Au|^2 + c_1'\|u\|^4|u|^{2\theta_5/(1-\theta_5)}, \\
|(Ru, Au)| &\le |Ru||Au| \\
&\le c_1\|u\|^{1-\theta_1}|Au|^{1+\theta_1} \\
&\le \text{(with Young's inequality)} \\
&\le \tfrac{1}{6}|Au|^2 + c_2'\|u\|^2.
\end{aligned}$$

Hence

$$\frac{d}{dt}a(u, u) + |Au|^2 \le 2|f|^2 + 2c_3'\|u\|^2. \tag{3.16}$$

This inequality, similar to (2.32), allows us to derive similar conclusions. In particular, we obtain the existence of an absorbing set $\mathscr{B}_1$ in V, and the fact that the operators $S(t)$ are uniformly compact.

Maximal Attractor

Theorem I.1.1 is applicable and gives (for the abstract equation (3.10)) a result totally similar to Theorem 3.2.

Theorem 3.2. *The dynamical system associated with equation* (3.10) *possesses an attractor $\mathscr{A}$ which is compact, connected, and maximal in H. $\mathscr{A}$ attracts the bounded sets of H and $\mathscr{A}$ is also maximal among the functional invariant sets bounded in H.*

3.2. Fluid Driven by Its Boundary

3.2.1. The General Case

The notations used here are those of Section 2. In particular, Ω denotes a bounded domain of $\mathbb{R}^2$ which is filled with an incompressible viscous liquid. If the boundary Γ of Ω is materialized and at rest, then the nonslip boundary condition for the velocity field is given by (2.4). If the boundary Γ is moving with a prescribed velocity φ, then the nonslip boundary condition is

$$u = \varphi \quad \text{on } \Gamma. \tag{3.17}$$

In order to have an autonomous dynamical system we assume that φ is independent of time. We assume that φ is given as the trace on Γ of a function Φ,

$$\Phi \in H^2(\Omega)^n, \qquad \operatorname{div} \Phi = 0, \qquad \int_\Gamma \Phi \cdot \mathbf{v}\, d\Gamma = 0, \tag{3.18}$$

where $\mathbf{v}$ is the unit outward normal on Γ.[1] It is then convenient to denote by $\tilde{u} = u + \Phi$ the velocity field in the fluid that was denoted by u in Section 2, while the pressure is still denoted by p. With these definitions (2.1) (with $\rho_0 = 1$) and (2.2) are rewritten as follows:

$$\frac{\partial u}{\partial t} + (u \cdot \nabla)u + (\Phi \cdot \nabla)u + (u \cdot \nabla)\Phi - \nu \Delta u + \nabla p = f + \nu \Delta \Phi - (\Phi \cdot \nabla)\Phi, \tag{3.19}$$

$$\operatorname{div} u = 0 \tag{3.20}$$

while (3.17) is now replaced by

$$u = 0 \quad \text{on } \Gamma. \tag{3.21}$$

The spaces H, V, being the same as in (2.8), (2.10), we obtain an analog of (2.14) by multiplying (3.19) by a test function $v \in V$, and integrating on Ω. We find

$$\frac{d}{dt}(u, v) + \nu((u, v)) + b(u, u, v) + b(u, \Phi, v) + b(\Phi, u, v) = (\bar{f}, v), \quad \forall v \in V, \tag{3.22}$$

where $\bar{f}$ is the right-hand side of (3.19).

The inequalities (2.16) on b are also valid for u, v, w in $H^2(\Omega)^n$ or $H^1(\Omega)^n$ in dimension 2, in the following form:

$$|b(u, v, w)| \le c_7 \times \begin{cases} |u|^{1/2}\|u\|_1^{1/2}\|v\|_1^{1/2}\|v\|_2^{1/2}|w|, & \forall u \in H^1(\Omega)^2,\ v \in H^2(\Omega)^2,\ w \in L^2(\Omega)^2, \\ |u|^{1/2}\|u\|_2^{1/2}\|v\|_1|w|, & \forall u \in H^2(\Omega)^2,\ v \in H^1(\Omega)^2,\ w \in L^2(\Omega)^2, \\ |u|\,\|v\|_1|w|^{1/2}\|w\|_2^{1/2}, & \forall u \in L^2(\Omega)^2,\ v \in H^1(\Omega)^2,\ w \in H^2(\Omega)^2, \\ |u|^{1/2}\|u\|_1^{1/2}\|v\|_1|w|^{1/2}\|w\|_1^{1/2}, & \forall u, v, w \in H^1(\Omega)^2, \end{cases} \tag{3.23}$$

where $\|\cdot\|_m$ is the norm in $H^m(\Omega)^2$, and as usual $|\cdot|$ is the norm in $L^2(\Omega)^2$ (see R. Temam [2], [3]). Thanks to these inequalities the operator B can be

[1] The physical data are φ while Φ is a (nonunique) extension of φ inside Ω. Concerning the hypotheses on Φ and φ we observe the following. The property $\Phi \in H^2(\Omega)^n$ is a regularity assumption; and $\int_\Gamma \Phi \cdot \mathbf{v}\, d\Gamma = 0$ is a consistency condition which is necessary because of (3.17) and $\operatorname{div} u = 0$. The condition $\operatorname{div} \Phi = 0$ (implying $\int_\Gamma \Phi \cdot \mathbf{v}\, d\Gamma = 0$) is added for convenience and is not restrictive at all (see (3.29) below).

extended as a bilinear continuous operator from $H^1(\Omega)^2 \times H^1(\Omega)^2$ into V' (and from $H^2(\Omega)^2 \times H^2(\Omega)^2$ into H) by setting

$$(B(u, v), w) = b(u, v, w), \qquad \forall u, v \in H^1(\Omega)^2, \quad \forall w \in V. \tag{3.24}$$

This allows us to rewrite (3.22) in a form similar to (3.10)

$$\frac{du}{dt} + \nu A u + B(u) + B(u, \Phi) + B(\Phi, u) = \bar{f}, \tag{3.25}$$

i.e.,

$$Ru = B(u, \Phi) + B(\Phi, u). \tag{3.26}$$

In order to apply Theorems 3.1 and 3.2 we check assumptions (3.3) to (3.9). Excepting (3.6) which is the same as (2.21), they follow from various applications of inequalities (3.23). The first and second inequality (3.23) give, for any $w \in H$,

$$\begin{aligned} |(Ru, w)| &= |b(u, \Phi, w) + b(\Phi, u, w)| \\ &\leq c_7 |u|^{1/2} \|u\|_1^{1/2} \|\Phi\|_1^{1/2} \|\Phi\|_2^{1/2} |w| \\ &\quad + c_7 |\Phi|^{1/2} \|\Phi\|_1^{1/2} \|u\|_1^{1/2} \|u\|_2^{1/2} |w|. \end{aligned} \tag{3.27}$$

Hence (3.3) with $\theta_1 = \frac{1}{2}$, since the norms $\|u\|$ and $\|u\|_1$ are equivalent on V and the norms $|Au|$ and $\|u\|_2$ are equivalent on $D(A)$. For (3.4) we use (2.21) and the last inequality (3.23)

$$|(Ru, u)| = |b(u, \Phi, u)| \leq c_7 |u| \, \|u\|_1 \, \|\Phi\|_1. \tag{3.28}$$

We obtain (3.4) with $\theta_2 = 0$. Then (3.7) with $\theta_3 = \frac{1}{2}$ follows from the last inequality (3.23); (3.8) with $\theta_4 = \frac{1}{2}$ follows from the first two inequalities (3.23). Finally, (3.9) with $\theta_5 = \frac{1}{2}$ is the same as the last inequality (3.23).

There remains to check (3.5). This follows from a technically difficult property:

For any given $\varphi \in H^{3/2}(\Gamma)^n$ such that $\int_\Gamma \varphi \cdot \mathbf{v} \, d\Gamma = 0$,[1] and for any $\varepsilon > 0$, there exists $\Phi \in H^2(\Omega)^n$ such that $\Phi = \varphi$ on Γ, $\operatorname{div} \Phi = 0$, and $|b(v, \Phi, v)| \leq \varepsilon \|v\|^2$, $\forall v \in V$. (3.29)

This property was proved by E. Hopf [1] (see also R. Temam [2] (Ch. II, Sec. 1.4 and App. I, Prop. 2.1)). Applying (3.29) with $\varepsilon = \nu/4$, we obtain (3.5) with $\alpha' = \nu/2$.

Remark 3.1. The function Φ given by (3.29) satisfies an estimate in terms of ε of the form

$$|\Phi| \leq \|\Phi\|_1 \leq c\varepsilon \exp\left(\frac{4}{\varepsilon}\right) |\varphi|_{H^{1/2}(\Gamma)^2} \qquad (n = 2) \tag{3.30}$$

[1] We assume here that Γ is connected. If Γ were not connected we must impose the stronger condition $\int_{\Gamma_i} \varphi \cdot \mathbf{v} \, d\Gamma = 0$, $i = 1, \ldots, N$, where $\Gamma_1, \ldots, \Gamma_N$ are the connected components of Γ (see, for instance, R. Temam [2]).

and hence for $\varepsilon = \nu/4$

$$|\Phi| \le \|\Phi\|_1 \le c'\nu \exp\left(\frac{16}{\nu}\right)|\varphi|_{H^{1/2}(\Gamma)^2}. \tag{3.30'}$$

This remark will be useful for later purposes (see Chapter VI, Section 4.2).

The assumptions of Theorems 3.1 and 3.2 are then satisfied and Theorem 3.2 gives the existence of a *maximal attractor* describing the flow in this case. As mentioned in the Introduction to Section 2 and in Remark 2.1 the physical significance of this attractor is, of course, the same as that of the attractor obtained in Section 2.

Remark 3.2. If we consider the Navier–Stokes equations (2.1), (2.2) with the space-periodicity boundary conditions (2.5), then, with the method used here, we can treat the case where the average flow does not vanish, i.e., when (2.6) is not satisfied:

Let us denote by $m(v)$ the average of a function v

$$m(v) = \frac{1}{|\Omega|}\int_\Omega v(x)\,dx,$$

where $|\Omega|$ = the measure (surface) of $\Omega = L_1 L_2$. The evolution of the average velocity $m(u(t))$ is fully determined by (2.1), (2.2), (2.5), (2.7). Indeed, by integration of (2.1) on $\Omega =]0, L_1[\times]0, L_2[$, we obtain ($\rho_0 = 1$)

$$\frac{d}{dt}m(u(t)) = m(f(t))$$

and thus

$$m(u(t)) = m(u_0) + \int_0^t m(f(s))\,ds.$$

Now we write $u = \bar{u} + m(u)$ and (2.1) becomes

$$\frac{\partial \bar{u}}{\partial t} + (\bar{u}\cdot\nabla)\bar{u} + m(u)\cdot\nabla\bar{u} - \nu\Delta\bar{u} + \nabla p = f. \tag{3.31}$$

This equation is similar to (3.19) (and $\bar{u}(t)$ belongs to the space V in (2.11), $\forall t$), and it can be treated similarly.

3.2.2. *An Improved Result*

We keep here the same notation as in Section 3.2.1. We assume that $\Omega \subset \mathbb{R}^2$ is smooth (at least $\mathscr{C}^3$) and that φ satisfies

$$\varphi \in \mathscr{C}^3(\Gamma)^2, \qquad \varphi\cdot\nu = 0 \quad \text{on } \Gamma. \tag{3.32}$$

As before we assume that φ is given as the trace on Γ of a function $\mathbf{\Phi}$ satisfying (at least) (3.18). We saw (Remark 3.1) that the L^2 and H^1 norms of the function $\mathbf{\Phi}$ considered in the general case contain an exponential of ε, leading to a very large bound on the diameter of the absorbing ball. By a modification of the Hopf construction we can improve (3.30) and (3.30′) when (3.32) is satisfied.

Indeed, we first have the following result (see A. Miranville and X. Wang [1]).

Lemma 3.1. *For every φ satisfying* (3.32) *and for every $\varepsilon > 0$, there exists $\mathbf{\Phi} \in H^2(\Omega)^2$ such that $\mathbf{\Phi} = \varphi$ on Γ, $\operatorname{div} \mathbf{\Phi} = 0$, and*

$$|b(v, \mathbf{\Phi}, v)| \leq c\varepsilon |\Omega|^{1/2} |\varphi|_{L^\infty(\Gamma)^2} \|v\|^2, \qquad \forall v \in V,$$

where c is independent of ε. Furthermore,

$$|\mathbf{\Phi}| \leq c'\varepsilon^{1/2} |\Omega|^{1/2} \operatorname{Max}(|\varphi|_{L^\infty(\Gamma)^2}, |\nabla\varphi|_{L^\infty(\Gamma)^4} \operatorname{diam} \Omega), \tag{3.33}$$

$$\|\mathbf{\Phi}\|_1 \leq \frac{c''}{\varepsilon^{1/2}} |\varphi|_{L^\infty(\Gamma)^2}, \tag{3.33′}$$

where c' and c'' are independent of ε.

Sketch of the Proof. Since Ω is bounded and regular, there exists $\delta_0 > 0$ such that the normals to Γ do not intersect in the neighborhood of Γ of width $2\delta_0$ (which we denote by $\mathcal{O}_{2\delta_0}(\Gamma)$). Moreover, for every $(x, y) \in \mathcal{O}_{\delta_0}(\Gamma)$, there exists a unique point $b(x, y) \in \Gamma$ such that

$$\operatorname{dist}((x, y), \Gamma) = \operatorname{dist}((x, y), b(x, y)).$$

Let $\tau_{b(x,y)}$ denote the clockwise tangent vector to Γ at the point $b(x, y)$. We consider a function $\rho \in \mathscr{C}^\infty([0, +\infty))$ such that

$$\begin{aligned}
&\operatorname{Supp} \rho \subset [0, 1], \\
&\rho(0) = 1, \\
&|\rho(s)| \leq 1, \qquad \forall s \in [0, +\infty), \\
&\int_0^1 \rho(s)\, ds = 0,
\end{aligned} \tag{3.34}$$

and we set

$$\begin{aligned}
\mathbf{\Psi} = \mathbf{\Psi}_\varepsilon &= \varphi(b(x, y)) \\
&\cdot \tau_{b(x,y)} \int_0^{\operatorname{dist}((x,y),b(x,y))} \rho\left(\frac{s}{|\Omega|^{1/2}\varepsilon}\right) ds \qquad \text{if} \quad (x, y) \in \mathcal{O}_{\delta_0}(\Gamma), \\
\mathbf{\Psi} = \mathbf{\Psi}_\varepsilon &= 0 \quad \text{elsewhere.}
\end{aligned}$$

We finally set

$$\mathbf{\Phi} = \mathbf{\Phi}_\varepsilon(x, y) = \operatorname{curl} \mathbf{\Psi} = \begin{pmatrix} -\dfrac{\partial \mathbf{\Psi}}{\partial y} \\ \dfrac{\partial \mathbf{\Psi}}{\partial x} \end{pmatrix}.$$

Now, for $v \in V$,

$$\begin{aligned} |b(v, \mathbf{\Phi}, v)| &\le \int_\Omega \left(\frac{v}{\operatorname{dist}((x, y), \Gamma)} \cdot \operatorname{dist}^2((x, y), \Gamma)\nabla \right) \mathbf{\Phi} \cdot \frac{v}{\operatorname{dist}((x, y), \Gamma)}\, dx\, dy \\ &\le \left| \frac{v}{\operatorname{dist}((x, y), \Gamma)} \right|^2 |\operatorname{dist}((x, y), \Gamma)^2 \nabla \mathbf{\Phi}|_{L^\infty}, \end{aligned}$$

and we find, using Hardy's inequality (see, e.g., R. Temam [2, p. 176])

$$|b(v, \mathbf{\Phi}, v)| \le c(\Omega) |\operatorname{dist}((x, y), \Gamma)^2 \nabla \mathbf{\Phi}|_{L^\infty} \|v\|^2. \qquad \square$$

We now proceed exactly as in Section 3.2.1 to check assumptions (3.3) to (3.9). Thus, Theorems 3.1 and 3.2 apply.

3.3. Magnetohydrodynamics (MHD)

We show that the equations of magnetohydrodynamics can be written in the form (3.10) so that Theorem 3.2 is applicable (as is Theorem 3.1).

Let Ω be an open bounded set of $\mathbb{R}^2$ with boundary Γ. We start from the MHD equations in a nondimensional form (see T.G. Cowling [1], L. Landau and E. Lifschitz [2])

$$\frac{\partial v}{\partial t} + (v \cdot \nabla)v - \frac{1}{R_e}\Delta v - S(B \cdot \nabla)B + \nabla\left(p + \frac{SB^2}{2}\right) = g, \tag{3.35}$$

$$\frac{\partial B}{\partial t} + (v \cdot \nabla)B + \frac{1}{R_m}\operatorname{curl}(\operatorname{curl} B) - (B \cdot \nabla)v = 0, \tag{3.36}$$

$$\operatorname{div} v = 0, \tag{3.37}$$

$$\operatorname{div} B = 0. \tag{3.38}$$

In these equations p is the pressure, v is the velocity vector (which was denoted u in Section 2 and 3.2), B is the magnetic field, and g (previously denoted f) represents the external volume forces applied to the fluid; all these quantities are nondimensionalized. The numbers R_e, R_m, $S > 0$, are classical nondimensional numbers: R_e is the Reynolds number defined as in (2.3), R_m is the magnetic Reynolds number, and $S = M^2/R_e R_m$ where M is the Hartman number.[1]

These equations are supplemented by the following boundary conditions:

$$v = 0 \quad \text{on } \Gamma, \tag{3.39}$$

$$B \cdot \mathbf{v} = 0 \quad \text{and} \quad \text{curl } B \times \mathbf{v} = 0 \quad \text{on } \Gamma, \tag{3.40}$$

where $\mathbf{v}$ is the unit outward normal on Γ. When initial-value problems are considered, we add the initial conditions

$$v(x, 0) = v_0(x), \qquad B(x, 0) = B_0(x), \qquad x \in \Omega. \tag{3.41}$$

Let us set this problem in the form of (3.10). The unknown $u = u(x, t)$ will be a vector taking its values in $\mathbb{R}^{2n}$, $u = \{v, B\}$, where $v(x, t)$ is the velocity vector and $B(x, t)$ is the magnetic field. We set $H = H_1 \times H_2$ and $V = V_1 \times V_2$, where H_i, V_i are the spaces appearing in (2.8), (2.10)

$$\begin{cases} H_1 = \{\varphi \in L^2(\Omega)^n, \text{div } \varphi = 0, \varphi \cdot \mathbf{v} = 0 \text{ on } \Gamma\}, \\ V_1 = \{\varphi \in H_0^1(\Omega)^n, \text{div } \varphi = 0\}, \end{cases} \tag{3.42}$$

and $H_2 = H_1$,

$$V_2 = \{\varphi \in H^1(\Omega)^n; \text{div } \varphi = 0, \varphi \cdot \mathbf{v} = 0 \text{ on } \Gamma\}. \tag{3.43}$$

We equip H_1 and H_2 with the usual scalar product and norm of $L^2(\Omega)^2$ (denoted $(\cdot, \cdot)$, $|\cdot|$), and we equip H with the scalar product and norm

$$[u_1, u_2] = (v_1, v_2) + S(B_1, B_2), \qquad [u] = \{[u, u]\}^{1/2},$$

$$u_i = \{v_i, B_i\} \in H, \qquad i = 1, 2,$$

that are equivalent to the "natural" scalar product and norm

$$(u_1, u_2) = (v_1, v_2) + (B_1, B_2), \qquad |u| = \{(u, u)\}^{1/2}.$$

The space V is endowed with the scalar product and norm

$$[\![\varphi, \psi]\!]_1 = \sum_{i=1}^{n} \left(\frac{\partial \varphi}{\partial x_i}, \frac{\partial \psi}{\partial x_i}\right), \qquad [\![\varphi]\!]_1 = \{[\![\varphi, \varphi]\!]_1\}^{1/2}, \tag{3.44}$$

while the space V_2 is endowed with the scalar product and norm

$$[\![\varphi, \varphi]\!]_2 = (\text{curl } \varphi, \text{curl } \psi), \qquad [\![\varphi]\!]_2 = \{[\![\varphi, \varphi]\!]_2\}^{1/2}. \tag{3.45}$$

According to a result in R. Temam [2] (Prop. 1.4, App. I), the norm

$$\{|\varphi|^2 + |\text{curl } \varphi|^2 + |\varphi \cdot \mathbf{v}|^2_{H^{1/2}(\Gamma)^n}\}^{1/2}$$

[1] $R_m = LU\sigma\mu$, where L is a typical length, U is a typical velocity, μ is the magnetic permeability, and σ is the conductivity of the fluid; $S = B_*^2/\mu\rho_* U^2$, where ρ_* is the density ($= 1$ here) and B_* is a reference magnetic field.

is, on $H^1(\Omega)^n$, a norm equivalent to that induced by $H^1(\Omega)^n$. Then $[\![\varphi]\!]_2$ is, on V_2, a norm equivalent to that induced by $H^1(\Omega)^n$ ($n = 2$ here).

Finally, we endow the space V with the scalar product and norm

$$[\![u_1, u_2]\!] = [\![v_1, v_2]\!]_1 + S[\![B_1, B_2]\!]_2, \qquad [\![u]\!] = \{[\![u, u]\!]\}^{1/2},$$
$$u_i = \{v_i, B_i\} \in V, \qquad i = 1, 2,$$

and we define the bilinear form a

$$a(u_1, u_2) = \frac{1}{R_e}[\![v_1, v_2]\!]_1 + \frac{S}{R_m}[\![B_1, B_2]\!]_2, \tag{3.46}$$

which is clearly continuous and coercive on V.

We define, as usual, the linear operator A associated with a, V, H (V, H endowed with the scalar product $[\![\cdot, \cdot]\!]$, $[\cdot, \cdot]$). Considered as a linear unbounded operator in H, its domain $D(A)$ is equal to $D(A_1) \times D(A_2)$ where $D(A_i)$ is the domain of A_i in H_i. For $i = 1$, as in Section 2,

$$D(A_1) = H^2(\Omega)^n \cap V_1,$$

and saying that $A_1 v = g, v \in D(A_1), g \in H_1$, amounts to saying that there exists $p \in H^1(\Omega)^n$ such that

$$\begin{cases} -\dfrac{1}{R_e}\Delta \cdot v + \operatorname{grad} p = g & \text{in } \Omega, \\ \operatorname{div} v = 0 & \text{in } \Omega, \\ v = 0 & \text{on } \Gamma, \end{cases} \tag{3.47}$$

For $i = 2$, the characterization of $D(A_2)$ was given in M. Sermange and R. Temam [1]. We have

$$D(A_2) = \{u \in H^2(\Omega)^n, \operatorname{div} u = 0 \text{ in } \Omega, u \cdot \mathbf{v} \text{ and } \operatorname{curl} u \times \mathbf{v} = 0 \text{ on } \Gamma\},$$

and saying that $A_2 B = g$, $B \in D(A_2)$, $g \in H_2$, amounts to say that

$$\begin{cases} \dfrac{1}{R_m} \operatorname{curl} \operatorname{curl} B = g & \text{in } \Omega, \\ \operatorname{div} B = 0 & \text{in } \Omega, \\ B \cdot \mathbf{v} = 0 & \text{on } \Gamma, \\ \operatorname{curl} B \times \mathbf{v} = 0 & \text{on } \Gamma. \end{cases} \tag{3.48}$$

Then we define on $V \times V \times V$ a trilinear form

$$b(u_1, u_2, u_3) = b_1(v_1, v_2, v_3) - Sb_1(B_1, B_2, v_3) + Sb_1(v_1, B_2, B_3)$$
$$- Sb_1(B_1, v_2, B_3), \qquad \forall u_i = \{v_i, B_i\} \in V, \tag{3.49}$$

where b_1 (defined on $H^1(\Omega)^n \times H^1(\Omega)^n \times H^1(\Omega)^n$) is the same as in (2.15)

$$b_1(\varphi, \psi, \theta) = \sum_{i,j=1}^{n} \int_\Omega \varphi_i \frac{\partial \psi_j}{\partial x_i} \theta_j \, dx. \tag{3.50}$$

The continuity properties (3.23) (and (2.16)) easily induce continuity properties for b_1. We can rewrite (2.21)

$$b_1(\varphi, \psi, \psi) = 0, \qquad \forall \varphi \in V_1, \quad \forall \psi \in H^1(\Omega)^n, \tag{3.51}$$

and this implies

$$b(u_1, u_2, u_2) = 0, \qquad \forall u_1, u_2 \in V. \tag{3.52}$$

We associate with the form b the bilinear continuous operator $\mathfrak{B}$ which maps $V \times V$ into V' (or $D(A) \times D(A)$ into H) and is defined by

$$(\mathfrak{B}(u_1, u_2), u_3) = b(u_1, u_2, u_3), \qquad \forall u_1, u_2, u_3 \in V. \tag{3.53}$$

We can now set the MHD problem in the form (3.10) (with $R = 0$ and B replaced by $\mathfrak{B}$). If $u = \{v, B\}$ is a solution of (3.35)–(3.41) and if $\varphi = \{w, C\}$ is a test function in V, we multiply (3.35) by w, (3.36) by SC, integrate over Ω, and add the resulting equations. After simplificiation, we find

$$\frac{d}{dt}[u, \varphi] + a(u, \varphi) + b(u, u, \varphi) = [f, \varphi], \qquad \forall \varphi \in V, \tag{3.54}$$

where $f = \begin{pmatrix} g \\ 0 \end{pmatrix} \in H$. We reinterpret (3.54) as

$$\frac{du}{dt} + Au + \mathfrak{B}(u) = f, \tag{3.55}$$

where we have set $\mathfrak{B}(u) = \mathfrak{B}(u, u)$.

We can verify that the assumptions of Theorems 3.1 and 3.2, and especially (3.3)–(3.9) are satisfied. The conditions (3.3)–(3.5) disappear since $R = 0$, (3.6) is the same as (3.52), and finally the conditions (3.7)–(3.9) are derived from (3.23) by elementary computations which are left to the reader ($\theta_3 = \theta_4 = \theta_5 = \frac{1}{2}$).

Theorems 3.1 and 3.2 are then applicable. Theorem 3.1 provides the existence (and some regularity properties) of the solution of the initial-value problem to (3.55). It allows us to define the dynamical system, i.e., the semigroup

$$S(t)\colon u(0) \in H \to u(t) \in H.$$

Theorem 3.2 then gives the existence of a global compact attractor (bounded in V, compact in H). This attractor is the mathematical object that describes the long-term behavior of the solutions of the MHD problem, and all the possible ones corresponding to all the possible initial data $u(0) = \{v(0), B(0)\}$.

Remark 3.3. For further details the reader is referred to M. Sermange and R. Temam [1]. See also J.M. Ghidaglia [1].

Remark 3.4. In a totally similar manner we can treat the case of space-periodic boundary conditions. We assume that $\Omega =]0, L_1[\times]0, L_2[$ and we replace (3.39)–(3.41) by

$$v, B, p, \text{ and the first derivatives of } v \text{ and } B \text{ are periodic on } \Gamma. \quad (3.55)$$

We also assume that the averages on v and B on Ω vanish at all time

$$\int_\Omega v\, dx = \int_\Omega B\, dx = 0. \quad (3.56)$$

In this case, we set $H_1 = H_2 =$ the space in (2.9)

$$\{\varphi \in \dot{L}^2(\Omega)^n, \operatorname{div} \varphi = 0, \varphi_{i|\Gamma_i} = -\varphi_{i|\Gamma_{i+n}}, i = 1, 2, n = 2\}$$

(see (2.9) and the footnote after (2.9)); we set $V_1 = V_2 =$ the space in (2.11)

$$\{\varphi \in \dot{H}^1_{\text{per}}(\Omega)^n, \operatorname{div} \varphi = 0\}.$$

We have $D(A_1) = D(A_2) = \dot{H}^2_{\text{per}}(\Omega)^n \cap V_1$. Everything then applies without modification (except the boundary conditions in (3.47), (3.48) which are replaced by the periodicity on Γ).

3.4. Geophysical Flows (Flows on a Manifold)

The next example, to which the general framework of Section 3.1 applies, is the flow of a fluid on a manifold. This could be, for instance, the flow of air around the earth, if the manifold is the surface of the earth. This kind of problem has of course some interest in meteorology. However, we consider for simplicity a fluid at constant temperature, and it would be necessary for a physically more significant result to introduce the equation of temperature (as in Section 3.5) and take into account some other effects, e.g., rotation and the resulting Coriolis force.

The Navier–Stokes equations on a manifold M can be found, for instance, in D. Ebin and J. Marsden [1] (see also A. Avez and Y. Bamberger [1]) for the flow around a two-dimensional sphere. When $M = T^n$ is the n-dimensional torus endowed with the flat metric ($g_{ij} = \delta_{ij}$ the Kronecker symbol, see below), we recover the Navier–Stokes equations with the space-periodic boundary condition as in (2.5).

3.4.1. Sobolev Spaces on a Riemannian Manifold

Let M be a finite-dimensional compact, connected, and oriented Riemann manifold without a boundary. We denote by g the Riemann metric on M: g

is a $\mathscr{C}^\infty$ field of symmetric bilinear forms on the tangent spaces

$$g(x; \xi, \eta) = g(x; \eta, \xi), \qquad \forall x \in M, \quad \forall \xi, \eta \in T_x M,$$

where $T_x M$ is the tangent space at M in x,

$$g(x; \xi, \xi) > 0 \qquad \forall x \in M, \quad \forall \xi \in T_x M, \quad \xi \neq 0.$$

If $x_1, \ldots, x_n$ is a coordinate system on M, we set

$$g_{ij} = g\left(x; \frac{\partial}{\partial x^i}, \frac{\partial}{\partial x^j}\right), \qquad G = \det(g_{ij}),$$

and we denote by

$$\xi \cdot \eta = g_{ij} \xi^i \eta^j, \qquad |\xi| = \{\xi \cdot \xi\}^{1/2}, \tag{3.57}$$

the scalar product and the norm induced by g on $T_x M$.

The inverse of the matrix g_{ij} is denoted by g^{ij}; hence, using the Einstein summation convention, $g_{ij} g^{jk} = \delta_i^k$. To a given vector ξ with components ξ^j, we associate the covector $\tilde{\xi}$ with components $\tilde{\xi}_i = g_{ij} \xi^j$, and to a covector η with component η_i, we associate the vector $\hat{\eta}$ with components $\hat{\eta}^i = g^{ij} \eta_j$.

We denote by $\mathscr{C}^\infty(M)$ (resp. $\mathscr{C}^\infty(TM)$) the space of $\mathscr{C}^\infty$ functions (resp. $\mathscr{C}^\infty$ vector field) on M. If $p \in \mathscr{C}^\infty(M)$, we define its gradient, $\nabla p \in \mathscr{C}^\infty(TM)$, by setting

$$(\nabla p)^i = g^{ij} \frac{\partial p}{\partial x^j}. \tag{3.58}$$

The total derivative ∇u of a vector field $u \in \mathscr{C}^\infty(TM)$ is defined by

$$(\nabla u)_i^j = D_i u^j, \tag{3.59}$$

where D is the Levi-Cività connection whose expression in a coordinate system $x_1, \ldots, x_n$ is given by the following formula using the Christoffel symbols.

$$D_i u^j = \frac{\partial u^j}{\partial x_i} + \begin{Bmatrix} & j & \\ i & & k \end{Bmatrix} u^k,$$

$$\begin{Bmatrix} & j & \\ i & & k \end{Bmatrix} = g^{kl}\{i \ j \ l\},$$

$$\{i \ j \ k\} = \frac{1}{2}\left(\frac{\partial g_{jk}}{\partial x^i} + \frac{\partial g_{ik}}{\partial x^j} - \frac{\partial g_{ij}}{\partial x^k}\right).$$

The divergence of a vector field $u \in \mathscr{C}^\infty(TM)$ is the contraction of its total derivative

$$\operatorname{div} u = D_i u^i; \tag{3.60}$$

The Laplacian of a scalar function $p \in \mathscr{C}^\infty(M)$ is

$$\Delta p = D_i\left(g^{ij}\frac{\partial p}{\partial x^j}\right), \tag{3.61}$$

and the Laplacian of a vector field $u \in \mathscr{C}^\infty(TM)$ is

$$(\Delta u)^i = g^{kl} D_k D_l u^i. \tag{3.62}$$

The canonical volume element on M is

$$dM = \sqrt{G}\, dx^1 \dots dx^n.$$

The space of real L^2 functions on M for the measure dM, $L^2(M, dM)$ is denoted by L^2; the space of vector fields that are L^2 on M for the measure dM is denoted $\mathbb{L}^2$. The scalar product on one of these spaces is written

$$(u, v) = \int u(x)\cdot v(x)\, dM(x)$$

and we set $|u| = \{(u, u)\}^{1/2}$. Finally, we define the scalar and vectorial Sobolev spaces H^m and $\mathbb{H}^m$ as the completed spaces of, respectively, $\mathscr{C}^\infty(M)$ and $\mathscr{C}^\infty(TM)$, for the norms

$$\|p\|_m = \left(\sum_{l=0}^m \int |\nabla^l p|^2\, dM\right)^{1/2}, \qquad p \in \mathscr{C}^\infty(M),$$

$$\|u\|_m = \left(\sum_{l=0}^m \int |\nabla^l u|^2\, dM\right)^{1/2}, \qquad u \in \mathscr{C}^\infty(TM).$$

These spaces are studied, for instance, in T. Aubin [1] and R.O. Wells [1]. We have, in particular, Sobolev embeddings and compactness results that are totally similar to those of Chapter II, Section 1.

When M is connected, which we assumed here, the seminorms

$$\|p\| = \left(\int |\nabla p|^2\, dM\right)^{1/2}, \qquad p \in H^1,$$

$$\|u\| = \left(\int |\nabla u|^2\, dM\right)^{1/2}, \qquad u \in \mathbb{H}^1,$$

are norms on the quotient spaces $H^1/\mathbb{R}$, $\mathbb{H}^1/\mathbb{R}$, and they are equivalent to the natural quotient norms. Furthermore, if M enjoys the following topological property

Every continuous vector field on M vanishes at one point at least, (3.63)

[1] In the case of a torus $\mathbb{T}^n$, $\chi(\mathbb{T}^n) = 0$ and (3.63) is not satisfied. However, (3.64) is satisfied on the set $\{u \in \mathbb{H}^1, \int_{\mathbb{T}^n} u\, dM = 0\}$.

then we have the following result which is equivalent to a Poincaré-like inequality

$$\|u\| \text{ induces on } \mathbb{H}^1 \text{ a norm equivalent to the norm } \|u\|_1. \tag{3.64}$$

Hypothesis (3.63), which will be assumed hereafter, is equivalent to the condition that the Euler–Poincaré characteristic $\chi(M)$ of M does not vanish and is satisfied, for instance, if M is a two-dimensional sphere or, more generally, a sphere of even dimension (see V. Arnold [1], Y. Choquet-Bruhat [1]).

3.4.2. Navier–Stokes Equations on a Manifold

Let M be given as above. The Navier–Stokes equations on M read

$$\frac{\partial u}{\partial t} - \nu\Delta u + u^l D_l u + \nabla p = f, \tag{3.65}$$

$$\operatorname{div} u = 0. \tag{3.66}$$

There are no boundary conditions since M is compact. If we consider an initial-value problem we supplement (3.65), (3.66) with the initial condition

$$u(x, 0) = u_0(x), \qquad x \in M. \tag{3.67}$$

Our aim now is to set the problem in the form (3.10), (3.11). Let

$$\mathscr{V} = \{v \in \mathscr{C}^\infty(TM), \operatorname{div} v = 0\}. \tag{3.68}$$

The spaces H and V are the closures of $\mathscr{V}$ in $\mathbb{L}^2$ and $\mathbb{H}^1$; H is endowed with the scalar product $(\cdot, \cdot)$ and, under assumption (3.63), V is equipped with the scalar product $((\cdot, \cdot))$. These spaces can be characterized as in the flat case (Section 2, see D. Ebin and J.E. Marsden [1])

$$H = \{u \in \mathbb{L}^2, \operatorname{div} u = 0\}, \tag{3.69}$$

$$V = \{u \in \mathbb{H}^1, \operatorname{div} u = 0\}. \tag{3.70}$$

We define the operator A associated in the usual manner with the form $a(u, v) = ((u, v))$, and the spaces V and H. The domain $D(A)$ of A in H is $V \cap \mathbb{H}^2$ and if $u \in D(A)$ and $f \in H$, saying that $Au = f$ amounts to saying that there exists $p \in H^1$ such that

$$-\Delta u + \nabla p = f.$$

We also define the trilinear form on $\mathscr{C}^\infty(TM)$

$$\begin{aligned} b(u, v, w) &= \int (u^l D_l v) \cdot w \, dM \\ &= \int g_{ij} u^l D_l v^i w^j \, dM \\ &= (u^l D_l v, w). \end{aligned} \tag{3.71}$$

For $n \leq 4$, this form can be extended as a trilinear continuous form on $\mathbb{H}^1$

and *for the case considered here where* $n = 2$ we have various estimates of b which are totally similar to (3.23) (and (2.16)). We have

$$b(u, v, v) = 0, \tag{3.72}$$

for $u, v \in \mathscr{C}^\infty(TM)$ if $\operatorname{div} u = 0$, and this relation extends by continuity for every u in V and v in $\mathbb{H}^1$.

The weak form of (3.65), (3.66) is obtained by taking the scalar product in H of (3.65) with a test function $v \in V$. We use an integration-by-parts formula on M which plays here the role of the Green formula, namely

$$\int (D_s \varphi_K^I) \psi_I^{sK}\, dM = -\int \varphi_k^I D_s \psi_I^{sK}\, dM, \tag{3.73}$$

where I and K are families of multi-indices. This formula, which is also necessary for the proof of (3.72), will be proved for the sake of completeness at the end of Section 3.4. We find

$$\frac{d}{dt}(u, v) + \nu((u, v)) + b(u, u, v) = (f, v), \qquad \forall v \in V. \tag{3.74}$$

We associate with b the bilinear continuous operator B mapping $V \times V$ into V' and $D(A) \times D(A)$ into H, defined by

$$(B(u, v), w) = b(u, v, w), \qquad \forall u, v, w \in V \tag{3.75}$$

and we set $B(u) = B(u, u)$. It is then clear that (3.74) is equivalent to

$$\frac{du}{dt} + \nu Au + B(u) = f. \tag{3.76}$$

This is exactly (3.10) (with $R = 0$ and A replaced by νA).

3.4.3. *The Main Results*

We now restrict ourselves to the space dimension 2, $\dim M = n = 2$. The assumptions of Theorems 3.1 and 3.2 are satisfied; (3.3)–(3.5) disappear, (3.6) is the same as (3.7), and (3.7)–(3.9) follow with $\theta_3 = \theta_4 = \theta_5 = \frac{1}{2}$ from the inequalities similar to (3.23) satisfied by b.

Theorem 3.1 proves the existence and uniqueness of solutions for the initial-value problem associated with (3.76) and gives some results of regularity on the solution. This allows us to define the semigroup of the dynamical systems, i.e., the operators

$$S(t)\colon u(0) \in H \to u(t) \in H.$$

Theorem 3.2 then shows the existence of a global attractor that describes all the long-term behavior of the solutions of (3.76) for all the initial data.

3.4.4. *Integration by Parts*

As promised, we show the integration-by-parts formula (3.73). It suffices to establish the result when I and K contain only one index.

Lemma 3.1. *Given $\mathscr{C}^\infty$-valued tensor functions φ_l^i, ψ_i^{kl}, we have*

$$\int (D_k \varphi_l^i)\psi_i^{kl}\, dM = -\int \varphi_l^i (D_k \psi_i^{kl})\, dM.$$

Proof. The general covariant differentiation is given by

$$D_i \theta_{i_1\ldots i_p}^{j_1\cdots j_q} = \frac{\partial}{\partial x_i}\theta_{i_1\ldots i_p}^{j_1\cdots j_q} - \sum_{r=1}^{p}\begin{Bmatrix} & k & \\ i & & i_r\end{Bmatrix}\theta_{i_1\ldots i_{r-1}k i_{r+1}\ldots i_p}^{j_1\cdots j_q}$$
$$+ \sum_{s=1}^{q}\begin{Bmatrix} & j_s & \\ i & & k\end{Bmatrix}\theta_{i_1\ldots i_p}^{j_1\cdots j_{s-1}k j_{s+1}\cdots j_q}$$

Hence

$$D_k \varphi_l^i = \frac{\partial \varphi_l^i}{\partial x_k} - \begin{Bmatrix} & q & \\ k & & l\end{Bmatrix}\varphi_q^i + \begin{Bmatrix} & i & \\ k & & p\end{Bmatrix}\varphi_l^p,$$

$$D_k \psi_i^{kl} = \frac{\partial \psi_i^{kl}}{\partial x_k} - \begin{Bmatrix} & p & \\ k & & i\end{Bmatrix}\psi_p^{kl} + \begin{Bmatrix} & k & \\ k & & q\end{Bmatrix}\psi_i^{ql} + \begin{Bmatrix} & l & \\ k & & q\end{Bmatrix}\psi_i^{kq}.$$

It suffices to consider functions φ and ψ supported by a chart, in which case,

$$I = \int (D_k \varphi_l^i)\psi_i^{kl}\, dM = \int_{\mathbb{R}^n} (D_k \varphi_l^i)\psi_i^{kl}\sqrt{G}\, dx$$
$$= \int_{\mathbb{R}^n} \frac{\partial \varphi_l^i}{\partial x_k}\psi_i^{kl}\sqrt{G}\, dx + \int_{\mathbb{R}^n}\begin{Bmatrix} & i & \\ k & & p\end{Bmatrix}\varphi_l^p\psi_i^{kl}\sqrt{G}\, dx$$
$$- \int_{\mathbb{R}^n}\begin{Bmatrix} & q & \\ k & & l\end{Bmatrix}\varphi_l^i\psi_i^{kl}\sqrt{G}\, dx.$$

By the usual integration by parts

$$\int_{\mathbb{R}^n} \frac{\partial \varphi_l^i}{\partial x_k}\psi_i^{kl}\sqrt{G}\, dx = -\int_{\mathbb{R}^n} \varphi_l^i \frac{\partial}{\partial x_k}(\psi_i^{kl}\sqrt{G})\, dx$$
$$= -\int_{\mathbb{R}^n} \varphi_l^i \frac{\partial \psi_i^{kl}}{\partial x_k}\sqrt{G}\, dx - \int_{\mathbb{R}^n} \varphi_l^i \psi_i^{kl}\begin{Bmatrix} & p & \\ p & & k\end{Bmatrix}\sqrt{G}\, dx$$

as

$$\begin{Bmatrix} & p & \\ p & & k\end{Bmatrix} = \frac{1}{\sqrt{G}}\frac{\partial\sqrt{G}}{\partial x_k}.$$

It is then clear that

$$-I = +J = \int \varphi_l^i (D_k \psi_i^{kl})\, dM = \int_{\mathbb{R}^n} \varphi_l^i (D_k \psi_i^{kl})\sqrt{G}\, dx. \qquad \square$$

3.5. Thermohydraulics

We conclude Section 3 by considering the equations of thermohydraulics, i.e., the coupled system of equations of fluid and temperature in the Boussinesq approximation. We restrict ourselves to the case of space dimension 2 (as for Navier–Stokes) so as to avoid the further mathematical difficulties encountered in dimension 3: we recall that in this case the semigroup $S(t)$ is not everywhere defined in H.[1]

In the case of thermohydraulics Theorems 3.1 and 3.2 do not apply directly, because assumption (3.5) is not satisfied. A subtle property of the semigroup and the attractor, related to the maximum principle in parabolic equations, allows us to overcome this difficulty and to obtain the desired results which are finally very similar to Theorems 3.1 and 3.2. We follow here the presentation in C. Foias, O. Manley, and R. Temam [1].

3.5.1. Equations of the Bénard Problem

We start from the Boussinesq equations in the nondimensional form as in C. Foias, O. Manley, and R. Temam [1]. The domain occupied by the fluid is $\Omega =]0, 1[\times]0, 1[$, e_1, e_2 is the canonical basis of $\mathbb{R}^2$, and $e_n = e_2$ is the unit vertical vector. We have

$$\frac{\partial v}{\partial t} + (v \cdot \nabla)v - \nu\Delta v + \nabla p = e_n(T - T_1), \tag{3.77}$$

$$\frac{\partial T}{\partial t} + (v \cdot \nabla)T - \kappa\Delta T = 0, \tag{3.78}$$

$$\operatorname{div} v = 0, \tag{3.79}$$

where v, p, T stand for the nondimensionalized velocity vector, pressure, and temperature, respectively; T_1 is the temperature at the top, $x_n = 1$, while $T_0 = T_1 + 1$ is the nondimensionalized temperature at the boundary below $x_n = 0$. We are considering a Bénard problem and therefore $T_0 > T_1$. The numbers ν and κ are nondimensional numbers which are related to the usual nondimensional numbers, i.e., Grashof (Gr), Prandtl (Pr), and Rayleigh (Ra).[2]

$$\mathrm{Gr} = \frac{1}{\nu^2}, \qquad \mathrm{Pr} = \frac{\nu}{\kappa}, \qquad \mathrm{Ra} = \frac{1}{\nu\kappa}. \tag{3.80}$$

[1] See Chapter VII for the three-dimensional Navier–Stokes equations.

[2] The reference length used in the nondimensionalization is the height of Ω, the typical temperature is the difference in temperatures at the top and bottom. This choice and the other physical data of the problem (gravity and volume expansion coefficient of the fluid) induce a typical time (see C. Foias, O. Manley, and R. Temam [1]).

We supplement (3.77)–(3.79) with the following boundary conditions:

$$v = 0 \quad \text{at} \quad x_n = 0 \quad \text{and} \quad x_n = 1 \tag{3.81}$$

$$T = T_0 \quad \text{at} \quad x_n = 0 \quad \text{and} \quad T = T_1 = T_0 - 1 \quad \text{at} \quad x_n = 1, \tag{3.82}$$

and

p, v, T and the first derivatives of v and T are periodic of period l in the direction x_1, (3.83)

which means that $\varphi|_{x_1=0} = \varphi|_{x_1=1}$ for the corresponding functions φ.

We substract from T the pure conduction solution and consider

$$\theta = T - T_0 - x_n(T_1 - T_0) = T - T_0 + x_n,$$

and we change p to

$$p - \left(x_n + \frac{x_n^2}{2}\right)(T_0 - T_1) = p - \left(x_n + \frac{x_n^2}{2}\right)$$

($n = 2$). We obtain the new equations

$$\frac{\partial v}{\partial t} + (v \cdot \nabla)v - \nu\Delta v + \nabla p = e_n\theta, \tag{3.84}$$

$$\frac{\partial \theta}{\partial t} + (v \cdot \nabla)\theta - v_n - \kappa\Delta\theta = 0, \tag{3.85}$$

$$\operatorname{div} v = 0, \tag{3.86}$$

with the boundary conditions

$$\theta = 0 \quad \text{at} \quad x_n = 0 \quad \text{and} \quad x_n = 1, \tag{3.87}$$

(3.81) and (3.83) hold with T replaced by θ. (3.88)

For the functional setting of the equation we consider the spaces H_1, H_2 and $H = H_1 \times H_2$

$$H_1 = \{v \in L^2(\Omega)^n, \operatorname{div} v = 0, v_{n|x_n=0} = v_{n|x_n=1}, v_{1|x_1=0} = v_{1|x_1=1}\},$$

$$H_2 = L^2(\Omega).$$

We use the same notation $(\cdot, \cdot)$ for the scalar products in H_1, H_2, H and the corresponding norms are denoted by $|\cdot|$. We also consider $V = V_1 \times V_2$, where V_2 is the space of functions in $H^1(\Omega)$ vanishing at $x_n = 0$ and $x_n = 1$ and periodic in the direction x_1. According to the Poincaré inequality[1]

$$|\varphi| \leq \|\varphi\|, \qquad \forall \varphi \in V_2, \tag{3.89}$$

[1] We use the version of the Poincaré inequality in II.(1.35), with

$$p(u) = \left(\int_{\Gamma_2 \cup \Gamma_4} |u|^2\, dx\right)^{1/2}, \qquad \Gamma_2 = \Gamma \cap \{x_2 = 0\}, \quad \Gamma_4 = \Gamma \cap \{x_2 = 1\}.$$

The fact that $c_0 = 1$ here follows from an elementary and easy computation.

and V_2 is a Hilbert space for the scalar product and the norm

$$((u, v)) = \int_\Omega \text{grad } u \text{ grad } u \, dx, \qquad \|u\| = \{((u, u))\}^{1/2}.$$

V_1 is the space

$$\{v \in V_2^n, \text{div } v = 0\}.$$

We also denote by $((\cdot, \cdot))$ and $\|\cdot\|$ the canonical scalar product and norm in V_1 and V; this should not be a source of confusion.

Let $D(A) = D(A_1) \times D(A_2)$,

$$D(A_i) = \left\{ v \in V_i \cap H^2(\Omega)^2, \left.\frac{\partial v}{\partial x_1}\right|_{x_1=0} = \left.\frac{\partial v}{\partial x_1}\right|_{x_1=1} \right\},$$

and let a be the linear operator from $D(A)$ into H and from V into V' defined by

$$(Au_1, u_2) = a(u_1, u_2), \qquad \forall u_i = \{v_i, \theta_i\} \in D(A), \qquad i = 1, 2,$$

with

$$a(u_1, u_2) = \nu((v_1, v_2)) + \kappa((\theta_1, \theta_2)). \tag{3.90}$$

As in the previous cases, Au is self-adjoint and positive and A^{-1} is a compact self-adjoint linear operator in H. We also consider the trilinear forms on V

$$b(u_1, u_2, u_3) = b_1(v_1, v_2, v_3) + b_2(v_1, \theta_2, \theta_3), \qquad \forall u_i = \{v_i, \theta_i\} \in V, \tag{3.91}$$

with

$$b_1(y, w, z) = \sum_{i,j=1}^n \int_\Omega y_i \frac{\partial w_j}{\partial x_i} z_j \, dx, \qquad \forall y, w, z \in H^1(\Omega)^n, \tag{3.92}$$

$$b_2(y, \varphi, \psi) = \sum_{i=1}^n \int_\Omega y_i \frac{\partial \varphi}{\partial x_i} \psi \, dx, \qquad \forall y \in H^1(\Omega)^n, \quad \forall \varphi, \psi \in H^1(\Omega). \tag{3.93}$$

The form b is trilinear continuous on V or even on $H^1(\Omega)^n \times H^1(\Omega)$, and further continuity properties follow from (3.23). We associate with the form b the bilinear continuous operator B which maps $V \times V$ into V' (= the dual of V) and $D(A) \times D(A)$ into H, defined by

$$(B(u_1, u_2), u_3) = b(u_1, u_2, u_3), \qquad \forall u_1, u_2, u_3 \in V. \tag{3.94}$$

Finally, we define the continuous operator in H

$$R: u = \{v, \theta\} \to Ru = \{-e_n\theta, -v_n\}$$

($n = 2$). Now we can set the thermohydraulics problem in the form (3.10). If $u = \{v, \theta\}$ is the solution of (3.84)–(3.88), and if $\varphi = \{w, \psi\}$ is a test function in V, we multiply (3.84) by w, (3.85) by ψ, integrate over Ω, and add the resulting equations. The pressure term disappears and after simplification we find

$$\frac{d}{dt}(u, \varphi) + a(u, \varphi) + b(u, u, \varphi) + (Ru, \varphi) = 0, \qquad \forall \varphi \in V. \tag{3.95}$$

Setting $B(u) = B(u, u)$, we can reinterpret (3.95) as

$$\frac{du}{dt} + Au + B(u) + Ru = 0. \tag{3.96}$$

This equation is the same as (3.10) (with $f = 0$). Most of the assumptions of Theorems 3.1 and 3.2 are satisfied: the properties (3.6)–(3.9) are proved as in the previous examples using (2.21) and (3.23), (3.3) and (3.4) are very easy ($\theta_1 = \theta_2 = 0$). There remains (3.5) which, unfortunately, is not satisfied when ν and κ are small. We will now see how we can overcome this difficulty by utilization of the maximum principle and obtain the same results. Let us observe that assumption (3.5) plays no role in the proof of Theorem 3.1; it only plays a role in the derivation of large time estimates (and the proof of existence of absorbing sets). Hence Theorem 3.1 is applicable and provides existence, uniqueness, and regularity properties for the solutions of the initial-value problem associated with (3.96). In particular, it allows us to define the semigroup $S(t)$ mapping H into itself (and even H into $D(A)$) which associates with any $u_0 \in H$, the value at time t, $u(t)$, of the solution of (3.96) satisfying $u(0) = u_0$.

3.5.2. *Maximum Principle*

We consider the truncation operators defined in Chapter II, (1.33)–(1.35), that associate with a function ψ, the functions ψ_+ and ψ_-

$$\psi_+(x) = \max(\psi(x), 0), \qquad \psi_-(x) = \max(-\psi(x), 0).$$

Using these operators we prove a variant of the maximum principle for θ

Lemma 3.2. *We assume that* v, θ satisfy (3.84)–(3.88) *and that*

$$-1 \le \theta(x, 0) \le 1, \quad a.e.\ x \in \Omega. \tag{3.97}$$

Then

$$-1 \le \theta(x, t) \le 1, \quad a.e.\ x \in \Omega, \quad a.e.\ t. \tag{3.98}$$

If $\{v, \theta\}$ *are defined for all* $t > 0$ *and* (3.97) *is not assumed, then*

$$\theta(\cdot, t) = \tilde{\theta}(\cdot, t) + \bar{\theta}(\cdot, t), \tag{3.99}$$

where $-1 \le \tilde{\theta}(x; t) \le 1$ *a.e., and*

$$\bar{\theta}(\cdot, t) \to 0 \quad \text{in } H_2\ (=L^2(\Omega)) \quad \text{as } t \to \infty. \tag{3.100}$$

Proof. The results are proved more naturally on the equation for T, i.e., (3.78). The existence and uniqueness results given by Theorem 3.1 are easily reinterpreted in terms of T. In terms of T, (3.97), (3.98) amount to

$$T_1 \le T(x, 0) \le T_0, \quad \text{a.e. } x \in \Omega, \tag{3.101}$$

$$T_1 \le T(x, t) \le T_0, \quad \text{a.e. } x \in \Omega, \quad \text{a.e. } t. \tag{3.102}$$

In order to establish the second inequality in (3.102), we consider the truncated function $(T - T_0)_+$ that belongs to $L^2(0, \tau, H^1(\Omega))$, $\forall \tau > 0$, thanks to (3.12) and II.(1.35), which vanishes at $x_n = 0$ and $x_n = 1$ and is space-periodic in the direction x_1; hence $(T - T_0)_+ \in L^2(0, \tau; V_2)$.

Multiplying (3.78) by $(T - T_0)_+$ and integrating over Ω we obtain, after utilization of Green's formula,

$$\frac{1}{2}\frac{d}{dt}|(T - T_0)_+|^2 + \kappa \|(T - T_0)_+\|^2 = 0.$$

Using the Poincaré inequality in V_2, i.e., (3.89), we find

$$\frac{1}{2}\frac{d}{dt}|(T - T_0)_+|^2 + \kappa |(T - T_0)_+|^2 \leq 0. \tag{3.103}$$

For (3.102) we observe that $|(T - T_0)_+(t)|$ is a decreasing function of t that vanishes at $t = 0$ and, therefore, it vanishes for all later time $t > 0$; thus $T(\cdot, t) \leq T_0, \forall t \geq 0$. For the proof of the first inequality in (3.102), we consider $(T - T_1)_-$ and proceed similarly.

If we do not assume (3.97), we conclude from (3.103) that $|(T - T_0)_+(t)|$ decreases exponentially

$$|(T - T_0)_+(t)| \leq |(T - T_0)_+(0)| \exp(-\kappa t). \tag{3.104}$$

Similarly, we can prove that

$$|(T - T_0)_-(t)| \leq |(T - T_0)_-(0)| \exp(-\kappa t). \tag{3.105}$$

Thus setting

$$T = \tilde{T} + \bar{T}, \qquad \bar{T} = (T - T_0)_+ - (T - T_1)_-$$

we see that $T_1 \leq \tilde{T}(x, t) \leq T_0$ almost everywhere and $\bar{T}(\cdot, t) \to 0$ in $L^2(\Omega)$ as $t \to \infty$

$$|\bar{T}(\cdot, t)| \leq \{|(T - T_0)_+(0)| + |(T - T_1)_-(0)|\} \exp(-\kappa t). \tag{3.106}$$

Then (3.99), (3.100) is just a rephrasing of this result in terms of θ, and (3.106) becomes

$$|\bar{\theta}(\cdot, t)| \leq \{|(\theta - 1)_+(0)| + |(\theta + 1)_-(0)|\} \exp(-\kappa t). \quad \square \tag{3.107}$$

3.5.3. *Absorbing Sets and Attractors*

We cannot apply the methods of Theorem 3.2 (using (3.5)) to derive uniform bounds on the solutions of (3.84)–(3.88). However, Lemma 3.2 already provides us with an uniform estimate for $|\theta(t)|$. Let us denote by $|\theta|_\infty$ the norm of θ in $L^\infty(0, \infty; H_2)$. From (3.99) and (3.107) it follows that

$$\begin{aligned} |\theta(t)| &\leq |\tilde{\theta}(t)| + |\bar{\theta}(t)| \\ &\leq |\Omega|^{1/2} + \{|(\theta - 1)|_+(0)| + |(\theta + 1)_-(0)| \exp(-\kappa t)\}, \end{aligned} \tag{3.108}$$

$$|\theta|_\infty \le |\Omega|^{1/2} + \{|(\theta - 1)_+(0)| + |(\theta + 1)_-(0)|\}, \tag{3.109}$$

$$\limsup_{t\to\infty} |\theta(t)| \le |\Omega|^{1/2}, \tag{3.110}$$

where $|\Omega|$ = the volume of $\Omega = \ell$ in the present nondimensional form of the equations.

The principle is then to treat (3.84) as we treated the Navier–Stokes equations in Section 2, considering $e_n\theta$ as a force f. The only differences lie in the boundary conditions for v which are not the same but are treated in exactly the same manner, and in the dependence of f on t with a bound at t infinite (namely (3.110)) which is better than the uniform bound (3.109).

We sketch the steps of the proof. First, we take the scalar product of (3.84) with v (i.e., we write (3.95) with $\varphi = \{v, 0\}$). The term involving b vanishes and we obtain

$$\begin{aligned}\frac{1}{2}\frac{d}{dt}|v|^2 + \nu\|v\|^2 = (\theta, v_n) &\le |\theta|\,|v_n| \le |\theta|\,|v| \\ &\le \text{(with (3.89))} \\ &\le |\theta|\,\|v\| \le \frac{\nu}{2}\|v\|^2 + \frac{1}{2\nu}|\theta|.\end{aligned} \tag{3.111}$$

Hence

$$\frac{d}{dt}|v|^2 + \nu\|v\|^2 \le \frac{|\theta|^2}{\nu}, \tag{3.112}$$

$$|v(t)|^2 \le |v_0|^2 \exp(-\nu t) + \frac{|\theta|_\infty^2}{\nu^2}(1 - \exp(-\nu t)), \tag{3.113}$$

and with (3.110)

$$\limsup_{t\to\infty} |v(t)|^2 \le \frac{|\Omega|}{\nu^2}. \tag{3.114}$$

From the previous results we can conclude the existence of an absorbing set in H. Because of (3.110), (3.114), for any $u_0 \in H$ and any $\varepsilon > 0$, there exists $t_0 = t_0(u_0, \varepsilon)$, such that for $t \ge t_0$

$$|v(t)| \le \frac{|\Omega|^{1/2}}{\nu} + \varepsilon, \tag{3.115}$$

$$|\theta(t)| \le |\Omega|^{1/2} + \varepsilon. \tag{3.116}$$

A perusal of the proof of (3.110), (3.114) shows that t_0 can be chosen uniformly for u_0 in a bounded set $\mathscr{B}$ of H, $t_0 = t_0(\mathscr{B}, \varepsilon)$; therefore, the set

$$\mathscr{B}_0 = \left\{(v, \theta) \in H, |v| \le \frac{|\Omega|^{1/2}}{\nu} + \varepsilon, |\theta| \le |\Omega|^{1/2} + \varepsilon\right\}$$

is absorbing in H for the semigroup $S(t)$.

Then by integration of (3.112) between t and $t+1$

$$|v(t)|^2 + \nu \int_t^{t+1} \|v(s)\|^2\, ds \le |v(t)|^2 + \frac{1}{\nu} \int_t^{t+1} |\theta(s)|^2\, ds,$$

and, for $t \ge t_0$,

$$\int_t^{t+1} \|v(s)\|^2\, ds \le \frac{1}{\nu}\left(\frac{|\Omega|^{1/2}}{\nu} + \varepsilon\right)^2 + \frac{1}{\nu^2}(|\Omega|^{1/2} + \varepsilon)^2. \tag{3.117}$$

An equation similar to (3.112) for θ is obtained by multiplying (3.25) by θ and integrating over Ω. After simplification we find

$$\begin{aligned}
\frac{1}{2}\frac{d}{dt}|\theta|^2 + \kappa\|\theta\|^2 &= (v_n, \theta) \\
&\le |v_n|\,|\theta| \le |v|\,|\theta| \\
&\le \text{(by (3.89))} \\
&\le |v|\,\|\theta\| \le \frac{\kappa}{2}\|\theta\|^2 + \frac{1}{2\kappa}|v|^2,
\end{aligned} \tag{3.118}$$

$$\frac{d}{dt}|\theta|^2 + \kappa\|\theta\|^2 \le \frac{1}{\kappa}|v|^2. \tag{3.119}$$

By integrating (3.119) between t and $t+1$, we obtain

$$|\theta(t+1)|^2 + \kappa \int_t^{t+1} \|\theta(s)\|^2\, ds \le |\theta(t)|^2 + \frac{1}{\kappa}\int_t^{t+1} |v(s)|^2\, ds.$$

Hence for $t \ge t_0$, using (3.115), (3.116), there results

$$\int_t^{t+1} \|\theta(s)\|^2\, ds \le \frac{1}{\nu}(|\Omega|^{1/2} + \varepsilon)^2 + \frac{1}{\kappa^2}\left(\frac{|\Omega|^{1/2}}{\nu} + \varepsilon\right)^2. \tag{3.120}$$

Absorbing Set in V

We now proceed and prove the existence of an absorbing set in V and the uniform compactness of the semigroup $S(t)$.

We take the scalar product of (3.84) with $-\Delta v$ in $L^2(\Omega)^2$

$$\begin{aligned}
\frac{1}{2}\frac{d}{dt}\|v\|^2 + \nu|\Delta v|^2 &= -(\theta, \Delta v_n) + b_1(v, v, \Delta v) \\
&\le \text{(with (3.23))} \\
&\le |\theta|\,|\Delta v| + c_7|v|^{1/2}\|v\|\,|\Delta v|^{3/2} \\
&\le \text{(with the Young inequality)} \\
&\le \frac{\nu}{2}|\Delta v|^2 + \frac{1}{\nu}|\theta|^2 + \frac{c_1'}{\nu^3}|v|^2\|v\|^4,
\end{aligned}$$

$$\frac{d}{dt}\|v\|^2 + \nu|\Delta v|^2 \leq \frac{2}{\nu}|\theta|^2 + \frac{2c_1'}{\nu^3}|v|^2\|v\|^4. \tag{3.121}$$

We apply the uniform Gronwall lemma (Lemma 1.1) with $r = 1$, $y = \|v\|^2$,

$$g = \frac{2c_1'}{\nu^3}|v|^2\|v\|^2, \qquad h = \frac{2}{\nu}|\theta|^2,$$

and with (3.115), (3.116), (3.117), (3.120) we find, for $t \geq t_0$,

$$a_1 = a_1(\varepsilon) = \frac{2c_1'}{\nu^3}\left(\frac{|\Omega|^{1/2}}{\nu} + \varepsilon\right)^2 a_3(\varepsilon),$$

$$a_2 = a_2(\varepsilon) = \frac{2}{\nu}(|\Omega|^{1/2} + \varepsilon)^2,$$

$$a_3 = a_3(\varepsilon) = \frac{1}{\nu}\left(\frac{|\Omega|^{1/2}}{\nu} + \varepsilon\right)^2 + \frac{1}{\nu}(|\Omega|^{1/2} + \varepsilon)^2.$$

We conclude from Lemma 1.1 that

$$\|v(t)\|^2 \leq (a_2(\varepsilon) + a_3(\varepsilon))\exp(a_1(\varepsilon)) \tag{3.122}$$

for $t \geq t_0(\mathscr{B}, \varepsilon) + 1$.

For the component θ, we take the scalar product of (3.85) with $-\Delta\theta$ and proceed in a totally parallel way; we omit the details. Finally, for any bounded set $\mathscr{B}$ of H, there exists t_1 depending on $\mathscr{B}$, $t_1 = t_0(\mathscr{B}, \varepsilon) + 1$ such that for $t \geq t_1$, $S(t)\mathscr{B}$ is included in the bounded set $\mathscr{B}_1$ of V that is defined by (3.122) and the similar relation for θ. We conclude that $\mathscr{B}_1$ is an absorbing set in V and that $S(t)$ is uniformly compact. Theorem I.1.1 is then applicable and leads to the same conclusions as those in Theorem 3.2; namely, there exists a global attractor $\mathscr{A}$ which is maximal among all the bounded invariant sets and the bounded attractors for $S(t)$. That global attractor is the mathematical object describing all the observable permanent flows in the present Bénard problem.

Remark 3.5. Since on the attractor $\mathscr{A}$ any trajectory $u(t) = \{v(t), \theta(t)\}$ is defined for all $t \in \mathbb{R}$ and bounded in H, we deduce easily from Lemma 3.2 and, in particular (3.107), that $-1 \leq \theta(x, t) \leq 1$, $\forall t \in \mathbb{R}$, a.e. $x \in \Omega$, for such a trajectory. Returning to the temperature function T, this means $T_1 \leq T(x, t) \leq T_0$, $\forall t \in \mathbb{R}$, a.e. $x \in \Omega$. Hence *all the orbits on the attractor satisfy the maximum principle.*

Remark 3.6. We can change the boundary condition in the direction x_1 and replace, for instance, (3.88) by a Dirichlet-type boundary condition

$$\begin{cases} v = 0, \text{ and} \\ T = T_0 - x_n = \text{the pure conduction solution} \\ \text{at } x_1 = 0 \text{ and } x_1 = 1. \end{cases} \tag{3.123}$$

Several other boundary conditions are also considered in C. Foias, O. Manley, and R. Temam [1]: free top surface, thermally driven cavity, and combinations of all these conditions. They all lead to the same results, with the only differences lying in the functional setting (spaces V, H, operator $A, \dots$).

4. Some Pattern Formation Equations

The equations under study in this section are related to various pattern formation phenomena accompanying the appearance of turbulence. They are related to weak forms of turbulence such as phase turbulence. In Section 4.1, we consider the Kuramoto–Sivashinsky equations related to turbulence phenomena in chemistry and combustion. Then Section 4.2 is devoted to the Cahn–Hilliard equation modeling pattern formation in phase transition phenomena.

Besides their physical interest, the motivation for studying these equations is that they contain differential operators of the fourth order in space, while the equations considered up to now involve only second-order elliptic operators. This section follows B. Nicolaenko, B. Scheurer, and R. Temam [1], [2], [3].

4.1. The Kuramoto–Sivashinsky Equation

The Kuramoto–Sivashinsky equation has been introduced by Kuramoto [1] in space dimension 1 for the study of phase turbulence in the Belousov–Zhabotinsky reactions. An extension of this equation to space dimension 2 (or more) has been introduced by G. Sivashinsky [1], [2] in studying the propagation of a flame front in the case of mild combustion. In space dimension 1, that equation is also encountered as a model for the Bénard problem is an elongated box; here many rolls appear which may not be in phase during their rotation, and the unknown u signifies the rotation angles of the rolls.

The mathematical theory of the initial-value problem of the Sivashinsky equation is not complete in space dimension $n \geq 2$ (i.e., the semigroup $S(t)$ is not defined everywhere in this case). Therefore, at this point, we cannot treat the higher-dimensional case and we restrict ourselves to space dimension one.[1]

4.1.1. The Equation and Its Mathematical Setting

Let $\Omega =]-L/2, L/2[$, $L > 0$ given. We consider the following nonlinear evolution equation for $u = u(x, t)$, $x \in \Omega$,

$$\frac{\partial u}{\partial t} + \nu \frac{\partial^4 u}{\partial x^4} + \frac{\partial^2 u}{\partial x^2} + \frac{1}{2}\left(\frac{\partial u}{\partial x}\right)^2 = 0, \tag{4.1}$$

[1] The higher-dimensional case can be treated by the methods of Chapter VII. The interested reader is referred to B. Nicolaenko, B. Scheurer, and R. Temam [1].

$\nu > 0$ given. This equation is supplemented with the space-periodicity boundary conditions

$$\frac{\partial^j u}{\partial x^j}\left(-\frac{L}{2}, t\right) = \frac{\partial^j u}{\partial x^j}\left(\frac{L}{2}, t\right), \qquad j = 0, \ldots, 3,^1 \tag{4.2}$$

and, for an initial-value problem, by the initial condition

$$u(x, 0) = u_0(x), \qquad x \in \Omega. \tag{4.3}$$

The nonlinear term is the same as in the eikonal equation. The dissipativity term is of the fourth order, $\nu\, \partial^4 u/\partial x^2$; note that the term $\partial^2 u/\partial x^2$ is anti-dissipative and, in fact, in the absence of an external excitation, this is the term which maintains the excitation, i.e., introduces energy into the system.

Alternatively, we can consider the space derivative of u, $v = \partial u/\partial x$ which satisfies the following equation obtained by differentiation of (4.1) with respect to x

$$\frac{\partial v}{\partial t} + \nu \frac{\partial^4 v}{\partial x^4} + \frac{\partial^2 v}{\partial x^2} + v\frac{\partial v}{\partial x} = 0. \tag{4.4}$$

The linear part is the same as in (4.1), the nonlinear term is now of the Burger type. We add the boundary conditions similar to (4.2)

$$\frac{\partial^j v}{\partial x^j}\left(-\frac{L}{2}, t\right) = \frac{\partial^j v}{\partial x^j}\left(+\frac{L}{2}, t\right), \qquad j = 0, \ldots, 3, \tag{4.5}$$

and the initial condition

$$v(x, 0) = v_0(x), \qquad x \in \Omega, \tag{4.6}$$

where obviously $v_0 = du_0/dx$. We also note that condition (4.2), with $j = 0$, implies

$$\int_\Omega v(x, t)dx = 0, \qquad \forall t. \tag{4.7}$$

Let us now write the functional setting of the problem in the form (4.4)–(4.7). We set

$$H = \dot{L}^2(\Omega) = \left(v \in L^2(\Omega), \int_{-L/2}^{L/2} v(x)\, dx = 0\right),$$

and (see Section II.1.3)

$$V = \dot{H}^2_{\text{per}}(\Omega) = H^2_{\text{per}}(\Omega) \cap H.$$

We endow H with the L^2 scalar product and norm (denoted $(\cdot, \cdot), |\cdot|$), while

[1] More derivatives may coincide depending on the space regularity of the function u.

V is endowed with the scalar product and norm

$$((v, w)) = \int_{-L/2}^{L/2} D^2 v D^2 w \, dx, \qquad D = \frac{\partial}{\partial x}, \tag{4.8}$$

$$\|v\| = \{((v, v))\}^{1/2}. \tag{4.9}$$

It follows from two successive applications of the Poincaré inequality II.(1.32) that $\|\cdot\|$ is a norm on V which is equivalent to that induced by $H^2(\Omega)$.

We denote by A the linear operator associated with V, H and the bilinear form $((\cdot, \cdot))$. It maps V onto V' and it can also be viewed as an unbounded self-adjoint linear operator in H with domain $D(A) = \dot{H}^4_{\text{per}}(\Omega)$. For every v in $D(A)$, $Av = D^4 v$. We denote by R the linear continuous operator from V into H defined by $Rv = D^2 v$. We also consider the trilinear form

$$b(\varphi, \psi, \theta) = \int_{-L/2}^{L/2} \varphi D\psi \theta \, dx, \tag{4.10}$$

which is trilinear continuous on V since $V \subset H^1(\Omega) \subset \mathscr{C}(\bar{\Omega})$ in space dimension 1 (see II.(1.12)). Further continuity properties of b will be given when needed. This form resembles the form b in (2.15) but instead of (2.21) we have the following easily verified orthogonality property

$$b(\varphi, \varphi, \varphi) = 0, \qquad \forall \varphi \in V. \tag{4.11}$$

We associate with b the bilinear continuous operator B from $V \times V$ into H defined by

$$(B(\varphi, \psi), \theta) = b(\varphi, \psi, \theta), \qquad \forall \varphi, \psi, \theta \in V,$$

and we write $B(\varphi) = B(\varphi, \varphi)$.

The weak form of the problem is obtained by multiplying (4.4) by a test function $w \in V$, integrating over Ω, and integrating by parts. We obtain

$$\frac{d}{dt}(v, w) + \nu((v, w)) + (Rv, w) + b(v, v, w) = 0, \qquad \forall w \in V. \tag{4.12}$$

This is equivalent to the abstract evolution equation

$$\frac{dv}{dt} + \nu A v + B(v) + Rv = 0, \tag{4.13}$$

with the initial condition

$$v(0) - v_0. \tag{4.14}$$

Equation (4.13) is similar to (3.10). It is elementary to show that the assumptions of Theorem 3.1 are satisfied, namely (3.3), (3.4) and (3.6)–(3.9). Theorem 3.1 is applicable (without any modification in the statement, except that $f = 0$ here) and provides existence, uniqueness, and regularity of solutions for (4.13), (4.14). In particular, this allows us to define the semigroup $S(t)$ corresponding to the mappings

$$v(0) \in H \to v(t) \in H.$$

We do not pursue the analogy with equation (3.10); (3.5) is not satisfied and Theorem 3.2 is not immediately applicable. A totally different method will be necessary to show the large-time stability of the solutions of (4.13), (4.14) and the existence of absorbing sets. That question will be investigated in Section 4.1.3, after we introduce some preliminary material in Section 4.1.2.

Remark 4.1. Once the evolution of $v = \partial u/\partial x$ is determined by (4.13), (4.14) we can determine the evolution of the average of u, and thus fully determine $u = u(x, t)$ for all time $t > 0$. We integrate (4.1) over Ω. Using the Green formula and (4.2), we see that the contribution from the linear terms vanishes and there remains

$$\frac{\partial}{\partial t}\int_\Omega u(x, t)\,dx = -\frac{1}{2}\int_\Omega\left(\frac{\partial u}{\partial x}(x, t)\right)^2 dx = -\frac{1}{2}\int_\Omega (v(x, t))^2\,dx. \tag{4.15}$$

Even when the equation for v is stable, i.e., $v(\cdot, t)$ is bounded as $t \to \infty$, according to (4.15) and to the physics of the problem, there is no reason for the average of u to remain bounded.

Remark 4.2. It is useful to compute the eigenvalues and eigenvectors of A and $\nu A + R$. The eigenvectors of both operators are the functions

$$\cos\frac{2k\Pi x}{L}, \qquad \sin\frac{2k\Pi x}{L}, \qquad k \in \mathbb{N}^*. \tag{4.16}$$

The corresponding eigenvalues of A are $2^4k^4\Pi^4/L^4$, and the associated eigenvalues of $\nu A + R$ are the numbers

$$\nu\left(\frac{2k\Pi}{L}\right)^4 - \left(\frac{2k\Pi}{L}\right)^2 = \left(\frac{2k\Pi}{L}\right)^2\left(\nu\left(\frac{2k\Pi}{L}\right)^2 - 1\right). \tag{4.17}$$

They are all positive if

$$\tilde{L} = \frac{L}{2\Pi\sqrt{\nu}} < 1, \tag{4.18}$$

but some are negative when (4.18) does not hold (and the number of negative modes is the integer part of $\tilde{L}$). The eigenvectors corresponding to the latter case are called the *unstable modes* of the equation.

We observe that if we remove the nonlinear term in (4.13) (and (4.18) is not satisfied) then the equation is unstable: $v(t)$ tends to infinity with t (for certain norms) for most initial values $v(0)$, i.e., those which have a nonzero projection on the space of unstable eigenmodes.

Remark 4.3. The previous remark indicates the nature of the difficulty in studying the stability of (4.4)–(4.7) for $t \to \infty$. If the solutions of this problem remain bounded at $t \to \infty$, then *it would be due in an essential manner to the presence of the nonlinear term.*

Another indication of the difficulty of the stability problem for this equation is as follows. We multiply (4.4) by v, integrate over Ω, and integrate by parts, using the boundary conditions. The cubic term disappears and we find

$$\frac{1}{2}\frac{d}{dt}|v|^2 + \nu|D^2 v|^2 = |Dv|^2, \tag{4.19}$$

$D = \partial/\partial x$; this energy-type equality is the same as for the linearized equation, i.e. when vDv is removed. We have the interpolation inequality

$$|Dv| \leq |v|^{1/2}|D^2 v|^{1/2}, \qquad \forall v \in H^2_{\text{per}}(\Omega), \tag{4.20}$$

which follows easily from the Parseval identity and the Cauchy–Schwarz inequality. Therefore

$$|Dv|^2 \leq \frac{\nu}{2}|D^2 v|^2 + \frac{1}{2\nu}|v|^2,$$

$$\frac{d}{dt}|v|^2 + \nu|D^2 v|^2 \leq \frac{1}{\nu}|v|^2, \tag{4.21}$$

$$\frac{d}{dt}|v|^2 \leq \frac{1}{\nu}|v|^2,$$

which indeed allows an exponential growth for $t \to \infty$,

$$|v(t)|^2 \leq |v(0)|^2 \exp\left(\frac{t}{\nu}\right).$$

4.1.2. *Eigenvalue of a Schrödinger Operator*

Let us consider the Schrödinger operator on Ω

$$\nu D^4 w - qw, \tag{4.22}$$

where q is in

$$\dot{\mathscr{C}}^\infty_{\text{per}} = \left\{\varphi \in \mathscr{C}^\infty(\mathbb{R}), \varphi(x+L) = \varphi(x), \int_{-L/2}^{L/2} \varphi(x)\,dx = 0\right\}.$$

We denote by K the operator (4.22) associated with the boundary conditions (4.2). It acts in $H = \dot{L}^2(\Omega)$ and its domain is $D(K) = \dot{H}^4_{\text{per}}(\Omega)$. Let H_0 be the subspace of *odd* functions of H. If q is an even function we observe that K maps $D(\text{K}) \cap H_0$ into H_0, and we denote by K_0 its restriction to H_0 with domain $D(K_0) = D(K) \cap H_0$. The first eigenvalue of K_0 is $\lambda_1 = \lambda_1(q)$

$$\lambda_1 = \inf_{w \in D(K_0)} \frac{(K_0 w, w)}{|w|^2}, \tag{4.23}$$

and the question we address here is whether this eigenvalue can be made arbitrarily large with an appropriate choice of q.

We have

Lemma 4.1. *Let K_0 and $D(K_0)$ be defined as above. Then, for any $\alpha > 0$, there exists an even function q of $\dot{\mathscr{C}}^\infty_{\text{per}}$ such that*

$$\lambda_1(q) \geq \alpha, \tag{4.24}$$

λ_1 defined by (4.23).

Proof. The functions of H_0 are expanded in Fourier series

$$w(x) = \sum_{k=1}^{\infty} w_k \sin \frac{2\Pi kx}{L}$$

and we write, for $s \geq 0$,

$$[w]_s = \left\{ \frac{L}{2} \sum_{k=1}^{\infty} \left(\frac{2k\Pi}{L} \right)^{2s} |w_k|^2 \right\}^{1/2}.$$

The L^2 norm of w is $|w| = [w]_0$, and $|w| + [w]_s$ is a norm on $H^s_{\text{per}}(\Omega)$ that is equivalent to the natural norm of $H^s(\Omega)$.

By application of the Parseval identity,

$$\begin{aligned}(K_0 w, w) &= \int_{-L/2}^{L/2} (\nu |D^2 w(x)|^2 - q(x)|w(x)|^2)\, dx \\ &= \nu [w]_2^2 + \alpha |w|^2 - \int_{-L/2}^{L/2} (\alpha + q(x))|w(x)|^2\, dx. \end{aligned} \tag{4.25}$$

The proof now consists in choosing q such that the last integral of (4.25) is arbitrarily small. Since $w(0) = 0$, we achieve this by choosing q such that $q + \alpha$ is a weak approximation of the Dirac measure at 0. Hence, we set

$$q(x) = \alpha \sum_{0 < |k| \leq M} \exp\left(2i\Pi \frac{kX}{L}\right) = 2\alpha \sum_{k=1}^{M} \cos \frac{2\Pi kx}{L}, \tag{4.26}$$

where M (to be chosen below) is sufficiently large. Of course, $q \in \dot{\mathscr{C}}^\infty_{\text{per}}$ and we have

$$\begin{aligned}\int_{-L/2}^{L/2} (\alpha + q(x))|w(x)|^2\, dx &= \alpha \sum_{|k| \leq M} \int_{-L/2}^{L/2} |w(x)|^2 \exp\left(2i\Pi \frac{kX}{L}\right) dx \\ &= \alpha L \sum_{|k| \leq M} f_k, \end{aligned} \tag{4.27}$$

where f_k is the kth Fourier coefficient of $f(x) = (w(x))^2$. Since $f(0) = \sum_{k\in\mathbb{Z}} f_k = (w(0))^2 = 0$, we have for any $s > \frac{1}{2}$

$$\begin{aligned}\left|\sum_{|k|\le M} f_k\right| &= \left|\sum_{|k|>M} f_k\right| \\ &\le \text{(by the Cauchy–Schwarz inequality)} \\ &\le \left(\sum_{|k|>M}\left(\frac{2k\Pi}{L}\right)^{2s}|f_k|^2\right)^{1/2}\left(\sum_{|k|>M}\left(\frac{2k\Pi}{L}\right)^{-2s}\right)^{1/2} \\ &\le c_1' L^{s-(1/2)}M^{-(s-(1/2))}[f]_s.^{1} \end{aligned} \tag{4.28}$$

In dimension 1, for $s > \frac{1}{2}$, the Sobolev space H^s is an algebra and

$$\|uv\|_{H^s} \le c_2'\|u\|_s\|v\|_s, \qquad \forall u, v \in H^s,$$

where c_2' depends only on L. Hence

$$[f]_s = [w^2]_s \le \|w^2\|_{H^s} \le c_2'\|w\|_{H^s}^2.$$

Since $w \in \dot{H}^s_{\text{per}}(\Omega)$ and, by the Poincaré inequality (see II.(1.32)), $[\cdot]_s$ is on $\dot{H}^s_{\text{per}}$ a norm equivalent to that induced by H^s, we also have

$$[w^2]_s \le c_3'[w]_s^2,$$

for an appropriate constant c_3' depending only on Ω and s. An elementary computation, based on the change of variable $x \to y = x/L$, shows that c_3' is of the form $c_4'L^{s-1/2}$, c_4' being an absolute constant.

With (4.27) and (4.28) we now obtain

$$\int_{-L/2}^{L/2} (\alpha + q(x))|w(x)|^2\,dx \le c_5'\alpha L^{2s}M^{-(s-1/2)}[w]_s^2 \qquad (c_5' = c_1'c_4').$$

Setting $s = 2$ and inserting this majorization into (4.25) we find

$$(K_0 w, w) \ge (\nu - c_5'\alpha L^4 M^{-3/2})[w]_2^2 + \alpha|w|^2. \tag{4.29}$$

We choose M so that $c_5'\alpha L^4 M^{-3/2} \le \nu/2$, i.e., M is the first integer larger than

$$\left(\frac{2c_5'\alpha}{\nu}\right)^{2/3} L^{8/3}. \tag{4.30}$$

Therefore

$$(K_0 w, w) \ge \frac{\nu}{2}[w]_2^2 + \alpha|w|^2 \tag{4.31}$$

and (4.24) follows. □

[1] The definition of $[f]_s$ is similar to that of $[w]_s$

$$[g]_s^2 = L\sum_{|k|=1}^{\infty}\left(\frac{2|k|\Pi}{L}\right)^{2s}|g_k|^2 \qquad \text{for} \quad g = \sum_k g_k \exp\left(\frac{2i\Pi kx}{L}\right).$$

Remark 4.4. Lemma 4.1 is valid only if we restrict ourselves to odd functions. Indeed, if we can find $q \in \dot{\mathscr{C}}^{\infty}_{\text{per}}$ such that

$$(Kw, w) = \nu[w]_2^2 - \int_{-L/2}^{L/2} q(x)|w(x)|^2\,dx \geq \alpha \int_{-L/2}^{L/2} |w(x)|^2\,dx \tag{4.32}$$

for every $w \in H^2_{\text{per}}(\Omega)$ or $\dot{H}^2_{\text{per}}(\Omega)$, then by translation, (4.32) will be valid for $q(x+\beta)$, $\forall \beta \in \mathbb{R}$, and after averaging with respect to β over a period, (4.32) will be valid with q replaced by its average over a period, i.e., 0 ($q \in \dot{\mathscr{C}}^{\infty}_{\text{per}}$). Inequality (4.32) with $q = 0$ is impossible in H^2_{per}, $\forall \alpha > 0$, and it is impossible in $\dot{H}^2_{\text{per}}$ for α sufficiently large.

4.1.3. Absorbing Sets and Attractors

We observe that if the initial condition $v(x, 0)$ is odd then the solution of (4.4)–(4.7) (or equivalently of (4.13), (4.14)) is odd for all time

$$v(x, t) = -v(-x, t), \qquad \forall x, t.$$

Because of this property and Lemma 4.1, from now on, we will consider only the restriction of the equation to the subspace H_0 of odd functions.

In order to study the stability of the solutions of (4.4)–(4.7) in the long-term, we introduce the change of the unknown function $v = w + \varphi$, where φ is a smooth odd function in $\dot{\mathscr{C}}^{\infty}_{\text{per}}$. The resulting equation for w reads

$$\frac{\partial w}{\partial t} + \nu D^4 w + D^2 w + \varphi D w + w D\varphi + w D w = g(\varphi), \tag{4.33}$$

with

$$g = g(\varphi) = -\nu D^4 \varphi - D^2 \varphi - \varphi D \varphi, \qquad D = \frac{\partial}{\partial x} \quad \text{or} \quad \frac{d}{dx}. \tag{4.34}$$

The boundary condition for w is again (4.2), while the initial condition is

$$w(x, 0) = w_0(x) = v_0(x) - \varphi(x). \tag{4.35}$$

There are two differences between (4.33) and (4.4): first is the appearance of a nonvanishing right-hand side, and second is the change of the linear part of the operator. If we multiply (4.33) by w, integrate over Ω, and integrate by parts, using the boundary condition, we obtain

$$\frac{1}{2}\frac{d}{dt}|w|^2 + \nu|D^2 w| - |Dw|^2 + \frac{1}{2}\int_{-L/2}^{L/2} w^2 D\varphi\,dx = (g, w). \tag{4.36}$$

This relation can be compared to (4.19), the new term $\frac{1}{2}(wD\varphi, w)$ being due to the change of the linear part of the operator. For (4.36) we can apply Lemma 4.1 with $q = -\frac{1}{2}D\varphi$: for every $\alpha > 0$ there exists $q \in \dot{\mathscr{C}}^{\infty}_{\text{per}}$, even, of the form

(4.26) such that (4.24) holds. By integration, we then find $\varphi \in \dot{\mathscr{C}}^\infty_{\text{per}}$ which is odd

$$\varphi(x) = -\frac{2\alpha L}{\Pi} \sum_{k=1}^{M} \frac{1}{k} \sin \frac{2\Pi k x}{L}. \tag{4.37}$$

As it appears in the proof of Lemma 4.1, q satisfies (4.31) which is stronger than (4.24)

$$\nu |D^2 w|^2 + \tfrac{1}{2}(wD\varphi, w) \geq \frac{\nu}{2}|D^2 w|^2 + \alpha |w|^2. \tag{4.38}$$

Then we infer from (4.36) that

$$\begin{aligned}\frac{1}{2}\frac{d}{dt}|w|^2 + \frac{\nu}{2}|D^2w|^2 + \alpha|w|^2 &\leq |Dw|^2 + (g, w) \\ &\leq \text{(by (4.20))} \\ &\leq |w|\,|D^2w| + |g|\,|w| \\ &\leq \frac{\nu}{4}|D^2w|^2 + \frac{2}{\nu}|w|^2 + \frac{\nu}{4}|g|^2.\end{aligned}$$

At this point we choose $\alpha = 3/\nu$ and we obtain

$$\frac{d}{dt}|w|^2 + \frac{\nu}{2}|D^2w|^2 + \frac{2}{\nu}|w|^2 \leq \frac{\nu}{2}|g|^2, \tag{4.39}$$

and M is then given by (4.30),

$$M - 1 \leq c_6' \tilde{L}^{8/3} < M, \qquad \tilde{L} = L/2\Pi\sqrt{\nu}. \tag{4.40}$$

With this choice of φ, we deduce from (4.39) that

$$\frac{d}{dt}|w|^2 + \frac{2}{\nu}|w|^2 \leq \frac{\nu}{2}|g|^2, \tag{4.41}$$

$$|w(t)|^2 \leq |w(0)|^2 \exp(-2t/\nu) + \frac{\nu^2}{4}|g|^2(1 - \exp(-2t/\nu)), \tag{4.42}$$

$$\limsup_{t\to\infty} |w(t)|^2 \leq \frac{\nu^2}{4}|g|^2. \tag{4.43}$$

This shows that $|w(t)|$ remains bounded for all time and establishes the nonlinear stability of equation (4.13) (i.e., (4.4)–(4.7)). Furthermore, we deduce from (4.42) in a straightforward manner the existence of an absorbing set for v in H; namely, the set $\mathscr{B}_0 = \mathscr{B}_{H_0}(\varphi, \rho_0')$ = the ball of H_0 centered at φ of radius $\rho_0' > \rho_0 = (\nu/2)|g|$.

Now we can prove the existence of an absorbing set in $V = \dot{H}^2_{\text{per}}(\Omega)$ with exactly the same methods as in the examples of Sections 1–3: we differentiate (4.4) with respect to x, multiply the equation by $\partial v/\partial x$, and integrate over Ω;

we obtain a differential inequality leading to the required conclusions by an application of the uniform Gronwall lemma. Theorem I.1.1 is then applicable and we obtain

Theorem 4.1. *We consider the dynamical system associated with odd solutions of the Kuramoto–Sivashinsky equations in space dimension* 1, (4.4)–(4.7). *This dynamical system possesses an attractor $\mathscr{A}$ which is maximal, connected, and compact in $H = \dot{L}^2(\Omega)$.*

Remark 4.5. Theorem 4.1 was proved in B. Nicolaenko, B. Scheurer, and R. Temam [1] where further results can be found; see also [3] for related results for more general equations:

(i) In particular, in [1], a better estimate of

$$R_0 = \limsup_{t \to \infty} |v(t)|,$$

than that given by (4.43), is obtained viz.

$$\limsup_{t \to \infty} |v(t)| \le |\varphi| + \frac{\nu}{2} |g(\varphi)|. \tag{4.44}$$

The improvement is due, in particular, to a better treatment of the right-hand side of (4.36): it involves the use of integrations by parts instead of the Cauchy–Schwarz inequality. The following bound is obtained

$$R_0 \le c_1 \nu^{-1/4} \tilde{L}^{5/2}. \tag{4.45}$$

Similarly, it is show that

$$Y_0 = \limsup_{t \to \infty} \frac{1}{t} \int_0^t |D^2 v(s)|^2 \, ds \le c_2 \nu^{-5/2} \tilde{L}^5, \tag{4.46}$$

where c_1, c_2 are appropriate absolute constants.

(ii) The generalizations of the Kuramoto–Sivashinsky equations considered in B. Nicolaenko, B. Scheurer, and R. Temam [1] include:

(a) the replacement of the operator $\nu D^4 + D^2$ by $\nu D^2 + \mathscr{H}$ where $\mathscr{H}$ is the Hilbert transform operator; or
(b) the replacement of D^4 by an elliptic pseudodifferential operator P of order $2m$, and the replacement of D^2 by R, where $-R$ is an elliptic pseudodifferential operator of order $< 2m$.

Equations of type (a) appear in combustion (as the Sivashinsky equation) when some hydrodynamical effects are taken into account (see Y. Pomeau and P. Manneville [1]). Equations of type (b) appear as a non-Hamiltonian version of the Benjamin–Ono equation in the modeling, under some circumstances, of plasma turbulence (see Y.C. Lee and H.H. Chen [1]).

4.2. The Cahn–Hilliard Equation

The nonlinear Cahn–Hilliard equation [1] was proposed as a continuum model for the description of the dynamics of pattern formation in phase transition. When a binary solution is cooled sufficiently, phase separation may occur and then proceed in two ways: either by nucleation in which nuclei of the second phase appear randomly and grow or, in the so-called spinodal decomposition, the whole solution appears to nucleate at once and then periodic or semiperiodic structures appear. Pattern formation resulting from phase transition has been observed in alloys, glasses, and polymer solutions. For further descriptions of the physical aspects the reader is referred to J.W. Cahn and J.E. Hilliard [1], J.S. Langer [1], A. Novick-Cohen and L.A. Segel [1]. For further results on the mathematical aspects, see A. Novick-Cohen and L.A. Segal [1], B. Nicolaenko and B. Scheurer [1], P. Constantin, C. Foias, B. Nicolaenko, and R. Temam [2], B. Nicolaenko, B. Scheurer, and R. Temam [3] where the existence of an attractor is proved.

4.2.1. The Equation and Its Mathematical Setting

Like the Kuramoto–Sivashinsky equation, this equation involves a fourth-order elliptic operator and it contains a negative viscosity term.

Let Ω denote an open bounded set of $\mathbb{R}^n$, $n = 1, 2$, or 3, with a smooth boundary Γ. The unknown function is a scalar $u = u(x, t)$, $x \in \Omega$, $t \in \mathbb{R}$, and the equation reads

$$\frac{\partial u}{\partial t} - \Delta K(u) = 0 \quad \text{in } \Omega \times \mathbb{R}_+, \tag{4.47}$$

$$K(u) = -\nu \Delta u + f(u), \qquad \nu > 0, \tag{4.48}$$

where f is a polynomial of order $2p - 1$

$$f(u) = \sum_{j=1}^{2p-1} a_j u^j, \qquad p \in \mathbb{N}, \quad p \geq 2. \tag{4.49}$$

We denote by g the primitive of f vanishing at $u = 0$,

$$g(u) = \sum_{j=2}^{2p} b_j u^j. \qquad j b_j = a_{j-1}, \qquad 2 \leq j \leq 2p, \tag{4.50}$$

and we assume that the leading coefficient of f (and g) is positive

$$a_{2p-1} - 2pb_{2p} > 0. \tag{4.51}$$

Thus the leading term is dissipative while the lower-order terms may not be dissipative and may produce a negative viscosity effect. Strictly speaking, the Cahn–Hilliard equation corresponds to the case where $p = 2$, $f(u) = -\alpha u + \beta u^3$, $\alpha, \beta > 0$, but the equation above permits more general expressions for f.

The equation is associated with boundary conditions which could be one of two types:

Either the Neumann boundary conditions

$$\frac{\partial u}{\partial \mathbf{v}} = -\frac{\partial}{\partial \mathbf{v}} K(u) = 0 \quad \text{on } \Gamma, \tag{4.52}$$

$\mathbf{v}$ the unit outward normal on Γ.

Or, assuming that $\Omega = \Pi_{i=1}^{n}]0, L_i[$, $L_i > 0$, the periodic boundary condition

$$\varphi|_{x_i=0} = \varphi|_{x_i=L_i}, \qquad i = 1, \ldots, n, \tag{4.53}$$

for u and the derivatives of u at least of order ≤ 3.[1]

Further, we observe that (4.52) is also equivalent to

$$\frac{\partial u}{\partial \mathbf{v}} = \frac{\partial \Delta u}{\partial \mathbf{v}} = 0 \quad \text{on } \Gamma. \tag{4.54}$$

When an initial-value problem is considered we supplement the equations with an initial condition

$$u(x, 0) = u_0(x), \qquad x \in \Omega. \tag{4.55}$$

For the mathematical setting of the problem we introduce $H = L^2(\Omega)$ (usual scalar product and norm) and V

$$V = \left\{\varphi \in H^2(\Omega), \frac{\partial \varphi}{\partial \mathbf{v}} = 0 \text{ on } \Gamma\right\} \quad \text{for b.c. (4.52)}, \tag{4.56}$$

$$V = H^2_{\text{per}}(\Omega) \quad \text{for b.c. (4.53)}. \tag{4.57}$$

The space V in (4.56) makes sense by the trace theorem (see II.(1.21)), and the space V in (4.57) was defined in Section II.1.3. In both cases, V is a closed subspace of $H^2(\Omega)$ and is equipped with the norm induced by $H^2(\Omega)$ denoted by $\|\cdot\|_2$.

The weak formulation of the problem is obtained by multiplying (4.47) by a test function $v \in V$, integrating over Ω, and using the Green formula and the boundary condition. We find

$$\frac{d}{dt}(u, v) + \nu(\Delta u, \Delta v) - (f(u), \Delta v) = 0, \qquad \forall v \in V. \tag{4.58}$$

Also, if we use the Green formula again on the last term and denote by f' the derivative of f, we have

[1] As usual, this condition for the derivatives of order ≤ 3 is a requirement and, for the derivatives of order ≥ 4, it is a regularity result which may or may not be satisfied.

$$\frac{d}{dt}(u, v) + \nu(\Delta u, \Delta v) + (f'(u)\nabla u, \nabla v) = 0, \qquad \forall v \in V. \tag{4.59}$$

The bilinear form associated with the linear operator is

$$a(u, v) = \nu(\Delta u, \Delta v), \tag{4.60}$$

which is continuous on V but not coercive; this difficulty will be overcome by using further properties of the equation. Although a is not coercive, we can still associate with a a linear unbounded operator A in H. Its domain is

$$D(A) = \left\{ v \in H^4(\Omega), \frac{\partial v}{\partial \mathbf{v}} = \frac{\partial \Delta v}{\partial \mathbf{v}} = 0 \text{ on } \Gamma \right\}$$

for the case of the Neumann boundary condition, and $D(A) = H_{\text{er}}(\Omega)$ for the space-periodic case. For $u \in D(A)$, Au is defined by

$$(Au, v) = a(u, v), \qquad \forall v \in H,$$

which amounts to saying that $Au = \Delta^2 u$.

A particular aspect of this problem is that the average of u is conserved, thus excluding the existence of an absorbing set in $L^2(\Omega)$. Indeed, when we replace v by 1 in (4.58) (which is possible), we find

$$\frac{\partial}{\partial t}\int_\Omega u(x, t)\, dx = 0,$$

$$\int_\Omega u(x, t)\, dx = \int_\Omega u_0(x)\, dx, \qquad \forall t > 0. \tag{4.61}$$

Another aspect of this problem is the existence of a Lyapunov function $J(u)$

$$J(u) = \frac{\nu}{2}|\nabla u|^2 + \int_\Omega g(u)\, dx. \tag{4.62}$$

Thus, assuming that u is a sufficiently regular solution of the problem, we multiply (4.47) by $K(u)$, integrate over Ω, and use the Green formula

$$\left(K(u), \frac{\partial u}{\partial t}\right) = -\nu \int_\Omega \Delta u \cdot \frac{\partial u}{\partial t}\, dx + \int_\Omega f(u)\frac{\partial u}{\partial t}\, dx$$

$$= \frac{d}{dt} J(u),$$

$$-(\Delta K(u), K(u)) = -\int_\Gamma \frac{\partial K(u)}{\partial \mathbf{v}} K(u)\, d\Gamma + \int_\Omega |\nabla K(u)|^2\, dx$$

$$= |\nabla K(u)|^2.$$

Therefore

$$\frac{d}{dt}J(u) + |\nabla K(u)|^2 = 0, \tag{4.63}$$

which shows that

$$J(u(t)) \leq J(u_0), \qquad \forall t \geq 0. \tag{4.64}$$

and if $J(u(t_1)) = J(u(t_2))$ for $t_1 < t_2$, then $u(t) = u^*, \forall t$, with $-\nu\Delta u^* + f(u^*) = 0$. The consequences of the existence of a Lyapunov function will be investigated further in Chapter VII. In particular, we will see that it can give complete information on the structure of the attractor.

Before stating Theorem 4.2, which gives the existence and uniqueness of solutions of the problem and thus defines the semigroup, we present the following simple lemma.

Lemma 4.2. *For every* $\eta > 0$,

$$\{|\Delta u|^2 + \eta|u|^2\}^{1/2} \quad \text{and} \quad \left\{|\Delta u|^2 + \eta\left(\int_\Omega u(x)\,dx\right)^2\right\}^{1/2} \tag{4.65}$$

are norms on V which are equivalent to the H^2*-norm. Similarly,*

$$\{|\Delta^2 u|^2 + \eta|u|^2\}^{1/2} \quad \text{and} \quad \left\{|\Delta^2 u|^2 + \eta\left(\int_\Omega u(x)\,dx\right)^2\right\}^{1/2} \tag{4.66}$$

are norms on $D(A)$ *which are equivalent to the* H^4-norm.

Proof. In the space-periodic case the results follow from elementary computations with the Fourier series. In the case of the Neumann boundary conditions, they follow from the regularity theory for elliptic boundary-value problems (see S. Agmon, A. Douglis, and L. Nirenberg [1]). For (4.65) we use the regularity theory of the Neumann problem

$$\Delta\varphi = h \quad \text{in } \Omega, \qquad \frac{\partial\varphi}{\partial\mathbf{v}} = 0 \quad \text{on } \Gamma, \tag{4.67}$$

which implies that

$$\|\varphi\|_{H^2(\Omega)/\mathbb{R}} \leq c(\Omega)|\Delta\varphi|.$$

For (4.66) we use the regularity theory for the Neumann biharmonic problem

$$\Delta^2\varphi = h \quad \text{in } \Omega, \qquad \frac{\partial\varphi}{\partial\mathbf{v}} = \frac{\partial\Delta\varphi}{\partial\mathbf{v}} = 0 \quad \text{on } \Gamma, \tag{4.68}$$

which implies that

$$\|\varphi\|_{H^4(\Omega)/\mathbb{R}} \leq c(\Omega)|\Delta^2\varphi|.$$

□

It follows, in particular from Lemma 4.2, that the bilinear forms

$$a(u, v) + \eta(u, v), \qquad a(u, v) + \eta\left(\int_\Omega u(x)\,dx\right)\left(\int_\Omega v(x)\,dx\right) \tag{4.69}$$

are coercive on V. □

We have

Theorem 4.2. *We assume that* (4.49), (4.51) *are satisfied. Then, for every u_0 given in H, the initial boundary-value problem* (4.47), (4.48), (4.52)–(4.55) *possesses a unique solution u which belongs to*

$$\mathscr{C}([0, T]; H) \cap L^2(0, T; V) \cap L^{2p}(0, T; L^{2p}(\Omega)), \qquad \forall T > 0. \tag{4.70}$$

The mapping $u_0 \to u(t)$ is continuous in H and the Lyapunov function $J(u(t))$, J defined by (4.62), *decays along the orbits.*[1]

Furthermore, if $p = 2$ when $n = 3$ (p arbitrary when $n = 1$ or 2) and $u_0 \in V$, then

$$u \in \mathscr{C}([0, T]; V) \cap L^2(0, T; D(A)), \qquad \forall T > 0. \tag{4.71}$$

Some indications on the proof of this result are given in Section 4.2.3. Theorem 4.2 gives the existence of the semigroup $S(t)$

$$S(t)\colon u_0 \in H \to u(t) \in H,$$

which clearly satisfies properties I.(1.1) and I.(1.4).

4.2.2. *Absorbing Sets and Attractors*

We denote by $m(\varphi)$ the average on Ω of a function φ in $L^2(\Omega)$ (or $L^1(\Omega)$)

$$m(\varphi) = \frac{1}{|\Omega|}\int_\Omega \varphi(x)\,dx \tag{4.72}$$

and we write $\bar{\varphi} = \varphi - m(\varphi)$.

For φ given in $L^2(\Omega)$, satisfying $m(\varphi) = 0$, we denote by $\psi = N(\varphi)$ the solution of the Poisson equation

$$-\Delta\psi = \varphi$$

associated to the Neumann boundary condition

$$\frac{\partial\psi}{\partial\nu} = 0 \quad \text{in the case (4.52),}$$

and to the periodicity boundary condition in the case (4.53).

[1] If u_0 does not belong to $H^1(\Omega) \cap L^{2p}(\Omega)$ then $J(u_0) = +\infty$. We can prove a regularizing property of $S(t)$ which implies that $u(t) \in H^1(\Omega) \cap L^{2p}(\Omega)$ (and even more), $\forall t > 0$. Thus $t \to J(u(t))$ decays along the orbit for $t > 0$ and $J(u(t)) \nearrow +\infty$ as $t \searrow 0$.

It is easily seen that $\{(N\varphi, \varphi)\}^{-1/2}$ is a continuous norm on $L^2(\Omega)$; we denote it by $\|\varphi\|_{-1}$; similarly, $((\varphi_1, \varphi_2))_{-1} = (N(\varphi_1), \varphi_2) = (\varphi_1, N(\varphi_2))$ is a (pre-Hilbertian) continuous scalar product on $L^2(\Omega)$.[1]

Because of (4.61) we can write (4.47) in the form

$$\frac{\partial \bar{u}}{\partial t} - \Delta K(u) = 0, \tag{4.73}$$

which is equivalent to

$$\frac{\partial}{\partial t} N\bar{u} + K(u) = 0. \tag{4.74}$$

We take the scalar product of this equation with $\bar{u}$ in $L^2(\Omega)$

$$\frac{1}{2}\frac{d}{dt}\|\bar{u}\|_{-1}^2 + (K(u), \bar{u}) = 0. \tag{4.75}$$

We then have

$$\begin{aligned}(K(u), \bar{u}) &= -\nu(\Delta u, \bar{u}) + (f(u), \bar{u}) \\ &= \text{(with the Green formula and (4.61))} \\ &= \nu|\nabla u|^2 + (f(u), u) - (f(u), m(u_0)).\end{aligned} \tag{4.76}$$

The leading term of $f(s)$ is $2pb_{2p}s^{2p-1}$ and that of $f(s)s$ is $2pb_{2p}s^{2p}$. Since $b_{2p} > 0$, it is easy to conclude that there exists a constant c_1 such that

$$f(s)s \geq pb_{2p}s^{2p} - c_1, \qquad \forall s \in \mathbb{R}, \tag{4.77}$$

and, for every $\varepsilon > 0$, there exists a constant $c_2 = c_2(\varepsilon)$ such that

$$|f(s)| \leq \varepsilon b_{2p}s^{2p} + c_2(\varepsilon), \qquad \forall s \in \mathbb{R}. \tag{4.78}$$

Similarly, the leading term of g being $b_{2p}s^{2p}$, there exists a constant c_3 such that

$$\tfrac{1}{2}b_{2p}s^{2p} - c_3 \leq g(s) \leq \tfrac{3}{2}b_{2p}s^{2p} + c_3, \qquad \forall s \in \mathbb{R}. \tag{4.79}$$

Assuming that $|m(u_0)| \leq \alpha$, we write (4.78) with $\varepsilon = (1/\alpha)(p - \frac{3}{2})$ and we deduce from (4.76) that

$$\begin{aligned}(K(u), \bar{u}) &\geq \nu|\nabla u|^2 + \tfrac{3}{2}b_{2p}\int_\Omega u^{2p}\,dx - (c_1 + \alpha c_2)|\Omega| \\ &\geq \text{(with (4.79))} \\ &\geq \nu|\nabla u|^2 + \int_\Omega g(u)\,dx - k_0(\alpha) \\ &\geq J(u) - k_0(\alpha)\end{aligned} \tag{4.80}$$

with

$$k_0(\alpha) = (c_1 + \alpha c_2 + c_3)|\Omega|. \tag{4.81}$$

[1] This scalar product is related to that of $D(A^{-1/2})$, but we will not develop this point here.

We observe that $\{(N\varphi, \varphi)\}^{1/2} = \|\varphi\|_{-1}$ is a continuous norm on $L^2(\Omega)$. Therefore there exists a constant c_1' depending only on Ω such that

$$\|\varphi\|_{-1} \leq c_1'|\varphi|, \qquad \forall \varphi \in L^2(\Omega). \tag{4.82}$$

By the generalized Poincaré inequality II.(1.35), applied with $p(u) = |m(u)|$, there exists a constant c_2' depending only on Ω such that

$$|\varphi - m(\varphi)| \leq c_2'|\nabla\varphi|, \qquad \forall \varphi \in H^1(\Omega) \tag{4.83}$$

and thus, setting $c_3' = c_1'c_2'$,

$$\|\varphi - m(\varphi)\|_{-1} \leq c_3'|\nabla\varphi|, \qquad \forall \varphi \in H^1(\Omega). \tag{4.84}$$

Using (4.79), (4.80), and (4.84) we infer from (4.75) the following inequalities

$$\frac{1}{2}\frac{d}{dt}\|\bar{u}\|_{-1}^2 + J(u) \leq k_0(\alpha), \tag{4.85}$$

$$\frac{1}{2}\frac{d}{dt}\|\bar{u}\|_{-1}^2 + \frac{\nu}{2(c_3')^2}\|\bar{u}\|_{-1}^2 \leq k_1(\alpha), \qquad k_1(\alpha) = k_0(\alpha) + c_3|\Omega|. \tag{4.86}$$

By the Gronwall lemma we find

$$\|\bar{u}(t)\|_{-1}^2 \leq \|\bar{u}(0)\|_{-1}^2 \exp(-\nu c_4' t) + \frac{2k_1(\alpha)}{c_4'}(1 - \exp(-\nu c_4' t),$$

$$\forall t,\ c_4' = 1/(c_3')^2. \tag{4.87}$$

Thus

$$\limsup_{t\to\infty} \|u(t) - m(u(t))\|_{-1}^2 \leq \rho_0^2, \qquad \rho_0(\alpha) = \left(\frac{2k_2(\alpha)}{\nu c_4'}\right). \tag{4.88}$$

Let us assume that

$$\|\bar{u}(0)\|_{-1} \leq R \quad \text{and} \quad m(u_0) \leq \alpha, \tag{4.89}$$

and let there be given $\rho > \rho_0$. Then there exists a time $t_0 = t_0(R, \alpha, \rho)$ that is easily computed with (4.87), such that

$$\|\bar{u}(t)\|_{-1} \leq \rho, \qquad \forall t \geq t_0. \tag{4.90}$$

This result resembles the existence of an absorbing set. More precisely, it shows the existence of an absorbing set for $S(t)$ on the affine (metric) space $H_\beta = \beta + H_0$, $\beta = m(u_0)$, endowed with the norm $\|\cdot\|_{-1}$; here we define H_0, H_β, and also $\mathscr{H}_\gamma$, $\beta \in \mathbb{R}$, $\gamma \geq 0$, by

$$H_\beta = \left\{\varphi \in L^2(\Omega), \frac{1}{|\Omega|}\int_\Omega \varphi(x)\,dx = \beta\right\}, \qquad \mathscr{H}_\gamma = \bigcup_{|\beta|\leq\gamma} H_\beta. \tag{4.91}$$

By integration of (4.85) with respect to t we find

$$\int_t^{t+r} J(u(s))\,ds \leq \frac{1}{\rho^2} + rk_1(\alpha), \qquad \forall t \geq t_0, \tag{4.92}$$

t_0 as above, $r > 0$. Since by Theorem 4.2, J decays along the orbits, we conclude from (4.92) that

$$J(u(t+r)) \leq \frac{1}{r\rho^2} + k_1(\alpha), \qquad \forall t \geq t_0. \tag{4.93}$$

By the definition (4.62) of J and the minorization (4.79) of g, that implies

$$\begin{gathered} \nu|\nabla u(t)|^2 + b_{2p}\int_\Omega u^{2p}(x,t)\,dx \leq k_2 \quad \text{for } t \geq t_0 + r, \\ k_2(\alpha) = 2c_3|\Omega| + \frac{2}{r\rho^2} + 2k_1(\alpha). \end{gathered} \tag{4.94}$$

Let $\mathscr{B}_1 = \mathscr{B}_1(\alpha)$ be the bounded set of $H^1(\Omega)$ defined by

$$|\nabla\varphi| \leq \sqrt{k_2(\alpha)}, \qquad |m(\varphi)| \leq \alpha; \tag{4.95}$$

the above results show that the image of any bounded set $\mathscr{B}$ of $\mathscr{H}_\alpha$ by $S(t)$ (and, in particular, of $\mathscr{H}_\alpha \cap H^1(\Omega)$) is included in $\mathscr{B}_1$ when $t \geq t_1(\mathscr{B})$ $(= t_0 + r, t_0$ as above, r arbitrary, say $= 1)$.

At the same time this shows the existence of an absorbing set in $\mathscr{H}_\alpha$ and in $\mathscr{H}_\alpha \cap H^1(\Omega)$, and the fact that $S(t)$ is uniformly compact in $\mathscr{H}_\alpha$.

We can then apply Theorem I.1.1 to the semigroup $S(t)$ *acting in* $\mathscr{H}_\alpha$, all the assumptions of this theorem being satisfied. We obtain

Theorem 4.3. *The space dimension is* $n = 1$, 2, *or* 3 *and we assume that* (4.49), (4.51) *are fulfilled;* $\mathscr{H}_\alpha$ *is defined by* (4.91). *For every* $\alpha \geq 0$, *the semigroup* $S(t)$ *associated with* (4.47), (4.52), (4.53) *maps* $\mathscr{H}_\alpha$ *into itself. It possesses in* $\mathscr{H}_\alpha$ *a maximal attractor* $\mathscr{A}_\alpha$ *that is compact and connected.*

Remark 4.6

(i) The set $\mathscr{A} = \bigcup_{\alpha\geq 0}\mathscr{A}_\alpha$ is a noncompact connected attractor in H. It is also clear that for each $\beta \in \mathbb{R}$, $X_\beta = \mathscr{A} \cap H_\beta$ is a compact connected attractor in the affine space H_β (see (4.91)); $S(t)$ maps H_β into itself, $\forall t \geq 0$, $\forall \beta \in \mathbb{R}$, and X_β attracts the bounded sets of H_β.

(ii) As previously indicated the structure of $\mathscr{A}$, $\mathscr{A}_\alpha$, X_α can be described in more detail using the results and methods of Section VII.4.

4.2.3. *Proof of Theorem 4.2*

The proof follows the same steps as that of Theorem 2.1; we present only the derivation of the a priori estimates that is made differently.

A first a priori estimate is obtained by replacing v by $u = u(t)$ in (4.59). We obtain

$$\frac{1}{2}\frac{d}{dt}|u|^2 + \nu|\Delta u|^2 + (f'(u)\nabla u, \nabla u) = 0. \tag{4.96}$$

The leading term of $f'(s)$ is $2p(2p-1)b_{2p}s^{2p-2}$ and, as in (4.77), we can show the existence of a constant $c_4 > 0$ such that

$$f'(s) \geq b_{2p}s^{2p-2} - c_4, \qquad \forall s \in \mathbb{R}. \tag{4.97}$$

Then (4.96) implies

$$\frac{1}{2}\frac{d}{dt}|u|^2 + |\Delta u|^2 + b_{2p}\int_\Omega u^{2p-2}|\nabla u|^2\,dx \leq c_4|\nabla u|^2. \tag{4.98}$$

By the interpolation inequality II.(1.24)

$$|\nabla u|^2 \leq c_1'|u|\,\|u\|_{H^2(\Omega)} \tag{4.99}$$

and, by Lemma 4.2,

$$\|u\|_{H^2(\Omega)} \leq c_2'(|\Delta u| + |m(u)|),$$

$$|\nabla u|^2 \leq c_3'|u|(|\Delta u| + \alpha) \leq \frac{\nu}{2}|\Delta u|^2 + \frac{1}{\nu c_4'}|u|^2 + \frac{\nu\alpha^2}{2}. \tag{4.100}$$

We obtain

$$\frac{d}{dt}|u|^2 + \nu|\Delta u|^2 + b_{2p}\int_\Omega u^{2p-2}|\nabla u|^2\,dx \leq k_3(\alpha)|u|^2 + k_4(\alpha). \tag{4.101}$$

Using the Gronwall lemma we derive from (4.98) an estimate of u in $L^\infty(0, T; L^2(\Omega))$ and $L^2(0, T; H^2(\Omega))$, $\forall T > 0$.

This is sufficient for the existence result (4.70) that is obtained by implementing a Galerkin method and passing to the limit. The passage to the limit, the continuity in H ($u \in \mathscr{C}([0, T]; H)$), and the uniqueness are proved as in Theorem 2.1.

In order to show that $J(u(t))$ decays along the trajectories we establish a relation similar to (4.63). The solution is not sufficiently regular to justify the computations leading to (4.63), and we proceed by approximation. We approximate f by the Lipschitz functions f_M

$$f_M(s) = f(s), \qquad |s| \leq M, \qquad = f(M)\operatorname{sgn} s, \qquad |s| \geq M.$$

For each M, we define a solution u_M of (4.47) (with f replaced by f_M), (4.52), (4.53). Since f_M is bounded, u_M is sufficiently regular for the computations leading to (4.63) to be valid. We obtain

$$\frac{d}{dt}J(u_M) + |\nabla K_M(u_M)|^2 = 0.$$

As $M \to \infty$, $u_M \to u$ in various spaces and we obtain by lower semicontinuity

$$\frac{d}{dt}J(u(t)) + |\nabla K(u(t))|^2 \leq 0,$$
$$\frac{d}{dt}J(u(t)) \leq 0. \tag{4.102}$$

It also follows from (4.102) and (4.79) that if $u_0 \in H^1(\Omega) \cap L^{2p}(\Omega)$, then

$$u \in L^\infty(0, T; H^1(\Omega) \cap L^{2p}(\Omega)), \qquad \forall T > 0. \tag{4.103}$$

Finally, the regularity results (4.71) are consequences of another a priori estimate that is obtained by multiplying (4.47) by $\Delta^2 u$ and integrating over Ω. The nonlinear terms must be estimated by delicate interpolation inequalities; the details are sketched hereafter (see also B. Nicolaenko, B. Scheurer, and R. Temam [3]). This gives an estimate of u in $L^\infty(0, T; V) \cap L^2(0, T; D(A))$, $\forall T > 0$, and we take advantage of this estimate by utilization of a basis made of the eigenfunctions of A in the Galerkin method; alternatively, we can also extend these estimates to the solution u_M of the truncated problem mentioned above (f replaced by f_M or a smoother truncation) and then let $M \to \infty$.

Another a priori Estimate

We conclude this section by sketching the proof of the a priori estimate leading to (4.71). We recall that $n = 1, 2$, or 3, p is finite arbitrary if $n = 1$ or 2, $p = 2$ if $n = 3$.

We multiply (4.47) by $\Delta^2 u$, integrate by parts using the Green formula and the boundary conditions, and this yields

$$\begin{aligned}\frac{1}{2}\frac{d}{dt}|\Delta u|^2 + \nu|\Delta^2 u|^2 &= (\Delta f(u), \Delta^2 u)\\ &\leq |\Delta f(u)|\,|\Delta^2 u|\\ &\leq \frac{1}{2\nu}|\Delta f(u)|^2 + \frac{\nu}{2}|\Delta^2 u|^2,\end{aligned}$$

$$\frac{d}{dt}|\Delta u|^2 + \nu|\Delta^2 u|^2 \leq \frac{1}{\nu}|\Delta f(u)|^2. \tag{4.104}$$

The rest of the proof consists of establishing an inequality

$$|\Delta f(u)|^2 \leq k(1 + |\Delta^2 u|^{2\sigma}), \tag{4.105}$$

where $0 \leq \sigma < 1$, the condition $p = 2$ when $n = 3$ being precisely necessary to ensure that $\sigma < 1$ in this case. Assuming (4.105) is proved, we infer from the Young inequality that

$$|\Delta f(u)|^2 \leq \frac{\nu}{2}|\Delta^2 u|^2 + k_1',$$

and (4.104) implies

$$\frac{d}{dt}|\Delta u|^2 + \frac{\nu}{2}|\Delta^2 u|^2 \leq k_1'. \tag{4.106}$$

Due to Lemma 4.2, there exists two constants c_7', c_8' depending only on Ω such

that

$$\begin{cases} \|u - m(u)\|_{H^2(\Omega)} \le c_7'|\Delta u|, \\ \|u - m(u)\|_{H^4(\Omega)} \le c_8'|\Delta^2 u|. \end{cases} \tag{4.107}$$

We readily infer from (4.106), (4.107) an a priori bound of $u - m(u)$ (and thus of u) in $L^\infty(0, \infty; H^2(\Omega))$ and in $L^2(0, T; H^4(\Omega))$, $\forall T > 0$.

To prove (4.105), we observe that

$$\Delta f(u) = f'(u)\Delta u + f''(u)\nabla u \cdot \nabla u,$$

and since $f(u)$ is a polynomial of order $2p - 1$

$$\begin{cases} |f'(s)| \le k_2'(1 + |s|^{2p-2}), \\ |f''(s)| \le k_3'(1 + |s|^{2p-3}), \end{cases} \tag{4.108}$$

so that

$$\begin{aligned} &|\Delta f(u)| \le |f'(u)|_{L^\infty(\Omega)}|\Delta u| + |f''(u)|_{L^\infty(\Omega)}|\nabla u|^2_{L^4(\Omega)}, \\ &|\Delta f(u)| \le k_4'((1 + |u|^{2p-2}_{L^\infty(\Omega)})|\Delta u| + (1 + |u|^{2p-3}_{L^\infty(\Omega)})|\nabla u|^2_{L^4(\Omega)}). \end{aligned} \tag{4.109}$$

Now, by interpolation (see II.(1.25), II.(2.24)), $H^2(\Omega) = [H^1(\Omega), H^4(\Omega)]_{2/3}$ and

$$\begin{aligned} \|u\|_{H^2(\Omega)} &\le c\|u\|^{2/3}_{H^1(\Omega)}\|u\|^{1/3}_{H^4(\Omega)}, \\ |\Delta u| &\le c|\nabla u|^{2/3}|\Delta^2 u|^{1/3}. \end{aligned} \tag{4.110}$$

Also, by the Sobolev embedding theorems, $H^{n/4}(\Omega) \subset L^4(\Omega)$ and

$$\begin{aligned} |\nabla u|_{L^4(\Omega)} &\le c\|u\|_{H^{1+n/4}} \\ &\le \text{(by interpolation)} \\ &\le c|\nabla u|^{1-n/12}|\Delta^2 u|^{n/12}. \end{aligned} \tag{4.111}$$

Since $J(u(t))$ decays along the orbits and g is bounded from below, we know that $|\nabla u(t)|$ is uniformly bounded for $t \ge 0$. The last two estimates then imply

$$\begin{cases} |\Delta u(t)| \le k_3(\alpha)|\Delta^2 u(t)|^{1/3}, \\ |\nabla u(t)|_{L^4(\Omega)} \le k_4(\alpha)|\Delta u(t)|^{n/12}. \end{cases} \tag{4.112}$$

In space dimension 1, $H^1(\Omega) \subset L^\infty(\Omega)$, and in space dimension 2, $H^{1+\varepsilon}(\Omega) \subset L^\infty(\Omega)$, $\forall \varepsilon > 0$. Hence

$$|\varphi - m(\varphi)|_{L^\infty(\Omega)} < c|\nabla\varphi|, \qquad \forall \varphi \in H^1(\Omega) \quad \text{if } n = 1, \tag{4.113}$$

$$|\varphi - m(\varphi)|_{L^\infty(\Omega)} \le c_\varepsilon|\nabla\varphi|^{1-\varepsilon}|\Delta^2\varphi|^\varepsilon, \qquad \forall \varphi \in H^4(\Omega) \quad \text{if } n = 2. \tag{4.114}$$

For $n = 3$, we use Agmon's inequality (see II.(1.40)) which implies

$$|\varphi - m(\varphi)|_{L^\infty(\Omega)} \le c|\nabla\varphi|^{5/6}|\Delta^2\varphi|^{1/6}. \tag{4.115}$$

We apply these relations to $\varphi = u(t)$; since $|m(u(t))| = |m(u_0)| \le \alpha$ and $|\nabla u(t)|$

is bounded for $t \geq 0$, we obtain

$$|u(t)|_{L^\infty(\Omega)} \leq k_{\varepsilon,n}(\alpha)\,|\Delta^2 u(t)|^{\beta_n},$$

$$\beta_n = 0, \qquad k_{\varepsilon,n} = k_{0,1} \quad \text{if } n = 1, \qquad \beta_n = \varepsilon \quad \text{if } n = 2,$$

$$\beta_n = \tfrac{1}{6}, \qquad k_{\varepsilon,n} = k_{0,3} \quad \text{if } n = 3. \tag{4.116}$$

Inserting (4.112), (4.113) into (4.109), we find precisely (4.105) ($p < 3$, i.e., $p = 2$ if $n = 3$).

The proof is complete. □

5. Semilinear Equations

In this section we consider nonlinear equations which contain a monotone term. A typical evolution equation that we consider contains the so-called "nonlinear Laplacian", i.e., the operator

$$-\sum_{i=1}^{n} \frac{\partial}{\partial x_i}\left(\left|\frac{\partial u}{\partial x_i}\right|^{p-2} \frac{\partial u}{\partial x_i}\right).$$

More general semilinear equations are considered as well as some equations of the Leray–Lions type ([1]). First we introduce the model equation, then we consider the evolution equation in its abstract form with the necessary assumptions, followed by further examples. We state the results concerning existence and uniqueness of solutions leading to the definition of the semigroup, and we prove the existence of the global attractor.

5.1. The Equations. The Semigroup

Let Ω denote an open bounded set of $\mathbb{R}^n$ with boundary Γ. We consider the evolution equation

$$\frac{\partial u}{\partial t} - \sum_{i=1}^{n} \frac{\partial}{\partial x_i}\left(\left|\frac{\partial u}{\partial x_i}\right|^{p-2} \frac{\partial u}{\partial x_i}\right) - \kappa u = f, \tag{5.1}$$

where $p > 2$, $\kappa \in \mathbb{R}$, and f is given. We associate with (5.1) the boundary condition

$$u = 0 \quad \text{on } \Gamma, \tag{5.2}$$

and when an initial-value problem is considered, we have the initial condition

$$u(x, 0) = u_0(x), \qquad u_0 \text{ given.} \tag{5.3}$$

For the functional setting we consider the Sobolev space

$$W^{1,p}(\Omega) = \{v \in L^p(\Omega),\ D_i v \in L^p(\Omega),\ i = 1, \ldots, n\},$$

which is Banach for the norm

$$\|v\|_{W^{1,p}(\Omega)} = \|v\|_{L^p(\Omega)} + \sum_{i=1}^{n} \|D_i v\|_{L^p(\Omega)}. \tag{5.4}$$

We also consider

$$\begin{cases} W_0^{1,p}(\Omega) = \text{the closure of } \mathscr{D}(\Omega) \text{ in } W^{1,p}(\Omega), \text{ or alternatively,} \\ W_0^{1,p}(\Omega) = \{v \in W^{1,p}(\Omega), v = 0 \text{ on } \Gamma\}, \\ W^{-1,p'}(\Omega) = \text{the dual of } W_0^{1,p}(\Omega). \end{cases} \tag{5.5}$$

We have

$$\begin{cases} f \in W^{-1,p'}(\Omega) \Leftrightarrow f = f_0 + \sum_{i=1}^{n} D_i f_i, \\ f_0, f_1, \ldots, f_n \in L^{p'}(\Omega), \qquad 1/p + 1/p' = 1. \end{cases} \tag{5.6}$$

We set $V = W_0^{1,p}(\Omega)$, $V' = W^{-1,p'}(\Omega)$, and for $\varphi \in V$,

$$A(\varphi) = -\sum_{i=1}^{n} \frac{\partial}{\partial x_i}\left(\left|\frac{\partial \varphi}{\partial x_i}\right|^{p-2} \frac{\partial \varphi}{\partial x_i}\right). \tag{5.7}$$

The operator $\varphi \mapsto A(\varphi)$ maps $W^{1,p}(\Omega)$ into $W^{-1,p'}(\Omega)$.

For φ, ψ in $W_0^{1,p}(\Omega)$, we have

$$(A(\varphi), \psi) = a(\varphi, \psi), \tag{5.8}$$

$$a(\varphi, \psi) = \sum_{i=1}^{n} \int_{\Omega} \left|\frac{\partial \varphi}{\partial x_i}\right|^{p-2} \frac{\partial \varphi}{\partial x_i} \frac{\partial \psi}{\partial x_i}\, dx. \tag{5.9}$$

Let H denote the space $L^2(\Omega)$ with the usual scalar product and norm $\{(\varphi, \psi), |\varphi| = \|\varphi\|_{L^2(\Omega)}\}$; V is included in H with a continuous embedding, V dense in H, and similarly H is included in V', with a continuous embedding, H dense in V'. Hence

$$V \subset H \subset V', \tag{5.10}$$

where the injections are continuous and each space is dense in the following one.

The initial-value problem (5.1)–(5.3) can now be written in the form

$$u' + Au + Cu = f \qquad (Cu = -\kappa u), \tag{5.11}$$

$$u(0) = u_0, \tag{5.12}$$

where we assume that f is given in V', u_0 in H, and we look for a solution of (5.11), (5.12) satisfying

$$u \in L^p(0, T; V), \qquad \forall T > 0. \tag{5.13}$$

Remark 5.1. If φ belongs to $L^p(0, T; W_0^{1,p}(\Omega))$, then we see that $A(\varphi)$ belongs to $L^{p'}(0, T; W^{-1,p'}(\Omega))$. Thus, if u satisfies (5.11) and (5.13), then

$$u' \in L^{p'}(0, T; V'), \qquad \forall T > 0. \tag{5.14}$$

Now it is a general fact (compare with Lemma II.3.2) that a function u satisfying (5.13), (5.14) is almost everywhere equal to a function in $\mathscr{C}([0, T]; H)$ $(\forall T > 0)$. This gives a regularity result for u and shows that (5.12) makes sense.

The operator A mapping V into V' enjoys the following two properties:

A is *hemicontinuous* from V into V', i.e., $\forall u, v, w \in V$, the function $\lambda \to (A(u + \lambda v), w)$ is continuous from $\mathbb{R} \to \mathbb{R}$. (5.15)

The operator A is *monotone*, i.e., $(A(u) - A(v), u - v) \geq 0$, $\forall u, v \in V$. (5.16)

The proof of (5.16) is elementary, while (5.15) is proved by using the Lebesgue dominated convergence theorem.

In order to be able to consider more general equations than (5.1), (5.2) we introduce an *abstract* framework and state

Theorem 5.1. *Let V be a separable reflexive Banach space with dual V', and let H be a Hilbert space, $V \subset H$, V dense in H, the injection being continuous* (*and thus* (5.10) *holds*).

Let A be a (*nonlinear*) *operator from V into V' which is monotone and hemicontinuous, i.e., it satisfies* (5.15), (5.16), *and furthermore*

$$(A(v), v) \geq \alpha \|v\|^p, \qquad \forall v \in V, \quad \alpha > 0, \quad 2 < p < \infty, \tag{5.17}$$

$$\|A(v)\|_{V'} \leq c_1 \|v\|^{p-1}, \qquad \forall v \in V. \tag{5.18}$$

Let there be given $\kappa \in \mathbb{R}$, f and u_0 satisfying

$$f \in V', \qquad u_0 \in H. \tag{5.19}$$

Then there exists a unique function u which satisfies (5.11)–(5.13), *and $u \in \mathbb{C}(\mathbb{R}_+; H)$. Furthermore, the mapping $u_0 \mapsto u(t)$ is continuous in H.*

This theorem is proved in J.L. Lions [2] (Theorem 1.2, Ch. II), when $\kappa = 0$ (except the continuity of $u_0 \to u(t)$). On letting $u(t) = \tilde{u}(t) \exp(\kappa t)$, and with a minor modification to the proof, the result of existence and uniqueness is valid for $\kappa \neq 0$ as well. Finally, the continuity of the mapping $u_0 \to u(t)$ is easily proved by using the property (5.16) of A.

Remark 5.2. When $\kappa \leq 0$ the dynamics of the system are trivial: there exists a unique solution u_* of $Au_* = f$ (a stationary solution for (5.11)) and, when $t \to \infty$,

$$u(t) \to u_* \quad \text{in } V.$$

The introduction of the "negative viscosity term" $-\kappa u$, $\kappa > 0$, is then necessary for a more complicated dynamics.

Theorem 5.1 is sufficient to define the semigroup $S(t)$

$$S(t)\colon u_0 \in H \to u(t) \in H,$$

which satisfies I.(1.1), I.(1.4).

A result more precise than Theorem 5.1 will however be necessary in the sequel. It is obtained under the following supplementary assumptions:

There exists a Gateaux differentiable function $J\colon V \to \mathbb{R}$ such that $J'(u) = A(u)$. (5.20)

There exist constants $c_2 > 0$, $c_3, c_4, c_5 \geq 0$, such that

$$c_2 \|\varphi\|^p - c_3 \leq J(\varphi) \leq c_4 \|\varphi\|^p + c_5, \qquad \forall \varphi \in V. \tag{5.21}$$

The first condition is equivalent to saying that, for every $u, v \in V$,

$$\frac{J(u + \lambda v) - J(u)}{\lambda} \to (A(u), v) \quad \text{as } \lambda \to 0, \quad \lambda > 0. \tag{5.22}$$

Such a function J is necessarily convex because of (5.16) and actually J convex is a necessary and sufficient condition for its differential $J'(u) = A(u)$ to be monotone.

We also define the domain of A in H, $D(A)$, as the set of v in V such that $A(v) \in H$ (instead of V'). We then have

Theorem 5.2. *The assumptions are those of Theorem* 5.1 *and, furthermore, we assume that* (5.20), (5.21) *hold and that* $u_0 \in D(A)$. *Then the solution* u *of* (5.11)–(5.13) *is in* $L^\infty(0, T; V)$, $\forall T > 0$, u' *and* Au *are in* $L^\infty(0, T; H)$, $\forall T > 0$.

A sketch of the proof is given in Section 5.3. We now give some examples.

Example 5.1. This is just (5.1), (5.2); assumptions (5.20), (5.21) are satisfied with

$$J(v) = \frac{1}{p} \sum_{i=1}^{n} \int_\Omega \left| \frac{\partial v}{\partial x_i} \right|^p dx.$$

Example 5.2. The spaces V, H are the same as in Example 5.1 and we set

$$A(\varphi) = -\sum_{i=1}^{n} \frac{\partial}{\partial x_i} a_i(x, \nabla u), \tag{5.23}$$

where $a_i = a_i(x, \xi)$ is a Carathéodory function on $\Omega \times \mathbb{R}^n$, i.e.,

For every $\xi \in \mathbb{R}^n$, $x \in \Omega \to a_i(x, \xi)$ is measurable and, for a.e. $x \in \Omega$, $\xi \mapsto a_i(x, \xi)$ is continuous. (5.24)

$$c_2' |\xi|^p \leq \sum_{i=1}^{n} a_i(x, \xi) \cdot \xi_i, \qquad c_2 > 0, \quad \forall \xi \in \mathbb{R}^n, \quad \text{a.e. } x \in \Omega, \tag{5.25}$$

$$|a_i(x, \xi)| \leq c_3' |\xi|^{p-1}, \qquad \forall \xi \in \mathbb{R}^n, \quad \text{a.e. } x \in \Omega. \tag{5.26}$$

The assumptions of Theorem 5.1 are satisfied. Those of Theorem 5.2 are satisfied if we assume, in addition, that

$$a_i(x, \xi) = \frac{\partial a(x, \xi)}{\partial \xi_i}, \qquad \text{a.e. } x, \quad \forall \xi, i = 1, \ldots, n, \tag{5.27}$$

where a is a Carathéodory function on $\Omega \times \mathbb{R}^n$, that is differentiable with respect to ξ for a.e. $x \in \Omega$, and such that there exist constants $c_4' > 0$, $c_5' - c_7'$, such that

$$c_4' |\xi|^p - c_5' \leq a(x, \xi) \leq c_6' |\xi|^p + c_7', \qquad \text{a.e. } x \in \Omega, \quad \forall \xi \in \mathbb{R}^n. \tag{5.28}$$

EXAMPLE 5.3. We could consider higher-order operators of a general form, but for the sake of simplicity we restrict ourselves to the following example involving the Laplace operator.

Let $2 < p < \infty$ and let

$$W^{m,p}(\Omega) = \{v \in L^p(\Omega), D^\alpha v \in L^p(\Omega), \forall \alpha, [\alpha] \leq m\},$$

which is Banach for the norm

$$\|v\|_{W^{m,p}(\Omega)} = \|v\|_{L^p(\Omega)} + \sum_{[\alpha]=1}^{m} \|D^\alpha v\|_{L^p(\Omega)}.$$

We set

$$V = W_0^{m,p}(\Omega) = \text{the closure of } \mathscr{D}(\Omega) \text{ in } W^{m,p}(\Omega).$$

For $u, v \in W_0^{m,p}(\Omega)$ we set

$$a(u, v) = \int_\Omega |\Delta u|^{p-2} \Delta u \Delta v \, dx, \tag{5.29}$$

and we define an operator A from $V = W_0^{m,p}(\Omega)$ into V' with

$$(A(u), v) = a(u, v), \qquad \forall u, v \in V. \tag{5.30}$$

Actually,

$$A(u) = \Delta(|\Delta u|^{p-2} \Delta u) \tag{5.31}$$

and $A(u) = J'(u)$ where

$$J(u) = \frac{1}{p} \int_\Omega |\Delta u|^p \, dx. \tag{5.32}$$

Theorems 5.1 and 5.2 are applicable. It is elementary to verify that assumptions (5.15), (5.16) (5.18), (5.20), (5.21) are satisfied. Assumption (5.17), which is not elementary, is satisfied if we observe that $\|\Delta u\|_{L^p(\Omega)}$ is a norm on $W_0^{m,p}(\Omega)$ that is equivalent to the natural norm: this follows from the L^p-regularity theory for the Dirichlet problem and for the Laplace operator, see S. Agmon, A. Douglis, and L. Nirenberg [1].

Remark 5.3. Although, for simplicity, we will not develop these aspects, let us mention that Theorems 5.1 and 5.2 can be extended to more general situations. The operator A can be of the form $A = \sum_{i=1}^{n} A_i$, where the operators A_i acting on spaces V_i satisfy the same assumptions as in Theorem 5.1. We can consider more general operators C, and in particular nonlinear operators satisfying some compactness properties. That allows us to add in (5.1) (or in Examples 5.2 and 5.3) a lower-order nonlinear term. More generally, with different assumptions, we could consider Leray–Lions operators (see [1]) or pseudo-monotone operators (see J.L. Lions [2]).

5.2. Absorbing Sets and Attractors

We assume that $\kappa \geq 0$ since otherwise the dynamics are trivial (see Remark 5.2). A first a priori estimate on the solution u of (5.11)–(5.13) is obtained by taking the scalar product of (5.11) with u. We obtain

$$\frac{1}{2}\frac{d}{dt}|u|^2 + (A(u), u) = (f, u) + \kappa|u|^2. \tag{5.33}$$

Since the embedding of V in H is continuous there exists a constant $c_6 > 0$ such that

$$|v| \leq c_6\|v\|, \qquad \forall v \in V. \tag{5.34}$$

With (5.34) and (5.17) we infer from (5.33) that

$$\begin{aligned}\frac{1}{2}\frac{d}{dt}|u|^2 + \alpha\|u\|^p &\leq \kappa c_6^2\|u\|^2 + c_6|f|\,\|u\| \\ &\leq \text{(with two applications of the Young inequality)} \\ &\leq \frac{\alpha}{2}\|u\|^p + c_1'|f|^{p'} + c_2'.\end{aligned}$$

Hence

$$\frac{d}{dt}|u|^2 + \alpha\|u\|^p \leq c_3'|f|^{p'} + c_4', \tag{5.35}$$

$$\frac{d}{dt}|u|^2 + \frac{\alpha}{c_2^p}|u|^p \leq c_3'|f|^{p'} + c_4'. \tag{5.36}$$

A uniform estimate for $|u(t)|$, and the existence of an absorbing set in H, then result from the following form of Gronwall's lemma (J.M. Ghidaglia):

Lemma 5.1. *Let y be a positive absolutely continuous function on $(0, \infty)$ which satisfies*

$$y' + \gamma y^p \leq \delta, \tag{5.37}$$

with $p > 1, \gamma > 0, \delta \geq 0$. *Then, for* $t \geq 0$,

$$y(t) \leq (\delta/\gamma)^{1/p} + (\gamma(p-1)t)^{-1/(p-1)}. \tag{5.38}$$

PROOF. If $y(0) \leq (\delta/\gamma)^{1/p}$, then $y(t) \leq (\delta/\gamma)^{1/p}, \forall t \geq 0$. If $y(0) > (\delta/\gamma)^{1/p}$, then there exists $t_0 \in]0, +\infty]$ such that

$$y(t) \geq (\delta/\gamma)^{1/p} \quad \text{for } 0 \leq t \leq t_0,$$

$$y(t) \leq (\delta/\gamma)^{1/p} \quad \text{for } t \geq t_0.$$

For $t \in [0, t_0]$ we write $z(t) = y(t) - (\delta/\gamma)^{1/p} \geq 0$ and since $a^p + b^p \leq (a+b)^p$ for $a, b \geq 0$, $p > 1$, we have

$$y^p = (z + (\delta/\gamma)^{1/p})^p \geq z^p + \delta/\gamma.$$

Hence

$$z' + \gamma z^p \leq y' + \gamma\left(y^p - \frac{\delta}{\gamma}\right) \leq 0,$$

and then by integration

$$z(t)^{p-1} \leq \frac{1}{z_0^{1-p} + \gamma(p-1)t} \leq \frac{1}{\gamma(p-1)t}.$$

This implies (5.38) for $t \in [0, t_0]$ and, since this inequality is obvious for $t \geq t_0$, the lemma is proved. □

We apply Lemma 5.1 with

$$y(t) = |u(t)|^2, \qquad \gamma = \frac{\alpha}{c_6^p}, \qquad \delta = c_3'|f|^{p'} + c_4'. \tag{5.39}$$

We obtain a uniform bound for $|u(t)|^2$, $t \geq 0$, and

$$\limsup_{t\to\infty} |u(t)|^2 \leq \left(\frac{\delta}{\gamma}\right)^{1/p}. \tag{5.40}$$

We also deduce from (5.38), (5.39) that the ball $\mathscr{B}_0(0, \rho_0')$ of H centered at 0 of radius $\rho_0' > \rho_0$, $\rho_0 = (\delta/\gamma)^{1/2p}$, is absorbing in H.

In fact, there exists t_0,

$$t_0 = t_0(\rho_0') = \frac{1}{\gamma(p-1)}\{(\rho_0')^2 - (\delta/\gamma)^{1/p}\}^{p-1},$$

such that for every $u_0 \in H$

$$|u(t)| \leq \rho_0', \qquad \forall t \geq t_0(\rho_0'); \tag{5.41}$$

this means that the entrance time of a set $\mathscr{B}$ in $\mathscr{B}_0$ ($S(t)\mathscr{B} \subset \mathscr{B}_0$ for $t \geq t_0$), is in fact independent of $\mathscr{B}$. As shown in most other examples this fact is not typical and it is due to the rapid growth of the nonlinear term ($p > 1$).

We then deduce from (5.35) that

$$\int_t^{t+r} \|u(s)\|^p\, ds \le \frac{1}{\alpha}(\rho_0')^2 + \frac{r}{\alpha}(c_3'|f|^2 + c_4'), \qquad \forall r > 0, \quad \forall t \ge t_0. \quad (5.42)$$

Absorbing Sets in V

A second uniform estimate for u is obtained by taking the scalar product of (5.11) with u'. We find, using (5.20),

$$(A(u), u') + |u'|^2 - \kappa(u, u') = (f, u'), \quad (5.43)$$

$$\frac{d}{dt}\left(J(u) - \frac{\kappa}{2}|u|^2 - (f, u)\right) + |u'|^2 = 0. \quad (5.44)$$

Hence

$$t \mapsto J(u(t)) - \frac{\kappa}{2}|u(t)|^2 - (f, u(t)) \text{ is decreasing for } t > 0. \quad (5.45)$$

In particular, given $t, s, r > 0$, $t < s < t + r$, we have

$$J(u(t+r)) - \frac{\kappa}{2}|u(t+r)|^2 - (f, u(t+r)) \le J(u(s)) - \frac{\kappa}{2}|u(s)|^2 - f(u(s)),$$

and after integration, with respect to s between t and $t + r$,

$$J(u(t+r)) - \frac{\kappa}{2}|u(t+r)|^2 - (f, u(t+r))$$

$$\le \frac{1}{r}\int_t^{t+r} (J(u(s)) - \frac{\kappa}{2}|u(s)|^2 - (f, u(s))\, ds, \qquad \forall t, r \ge 0.$$

Using (5.21) and (5.34) we then obtain

$$c_2\|u(t+r)\|^p \le c_3 + c_5 + |f|\,|u(t+r)|$$

$$+ \frac{1}{r}\int_t^{t+r} (c_4\|u(s)\|^p + |f|\,|u(s)|)\, ds, \qquad \forall t, r > 0. \quad (5.46)$$

Now, if we assume that u_0 belongs to a bounded set $\mathscr{B}$ of H, using (5.41), (5.42), we obtain

$$\|u(t+r)\| \le \rho_1 \quad \text{for } t \ge t_0(\mathscr{B}, \rho_0') + r,$$

$$\rho_1^p = \frac{1}{c_2}\left(c_3 + c_5 + 2|f|\rho_0' + \frac{1}{r_\alpha}(\rho_0')^2 + \frac{c_3'}{\alpha}|f|^2 + \frac{c_4'}{\alpha}\right). \quad (5.47)$$

This shows that $\|u(t)\|$ is uniformly bounded for $t \ge r + t_0(\mathscr{B}, \rho_0')$, $u_0 \in \mathscr{B}$. We obtain the existence of an *absorbing set in V*. Note also that this result (like the uniform Gronwall lemma) is valid even if $J(u_0) = +\infty$, i.e., $u_0 \in H\backslash V$, in which case $u(t) \in V$, $\forall t > 0$. Thus the image by $S(t)$ of the bounded set $\mathscr{B}$ is included in a bounded set of V for $t \geq t_0(\mathscr{B}) + r$. If we assume that

$$\text{the injection of } V \text{ in } H \text{ is compact,} \tag{5.48}$$

we see that the semigroup $S(t)$ is uniformly compact, i.e., satisfies I.(1.12).
Theorem I.1.1 is applicable.

Theorem 5.3. *The assumptions are those of Theorems* 5.1, 5.2, *and* (5.48). *The corresponding semigroup* $S(t)$ *possesses a global attractor* $\mathscr{A}$ *whose basin of attraction is* H, *and which is maximal connected, compact in* H.

5.3. Proof of Theorem 5.2

We give a sketch of the proof of Theorem 5.2 that contains an alternative proof of the existence in Theorem 5.1.

Under assumptions (5.15)–(5.18),[1] A and $A + \lambda I$, $\forall \lambda > 0$, are surjective operators from V (in fact $D(A)$) onto H. Hence the operator $R_\lambda = (I + \lambda A)^{-1}$, the resolvent of H, is well defined in H, and it is easy to see with (5.16) that this is a Lipschitz operator with constant 1, $\forall \lambda > 0$,

$$|R_\lambda \varphi - R_\lambda \psi| \leq |\varphi - \psi|, \qquad \forall \varphi, \psi \in H, \quad \lambda > 0. \tag{5.49}$$

We consider then the Yosida regularization of A, namely

$$A_\lambda = \frac{I - R_\lambda}{\lambda}.$$

The operators A_λ are monotone Lipschitz operators in H with constant $1/\lambda$, $\forall \lambda > 0$, see K. Yosida [1], H. Brézis [1], and as $\lambda \to \infty$,

$$A_\lambda \varphi \to A\varphi \quad \text{in } H \quad \text{if } \varphi \in D(A), \qquad |A_\lambda \varphi| \to +\infty \quad \text{if not.} \tag{5.50}$$

For every $\lambda > 0$ we solve the following differential equation that approximates (5.11), (5.12)

$$\frac{du_\lambda}{dt} + A_\lambda(u_\lambda) + Cu_\lambda = f, \tag{5.51}$$

$$u_\lambda(0) = u_0. \tag{5.52}$$

Since A_λ is a Lipschitz operator, the existence and uniqueness of A_λ on some interval $(0, T_\lambda)$ follow from standard theorems, and $T_\lambda = +\infty$ is a consequence of the following a priori estimates.

The first a priori estimates are obtained by taking the scalar product of (5.46) with u_λ in H

$$\frac{1}{2}\frac{d}{dt}|u_\lambda|^2 + (A_\lambda(u_\lambda), u_\lambda) + (Cu_\lambda, u_\lambda) = (f, u_\lambda). \tag{5.53}$$

[1] Assumptions (5.17), (5.18) can be weakened here.

Since

$$(A_\lambda(\varphi), \varphi) = (A(\varphi), \psi) + \lambda |A\psi|^2 \geq 0, \qquad \forall \varphi \in H, \quad \psi = (I + \lambda A)^{-1}\varphi, \quad (5.54)$$

using the Gronwall lemma, we easily deduce from (5.53) that

$$u_\lambda \text{ is bounded in } L^\infty(0, T; H) \text{ independently of } \lambda, \qquad \forall T > 0. \quad (5.55)$$

Then we write (5.49) as

$$\frac{du_\lambda(t)}{dt} + \frac{1}{\lambda}(u_\lambda - R_\lambda(u_\lambda(t))) = g(t), \qquad g(t) = f + \kappa u(t),$$

and deduce from (5.49) and the estimate (12), Theorem 1.6 in H. Brézis [1], that

$$\left|\frac{du_\lambda(t)}{dt}\right| \leq \left|\frac{du_\lambda(0)}{dt}\right| + \int_0^t |g(s)|\, ds. \quad (5.56)$$

We have

$$\frac{du_\lambda(0)}{dt} = f - A_\lambda(u_0) - \kappa u_0,$$

and since $u_0 \in D(A)$, $A_\lambda(u_0) \to Au_0$ in H as $\lambda \to 0$ (see (5.50)) and $du_\lambda(0)/dt$ is bounded as $\lambda \to 0$. With (5.55), (5.56) we conclude that

$$\frac{du_\lambda}{dt} \text{ remains bounded in } L^\infty(0, T; H), \text{ when } \lambda \to \infty, \forall T > 0. \quad (5.57)$$

Since $R_\lambda(0) = 0$, $|R_\lambda(u_\lambda)| \leq |u_\lambda|$, we see that $R_\lambda u_\lambda$ is bounded in $L^\infty(0, T; H)$, $\forall T > 0$, and then returning to (5.51) we see that

$$A_\lambda(u_\lambda) = A(R_\lambda u_\lambda)$$

is bounded in $L^\infty(0, T; H)$, $\forall T > 0$. With (5.17) we conclude that $R_\lambda(u_\lambda)$ is bounded in $L^p(0, T; V)$, $\forall T > 0$.

We can then pass to the limit in (5.51) and (5.52) using the classical methods for this type of equation and which, in an essential manner, are based on (5.16) (the *monotony* method). We find that u_λ and $R_\lambda(u_\lambda)$ converge to some function u that satisfies all the requirements in Theorems 5.1 and 5.2.

6. Backward Uniqueness

In this, the last section of Chapter III, we prove a result of backward uniqueness for evolution equations of the first order.

The question of backward uniqueness for a general evolution equation

$$\frac{du(t)}{dt} + N(u(t)) = 0 \quad (6.1)$$

is the following: assuming that two solutions u, v of equation (6.1) coincide at some time t_1

$$\frac{du}{dt} + N(u) = \frac{dv}{dt} + N(v) = 0 \quad \text{for} \quad t_1 - \varepsilon < t < t_1, \tag{6.2}$$

and

$$u(t_1) = v(t_1), \tag{6.3}$$

is it true that $u(t) = v(t)$ for all time $t < t_1$ for which u and v are defined? Let $S(t)$ denote the semigroup associated with the solution of the evolution equation (6.1),

$$S(t)u(s) = u(t + s), \qquad \forall t > 0, \quad \forall s \in \mathbb{R},$$

then for $0 < \tau < \varepsilon$, the question is whether

$$S(\tau)u(t_1 - \tau) = S(\tau)v(t_1 - \tau)$$
$$\Rightarrow u(t_1 - \tau) = v(t_1 - \tau).$$

This is exactly the injectivity property of $S(\tau)$ and, as indicated before, the backward uniqueness for all time is equivalent to the injectivity of $S(t)$, $\forall t > 0$.

In Section 6.1 we prove an abstract result and then in Section 6.2 we show how this result produces the backward uniqueness for most of the equations studied in Chapter III.

6.1. An Abstract Result

Let H be a Hilbert space (scalar product $(\cdot, \cdot)$, norm $|\cdot|$), and let A be a linear positive self-adjoint unbounded operator in H with domain $D(A) \subset H$. We denote by V the domain $D(A^{1/2})$ of $A^{1/2}$ that is endowed with the norm

$$\|v\| = |A^{1/2}v|, \qquad \forall v \in D(A^{1/2}),$$
$$= \{(Av, v)\}^{1/2}, \qquad \forall v \in D(A).$$

These assumptions are satisfied by all the examples in Sections 1–4.

We now consider a function w

$$w \in L^\infty(0, T; V) \cap L^2(0, T; D(A)), \tag{6.4}$$

that satisfies a relation

$$\frac{dw(t)}{dt} + Aw(t) = h(t, w(t)), \qquad t \in (0, T). \tag{6.5}$$

We assume that h is a function from $(0, T) \times V$ into H such that for any function w satisfying (6.4)

$$|h(t, w(t))| \leq k(t)\,\|w(t)\| \quad \text{for a.e.} \quad t \in (0, T), \tag{6.6}$$

with

$$k \in L^2(0, T), \tag{6.7}$$

the function $t \to h(t, w(t))$ also being measurable from $(0, T)$ into H.

We denote by $\Lambda = \Lambda(t)$ the quotient of norms

$$\Lambda(t) = \frac{\|w(t)\|^2}{|w(t)|^2}, \tag{6.8}$$

and we prove

Lemma 6.1. *Under the above assumptions*

$$\Lambda' \leq 2k^2\Lambda \quad on\ (0, T), \tag{6.9}$$

$$\Lambda(t) \leq \Lambda(0) \exp\left(2 \int_0^t k^2(s)\, ds\right), \qquad t \in (0, T). \tag{6.10}$$

Proof. By differentiation we write

$$\begin{aligned}
\frac{1}{2}\frac{d\Lambda}{dt} &= \frac{((w', w))}{|w|^2} - \frac{\|w\|^2}{|w|^4}(w', w) \\
&= \frac{1}{|w|^2}(w', Aw - \Lambda w) \\
&= \frac{1}{\|w\|^2}(h - Aw, Aw - \Lambda w) \\
&= (\text{since } (Aw - \Lambda w, w) = 0) \\
&= -\frac{|Aw - \Lambda w|^2}{|w|^2} + \frac{1}{|w|^2}(Aw - \Lambda w, h) \\
&\leq -\frac{|Aw - \Lambda w|^2}{2|w|^2} + \frac{1}{2}\frac{|h|^2}{|w|^2} \\
&\leq (\text{with } (6.6)) \\
&\leq -\frac{|Aw - \Lambda w|^2}{2|w|^2} + k^2\Lambda.
\end{aligned}$$

Therefore

$$\Lambda' + \frac{|Aw - \Lambda w|^2}{|w|^2} \leq 2k^2\Lambda \tag{6.11}$$

and (6.9), (6.10) follow. □

Next we prove the "abstract" backward uniqueness result.

Lemma 6.2. *Under the above assumptions, if a function w satisfies conditions* (6.4)–(6.7) *and*

$$w(\tau) = 0, \tag{6.12}$$

then

$$w(t) = 0, \qquad 0 \le t \le T. \tag{6.13}$$

PROOF. We argue by contradiction and assume that $|w(t_0)| \neq 0$ for some $t_0 \in [0, T[$. Then, by continuity, $|w(t)| \neq 0$ on some interval $(t_0, t_0 + \varepsilon)$ and we denote by $t_1 \le T$ the largest time for which

$$|w(t)| \neq 0 \quad \text{on } [t_0, t_1[.$$

Necessarily, $|w(t_1)| = 0$.

On $[t_0, t_1[$ the function $t \to \log|w(t)|$ is well defined and, by differentiation, we see that

$$\begin{aligned} \frac{d}{dt}\log\frac{1}{|w|} &= -\frac{1}{2}\frac{d}{dt}\log|w|^2 = -\frac{(w', w)}{|w|^2} \\ &= -\frac{(h - Aw, w)}{|w|^2} = +\Lambda - \frac{(h, w)}{|w|^2} \\ &\le (\text{by } (6.6)) \\ &\le \Lambda + k\Lambda^{1/2}, \end{aligned} \tag{6.14}$$

$$\frac{d}{dt}\log\frac{1}{|w|} \le 2\Lambda + k^2.$$

Then, by integration between t_0 and $t \in [t_0, t_1[$,

$$\begin{aligned} \log\frac{1}{|w(t)|} &\le \log\frac{1}{|w(t_0)|} + \int_{t_0}^{t} (2\Lambda(s) + k^2(s))\, ds \\ &\le \log\frac{1}{|w(t_0)|} + \int_0^T (2\Lambda(s) + k^2(s))\, ds. \end{aligned}$$

This inequality shows that $1/|w(t)|$ is bounded from above as $t \to t_1 - 0$ and produces the contradiction. □

Remark 6.1. A proof of backward uniqueness, based on an inequality of Carleman, is due to J.L. Lions and B. Malgrange [1]; the proof presented here uses the log-convexity method of S. Agmon and L. Nirenberg [1] and is based on C. Bardos and L. Tartar [1], J.M. Ghidaglia [3].

6.2. Applications

All the equations considered in Sections 1–4 of this chapter are of the form

$$\frac{du}{dt} + Au + G(u) = 0, \tag{6.15}$$

with $A, D(A), V, H$ as in Section 6.1. Consider another solution v, set $w = u - v$, then

$$\begin{gathered} \frac{dv}{dt} + Av + G(v) = 0, \\ \frac{dw}{dt} + Aw = h, \qquad h = G(v) - G(u). \end{gathered} \tag{6.16}$$

Now it suffices to show that if u, v are in $L^\infty(0, T; V) \cap L^2(0, T; D(A))$, we have the majorization (6.6), (6.7) that is equivalent to

$$|G(u(t)) - G(v(t))| \leq k(t)\,\|u(t) - v(t)\|, \qquad k \in L^2(0, T). \tag{6.17}$$

We conclude by checking (6.17) for all the equations of Sections 1–4. In each case the notations are those used in the corresponding section.

Example in Section 1.1

In order to establish (6.17) we need to know that the solutions of the problem belong to $L^{4p-4}(0, T; L^{4p-4}(\Omega))$, $\forall T > 0$. The result is valid without further assumptions if p is not too large (when the dimension n is fixed) but, for arbitrary p, a further assumption is needed on f and u_0

$$f \in L^q(\Omega), \qquad q = \frac{4p-4}{2p-1}, \qquad u_0 \in L^{2p-2}(\Omega). \tag{6.18}$$

In order to show that this implies

$$u \in L^{4p-4}(\Omega)), \qquad \forall T > 0, \tag{6.19}$$

we proceed *formally* as follows: we multiply (1.2) by u^{2p-3} and integrate over Ω; using (1.3) and Green's formula we obtain

$$\begin{aligned} &\frac{1}{2p-2}\frac{d}{dt}\int_\Omega u^{2p-2}(x, t)\,dx + (2p-3)\int_\Omega u^{2p-4}(x, t)\,|\nabla u(x, t)|^2\,dx \\ &\quad + \int_\Omega g(u(x, t))u^{2p-3}(x, t)\,dx = \int_\Omega f(x)u^{2p-3}(x, t)\,dx. \end{aligned} \tag{6.20}$$

As in (1.8), (1.18) we show that there exists a constant c_1' such that

$$g(s)s^{2p-3} \geq \tfrac{1}{2}b_{2p-1}s^{4p-4} - c_1', \qquad \forall s \in \mathbb{R},$$

and therefore

$$\int_\Omega g(u)u^{2p-3}\,dx \geq \tfrac{1}{2}b_{2p-1}\int_\Omega u^{4p-4}\,dx - c_1'|\Omega|.$$

Then by the Holder and Young inequalities

$$\left|\int_\Omega f^{2p-3}\,dx\right| \leq \left(\int_\Omega |f|^{(4p-4)/(2p-1)}\,dx\right)^{(2p-1)/(4p-4)}\left(\int_\Omega |u|^{4p-4}\,dx\right)^{(2p-3)/(4p-4)}$$
$$\leq \tfrac{1}{4}b_{2p-1}\int_\Omega u^{4p-4}\,dx + c_2'\|f\|^q_{L^q(\Omega)},$$

and we obtain finally

$$\frac{1}{2p-2}\frac{d}{dt}\int_\Omega u^{2p-2}\,dx + \tfrac{1}{4}b_{2p-1}\int_\Omega u^{4p-4}\,dx \leq c_1'|\Omega| + c_2'\|f\|^q_{L^q(\Omega)}. \quad (6.21)$$

By integration of (6.21) in t between 0 and T, we obtain (6.19). The derivation above of (6.21) is heuristical. We make it rigorous as follows: let $\theta_M = \theta_M(s)$ be a smooth increasing $\mathscr{C}^\infty$ function from $\mathbb{R}$ into $\mathbb{R}$, $\theta_M(s) = s$ for $|s| \leq M$, $\theta_M(s) = 2M$ for $|s| \geq 3M$. We multiply (1.2) by $\theta_M(u^{2p-3})$ and obtain an inequality similar to (6.21). When we let $M \to +\infty$ in this inequality we obtain precisely (6.21).

Remark 6.2. The assumption $u_0 \in L^{2p-2}(\Omega)$ is not essential on the attractor. The regularity (6.19) and more regularity results can be proved on the attractor; see Chapter IV, Section 6.

We complete the proof of (6.17) in this case. We write

$$G(u) - G(v) = \sum_{j=1}^{2p-1} b_j(u^j - v^j)$$
$$= \sum_{j=1}^{2p-1}\sum_{i=1}^{j-1} b_j(u-v)u^i v^{j-i}.$$

There exists a constant $c_2' > 0$ such that

$$\left|\sum_{j=1}^{2p-1}\sum_{i=1}^{j-1} b_j r^i s^{j-i}\right| \leq c_1'(1 + |r|^{2p-2} + |s|^{2p-2}), \qquad \forall r, s \in \mathbb{R}.$$

Then

$$|G(u) - G(v)| \leq c_1'\int_\Omega |w(x)|(1 + |u(x)|^{2p-2} + |v(x)|^{2p-2})\,dx$$
$$\leq \text{(with the Schwarz inequality)}$$
$$\leq c_2'\|w\|_{L^2(\Omega)}(1 + \|u\|^{2p-2}_{L^{4p-4}(\Omega)} + \|v\|^{2p-2}_{L^{4p-4}(\Omega)}.$$

We then obtain (6.17) with

$$k(t) = c_3'(1 + \|u(t)\|_{L^{4p-4}(\Omega)}^{2p-2} + \|v(t)\|_{L^{4p-4}(\Omega)}^{2p-2}).$$

Since u, v are in $L^{4p-4}(0, T; L^{4p-4}(\Omega))$, the function k is in $L^2(0, T)$, which is precisely the desired result.

Examples in Section 1.2

In order to prove (6.17), we supplement (1.44), (1.46) with the following assumption:

For every $x \in \Omega$, g is differentiable with respect to u in $\mathscr{D}$ and the differential $g_u'(u, x)$ is uniformly bounded in $\mathscr{D} \times \bar{\Omega}$. (6.22)

This assumption is satisfied in Examples 1.1–1.5.

With (6.22) we write

$$\begin{aligned} |G(u) - G(v))(x, t)| &= |g(v(x, t), x) - g(u(x, t), x)| \\ &\leq \eta |v(x, t) - u(x, t)|, \end{aligned}$$

$$|G(u) - G(v)| \leq \eta |w| \leq c_4' \|w\|, \tag{6.23}$$

where $\eta = \operatorname{Sup}_{\substack{u \in \mathscr{D} \\ x \in \bar{\Omega}}} |g_u'(u, x)|$; (6.17) follows.

Examples in Sections 2, 3, and 4.1

We now consider equation (3.10), with the assumptions in Section 3.1, especially (3.8). In this way, we prove the backward uniqueness in the Navier–Stokes equations, and in the examples of Sections 3.1–3.4. The result will also be valid for the examples in Section 3.5 (thermohydraulics) and Section 4.1 (the Kuramoto–Sivashinsky equation) for which (3.5) is not satisfied, because we do not use (3.5) here (nor (3.3), (3.4), (3.6), (3.7), (3.9)).

To verify the validity of (6.17) we write

$$\begin{aligned} G(u) - G(v) &= -B(u, u) + B(v, v) - Rw, \\ &= -B(u, w) - B(w, v) - Rw, \end{aligned}$$

$$\begin{aligned} |G(u) - G(v)| &\leq |B(u, w)| + |B(w, v)| + |Rw| \\ &\leq \text{(with (3.8))} \\ &\leq c_4(\|u\|^{1-\theta_4}|Au|^{\theta_4} + \|v\|^{1-\theta_4}|Au|^{\theta_4})\,\|w\| + |Rw|. \end{aligned}$$

When u, v are solutions in $L^2(0, T; D(A))$ and $\theta_4 < 1$, the function

$$k_1(t)\colon t \to c_4(\|u(t)\|^{1-\theta_4}|Au(t)|^{\theta_4} + \|v(t)\|^{1-\theta_4}|Av(t)|^{\theta_4})$$

is in $L^2(0, T)$. For the term Rw, we could prove the desired result if (3.3) holds with $\theta_1 = 0$. This is obviously true for the Navier–Stokes equations (Section 2), and the examples in Sections 3.3 and 3.4 where $R = 0$. For thermohydraulic equations R is continuous in H and (3.3) is satisfied with $\theta_1 = 0$, and this is also true for the Kuramoto–Sivashinsky equation for which R is continuous from V into H. There remains to consider the equations in Section 3.2 (Navier–Stokes equations with a nonhomogeneous boundary condition Φ). There we have

$$Rw = B(w, \Phi) + B(\Phi, w).$$

It is easy to see with the definition (3.24) of B that

$$|B(w, \Phi)| \leq \|w\|_{L^2(\Omega)} \|\text{grad } \Phi\|_{L^\infty(\Omega)},$$

$$|B(\Phi, w)| \leq \|\Phi\|_{L^\infty(\Omega)} \|\text{grad } w\|_{L^2(\Omega)},$$

$$|Rw| \leq c_4'(\|\Phi\|_{L^\infty(\Omega)} + \|\text{grad } \Phi\|_{L^\infty(\Omega)}) \|w\|,$$

and (3.3) holds with $\theta_1 = 1$ provided we supplement (3.18) with

$$\Phi \in W^{1,\infty}(\Omega)^n, \tag{6.24}$$

which, for $n = 2$, is slightly more than (3.18) ($H^2(\Omega) \subset W^{1,p}(\Omega)$, $\forall p < \infty$).

In conclusion, (6.17) is valid for equation (3.10) provided

$$\text{(3.8) holds } (0 \leq \theta_4 < 1) \text{ and (3.3) holds with } \theta_1 = 0. \tag{6.25}$$

And we have proved that (6.25) is satisfied in all of the examples in Sections 2, 3, and 4.1.

Remark 6.3.

(i) The backward uniqueness result (again based on (6.17)) is also valid for the example in Section 4.2 (the Cahn–Hilliard equation). However, the proof necessitates some delicate interpolation inequalities and it will not be given here.

(ii) We do not know if there is backward uniqueness for the equations of Section 5.

CHAPTER IV

Attractors of Dissipative Wave Equations

Introduction

Our aim in this chapter is to study some nonlinear wave equations and a nonlinear Schrödinger equation, the Ginzburg–Landau equation. The wave equations that we consider are the sine–Gordon equation, a nonlinear wave equation of relativisitic quantum mechanics, and some nonlinear vibration equations in solid mechanics that involve fourth-order differential operators in space variables. Strictly speaking, the nonlinear Schrödinger equation is an evolution equation of the first order in time, and it is studied by the methods of Chapter III; however, this equation is related to wave phenomena and from the physical point of view has some properties in common with wave equations.

The questions that we address are the same as those developed in Chapter III, of which this chapter is a continuation:

definition and properties of the semigroup, i.e., existence and uniqueness of solution;
nonlinear stability of the equation;
existence of absorbing sets (dissipativity);
existence of a global attractor.

In Section 1 we present some complementary results concerning abstract linear equations of the second order in time which include, of course, the linear wave equation. Those results are related, in particular, to the exponential decay of the solutions in the absence of external excitation, and also include the study of bounded solutions on the real line. Section 2 is devoted to the sine–Gordon equation subject to various boundary conditions and a time-independent force. In Sections 2.1 and 2.2 we consider the Dirichlet boundary condition which, due to the existence of a Lyapunov function, actually leads to phenomena of limited physical interest when the forces are independent of time.[1] In Section 2.3 we consider the sine–Gordon equation with a Neumann or space-periodicity boundary condition. This leads to a physically more interesting dynamical situation; we prove the existence of a maximal attractor, but in this case a slight extension of the general theorem (Theorem 1.1) of Chapter I is necessary. Section 3 is devoted to the study of a nonlinear wave equation of relativistic quantum mechanics. In Section 4 we study a general abstract equation of the second order in time which is representative of the examples of Sections 2 and 3 and is suitable for many more situations. Some further new examples are given, including vibration equations drawn from solid mechanics that involve differential operators of the fourth order in the space variables. Finally, in Section 5, we treat the Ginzburg–Landau equation.

Following those results that can be viewed as a continuation of Chapter III, this chapter ends with two sections containing results of a different nature. Section 6 deals with the regularity of the attractors. The regularity question is understood here in the sense of the theory of partial differential equations: if the data are sufficiently regular then the attractors (or functional invariant sets) are contained in spaces of more (spatially) regular functions. Two examples of such results are derived in Section 6 without any attempt at generality. Finally, Section 7 contains a result on stability of attractors which illustrates Theorem I.1.2: again, without any attempt at generality, it is shown for a specific example, how the maximal attractor associated with the Galerkin approximation of a given equation approximates the maximal attractor of the exact equation.

1. Linear Equations: Summary and Additional Results

We consider an abstract damped linear equation of the second order in time

$$\frac{d^2u}{dt^2} + \alpha\frac{du}{dt} + Au = f, \qquad \alpha > 0. \tag{1.1}$$

In Section 1.1 we describe the appropriate mathematical framework and recall the classical results of existence and uniqueness of solutions. In Section 1.2 we show that the solutions of (1.1) decay exponentially as $t \to \infty$ when $f = 0$. Finally, in Section 1.3, we study the solutions of (1.1) which are bounded on the real line.

[1] This will be clarified in Chapter VII.

1.1. The General Framework

We are given two Hilbert spaces V and H, $V \subset H$, V dense in H, the injection of V in H being continuous. The scalar product and the norm in V and H are, respectively, denoted $((\cdot, \cdot))$, $\|\cdot\|$, $(\cdot, \cdot)$, $|\cdot|$. We identify H with its dual H', and H' with a dense subspace of the dual V' of V (norm $\|\cdot\|_*$); thus

$$V \subset H \subset V', \tag{1.2}$$

where the injections are continuous and each space is dense in the following one.

Let $a(u, v)$ be a bilinear continuous form on V which is symmetric and coercive

$$\exists \alpha_0 > 0, \qquad a(u, u) \geq \alpha_0 \|u\|^2, \qquad \forall u \in V. \tag{1.3}$$

As usual, we associate with this form the linear operator A from V into V' defined by

$$(Au, v) = a(u, v), \qquad \forall u, v \in V;$$

A is an isomorphism from V onto V' and it can also be considered as a self-adjoint unbounded operator in H with domain $D(A) \subset V$,

$$D(A) = \{v \in V, Av \in H\}.$$

As indicated in Section II.2 we can define the powers A^s of A for $s \in \mathbb{R}$, which operate on the spaces $D(A^s)$. We have $D(A^0) = H, D(A^{1/2}) = V, D(A^{-1/2}) = V'$, and we write

$$V_{2s} = D(A^s), \qquad s \in \mathbb{R}. \tag{1.4}$$

This is a Hilbert space for the scalar product and the norm

$$(u, v)_{2s} = (A^s u, A^s v), \qquad |u|_{2s} = \{(u, u)_{2s}\}^{1/2}, \qquad \forall u, v \in D(A^s), \tag{1.5}$$

and A^r is an isomorphism for $D(A^s)$ onto $D(A^{s-r})$, $\forall s, r \in \mathbb{R}$. In particular, A is an isomorphism from $D(A^s)$ onto $D(A^{s-1})$ and *in many cases, once A is defined, the spaces V, H do not play any particular role and they can be replaced by any pair V_s, V_{s-1} in the scale of spaces V_r*.

Given $T > 0$ and f, u_0, u_1 satisfying

$$f \in L^2(0, T; H), \qquad u_0 \in V, \qquad u_1 \in H, \tag{1.6}$$

we consider the initial-value problem

$$\frac{d^2u}{dt^2} + \alpha \frac{du}{dt} + Au = f \quad \text{on } (0, T), \tag{1.7}$$

$$u(0) = u_0, \qquad \frac{du}{dt}(0) = u_1, \tag{1.8}$$

where, for the moment, $\alpha \in \mathbb{R}$.

We have the following theorem which recapitulates the results given in Section II.4 (Theorems 4.1 and 4.2, Lemma 4.1).

Proposition 1.1. *Under the above assumptions, let there be given* u_0, u_1, f *satisfying*

$$u_0 \in V, \qquad u_1 \in H, \qquad f \in L^2(0, T; H). \tag{1.9}$$

Then there exists a unique function u satisfying

$$u \in \mathscr{C}([0, T]; V), \qquad u' \in \mathscr{C}([0, T]; H), \tag{1.10}$$

and (1.7), (1.8) ($u' = du/dt$). *Furthermore, the following equation of energy holds*

$$\frac{1}{2}\frac{d}{dt}(\|u\|^2 + |u'|^2) + \alpha|u'|^2 = (f, u'). \tag{1.11}$$

If we assume that

$$f \in \mathscr{C}([0, T]; H), \qquad f' \in L^2(0, T; H), \qquad u_0 \in D(A), \quad u_1 \in V, \tag{1.12}$$

then

$$\{u, u', u''\} \in \mathscr{C}([0, T]; D(A) \times V \times H). \tag{1.13}$$

Remark 1.1. If we perform the change of variable $t \to -t$ in (1.7) we obtain the same equation with α changed into $-\alpha$. Since the sign of α is arbitrary in Theorem 1.1, we see that if f is defined on $[-T, 0]$, then we can solve the initial-value problem backward in time on $[-T, 0]$. Proposition 1.1 is easily adapted to this situation.

Remark 1.2. As indicated in Section II.4, Proposition 1.1 and its extension in Remark 1.1 can be generalized to other spaces V_s, $D(A)$, V, H being replaced by V_{s+2}, V_{s+1}, V_s, $\forall s \in \mathbb{R}$. This also applies to most of the following results in the present section.

The group $\Sigma(t)$

When f is independent of t, $f(t) = f \in H$, $\forall t$, Proposition 1.1 and Remark 1.1 show that the mapping

$$\Sigma_f(t)\colon \{u_0, u_1\} \to \{u(t), u'(t)\}, \qquad t \in \mathbb{R},$$

is well defined from $E_0 = V \times H$ into itself (or from $E_1 = V_2 \times V_1$ into itself ($V_2 = D(A)$, $V_1 = V$). These mappings form a group of operators in E_0 or E_1, i.e., I.(1.1) is satisfied for every $s, t \in \mathbb{R}$. When $f = 0$ we write $\Sigma_f(t) = \Sigma(t)$.

Remark 1.3. When $\alpha = 0$, equation (1.7) is conservative. It follows from (1.11), with say $s = 0$, that

$$|u(t)|_1^2 + |u'(t)|^2 = |u_0|_1^2 + |u_1|^2 + 2\int_0^t (f(s), u'(s))\, ds,$$

and if $f = 0$, $|u(t)|_1^2 + |u'(t)|^2$ is conserved. In the case of interest to us, $\alpha > 0$ and the equation is dissipative; for instance, if $f = 0$ then (1.11) with $s = 0$ shows that the "energy"

$$\|u(t)\|^2 + |u'(t)|^2 \tag{1.14}$$

decays as $t \to +\infty$. We note that (1.11) does not specify the rate of decay of (1.14), nor its limit. This point will be clarified in Section 1.2.

1.2. Exponential Decay

From now on *we assume that* $\alpha > 0$ unless otherwise specified.

Let f be given, continuous and bounded from $\mathbb{R}_+$ into H,

$$f \in \mathscr{C}_b(\mathbb{R}_+; H). \tag{1.15}$$

We want to show that the solution of (1.7), (1.8), which is then defined for all $t > 0$, remains bounded as $t \to \infty$ and decays exponentially if $f \equiv 0$; as is stated in Remark 1.3, this does not follow from (1.11).

For that purpose let $\varepsilon > 0$ be fixed; ε will be specified below. We set $v = u' + \varepsilon u$ and rewrite (1.7) as follows ($\varphi' = d\varphi/dt$):

$$v' + (\alpha - \varepsilon)v + (A - \varepsilon(\alpha - \varepsilon))u = f. \tag{1.16}$$

Assume that u is sufficiently regular, for instance, that it is twice continuously differentiable from $\mathbb{R}_+$ in $D(A)$. Then we take the scalar product in H of (1.16) with v and we obtain

$$\frac{1}{2}\frac{d}{dt}(\|u\|^2 + |v|^2) + \varepsilon\|u\|^2 + (\alpha - \varepsilon)|v|^2 - \varepsilon(\alpha - \varepsilon)(u, v) = (f, v). \tag{1.17}$$

Under the present assumptions, u satisfying (1.10) for every $T > 0$, (1.17) makes sense and is valid. This can easily be deduced from (1.11) which in turns follows from the approximation lemma in Chapter II: Lemma II.4.1.

Now let λ_1 denote the first eigenvalue of A,

$$\lambda_1 = \operatorname*{Inf}_{\substack{v \in V \\ v \neq 0}} \frac{\|v\|^2}{|v|^2}. \tag{1.18}$$

From (1.17) we infer the following:

Proposition 1.2. *The hypotheses are those of Proposition* 1.1 *and we assume that* $f \in \mathscr{C}_b(\mathbb{R}_+; H)$, $u_0 \in V$, $u_1 \in H$. *Then if* u *is the solution of* (1.7), (1.8), $\{u, u'\} \in \mathscr{C}_b(\mathbb{R}_+; V \times H)$ *and if*

$$0 < \varepsilon \leq \varepsilon_0, \qquad \varepsilon_0 = \min\left(\frac{\alpha}{4}, \frac{\lambda_1}{2\alpha}\right), \tag{1.19}$$

and $\alpha_1 = \varepsilon/2$, *u satisfies the following relations*

$$\|u(t)\|^2 + |u'(t) + \varepsilon u(t)|^2 \le \{\|u_0\|^2 + |u_1 + \varepsilon u_0|^2\} \exp(-\alpha_1 t) + \frac{1}{\alpha_1 \varepsilon} |f|^2_{L^\infty(\mathbb{R}_+; H)} (1 - \exp(-\alpha_1 t)), \tag{1.20}$$

$$|u(t)|^2 + |u'(t) + \varepsilon u(t)|^2_{-1} \le \{|u_0|^2 + |u_1 + \varepsilon u_0|^2_{-1}\} \exp(-\alpha_1 t) + \frac{1}{\alpha_1^2} |f|^2_{L^\infty(\mathbb{R}_+; V')} (1 - \exp(-\alpha_1 t)). \tag{1.21}$$

Proof. We start from (1.17) and write

$$\begin{aligned}
\varepsilon \|u\|^2 + (\alpha - \varepsilon)|v|^2 - \varepsilon(\alpha - \varepsilon)(u, v) &\ge \varepsilon \|u\|^2 + (\alpha - \varepsilon)|v|^2 - \frac{\varepsilon(\alpha - \varepsilon)}{\sqrt{\lambda_1}} \|u\| \, |v| \\
&\ge \text{(by the assumptions on } \varepsilon) \\
&\ge \varepsilon \|u\|^2 + \frac{3\alpha}{4}|v|^2 - \frac{\alpha\varepsilon}{4\sqrt{\lambda_1}} \|u\| \, |v| \\
&\ge \frac{\varepsilon}{2} \|u\|^2 + \frac{\alpha}{2} |v|^2 \\
&\ge \alpha_1(\|u\|^2 + |v|^2). \qquad (1.22)
\end{aligned}$$

We also write

$$(f, v) \le |f||v| \le \frac{\alpha_1}{2}|v|^2 + \frac{1}{2\alpha_1}|f|^2,$$

and we obtain

$$\frac{d}{dt}(\|u\|^2 + |v|^2) + \alpha_1(\|u\|^2 + |v|^2) \le \frac{1}{\alpha_1}|f|^2. \tag{1.23}$$

Using the Gronwall lemma we easily deduce (1.20) from (1.23).

To prove (1.21) we proceed exactly in the same manner, starting from the analog of (1.17) which is obtained by replacing $s = 0$ by $s = -1$ in (1.11) and $V = V_1$, $H = V_0$ by H and $V' = V_{-1}$ (see Remark 1.1)

$$\frac{1}{2}\frac{d}{dt}(|u|^2 + |v|^2_{-1}) + \varepsilon |u|^2 + (\alpha - \varepsilon)|v|^2_{-1} - \varepsilon(\alpha - \varepsilon)(u, v)_{-1} = (f, v)_{-1}. \tag{1.24}$$

Of course, we can derive similar relations with norms $|\cdot|_{s+1}, |\cdot|_s$, $\forall s \in \mathbb{R}$.

The proposition is proved. □

Sometimes, in the sequel, it will be convenient to reduce (1.7) to an evolution equation of the first order in time in the following manner. We set $\varphi = \{u, v\}$, $v = u' + \varepsilon u$, $0 < \varepsilon \le \varepsilon_0$, and we write (1.7) as

$$\varphi' + \Lambda_\varepsilon \varphi = \tilde{f}, \tag{1.25}$$

where $\tilde{f} = \{0, f\}$ and

$$\Lambda_\varepsilon = \begin{pmatrix} \varepsilon I & -I \\ A - \varepsilon(\alpha - \varepsilon)I & (\alpha - \varepsilon)I \end{pmatrix}. \tag{1.26}$$

We can then consider (1.7), (1.8) as an evolution problem in the product spaces $E_i = V_{i+1} \times V_i$, $i \in \mathbb{R}$, which are Hilbert spaces for the natural scalar products and norms. Proposition 1.1 and Remark 1.1 are easily rewritten as existence and uniqueness results for (1.25). In particular, when $f \equiv 0$, we can define the linear operators $\Sigma_\varepsilon(t)$, $t \in \mathbb{R}$,

$$\begin{gathered} \Sigma_\varepsilon(t)\colon E_i \to E_i, \\ \varphi_0 = \{u_0, v_0 = u_1 + \varepsilon u_0\} \to \varphi(t) = \{u(t), v(t) = u'(t) + \varepsilon u(t)\}, \end{gathered} \tag{1.27}$$

where φ is the solution of (1.25) which satisfies $\varphi(0) = \varphi_0$. Alternatively,

$$\Sigma_\varepsilon(t) = R_\varepsilon \Sigma(t) R_{-\varepsilon}, \tag{1.28}$$

where $\Sigma(t)$ is the linear operator in E_i introduced in Section 1.1

$$\{u_0, u_1\} \to \{u(t), u'(t)\}$$

(u is the solution of (1.7), (1.8) when $f = 0$), and R_γ, $\gamma \in \mathbb{R}$, is the isomorphism of E_i

$$R_\gamma\colon \{a, b\} \to \{a, b + \gamma a\}. \tag{1.29}$$

These operators Σ_ε form a group on E_i, $i \in \mathbb{R}$. Furthermore, we infer from (1.20), (1.21) that, for $t \geq 0$,

$$\|\Sigma_\varepsilon(t)\varphi_0\|_{E_i}^2 \leq \|\varphi_0\|_{E_i}^2 \exp(-\alpha_1 t), \qquad \forall t \geq 0, \quad i = 0, -1, \tag{1.30}$$

with $\|\varphi\|_{E_0} = (\|u\|^2 + |v|^2)^{1/2}$, $\|\varphi\|_{E_{-1}} = (|u|^2 + |v|_{-1}^2)^{1/2}$. This implies the following bounds on the norms of $\Sigma_\varepsilon(t)$ as a linear operator in E_0, E_{-i}:

$$\|\Sigma_\varepsilon(t)\|_{\mathscr{L}(E_i)} \leq \exp\left(-\frac{\alpha_1 t}{2}\right), \qquad \forall t \geq 0, \quad i = 0, -1. \tag{1.31}$$

Similarly, for $\Sigma(t) = R_{-\varepsilon}\Sigma_\varepsilon(t)R_\varepsilon$,

$$\begin{gathered} \|\Sigma(t)\|_{\mathscr{L}(E_i)} \leq c_0 \exp\left(-\frac{\alpha_1 t}{2}\right), \qquad \forall t \geq 0, \quad i = 0, -1, \\ c_0 = (1 + \varepsilon\lambda_1^{-1/2})^2, \end{gathered} \tag{1.32}$$

where λ_1 is the first eigenvalue of A (see (1.18)). We derive (1.31) from (1.30) by simply observing that

$$\|R_\gamma\|_{\mathscr{L}(E_i)} \leq 1 + |\gamma|\lambda_1^{-1/2}, \qquad \forall \gamma \in \mathbb{R}. \tag{1.33}$$

Remark 1.4. This proof of the exponential decay of the solutions of the linear homogeneous equation is due to P.H. Rabinowitz [1]; see also J.M. Ghidaglia and R. Temam [1].

1.3. Bounded Solutions on the Real Line

Another property of equation (1.7), in the dissipative case ($\alpha > 0$), is that for f given continuous and bounded from $\mathbb{R}$ to H,

$$f \in \mathscr{C}_b(\mathbb{R}; H), \tag{1.34}$$

we can solve (1.7) with the boundary condition

$$\limsup_{t \to \infty} \{\|u(t)\| + |u(t)|\} < +\infty. \tag{1.35}$$

More precisely,

Proposition 1.3. *The hypotheses are those of Proposition* 1.1. *Let f be given in $\mathscr{C}_b(\mathbb{R}; H)$. Then equation* (1.7) *possesses a unique solution u which satisfies*

$$u \in \mathscr{C}_b(\mathbb{R}_+; V), \qquad u' \in \mathscr{C}_b(\mathbb{R}; H), \tag{1.36}$$

and this solution is given by

$$\{u(t), u'(t) + \varepsilon u(t)\} = \int_{-\infty}^{t} \Sigma_\varepsilon(t - \tau)\{0, f(\tau)\}\, d\tau, \tag{1.37}$$

where $0 < \varepsilon \le \varepsilon_0$.

If, furthermore, $f' \in \mathscr{C}_b(\mathbb{R}; H)$, *then*

$$u \in \mathscr{C}_b(\mathbb{R}; D(A)), \qquad u' \in \mathscr{C}_b(\mathbb{R}; V), \qquad u'' \in \mathscr{C}_b(\mathbb{R}; H). \tag{1.38}$$

Proof. We first prove uniqueness. We must show that if u satisfies (1.7), with $f = 0$, and (1.36), then $u = 0$. Actually, we can prove a slightly stronger result which will be used below, namely, if u satisfies (1.7) with $f = 0$ and

$$u \in \mathscr{C}_b(\mathbb{R}; H), \qquad u' \in \mathscr{C}_b(\mathbb{R}; V'), \tag{1.39}$$

then $u = 0$.

Indeed, for $t \ge s$, we infer from (1.21), (1.39) that

$$\begin{aligned} |u(t)|^2 + |u'(t) + \varepsilon u(t)|_{-1}^2 &\le \{|u(s)|^2 + |u'(s) + \varepsilon u(s)|_{-1}^2\} \exp(-\alpha_1(t - s)) \\ &\le c_1' \exp(-\alpha_1(t - s)). \end{aligned}$$

Letting $s \to -\infty$, we obtain $u(t) = 0$.

We then prove the existence. We set $\psi(t, \tau) = \Sigma_\varepsilon(t - \tau)\{0, f(\tau)\}$ and we infer from (1.29) that for every $t, \tau \in \mathbb{R}$,

$$\begin{aligned} \|\psi(t, \tau)\|_{E_0} &\le |f(\tau)| \exp\left(-\frac{\alpha_1}{2}(t - \tau)\right), \\ &\le |f|_{L^\infty(\mathbb{R}, H)} \exp\left(-\frac{\alpha_1}{2}(t - \tau)\right), \qquad \tau \le t, \end{aligned}$$

and therefore $\psi(t, \cdot) \in L^1(-\infty, t; E_0)$. Hence we can define $\varphi \in \mathscr{C}(\mathbb{R}; E_0)$ by

$$\varphi(t) = \int_{-\infty}^{t} \psi(t, \tau)\, d\tau, \tag{1.40}$$

and

$$\begin{aligned} \|\varphi(t)\|_{E_0} &\leq \left(\int_{-\infty}^{t} \exp\left(-\frac{\alpha_1}{2}(t-\tau)\right) d\tau\right) |f|_{L^\infty(\mathbb{R}; H)}, \\ \|\varphi\|_{L^\infty(\mathbb{R}; E_0)} &\leq \frac{2}{\alpha_1} |f|_{L^\infty(\mathbb{R}; H)}. \end{aligned} \tag{1.41}$$

On the other hand, the function $\psi(\cdot, \tau)$ is the unique solution in $\mathscr{C}(\mathbb{R}; E_0)$ of the Cauchy problem

$$\frac{d}{dt}\psi(t, \tau) + \Lambda_\varepsilon \psi(t, \tau) = 0 \tag{1.42}$$

$$\psi(\tau, \tau) = \{0, f(\tau)\}. \tag{1.43}$$

We observe that Λ_ε is linear continuous from $E_0 = V \times H$ into $E_{-1} = H \times V'$. Then by (1.41), $(d/dt)\psi(t, \tau)$ is in $\mathscr{C}(\mathbb{R}; E_{-1})$, and if β denotes the norm of Λ_ε in $\mathscr{L}(E_0, E_{-1})$,

$$\begin{aligned} \left\| \frac{d}{dt}\psi(t, \tau) \right\|_{E_{-1}} &\leq \beta \|\psi(t, \tau)\| \\ &\leq \beta \exp\left(-\frac{\alpha_1}{2}(t-\tau)\right) |f|_{L^\infty(\mathbb{R}; H)}. \end{aligned}$$

This last estimate shows that φ defined by (1.40) belongs to $\mathscr{C}^1(\mathbb{R}; E_{-1})$ and, in E_{-1},

$$\begin{aligned} \varphi'(t) &= \psi(t, t) + \int_{-\infty}^{t} \frac{d\psi}{dt}(t, \tau)\, d\tau \\ &= \{0, f(t)\} - \Lambda_\varepsilon \int_{-\infty}^{t} \psi(t, \tau)\, d\tau \\ &= \{0, f(t)\} - \Lambda_\varepsilon \varphi(t). \end{aligned}$$

If we set $\varphi(t) = \{u(t), v(t)\}$ we deduce from the last relation that $v = u' + \varepsilon u$, and (1.36), (1.37) are proved.

We must then prove that $\{u, u', u''\} \in \mathscr{C}_b(\mathbb{R}; D(A) \times V \times H)$ if, furthermore, $f' \in \mathscr{C}_b(\mathbb{R}; H)$. We observe that, by (1.7) and (1.36),

$$u'' = -\alpha u' - Au + f \in \mathscr{C}_b(\mathbb{R}; V'),$$

and, therefore, if we set $w = u'$, then $\{w, w'\} \in \mathscr{C}_b(\mathbb{R}; E_{-1})$ and by time differentiation of (1.7),

$$w'' + \alpha w' + Aw = f'. \tag{1.44}$$

We have just shown above that (1.43) possesses a solution $\{w, w'\} \in \mathscr{C}_b(\mathbb{R}; E_0)$ which is unique in $\mathscr{C}_b(\mathbb{R}; E_{-1})$. Thus $\{u', u''\} \in \mathscr{C}_b(\mathbb{R}; E_0)$, i.e., $u' \in \mathscr{C}_b(\mathbb{R}; V)$, $u'' \in \mathscr{C}_b(\mathbb{R}; H)$. Then (1.7) implies that $Au \in \mathscr{C}_b(\mathbb{R}; H)$, so that $u \in \mathscr{C}_b(\mathbb{R}; D(A))$. The proof is complete. □

Remark 1.5. As in Remark 1.2, we observe that the spaces V, H do not play a particular role in Proposition 1.3, and the proposition can easily be generalized to the case where $D(A)$, V, H are replaced by V_{s+2}, V_{s+1}, V_s, $s \in \mathbb{R}$.

In that respect, we conclude Section 1 with a complement to Propositon 1.3.

Proposition 1.4. *The hypotheses are those of Proposition* 1.1.

If $\{u, u'\} \in \mathscr{C}_b(\mathbb{R}; V_{s+1} \times V_s)$, $s \in \mathbb{R}$, *is a solution of* (1.7) *with* $f \in \mathscr{C}_b(\mathbb{R}; V_\sigma)$, $\sigma > s$, *then* $\{u, u'\} \in \mathscr{C}_b(\mathbb{R}; V_{\sigma+1} \times V_\sigma)$.

PROOF. We apply $A^{\sigma/2}$ to equation (1.7) and set $w = A^{\sigma/2}u$, $g = A^{\sigma/2}f$. Hence

$$w'' + \alpha w' + Aw = g, \qquad t \in \mathbb{R}, \tag{1.45}$$

where $g \in \mathscr{C}_b(\mathbb{R}; H)$ and $w \in \mathscr{C}_b(\mathbb{R}; V_{s-\sigma+1})$, $w' \in \mathscr{C}_b(\mathbb{R}; V_{s-\sigma})$. Due to the uniqueness result established in Proposition 1.3,[1] we have $\{w, w'\} \in \mathscr{C}_b(\mathbb{R}; V \times H)$ and $\{u, u'\} = A^{-\sigma/2}\{w, w'\} \in \mathscr{C}_b(\mathbb{R}; V_{\sigma+1} \times V_\sigma)$.

2. The Sine–Gordon Equation

In physics the sine–Gordon equation is used to model, for instance, the dynamics of a Josephson junction driven by a current source. For a single junction the governing equation is an ordinary differential equation similar to the pendulum equation described in Section 2.1, with θ here being replaced by u, representing the time derivative of the junction voltage (up to a multiplicative factor). A coupled system of such equations appears when we consider a family of coupled junctions, and the continuous case is modeled by the sine–Gordon equation (see (2.1)).

In Section 2.1 we describe the equation and its mathematical setting in the case of a Dirichlet boundary condition. In Section 2.2 we show the nonlinear stability of the equation (boundedness of trajectories), the existence of absorbing sets, and the existence of a maximal attractor. In Section 2.3 we consider other boundary conditions, viz. a Neumann boundary condition and the space-periodicity boundary condition. In those cases which are physically the most interesting, the average value of the function u is not expected to remain bounded and that actually leads to nontrivial dynamics (see A.R. Bishop *et al.* [1], D. McLaughlin [1]). From the mathematical point of view the

[1] In Proposition 1.3, $V_{s-\sigma+1}$ and $V_{s-\sigma}$ were replaced by H and V_{-1}, i.e., $\sigma - s = 1$. We easily generalize the result to arbitrary values of $\sigma - s$ (using, for instance, Remark 1.5 and the uniqueness result of Proposition 1.3 repeatedly).

situation is also interesting: Theorem I.1.1 does not apply directly and a slight generalization of it is necessary for the existence of a global attractor.

2.1. The Equation and Its Mathematical Setting

Let Ω be an open bounded set of $\mathbb{R}^n$ with a boundary Γ sufficiently regular. In the sine–Gordon equation the unknown is a scalar function $u = u(x, t)$, $x \in \Omega$, $t \in \mathbb{R}$ (or some interval of $\mathbb{R}$) and in its nondimensionalized form the equations reads

$$\frac{\partial^2 u}{\partial t^2} + \alpha \frac{\partial u}{\partial t} - \Delta u + \beta \sin u = f \quad \text{in } \Omega \times \mathbb{R}_+, \tag{2.1}$$

where f and α are given, $\alpha > 0$, and, in the physical problem, f is proportional to the current intensity applied to the function. The boundary condition considered in Section 2.1 is the Dirichlet boundary condition

$$u = 0 \qquad \text{on} \quad \Gamma \times \mathbb{R}_+, \tag{2.2}$$

and when an initial-value problem is considered we provide u_0, u_1, such that

$$u(x, 0) = u_0(x), \qquad \frac{\partial u}{\partial t}(x, 0) = u_1(x), \qquad x \in \Omega. \tag{2.3}$$

The functional setting corresponding to the linear part has already been described several times. We set $H = L^2(\Omega)$, $V = H_0^1(\Omega)$, and endow these spaces with the usual scalar products and norms

$$(u, v) = \int_\Omega u(x)v(x)\,dx, \qquad |u| = \{u, u\}^{1/2}, \qquad \forall u, v \in L^2(\Omega),$$

$$((u, v)) = \sum_{i=1}^n (D_i u, D_i v), \qquad \|u\| = \{((u, u))\}^{1/2}, \qquad \forall u, v \in H_0^1(\Omega).$$

We write $D(A) = H_0^1(\Omega) \cap H^2(\Omega)$ and for $u \in D(A)$, $Au = -\Delta u$.

Then (2.1)–(2.3) is equivalent to the following second-order differential equation in H ($\varphi' = d\varphi/dt$):

$$u'' + \alpha u' + Au + g(u) = f, \tag{2.4}$$

$$u(0) = u_0, \qquad u'(0) = u_1, \qquad g(u) = \beta \sin u. \tag{2.5}$$

The existence and uniqueness of solution of (2.4), (2.5) is given by the following result which is a particular case of Theorem 4.1 (whose proof is sketched in Section 4.4). Although we are interested in the case $\alpha > 0$, we state Theorem 2.1 in the general case, $\alpha \in \mathbb{R}$.

Theorem 2.1. *Let $\alpha \in \mathbb{R}$ and let f, u_0, and u_1 be given satisfying*

$$f \in \mathscr{C}([0, T]; H), \qquad u_0 \in V, \qquad u_1 \in H. \tag{2.6}$$

Then there exists a unique solution u of (2.4), (2.5) *such that*

$$u \in \mathscr{C}([0, T]; V), \qquad u' \in \mathscr{C}([0, T]; H). \tag{2.7}$$

If, furthermore,

$$f' \in \mathscr{C}([0, T]; H), \qquad u_0 \in D(A), \qquad u_1 \in V, \tag{2.8}$$

then u satisfies

$$u \in \mathscr{C}([0, T]; D(A)), \qquad u' \in \mathscr{C}([0, T]; V). \tag{2.9}$$

As in the linear case, the only effect of the change of variable $t/-t$ on equation (2.4) is to change α into $-\alpha$. Thus, if f is defined on $[-T, 0]$ or on $[-T, T]$ ($f \in \mathscr{C}([-T, 0); H])$ or $f \in \mathscr{C}([-T, T]; H)$), then Theorem 2.1 allows us to solve (2.4), (2.5) backward in time or both backward and forward.

In the autonomous case, i.e., when f is given independent of time ($f(t) = f \in H, \forall t$), we define the mappings

$$S(t)\colon \{u_0, u_1\} \to \{u(t), u'(t)\} \quad \textit{for all } t \in \mathbb{R}. \tag{2.10}$$

They map $E_0 = V \times H$ into itself and $E_1 = D(A) \times V$ into itself, and they enjoy the usual *group properties* similar to I.(1.1).

$$\begin{cases} S(t + s) = S(t)S(s), & \forall s, t \in \mathbb{R}, \\ \quad S(0) = I. \end{cases} \tag{2.11}$$

It follows from Remark 2.2 below that the operators $S(t)$ are continuous in E_0 and E_1, $\forall t \in \mathbb{R}$, and $S(t)$ is then an homeomorphism of E_0 or E_1, $\forall t \in \mathbb{R}$ (see Proposition 2.1 below).

Remark 2.1. It is clear that $g(\varphi) \in L^2(\Omega)$ for any $\varphi \in L^2(\Omega)$. Thus, by using the results proved in the linear case, we see that the solution u of (2.4), (2.5) satisfies the energy equation (1.11) with f replaced by $f - g(u)$. It also satisfies the more general energy equations II.(4.23) with $0 \le s \le 1$, and the analog of (1.20), (1.21) (see below): we present explicitly the equations resulting from (1.11), II.(4.23), while those similar to (1.20), (1.21) will be presented and used in Section 2.2:

$$\frac{1}{2}\frac{d}{dt}(\|u\|^2 + |u'|^2) + \alpha|u'|^2 + (g(u), u') = (f, u'), \tag{2.12}$$

$$\frac{1}{2}\frac{d}{dt}(|u|_{s+1}^2 + |u'|_s^2) + \alpha|u'|_s^2 + (g(u), u')_s = (f, u')_s \qquad (0 \le s \le 1). \tag{2.13}$$

Remark 2.2. The continuity of the mappings $S(t)$ in E_0 or E_1 follows easily from Remark 2.1. Let u, v be the solutions of (2.4) associated with the initial data u_0, u_1, and v_0, v_1, and let $w = u - v$. Then

$$w'' + \alpha w' + Aw = -\beta(\sin u - \sin v), \tag{2.14}$$

and by (2.12)

$$
\begin{aligned}
\frac{1}{2}\frac{d}{dt}(\|w\|^2 + |w'|^2) + \alpha|w'|^2 &= -\beta(\sin u - \sin v, w') \\
&\leq \beta|w||w'| \leq \frac{\beta}{\sqrt{\lambda_1}}\|w\|\,|w'| \\
&\leq \frac{1}{2}\|w\|^2 + \frac{1}{2}\frac{\beta^2}{\lambda_1}|w'|^2,
\end{aligned}
\tag{2.15}
$$

$$\|w(t)\|^2 + |w'(t)|^2 \leq (\|w(0)\|^2 + |w'(0)|^2)\exp(\gamma t), \qquad t \geq 0,$$

where λ_1 is the first eigenvalue of A (see (1.18)), and $\gamma = \max(1, \beta^2/\lambda_1 + 2\alpha_-)$. The continuity of $S(t)$ in E_0 follows for $t > 0$. For $t < 0$, we just replace α by $-\alpha$, i.e., we replace α_- by α_+ in the expression of γ. Finally, the continuity of $S(t)$ in E_1 is proved in a similar manner starting from relation (2.13) written with $s = 1$.

The properties of the operators $S(t)$ are slightly different than in the first-order case. We recapitulate them in the

Proposition 2.1. *When $f(t) \equiv f \in H$, we can associate with equation* (2.5) *the group of operators $S(t)$ defined by* (2.10), *α fixed in $\mathbb{R}$, $t \in \mathbb{R}$. For every t, $S(t)$ is an homeophormism from E_i onto E_i, $i = 0, 1$.*

All the statements in this proposition have been proved, except the invertibility which is a consequence of (2.11), and the continuity of the inverse which results from the change of α into $-\alpha$.

Remark 2.3. We can write (2.4), (2.5) in the form of a system similar to (1.25), (1.26). For ε fixed (arbitrary at the moment) we set $\varphi = \{u, v = u' + \varepsilon u\}$, $\varphi_0 = \{u_0, v_0 = u_1 + \varepsilon u_0\}$ and we see that (2.4), (2.5) is equivalent to

$$\varphi' + \Lambda_\varepsilon \varphi + \mathscr{G}_\varepsilon(\varphi) = \mathscr{F}, \tag{2.16}$$

$$\varphi(0) = \varphi_0, \tag{2.17}$$

Λ_ε as in (1.26), $\mathscr{G}_\varepsilon(\varphi) = \{0, \beta \sin u\}$, $\mathscr{F} = \{0, f\}$.

2.2. Absorbing Sets and Attractors

We return to the hypothesis $\alpha > 0$ and as in Section 1 we consider ε fixed, $0 < \varepsilon \leq \varepsilon_0$,

$$\varepsilon_0 = \min\left(\frac{\alpha}{4}, \frac{\lambda_1}{2\alpha}\right). \tag{2.18}$$

As indicated above, a relation similar to (1.17) is valid and is obtained by taking the scalar product of (2.4) with $v = u' + \varepsilon u$ in H

$$\frac{1}{2}\frac{d}{dt}\{\|u\|^2 + |u|^2\} + \varepsilon\|u\|^2 + (\alpha - \varepsilon)|v|^2 - \varepsilon(\alpha - \varepsilon)(u, v) + (g(u), v) = (f, v). \tag{2.19}$$

We will show how we can derive uniform estimates for u from (2.19). First, as in (1.22), we write

$$\varepsilon\|u\|^2 + (\alpha - \varepsilon)|v|^2 - \varepsilon(\alpha - \varepsilon)|v|^2 \geq \frac{\varepsilon}{2}\|u\|^2 + \frac{\alpha}{2}|v|^2 \geq \alpha_1(\|u\|^2 + |v|^2), \tag{2.20}$$

($\alpha_1 = \varepsilon/2$). Then

$$\frac{d}{dt}(\|u\|^2 + |v|^2) + 2\alpha_1(\|u\|^2 + |v|^2) \leq 2(f - g(u), v) \leq 2|f||v| + 2|\beta||\Omega|^{1/2}|v|$$

$$\leq \alpha_1|v|^2 + \frac{2}{\alpha_1}(|f|^2 + \beta^2|\Omega|),$$

$$\frac{d}{dt}(\|u\|^2 + |v|^2) + \alpha_1(\|u\|^2 + |v|^2) \leq \frac{2}{\alpha_1}(|f|^2 + \beta^2|\Omega|), \tag{2.21}$$

where $|\Omega|$ is the (n-dimensional) measure of Ω.

Using Gronwall's lemma we deduce from (2.21)

$$\|u(t)\|^2 + |v(t)|^2 \leq (\|u_0\|^2 + |v_0|^2)\exp(-\alpha_1 t) + \frac{2}{\alpha_1^2}(|f|^2 + \beta^2|\Omega|)(1 - \exp(-\alpha_1 t)), \qquad t \geq 0, \tag{2.22}$$

and

$$\limsup_{t\to\infty}(\|u(t)\|^2 + |v(t)|^2) \leq \frac{2}{\alpha_1^2}(|f|^2 + \beta^2|\Omega|). \tag{2.23}$$

We consider the space E_0 normed by $\|\varphi\|_{E_0} = \{\|u\|^2 + |v|^2\}^{1/2}$, $\forall\varphi = \{u, v\}$ and set

$$\rho_0^2 = \frac{2c_0}{\alpha_1}(|f|^2 + \beta^2|\Omega|), \tag{2.24}$$

c_0 as in (1.32), (1.33). We deduce from (1.33), (2.22), and (2.23) that

The balls $B_{E_0}(0, \rho)$ of E_0 centered at 0 of radius $\rho > \rho_0$ are absorbing in E_0 for the semigroup $S(t)$, $t \geq 0$. (2.25)

We choose $\rho_0' > \rho_0$ and set $\mathscr{B}_0 = B_{E_0}(0, \rho_0')$. If $\mathscr{B}$ is any bounded set of E_0, $\mathscr{B} \subset B_{E_0}(0, R_0)$, then $S(t)\mathscr{B} \subset \mathscr{B}_0$ for $t \geq t_0(\mathscr{B}, \rho_0')$; the time t_0 is easily computed from (2.22) and (1.33)

$$t_0 = \frac{1}{\alpha_1}\log\frac{c_0^2 R_0^2}{(\rho_0')^2 - \rho_0^2}. \tag{2.26}$$

Absorbing Set in E_1

Although it will not be used here, we prove the existence of an absorbing set in E_1.

We proceed as for E_0. We begin with the analog of (2.19) which is obtained by taking the scalar product of (2.4) with $A(u' + \varepsilon u)$ in H. Assuming that u is $\mathscr{C}^2$ in time with values in $D(A)$ we obtain

$$\frac{1}{2}\frac{d}{dt}(|Au|^2 + \|v\|^2) + \varepsilon|Au|^2 + (\alpha - \varepsilon)\|v\|^2 - \varepsilon(\alpha - \varepsilon)(Au, v) + ((g(u), v)) = (f, Av). \tag{2.27}$$

This relation makes sense when u_0, u_1 satisfy (2.8) and u possesses the regularity level provided by (2.9) ($\forall T > 0, f \in H$). As in the case of (2.19), (2.27) can be rigorously proved by using II.(4.23) and Lemma II.4.1.

Exactly as in (1.22), (2.20), for $0 < \varepsilon \leq \varepsilon_0$, we write

$$\varepsilon|Au|^2 + (\alpha - \varepsilon)\|v\|^2 - \varepsilon(\alpha - \varepsilon)((u, v)) \geq \frac{\varepsilon}{2}|Au|^2 + \frac{\alpha}{2}\|v\|^2. \tag{2.28}$$

Also

$$(f, Av) = \frac{d}{dt}(f, Au) + \varepsilon(f, Au) \leq \frac{d}{dt}(f, Au) + \frac{\varepsilon}{4}|Au|^2 + \varepsilon|f|^2, \tag{2.29}$$

$$|((g(u), v))| = \left|\int_\Omega \beta \cos u \operatorname{grad} u \operatorname{grad} v\, dx\right| \leq |\beta|\,\|u\|\,\|v\| \leq \frac{\alpha}{4}\|v\|^2 + \frac{|\beta|}{\alpha}\|u\|^2. \tag{2.30}$$

Hence, again setting $\alpha_1 = \varepsilon/2$, we have

$$\frac{d}{dt}(|Au|^2 + \|v\|^2 - 2(f, Au)) + \alpha_1(|Au|^2 + \|v\|^2) \leq 2\varepsilon|f|^2 + \frac{2}{\alpha}|\beta|\,\|u\|^2, \tag{2.31}$$

$$\frac{dy}{dt} + \frac{\alpha_1}{2}y \leq (2\varepsilon + \alpha_1)|f|^2 + \frac{2}{\alpha}|\beta|\,\|u\|^2, \tag{2.32}$$

$$y = |Au - f|^2 + \|v\|^2. \tag{2.33}$$

If $\{u_0, u_1\}$ belongs to a bounded set $\mathscr{B}$ of E_1, then since $\mathscr{B}$ is also bounded in E_0 there exists a time t_0 given by (2.26) such that for $t \geq t_0$, $S(t)\mathscr{B} \subset \mathscr{B}_0$, which implies that $\|u(t)\| \leq \rho_0'$. Thanks to the Gronwall lemma we infer from

(2.32) that

$$y(t) \leq y(t_0) \exp\left(-\frac{\alpha_1}{2}(t-t_0)\right)$$
$$+\frac{1}{\alpha_1}\left\{(2\varepsilon+\alpha_1)|f|^2+\frac{2}{\alpha}|\beta|\rho_0'^2\right\}\left\{1-\exp\left(-\frac{\alpha_1}{2}(t-t_0)\right)\right\} \quad \text{for } t \geq t_0. \tag{2.34}$$

Defining ρ_1 by

$$\rho_1^2 = \frac{1}{\alpha_1}\left\{(2\varepsilon+\alpha_1)|f|^2+\frac{2}{\alpha}|\beta|\rho_0'^2\right\}, \tag{2.35}$$

we see that

$$\limsup_{t\to\infty} y(t) \leq \rho_1^2 \tag{2.36}$$

and we conclude that

The ball of E_1, $\mathscr{B}_1 = B_{E_1}((A^{-1}f, 0), \rho_1')$, centered at $(A^{-1}f, 0)$ of radius $\rho_1' > \rho_1$, is absorbing in E_1 for the (semi)group $S(t)$, $t \geq 0$. (2.37)

The time $t_1 = t_1(\mathscr{B}, \rho_1')$ after which $S(t)\mathscr{B}$ is included in $\mathscr{B}_1$ cannot be written explicitly: let us set

$$R_1 = \operatorname*{Sup}_{(u(0), v(0)) \in \mathscr{B}} (|Au(t_0) - f|^2 + \|v(t_0)\|^2), \tag{2.38}$$

then $t_1 \geq t_0 + t_1'$

$$t_1' = \frac{2}{\alpha_1} \log \frac{R_1}{(\rho_1')^2 - \rho_1^2} \tag{2.39}$$

Hypothesis I.(1.13)

A major difference between the equations of the first order in time, considered in Chapter III, and those considered here is that the compactness assumption I.(1.12) is not satisfied; this is obvious from Proposition 2.1 since the operators $S(t)$ are homeomorphisms of the spaces E_0, E_1, in which they operate. We will see that in all the examples of Sections 2, 3, and 4, the assumption satisfied instead is I.(1.13) with H replaced by E_0.

Lemma 2.1. *We can write $S(t)$ in the form $S_1(t) + S_2(t)$, $\forall t > 0$, where for each $t \geq 0$, $S_1(t)$ maps E_0 into itself, the operators $S_1(t)$ being uniformly compact in E_0 in the sense of* I.(1.12) *and $S_2(t) \in \mathscr{L}(E_0)$ with*

$$\|S_2(t)\|_{\mathscr{L}(E_0)} \leq c_0 \exp\left(-\frac{\alpha_1}{2}t\right), \qquad \forall t \geq 0, \tag{2.40}$$

c_0 as in (1.32).

Proof. We write $S_2(t) = \Sigma(t)$, $S_1(t) = S(t) - \Sigma(t)$; (2.40) is nothing other than (1.32) and S_2 satisfies the required properties.

In order to prove that the operators $S_1(t)$ are uniformly compact in E_0 we observe that $S_1(t)\cdot\{u_0, u_1\} = \{\tilde{u}(t), \tilde{u}'(t)\}$ where $\tilde{u}$ is solution of the linear evolution problem

$$\tilde{u}'' + \alpha\tilde{u}' + A\tilde{u} = f - g(u(t)), \tag{2.41}$$

$$\tilde{u}(0) = 0, \qquad \tilde{u}'(0) = 0, \tag{2.42}$$

u being of course the solution of (2.4), (2.5).

We can obtain some estimates of $\tilde{u}$. The computations leading to (2.36) are valid when f depends on t provided we replace, in (2.36), $|f|$ by $|f|_{L^\infty(\mathbb{R};H)}$. We can then apply (2.36) to $\bar{u}$, f being replaced by

$$f - \beta \sin u,$$

with

$$|f - \beta \sin u|_{L^\infty(\mathbb{R};H)} \leq |f| + |\beta||\Omega|^{1/2}.$$

Hence

$$\begin{aligned}|A\tilde{u}(t)|^2 + |\tilde{u}(t) + \varepsilon\tilde{u}'(t)|^2 &\leq 8(|f|^2 + \beta^2|\Omega| + |\beta||\Omega|^{1/2}) \exp\left(-\frac{\alpha_1 t}{2}\right) \\ &\quad + 4\frac{\kappa_1}{\alpha_1}\left\{1 - \exp\left(-\frac{\alpha_1 t}{2}\right)\right\} + 4(|f|^2 + \beta^2|\Omega|) \\ &\leq 12(|f|^2 + \beta|\Omega| + |\beta||\Omega|^{1/2}) + \frac{\kappa_1}{\alpha}, \qquad \forall t \geq 0.\end{aligned} \tag{2.43}$$

Thus when $\mathscr{B}$ is a bounded set of E_0, $\bigcup_{t\geq 0} S_1(t)\mathscr{B}$ is included in a bounded set of E_1 and hence in a compact set of E_0 (which is independent of $\mathscr{B}$ owing to (2.43)). The uniform compactness of S_1 is proved and the proof of Lemma 2.1 is complete. □

Maximal Attractor

We have the following

Theorem 2.2. *The dynamical system associated with the sine–Gordon equation* (2.1) *supplemented by the boundary condition* (2.2) *possesses a global attractor* $\mathscr{A}$ *which is compact, connected, and maximal in* E_0. $\mathscr{A}$ *is included in* E_1. *It attracts the bounded sets of* E_0 *and* $\mathscr{A}$ *is also maximal among the functional invariant sets bounded in* E_0.

Proof. We apply Theorem I.1.1 with H replaced by E_0 and obtain all the results stated in Theorem 1.2 except the inclusion of $\mathscr{A}$ in E_1. The necessary assumptions of Theorem I.1.1 have been proved above, namely I.(1.1), I.(1.4), and I.(1.13).

In order to prove that $\mathscr{A} \subset E_1$ we observe that if $\{u_0, u_1\} \in \mathscr{A}$ then, since $S(t)\mathscr{A} = \mathscr{A}$, $\forall t > 0$, $\{u_0, u_1\}$ is on an orbit belonging to $\mathscr{C}_b(\mathbb{R}; V \times H)$ of solutions of

$$u'' + \alpha u' + Au = f - g(u).$$

Obviously, $f - g(u) = f - \beta \sin u \in \mathscr{C}_b(\mathbb{R}; H)$ and $(f - g(u))' = -\beta u' \cos u \in \mathscr{C}_b(\mathbb{R}; H)$. We conclude with Proposition 1.3 that the orbit belongs to $\mathscr{C}_b(\mathbb{R}; D(A) \times V)$ and, in particular, $\{u_0, u_1\} \in D(A) \times V = E_1$. □

Remark 2.4. More information on the structure of $\mathscr{A}$ and on the convergence of the trajectories towards $\mathscr{A}$ will be given in Chapter VII.

Remark 2.5. In the present case the operators $S(t)$ are homeomorphisms of E_0 (or E_1) onto itself, $\forall t \in \mathbb{R}$, and the injectivity property of $S(t)$ is obvious.

2.3. Other Boundary Conditions

We again consider the sine–Gordon equation (2.1), the assumptions on α, β, f, Ω being the same as in Section 2.1, and we replace the Dirichlet boundary condition by either a Neumann boundary condition or a space-periodicity condition, i.e.,

$$\frac{\partial u}{\partial \mathbf{v}} = 0 \quad \text{on } \Gamma \times \mathbb{R}_+, \tag{2.44}$$

or

$$\Omega \text{ is a product } \prod_{i=1}^{n} \,]0, L_1[\text{ and } u \text{ is } \Omega\text{-periodic.} \tag{2.45}$$

With these boundary conditions the situation is similar to the Dirichlet boundary condition case, except that we are unable to bound the average value of u on Ω for large values of t. In fact, this difficulty leads to more interesting physical situations. As we will see in Chapter VII, in the case of the Dirichlet boundary condition, the dynamics are in a sense trivial since, due to the existence of a Lyapunov function, $u(t)$ converges to a stationary solution as $t \to \infty$. In the Neumann and space-periodic cases the "uncontrolled" space average of u sustains the motion and produces more complex dynamics.

For the mathematical study we set $H = L^2(\Omega)$, $V = H^1(\Omega)$, and we endow these spaces with the usual scalar product and norms

$$(u, v) = \int_\Omega u(x)v(x)\,dx, \qquad |u| = \{(u, u)\}^{1/2}, \qquad \forall u, v \in L^2(\Omega), \tag{2.46}$$

$$((u, v))_1 = (u, v) + \sum_{i=1}^{n} (D_i u, D_i v), \qquad \|u\|_1 = \{((u, u))_1\}^{1/2},$$
$$\forall u, v \in H^1(\Omega). \tag{2.47}$$

We define an unbounded linear operator A in H. Its domain $D(A)$ is

$$D(A) = \left\{ v \in H^2(\Omega), \frac{\partial v}{\partial \mathbf{v}} = 0 \text{ on } \Gamma \right\} \qquad \text{in the case (2.44)}, \tag{2.48}$$

$$D(A) = H^2_{\text{per}}(\Omega) \qquad \text{in the case (2.45)}, \tag{2.49}$$

and for $u \in D(A)$, $Au = -\Delta u$. It follows from the results of Section II.2 that for every $\varepsilon > 0$, $A + \varepsilon I$ is an isomorphism from $D(A)$ (endowed, for instance, with the norm of $H^2(\Omega)$) onto H.

We can then write the initial boundary-value problem consisting of (2.1), (2.3) and (2.44) or (2.45) in the form of a second-order differential equation in H totally similar to (2.4), (2.5)

$$u'' + \alpha u' + Au + g(u) = f, \tag{2.50}$$

$$u(0) = u_0, \qquad u'(0) = u_1, \tag{2.51}$$

$g(u) = \beta \sin u$. Theorem 2.1 concerning existence, uniqueness, and regularity of solution applies without any modification. The comments and remarks following Theorem 2.1 are also valid with, however, some slight modifications that we now describe.

When $f(t) = f \in H$ is independent of t, we define the operators $S(t)$, $t \in \mathbb{R}$, as in (2.10)

$$S(t)\colon \{u_0, u_1\} \to \{u(t), u'(t)\}. \tag{2.52}$$

They map $E_0 = V \times H$ into itself and $E_1 = D(A) \times V$ into itself. They enjoy the group properties (2.11) and for all t they are homeomorphisms from E_0 (or E_1) onto itself. Relation (2.14) is valid with

$$((u, v)) = \sum_{i=1}^{n} (D_i u, D_i v), \qquad \|u\| = \{((u, u))\}^{1/2}, \qquad \forall u \in V, \tag{2.53}$$

and (2.13), for $s = 1$, is replaced by

$$\frac{1}{2}\frac{d}{dt}(|Au|^2 + \|u'\|^2) + \alpha\|u'\|^2 + ((g(u), u')) = ((f, u')). \tag{2.54}$$

The computations in Remark 2.2, showing that $S(t)$ is bicontinuous, are based on (2.12), (2.13) and are derived here from (2.12), (2.56) in a similar manner.

Since the average value of u is not expected to remain bounded in this case, new difficulties arise when we try to derive time uniform estimates for the solutions of (2.50), (2.51).

Absorbing Set in E_0

If $\varphi \in L^2(\Omega)$ we denote by $m(\varphi)$ its average on Ω

$$m(\varphi) = \frac{1}{|\Omega|}\int_\Omega \varphi(x)\, dx, \tag{2.55}$$

and we set $\bar{\varphi} = \varphi - m(\varphi)$. When $u = u(t)$ is the solution of (2.50), (2.51) we write, for simplicity, $m(u(t)) = m(t)$ and thus

$$u(t) = \bar{u}(t) + m(t), \qquad \forall t. \tag{2.56}$$

To study the stability of the solutions of (2.50) for t large, it is useful to rewrite this equation as a system for $\bar{u}$ and m. The first equation is obtained by taking the scalar product of (2.50) with $|\Omega|^{-1}$; this amounts to taking the average of (2.1), using Green's formula and the boundary condition (2.44), (2.45). We obtain for $m = m(t)$

$$m'' + \alpha m' + m(g(\bar{u} + m)) = m(f). \tag{2.57}$$

The second equation is obtained by substracting (2.57) from (2.50)

$$\bar{u}'' + \alpha\bar{u}' + A\bar{u} + \overline{g(\bar{u} + m)} = \bar{f}. \tag{2.58}$$

The system (2.57), (2.58) is supplemented by the initial conditions

$$m(0) = m(u_0), \qquad m'(0) = m(u_1), \tag{2.59}$$

$$\bar{u}(0) = \bar{u}_0, \qquad \bar{u}'(0) = \bar{u}_1. \tag{2.60}$$

If we assume that the function $t \to m(t)$ is known, then (2.58), (2.60) produce a well-posed initial-value problem with a proof of existence, uniqueness, and regularity similar to that of Theorem 2.1: the natural spaces for this problem are

$$\dot{L}^2(\Omega) = \{v \in L^2(\Omega), m(v) = 0\}, \qquad \dot{H}^1(\Omega) = H^1(\Omega) \cap \dot{L}^2(\Omega),$$

and we recall that because of the generalized Poincaré inequality (see Section II.1.4) the following infimum denoted by λ_1' is strictly positive

$$\operatorname*{Inf}_{\substack{\varphi \in \dot{H}^1(\Omega)\\ \varphi \neq 0}} \frac{\|\varphi\|^2}{|\varphi|^2} = \lambda_1' > 0. \tag{2.61}$$

With this remark in mind we can repeat for $\bar{u}$ the computations (2.18)–(2.23) carried out for u in Section 2.2. Assume that $0 < \varepsilon \leq \varepsilon_0'$ where (compare to (2.18))

$$\varepsilon_0' = \min\left(\frac{\alpha}{4}, \frac{\lambda_1'}{2\alpha}\right), \tag{2.62}$$

and let $\bar{v} = \bar{u}' + \varepsilon\bar{u}$. Taking the scalar product of (2.58) with $\bar{v}$ we obtain the analog of (2.19)

$$\frac{1}{2}\frac{d}{dt}\{\|\bar{u}\|^2 + |\bar{v}|^2\} + \varepsilon\|\bar{u}\|^2 + (\alpha - \varepsilon)|\bar{v}|^2 - \varepsilon(\alpha - \varepsilon)(\bar{u}, \bar{v}) + (\bar{g}(\bar{u} + m), \bar{v})$$
$$= (\bar{f}, \bar{v}). \tag{2.63}$$

The computations leading to (2.21) do not use the explicit form of g, but only the fact that $|g(\varphi)| \leq |\beta| \, \|\varphi\|^{1/2}$, $\forall \varphi \in L^2(\Omega)$. Here we have

$$|\bar{g}(\varphi)| = |\beta \sin \varphi - \beta m(\sin \varphi)| \leq 2|\beta||\Omega|^{1/2}. \tag{2.64}$$

Thus, instead of (2.21),

$$\frac{d}{dt}(\|\bar{u}\|^2 + |\bar{v}|^2) + \alpha_1(\|\bar{u}\|^2 + |\bar{v}|^2) \leq \frac{2}{\alpha_1}(\|\bar{f}\|^2 + 4\beta^2|\Omega|), \tag{2.65}$$

$\alpha_1 = \varepsilon/2$, and

$$\begin{gathered}\|\bar{u}(t)\|^2 + |\bar{v}(t)|^2 \leq (\|\bar{u}_0\|^2 + |\bar{v}_0|^2)\exp(-\alpha_1 t) + \rho_0^2(1 - \exp(-\alpha_1 t)), \\ \rho_0^2 = \frac{2}{\alpha_1^2}(|\bar{f}|^2 + 4\beta^2|\Omega|).\end{gathered} \tag{2.66}$$

We infer from (2.66) the following limited result similar to that concerning the existence of an absorbing set.

Lemma 2.2. *Assume that $\rho_0' > \rho_0$ and that u_0, u_1 are given such that*

$$\|\bar{u}_0\|^2 + |\bar{v}_0|^2 \leq R_0^2.$$

Then there exists $t_0' = t_0'(R_0, \rho_0')$,

$$t_0' = \frac{1}{\alpha_1}\log\frac{R_0^2}{(\rho_0')^2 - \rho_0^2}, \tag{2.67}$$

such that for $t \geq t_0'$, $\{\bar{u}(t), \bar{v}(t)\}$ belongs to the ball $(B_{E_0}(0, \rho_0')$ of E_0.

A result concerning m' can also be derived from (2.57), (2.60). The average value of $g(\bar{u} + m)$ is bounded pointwise by $|\beta|$. Hence

$$\begin{gathered}-c_1 \leq m'' + \alpha m' \leq c_1, \qquad c_1 = |m(f)| + |\beta|, \\ |m'(u(t)) - m(u_1)\exp(-\alpha t)| \leq \frac{c_1}{\alpha}(1 - \exp(-\alpha t)),\end{gathered} \tag{2.68}$$

$$\limsup_{t\to\infty} |m'(u(t))| \leq \frac{c_1}{\alpha}. \tag{2.69}$$

If $a_0 = c_1/\alpha$, $a_0' > a_0$, and u_0, u_1 are given such that $|m(u_1)| \leq b_0$, then for $t \geq t_0'' = t_0''(b_0, a_0')$

$$t_0''(b_0, a_0') = \frac{1}{\alpha}\log\frac{b_0}{a_0' - a_0},$$

we have

$$|m'(u(t))| \leq a_0'. \tag{2.70}$$

We combine Lemma 2.2 and (2.70) and we obtain a partial result concerning the existence of an absorbing set. We denote by P the projector in E_0 defined as follows:

$$\begin{gathered}P\colon E_0 \to E_0,\\ \varphi\colon \{\bar{u} + m(u), w = \bar{w} + m(w)\} \to P\varphi = \{\bar{u}, w = \bar{w} + m(w)\},\end{gathered} \tag{2.71}$$

and we select in PE_0 a bounded set $\mathscr{B}_0$ defined by[1]

$$\|\{\bar{u}, \bar{w} + \varepsilon\bar{u}\}\|_{E_0} \leq \rho_0', \qquad |m(\bar{w})| \leq a_0', \tag{2.72}$$

ρ_0', a_0' as in Lemma 2.2 and (2.70). We then have

Lemma 2.3. *Assume that $\mathscr{B} \subset E_0$ is such that $P\mathscr{B}$ is bounded in E_0 and that $\{u_0, w_0 = u_1\}$ is given in $\mathscr{B}$. Then there exists $t_0 = t_0(\mathscr{B}, \mathscr{B}_0)$[2] depending on $\mathscr{B}$ and $\mathscr{B}_0$ such that for $t \geq t_0$, $P\{u(t), u'(t)\}$* belongs to $\mathscr{B}_0$.

We can reinterpret Lemma 2.3 in a different manner. We observe that the *change* of u *into* $u + 2k\Pi$, $k \in \mathbb{Z}$, *leaves equation* (2.58) *unchanged* and it is then natural to consider the function u modulo 2Π. For that purpose we write, for $i = 0, 1$,

$$E_i \equiv PE_i \times \mathbb{R}, \tag{2.73}$$

so that if $\varphi = \{u = \bar{u} + m(u), w = \bar{w} + m(w)\} \in E_i$, then $\varphi \equiv \{P\varphi, m(u)\}$, $P\varphi$ as in (2.71). Then we consider

$$\tilde{E}_i = PE_i \times \mathbb{T}^1, \tag{2.74}$$

where $\mathbb{T}^1 = \mathbb{R}/2\Pi\mathbb{Z}$ is the one-dimensional torus. We observe that the group $S(t)$ induces a group on $\tilde{E}_i$, $i = 0, 1$,

$$\{\bar{u}_0, u_1, m(u_0) \ (\text{mod } 2\Pi)\} \to \{\bar{u}(t), \bar{u}'(t), m(u(t)) \ (\text{mod } 2\Pi)\}. \tag{2.75}$$

Then Lemma 2.3 implies simply

Lemma 2.4. *The sets $\mathscr{B}_0 \times \mathbb{T}^1$ and $\mathscr{B}_1 \times \mathbb{T}^1$ are absorbing sets for the (semi)group $\tilde{S}(t)$ in $\tilde{E}_0$ and $\tilde{E}_1$, respectively.*

Hypothesis I.(1.13)

We intend to derive a suitable generalization of Theorem I.1.1 for $\tilde{S}(t)$. For that purpose we use the decomposition of $S(t)$ in the form $S_1(t) + S_2(t)$ and show that this decomposition induces a decomposition of $\tilde{S}(t)$. For $\{u_0, u_1\} \in E_i$, $\tilde{S}_2(t)\{u_0, u_1\} = \{u_2(t) \ (\text{mod } 2\Pi), u_2'(t)\} \in E_i$, where u_2 is the solution of

$$\begin{cases} u_2'' + \alpha u_2' + Au_2 = 0,\\ u_2(0) = \bar{u}_0, \qquad u_2'(0) = u_1.\end{cases} \tag{2.76}$$

[1] In Lemma 2.3 the second component of φ in E_0 will be u' and not $v = u' + \varepsilon u$.

[2] $t_0(\mathscr{B}, \mathscr{B}_0) = \max(t_0'(R_0, \rho_0'), t_0''(b_0, a_0'))$, t_0', t_0'' as in Lemma 2.2 and (2.70) provided $\|\bar{u}\|^2 + |\bar{w} + \varepsilon\bar{u}|^2 \leq R_0^2$ and $|m(w)| \leq b_0$, $\forall \{u = \bar{u} + m(u), w = \bar{w} + m(w)\} \in \mathscr{B}$.

In (2.76), $\bar{u}_0 = u_0 - m(u_0)$, $m(u_0)$ being the average of u_0 on Ω. Of course, u_0 is defined mod 2Π but $\bar{u}_0$ is independent of the element chosen in the equivalence class. We observe that $\tilde{S}_2(t)$ is linear continuous in PE_i, $i = 0, 1$, and with the analog of (2.27) and (2.65) we prove exactly, as in (1.32), that the norm of $\tilde{S}_2(t)$ in $\mathscr{L}(PE_i)$, $i = 0, 1$, decays exponentially,

$$\|\tilde{S}_2(t)\|_{\mathscr{L}(PE_i)} \leq c_0 \exp\left(-\frac{\alpha_1 t}{2}\right), \tag{2.77}$$

α_1 as before, c_0 as in (1.32). Then, as in Lemma 2.1, we notice that $\tilde{S}_1(t)\cdot\{u_0, u_1\} = \{\tilde{u}(t) + m(u_0) \pmod{2\Pi}, \tilde{u}'(t)\}$ where $\tilde{u}$ is the solution of (2.41), (2.42), and we show exactly, as in (2.43), that if $\mathscr{B}$ is a bounded set of E_0, then $\bigcup_{t\geq 0} S_1(t)\mathscr{B}$ is a bounded set of E_1 and thus a relatively compact set of E_0. We have therefore proved that

$\tilde{S}(t) = \tilde{S}_1(t) + \tilde{S}_2(t)$, $\forall t$, where the operators $\tilde{S}_i(t)$ are continuous in $\tilde{E}_0$, $\forall t$, $i = 1, 2$, the $\tilde{S}_1$ are uniformly compact in $\tilde{E}_0$, and the operators $P\tilde{S}_2(t)$ are linear continuous in $\mathscr{L}(PE_i)$ and satisfy (2.77). (2.78)

Maximal Attractor

We arrive at the final result

Theorem 2.3. *We consider the sine–Gordon equation* (2.1) *supplemented with the boundary condition*

(2.44) (*Neumann*) *or* (2.45) (*space-periodicity*).

We can associate with this boundary-value problem a continuous group $\tilde{S}(t)$ *which operates in*

$$\tilde{E}_0 = (V \cap \dot{L}^2(\Omega)) \times L^2(\Omega) \times \mathbb{T}^1.$$

The corresponding dynamical system possesses a global attractor $\tilde{\mathscr{A}}$ *which is compact, connected, and maximal in* $\tilde{E}_0$, $\tilde{\mathscr{A}}$ *is included in* $\tilde{E}_i$. *It attracts the bounded sets of* $\tilde{E}_0$, *and* $\tilde{\mathscr{A}}$ *is also maximal among the functional invariant sets bounded in* $\tilde{E}_0$.

Proof. Since $\tilde{E}_0 = PE_0 \times \mathbb{T}^1$ is not a vector space, Theorem I.1.1 with assumption I.(1.13) is not directly applicable. However, PE_0 is a Hilbert space and $\mathbb{T}^1$ is a compact additive group. Instead of I.(1.13) we have the following:

$\tilde{E}_0$ is an additive (metric) group;

$\tilde{S}(t) = \tilde{S}_1(t) + \tilde{S}_2(t)$ where $\tilde{S}_1(t)$ is uniformly compact in $\tilde{E}_0$, and $\tilde{S}_2(t)$ is a Lipschitz operator in $\tilde{E}_0$ such that $\tilde{S}_2(t)0 = 0$, and

$$d(\tilde{S}_2(t)\varphi, \tilde{S}_2(t)\psi) \leq r(t)\, d(\varphi, \psi), \qquad \forall \varphi, \psi \in \tilde{E}_0,$$

where $r(t)$ is a decreasing function converging to 0 as $t \to \infty$.

We leave as an exercise to the reader the extension of the proof of Theorem I.1.1 to the present situation. We obtain in this way all the statements in Theorem 2.3 except the inclusion of $\tilde{\mathscr{A}}$ in $\tilde{E}_1$.

The inclusion of $\tilde{\mathscr{A}}$ in $\tilde{E}_1$ follows from Proposition 1.3 as in Theorem 2.2. Indeed, if $\{u_0, u_1\} \in \tilde{\mathscr{A}}$, then a representative of this point belongs to a solution in $\mathscr{C}_b(\mathbb{R}; V \times H)$ of

$$u'' + \alpha u' + Au + u = f - \beta \sin u + u.$$

Clearly, $f - \beta \sin u + u \in \mathscr{C}_b(\mathbb{R}; H)$ and its time derivative $(-\beta \cos u + 1)u'$ is in $\mathscr{C}_b(\mathbb{R}; H)$ too. Thus, by Proposition 1.3, the orbit is in $\mathscr{C}_b(\mathbb{R}; D(A) \times V)$ and in particular the considered point is in $D(A) \times V$ and $\{u_0, u_1\} \in \tilde{E}_1$. $\square$

Remark 2.6. We can return to E_0 and associate with $\tilde{\mathscr{A}}$ the set $\mathscr{A} = P\tilde{\mathscr{A}} \times \mathbb{R}$ which is obviously not compact in E_0. This set attracts the sets $\mathscr{B}$ of E_0, such that $P\mathscr{B}$ is bounded in the semidistance associated with

$$d(\varphi, \varphi^*) = \|\bar{u} - \bar{u}^*\| + |w - w^*| + \operatorname*{Inf}_{k \in \mathbb{Z}} |m(u - u^*) - 2k\Pi|,$$

$$\forall \varphi = \{u, w\}, \quad \varphi^* = \{u^*, w^*\} \in E_0.$$

For a single orbit $u(t)$ we have

$$\operatorname*{Inf}_{\substack{\{u^*, w^*\} \in \mathscr{A} \\ k \in \mathbb{Z}}} \{\|\bar{u}(t) - \bar{u}^*\| + |u'(t) - w^*| + |m(u(t) - u^*) - 2k\Pi|\} \to 0, \quad t \to \infty.$$

3. A Nonlinear Wave Equation of Relativistic Quantum Mechanics

In this section we consider a nonlinear wave equation arising in relativistic quantum mechanics, see K. Jörgens [1], L.I. Schiff [1], I.E. Segal [1], [2]. The methods are similar to those used in Sections 2.1 and 2.2 for the sine–Gordon equation but further difficulties arise here since g is unbounded.

3.1. The Equation and Its Mathematical Setting

Let Ω be an open bounded set of $\mathbb{R}^n$ with a boundary Γ sufficiently regular. The unknown function $u = u(x, t)$, $x \in \Omega$, $t \in \mathbb{R}$ (or some interval of $\mathbb{R}$) is solution of the equation

$$\frac{\partial^2 u}{\partial t^2} + \alpha \frac{\partial u}{\partial t} - \Delta u + g(u) = f \quad \text{in} \quad \Omega \times \mathbb{R}_+, \tag{3.1}$$

where f and α are given, $\alpha > 0$ in the damped case of interest to us, and g is a $\mathscr{C}^2$ function from $\mathbb{R}$ into $\mathbb{R}$ satisfying some assumptions specified hereafter.

The equation is supplemented with the boundary condition

$$u = 0 \qquad \text{on} \quad \Gamma \times \mathbb{R}_+, \tag{3.2}$$

and when an initial-value problem is considered we provide the initial conditions u_0, u_1,

$$u(x, 0) = u_0(x), \qquad \frac{\partial u}{\partial t}(x, 0) = u_1(x), \qquad x \in \Omega. \tag{3.3}$$

We denote by G the function $G(s) = \int_0^s g(r)\, dr$ and we make the following assumptions on g, G:

$$\liminf_{|s|\to\infty} \frac{G(s)}{s^2} \geq 0. \tag{3.4}$$

There exists $c_1 > 0$ such that

$$\liminf_{|s|\to\infty} \frac{sg(s) - c_1 G(s)}{s^2} \geq 0, \tag{3.5}$$

and

$$|g'(s)| \leq c_2(1 + |s|^\gamma) \quad \text{with} \quad \begin{cases} 0 \leq \gamma < \infty & \text{when } n = 1, 2, \\ 0 \leq \gamma < 2 & \text{when } n = 3, \\ \gamma = 0 \text{ (i.e., } g' \text{ bounded)} & \text{when } n \geq 4. \end{cases} \tag{3.6}$$

In the equation of relativistic quantum mechanics

$$g(u) = |u|^\gamma u, \tag{3.7}$$

and if γ satisfies the conditions in (3.6), then (3.4)–(3.6) hold with $c_1 = 1 + \gamma$.[1]

We infer from (3.4), (3.5) that, for every $\eta > 0$, there exists C_η, C'_η such that

$$G(s) + \eta s^2 \geq -C_\eta, \tag{3.8}$$

$$sg(s) - c_1 G(s) + \eta s^2 \geq -C'_\eta, \qquad \forall s \in \mathbb{R}, \tag{3.9}$$

and a proper choice of η will be made when necessary.

Let us write problem (3.1)–(3.3) as a second-order evolution equation in infinite dimension. For that purpose, we set $H = L^2(\Omega)$, $V = H_0^1(\Omega)$, and we endow these spaces with the usual scalar products and norms, $(\cdot, \cdot)$, $|\cdot|$, $((\cdot, \cdot))$, $\|\cdot\|$. We write $D(A) = H_0^1(\Omega) \cap H^2(\Omega)$ and for every $u \in D(A)$, $Au = -\Delta u$. We have

$$D(A) \subset V \subset H \subset V',$$

V' the dual of V, where each space is dense in the following one and the injections are continuous.

[1] The case $\gamma = 2$, $n = 3$, of interest in quantum mechanics is not contained in (3.6). See E. Fereisl [2], J. Arrieta, A. Cavarlho and J.K. Hale [1].

By the Sobolev embeddings (see II.(1.10)),

$$H_0^1(\Omega) \subset L^q(\Omega), \quad \forall q < \infty \text{ if } n = 2, \quad q = 2n/(n-2), \quad \forall n \geq 3\ (= 6 \text{ if } n = 3). \tag{3.10}$$

Thanks to the assumptions (3.6) on g, $g \circ u$ and $G \circ u$ are then integrable for every $u \in H_0^1(\Omega)$ and we will use g and G as well to denote the nonlinear operators

$$\begin{aligned} g&: H_0^1(\Omega) \to L^2(\Omega), \qquad u \to g \circ u, \\ G&: H_0^1(\Omega) \to L^2(\Omega), \qquad u \to G \circ u. \end{aligned} \tag{3.11}$$

Then (3.1)–(3.3) is equivalent to the following equation in H ($\varphi' = d\varphi/dt$):

$$u'' + \alpha u' + Au + g(u) = f, \tag{3.12}$$

$$u(0) = u_0, \qquad u'(0) = u_1. \tag{3.13}$$

The existence and uniqueness of solution of (3.12), (3.13) is proved in J.L. Lions [2] and also follows from Theorem 4.1 below.

Theorem 3.1. *Let $\alpha \in \mathbb{R}$, and let f, u_0, u_1 be given satisfying*

$$f \in \mathscr{C}([0, T]; H), \qquad u_0 \in V, \quad u_1 \in H. \tag{3.14}$$

Then there exists a unique solution u of (3.12), (3.13) *such that*

$$u \in \mathscr{C}([0, T]; V), \qquad u' \in \mathscr{C}([0, T]; H). \tag{3.15}$$

If, furthermore,

$$f' \in \mathscr{C}([0, T]; H), \qquad u_0 \in D(A), \quad u_1 \in V, \tag{3.16}$$

then u satisfies

$$u \in \mathscr{C}([0, T]; D(A)), \qquad u' \in \mathscr{C}([0, T]; V). \tag{3.17}$$

As in the previous cases, by changing t into $-t$, we see that we can also solve equation (3.12) with initial data (3.13) on $[-T, 0]$ provided f is defined on $[-T, 0]$.

In the autonomous case f is independent of t, $f(t) = f \in H$, $\forall t$, then u is defined for all time,

$$u \in \mathscr{C}(\mathbb{R}; V), \qquad u' \in \mathscr{C}(\mathbb{R}; H), \tag{3.18}$$

if $u_0 \in V$, $u_1 \in H$, whereas if $u_0 \in D(A)$, $u_1 \in V$,

$$u \in \mathscr{C}(\mathbb{R}; D(A)), \qquad u' \in \mathscr{C}(\mathbb{R}; V). \tag{3.19}$$

We define for every $t \in \mathbb{R}$, the mapping

$$S(t): \{u_0, u_1\} \to \{u(t), u'(t)\}. \tag{3.20}$$

They map $E_0 = V \times H$ into itself and $E_1 = D(A) \times V$ into itself and they enjoy the *group properties*

$$\begin{cases} S(t+s) = S(t)S(s), & \forall s, t \in \mathbb{R}, \\ S(0) = I. \end{cases} \tag{3.21}$$

Hence $S(-t)$ is the inverse of $S(t)$, $\forall t \in \mathbb{R}$, and these operators are one-to-one. We have

Proposition 3.1. *When $f(t) = f \in H$, we can associate with equation* (3.12) *the group of operators $S(t)$ defined by* (3.20), *α fixed in $\mathbb{R}$, $t \in \mathbb{R}$. For every t, $S(t)$ is an homeomorphism from E_0 onto E_0.*

Proof. There only remains to prove that $S(t)$ is continuous in E_0, $\forall t > 0$, since $S(-t) = \{S(t)\}^{-1}$, and the change of orientation in time has only for effect to change α into $-\alpha$.

The solution u of (3.12), (3.13) satisfies the following energy equality if $\{u_0, u_1\} \in V \times H$ (compare to (2.12), (2.13))

$$\frac{1}{2}\frac{d}{dt}(\|u\|^2 + |u'|^2) + \alpha|u'|^2 + (g(u), u') = (f, u'), \tag{3.22}$$

and if $\{u_0, u_1\} \in D(A) \times V$

$$\frac{1}{2}\frac{d}{dt}(|Au|^2 + \|u\|^2) + \alpha\|u'\|^2 + ((g(u), u')) = (f, Au'). \tag{3.23}$$

They are formally obtained by taking the scalar product in H of (3.12) with, respectively, u' and Au'; the rigorous proof relies on Lemma II.4.1.

Let u, v be two solutions of (3.12) associated with initial data u_0, u_1 and v_0, v_1, and let $w = u - v$. Then

$$w'' + \alpha w' + Aw = -(g(u) - g(v)), \tag{3.24}$$

and by taking the scalar product with w' in H we find

$$\frac{1}{2}\frac{d}{dt}(\|w\|^2 + |w'|^2) + \alpha|w'|^2 = -(g(u) - g(v), w).$$

For $t \in [0, T]$, $u, v \in \mathscr{C}([0, T]; V)$, and we infer from Lemma 3.1 below that there exists $c_1' = c_1'(T)$ such that

$$\begin{aligned} &|(g(u(t)) - g(v(t)), u'(t) - v'(t))| \\ &\quad \le c_1' \|u(t) - v(t)\| \, |u'(t) - v'(t)|, \qquad \forall t \in [0, T]. \end{aligned} \tag{3.25}$$

Hence we have

$$\begin{aligned} \frac{d}{dt}(\|w\|^2 + |w'|^2) &\le 2|\alpha||w'|^2 + 2c_1' \|w\| \, |w'| \\ &\le \beta(\|w\|^2 + |w'|^2), \qquad \beta = (2|\alpha| + c_1'), \end{aligned} \tag{3.26}$$

and the Gronwall lemma implies

$$\|w(t)\|^2 + |w'(t)|^2 \leq (\|w(0)\|^2 + |w'(0)|^2)\exp(\beta t), \tag{3.27}$$

which proves the continuity of $S(t)$ in E_0.

Remark 3.1. $S(t)$ is also an homeomorphism from E_1 onto itself, $\forall t \in \mathbb{R}$, but the continuity of $S(t)$ in E_1, which is not essential here, is more delicate to establish and will not be proved here.

We establish Lemma 3.1 used in the proof of Proposition 3.1

Lemma 3.1. *The operator g is locally Lipschitz from V into H.*

Proof. The proof consists of showing the existence of a constant c_2' such that

$$|g(\varphi) - g(\psi)| \leq c_2'(1 + \|\varphi\| + \|\psi\|)^\gamma \|\varphi - \psi\|, \qquad \forall \varphi, \psi \in V. \tag{3.28}$$

We have

$$\begin{aligned}|g(\varphi) - g(\psi)|^2 &= \int_\Omega (g(\varphi(x)) - g(\psi(x)))^2\,dx \\ &\leq (\text{by (3.6) and for } \theta = \theta(x) \in [0, 1]) \\ &\leq c_2^2 \int_\Omega (1 + |\theta\varphi + (1-\theta)\psi|^\gamma)^2(\varphi - \psi)^2\,dx \\ &\leq c_3' \int_\Omega (1 + |\varphi|^\gamma + |\psi|^\gamma)^2(\varphi - \psi)^2\,dx\end{aligned}$$

For $n \geq 4$, $\gamma = 0$, and we simply majorize this expression by

$$c_4'|\varphi - \psi|^2 \leq (\text{with (1.18)}) \leq c_4'\lambda_1^{-1}\|\varphi - \psi\|^2.$$

For $n \leq 3$ we use the Hölder inequality and majorize the last expression by

$$\begin{aligned}&c_3'\left(\int_\Omega (1 + |\varphi|^\gamma + |\psi|^\gamma)^3\,dx\right)^{2/3}\left(\int_\Omega (\varphi - \psi)^6\,dx\right)^{1/3} \\ &\leq (\text{by the hypothesis (3.6) on } \gamma \text{ and (3.10)}) \\ &\leq c_5'(1 + \|\varphi\| + \|\psi\|)^{2\gamma}\|\varphi - \psi\|^2,\end{aligned}$$

and (3.28) follows in all cases. □

3.2. Absorbing Sets and Attractors

We now proceed with the existence of absorbing sets. From now on $\alpha > 0$ is fixed and we choose ε as in Section 1

$$0 < \varepsilon \leq \varepsilon_0, \qquad \varepsilon_0 = \min\left(\frac{\alpha}{4}, \frac{\lambda_1}{2\alpha}\right). \tag{3.29}$$

Absorbing Sets in E_0

We formally take the scalar product in H of equation (3.12) with $v = u' + \varepsilon u$; after a computation, which can be rigorously justified with Lemma II.4.1, we find (compare to (1.17))

$$\frac{1}{2}\frac{d}{dt}\{\|u\|^2 + |v|^2\} + \varepsilon\|u\|^2 + (\alpha - \varepsilon)|v|^2 - \varepsilon(\alpha - \varepsilon)(u, v) + (g(u), v) = (f, v). \tag{3.30}$$

Thanks to assumption (3.29) on ε and (1.33)

$$\begin{aligned} \varepsilon\|u\|^2 + (\alpha - \varepsilon)|v|^2 - \varepsilon(\alpha - \varepsilon)|v|^2 &\geq \frac{\varepsilon}{2}\|u\|^2 + \frac{\alpha}{2}|v|^2, \\ (g(u), v) = (g(u), u') + \varepsilon(g(u), u) &= \frac{d}{dt}G(u) + \varepsilon(g(u), u), \end{aligned} \tag{3.31}$$

and we deduce from (3.30)

$$\frac{1}{2}\frac{d}{dt}\{\|u\|^2 + |v|^2 + 2G(u)\} + \frac{\varepsilon}{2}\|u\|^2 + \frac{\alpha}{2}|v|^2 + \varepsilon(g(u), u) \leq (f, v).$$

We infer from (3.8), (3.9), and (1.18) that there exist two constants κ_1, κ_2 such that

$$G(\varphi) + \frac{1}{8(1 + c_1)}\|\varphi\|^2 + \kappa_1 \geq 0, \qquad \forall \varphi \in V, \tag{3.32}$$

$$(\varphi, g(\varphi)) - c_1 G(\varphi) + \tfrac{1}{8}\|\varphi\|^2 + \kappa_2 \geq 0, \qquad \forall \varphi \in V. \tag{3.33}$$

Thus

$$\begin{aligned} (g(u), u) &\geq c_1 G(u) - \tfrac{1}{8}\|u\|^2 - \kappa_2 \\ &\geq c_1 G(u) - \tfrac{1}{4}\|u\|^2 - (\kappa_2 + c_1\kappa_1), \end{aligned}$$

and

$$\begin{aligned} &\frac{1}{2}\frac{d}{dt}\{\|u\|^2 + |v|^2 + 2G(u)\} + \frac{\varepsilon}{4}\|u\|^2 + \frac{\alpha}{2}|v|^2 + \varepsilon c_1 G(u) \\ &\leq \varepsilon(\kappa_2 + c_1\kappa_1) + (f, v) \\ &\leq \varepsilon(\kappa_2 + c_1\kappa_1) + \frac{\alpha}{4}|v|^2 + \frac{1}{\alpha}|f|^2. \end{aligned}$$

Setting $\alpha_2 = \min(\alpha/2, \varepsilon c_1)$, we obtain using again (3.32)

$$\frac{d}{dt}y + \alpha_2 y \leq c_6' + \frac{2}{\alpha}|f|^2, \tag{3.34}$$

$$y = \|u\|^2 + |v|^2 + 2G(u) + 2\kappa_1 \geq \frac{3}{4}\|u\|^2 + |v|^2 \geq 0,$$

$$c_6' = 2\varepsilon(\kappa_2 + c_1\kappa_1) + 2\alpha_2\kappa_1.$$

With the Gronwall lemma

$$y(t) \leq y(0)\exp(-\alpha_2 t) + \left(\frac{c_6'}{\alpha_2} + \frac{2}{\alpha\alpha_2}|f|^2\right)\{1 - \exp(-\alpha_2 t)\}, \quad \forall t \geq 0, \quad (3.35)$$

and

$$\limsup_{t\to\infty} y(t) \leq \mu_0^2, \qquad \mu_0^2 = \left(\frac{c_6'}{\alpha_2} + \frac{2}{\alpha\alpha_2}|f|^2\right). \quad (3.36)$$

Let $\mu_0' > \mu_0$ be fixed and assume that $y(0) \leq R$. It readily follows from (3.35) that for $t \geq t_0 = t_0(R, \mu_0')$,

$$t_0(R, \mu_0') = \frac{1}{\alpha_2}\log\frac{R}{(\mu_0')^2 - \mu_0^2}, \quad (3.37)$$

we have $y(t) \leq \mu_0'$ and

$$\|u(t)\|^2 + |u'(t)|^2 \leq (1 + \varepsilon\lambda_1^{-1/2})(\|u(t)\|^2 + |u'(t) + \varepsilon u(t)|^2)$$
$$\leq (1 + \varepsilon\lambda_1^{-1/2})y(t) \leq (1 + \varepsilon\lambda_1^{-1/2})\mu_0'.$$

We observe that G is a bounded operator from V into H and therefore if $\mathscr{B}$ is a bounded set of $E_0 = V \times H$, then

$$R = R(\mathscr{B}) = \operatorname*{Sup}_{\varphi=(\varphi_0,\varphi_1)\in\mathscr{B}} \{\|\varphi_0\|^2 + |\varphi_1 + \varepsilon\varphi_0|^2 + 2G(\varphi_0) + 2\kappa_1\} < \infty. \quad (3.38)$$

We set $\rho_0 = (1 + \varepsilon\lambda_1^{-1/2})\mu_0'$ and we conclude with

Lemma 3.2. *The ball of E_0, $\mathscr{B}_0 = B_{E_0}(0, \rho_0)$, centered at 0 of radius ρ_0, is an absorbing set in E_0 for the (semi)group $S(t)$. For many bounded set $\mathscr{B}$ of E_0, $S(t)\mathscr{B} \subset \mathscr{B}_0$ for $t \geq t_0$, t_0 given by* (3.37), *R given by* (3.38).

Hypothesis I.(1.13)

Our next aim is to check assumption I.(1.13) for the operators $S(t)$ with H replaced by E_0. For that purpose we need the following technical result on g:

Lemma 3.3. *There exists $\sigma_2 > 0$ such that for every $\varphi \in V$, the differential $g'(\varphi)$ belongs to $\mathscr{L}(H, V_{-1+\sigma_2})$ and, for every $R > 0$,*

$$\operatorname*{Sup}_{\|\varphi\|\leq R} |g'(\varphi)|_{\mathscr{L}(H,V_{-1+\sigma_2})} < \infty. \quad (3.39)$$

Proof. Due to (3.6), if $n \geq 4$, $g'(\varphi)$ is in $L^\infty(\Omega)$ for any φ in $H = L^2(\Omega)$ and the result is obvious in this case. For $n = 3$, due to the Sobolev embedding theorems (see II.(1.10) and Section II.1.3)

$$H^1(\Omega) \subset L^6(\Omega) \quad \text{and} \quad H^s(\Omega) \subset L^{6/(3-2s)}(\Omega) \quad \text{for } 0 \leq s < \tfrac{3}{2}.$$

If $\varphi \in V = H_0^1(\Omega)$ with $\|\varphi\| \leq R$, then by (3.6)

$$|g'(\varphi)|_{L^{6/\gamma}(\Omega)} \leq c_1'(R). \quad (3.40)$$

Let $\sigma_2 = 1 - \gamma/2$. By interpolation (see Section II.1.3) $V_{1-\sigma_2} \subset H^{1-\sigma_2}(\Omega)$ and by the Sobolev embedding this space is included in $L^{6/(3-\gamma)}(\Omega)$, all the injections being continuous. Let $\psi \in V_{1-\sigma_2}$ and $\theta \in H$. We apply the Holder inequality with exponents $6/\gamma$, $6/(3-\gamma)$, 2 and find

$$\left|\int_\Omega g'(\varphi)\psi\theta\,dx\right| \le |g'(\varphi)|_{L^{6/\gamma}(\Omega)}|\psi|_{L^{6/(3-\gamma)}(\Omega)}|\theta|$$
$$\le c_1'(R)c_2'|\psi|_{V_{1-\sigma_2}}|\theta|.$$

This shows that $g'(\varphi)\theta$ is in the dual V_{σ_2-1} of $V_{1-\sigma_2}$ and that its norm in V_{σ_2-1} is bounded by $c_1'(R)c_2'|\theta|$, and the lemma is proved, the supremum in (3.39) being bounded by $c_1'(R)c_2'$ ($c_2' \ge$ the norm of the injection of $V_{1-\sigma_2}$ in $L^{6/(3-\gamma)}(\Omega)$).

For $n = 1, 2$, since $H_0^1(\Omega) \subset L^q(\Omega)$, $\forall q < \infty$, we have instead of (3.40) that $g'(\varphi)$ belongs to $L^{q/\gamma}(\Omega)$, $\forall q < \infty$, and $|g'(\varphi)|_{L^{q/\gamma}(\Omega)} \le c_1'(R, q)$ when $\|\varphi\| \le R$. We can take $q = 4\gamma$, proceed exactly as above, and obtain the result with $\sigma_2 = 1/2$. We can also improve the result by taking q arbitrarily large ($q = \infty$ if $n = 1$) and obtain σ_2 arbitrarily close to 1, $\sigma_2 < 1$ ($\sigma_2 = 1$ if $n = 1$). □

We now check hypothesis I.(1.13) for this problem

Lemma 3.4. *We can write $S(t)$ in the form $S_1(t) + S_2(t)$, $\forall t > 0$, where for each $t \ge 0$, $S_i(t)$ maps E_0 into itself, the operators $S_1(t)$ being uniformly compact in E_0 in the sense of* I.(1.12) *and $S_2(t) \in \mathscr{L}(E_0)$ with*

$$\|S_2(t)\|_{\mathscr{L}(E_0)} \le c_0 \exp\left(-\frac{\alpha_1 t}{2}\right), \qquad \forall t \ge 0, \tag{3.41}$$

c_0 as in (1.32).

Proof. We write $S_2(t) = \Sigma(t)$, $S_1(t) = S(t) - \Sigma(t)$; (3.41) is nothing other than (1.32) and S_2 satisfies the required properties. To prove that the operators $S_1(t)$ are uniformly compact in E_0 we observe, as in Lemma 2.1, that $S_1(t)\cdot\{u_0, u_1\} = \{\tilde u(t), \tilde u'(t)\}$ where $\tilde u$ is a solution of the linear problem

$$\tilde u'' + \alpha\tilde u' + A\tilde u = f - g(u), \tag{3.42}$$

$$\tilde u(0) = 0, \qquad \tilde u'(0) = 0, \tag{3.43}$$

u being, of course, the solution of (2.4), (2.5). If $\{u_0, u_1\}$ belongs to a bounded set $\mathscr{B}$ in E_0, then for $t \ge t_0$ (given by (3.37), (3.38)), Lemma 3.2 implies that $\{u(t), u'(t)\}$ is in $\mathscr{B}_0$ and $\|u(t)\|^2 + |u'(t)|^2 < \rho_0^2$, $\forall t > t_0$. Since $\{u, u'\}$ is in $\mathscr{C}([0, T]; V \times H)$, $\forall T > 0$, u is bounded from $\mathbb{R}_+$ into V, $u \in \mathscr{C}_b(\mathbb{R}_+; V)$, and u' is bounded from $\mathbb{R}_+$ into H, $u' \in \mathscr{C}_b(\mathbb{R}_+; H)$,

$$\|u(t)\| \le R, \qquad |u'(t)| \le R, \qquad \forall t \ge 0. \tag{3.44}$$

It is clear, with (3.6) and (3.10), that g is a bounded-continuous map from V into H and therefore $g(u(\cdot))$ is in $\mathscr{C}_b(\mathbb{R}_+; H)$. It follows from Proposition 1.2 that $\{\tilde u, \tilde u'\}$ is in $\mathscr{C}_b(\mathbb{R}_+; V \times H)$. Then, by differentiation of (3.42), we find that

$w = \tilde{u}'$ is a solution of

$$w'' + \alpha w' + Aw = -g'(u)u', \tag{3.45}$$

$$w(0) = 0, \qquad w'(0) = f - g(u_0). \tag{3.46}$$

Due to Lemma 3.3 and (3.44), $g'(u)u'$ is in $\mathscr{C}_b(\mathbb{R}_+; V_{-1+\sigma_2})$ and we infer, from Proposition 1.2 applied to $A^{\sigma_2-1}w$, that $\{w, w'\} = \{\tilde{u}', \tilde{u}''\} \in \mathscr{C}_b(\mathbb{R}_+; V_{\sigma_2} \times V_{\sigma_2-1})$. We then return to (3.42); since $f - g(u) \in \mathscr{C}_b(\mathbb{R}_+; H)$ we find that $A\tilde{u} \in \mathscr{C}_b(\mathbb{R}_+; V_{\sigma_2-1})$, i.e., $\tilde{u} \in \mathscr{C}_b(\mathbb{R}_+; V_{\sigma_2+1})$. Finally, $\{\tilde{u}, \tilde{u}'\} \in \mathscr{C}_b(\mathbb{R}_+; V_{\sigma_2+1} \times V_{\sigma_2})$ and $\bigcup_{t\geq 0} S_1(t)\mathscr{B}$ is included in a bounded set of $V_{\sigma_2+1} \times V_{\sigma_2}$. Since the injection of V_{s_1} into V_{s_2} is compact $\forall s_1 > s_2$, this set is compact in $E_0 = V \times H$.

The proof is complete. □

Maximal Attractor

The hypotheses I.(1.1), I.(1.4), and I.(1.13) of Theorem I.1.1 have been verified (with H replaced by E_0). By application of this theorem we now obtain

Theorem 3.2. *The dynamical system associated with the wave equation* (3.1), *supplemented by the boundary condition* (3.2), *possesses a global attractor $\mathscr{A}$ which is compact, connected, and maximal in E_0. $\mathscr{A}$ is included and bounded in E_1. It attracts the bounded sets of E_0, and $\mathscr{A}$ is also maximal among the functional-invariant sets bounded in E_0.*

Proof. All the results are contained in Theorem I.1.1 except the properties $\mathscr{A} \subset E_1$, $\mathscr{A}$ bounded in E_1. The principle of the proof of these properties is the same as in Theorem 2.2 but the technical details are much more complicated.

Let $\varphi_0 = \{u_0, u_1\} \in \mathscr{A}$. Then since $S(t)\mathscr{A} = \mathscr{A}$, $\forall t > 0$, φ_0 is on an orbit of solutions of (3.12) which belongs to $\mathscr{C}_b(\mathbb{R}; V \times H)$. In Lemma 3.5 below we show that such an orbit actually belongs to $\mathscr{C}_b(\mathbb{R}; D(A) \times V)$ with a bound depending only on $\mathscr{A}$. Thus $\varphi_0 \in E_1$ and $\mathscr{A}$ is bounded in E_1 since $\mathscr{A}$ is the union of such orbits. □

The theorem is then proved after we establish

Lemma 3.5. *For f, f' given in $\mathscr{C}_b(\mathbb{R}; H)$,*[1] *any solution $\{u, u'\}$ in $\mathscr{C}_b(\mathbb{R}; E_0)$ of equation* (3.12) *belongs to $\mathscr{C}_b(\mathbb{R}; E_1)$, and the norm of $\{u, u'\}$ in $\mathscr{C}_b(\mathbb{R}; E_1)$ is majorized by a bounded function of*

$$|f|_{\mathscr{C}_b(\mathbb{R}; H)} + |f'|_{\mathscr{C}_b(\mathbb{R}; H)} + |\{u, u'\}|_{\mathscr{C}_b(\mathbb{R}; E_0)}. \tag{3.47}$$

Proof. Let u be as indicated in the assumptions; we write $w = u'$ and by differentiation of (3.12) we see that

$$w'' + \alpha w' + Aw = f' - g'(u)u'. \tag{3.48}$$

[1] We assumed that f is independent of t, so that $f' = 0$, but we prove this lemma with a more general f.

By Lemma 3.3, $h = f' - g'(u)u'$ belongs to $\mathscr{C}_b(\mathbb{R}; V_{-1+\sigma_2})$, and by application of Proposition 1.4 we find that $\{w, w'\} \in \mathscr{C}_b(\mathbb{R}; V_{\sigma_2}, V_{-1+\sigma_2})$. In particular, $u'' \in \mathscr{C}_b(\mathbb{R}; V_{-1+\sigma_2})$ and since f, u', and $g'(u)$ are in $\mathscr{C}_b(\mathbb{R}; H)$, equation (3.12) shows that $Au \in \mathscr{C}_b(\mathbb{R}; V_{-1+\sigma_2})$, i.e., $u \in \mathscr{C}_b(\mathbb{R}; V_{1+\sigma_2})$ and its norm in this space is majorized by a bounded function of expression (3.47).

We consider the most complicated case where $n = 3$. The proof of Lemma 3.3 shows that $\sigma_2 = 1 - \gamma/2$, and thus $1 + \sigma_2 = 2 - (\gamma/2)$. If $2 - (\gamma/2) > \frac{3}{2}$ (i.e., if $0 \le \gamma < 1$), then $V_{2-(\gamma/2)} \subset H^{2-(\gamma/2)}(\Omega) \subset L^\infty(\Omega)$, and $u \in \mathscr{C}_b(\mathbb{R}; L^\infty(\Omega))$. Consequently, $g'(u) \in \mathscr{C}_b(\mathbb{R}; L^\infty(\Omega))$ and $h = f' - g'(u)u' \in \mathscr{C}_b(\mathbb{R}; H)$. Due to (3.48) and Proposition 1.4, $\{u', u''\} = \{w, w'\} \in \mathscr{C}_b(\mathbb{R}; V \times H)$, hence by (3.12), $Au \in \mathscr{C}_b(\mathbb{R}; H)$ and $u \in \mathscr{C}_b(\mathbb{R}; D(A))$ and the lemma is proved.

If $1 \le \gamma < 2$ we iterate the argument as many times as necessary until we obtain $g'(u)u' \in \mathscr{C}_b(\mathbb{R}; H)$. We start from $\{u, u'\} \in \mathscr{C}_b(\mathbb{R}; V_{1+\delta} \times V_\delta)$, $\delta = 1 - \gamma/2 \in]0, \frac{1}{2}]$. Since $V_{1+\delta} \subset H^{1+\delta}(\Omega)$ and $H^{1+\delta}(\Omega) \subset L^{6/(1-2\delta)}(\Omega)$, we see with (3.6) that $g'(u) \in \mathscr{C}_b(\mathbb{R}; L^{6/\gamma(1-2\delta)}(\Omega))$, whereas $u' \in \mathscr{C}_b(\mathbb{R}; L^{6/(3-2\delta)}(\Omega))$ since $V_\delta \subset H^\delta(\Omega) \subset L^{6/(3-2\delta)}(\Omega)$. Hence,

$$g'(u)u' \in \mathscr{C}_b(\mathbb{R}; L^q(\Omega)), \qquad q = 6/(3 - 2\delta + \gamma - 2\delta\gamma). \tag{3.49}$$

If $q \ge 2$, $g'(u)u'$ is in $\mathscr{C}_b(\mathbb{R}; H)$ and we conclude as above that $\{u', u''\} \in \mathscr{C}_b(\mathbb{R}; V \times H)$, $\{u, u'\} \in \mathscr{C}_b(\mathbb{R}; D(A) \times V)$. If $q < 2$, then we set $s = -\delta - [(2\delta - 1)/2]\gamma > 0$, and observe that $V_s \subset H^s(\Omega) \subset L^{q'}(\Omega)$ where q' is the conjugate exponent of q, $1/q' = 1 - 1/q$, and by duality this gives

$$L^q(\Omega) \subset V_{-s},$$

the injection being continuous. Therefore $g'(u)u' \in \mathscr{C}_b(\mathbb{R}; V_{-s})$. Using (3.48), (3.12), and Proposition 1.4, $\{u'', u'\} \in \mathscr{C}_b(\mathbb{R}; V_{1-s} \times V_{-s})$, $\{u, u'\} \in \mathscr{C}_b(\mathbb{R}; V_{2-s} \times V_{1-s})$. The situation is the same as above with δ replaced by $\bar{\delta} = 1 - s = 1 + \delta + [(2\delta - 1)/2]\gamma = 1 - \gamma/2 + \delta(1 + \gamma)$. As long as we reiterate, we define a sequence

$$\delta_0 = \delta, \qquad \delta_1 = \bar{\delta}, \qquad \delta_2, \ldots,$$

with

$$\delta_{m+1} = \left(1 - \frac{\gamma}{2}\right) + (1 + \gamma)\delta_m.$$

Since $\gamma \ge 1$, in the present case, the sequence δ_m converges to $+\infty$ as $m \to \infty$; thus $\delta_m \ge \frac{1}{2}$ for some m, in which case the corresponding value of q is ≥ 2 and we can conclude.

If $n \ge 4$, g' is bounded and we immediately find that $h = f' - g'(u)u'$ is in $\mathscr{C}_b(\mathbb{R}; H)$. If $n = 1$, $u \in \mathscr{C}_b(\mathbb{R}; V)$ implies $u \in \mathscr{C}_b(\mathbb{R}; L^\infty(\Omega))$ and $g'(u)u' \in \mathscr{C}_b(\mathbb{R}; H)$. If $n = 2$, $u \in \mathscr{C}_b(\mathbb{R}; V)$ implies $u \in \mathscr{C}_b(\mathbb{R}; L^r(\Omega))$, $\forall r < \infty$, $g'(u)u' \in \mathscr{C}_b(\mathbb{R}; L^q(\Omega))$, $q < 2$, q arbitrarily close to 2 and one single step of the iteration (3.49) is sufficient to obtain the desired result. □

4. An Abstract Wave Equation

Our aim in this section is to generalize the previous results to an abstract equation which includes, as a particular case, the sine–Gordon equation and the nonlinear wave equation of relativistic quantum mechanics. Many types of equations and boundary conditions can be treated with this setting. After the adequate treatment of the equation we briefly show how the general framework applies to the sine–Gordon and relativistic quantum mechanics equations with Dirichlet boundary conditions[1] and we then present some new examples, including an equation of fourth order in space variables which is related to vibrating beams.

In Section 4.1 we present the abstract setting and the assumptions, and we define the corresponding (semi)group, i.e., we show the existence, uniqueness, and regularity of solutions for all time. In Section 4.2 we prove the existence of an absorbing set and the existence of a global attractor. Section 4.3 is devoted to examples, and Section 4.4 contains a sketch of the proof of existence and uniqueness of solutions. In Sections 4.1 and 4.2 we will refer very often to Section 3 where the derivation of the stated results is very similar.

4.1. The Abstract Equation. The Group of Operators

The abstract equation that we consider is formally written as (2.4) or (3.12)

$$u'' + \alpha u' + Au + g(u) = f. \tag{4.1}$$

For the linear part the setting is the same as in Section 1 (or Section II.4). We are given two Hilbert spaces V and H, $V \subset H$, V dense in H, and

$$\text{The injection of } V \text{ in } H \text{ is compact.} \tag{4.2}$$

The scalar product and the norm in V and H are, respectively, denoted $((\cdot, \cdot))$, $\|\cdot\|$, $(\cdot, \cdot)$, $|\cdot|$. We identify H with its dual H', and H' with a dense subspace of the dual V' of V (norm $\|\cdot\|_*$); thus

$$V \subset H \subset V', \tag{4.3}$$

where the injections are continuous and each space is dense in the following one.

Let $a(u, v)$ be a bilinear continuous form on V which is symmetric and coercive

$$\exists \alpha_0 > 0, \qquad a(u, u) \geq \alpha_0 \|u\|^2, \qquad \forall u \in V. \tag{4.4}$$

With this form we associate the linear operator A from V into V' defined by

$$(Au, v) = a(u, v), \qquad \forall u, v \in V;$$

[1] The setting includes the sine–Gordon and the relativistic quantum mechanics equations with the Dirichlet boundary condition. It does not include ...

A is an isomorphism from V onto V' and it can also be considered as a self-adjoint unbounded operator in H with domain $D(A) \subset V$,

$$D(A) = \{v \in V, Av \in H\}.$$

Due to (4.2) there exists an orthonormal basis of H, $\{w_j\}_{j \in \mathbb{N}}$ which consists of eigenvectors of A,

$$\begin{cases} Aw_j = \lambda_j w_j, & \forall j, \\ 0 < \lambda_1 \le \lambda_2 \le \dots, & \lambda_j \to \infty \quad \text{as } j \to \infty. \end{cases} \tag{4.5}$$

Using this basis we define easily the powers A^s of A for $s \in \mathbb{R}$ as indicated in Section II.2; these operators are isomorphisms from their domain $D(A^s)$ onto H. We have $D(A^0) = H$, $D(A^{1/2}) = V$, $D(A^{-1/2}) = V'$, and we write

$$V_{2s} = D(A^s), \tag{4.6}$$

which is a Hilbert space for the scalar product and the norm

$$\begin{aligned} (u, v)_{2s} &= \sum_{j=1}^{\infty} \lambda_j^{2s}(u, w_j)(v, w_j), \\ |u|_{2s} &= \left\{ \sum_{j=1}^{\infty} \lambda_j^{2s}(u, w_j)^2 \right\}^{1/2}, \qquad \forall u, v \in D(A^s). \end{aligned} \tag{4.7}$$

The operator A^r is an isomorphism from $D(A^s)$ onto $D(A^{s-r})$, $\forall s, r \in \mathbb{R}$.

We then consider the nonlinear operator g in (4.1) for which we assume

g is a $\mathscr{C}^1$ bounded operator from V into H, Fréchet differentiable with differential g'. (4.8)

(i) g maps $D(A)$ into V and is Lipschitzian from the bounded sets of $D(A)$ into V and from the bounded sets of V into H.

(ii) There exists $\sigma_1 > 0$, and for every $R \ge 0$ there exists $c_0 = c_0(R)$ such that

$$\forall \varphi \in D(A), \qquad \|\varphi\| \le R, \qquad \|g(\varphi)\| \le c_0(R)(1 + |A\varphi|)^{1-\sigma_1}, \tag{4.9}$$

g is continuous from $L^\infty_{\mathrm{ws}}(0, T; V) \cap L^2(0, T; H)$ into $L^2_{\mathrm{w}}(0, T; H)$ (ws = weak-star topology, w = weak topology). (4.10)

We make on g' the following assumptions:

(i) g' is a bounded continuous mapping from V into $\mathscr{L}(V, H)$ and a bounded mapping from $D(A)$ into $\mathscr{L}(V_s, H)$, for some $s \in [0, 1[$ and from V in $\mathscr{L}(H, V_{-1+\sigma_2})$ for some $\sigma_2 > 0$.

(ii) There exists $0 < \delta < 1$, and for every R there exists $c_0' = c_0'(R)$ such that

$$\begin{aligned} &|g'(\xi) - g'(\eta)|_{\mathscr{L}(V,H)} \le c_0' \|\xi - \eta\|^\delta, \\ &\forall \xi, \eta \in V, \ \|\xi\| \le R, \ \|\eta\| \le R. \end{aligned} \tag{4.11}$$

We then formulate the following assumptions which express that g is a perturbation of a gradient G', and relate g and G:

There exists $G \in \mathscr{C}^1(V; \mathbb{R})$, $G(0) = 0$, and $p \in \mathscr{C}(V; H)$ such that $g(\varphi) = G'(\varphi) + p(\varphi)$, $\forall \varphi \in V$, G (resp. p) being bounded from V into $\mathbb{R}$ (resp. into H). (4.12)

$$\liminf_{\|\varphi\| \to \infty} \frac{G(\varphi)}{\|\varphi\|^2} \geq 0. \tag{4.13}$$

There exists $c_1 > 0$ such that

$$\liminf_{\|\varphi\| \to \infty} \frac{(\varphi, g(\varphi)) - c_1 G(\varphi)}{\|\varphi\|^2} \geq 0. \tag{4.14}$$

There exists $\sigma_3 > 0$ and a constant c_3 such that

$$|p(\varphi)| \leq c_3(1 + |G(\varphi)|)^{1/2 - \sigma_3}, \qquad \forall \varphi \in V. \tag{4.15}$$

As a consequence of (4.12)–(4.14) for every $\eta > 0$, there exist two constants C_η, C'_η such that

$$\begin{aligned} &\mathrm{G}(\varphi) + \eta \|\varphi\|^2 \geq -C_\eta, \\ &(\varphi, g(\varphi)) - c_1 G(\varphi) + \eta \|\varphi\|^2 \geq -C'_\eta, \qquad \forall \varphi \in V. \end{aligned}$$

In particular, there exist two constants κ_1, κ_2 such that

$$G(\varphi) + \frac{1}{8(1 + c_1)} \|\varphi\|^2 + \kappa_1 \geq 0, \tag{4.16}$$

$$(\varphi, g(\varphi)) - c_1 G(\varphi) + \tfrac{1}{8} \|\varphi\|^2 + \kappa_2 \geq 0, \qquad \forall \varphi \in V. \tag{4.17}$$

First we consider the initial-value problem for (4.1)

$$u(0) = u_0, \qquad u'(0) = u_1. \tag{4.18}$$

Theorem 4.1. *The hypotheses are the general hypotheses above on V, H, a, A, and we assume that g satisfies the hypotheses* (4.8) *and* (4.10)–(4.15).

Let $\alpha \in \mathbb{R}$ and let f, u_0, u_1, be given such that

$$f \in \mathscr{C}([0, T]; H), \qquad u_0 \in V, \quad u_1 \in H. \tag{4.19}$$

Then there exists a unique solution u of (4.1), (4.18) *such that*

$$u \in \mathscr{C}([0, T]; V), \qquad u' \in \mathscr{C}([0, T]; H). \tag{4.20}$$

If, furthermore, g satisfies (4.9) *and*

$$f' \in \mathscr{C}([0, T]; H), \qquad u_0 \in D(A), \quad u_1 \in V, \tag{4.21}$$

then u satisfies

$$u \in \mathscr{C}([0, T]; D(A)), \qquad u' \in \mathscr{C}([0, T]; V). \tag{4.22}$$

The proof of Theorem 4.1 (which includes Theorems 2.1 and 3.1) is sketched in Section 4.3. Since changing the sense of t amounts to changing α into $-\alpha$ in (4.1), Theorem 4.1 shows that (4.1), (4.18) can also be solved on $[-T, 0]$ (resp. $[-T, T]$) if f is defined on $[-T, 0]$ (resp. $[-T, T]$).

In the autonomous case, f is independent of t, $f(t) = f \in H$, $\forall t$, then u is defined for all time

$$u \in \mathscr{C}(\mathbb{R}; V), \qquad u' \in \mathscr{C}(\mathbb{R}; H), \tag{4.23}$$

if $u_0 \in V$, $u_1 \in H$, and

$$u \in \mathscr{C}(\mathbb{R}; D(A)), \qquad u' \in \mathscr{C}(\mathbb{R}; V), \tag{4.24}$$

if $u_0 \in D(A)$, $u_1 \in V$. We define, for every $t \in \mathbb{R}$, the mapping

$$S(t)\colon \{u_0, u_1\} \to \{u(t), u'(t)\}. \tag{4.25}$$

These operators map $E_0 = V \times H$ into itself and $E_1 = D(A) \times V$ into itself and they enjoy the *group properties*

$$\begin{cases} S(t + s) = S(t)S(s), & \forall s, t \in \mathbb{R}, \\ S(0) = I. \end{cases} \tag{4.26}$$

Hence $S(-t)$ is the inverse of $S(t)$ and these operators are one-to-one. We also have

Proposition 4.1. *When* $f(t) = f \in H$, *we can associate with equation* (4.1) *the group of operators* $S(t)$ *defined by* (4.25), α *fixed in* $\mathbb{R}$, $t \in \mathbb{R}$. *For every* t, $S(t)$ *is an isomorphism from* E_0 *onto* E_0.

PROOF. It suffices to prove that $S(t)$ is continuous in E_0, $\forall t > 0$, for $\alpha \in \mathbb{R}$. The proof is exactly that of Proposition 3.1; (3.25) which is a consequence of Lemma 3.1 is valid: g is locally Lipschitz, from V into H as asserted in Lemma 3.1, because of (4.11). □

4.2. Absorbing Sets and Attractors

We assume from now on that $\alpha > 0$ and we choose ε,

$$0 < \varepsilon \leq \varepsilon_0, \qquad \varepsilon_0 = \min\left(\frac{\alpha}{4}, \frac{\lambda_1}{2\alpha}\right). \tag{4.27}$$

Absorbing Sets in E_0

We take the scalar product in H of equation (4.1) with $v = u' + \varepsilon u$. After a formal computation, which can be rigorously justified by Lemma II.4.1, we find the same relation as (3.30), and (3.31) is valid too. The computation of $(g(u), v)$ is different here since we had $p = 0$ in Section 3: thanks to (4.12),

(4.15)–(4.17)

$$
\begin{aligned}
(g(u), v) &= (G'(u) + p(u), u') + \varepsilon(g(u), u) \\
&= \frac{d}{dt} G(u) + (p(u), v) - \varepsilon(p(u), u) + \varepsilon(g(u), u) \\
&\geq \frac{d}{dt} G(u) - c_3(1 + |G(u)|)^{1/2-\sigma_3}(|v| + \varepsilon|u|) \\
&\quad + \varepsilon c_1 G(u) - \frac{\varepsilon}{8}\|u\|^2 - \varepsilon\kappa_2 \\
&\geq \frac{d}{dt} G(u) - c_3(1 + |G(u)|)^{1/2-\sigma_3}(|v| + \varepsilon|u|) \\
&\quad + \varepsilon c_1 G(u) - \frac{\varepsilon}{4}\|u\|^2 - \varepsilon(\kappa_2 + c_1\kappa_1).
\end{aligned} \tag{4.28}
$$

With (3.31) and this inequality, (3.30) leads to

$$
\begin{aligned}
&\frac{1}{2}\frac{d}{dt}\{\|u\|^2 + |v|^2 + 2G(u)\} + \frac{\varepsilon}{4}\|u\|^2 + \frac{\alpha}{2}|v|^2 + \varepsilon c_1 G(u) \\
&\quad \leq \varepsilon(\kappa_2 + c_1\kappa_1) + (f, v) + c_3(1 + |G(u)|)^{1/2-\sigma_3}(|v| + \varepsilon|u|).
\end{aligned} \tag{4.29}
$$

But

$$
(f, v) \leq \frac{\alpha}{8}|v|^2 + \frac{2}{\alpha}|f|^2, \tag{4.30}
$$

$$
c_3(1 + |G(u)|)^{1/2-\sigma_3}(|v| + \varepsilon|u|) \leq \frac{\alpha}{8}|v|^2 + c_1'(1 + |G(u)|)^{1/2-\sigma_3} + \frac{\varepsilon}{16}\|u\|^2, \tag{4.31}
$$

where we have used

$$
\operatorname*{Inf}_{\substack{\varphi \in V \\ \varphi \neq 0}} \frac{\|\varphi\|^2}{|\varphi|^2} = \lambda_1 > 0. \tag{4.32}
$$

Due to (4.16)

$$
|G(\varphi)| \leq G(\varphi) + \frac{1}{4c_1}\|\varphi\|^2 + 2\lambda_1, \qquad \forall \varphi \in V, \tag{4.33}
$$

and thus

$$
\begin{aligned}
c_1'(1 + |G(u)|)^{1-2\sigma_3} &\leq c_1'\left(1 + 2\kappa_1 + G(\varphi) + \frac{1}{4c_1}\|\varphi\|^2\right)^{1-2\sigma_3} \\
&\leq \text{(with the Young inequality and since } \sigma_3 > 0) \\
&\leq \frac{\varepsilon}{16}\|u\|^2 + \frac{\varepsilon c_1}{4} G(u) + c_2'.
\end{aligned}
$$

Setting $\alpha_2 = \varepsilon/4$ we infer from (4.29)–(4.33) the following inequality similar to (3.34):

$$\begin{aligned}\frac{dy}{dt} + \alpha_2 y \le c_3' + \frac{4}{\alpha}|f|^2,&\\ y = \|u\|^2 + |v|^2 + 2G(u) + 2\kappa_1 \ge 0,&\\ c_3' = 2\varepsilon(\kappa_2 + c_1\kappa_1) + 2\alpha_2\kappa_1 + 2c_2'.&\end{aligned} \tag{4.34}$$

We deduce from (4.34) the analog of (3.35)–(3.38) and Lemma 3.2: assume that $\{u_0, u_1\}$ belongs to a bounded set $\mathscr{B}$ of E_0 and let

$$R = R(\mathscr{B}) = \sup_{\varphi=\{u_0,u_1\}\in\mathscr{B}} \{\|u_0\|^2 + |u_1 + \varepsilon u_0|^2 + 2G(u_0) + 2\kappa_1\} < +\infty, \tag{4.35}$$

We set $\mu_0^2 = (1/\alpha_2)(c_3' + (4/\alpha)|f|^2)$ and fix $\mu_0' > \mu_0$. Then, for $t \ge t_0(R, \mu_0')$,

$$t_0(R, \mu_0') = \frac{1}{\alpha_2}\log\frac{R}{(\mu_0')^2 - \mu_0^2}, \tag{4.36}$$

We have $y(t) \le \mu_0'$ and by (1.32)

$$\begin{aligned}\|u(t)\|^2 + |u'(t)|^2 &\le (1 + \varepsilon\lambda_1^{-1/2})(\|u(t)\|^2 + |u'(t) + \varepsilon u(t)|^2)\\ &\le (1 + \varepsilon\lambda_1^{-1/2})y(t) \le (1 + \varepsilon\lambda_1^{-1/2})\mu_0'.\end{aligned}$$

Setting $\rho_0 = (1 + \varepsilon\lambda_1^{-1/2})\mu_0'$ we are able to state

Lemma 4.1. *The ball of E_0, $\mathscr{B}_0 = B_{E_0}(0, \rho_0)$, centered at 0 of radius ρ_0 is an absorbing set in E_0 for the (semi)group $S(t)$. If $\{u_0, u_1\}$ belongs to a bounded set $\mathscr{B}$ of E_0, then $S(t)\mathscr{B} \subset \mathscr{B}_0$ for $t \ge t_0$, t_0 given by* (4.36), (4.35).

Hypothesis I.(1.13)

Lemma 3.4 applies here without any modification. The two properties of g needed in the proof of Lemma 3.4 are that:

(i) g is bounded continuous from V into H;
(ii) g' satisfies the properties in Lemma 3.3.

Here (i) is part of assumption (4.8), whereas the properties in Lemma 3.3 are exactly the same as those in assumption (4.11).

Maximal Attractors

We have the analog of Theorem 3.2:

Theorem 4.2. *The hypotheses are the general hypotheses in Section* 4.1 *on V, H, a, A, and we assume that $f \in H$ and g satisfies the hypotheses* (4.8) *and* (4.10)–(4.15).

The abstract evolution equation (4.1) *defines a dynamical system which* *possesses a global attractor $\mathscr{A}$ compact, connected, and maximal in E_0. $\mathscr{A}$ attracts the bounded sets of E_0 and $\mathscr{A}$ is also maximal among the functional-invariant sets bounded in E_0.*

The theorem is a direct consequence of Theorem I.1.1, the assumptions of which have been verified, namely I.(1.1), I.(1.4), I.(1.13).

We can obtain supplementary information on $\mathscr{A}$ after we make another assumption

For f, f' given in $\mathscr{C}_b(\mathbb{R}; H)$, for any solution u of (4.1), such that $\{u, u'\} \in \mathscr{C}_b(\mathbb{R}; E_0)$ we have $\{u, u'\} \in \mathscr{C}_b(\mathbb{R}; E_1)$, and the norm of $\{u, u'\}$ in $\mathscr{C}_b(\mathbb{R}; E_1)$ is majorized by a bounded function of

$$|f|_{L^\infty(\mathbb{R}; H)} + |f'|_{L^\infty(\mathbb{R}; H)} + |\{u, u'\}|_{L^\infty(\mathbb{R}; E_0)}. \tag{4.37}$$

Then

Theorem 4.3. *The hypotheses are those of Theorem* 4.2 *and* (4.37). *Then $\mathscr{A}$ is included and bounded in E_1.*

This result follows immediately from (4.37), as in the proof of Theorem 3.2 where this result was a consequence of Lemma 3.5 (which proves precisely (4.37) for the equation considered in Section 3).

Absorbing Sets in E_1

Although this is not of direct use here we can prove the existence of an absorbing set in E_1 for the (semi)group $S(t)$.[1]

We take the scalar product in H of (4.1) with $Av = Au' + \varepsilon Au$, ε as in (4.27), and we obtain a relation similar to (2.27)

$$\frac{1}{2}\frac{d}{dt}(|Au|^2 + \|v\|^2) + \varepsilon|Au|^2 + (\alpha - \varepsilon)\|v\|^2 - \varepsilon(\alpha - \varepsilon)(Au, v) + ((g(u), v))$$
$$= (f, Av). \tag{4.38}$$

For $0 < \varepsilon \leq \varepsilon_0$, we have

$$\varepsilon|Au|^2 + (\alpha - \varepsilon)\|v\|^2 - \varepsilon(\alpha - \varepsilon)((u, v)) \geq \frac{\varepsilon}{2}|Au|^2 + \frac{\alpha}{2}\|v\|^2. \tag{4.39}$$

Also

$$(f, Av) = \frac{d}{dt}(f, Au) + \varepsilon(f, Au)$$
$$\leq \frac{d}{dt}(f, Au) + \frac{\varepsilon}{8}|Au|^2 + 2\varepsilon|f|^2. \tag{4.40}$$

[1] The following computations are also used in the proof of Theorem 4.1 that is given below in Section 4.4.

If $\{u_0, u_1\}$ belongs to a bounded set $\mathscr{B}$ of E_1, then $\mathscr{B}$ is also bounded in E_0 and, for $t \geq t_0(\mathscr{B})$, we have

$$\|u(t)\|^2 + |u'(t)|^2 \leq \rho_0^2,$$

t_0, ρ_0 given in Lemma 4.1.

Using (4.9) with $R = \rho_0$ we majorize $\|g(u(t))\|$, for $t \geq t_0$, as follows:

$$\|g(u(t))\| \leq c_0(\rho_0)(1 + |Au(t)|)^{1-\sigma_1}.$$

Then

$$\begin{aligned} |((g(u), v))| &\leq c_0(1 + |Au(t)|)^{1-\sigma_1}\|v\| \\ &\leq \frac{\alpha}{4}\|v\|^2 + \frac{c_0^2}{\alpha}(1 + |Au|)^{2(1-\sigma_1)} \\ &\leq \text{(with the Young inequality)} \\ &\leq \frac{\alpha}{4}\|v\|^2 + \frac{\varepsilon}{8}|Au|^2 + c_1'. \end{aligned} \tag{4.41}$$

Inserting all these inequalities in (4.38) we obtain, setting $\alpha_1 = \varepsilon/2$,

$$\frac{d}{dt}(|Au|^2 + \|v\|^2 - 2(f, Au)) + \alpha_1(|Au|^2 + \|v\|^2) \leq 2c_1' + 4\varepsilon|f|^2 \quad \text{for } t \geq t_0, \tag{4.42}$$

$$\frac{d}{dt}(|Au - f|^2 + \|v\|^2) + \frac{\alpha_1}{2}(|Au - f|^2 + \|v\|^2) \leq 2c_1' + (4\varepsilon + \alpha_1)|f|^2 \quad \text{for } t \geq t_0. \tag{4.43}$$

Hence

$$\begin{aligned} |Au(t) - f|^2 + \|v(t)\|^2 \leq {} & (|Au(t_0) - f|^2 + \|v(t_0)\|^2)\cdot\exp\left(-\frac{\alpha_1}{2}(t - t_0)\right) \\ & + \frac{\rho_1^2}{c_0}\left(1 - \exp\left(-\frac{\alpha_1}{2}(t - t_0)\right)\right) \quad \text{for } t \geq t_0, \end{aligned} \tag{4.44}$$

$$\rho_1^2 = \frac{2c_0}{\alpha_1}(2c_1' + (4\varepsilon + \alpha_1)|f|^2),$$

c_0 as in (1.32). Using (1.33) we deduce from (4.44) that

$$\begin{aligned} |Au(t) - f|^2 + \|u'(t)\|^2 \leq {} & c_0^2(|Au(t_0) - f|^2 + \|u'(t_0)\|^2)\cdot\exp\left(-\frac{\alpha_1}{2}(t - t_0)\right) \\ & + \rho_1^2\left(1 - \exp\left(-\frac{\alpha_1}{2}(t - t_0)\right)\right), \qquad t \geq t_0. \end{aligned} \tag{4.45}$$

In conclusion,

The ball of E_1, $\mathscr{B}_1 = B_{E_1}((A^{-1}f, 0), \rho_1'))$, centered at $\{A^{-1}f, 0\}$ of radius $\rho_1' > \rho_1$ is absorbing in E_1 for $S(t)$. (4.46)

If $\{u_0, u_1\}$ belongs to a bounded set $\mathscr{B}$ of E_1, then $\mathscr{B}$ is bounded in E_0 and we define $t_0(\mathscr{B})$ as in Lemma 4.1. The time $t_1 = t_1(\mathscr{B})$ after which $S(t)\mathscr{B} \subset \mathscr{B}_1$ is of the form $t_0 + t_1'$, $t_1' = t_1'(\mathscr{B}, \rho_1')$,

$$\begin{aligned} t_1' &= \frac{2}{\alpha_1} \log \frac{c_0^2 R_1^2}{(\rho_1')^2 - \rho_1^2}, \\ R_1 &= \sup_{\{u(t_0), u'(t_0)\} \in S(t_0)\mathscr{B}} (|Au(t_0) - f|^2 + \|u'(t_0)\|^2). \end{aligned} \tag{4.47}$$

4.3. Examples

Examples of Sections 2 and 3

As indicated above, the sine–Gordon equation and the nonlinear wave equation of relativistic quantum mechanics associated with the Dirichlet boundary conditions are particular cases of (4.1). The necessary assumptions (4.8)–(4.15) and (4.37) have already been proved in Sections 2 and 3 or are very easy to prove ($p = 0$ in (4.12)–(4.15)). We note that some of the assumptions have not yet been used and will be needed only in Chapter VI (see Section VI.6). Also, for the wave equation of Section 3, assumption (4.11)(ii), which will be used only subsequently, necessitates a strengthening of assumption (3.6)

$$\begin{cases} |g''(s)| \le c(1 + |s|^{\gamma - 1}), & \gamma \text{ as in (3.6) if } n = 1, 2, 3, \\ g'' \text{ bounded if } n \ge 4. \end{cases} \tag{4.48}$$

With this assumption on g'' and the Sobolev embeddings, it is easy to verify (4.11)(ii); this is left as an exercise for the reader

We now describe some other examples, including examples where p does not vanish, i.e., the nonlinear term is not a gradient which is important as far as the large-time dynamics is concerned (see Chapter VII). In all the examples, Ω is an open bounded set of $\mathbb{R}^n$ with boundary Γ.

EXAMPLE 4.1. We consider the wave equation (3.1) with a lower-order term ηu, $\eta > 0$. We look for a u solution of

$$\frac{\partial^2 u}{\partial t^2} + \alpha \frac{\partial u}{\partial t} - \Delta u + \eta u + g(u) = f \quad \text{in } \Omega \times \mathbb{R}_+. \tag{4.49}$$

The assumptions on g are the same as in Section 3 (or in Section 2: $g(u) = \beta \sin u$ satisfies the assumptions on g in Section 3). We supplement (4.49) with one of the boundary conditions:

Dirichlet: $u = 0$ on $\Gamma \times \mathbb{R}_+$;
Neumann: $\partial u/\partial \mathbf{v} = 0$ on $\Gamma \times \mathbb{R}_+$;
Space-periodicity, $\Omega = \prod_{i=1}^{n}]0, L_i[$, and u is Ω-periodic.

The functional setting of the equation is clear: $H = L^2(\Omega)$, $V =$, respectively, $H_0^1(\Omega)$, $H^1(\Omega)$, $H_{\text{per}}^1(\Omega)$, $D(A) =$, respectively, $H_0^1(\Omega) \cap H^2(\Omega)$, $\{v \in H^2(\Omega), \partial v/\partial \mathbf{v} = 0 \text{ on } \Gamma\}$, $H_{\text{per}}^2(\Omega)$, $A\varphi = -\Delta\varphi + \eta\varphi$, $\forall \varphi \in D(A)$, $g(\varphi) = g \circ \varphi$, $G(\varphi) = \int_\Omega G(\varphi(x))\, dx$, where $G(s) = \int_0^s g(t)\, dt$, $p \equiv 0$.

Due to the adjunction of the term ηu, the coercivity condition (4.4) is satisfied and all the other hypotheses are checked as in Section 3.

Example 4.2. We consider a *system* of sine–Gordon equations occurring in the Josephson junctions (see M. Levi [1]). The unknown function $u = \{u_1, u_2\}$ is a vector. It satisfies ($k \geq 0$)

$$\begin{cases} \dfrac{\partial^2 u_1}{\partial t^2} + \dfrac{\partial u_1}{\partial t} - \Delta u_1 + \sin u_1 + k(u_1 - u_2) = f_1, \\[2ex] \dfrac{\partial^2 u_2}{\partial t^2} + \dfrac{\partial u_2}{\partial t} - \Delta u_2 + \sin u_2 + k(u_2 - u_1) = f_2, \end{cases} \tag{4.50}$$

$$u_i(x, t) = 0 \quad \text{on } \Gamma \times \mathbb{R}_+, \qquad i = 1 \text{ or } 2. \tag{4.51}$$

We take $H = L^2(\Omega)^2$, $V = H_0^1(\Omega)^2$,

$$D(A) = (H_0^1(\Omega) \cap H^2(\Omega))^2,$$

$$A\{\varphi_1, \varphi_2\} = \{-\Delta\varphi_1 - \Delta\varphi_2\},$$

$$g(\varphi_1, \varphi_2) = \{\sin \varphi_1 + k(\varphi_1 - \varphi_2), \sin \varphi_2 + k(\varphi_2 - \varphi_1)\},$$

$$G(\varphi_1, \varphi_2) = \int_\Omega \left\{\cos \varphi_1(x) + \cos \varphi_2(x) + \frac{k}{2}(\varphi_1(x) - \varphi_2(x))^2\right\} dx, \quad p \equiv 0.$$

The assumptions are checked as in Section 2.

Example 4.3. This is again a system much like the sine–Gordon equations, but in this case p *does not vanish*, i.e., *the nonlinear term is not a gradient*.

The equations read

$$\begin{cases} \dfrac{\partial^2 u_1}{\partial t^2} + \dfrac{\partial u_1}{\partial t} - \Delta u_1 + \sin(u_1 + u_2) = f_1, \\[2ex] \dfrac{\partial^2 u_2}{\partial t^2} + \dfrac{\partial u_2}{\partial t} - \Delta u_2 + \sin(u_1 - u_2) = f_2. \end{cases} \tag{4.52}$$

$$u_i = 0 \qquad \text{on} \quad \Gamma \times \mathbb{R}_+. \tag{4.53}$$

We choose, as in Example 4.2, $H = L^2(\Omega)^2$, $V = H_0^1(\Omega)^2$,

$$D(A) = (H_0^1(\Omega) \cap H^2(\Omega))^2,$$

$$A\{\varphi_1, \varphi_2\} = \left\{-\Delta\varphi_1 - \frac{\lambda_1'}{2}\varphi_1, -\Delta\varphi_2 - \frac{\lambda_1'}{2}\varphi_2\right\},$$

$$g(\{\varphi_1, \varphi_2\}) = \left\{\frac{\lambda_1'}{2}\varphi_1, \frac{\lambda_1'}{2}\varphi_2\right\},$$

$$G(\{\varphi_1, \varphi_2\}) = \frac{\lambda_1'}{4}\int_\Omega (\varphi_1^2 + \varphi_2^2)\,dx,$$

$$p(\{\varphi_1, \varphi_2\}) = \{\sin(\varphi_1 + \varphi_2), \sin(\varphi_1 - \varphi_2)\},$$

where λ_1' is the first eigenvalue of the Dirichlet problem in Ω.

There is no new difficulty in checking assumptions (4.8)–(4.15) and (4.37).

Example 4.4. We give now an example involving a higher (fourth)-order operator in the space variables.

The function u is a scalar function which is a solution of

$$\frac{\partial^2 u}{\partial t^2} + \alpha\frac{\partial u}{\partial t} + \Delta^2 u + |u|^\gamma u = f \quad \text{in } \Omega \times \mathbb{R}_+, \tag{4.54}$$

$$u = \frac{\partial u}{\partial \mathbf{v}} = 0 \quad \text{on } \Gamma \times \mathbb{R}_+, \tag{4.55}$$

where $\gamma > 0$ and $\Omega \subset \mathbb{R}^n$, $n \leq 3$, for simplicity. For the functional setting, we write $H = L^2(\Omega)$, $V = H_0^2(\Omega)$,

$$D(A) = H_0^2(\Omega) \cap H^4(\Omega) = \left\{v \in H^4(\Omega),\ v = \frac{\partial v}{\partial \mathbf{v}} = 0 \text{ on } \Gamma\right\},$$

$$g(\varphi) = |\varphi|^\gamma \varphi,$$

$$G(\varphi) = \frac{1}{\gamma + 2}\int_\Omega |\varphi|^{\gamma+2}\,dx, \qquad p = 0.$$

In dimension $n \leq 3$, as a consequence of the Sobolev embeddings, $H_0^2(\Omega) \subset \mathscr{C}(\bar{\Omega})$ and the necessary properties of g are easily checked.

Example 4.5. This example and the next one are borrowed from nonlinear elasticity (see, for instance, P. Germain [1]). They are related to vibration of bars and they involve a fourth-order operator in one-space dimension.

The function u is defined on $\mathbb{R} \times \mathbb{R}_+$ and satisfies

$$\frac{\partial^2 u}{\partial t^2} + \alpha\frac{\partial u}{\partial t} + \frac{\partial^4 u}{\partial x^4} + u + g(u) = f, \tag{4.56}$$

$$u(x + L, t) = u(x, t), \qquad x \in \mathbb{R}, \quad t \geq 0, \tag{4.57}$$

where $L > 0$ is given. As in all problems involving a space-periodicity boundary condition it suffices to consider the restriction of u to the periodicity interval $\Omega =]0, L[$, in which case (4.57) is replaced by

$$\frac{\partial^j u}{\partial x^j}(L, t) = \frac{\partial^j u}{\partial x^j}(0, t), \qquad j = 0, \ldots, 3, \quad t \geq 0. \tag{4.58}$$

The nonlinear term g is defined as follows. Let

$$\varphi = \sum_{k \in \mathbb{Z}} \varphi_k \exp\left(2i\frac{kx}{L}\right)$$

denote the Fourier series expansion of a function φ in $L^2(\Omega)$. Then

$$g(\varphi)_k = \left(\sum_{m \in \mathbb{Z}} |m|^\gamma |\varphi_k|\right)^\delta |k|^{2\gamma} \varphi_k,$$

$\delta > 0$, i.e., $g(\varphi)$ is proportional to

$$\left(\int_0^L |D^\gamma \varphi|^2 \, dx\right)^\delta D^{2\gamma}\varphi,$$

where D is the square root of $-\Delta$.

We set $H = L^2(0, L)$, $V = H^2_{\text{per}}(]0, L[)$,

$$D(A) = H^4_{\text{per}}(]0, L[),$$

and for $\varphi \in D(A)$,

$$A\varphi = \frac{\partial^4 \varphi}{\partial x^4} + \varphi,$$

$$G(\varphi) = \frac{1}{\delta + 2}\left(\sum_{m \in \mathbb{Z}} |m|^\gamma |\varphi_m|^2\right)^\delta, \qquad p = 0.$$

For $0 \leq \gamma < 1$, the necessary assumptions are satisfied and Theorems 4.1–4.3 apply.

This model, with $\gamma = 1$ instead of $0 \leq \gamma < 1$, has been proposed by S. Woinowsky-Krieger [1] for vibrating bars.

EXAMPLE 4.6. This last example describes plan oscillations of a thin elastic rod with fixed ends in the Kirchhoff model. The existence of a maximal attractor for the corresponding dynamical system has been proved by V.S. Stepanov [1]. The results obtained by application of Theorems 4.2 and 4.3 improve those of V.S. Stepanov.

The vector function u defined on $\Omega \times \mathbb{R}_+$, $\Omega =]0, L[$, $L > 0$, with values in $\mathbb{R}^2$ is a solution of the following system:

$$\frac{\partial^2 u}{\partial t^2} + \alpha\frac{\partial u}{\partial t} + \frac{\partial^4 u}{\partial x^4} - 2\frac{\partial}{\partial x}\left\{\left(\left|\frac{\partial}{\partial x}(u + b)\right|^2 - 1\right)\frac{\partial}{\partial x}(u + b)\right\} = f \quad \text{in } \Omega \times \mathbb{R}_+, \tag{4.59}$$

where b and f are given vector functions on $(0, L)$, b sufficiently regular, $f \in L^2(0, L)$, as in Theorems 4.1–4.3. We supplement (4.59) with the boundary conditions

$$u(0, t) = u(L, t) = \frac{\partial u}{\partial x}(0, t) = \frac{\partial u}{\partial x}(L, t) = 0, \qquad t \geq 0. \tag{4.60}$$

We set $H = L^2(\Omega)$, $V = H_0^2(\Omega)$,

$$D(A) = H_0^2(\Omega) \cap H^4(\Omega),$$

$$A\varphi = +\frac{\partial^4 \varphi}{\partial x^4} - \frac{\lambda_1}{2}\varphi,$$

$$g(\varphi) = -2\frac{\partial}{\partial x}\left\{\left(\left|\frac{\partial}{\partial x}(\varphi + b)\right|^2 - 1\right)\frac{\partial}{\partial x}(\varphi + b)\right\},$$

$$G(\varphi) = \int_0^L \left(\left|\frac{\partial}{\partial x}(\varphi + b)\right|^2 - 1\right)^2 dx, \qquad p = 0.$$

The assumptions (4.8)–(4.15) and (4.37) are readily checked.

4.4. Proof of Theorem 4.1 (Sketch)

The principle of the proof is classical. We implement a Faedo–Galerkin method using as a basis the eigenfunctions w_j of A (see (4.5)). For each m we look for an approximate solution u_m of the form

$$u_m(t) = \sum_{i=1}^{m} g_{im}(t) w_i \tag{4.61}$$

satisfying

$$\left(\frac{d^2 u_m}{dt^2}, w_j\right) + \alpha\left(\frac{du_m}{dt}, w_j\right) + a(u_m, w_j) + (g(u_m), w_j) = (f, w_j), \qquad j = 1, \ldots, m, \tag{4.62}$$

$$u_m(0) = P_m u_0, \qquad u_m'(0) = P_m u_1, \tag{4.63}$$

where P_m is the orthogonal projector in H (or V, or any space V_s) onto the space spanned by $w_1, \ldots, w_m$. Since A and P_m commute, equation (4.62) is also equivalent to

$$\frac{d^2 u_m}{dt^2} + \alpha\frac{du_m}{dt} + Au_m + P_m g(u_m) = P_m f. \tag{4.64}$$

The existence and uniqueness of u_m on some interval $[0, T_m[$ is elementary and then $T_m = +\infty$, because of the a priori estimates that we obtain for u_m. A first energy equality is obtained by multiplying (4.62) by $g_{jm}' + \varepsilon g_{jm}$, ε as in (4.27), and summing these relations for $j = 1, \ldots, m$. We obtain (3.30) with u

replaced by u_m.[1] We repeat on (3.30) the computations made in Section 4.1, (4.28)–(4.33) and we obtain exactly (4.34) in which u is replaced by u_m. We easily infer from (4.34) that

$$\{u_m, u_m'\} \text{ remains in a bounded set of } L^\infty(0, T; V \times H) \quad \text{as } m \to \infty. \tag{4.65}$$

Thanks to (4.65) we can extract a subsequence, still denoted m, such that

$$\begin{cases} u_m \to u & \text{in } L^\infty(0, T; V) \text{ weak-star}, \\ u_m' \to u' & \text{in } L^\infty(0, T; H) \text{ weak-star} \quad \text{as } m \to \infty. \end{cases} \tag{4.66}$$

Thanks to a classical compactness theorem (see, for instance, R. Temam [1] (Theorem 2.3, Chap. III)), (4.66) implies

$$u_m \to u \quad \text{in } L^2(0, T; H) \text{ strongly}. \tag{4.67}$$

Due to (4.10), $g(u_m)$ converges to $g(u)$ weakly in $L^2(0, T; V)$. It is then easy to pass to the limit in (4.62), (4.63) and we find that u is a solution of (4.1), (4.18) such that

$$u \in L^\infty(0, T; V), \qquad u' \in L^\infty(0, T; H).$$

The continuity properties

$$u \in \mathscr{C}([0, T]; V), \qquad u' \in \mathscr{C}([0, T]; H),$$

are proved with the methods indicated in Sections II.3 and II.4. The uniqueness is implicitly proved in Proposition 4.1 and relies on (3.27) (here we use the assumption (4.11) which is necessary for (3.25)).

It remains to show the supplementary properties (4.22) when conditions (4.21) are satisfied. These properties are consequences of another energy equality obtained as follows:[2] we multiply (4.62) by $\lambda_j(g_{jm}' + \varepsilon g_{jm}), j = 1, \dots, m$, and add the resulting relations. Since

$$\begin{cases} \lambda_j(\varphi, w_j) = (\varphi, Aw_j) = ((\varphi, w_j)), & \forall \varphi \in V, \\ \lambda_j a(\varphi, w_j) = \lambda_j(A\varphi, w_j) = (A\varphi, Aw_j), & \forall \varphi \in D(A), \end{cases} \tag{4.68}$$

we obtain exactly (4.38) with u, v replaced by u_m, v_m, $v_m = u_m' + \varepsilon u_m$, ε as in (4.27). The computations leading from (4.38) to (4.43) are valid with the following modifications:

We are interested in the interval $[0, T]$ instead of $[t_0, \infty[$; hence we use (4.9) with R replaced by the bound of u_m in $\mathscr{C}([0, T]; V)$ given by (4.65).

We replace (4.40) by

$$\begin{aligned} (f, Av) &= (f, Au') + \varepsilon(f, Au) \\ &= \frac{d}{dt}(f, Au) + (\varepsilon f - f', Au), \end{aligned}$$

[1] Here we could take $\varepsilon = 0$ since we are only interested in a finite interval of time, $[0, T]$. We choose $\varepsilon > 0$ to avoid repeating the computations following (3.30) and (4.28).

[2] This equality is close to that used in the demonstration of the existence of an absorbing set for $S(t)$ in E_1.

and then (4.43) takes the form

$$\frac{d}{dt}(|Au_m - f|^2 + \|v_m\|^2) + \frac{\alpha_1}{2}(|Au_m - f|^2 + \|v_m\|^2)$$

$$\leq 2c_1' + (4\varepsilon + \alpha_1)\left|f - \frac{1}{\varepsilon}f'\right|^2_{L^\infty(0,T;H)} \qquad \text{for } t \in [0, T]. \tag{4.69}$$

It follows from (4.69) and the Gronwall lemma that

$$\{u_m, u_m'\} \text{ remains in a bounded set of } L^\infty(0, T; D(A) \times V) \quad \text{as } m \to \infty. \tag{4.70}$$

Thus, with (4.66)

$$u \in L^\infty(0, T; D(A)), \qquad u' \in L^\infty(0, T; V), \tag{4.71}$$

$$\begin{cases} u_m \to u & \text{in } L^\infty(0, T; D(A)) \text{ weak-star,} \\ u_m' \to u' & \text{in } L^\infty(0, T; V) \text{ weak-star.} \end{cases} \tag{4.72}$$

We conclude the proof by showing that

$$u \in \mathscr{C}([0, T]; D(A)), \qquad u' \in \mathscr{C}([0, T]; V),$$

usng the methods in Sections II.4 and II.3.

Remark 4.1. Parts of the results in Sections 2, 3, and 4, follow J.M. Ghidaglia and R. Temam [1]. Existence of attractors for dissipative wave equations is also derived in J. Hale [2], A. Haraux [1], and A.V. Babin and M.I. Vishik [2].

5. The Ginzburg–Landau Equation

This section is devoted to the study of a Schrödinger equation with a nonlinear term, the Ginzburg–Landau equation. This equation governs the finite amplitude evolution of instability waves in a large variety of dissipative systems which are close to criticality. Various forms of the Ginzburg–Landau equation arise, for instance, in hydrodynamic instability theory: the development of Tollmien–Schlichting waves in plane Poiseuille flows, the nonlinear growth of convection rolls in the Rayleigh–Bénard problem, and the appearance of Taylor vortices in the flow between counterrotating circular cylinders: see P.J. Blennerhassett [1], H.T. Moon, P. Huerre, and L.G. Redekopp [1], [2], A.C. Newell and J.A. Whitehead [1], J.T. Stuart and R.C. Di Prima [1]. The equation also arises in the study of chemical systems governed by reaction–diffusion equations. It has been shown by Y. Kuramoto and T. Tsuzuki [1], [3] that the perturbation concentration $c - c^*$, away from a steady state solution c^*, also satisfies the Ginzburg–Landau equation.

The equation that we study contains, of course, a dissipative term and is presented in Section 5.1 where its functional setting is also described. Since the unknown function is complex valued we are naturally led here to introduce complex-valued spaces; this is the only occasion in this book in which such spaces appear. Section 5.2 contains the proofs of existence of absorbing sets and of a global attractor.

5.1. The Equation and Its Mathematical Setting

We denote by Ω an open bounded set of $\mathbb{R}^n$, $n = 1$ or 2, with a boundary Γ sufficiently regular. We consider the Ginzburg–Landau equation where the unknown u is a complex-valued function defined on $\Omega \times \mathbb{R}_+$

$$\frac{\partial u}{\partial t} - (\lambda + i\alpha)\Delta u + (\kappa + i\beta)|u|^2 u - \gamma u = 0. \tag{5.1}$$

The parameters λ, α, β, γ, κ are real numbers and, as appears below, the conditions

$$\lambda > 0, \qquad \kappa > 0, \tag{5.2}$$

render equation (5.1) dissipative. In Section 6 we shall consider the case where $\lambda = \kappa = 0$, in which case equation (5.1) becomes the nonlinear Schrödinger equation (and $-\gamma = \alpha > 0$ ensures mild dissipativity).

The equation will be supplemented with one of the following usual boundary conditions:

the Dirichlet boundary condition

$$u = 0 \quad \text{on } \Gamma \times \mathbb{R}_+; \tag{5.3a}$$

the Neumann boundary condition

$$\frac{\partial u}{\partial \mathbf{v}} = 0 \quad \text{on } \Gamma \times \mathbb{R}_+, \tag{5.3b}$$

where $\mathbf{v}$ is the unit outward normal on Γ;

space-periodicity, in which case,

$$\Omega =]0, L[\quad (n = 1) \quad \text{or} \quad]0, L_1[\times]0, L_2[\quad (n = 2) \quad \text{and}$$

$$u \text{ is } \Omega\text{-periodic.} \tag{5.3c}$$

Below we will give a unified treatment for these three cases.

For an initial-value problem we also provide the initial value of u:

$$u(x, 0) = u_0(x), \qquad x \in \Omega. \tag{5.4}$$

For the mathematical setting we introduce complex Sobolev spaces. In general, we denote in this section by $\mathbb{X}$, $\mathbb{Y}$, ..., the complexified space of a function space X, Y, ... For example $\mathbb{L}^2(\Omega)$ is the complexified space of $L^2(\Omega)$; we denote by $(\cdot, \cdot)$ and $|\cdot|_{L^2(\Omega)}$[1] the scalar product and the norm in either $L^2(\Omega)$ or $\mathbb{L}^2(\Omega)$. Hence, if $u \in \mathbb{L}^2(\Omega)$, then $u = \{u_1, u_2\}$, $u_j \in L^2(\Omega)$, $j = 1$,

[1] The norm on L^2 spaces is denoted here by $|\cdot|_{L^2(\Omega)}$, to avoid any confusion with the modulus $|\cdot|$ which is frequently used. If z is a complex number or a complex-valued function we denote by Re z, Im z, $|z|$, $\bar{z}$, its real or imaginary part and its modulus or conjugate.

2, and

$$|u|_{L^2} = \{|u_1|_{L^2}^2 + |u_2|_{L^2}^2\}^{1/2}.$$

If $u = u_1 + iu_2$, $v = v_1 + iv_2$ are in $\mathbb{L}^2(\Omega)$,

$$(u, v) = \{(u_1, v_1) + (u_2, v_2)\} + i\{(u_2, v_1) - (u_1, v_2)\}.$$

The space $\mathbb{H}^1(\Omega)$ is the complexified space of $H^1(\Omega)$. We denote by $((\cdot, \cdot))_1$, $\|\cdot\|_1$ the scalar product and the norm on either $\mathbb{H}^1(\Omega)$ or $H^1(\Omega)$ and for $u, v \in \mathbb{H}^1(\Omega)$ we write

$$((u, v)) = \sum_{i=1}^{n} (D_i u, D_i v), \qquad \|u\| = \{((u, u))\}^{1/2}.$$

Hence

$$((u, v))_1 = (u, v) + ((u, v)), \qquad \|u\|_1 = \{|u|_{L^2}^2 + \|u\|^2\}^{1/2}.$$

For the functional setting of (5.1)–(5.5) we choose $H = \mathbb{L}^2(\Omega)$, and

$V = \mathbb{H}_0^1(\Omega)$, $\quad D(A) = \mathbb{H}_0^1(\Omega) \cap \mathbb{H}^2(\Omega)$ in case (5.3a),

$V = \mathbb{H}^1(\Omega)$, $\quad D(A) = \{v \in \mathbb{H}^2(\Omega), \partial v/\partial \mathbf{v} = 0 \text{ on } \Gamma\}$ in case (5.3b),

$V = \mathbb{H}_{\text{per}}^1(\Omega)$, $\quad D(A) = \mathbb{H}_{\text{per}}^2(\Omega)$ in case (5.3c).

In the three cases, $Au = -\Delta u$. For every $\eta > 0$, $A + \eta I$ is an isomorphism from V onto its dual V' or from $D(A)$ (endowed with the $\mathbb{H}^2$ norm) onto H. Due to the Poincaré inequality the same is also true for $\eta = 0$ in case (5.3a) (see Section II.2).

We denote by w_j and λ_j the eigenvectors (orthonormal in H) and eigenvalues of A in H

$$\begin{gathered} Aw_j = \lambda_j w_j, \qquad j \geq 1, \\ 0 \leq \lambda_1 \leq \lambda_2 \leq \ldots, \qquad \lambda_j \to \infty \quad \text{as } j \to \infty. \end{gathered} \tag{5.5}$$

In case (5.3a), $\lambda_1 > 0$, whereas in cases (5.3b) and (5.3c), $\lambda_1 = 0$, $w_1 = |\Omega|^{-1/2}$, $\lambda_2 > 0$. The powers A^s of A, $s \in \mathbb{R}$, are defined as in Section II.2, $V = D(A^{1/2})$, $V' = D(A^{-1/2})$, $H = D(A^0)$.

Problem (5.1), (5.3) is now equivalent to the following functional evolution equation:

$$u' + (\lambda + i\alpha)Au + (\kappa + i\beta)|u|^2 u - \gamma u = 0, \tag{5.6}$$

and (5.4) is written

$$u(0) = u_0. \tag{5.7}$$

Concerning the existence and uniqueness of solutions of the initial-value problem (5.6), (5.7) we can state

Theorem 5.1. *We assume that $n = 1$ or 2 and that* (5.2) *holds. For u_0 given in H, there exists a unique solution u of* (5.6), (5.7)

$$u \in \mathscr{C}([0, T]; H) \cap L^2(0, T; V), \qquad \forall T < \infty. \tag{5.8}$$

Furthermore, the mapping

$$u_0 \to u(t)$$

is continuous from H into itself, $\forall t > 0$.

If $u_0 \in V$, *then*

$$u \in \mathscr{C}([0, T]; V) \cap L^2(0, T; D(A)), \qquad \forall T < \infty. \tag{5.9}$$

PROOF. We apply the methods of Theorem III.3.1 to equation (5.6), considered as a real system for the real and imaginary parts u_1, u_2 of u ($u = u_1 + iu_2$). We choose H, V as above ($\mathbb{L}^2(\Omega) \equiv L^2(\Omega)^2$, $\mathbb{H}^1(\Omega) \equiv H^1(\Omega)^2$, etc...); $Ru = -(\gamma + 1)u = -(\gamma + 1)\{u_1, u_2\}$ and A is replaced by $(\lambda + i\alpha)(Au + u)$, i.e., for $u = \{u_1, u_2\}$,

$$Au = \begin{cases} \lambda(-\Delta u_1 + u_1) - \alpha(-\Delta u_2 + u_2), \\ \alpha(-\Delta u_1 + u_1) + \lambda(-\Delta u_2 + u_2). \end{cases}$$

We have for $u, v \in D(A)$ (or V),

$$a(u, v) = (Au, v) = \lambda\{((u_1, v_1))_1 + ((u_2, v_2))_1\} - \alpha\{((u_2, v_1)) - ((u_1, v_2))\}.$$

The nonlinear term B is defined as follows: $B(u) = B(u; u)$

$$(B(u; v), w) = \int_\Omega |u|^2\{(\kappa v_1 - \beta v_2)w_1 + (\beta v_1 + \kappa v_2)w_2\}\, dx, \qquad \forall u, v \in V,$$

or, alternatively,

$$B(u; v) = \begin{cases} |u|^2(\kappa v_1 - \beta v_2), \\ |u|^2(\beta v_1 + \kappa v_2). \end{cases}$$

Hypothesis III.(3.1) is satisfied (with α replaced by λ); a is not symmetric but this is not necessary for Theorem III.3.1. The hypotheses III.(3.3) and III.(3.4) are obvious, with $\theta_1 = \theta_2 = 0$, since R is continuous in H. The operator B maps $V \times V$ into V' and even into H since by the Sobolev embedding theorems $H^1(\Omega) \subset L^s(\Omega)$, $\forall s < \infty$ if $n = 2$, and $H^1(\Omega) \subset \mathscr{C}(\bar{\Omega})$ if $n = 1$. The operator $B(\cdot, \cdot)$ is not linear with respect to its first argument and satisfies, instead of II.(3.6),

$$(B(u; v), v) = \kappa \int_\Omega |u|^2 |v|^2\, dx \geq 0, \qquad \forall u, v \in V, \tag{5.10}$$

but this is sufficient for Theorem III.3.1. We also have

$$\begin{aligned} |B(u; v)|_{L^2}^2 &\leq 2(\kappa^2 + \beta^2) \int_\Omega |u|^4 |v|^2\, dx, \\ &\leq \text{(by the Schwarz inequality)}, \\ &\leq 2(\kappa^2 + \beta^2)\left(\int_\Omega |u|^8\, dx\right)^{1/2}\left(\int_\Omega |v|^4\, dx\right)^{1/2}, \end{aligned}$$

$$\leq \text{(by the Sobolev embeddings)},$$

$$\leq c_1'(\kappa^2 + \beta^2)|u|^4_{\mathbb{H}^{3/4}(\Omega)}|v|^2_{\mathbb{H}^{1/2}(\Omega)}.$$

Finally, by interpolation,

$$|B(u; v)|^2_{L^2} \leq (\kappa^2 + \beta^2)|u|_{L^2}\|u\|^3_1|v|^{1/2}_{L^2}\|v\|^{1/2}_1. \tag{5.11}$$

This inequality replaces III.(3.7)–III.(3.9) in the proof of Theorem III.3.1.

We leave as an exercise for the reader the repetition of the proof of Theorem III.3.1 after these modifications in the assumptions. □

Theorem 5.1 allows us to define the semigroup $S(t)$. For every $t \geq 0$, we define the operator $S(t)$ mapping H into itself by

$$S(t)\colon u_0 \to u(t). \tag{5.12}$$

It is clear that these operators enjoy the semigroup properties I.(1.1) and that they are continuous from H into itself and even from H into $D(A)$; I.(1.4) is thus satisfied.

5.2. Absorbing Sets and Attractors

Absorbing Sets in H

We first prove the existence of an absorbing set in H. The necessary energy equation is obtained by multiplying (5.1) by $\bar{u}$ (= the conjugate of u), integrating over Ω, using Green's formula, and taking the real part of the equation that we obtain

$$\frac{1}{2}\frac{d}{dt}\int_\Omega |u|^2\,dx + \lambda\int_\Omega |u|^2\,dx + \kappa\int_\Omega |u|^4\,dx - \gamma\int_\Omega |u|^2\,dx = 0,$$

or, equivalently,

$$\frac{1}{2}\frac{d}{dt}|u|^2_{L^2} + \lambda\|u\|^2 + \kappa|u|^4_{L^4} - \gamma|u|^2_{L^2} = 0. \tag{5.13}$$

Case $\gamma \leq 0$

The case $\gamma \leq 0$ leads to trivial dynamics. If $\gamma < 0$, we infer from (5.13) that

$$\frac{d}{dt}|u|^2_{L^2} - \gamma|u|^2_{L^2} \leq 0, \tag{5.14}$$

$$|u(t)|^2_{L^2} \leq |u_0|^2_{L^2}\exp(\gamma t), \tag{5.15}$$

and therefore

$$|u(t)|^2_{L^2} \to 0 \quad \text{as } t \to \infty, \qquad \forall u_0 \in \mathbb{L}^2(\Omega). \tag{5.16}$$

If $\gamma = 0$, (5.16) is still valid. Indeed, by Holder's inequality,

$$|u|^2_{L^2} \leq |\Omega|^{1/2}|u|^2_{L^4},$$

and we see with (5.13) that

$$y' + \frac{2\kappa}{|\Omega|} y^2 \leq 0, \qquad y(t) = |u(t)|_{L^2}^2.$$

Hence

$$\frac{1}{y(0)} + \frac{2\kappa}{|\Omega|} t \leq \frac{1}{y(t)}$$

and (5.16) follows.

The conclusion (5.16) is still valid if $\gamma < \lambda\lambda_1$ in the case of the boundary condition (5.3a), λ_1 denoting the first eigenvalue of $-\Delta$ in Ω, with the Dirichlet boundary condition. For cases (5.3b), (5.3c) there are stationary solutions of (5.1), (5.3b, c) which are constant in space and time (i.e., when $\beta = 0$ and $|u|^2 = \gamma\kappa^{-1}$ and (5.16) is not no longer valid.

Case $\gamma > 0$

From now on we assume that $\gamma > 0$. We have

$$\frac{\kappa}{2} s^4 - 2\gamma s^2 \geq -\frac{2}{\kappa}\gamma^2, \qquad \forall s \in \mathbb{R}, \tag{5.17}$$

$$\frac{\kappa}{2}\int_\Omega |\varphi|^4 \, dx - 2\gamma \int_\Omega |\varphi|^2 \, dx \geq -\frac{2}{\kappa}\gamma^2 |\Omega|, \qquad \forall \varphi \in \mathbb{L}^4(\Omega), \tag{5.18}$$

and with this last inequality (5.13) yields

$$\frac{d}{dt}|u|_{L^2}^2 + 2\lambda \|u\|^2 + \kappa |u|_{L^4}^4 + 2\gamma |u|_{L^2}^2 \leq \frac{2\gamma^2}{\kappa}|\Omega|. \tag{5.19}$$

Then with the Gronwall lemma

$$|u(t)|_{L^2}^2 \leq |u(0)|_{L^2}^2 \exp(-\gamma t) + \frac{\gamma}{\kappa}|\Omega|(1 - \exp(-\gamma t)), \quad \forall t \geq 0, \tag{5.20}$$

$$\limsup_{t\to\infty} |u(t)|_{L^2}^2 \leq \rho_0^2, \qquad \rho_0^2 = \frac{\gamma}{\kappa}|\Omega|. \tag{5.21}$$

Therefore

The ball of H, $B_H(0, \rho_0')$ centered at 0 of radius $\rho_0' \geq \rho_0$, is positively invariant for the semigroup $S(t)$.

The ball of H, $\mathscr{B}_0 - B_H(0, \rho_0')$ centered at 0 of radius $\rho_0' > \rho_0$, is absorbing in H for the semigroup $S(t)$. (5.22)

If $\mathscr{B}$ is a bounded set of H, included say in the ball $B_H(0, R)$ of H centered at 0 of radius R, then $S(t)\mathscr{B} \subset \mathscr{B}_0$ for $t \geq t_0 = t_0(\mathscr{B}, \mathscr{B}_0)$,

$$t_0 = \frac{1}{\gamma}\log\frac{R^2}{(\rho_0')^2 - \rho_0^2}. \tag{5.23}$$

We then integrate (5.19) between t and $t+r$ $(r>0)$. If $u_0 \in \mathscr{B}$ and $t \geq t_0(\mathscr{B}, \mathscr{B}_0)$, $\mathscr{B}$, t_0 as above, we obtain

$$\int_t^{t+r} \{2\lambda \|u\|^2 + \kappa |u|_{L^4}^4 + 2\gamma |u|_{L^2}^2\}\, ds \leq \rho_0'^2 + \frac{2r\gamma^2}{\kappa}|\Omega|,$$
$$\forall t \geq t_0, \quad \forall r > 0. \tag{5.24}$$

Absorbing Sets in V

We continue and show the existence of an absorbing set in V. For that purpose we obtain another energy equation as follows: we multiply (5.1) by $-\Delta\bar{u}$, integrate over Ω, use the Green formula, and take the real part of the equation. At the end of these operations we obtain

$$\begin{aligned}\frac{1}{2}\frac{d}{dt}\|u\|^2 + \lambda|\Delta u|_{L^2}^2 - \gamma\|u\|^2 &= \operatorname{Re}(\kappa + i\beta)\int_\Omega |u|^2 u \Delta\bar{u}\, dx \\ &= \operatorname{Re}(\kappa + i\beta)\int_\Omega (\nabla(|u|^2 u))\nabla\bar{u}\, dx \\ &\leq 3(\kappa^2 + \beta^2)^{1/2}\int_\Omega |u|^2|\nabla u|^2\, dx. \end{aligned} \tag{5.25}$$

Using the Schwarz inequality, this last expression on the right-hand side of (5.25) is bounded by

$$3(\kappa^2 + \beta^2)^{1/2}|u|_{L^4}^2|\nabla u|_{L^4}^2. \tag{5.26}$$

By the Sobolev embedding (see Sections II.1.1 and II.1.3) and the interpolation inequality II.(1.24), we have $H^{1/2}(\Omega) \subset L^4(\Omega)$ (for $n = 2$ or 1), and

$$|\varphi|_{L^4(\Omega)} \leq c_1'|\varphi|_{L^2}{}^{1/2}(\|\varphi\|^2 + |\varphi|_{L^2}^2)^{1/4}, \qquad \forall\varphi \in H^1(\Omega). \tag{5.27}$$

Also, the norm $(|u|_{L^2}^2 + |\Delta u|_{L^2}^2)$ is on H^2 a norm equivalent to the natural one. Thus, also using (5.27) for $\nabla\varphi$, we see that there exists a constant c_2' depending only on Ω such that

$$|\nabla\varphi|_{L^4} \leq c_2'\|\varphi\|^{1/2}(|\varphi|_{L^2}^2 + |\Delta\varphi|_{L^2}^2)^{1/4}. \tag{5.28}$$

This allows us to majorize (5.26) by

$$\begin{aligned} &3(c_2')^2(\kappa^2 + \beta^2)^{1/2}|u|_{L^4}^2\|u\|(|u|_{L^2}^2 + |\Delta u|_{L^2}^2)^{1/2} \\ &\leq \frac{\lambda}{2}|\Delta u|_{L^2}^2 + \frac{\lambda}{2}|u|_{L^2}^2 + c_3'|u|_{L^4}^4\|u\|^2, \end{aligned} \tag{5.29}$$

$$c_3' = \frac{9}{2\lambda}(c_2')^4(\kappa^2 + \beta^2),$$

and inserting the bounds (5.26), (5.29) in (5.25) we obtain

$$\frac{d}{dt}\|u\|^2 + \lambda|\Delta u|_{L^2}^2 \leq 2(\gamma + c_3'|u|_{L^4}^4)\|u\|^2 + \lambda|u|_{L^2}^2. \tag{5.30}$$

At this point we apply the uniform Gronwall lemma (Lemma II.1.1), with

$$y = \|u\|^2, \qquad g = 2(\gamma + c_3'|u|_{L^4}^4), \qquad h = \lambda |u|_{L^2}^2,$$

$$a_1 = 2\gamma r + \frac{2c_3'}{\kappa}\left(\rho_0'^2 + \frac{2r\gamma^2}{\kappa}|\Omega|\right) \quad \text{(thanks to (5.24))},$$

$$a_2 = \lambda \rho_0'^2 \quad \text{(thanks to (5.22), (5.23))},$$

$$a_3 = \frac{1}{2\lambda}\left(\rho_0'^2 + \frac{2r\gamma^2}{\kappa}|\Omega|\right) \quad \text{(thanks to (5.24))}.$$

Lemma II.1.1 allows us to assert that

$$\|u(t)\|^2 \leq \left(\frac{a_3}{r} + a_2\right)\exp(a_1) \quad \text{for} \quad t \geq t_0 + r, \tag{5.31}$$

where $r > 0$ is arbitrarily chosen, $u_0 \in \mathscr{B}$, and $t_0 = t_0(\mathscr{B})$ as in (5.23).

We recall that $(|\varphi|_{L^2}^2 + \|\varphi\|^2)^{1/2}$ is the norm on V (= the H^1 norm). Combining (5.31) with (5.22), (5.23) we obtain the existence of an absorbing set in V for $S(t)$ and the uniform compactness property I.(1.12). Indeed, if $\mathscr{B}$ is a bounded set of V, then it is also a bounded set of H, (5.22), (5.23) apply, $S(t)\mathscr{B} \subset \mathscr{B}_0$ for $t \geq t_0(\mathscr{B}, \mathscr{B}_0)$, and then with (5.31), $S(t)\mathscr{B} \subset \mathscr{B}_1$ for $t \geq t_0 + r$, where $\mathscr{B}_1$ is the ball of V centered at 0 of radius ρ_1,

$$\rho_1^2 = (\rho_0')^2 + \left(\frac{a_3}{r} + a_2\right)\exp(a_1), \tag{5.32}$$

a_1, a_2, a_3 as in (5.31). Thus

> The ball of V, $\mathscr{B}_1 = B_V(0, \rho_1)$ centered at 0 of radius ρ_1, is absorbing in V for the semigroup $S(t)$. (5.33)

If $u_0 \in \mathscr{B}$ where $\mathscr{B}$ is only bounded in H, the above analysis still applies and $S(t)\mathscr{B} \subset \mathscr{B}_1$ for $t \geq t_0(\mathscr{B}) + r$. Since $\mathscr{B}_1$ is bounded in V and the injection of V in H is compact, we conclude that

$$\bigcup_{t \geq t_0 + r} S(t)\mathscr{B} \text{ is relatively compact in } H, \tag{5.34}$$

i.e., I.(1.12) holds.

Maximal Attractor

All the assumptions of Theorem I.1.1 are satisfied: I.(1.1), I.(1.4), I.(1.12), and the existence of an absorbing set in H. Theorem I.1.1 then implies the existence of a global attractor in $\mathbb{L}^2(\Omega)$ for the Ginzburg–Landau equation.

Theorem 5.2. *We consider the dynamical system associated with the Ginzburg–Landau equation* (5.1) *supplemented by one of the boundary conditions* (5.3), *with* $\lambda > 0$, $\kappa > 0$. *This dynamical system possesses an attractor* $\mathscr{A}$ *which is* *compact, connected, and maximal in* $\mathbb{L}^2(\Omega)$. $\mathscr{A}$ *attracts the bounded sets of* $\mathbb{L}^2(\Omega)$ *and* $\mathscr{A}$ *is also maximal among the functional-invariant sets bounded in* $\mathbb{L}^2(\Omega)$.

Remark 5.1. The analysis above (see (5.16)) shows that, if $\gamma \leq 0$, all the orbits converge to 0 as $t \to \infty$. More precisely, (5.15) and the proof of (5.16) show that

$$\operatorname{dist}(S(t)B, \{0\}) \to 0 \quad \text{as } t \to \infty,$$

for every bounded set $B \subset H$. Hence the universal attractor $\mathscr{A}$ is reduced to $\{0\}$.

Remark 5.2. The attractors for the Ginzburg–Landau equation have been studied by J.M. Ghidaglia and B. Héron [1], C.R. Doering, J.D. Gibbon, D.D. Holm, and B. Nicolaenko [1], C.R. Doering, J. Gibbon and C.D. Levermore [1]. For other (less dissipative) Schrödinger equations, see next section and the references therein.

6. Weakly Dissipative Equations I. The Nonlinear Schrödinger Equation

In this section, and the next two, we consider equations presenting some new difficulties for which specific techniques are needed. In Section 8 we will address the case of unbounded domains leading to a lack of compactness. In this section, and in Section 7, we consider weakly dissipative equations, that is, equations for which the dissipation occurs only on the lowest-order terms (zeroth-order term), i.e., terms which do not contain derivatives of the unknown function u; in all the examples previously considered in Chapters II, III, and this chapter, the dissipativity was strong, occurring on some of the spatial derivatives of u if not on its highest derivatives.

Two examples of weakly dissipative equations are considered: the nonlinear Schrödinger equation in the present section, and the weakly damped Korteweg–de Vries equation in Section 7. The questions that we address, beside that of existence and uniqueness of solution, are the question of the existence of a compact attractor and the question of regularity of the attractor. The latter problem is the following: assuming that, for a given equation, well-posedness and the existence of the attractor are established in two different spaces, one smaller than the other, is it true that the attractor is the same (as a set) in both spaces? This question is addressed in Section VI.3.1 in the particular case of the two-dimensional Navier–Stokes equations with space-periodic boundary conditions and in a more general context, from a slightly different point of view, in Section 9 of this chapter. However, in the case of the weakly dissipative equations, as we shall see, the question of the regularity of the attractor is closely related to that of its existence, and we shall study both problems at the same time.

Technically, the existence and uniqueness of the solutions are proved by standard methods and will be very briefly addressed. The existence of the attractor is proved using Theorem I.1.1; for the nonlinear Schrödinger equation we rely on I.(1.13) whose verification is, however, new and involved; for the Korteveg–de Vries equation the attractor is obtained with Theorem I.1.1 supplemented by Remark I.1.4, with an appropriate technique for the verification of I.(1.17). Finally, the energy equalities used for the nonlinear Schrödinger equations are (nearly) standard, whereas for the Korteweg–de Vries equation, they are based on the classical infinite sequence of nonquadratic invariants obtained by multiplying the equation by nonlinear appropriate quantities; the computations are quite involved in this case.

For this section we follow essentially the article by O. Goubet [1]; further bibliographical comments are given in Remark 6.3 at the end of this section.

6.1. The Nonlinear Schrödinger Equation

We restrict ourselves to space dimension one. The nonlinear Schrödinger equation is similar to the Ginzburg–Landau equation (5.1), except that $\lambda = \kappa = 0$ here (see (5.2)). It occurs in similar phenomena as the Ginzburg–Landau equation but it is also, in nonlinear optics and the propagation of laser beams, an important equation for the study of the propagation of solitons (see, e.g., G.P. Agrawal [1], G.P. Agrawal and W. Boyd [1], J. Ablowitz and H. Segur [1], and A.C. Newell [1]).

The equations reads

$$\frac{\partial u}{\partial t} + i\frac{\partial^2 u}{\partial x^2} + i|u|^2 u + \alpha u - iu = f, \tag{6.1}$$

where $u = u(x, t)$ is a complex-valued function. Here α and f are given, with

$$\alpha > 0 \tag{6.2}$$

the term αu producing the weak damping. We shall generally assume that f is C^∞ and one-periodic except when otherwise stated. We assume that u is space-periodic

$$u(x + 1, t) = u(x, t), \qquad x \in \mathbb{R}, \quad t > 0, \tag{6.3}$$

and we set $\Omega = (0, 1)$. We supplement (6.1) and (6.3) with the initial condition

$$u(x, 0) = u_0(x), \qquad x \in \Omega. \tag{6.4}$$

For the mathematical setting we introduce, as in Section 5, complex

Sobolev spaces and we set (the notation is not the same as in Section 5)

$$H^k = \{u\colon [0, 1] \mapsto \mathbb{C},\ u \text{ and a } \frac{\partial^j u}{\partial x^j} \text{ belong to } L^2(0, 1), j = 1, \ldots, k;$$

$$x \mapsto u(x), \text{ and } x \mapsto \frac{\partial^j u}{\partial x^j}(x) \text{ are one-periodic}, j = 1, \ldots, k - 1\}, \quad (6.5)$$

$L^2 = L^2(0, 1)$ is the space of complex-valued L^2-functions. Of course, all these spaces are endowed with their natural Hilbert structure, scalar products and norms denoted $(\cdot, \cdot)_{L^2}$, $|\cdot|_{L^2}$ for L^2 and $((\cdot, \cdot))_{H^k}$, $\|\cdot\|_{H^k}$ for H^k.

6.2. Existence and Uniqueness of Solution. Absorbing Sets

For the existence of solution, we have the following result.

Theorem 6.1. *For u_0 given in H^1 and f given in L^2, there exists a unique solution u of* (6.1)–(6.4),

$$u \in \mathscr{C}_b(\mathbb{R}_+; H^1), \qquad u_t = \frac{\partial u}{\partial t} \in \mathscr{C}_b(\mathbb{R}_+; H^{-1}), \quad (6.6)$$

where H^{-1} is the dual space of H^1.

Proof (Sketch). This theorem is proved essentially as is Theorem 5.1, using the methods of Theorem III.3.1. The main difference occurs in the a priori estimates on which the proof is based. Let us make explicit the a priori estimates which will be used again several times in the sequel, in particular, for the absorbing sets.

We write $u_t = \partial t/\partial t$, $u_x = \partial u/\partial x$; we multiply equation (6.1) by $\bar{u}$ (= the complex conjugate of u), and integrate over Ω; using Green's formula, and taking the real part of the resulting equation, we find (compare to (5.13)):

$$\frac{1}{2}\frac{d}{dt}|u|_{L^2}^2 + \alpha|u|_{L^2}^2 = \operatorname{Re}(f, u)_{L^2}.$$

Writing

$$\operatorname{Re}(f, u)_{L^2} \leq |f|_{L^2}|u|_{L^2} \leq \frac{\alpha}{2}|u|_{L^2}^2 + \frac{1}{2\alpha}|f|_{L^2}^2,$$

we obtain

$$\frac{d}{dt}|u|_{L^2}^2 + \alpha|u|_{L^2}^2 \leq \frac{1}{\alpha}|f|_{L^2}^2, \quad (6.7)$$

from which we easily derive an a priori estimate of u in $L^\infty(\mathbb{R}_+; L^2)$. Namely,

$$|u(t)|_{L^2}^2 \le |u_0|_{L^2}^2 \exp(-\alpha t) + \frac{1}{\alpha^2}|f|_{L^2}^2\{1 - \exp(-\alpha t)\}. \tag{6.8}$$

Then we multiply (6.1) by $-(\bar{u}_t + \alpha\bar{u})$, integrate over Ω, use Green's formula, and take the imaginary part of the resulting equation. This yields

$$\begin{aligned}\frac{d}{dt}\left\{\frac{1}{2}|u_x|_{L^2}^2 + \operatorname{Im}\int_\Omega f\bar{u}\,dx\right\} &+ \alpha|u_x|_{L^2}^2 \\ &- \alpha\int_\Omega (|u|^2 - 1)|u|^2\,dx + \alpha\operatorname{Im}\int_\Omega f\bar{u}\,dx \\ &- \operatorname{Im} i\int_\Omega (|u|^2 - 1)u\bar{u}_t\,dx - \operatorname{Im}\int_\Omega f_t\bar{u}\,dx = 0. \end{aligned}\tag{6.9}$$

After expanding, we obtain ($f_t = 0$)

$$\frac{1}{2}\frac{d}{dt}\varphi_1(u) + \alpha\psi_1(u) = 0, \tag{6.10}$$

with

$$\varphi_1(u) = |u_x|_{L^2}^2 + |u|_{L^2}^2 - \frac{1}{2}\int_\Omega |u|^4\,dx + 2\operatorname{Im}\int_\Omega f\bar{u}\,dx,$$

$$\psi_1(u) = |u_x|_{L^2}^2 + |u|_{L^2}^2 - \int_\Omega |u|^4\,dx + \operatorname{Im}\int_\Omega f\bar{u}\,dx.$$

We observe that

$$\operatorname*{Sup}_{0<x<1} |u(x)|^2 = |u|_{L^\infty}^2 \le |u|_{L^2}(2|u_x|_{L^2} + |u|_{L^2}), \qquad \forall u \in H^1, \tag{6.11}$$

and we deduce that

$$\begin{aligned}&\varphi_1(u) \le |u_x|_{L^2}^2 + \tfrac{3}{2}|u|_{L^2}^2 + 2|f|_{L^2}^2,\\ &\varphi_1(u) \ge \tfrac{1}{2}|u_x|_{L^2}^2 + \tfrac{1}{2}|u|_{L^2}^2 - 2|f|_{L^2}^2 - \tfrac{1}{2}|u|_{L^2}^4 - \tfrac{1}{2}|u|_{L^2}^6.\\ &\psi_1(u) \le |u_x|_{L^2}^2 + \tfrac{3}{2}|u|_{L^2}^2 + \tfrac{1}{2}|f|_{L^2}^2,\\ &\psi_1(u) \ge \tfrac{1}{2}|u_x|_{L^2}^2 + \tfrac{1}{2}|u|_{L^2}^2 - \tfrac{1}{2}|f|_{L^2}^2 - |u|_{L^2}^4 - 2|u|_{L^2}^6.\end{aligned}\tag{6.12}$$

Hence (6.10) yields

$$\begin{aligned}&\frac{d}{dt}\varphi_1(u) + \alpha\varphi_1(u) \le L_1(u),\\ &\qquad L_1(u) = 3\alpha|f|_{L^2}^2 + 2\alpha|u|_{L^2}^4 + 4\alpha|u|_{L^2}^6 + \frac{\alpha}{2}|u|_{L^2}^2\end{aligned}\tag{6.13}$$

By Gronwall's inequality

$$\varphi_1(u(t)) \leq \varphi_1(u(0)) \exp(-\alpha t) + \frac{\kappa_1}{\alpha}\{1 - \exp(-\alpha t)\}, \tag{6.14}$$

where κ_1 is a time uniform bound of $L_1(u)$ which can be found since u is a priori bounded in $L^\infty(\mathbb{R}_+; L^2)$.

After straightforward calculations we obtain

$$\|u(t)\|_{H^1}^2 = |u_x(t)|_{L^2}^2 + |u(t)|_{L^2}^2 \leq 2\kappa_2 + \frac{2\kappa_1}{\alpha} + \kappa_3, \qquad t \geq 0, \tag{6.15}$$

where κ_2 is a time uniform bound of

$$L_2(u) = 2|f|_{L^2}^2 + \tfrac{1}{2}|u|_{L^2}^2 + \tfrac{1}{2}|u|_{L^2}^2,$$

and

$$\kappa_3 = 2|u_{0x}|_{L^2}^2 + 3|u_0|_{L^2}^2 + 4|f|_{L^2}^2.$$

We derive from (6.15) an a priori bound for u in $L^\infty(\mathbb{R}_+; H^1)$. Returning to equation (6.1) we obtain then an a priori estimate of $u_t = \partial u/\partial t$ in $L^\infty(\mathbb{R}_+; H^{-1})$. □

From Theorem 6.1 we derive the existence of the semigroup $\{S(t)\}_{t\geq 0}$, where

$$S(t)\colon u_0 \mapsto u(t), \tag{6.16}$$

is continuous from H^1 into itself; equations I.(1.1)–I.(1.3) are satisfied and it can be shown that I.(1.4) is also satisfied.

Remark 6.1. Changing t into $-t$ amounts to changing α into $-\alpha$ and i into $-i$. Hence Theorem 6.1 is valid as well for $t < 0$, except that (6.8) and (6.14) now allow exponential growth of the norms as $t \to -\infty$: $u \in \mathscr{C}(\mathbb{R}; H^1)$, $u_t \in \mathscr{C}(\mathbb{R}; H^{-1})$ are not bounded as $t \to -\infty$. The operators $S(t)$ in (6.16) defined for $t \in \mathbb{R}$ form a group.

Absorbing Sets

It is easy to derive the existence of absorbing sets from the above estimates. The absorbing set in L^2 is derived from (6.7) and (6.8). Let

$$\rho_0^2 = \frac{1}{\alpha^2}|f|^2,$$

and let ρ_0' be any number, $\rho_0' > \rho_0$. Then

The ball $\mathscr{B}_0$ of L^2 centered at 0 of radius ρ_0' is an absorbing ball for the semigroup $S(t)$. (6.17)

If $\mathscr{B}$ is a bounded set of L^2, included say in the ball of L^2 centered at 0 of radius R, then $S(t)\mathscr{B} \subset \mathscr{B}_0$ for $t \geq t_0(\mathscr{B}, \mathscr{B}_0)$,

$$t_0 = \frac{1}{\alpha} \log \frac{R^2}{(\rho_0')^2 - \rho_0^2}. \tag{6.18}$$

The absorbing ball in H^1 is derived from (6.15). For $t \geq t_0$, (6.13) and (6.14) are valid with $|u(t)| \leq \rho_0'$ giving

$$L_1(u(t)) \leq \kappa_1 = 3\alpha |f|_{L^2}^2 + 2\alpha(\rho_0')^4 + 4\alpha(\rho_0')^6 + \frac{\alpha}{2}(\rho_0')^2,$$

$$L_2(u(t)) \leq \kappa_2 = 2|f|_{L^2}^2 + \tfrac{1}{2}(\rho_0')^4 + \tfrac{1}{2}(\rho_0')^6.$$

Integrating (6.14) between t_0 and t, $t \geq t_0$, we obtain, by a relation similar to (6.15),

$$\|u(t)\|_{H^1}^2 = |u_x(t)|_{L^2}^2 + |u(t)|_{L^2}^2 \leq \rho_1^2, \tag{6.19}$$

for $t \geq t_1$, where ρ_1 depends on the data but not on u_0, while $t_1 = t_1(R_1)$ depends on the data and on u_0 through R_1, where

$$\|u_0\|_{H^1}^2 = |u_{0x}|_{L^2}^2 + |u_0|_{L^2}^2 \leq R_1^2. \tag{6.20}$$

Returning to equation (6.1) we then infer from (6.19) that

$$\|u_t(t)\|_{H^{-1}} = \left\| \frac{du}{dt}(t) \right\|_{H^{-1}} \leq \rho' \qquad \text{for} \quad t \geq t_1 \tag{6.21}$$

for some suitable ρ'.

We skip the details of these calculations, very similar to calculations we have done many times, and we now emphasize the technics specifically needed for the equation under consideration.

6.3. Decomposition of the Semigroup

The proof of existence of the attractor relies on Theorem I.1.1 used with hypothesis I.(1.13) corresponding to a very specific decomposition of $S(t)$ in the form $S_1(t) + S_2(t)$ which we now present and study. Note that by Theorem 6.1 and Remark 6.1 the (semi)group does not possess any smoothing effect. In fact, S_1 will correspond to a smoothing operator and we will identify a suitable S_2 corresponding to a nonsmoothing, high-frequency part.

6.3.1. The Decomposition

Let $u = u(t)$ be the solution of (6.1)–(6.4). We expand $u(t)$ into its Fourier series

$$u(t) = \sum_{k \in \mathbb{Z}} u_k(t) e^{2i\pi kx}. \tag{6.22}$$

For a given level $N \in \mathbb{N}$, we denote by $y(t)$ the low-frequency part of u,

$$y(t) = \sum_{|k| \leq N} u_k(t) e^{2i\pi kx}, \tag{6.23}$$

which is a smooth function with respect to the x variable (an analytical one). Then the regularity of u with respect to x depends on its high-frequency part, that is,

$$z(t) = \sum_{|k| > N} u_k(t) e^{2i\pi kx}. \tag{6.24}$$

By projecting (6.1) on the high modes we see that z is the solution of the nonautonomous partial differential equation

$$z_t + \alpha z + i z_{xx} - iz + iQ(|y + z|^2(y + z)) = Qf, \tag{6.25}$$

with the initial condition,

$$z(0) = Qu_0 = z_0, \tag{6.26}$$

where Q denotes the orthogonal projector onto

$$QH^1 = \left\{ z \in H^1;\, z = \sum_{|k| > N} u_k e^{2i\pi kx} \right\}. \tag{6.27}$$

Since we are interested in the long-time behavior of $z(t)$, we may focus on $z(t)$ for $t \geq t_1$, t_1 being as in (6.19).

Hence, for $t \geq t_1$, z is the solution of (6.25) with the initial condition

$$z(t_1) = Qu(t_1). \tag{6.28}$$

We also introduce $Z\colon [t_1, +\infty) \to QH^1$ which is the solution of

$$\begin{cases} Z_t + \alpha Z + iZ_{xx} - iZ + iQ(|y + Z|^2(y + Z)) = Qf, \\ Z(t_1) = 0; \end{cases} \tag{6.29}$$

here $y = Pu = (Id - Q)u$ is as above.

The fact that Z is well defined by (6.29) will be proven in Section 6.1.2 below. Actually, we shall prove that for N given, large enough, depending on the data, $Z(t)$ exists for all $t \geq t_1$ and takes its values in QH^k, for any integer k.

Then z splits into

$$z = Z + (z - Z), \tag{6.30}$$

where Z is smooth, and where $z - Z$ converges toward 0 in H^1 when t goes to infinity (this last point will be proven in Section 6.3.3). Hence we shall eventually set $S_1(t)u_0 = y(t) + Z(t)$, $S_2(t)u_0 = z(t) - Z(t)$, i.e., we shall decompose $u(t)$ into $u(t) = v(t) + \chi(t)$, $v(t) = y(t) + Z(t)$, $\chi(t) = z(t) - Z(t)$, where $v(t)$ is smooth for $t \geq t_1$ and $\chi(t) \to 0$ as $t \to \infty$ (see Sections 6.4 and 6.5).

6.3.2. An Existence Result for Z

For the convenience of the notations we shift time and we assume in Sections 6.3 and 6.4 that $t_1 = 0$ in (6.19). Hence $u(t)$ remains for $t \geq 0$ in the absorbing ball in H^1 whose radius is ρ_1. We shall go back (in Section 6.5) to the general case by translating time.

Let N be fixed large enough (the condition on N will be subsequently specified). Let $y(t) = Pu(t)$ as above. For $m > N$, let $Z^m(t)$ be the solution in

$$P_m QH^1 = \left\{ \sum_{N < |k| \leq m} u_k e^{2i\pi kx} \right\}, \tag{6.31}$$

of the nonautonomous *ordinary differential equation*

$$\begin{cases} Z_t^m + \alpha Z^m + iZ_{xx}^m - iZ^m + iP_m Q(|y + Z^m|^2(y + Z^m)) = P_m Qf, \\ Z^m(0) = 0. \end{cases} \tag{6.32}$$

This equation is a Galerkin approximation at order m of problem (6.29) (with $t_1 = 0$). Due to the Cauchy–Lipschitz theorem, Z^m exists as an application from $[0, T_m)$ into $P_m QH^1$ for some $T_m > 0$. We shall prove below a priori estimates which show that in fact $T_m = +\infty$ and which allow us to let m go to infinity. Further a priori estimates will provide us with further regularity results.

We first prove the existence, global in time, of Z^m and we then let $m \to \infty$ to obtain the existence of Z.

Proposition 6.1. *There exists N_0 that depends on the data, α and f, such that, for any $N \geq N_0$, the solution Z^m of (6.32) satisfies*

$$\sup_{t \geq 0} \|Z^m(t)\|_{H^1} \leq K_1, \tag{6.33}$$

where K_1 is a constant which depends only on the data, α and f.

PROOF. For the sake of convenience, we drop (in this proof) the superscript m and write $Z = Z^m$, $v = v^m = y + Z^m$. This will not introduce any confusion.

Multiplying (6.32) by $-\bar{Z}_t - \alpha\bar{Z}$, taking the imaginary part, and integrating on (0, 1) (compare to (6.9)), we obtain

$$\begin{aligned} \frac{1}{2}\frac{d}{dt}J(Z) + \alpha J(Z) = {} & \alpha \operatorname{Im} \int f\bar{Z} + \frac{\alpha}{2}\int |Z|^4 - \operatorname{Re} \int (|y|^2 y)_t \bar{Z} \\ & \alpha \operatorname{Re} \int |y|^2 y\bar{Z} - \int \operatorname{Re}(\bar{y}_t Z)|Z|^2 \\ & + \alpha \int \operatorname{Re}(\bar{y}Z)|Z|^2 - \operatorname{Re} \int \bar{y}y_t|Z|^2 \\ & - 2\int \operatorname{Re}(\bar{y}_t Z)\operatorname{Re}(\bar{y}Z), \end{aligned} \tag{6.34}$$

where

$$J(Z) = \|Z\|_{H^1}^2 + 2\,\mathrm{Im}\int f\bar{Z} - \frac{1}{2}\int |Z|^4 - 2\,\mathrm{Re}\int |y|^2 y\bar{Z}$$
$$- 2\,\mathrm{Re}\int |Z|^2 Z\bar{y} - \int |y|^2|Z|^2 - 2\int (\mathrm{Re}(\bar{y}Z))^2. \tag{6.35}$$

As before, Re and Im denote, respectively, the real and the imaginary parts of a complex number, and $|Z|^2 = Z\bar{Z}$. We have also set $\int g = \int_0^1 g(x)\,dx$.

We easily derive from (6.35) a lower bound for $J(Z)$ that reads

$$J(Z) \geq \|Z\|_{H^1}^2 - 2|f|_{L^{4/3}}|Z|_{L^4} - \tfrac{1}{2}|Z|_{L^4}^4 - 2|y|_{L^4}^3|Z|_{L^4}$$
$$- 2|Z|_{L^4}^3|y|_{L^4} - 3|y|_{L^4}^2|Z|_{L^4}^2. \tag{6.36}$$

Then, due to the Sobolev embedding $H^1 \subset L^4$ and to the fact that $y = Pu$, we have

$$|y|_{L^4} \leq c\|y\|_{H^1} \leq c\|u\|_{H^1} \leq c\rho_1, \tag{6.37}$$

where ρ_1 is as in (6.19). We recall that $u(t)$ remains in the absorbing ball of H^1 whose radius is ρ_1.

Remark 6.2. In the sequel c denotes a numerical constant, that may vary from one line to another. We also denote by K a constant that depends on the data of the equation, like ρ_1, for instance. We allow K to vary from one line to another in the computations.

We then easily infer from (6.36) and (6.37) that

$$J(Z) \geq \|Z\|_{H^1}^2 - |Z|_{L^4}^4 - c[|f|_{L^{4/3}}^{4/3} + M_1^4] = \|Z\|_{H^1}^2 - |Z|_{L^4}^4 - K. \tag{6.38}$$

We now majorize the right-hand side (r.h.s.) of (6.34)

$$\text{r.h.s. of (6.34)} \leq \alpha|f|_{L^{4/3}}|Z|_{L^4} + \frac{\alpha}{2}|Z|_{L^4}^4 + \|y_t\|_{H^{-1}}\||y|^2 Z\|_{H^1}$$
$$+ 2\|y_t\|_{H^{-1}}\|y\,\mathrm{Re}(\bar{y}Z)\|_{H^1} + \alpha|y|_{L^4}^3\|Z\|_{H^1}$$
$$+ \|y_t\|_{H^{-1}}\|Z|Z|^2\|_{H^1} + \alpha|y|_{L^4}|Z|_{L^4}^3$$
$$+ \|y_t\|_{H^{-1}}\|\bar{y}|Z|^2\|_{H^1} + 2\|y_t\|_{H^{-1}}\|2\,\mathrm{Re}(\bar{y}Z)\|_{H^1}. \tag{6.39}$$

Let us first majorize the third term in the right-hand side of (6.39). On the one hand, due to (6.21),

$$\|y_t\|_{H^{-1}} \leq \|u_t\|_{H^{-1}} \leq \rho'. \tag{6.40}$$

On the other hand,

$$\||y|^2 Z\|_{H^1} \leq c[|y|_{L^\infty}^2\|Z\|_{H^1} + |y|_{L^\infty}|Z|_{L^\infty}\|y\|_{H^1}], \tag{6.41}$$

due to formula

$$\|u_1 u_2 u_3\|_{H^1} \leq c \sum_{(i,j,k)=(1,2,3)} |u_i|_{L^\infty} |u_j|_{L^\infty} \|u_k\|_{H^1}. \tag{6.42}$$

Using then the embedding $H^1 \subset L^\infty$, (6.37), (6.40), and (6.41), we see that

$$\begin{aligned}\|y_t\|_{H^{-1}} \| |y|^2 Z\|_{H^1} &\leq c\|y_t\|_{H^{-1}} \|y\|_{H^1}^2 \|Z\|_{H^1} \\ &\leq K\|Z\|_{H^1} \\ &\leq \frac{\alpha}{4}\|Z\|_{H^1}^2 + K. \end{aligned}\tag{6.43}$$

Similarly, we handle the sixth term of the right-hand side of (6.39) as follows

$$\begin{aligned}\|y_t\|_{H^{-1}} \| |Z|^2 Z\|_{H^1} &\leq c\|Z\|_{L^\infty}^2 \|Z\|_{H^1} \|y_t\|_{H^{-1}} \\ &\leq \frac{K}{N}\|Z\|_{H^1}^3, \end{aligned}\tag{6.44}$$

using (6.40) and the following enhanced Poincaré inequality valid on QH^1

$$|Z|_{L^\infty} \leq c|Z|_{L^2}^{1/2} \|Z\|_{H^1}^{1/2} \leq \frac{c}{\sqrt{N}}\|Z\|_{H^1}. \tag{6.45}$$

The first inequality in (6.45) is Agmon's inequality (see I.(1.40)). The second inequality follows easily from the Fourier expansion of Z.

The last term in the right-hand side of (6.39) can also be majorized in the same way

$$\begin{aligned}\|y_t\|_{H^{-1}} \|Z \operatorname{Re}(\bar{y}Z)\|_{H^1} &\leq c\|y_t\|_{H^{-1}} \|y\|_{H^1} |Z|_{L^\infty} \|Z\|_{H^1} \\ &\leq \frac{K}{\sqrt{N}}\|Z\|_{H^1}^2. \end{aligned}\tag{6.46}$$

On the other hand, we majorize the fifth and seventh terms in the right-hand side of (6.39) by, due to (6.37),

$$\frac{\alpha}{4}|Z|_{L^4}^4 + c\alpha|y|_{L^4}^4 \leq \frac{\alpha}{4}|Z|_{L^4}^4 + K. \tag{6.47}$$

From all these computations, we infer that the

$$\text{r.h.s. of (6.39)} \leq \left(\frac{\alpha}{4} + \frac{K}{\sqrt{N}}\right)\|Z\|_{H^1}^2 + \alpha|Z|_{L^4}^4 + \frac{K'}{N}\|Z\|_{H^1}^3 + K'', \tag{6.48}$$

where, as indicated before, K, K', K'' depend only on the data.

We deduce from classical interpolation results and from the enhanced Poincaré inequality (6.45)

$$|Z|_{L^4} \leq c|Z|_{L^2}^{3/4} \|Z\|_{H^1}^{1/4} \leq \frac{c}{N^{3/4}}\|Z\|_{H^1}. \tag{6.49}$$

Then we derive from (6.48), (6.49), and from

$$\frac{K'}{N}\|Z\|_{H^1}^3 \le \frac{\alpha}{8}\|Z\|_{H^1}^2 + \frac{K}{N^2}\|Z\|_{H^1}^4, \tag{6.50}$$

that the

$$\text{r.h.s. of (6.39)} \le \left(\frac{3\alpha}{8}\|Z\|_{H^1}^2\right) + \left(\frac{K}{\sqrt{N}}\|Z\|_{H^1}^2\right) + \frac{K'}{N^2}\|Z\|_{H^1}^4 + K''. \tag{6.51}$$

At that stage appears the first condition on N_0 (and on N). We assume in the sequel that

$$\frac{K}{\sqrt{N}} \le \frac{K}{\sqrt{N_0}} \le \frac{\alpha}{8}, \qquad \text{i.e.,}\quad N_0 \ge 64\frac{K^2}{\alpha^2}, \tag{6.52}$$

where K is as in (6.51). Then the

$$\text{r.h.s. of (6.39)} \le \frac{\alpha}{2}\|Z\|_{H^1}^2 + \frac{K}{N^2}\|Z\|_{H^1}^4 + K'. \tag{6.53}$$

We finally infer from (6.39), (6.34), (6.38), (6.49), and (6.53) that

$$\frac{1}{2}\frac{d}{dt}J(Z) + \alpha J(Z) \le \frac{\alpha}{2}J(Z) + \left(\frac{c\alpha}{2N^3} + \frac{K}{N^2}\right)\|Z\|_{H^1}^4 + K, \tag{6.54}$$

and then

$$\frac{d}{dt}J(Z) + \alpha J(Z) \le \frac{K}{N^2}\|Z\|_{H^1}^4 + K'. \tag{6.55}$$

We now integrate (6.55) for t between 0 and t, observing that $J(Z(0)) = J(0) = 0$. This leads to

$$J(Z(t))e^{\alpha t} \le \frac{K}{N^2}\int_0^t \|Z(s)\|_{H^1}^4 e^{\alpha s}\,ds + \frac{K'e^{\alpha t}}{\alpha}. \tag{6.56}$$

We then easily deduce from (6.38) and (6.56) that

$$\|Z(t)\|_{H^1}^2 \le \frac{c}{N^3}\|Z(t)\|_{H^1}^4 + \frac{K}{N^2}\int_0^t \|Z(s)\|_{H^1}^4 e^{\alpha(s-t)}\,ds + K'. \tag{6.57}$$

Let us now introduce

$$\zeta(t) = \sup_{s\in[0,t]} \|Z(s)\|_{H^1}^2, \tag{6.58}$$

which is a continuous function of t satisfying

$$\zeta(0) = 0.$$

We infer from (6.57) that $\zeta(t)$ satisfies

$$\zeta(t) \le \frac{K}{N^2}\zeta^2(t) + K', \tag{6.59}$$

where K and K' are as in Remark 6.2.

Let us set

$$\phi(\zeta) = \zeta - \frac{K}{N^2}\zeta^2 - K', \tag{6.60}$$

and let us assume that $\phi(2K') > 0$, namely (this is the second assumption on N_0),

$$\frac{4KK'}{N^2} \leq \frac{4KK'}{N_0^2} < 1, \qquad \text{i.e.,} \quad N_0 \geq 2\sqrt{KK'}. \tag{6.61}$$

The graph of ϕ is shown in Figure 6.1, under assumption (6.61). Therefore, since $\phi(\zeta(0)) = \phi(0) < 0$ and since $t \mapsto \zeta(t)$ is a continuous nonnegative function of t, $\zeta(t)$ remains bounded in $[0, \alpha_1]$, where α_1 is the first root of ϕ. Hence, a fortiori, the following holds true

$$\zeta(t) = \sup_{s \in [0,t]} \|Z^m(t)\|_{H^1}^2 \leq 2K', \tag{6.62}$$

where K' is as above. The proof of Proposition 6.1 is complete. □

We deduce from Proposition 6.1 the following.

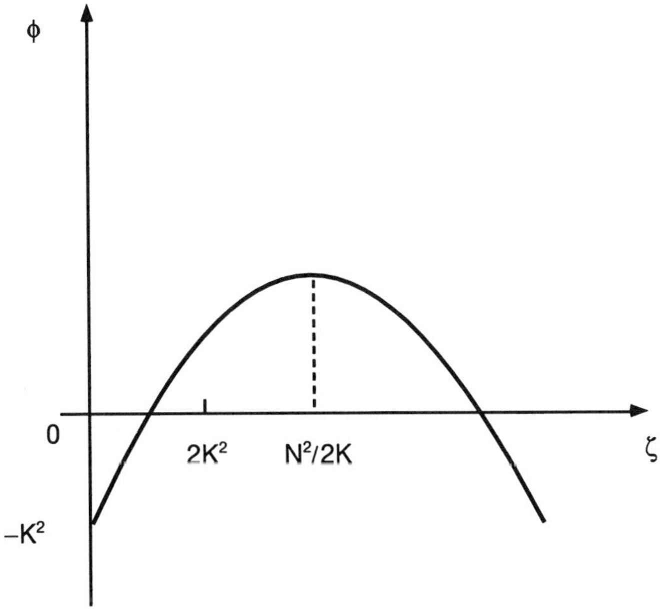

Figure 6.1. Graph of ϕ.

Corollary 6.1. *Let N_0 be as in Proposition* 6.1. *Let $N \geq N_0$ be fixed. Then there exists $Z = Z(t) \in C_b(\mathbb{R}_+; QH^1)$ continuous and bounded from $\mathbb{R}_+$ into QH^1 satisfying*

$$\begin{cases} Z_t + \alpha Z + iZ_{xx} - iZ + iQ(|y + Z|^2(y + Z)) = Qf, \\ Z(0) = 0. \end{cases} \tag{6.63}$$

Proof. The proof is classical and left as an exercise to the reader. It is based on the fact that the bound (6.62) for Z^m is independent of m. We just have to pass to the limit $m \to +\infty$ in (6.32), and to use Proposition 6.1. Moreover, stronger convergence results (on bounded time intervals $[0, T]$) may be used; the results are derived from the a priori estimates described below in Section 6.3.3.

Of course Z satisfies (6.62) as well.

We recall that $y = Pu$ where $u = u(t)$ is a trajectory included in the absorbing ball (6.19) for all $t \geq 0$. □

6.3.3. *Further a priori Estimates*

We shall prove, recursively in k, the following:

Proposition 6.2. *Let N_0 be as in Proposition* 6.1. *Let $N \geq N_0$ be fixed. For any $k \geq 2$, there exists $K(N, k)$ which depends on N and on k, such that the solution Z of* (6.63) *satisfies*

$$\sup_{t \geq 0} \|Z(t)\|_{H^k} \leq K(N, k). \tag{6.64}$$

Proof. We proceed by induction on k.

(i) We first start with $k = 2$. In all cases, we actually first prove the result for Z^m, and then let m go to infinity. Nevertheless, for the sake of simplicity, we just prove below the a priori estimates for Z^m omitting the details of the convergence $m \to \infty$. We also drop the subscript m so that we write $Z = Z^m$, $v = v^m = y + Z^m$, and we write Q for $P_m Q$. This simplifies the reading and does not introduce any confusion.

Let us differentiate (6.32) with respect to x. We obtain that Z_x is the solution of

$$\begin{cases} (Z_x)_t + \alpha Z_x + Q(A - F'(v))Z_x = G, \\ Z_x(0) = 0, \end{cases} \tag{6.65}$$

where we have set

$$AZ_x = i(Z_x)_{xx} - iZ_x, \tag{6.66}$$

and

$$F'(v)Z_x = -i|v|^2 Z_x - 2i\,\mathrm{Re}(\bar{v}Z_x)v, \tag{6.67}$$

where F' is the derivative of F. We have also set

$$G = Q[f_x + F'(v)y_x]. \tag{6.68}$$

We now apply Proposition 6.3 below to Z_x which solves (6.65). For that purpose, we observe that $v = y + Z = y + Z^m$ satisfies (6.85) with

$$K_0 = 2\max(\rho_1^2, (\rho')^2, K_1^2, K'^2), \tag{6.69}$$

where ρ_1, ρ', K' are as in (6.19), (6.21), and (6.33) and where

$$K' = \sup_{t\geq 0} \|Z_t^m(t)\|_{H^{-1}} \tag{6.70}$$

can be computed from ρ_1, ρ', K_1 using (6.32). On the other hand, we have to bound G (defined in (6.68)) in $C_b^1(\mathbb{R}_+; H^{-1})$. Let us first observe that the term Qf_x does not introduce any difficulty, since f is assumed to be as smooth as we want with respect to x, and since f is time-independent. Let us now focus on

$$QF'(v)y_x = -iQ|v|^2 y_x - 2iQ\,\mathrm{Re}(\bar{v}y_x)v. \tag{6.71}$$

Let us consider the first term in the right-hand side of (6.71); the other one can be handled exactly in the same manner. On the one hand, v belongs to $C_b(\mathbb{R}_+; H^1)$ (see (6.85)). On the other hand, y_x belongs to $C_b(\mathbb{R}_+; H^k)$, for any k in $\mathbb{N}$. This last point is a consequence of the usual inverse inequality valid on PH^1:

$$\|y_x\|_{H^k} \leq (2\pi N)^{k-1}\|y\|_{H^1}, \tag{6.72}$$

and of (6.37). Hence we have bounded G in $C_b(\mathbb{R}_+; H^{-1})$. It remains to consider its time derivative. The derivative of the first term in the right-hand side of (6.71) reads

$$(-iQ|v|^2 yx)_t = Q[-2i\,\mathrm{Re}(\bar{v}v_t)y_x - i|v|^2 y_{xt}]. \tag{6.73}$$

Let φ be a test function in H^1 that satisfies $\|\varphi\|_{H^1} = 1$. We write

$$\begin{aligned}(-2i\,\mathrm{Re}(\bar{v}v_t)y_x, \varphi)_{L^2} &= \mathrm{Re}\int -2i\,\mathrm{Re}(\bar{v}v_t)y_x\bar{\varphi}\\ &= 2\,\mathrm{Re}\int v_t\bar{v}\;\;\mathrm{Im}(y_x\bar{\varphi})\\ &\leq 2\|v_t\|_{H^{-1}}\|\bar{v}\,\mathrm{Im}(y_x\bar{\varphi})\|_{H^1}.\end{aligned} \tag{6.74}$$

Since $\bar{v}$, y_x, and $\bar{\varphi}$ are bounded in H^1 (see (6.85) and (6.72) and since H^1 is a multiplicative algebra (see (6.42)), $\bar{v}\,\mathrm{Im}(y_x\bar{\varphi})$ is bounded in H^1. On the other hand, v_t is bounded in H^{-1} (see (6.69)). Therefore

$$\sup_{\|\varphi\|_{H^1}=1} (-2i\,\mathrm{Re}(\bar{v}v_t)y_x, \varphi)_{L^2} \leq K(N), \tag{6.75}$$

and we have an upper bound for $-2i\,\mathrm{Re}(\bar{v}v_t)y_x$ in $C_b(\mathbb{R}_+; H^{-1})$.

Let us now majorize the second term in the right-hand side of (6.73) as follows: let φ be as above. Then

$$(-i|v|^2 y_{xt}, \varphi)_{L^2} \leq \|y_{xt}\|_{H^{-1}} \| |v|^2 \varphi \|_{H^1}. \tag{6.76}$$

Applying the inverse inequality

$$\|y_{xt}\|_{H^{-1}} \leq (2\pi N) \|y_t\|_{H^{-1}}, \tag{6.77}$$

and (6.40), leads to the boundedness for $\|y_{xt}\|_{H^{-1}}$. On the other hand, since H^1 is an algebra, then $\| |v|^2 \phi \|_{H^1}$ is also majorized as above. Therefore, at this stage, we have proven the boundedness of $-2i \operatorname{Re}(\bar{v}v_t) y_x - i|v|^2 y_{xt}$ in H^{-1}. Since the term $(-2i \operatorname{Re}(\bar{v}y_x)v)_t$ can be handled in exactly the same manner and since the projector Q is (uniformly) bounded from H^{-1} into H^{-1}, we obtain that G_t is bounded in $C_b(\mathbb{R}_+; H^{-1})$. Let us point out, however, that this bound on G and G_t depends on N. Actually, a finer analysis shows that this bound behaves like $K(1 + N)$, where K is independent of N.

We are now able to apply Proposition 6.3 to Z_x which solves (6.65). Therefore, the proof of (6.64) for $k = 2$ is completed.

This result implies that $v = y + Z = y + Z^m$ belongs to $C_b(\mathbb{R}_+; H^2) \cap C_b^1(\mathbb{R}; L^2)$. For y, it is a consequence of the inverse inequalities (6.72) and (6.77), and for Z, this is a consequence of (6.64) for $k = 2$.

(ii) Let us now prove further a priori estimates. Let us differentiate (6.65) with respect to x. This leads to

$$\begin{cases} (Z_{xx})_t + \alpha Z_{xx} + Q(A - F'(v))Z_{xx} = G_x + QF''(v)(v_x, Z_x), \\ Z_{xx}(0) = 0, \end{cases} \tag{6.78}$$

where we have set

$$F''(v)(v_x, Z_x) = -2i[\operatorname{Re}(\bar{v}v_x)Z_x + \operatorname{Re}(\bar{v}_x Z_x)v + \operatorname{Re}(\bar{v}Z_x)v_x], \tag{6.79}$$

and

$$G_x = Q[f_{xx} + F'(v)y_{xx} + F''(v)(v_x, y_x)]. \tag{6.80}$$

We now plan to apply Proposition 6.3 to the Z_{xx} solution of (6.79) which features the same properties as (6.65). For that purpose, we just have to check that

$$\tilde{G} = G_x + QF''(v)(v_x, Z_x) \tag{6.81}$$

belongs to $C_b^1(\mathbb{R}_+; H^{-1})$. This is technical but straightforward. In fact, we use that v and v_x belong to $C_b^1(\mathbb{R}_+; H^{-1}) \cap C_b(\mathbb{R}_+; H^1)$; this is a consequence, for the y term, of the inverse inequalities; and this is a consequence of (6.64) with $k = 2$ for the Z term. We also use that y_{xx}, y_{xxt} are as smooth as we want. Details are omitted. We may now claim that (6.64) is proved for $k = 3$.

Hence v, v_x, *and* v_{xx} belong to $C_b^1(\mathbb{R}_+; H^{-1}) \cap C_b(\mathbb{R}_+; H^1)$. This allows us to differentiate (6.78) with respect to x, and to apply Proposition 6.3.

The proof can be continued in this way by induction on k; we first differentiate with respect to x, then apply Proposition 6.3, and so on.

Hence the proof of Proposition 6.2 is completed. The conditions on N_0 are given in (6.52), (6.61), Lemma 6.1, and (6.91). A finer analysis shows that $C(N, k) \sim K(1 + N)^{k-1}$ in (6.64). □

Before stating and proving Proposition 6.3, we first state an important lemma which asserts the coercivity of the linear operator $iQ(A - F'(v))$ on QH^1.

Lemma 6.1. *There exists N_1 that depends on K_0 such that if $N \geq N_1$, then for any Ψ in QH^1,*

$$(i(A - F'(v))\Psi, \Psi)_{L^2} \geq \tfrac{3}{4}\|\Psi\|_{H^1}^2. \tag{6.82}$$

Proof. Let us first write

$$(i(A - F'(v))\Psi, \Psi)_{L^2} = \|\Psi\|_{H^1}^2 - \int |v|^2|\Psi|^2 - 2\int \operatorname{Re}(\bar{v}\Psi)^2. \tag{6.83}$$

Now we observe that (see (6.45))

$$\int |v|^2|\Psi|^2 + 2\int \operatorname{Re}(\bar{v}\Psi)^2 \leq 3|v|_{L^\infty}^2|\Psi|_{L^2}^2 \leq \frac{cK_0^2}{N^4}\|\Psi\|_{H^1}^2. \tag{6.84}$$

Hence (6.84) holds when $cK_0^2/N_1^4 \leq \frac{1}{4}$. □

We now state Proposition 6.3 which has been used several times.

Proposition 6.3. *Let v be in $C_b(\mathbb{R}_+, H^1) \cap C_b^1(\mathbb{R}_+; H^{-1})$ such that*

$$\sup_{t\geq 0} (\|v(t)\|_{H^1} + \|v_t(t)\|_{H^{-1}}) \leq K_0. \tag{6.85}$$

Let G be in $C_b^1(\mathbb{R}_+; H^{-1})$. Then there exists N_2 which depends on K_0 (but is independent of G) such that for each $N \geq N_2$ fixed, the solution Z_x of (6.65) *belongs to $C_b(\mathbb{R}_+, QH^1)$ and satisfies*

$$\sup_{t\geq 0} \|Z_x(t)\|_{H^1} \leq \frac{c}{\alpha} \sup_{t\geq 0} [\|G(t)\|_{H^{-1}} + \|G_t(t)\|_{H^{-1}}]. \tag{6.86}$$

Proof. Let us multiply (6.65) by $-(\bar{Z}_x)_t - \alpha\bar{Z}_x$, take the imaginary part and then integrate on (0, 1). This leads to

$$\begin{aligned}\frac{1}{2}\frac{d}{dt}J(Z_x) + \alpha J(Z_x) = &-\int \operatorname{Re}(\bar{v}v_t)|Z_x|^2 - 2\int \operatorname{Re}(\bar{v}_t Z_x)\cdot\operatorname{Re}(\bar{v}Z_x)\\ &+ \int \operatorname{Im}(G_t\bar{Z}_x) + \alpha\int \operatorname{Im}(G\bar{Z}_x),\end{aligned} \tag{6.87}$$

where

$$J(Z_x) = \|Z_x\|_{H^1}^2 - \int |v|^2|Z_x|^2 - 2\int \mathrm{Re}(\bar{v}Z_x)^2 + 2\int \mathrm{Im}(G\bar{Z}_x). \tag{6.88}$$

On the one hand, due to Lemma 6.1, and assuming $N \geq N_1$, N_1 as in Lemma 6.1,

$$\begin{aligned} J(Z_x) &\geq \tfrac{3}{4}\|Z_x\|_{H^1}^2 - 2\|G\|_{H^{-1}}\|Z_x\|_{H^1} \\ &\geq \tfrac{1}{2}\|Z_x\|_{H^1}^2 - c\|G\|_{H^{-1}}^2. \end{aligned} \tag{6.89}$$

On the other hand, the right-hand side of (6.87) can be handled exactly as in Section 6.3.2. This yields the

$$\begin{aligned} \text{r.h.s. of (6.87)} &\leq \frac{c}{\sqrt{N}}\|v_t\|_{H^{-1}}\|v\|_{H^1}\|Z_x\|_{H^1}^2 + (\|G_t\|_{H^{-1}} + \|G\|_{H^{-1}})\|Z_x\|_{H^1} \\ &\leq \left(\frac{\alpha}{8} + \frac{cK_0^2}{\sqrt{N}}\right)\|Z_x\|_{H^1}^2 + \frac{c}{\alpha}[\|G\|_{H^{-1}}^2 + \|G_t\|_{H^{-1}}^2]. \end{aligned} \tag{6.90}$$

Assuming that $N \geq N_2$ is large enough,

$$\frac{\alpha}{8} + \frac{cK_0^2}{\sqrt{N}} \leq \frac{\alpha}{8} + \frac{cK_0^2}{\sqrt{N_2}} \leq \frac{\alpha}{4}, \qquad \text{i.e.,} \quad N_2 \geq \frac{64K_0^4}{\alpha^2}, \tag{6.91}$$

we infer from (6.87), (6.89), and (6.90) that

$$\frac{1}{2}\frac{d}{dt}J(Z_x) + \frac{\alpha}{2}J(Z_x) \leq \frac{c}{\alpha}(\|G\|_{H^{-1}}^2 + \|G_t\|_{H^{-1}}^2). \tag{6.92}$$

Deriving (6.86) from (6.89) and (6.92) is just a simple application of the usual Gronwall lemma, observing that $J(Z_x(0)) = J(0) = 0$. □

6.4. Comparison of z and Z for Large Times

For the sake of convenience, we again assume in this section that $t_1 = 0$ in (6.29), and that hence $u(t)$ remains in the absorbing ball in H^1 whose radius is ρ_1. The general case will be handled in Section 6.5 by a time translation.

Let $z = Qu$ be the solution of (6.25) and let Z be the solution of (6.29). We state

Proposition 6.4. *Let N_0 be as in Proposition* 6.1. *Let $N \geq N_0$ be fixed. We then have*

$$\|Z(t) - z(t)\|_{H^1} \leq c_0 e^{-\alpha t}, \tag{6.93}$$

where c_0 depends on $\|u_0\|_{H^1}$ and the data α, f.

PROOF. Let $v = y + Z$ and let $\chi = z - Z = u - v$. Then χ is the solution of

$$\chi_t + \alpha\chi + i\chi_{xx} - i\chi = iQ(|v|^2 v - |u|^2 u) = iQ[(|u|^2 + |v|^2)\chi + uv\bar{\chi}]. \tag{6.94}$$

Multiplying (6.94) by $-\bar{\chi}_t - \alpha\bar{\chi}$, taking the imaginary part and integrating on (0, 1), we obtain

$$\frac{1}{2}\frac{d}{dt}J(\chi) + \alpha J(\chi) = -\frac{1}{2}\int(|u|^2 + |v|^2)_t|\chi|^2 - \frac{\text{Re}}{2}\int(uv)_t\bar{\chi}^2, \tag{6.95}$$

where

$$J(\chi) = \|\chi\|_{H^1}^2 - \int(|u|^2 + |v|^2)|\chi|^2 - \text{Re}\int uv\bar{\chi}^2. \tag{6.96}$$

Since Z satisfies (6.62) (see Corollary 6.1) and since $\|u\|_{H^1} \leq \rho_1$, we have

$$J(\chi) \geq \|\chi\|_{H^1}^2\left(1 - \frac{c(\rho_1^2 + K_1^2)}{N^2}\right). \tag{6.97}$$

Hence if N_0 is large enough as required by Proposition 6.1, we obtain the coerciveness of J on H^1, that is,

$$J(\chi) \geq \tfrac{1}{2}\|\chi\|_{H^1}^2. \tag{6.98}$$

We can derive an upper bound for the right-hand side of (6.95) as we did in the proof of Proposition 6.1. This yields, due to (6.42) and (6.45), the

$$\begin{aligned}\text{r.h.s. of (6.95)} &\leq c(\|u_t\|_{H^{-1}} + \|v_t\|_{H^{-1}})(\|u\|_{H^1} + \|v\|_{H^1})\cdot\|\chi\|_{H^1}|\chi|_{L^\infty}\\ &\leq \frac{K}{\sqrt{N}}\|\chi\|_{H^1}^2,\end{aligned} \tag{6.99}$$

since u, v are bounded in $C_b(\mathbb{R}_+; H^1) \cap C_b^1(\mathbb{R}_+; H^{-1})$: this follows from (6.19) and (6.21) for u and y, and from (6.62) and (6.63) for Z, and then for v.

We then easily deduce from (6.95), (6.98), and (6.99) that

$$\frac{1}{2}\frac{d}{dt}J(\chi) + \alpha J(\chi) \leq \frac{K}{\sqrt{N}}\|\chi\|_{H^1}^2 \leq \frac{2K}{\sqrt{N}}J(\chi). \tag{6.100}$$

Therefore, since $2K/\sqrt{N} \leq \alpha/2$ (as in Proposition 6.1), the classical Gronwall lemma leads to

$$J(\chi(t)) \leq J(\chi(0))e^{-\alpha t}. \tag{6.101}$$

Using $\chi(0) = z_0$ and the coerciveness of J on QH^1 (see (6.98)), we easily complete the proof of (6.93). □

6.5. Application to the Attractor. The Main Result

6.5.1. Existence of the Attractor in H^1

We now aim to prove

Theorem 6.2. *We assume that f is given in L^2. The semigroup $\{S(t)\}_{t\geq 0}$ defined by* (6.1) *possesses a compact global attractor $\mathscr{A}$ in H^1.*

PROOF. Let $u(t) = S(t)u_0$, where u_0 belongs to a given bounded set $\mathscr{B}$. Let t_1 be as in (6.19); hence $u(t)$ remains, for $t \geq t_1$ in the ball $\mathscr{B}_1$ of H^1, centered at 0 of radius ρ_1. Let N_0 be large enough (as in Proposition 6.1), and let $N \geq N_0$ be fixed. We now set

$$\begin{cases} S_1(t)u_0 = y(t) = Pu(t), & \text{for} \quad t \leq t_1, \\ \qquad\quad\; = v(t) = y(t) + Z(t), & \text{for} \quad t \geq t_1, \end{cases} \tag{6.102}$$

where $Z(t)$ solves (6.29) and

$$\begin{cases} S_2(t)u_0 = z(t) = Qu(t), & \text{for} \quad t \leq t_1, \\ \qquad\quad\; = \chi(t) = z(t) - Z(t), & \text{for} \quad t \geq t_1. \end{cases} \tag{6.103}$$

Because of Propositon 6.3, $Z(t)$ remains, for $t \geq t_1$, in a bounded set of H^2. Since $y(t)$, due to (6.72), also remains in a bounded set of H^2, $\bigcup_{t\geq t_1} S_1(t)\mathscr{B}$ is bounded in H^2, and hence relatively compact in H^1.

Furthermore, applying Proposition 6.4 leads to

$$\|Z(t) - z(t)\|_{H^1} \leq c_0 \exp(-\alpha(t - t_1)) \qquad \text{for} \quad t \geq t_1, \tag{6.104}$$

where c_0 depends boundedly on $\|u(t_1)\|_{H^1}$ ($\leq \rho_1$). Hence I.(1.13) is satisfied.

It is easy to prove that, for a given t, $S_1(t)$ and $S_2(t)$ are continuous mappings on H^1. In fact, $Z(t)$ depends continuously on $y(t)$, and $y(t) = Pu(t)$ depends continuously on u_0 (on bounded intervals of time).

All the other hypotheses of Theorem I.1.1 are easy to check and therefore Theorem 6.2 follows. By perusing the proofs we see that $f \in \mathscr{C}^\infty$ is not needed, and $f \in L^2$ is sufficient. □

6.5.2. Regularity of the Attractor

The aim of this section is to prove

Theorem 6.3.

(i) *We assume that $f \in \mathscr{C}^\infty$, the space of one-periodic, $\mathscr{C}^\infty$ complex-valued functions on $\mathbb{R}$. Then the global attractor $\mathscr{A}$ defined in Theorem 6.2 is included and bounded in H^k, for any $k \geq 2$.*

(ii) $\mathscr{A} \subset \mathscr{C}^\infty$.

(iii) For each t, and *for each integer* $m \geq 1$, $S(t)$ *maps* H^m *into itself and* $\mathscr{A}$ *is the global attractor for* $S(t)$ *in* H^m.

PROOF. For (i), let u_0 be in $\mathscr{A}$. We want to show that $u_0 \in H^k$, $\forall k \geq 1$, and we observe that

$$u_0 \in H^k \quad \Leftrightarrow \quad z_0 = Qu_0 \in H^k. \tag{6.105}$$

Here $Q = Q_N$ and $N \geq N_0$, N_0 as in Proposition 6.1. It is well known that there exists a complete orbit $u(t) \in C_b(\mathbb{R}; H^1)$ such that $u(0) = u_0$; actually, $\mathscr{A}$ is the union of such complete orbits (see Section I.1.3 and Definition I.1.2(i)). Moreover, we may assume that (6.19) and (6.21) are valid for such a complete orbit $u(t)$, for any $t \in \mathbb{R}$.

We now plan to approximate z_0 as follows. Let m be an integer. Let $Z^m(t)$ be the solution in $C_b([-m, +\infty[, QH^1)$ of

$$\begin{cases} Z_t^m + \alpha Z^m + iZ_{xx}^m - iZ^m + iQ(|y + Z^m|^2(y + Z^m)) = Qf, \\ Z^m(-m) = 0. \end{cases} \tag{6.106}$$

The (global) existence result for Z^m can be proven exactly as in Section 6.3. Moreover, the analog of Proposition 6.2 is also valid, and Z^m satisfies: for any $k \geq 1$, there exists $C(N, k)$ that depends only on N, k, and on the data, such that

$$\sup_{t \geq -m} \|Z^m(t)\|_{H^k} \leq C(N, k). \tag{6.107}$$

We now set

$$Z_m = Z^m(0). \tag{6.108}$$

The sequence $(Z_m)_{m \in N}$ is bounded in H^k, for any k, and therefore strongly converges in any H^k. We shall prove below that

$$Z_m \to z_0 \quad \text{strongly in } H^1, \tag{6.109}$$

and this will complete the proof of (i).

To prove (6.109), we observe that the analog of Proposition 6.4 is valid (the proof is the same as in Section 6.4); we obtain

$$\|Z^m(t) - z(t)\|_{H^1} \leq c_m e^{-\alpha(t+m)}, \tag{6.110}$$

for any $t \geq -m$, and where c_m depends boundedly on $\|u(-m)\|_{H^1}$. Due to (6.19), $\|u(-m)\|_{H^1}$ is bounded independently of m by ρ_1. Hence setting $t = 0$ in (6.110) leads to

$$\|Z_m - z_0\|_{H^1} \leq Ke^{-\alpha m}, \tag{6.111}$$

and, as $m \to \infty$, we obtain (6.109) and (6.105); (i) is proved.

The proof of (ii) is a mere consequence of the Sobolev embeddings (see

II.(1.12))

$$\bigcap_{k=1}^{\infty} H^k = \mathscr{C}^\infty.$$

For (iii) we observe that if $u_0 \in H^m$, where m is an integer ≥ 1, then $u(t) = S(t)u_0$ is well defined. Also it can be shown by space differentiation of equation (6.1), that $\{t \mapsto S(t)u_0\}$ belongs to $\mathscr{C}_b(\mathbb{R}_+; H^m)$. The proof conducted by induction on m is based on a priori estimates similar to those used for Theorem 6.1; namely, we differentiate (6.1) k times in x, $k \leq m$, and we successively multiply the corresponding equation by $\partial^k \bar{u}/\partial x^k$ and by $-(\partial^{k+1}\bar{u}/\partial x^k \partial t + \alpha \partial^k \bar{u}/\partial x^k)$; we omit the calculations. Similarly, we show that the operators $S(t)$ satisfy I.(1.1)–I.(1.4) in H^m, and, based on Theorem I.1.1, we conclude that $S(t)$ possesses an attractor $\mathscr{A}_m$ in H^m. It is clear by the Sobolev embedding theorems and by the invariance property ($S(t)\mathscr{A}_m = \mathscr{A}_m$, $\forall t \in \mathbb{R}$), that

$$\mathscr{A}_m \subset \mathscr{A}_{m-1} \subset \cdots \subset \mathscr{A}_2 \subset \mathscr{A}_1 = \mathscr{A}.$$

Now $\mathscr{A}_m = \mathscr{A}$; indeed, since $\mathscr{A}$ is bounded in H^m by (i), we have

$$\operatorname{dist}_{H^m}(S(t)\mathscr{A}, \mathscr{A}_m) = \sup_{a \in \mathscr{A}} \inf_{a_m \in \mathscr{A}_m} \|S(t)a - a_m\|_{H^m} \to 0,$$

when $t \to \infty$. But since $\mathscr{A}$ is invariant, $S(t)\mathscr{A} = \mathscr{A}$ and thus

$$\operatorname{dist}_{H^m}(\mathscr{A}, \mathscr{A}_m) = 0, \qquad \text{i.e.,} \quad \mathscr{A} \subset \mathscr{A}_m. \tag{6.112}$$

Theorem 6.3 is proved. □

6.6. Determining Modes

We conclude this section by showing a result related to the concept of determining modes introduced in C. Foias and G. Prodi [1] and generalized in C. Foias, O. Manley, R. Temam, and Y. Treve [1]. It was shown for the two-dimensional Navier–Stokes equations that the large time behavior of the orbits is completely determined by its projection on a spectral finite-dimensional space. This result was subsequently extended to numerous (strongly) dissipative equations. Following O. Goubet [1] we want to show that the orbits on $\mathscr{A}$ are fully determined by their projections P_N, where N is sufficiently large.

Theorem 6.4. *Let $\{u_1(t)\}_{t \in \mathbb{R}}$ and $\{u_2(t)\}_{t \in \mathbb{R}}$ be two complete orbits of $\mathscr{A}$ such that*

$$P_N u_1(t) = P_N u_2(t), \qquad \forall t \in \mathbb{R}, \tag{6.113}$$

where $N \geq N_0$, N_0 as in Theorem 6.1. Then

$$u_1(t) = u_2(t), \qquad \forall t \in \mathbb{R}. \tag{6.114}$$

Proof. It is equivalent to prove the following statement: let $y(t) = P_{u1}(t) = P_{u2}(t)$, $P = P_N$, be such that

$$\|y(t)\|_{H^1} \le \rho_1 \quad \text{and} \quad \|y_t(t)\|_{H^{-1}} \le \rho', \quad \forall t \in \mathbb{R}; \tag{6.115}$$

then there exists only one solution $z = z(t) \in \mathscr{C}_b(\mathbb{R}, QH^1)$ of the nonlinear autonomous equation

$$\begin{cases} z_t + \alpha z + iz_{xx} - iz + iQ(|y + z|^2(y + z)) = Qf, \\ \limsup_{t \to -\infty} \|z(t)\|_{H^1} < +\infty. \end{cases} \tag{6.116}$$

Let $z^1(t)$ and $z^2(t)$ be two solutions of (6.116). We now follow step by step the proof of Proposition 6.4. We set $v^1 = y + z^1$ and $v^2 = y + z^2$. Let us also set $\chi = z^2 - z^1 = v^2 - v^1$. Then χ satisfies

$$\chi_t + \alpha\chi + i\chi_{xx} - i\chi = -iQ[|v_1|^2 + |v_2|^2]\chi - iQv_1v_2\bar{\chi}. \tag{6.117}$$

Multiplying (6.117) by $-\bar{\chi}_t - \alpha\bar{\chi}$, taking the imaginary part, and integrating on (0, 1), we obtain, as in (6.100),

$$\frac{1}{2}\frac{d}{dt}J(\chi) + \frac{\alpha}{2}J(\chi) \le 0, \tag{6.118}$$

where

$$J(\chi) = \|\chi\|_{H^1}^2 - \int (|v_1|^2 + |v_2|^2)|\chi|^2 - \operatorname{Re}\int v_1v_2\bar{\chi}^2 \ge \tfrac{1}{2}\|\chi\|_{H^1}^2, \tag{6.119}$$

provided $N \ge N_0$. Integrating (6.118) between t' and $t > t'$, and also using (6.119), leads to

$$\|\chi(t)\|_{H^1}^2 \le J(\chi(t'))e^{-\alpha(t-t')} \quad \text{for} \quad t \ge t'. \tag{6.120}$$

We now let t' go to $-\infty$. Observing that $J(\chi(t'))$ remains bounded, we then obtain the result. □

Remark 6.3. The presentation of this section follows essentially O. Goubet [1]. Partial results for attractors for the weakly damped nonlinear Schrödinger equations were previously derived by J.M. Ghidaglia [5] and by X. Wang [1]. In the former article, the existence of a *weak* global attractor in H^1 is proven, i.e., the existence of a compact attractor for the weak topology of H^1. In the latter article the result is improved, following an idea of J. Ball [1] (also used in J.M. Ghidaglia [?]), and the weak attractor is shown to be a strong attractor in H^1; this is the attractor called $\mathscr{A}$ in Theorem 6.2. In our presentation here, the existence of the strong attractor in H^1 is obtained at the end, but the proof also yields a number of other important results.

As indicated before, we study in Section 7 another weakly dissipative equation, namely, the (weakly damped) Korteweg–de Vries equation.

7. Weakly Dissipative Equations II. The Korteweg–de Vries Equation

In this section, we consider a second example of weakly dissipative equations, namely, the Korteweg–de Vries (KdV) equation with a weak dissipation and an external forcing term. The KdV equation (without dissipation or forcing) was initially derived by D.J. Korteweg and G. de Vries [1] as a model for one-directional water waves of small amplitude in shallow water, and it was later shown to model a number of other physical systems related to the propagation of one-directional waves in nonlinear dispersive media, when the nonlinearity and dispersion have comparable effects, the dispersion relation has a particular form, and the nonlinearity is weak and quadratic (e.g., magnetosonic waves and ion-sound waves; see, for instance, J.W. Miles [1] and A.C. Newell [1]). In many real situations, however, one cannot neglect external excitation and energy dissipation mechanisms, especially for the long-time behavior. Several energy dissipation mechanisms were derived by E. Ott and R.N. Sudan [1] depending on the physical situation. We consider here a weak dissipation on the lower-order (zero derivative) term which has no smoothing effect. Hence, from the dynamical system point of view the main difficulty is the lack of compactness of the semigroup. We overcome this difficulty and derive the existence of the attractor by using energy equations and an idea of J. Ball [1] (see J.M. Ghidaglia [6]) and this essentially amounts to an application of Theorem I.1.1 supplemented by Remark I.1.4.

As in Section 6, the other problem that we address is the regularity of the attractor; it is proved that the global attractor is as smooth as the forcing term. As in Section 6, this is obtained through a nonobvious decomposition of the solutions of the equation in order to overcome the lack of a suitable smoothing effect of the equation.

Despite the similarities in the results, this section is technically very different from Section 6. Indeed, we know that the energy equalities for the Korteweg–de Vries equation are obtained by using the infinite sequence of polynomial invariants which follow, in particular, from its exact integrability via the Inverse Scattering Theory (see, e.g., R.M. Miura, C.S. Gardner, and M.D. Kruskal [1], M.D. Kruskal, R.M. Miura, C.S. Gardner, and N.J. Zabusky [1], and A.C. Newell [1]). Contrarily to what was done for all the other equations in this book, these invariants are obtained by multiplying the equation by *nonlinear* (and nonobvious) functions of u and its spatial derivatives; see Section 7.2 for the details.

In Section 7.1 we describe the equation and its mathematical setting and state some preliminary results. Section 7.2 contains the proof of the existence of the attractor in higher-order Sobolev spaces. In Section 7.3 the regularity of the attractor is obtained. Finally, Sections 7.4 and 7.5 contain the proofs of technical results; Section 7.4 contains a sketch of the proofs of the results stated in Section 7.1; Section 7.5 contains the proof of Proposition 7.2.

Section 7 follows essentially the article by I. Moise and R. Rosa [1].

7.1. The Equation and its Mathematical Setting

We consider the following nonlinear evolution equation for an unknown scalar function $u = u(x, t)$, $x \in \mathbb{R}$, $t \in \mathbb{R}$:

$$\frac{\partial u}{\partial t} + u\frac{\partial u}{\partial x} + \frac{\partial^3 u}{\partial x^3} + \gamma u = f, \tag{7.1}$$

where $f = f(x)$ and γ are given, with $\gamma > 0$ in the case of interest to us. We supplement this equation with the space-periodicity boundary condition

$$u(x + L, t) = u(x, t), \qquad \forall x \in \mathbb{R}, \quad \forall t \in \mathbb{R}, \tag{7.2}$$

$L > 0$ given, and with the initial condition

$$u(x, 0) = u_0(x), \qquad \forall x \in \mathbb{R}. \tag{7.3}$$

For the functional setting of the problem we consider the Sobolev spaces $H^k_{\text{per}}(\Omega)$ of periodic functions, locally in H^k, where k is a nonnegative integer and $\Omega = (0, L)$; these spaces are endowed with the following norm:

$$\|u\|_k = \left(\sum_{j=0}^{k} L^{2j}|u|_j^2\right)^{1/2}, \tag{7.4}$$

where

$$|u|_j = \left(\int_\Omega |D^j u|^2\, dx\right)^{1/2}, \qquad D = \frac{\partial}{\partial x}. \tag{7.5}$$

The space $H^0_{\text{per}}(\Omega)$ is simply $L^2(\Omega)$, and for $k \geq 1$, functions in $H^k_{\text{per}}(\Omega)$ are bounded continuous functions on $[0, L]$. Indeed, it is easy to deduce the following explicit form of the Agmon inequality:

$$\|\omega\|_\infty = \sup_{0 \leq x \leq L} |u(x)| \leq |u|_0^{1/2}(2|u|_1 + L^{-1}|u|_0)^{1/2}, \qquad \forall u \in H^k_{\text{per}}(\Omega), \tag{7.6}$$

for $k \geq 1$. In case the function has zero average, we have simply

$$\|u\|_\infty \leq \sqrt{2}|u|_0^{1/2}|u|_1^{1/2}, \qquad \forall u \in H^k_{\text{per}}(\Omega), \qquad \int_\Omega u(x)\, dx = 0,$$

for $k \geq 1$. We also write $(\cdot, \cdot)_{L^2}$ for the usual scalar product in $L^2(\Omega)$.

We assume that $u_0, f \in H^k_{\text{per}}(\Omega)$, $k \geq 2$, and study the problem (7.1), (7.2), and (7.3) as a dynamical system in $H^k_{\text{per}}(\Omega)$. The existence and uniqueness of solutions for this problem in $H^2_{\text{per}}(\Omega)$ for the case $\gamma = 0$ and $f = 0$ was proved by R. Temam [6] and later by J.L. Bona and R. Smith [1] with the help of three polynomial invariants. When γ and f are not zero, these polynomials are no longer invariant, but satisfy some energy-like equations and are important for the study of the long-time behavior of the solutions, as well as for the proof of the existence of solutions (see Section 7.2). The original KdV equation (equation (7.1) with $\gamma = 0$ and $f = 0$) possesses, in fact, an infinite number of polynomial invariants, which follow, in particular, from its exact integrability via the Inverse Scattering Theory (see, e.g., R.M. Miura, C.S. Gardner, and M.D. Kruskal [1], M.D. Kruskal, R.M. Miura, C.S. Gardner, and N.J. Zabusky [1], and A.C. Newell [1]). The polynomial invariants are of the form

$$I_m(u) = \int_\Omega [(D^m u)^2 - \alpha_m u(D^{m-1}u)^2 + Q_m(u, Du, \ldots, D^{m-2}u)]\, dx,$$

$$m = 0, 1, 2, \ldots, \tag{7.7}$$

where $D = \partial/\partial x$ and $Q_m = Q_m(u, Du, \ldots, D^{m-2}u)$ is a polynomial composed only of monomials of rank $m + 2$; here the rank of a monomial $u^{a_0}(Du)^{a_1} \cdots (D^l u)^{a_l}$ ($a_j \geq 0$ integer, $j = 0, 1, \ldots, l$) is defined as $\sum_{j=0}^{l} (1 + j/2)a_j$. For $m = 0$, 1, we simply set $Q_m \equiv 0$ in formula (7.7). Because of the rapid proliferation of terms with increasing rank, it is very difficult to find more explicit recursive formulas for expressing these invariants $I_m(u)$.

We write $Q_m = Q_m(u, Du, \ldots, D^{m-2}u)$ as a linear combination of all possible terms of rank $m + 2$, but consider only the *irreducible terms*, i.e., the terms which do not have the highest-derivative factor occurring linearly. The reason is as follows: if a term is not irreducible, then one can show that it can be integrated by parts and expressed as a combination of irreducible terms. The coefficients of Q_m, as well as α_m, can be obtained through the method of undetermined coefficients (see M.D. Kruskal, R.M. Miura, C.S. Gardner, and N.J. Zabusky [1]), which is as follows: we consider

$$\begin{aligned}
\frac{d}{dt} I_m(u) &= \frac{d}{dt} \int_\Omega [(D^m u)^2 - \alpha_m u(D^{m-1}u)^2 + Q_m(u, Du, \ldots, D^{m-2}u)]\, dx \\
&= \int_\Omega \Bigg[2D^m u D^m u_t - \alpha_m u_t (D^{m-1}u)^2 - 2\alpha_m u D^{m-1} u D^{m-1} u_t \\
&\qquad + \sum_{j=0}^{m-2} \frac{\partial Q_m}{\partial y_j}(u, Du, \ldots, D^{m-2}u) D^j u_t \Bigg] dx \\
&= \int_\Omega \Bigg[2(-1)^m D^{2m} u - \alpha_m (D^{m-1}u)^2 + 2\alpha_m (-1)^m D^{m-1}(u D^{m-1} u) \\
&\qquad + \sum_{j=0}^{m-2} (-1)^j D^j \left(\frac{\partial Q_m}{\partial y_j}(u, Du, \ldots, D^{m-2}u) \right) \Bigg] u_t \, dx,
\end{aligned}$$

and denote

$$L_m(u) = 2(-1)^m D^{2m} u - \alpha_m (D^{m-1}u)^2 + 2\alpha_m (-1)^m D^{m-1}(u D^{m-1} u) + \sum_{j=0}^{m-2} (-1)^j D^j \left(\frac{\partial Q_m}{\partial y_j}(u, Du, \dots, D^{m-2}u) \right). \tag{7.8}$$

Then, we have

$$\frac{d}{dt} I_m(u) = \int_\Omega L_m(u) u_t \, dx. \tag{7.9}$$

For $I_m(u)$ to be invariant for the KdV equation, we must impose

$$(L_m(u), uDu + D^3 u)_{L^2} = 0. \tag{7.10}$$

After making several integrations by parts in order to obtain only irreducible terms in (7.10), we demand that this final expression vanish identically, and we find α_m and the polynomial Q_m.

In case f and γ do not vanish, we obtain instead the following energy equation:

$$\frac{d}{dt} I_m(u) + \gamma(u, L_m(u))_{L^2} = (f, L_m(u))_{L^2}.$$

Using (7.7) and (7.8), we obtain after integration by parts

$$\frac{d}{dt} I_m(u) + 2\gamma I_m(u) = K_m(u), \tag{7.11}$$

where

$$K_m(u) = \int_\Omega \Big[2D^m f D^m u - \alpha_m f (D^{m-1}u)^2 - 2\alpha_m u D^{m-1} u D^{m-1} f + \gamma \alpha_m u (D^{m-1}u)^2 + 2\gamma Q_m(u, Du, \dots, D^{m-2}u) + \sum_{j=0}^{m-2} (D^j f - \gamma D^j u) \frac{\partial Q_m}{\partial y_j}(u, Du, \dots, D^{m-2}u) \Big] dx. \tag{7.12}$$

With the help of the energy equations (7.11) we have the following result (see also J.M. Ghidaglia [4], when $k = 2$):

Theorem 7.1. *For $\gamma \in \mathbb{R}$, $f \in H^k_{\text{per}}(\Omega)$, and $u_0 \in H^k_{\text{per}}(\Omega)$, $k \geq 2$, there exists a unique solution u of* (7.1)–(7.3) satisfying

$$u \in L^\infty(0, T; H^k_{\text{per}}(\Omega)) \cap \mathscr{C}([0, T], L^2(\Omega)), \qquad \forall T > 0. \tag{7.13}$$

Note that replacing simultaneously x and t by $-x$ and $-t$, we obtain the same equation (7.1) with γ and f replaced by $-\gamma$ and $-f$, respectively. Therefore, Theorem 7.1 says that the solutions exist for all $t \in \mathbb{R}$. Then, we have

Theorem 7.2. *Let $\gamma \in \mathbb{R}$ and $f \in H^k_{\text{per}}(\Omega)$, $k \geq 2$. Then, for every $u_0 \in H^k_{\text{per}}(\Omega)$, the solution $u = u(t)$ of (7.1)–(7.3) restricted to $[-T, T]$ belongs to $\mathscr{C}([-T, T], H^k_{\text{per}}(\Omega))$, for all $T > 0$. Hence, we can define on $H^k_{\text{per}}(\Omega)$ the group $\{S^{(k)}(t)\}_{t \in \mathbb{R}}$ given by*

$$S^{(k)}(t)u_0 = u(t), \qquad \forall t \in \mathbb{R}, \tag{7.14}$$

for all $u_0 \in H^k_{\text{per}}(\Omega)$, where $u(t)$ is the solution of (7.1)–(7.3) at time t. For each $t \in \mathbb{R}$, $S^{(k)}(t)$ is weakly and strongly continuous on $H^k_{\text{per}}(\Omega)$; moreover, the following estimate holds:

$$\sup_{\substack{|t| \leq T \\ \|u_0\|_k \leq R}} \|S^{(k)}(t)u_0\|_k \leq C(R, T), \qquad \forall T, R > 0, \tag{7.15}$$

where $C(R, T)$ is a constant depending on R and T, as well as on L, γ, and $\|f\|_k$.

As a corollary of the proof of Theorem 7.2, we have the following result:

Corollary 7.1. *For $\gamma \in \mathbb{R}$, $f \in H^k_{\text{per}}(\Omega)$, and $u_0 \in H^k_{\text{per}}(\Omega)$, $k \geq 2$, the solution $u = u(t) = S^{(k)}(t)u_0$, $t \in \mathbb{R}$, satisfies the energy equations (7.11) for $m = 0, 1, \ldots, k$.*

One way to prove Theorems 7.1 and 7.2 is to use the energy equation (7.11) with $m = 0, 1, \ldots, k$. However, a simpler way is to use (7.11) with $m = 0, 1, 2$ to prove the result *for* $k = 2$, and then to use some commutator estimates as done by J.C. Saut and R. Temam [2] for the original KdV equation to obtain the result for $k > 2$. These estimates will also be useful for our study of the regularity of the global attractor and for this reason we state them below.

Lemma 7.1. *Let u, v belong to $H^s_{\text{per}}(\Omega)$, $s \in \mathbb{N}$, $s > 1$, and $\beta \in \mathbb{R}$, $\beta > \frac{1}{2}$. Then*

$$|D^s(uv) - uD^s v|_0 \leq c(\beta, s, L)\{\|u\|_s \|v\|_\beta + \|u\|_{1+\beta}\|v\|_{s-1}\}, \tag{7.16}$$

where, for $\beta \in \mathbb{R}$, $\beta \geq 0$,

$$\|u\|_\beta = \left\{L \sum_{j \in \mathbb{Z}} (1 + (2\pi)^2 |j|^2)^\beta |\hat{u}_j|^2\right\}^{1/2} \qquad \text{for every} \quad u = \sum_{j \in \mathbb{Z}} \hat{u}_j e^{2\pi i j x/L}.$$

For the proof, see J.C. Saut and R. Temam [2].

7.2. Absorbing Sets and Attractors

In this section, we assume $\gamma > 0$, for dissipativity, and consider $f \in H^k_{\text{per}}(\Omega)$ given, with $k \geq 2$ fixed. First, we show in the following proposition that there exists a bounded absorbing set in $H^k_{\text{per}}(\Omega)$ for $\{S^{(k)}(t)\}_{t \in \mathbb{R}}$.

Proposition 7.1. *There exists a constant $\rho_k = \rho_k(\gamma, L, \|f\|_k)$ and for every $R > 0$ there exists $T_k = T_k(R, \gamma, L, \|f\|_k)$ such that*

$$\|S^{(k)}(t)u_0\|_k \leq \rho_k, \qquad \forall t \geq T_k, \quad \forall u_0 \in H^k_{\text{per}}(\Omega), \quad \|u_0\|_k \leq R.$$

Hence, the ball $\mathscr{B}_k$ of $H^k_{\text{per}}(\Omega)$ centered at 0 and of radius ρ_k is absorbing in $H^k_{\text{per}}(\Omega)$ for the semigroup $\{S^{(k)}(t)\}_{t\geq 0}$.

PROOF. The case $k = 2$ was done by J.M. Ghidaglia [4], but we include it here for completeness. The proof is obtained using the energy equations (7.11) for $m = 0, 1, \ldots, k$. We first need some intermediate results.

Step I. *Time-uniform a priori estimates in $L^2(\Omega)$ and $H^1_{\text{per}}(\Omega)$.*

For $m = 0$, we have $L_0(u) = 2u$, and thus the energy equation (7.11) with $m = 0$ becomes

$$\frac{d}{dt}|u|_0^2 + 2\gamma|u|_0^2 = 2\int_\Omega fu\,dx. \tag{7.17}$$

Therefore,

$$\frac{d}{dt}|u|_0^2 + 2\gamma|u|_0^2 \leq 2|f|_0|u|_0 \leq \gamma|u|_0^2 + \frac{1}{\gamma}|f|_0^2,$$

which gives

$$|u(t)|_0^2 \leq |u_0|_0^2 e^{-\gamma t} + \frac{|f|_0^2}{\gamma^2}, \qquad \forall t \geq 0. \tag{7.18}$$

Consequently,

$$|u(t)|_0 \leq \frac{2|f|_0}{\gamma}, \qquad \forall t \geq T_0, \tag{7.19}$$

where

$$T_0 = T_0(|u_0|_0, \gamma, |f|_0) = \frac{1}{\gamma}\log\left(\frac{\gamma^2|u_0|_0^2}{3|f|_0^2}\right).$$

For $m = 1$, a simple computation using the method of undetermined coefficients described in Section 7.1 yields

$$I_1(u) = \int_\Omega [(Du)^2 - \tfrac{1}{3}u^3]\,dx \qquad (\alpha_1 = \tfrac{1}{3}), \tag{7.20}$$

and

$$L_1(u) = -2D^2u - u^2. \tag{7.21}$$

Hence, the energy equation (7.11) for $m = 1$ becomes

$$\frac{d}{dt} I_1(u) + 2\gamma I_1(u) = \frac{\gamma}{3} \int_\Omega u^3 \, dx + \int_\Omega [2DfDu - fu^2] \, dx. \tag{7.22}$$

We estimate

$$\left| \int_\Omega u^3 \, dx \right| \le |u|_0^2 \|u\|_\infty \le \text{(using the Agmon inequality (7.6))}$$

$$\le |u|_0^{5/2} \left(2|u|_1 + \frac{1}{L} |u|_0 \right)^{1/2} \le \sqrt{2} |u|_0^{5/2} |u|_1^{1/2} + L^{-1/2} |u|_0^3$$

$$\le \text{(by the Young inequality)} \le |u|_1^2 + \tfrac{3}{4} |u|_0^{10/3} + L^{-1/2} |u|_0^3. \tag{7.23}$$

From (7.20) and (7.23) we deduce that

$$|u|_1^2 \le \tfrac{3}{2} I_1(u) + \tfrac{3}{8} |u|_0^{10/3} + \frac{1}{2L^{1/2}} |u|_0^3, \tag{7.24}$$

and also

$$I_1(u) \le \tfrac{4}{3} |u|_1^2 + \tfrac{1}{4} |u|_0^{10/3} + \frac{1}{3L^{1/2}} |u|_0^3. \tag{7.25}$$

We then estimate (7.22) as follows:

$$\frac{d}{dt} I_1(u) + 2\gamma I_1(u) \le \frac{\gamma}{3} \left[|u|_1^2 + \tfrac{3}{4} |u|_0^{10/3} + \frac{1}{L^{1/2}} |u|_0^3 \right] + 2|f|_1 |u|_1 + \|f\|_\infty |u|_0^2$$

$$\le \frac{2\gamma}{3} |u|_1^2 + \frac{\gamma}{4} |u|_0^{10/3} + \frac{\gamma}{3L^{1/2}} |u|_0^3 + \frac{3}{\gamma} |f|_1^2 + \|f\|_\infty |u|_0^2$$

$$\le \text{(using (7.24))}$$

$$\le \gamma I_1(u) + \frac{\gamma}{2} |u|_0^{10/3} + \frac{2\gamma}{3L^{1/2}} |u|_0^3 + \frac{3}{\gamma} |f|_1^2 + \|f\|_\infty |u|_0^2.$$

Therefore,

$$\frac{d}{dt} I_1(u) + \gamma I_1(u) \le \frac{\gamma}{2} |u|_0^{10/3} + \frac{2\gamma}{3L^{1/2}} |u|_0^3 + \frac{3}{\gamma} |f|_1^2 + \|f\|_\infty |u|_0^2. \tag{7.26}$$

Using now (7.18) and (7.26), we find that

$$\frac{d}{dt} I_1(u) + \gamma I_1(u) \le c\gamma \left[|u_0|_0^{10/3} e^{-(5/3)\gamma t} + \frac{|f|_0^{10/3}}{\gamma^{10/3}} \right]$$

$$+ \frac{c\gamma}{L^{1/2}} \left[|u_0|_0^3 e^{-(3/2)\gamma t} + \frac{|f|_0^3}{\gamma^3} \right]$$

$$+ \frac{3}{\gamma} |f|_1^2 + \|f\|_\infty \left[|u_0|_0^2 e^{-\gamma t} + \frac{|f|_0^2}{\gamma^2} \right], \qquad \forall t \ge 0, \tag{7.27}$$

for a numerical constant $c > 0$.

We deduce from (7.27) that

$$\frac{d}{dt} I_1(u) + \gamma I_1(u) \leq c_1 e^{-(\gamma/2)t} + c_2, \qquad \forall t \geq 0, \tag{7.28}$$

where

$$c_1 = c[\gamma |u_0|_0^{10/3} + \gamma L^{-1/2} |u_0|_0^3 + \|f\|_\infty |u_0|_0^2]$$

and

$$c_2 \; c[\gamma^{-7/3} |f|_0^{10/3} + \gamma^{-2} L^{-1/2} |f|_0^3 + \gamma^{-1} |f|_1^2 + \gamma^{-2} \|f\|_\infty |f|_0^2].$$

Then, by the Gronwall lemma we obtain

$$I_1(u(t)) \leq I_1(u_0) e^{-\gamma t} + \frac{2c_1}{\gamma} e^{-(\gamma/2)t} + \frac{c_2}{\gamma}, \qquad \forall t \geq 0. \tag{7.29}$$

Using (7.24), (7.25), and (7.18) in (7.29), we deduce that

$$\begin{aligned}
|u(t)|_1^2 &\leq \tfrac{3}{2} I_1(u(t)) + \tfrac{3}{8} |u(t)|_0^{10/3} + \frac{1}{2L^{1/2}} |u(t)|_0^3 \\
&\leq \tfrac{3}{2} I_1(u_0) e^{-\gamma t} + \frac{3c_1}{\gamma} e^{-(\gamma/2)t} + \frac{3c_2}{2\gamma} + c\left[|u_0|_0^{10/3} e^{-(5/3)\gamma t} + \frac{|f|_0^{10/3}}{\gamma^{10/3}} \right] \\
&\quad + cL^{-1/2} \left[|u_0|_0^3 e^{-(3/2)\gamma t} + \frac{|f|_0^3}{\gamma^3} \right] \\
&\leq \tfrac{3}{2} \left[\tfrac{4}{3} |u_0|_1^2 + \tfrac{1}{4} |u_0|_0^{10/3} + \frac{1}{3L^{1/2}} |u_0|_0^3 \right] e^{-\gamma t} + \frac{3c_1}{\gamma} e^{-(\gamma/2)t} + \frac{3c_2}{2\gamma} \\
&\quad + c\left[|u_0|_0^{10/3} e^{-(5/3)\gamma t} + \frac{|f|_0^{10/3}}{\gamma^{10/3}} \right] \\
&\quad + cL^{-1/2} \left[|u_0|_0^3 e^{-(3/2)\gamma t} + \frac{|f|_0^3}{\gamma^3} \right], \qquad \forall t \geq 0,
\end{aligned}$$

where, as before, c is a numerical constant. Hence,

$$|u(t)|_1^2 \leq \chi_1 e^{-(\gamma/2)t} + \tilde{\chi}_1, \qquad \forall t \geq 0, \tag{7.30}$$

with

$$\begin{aligned}
\chi_1 &= 2|u_0|_1^2 + c|u_0|_0^{10/3} + cL^{-1/2} |u_0|_0^3 + \frac{3c_1}{\gamma} \\
&\leq c(|u_0|_1^2 + |u_0|_0^{10/3} + L^{-1/2} |u_0|_0^3 + \gamma^{-1} \|f\|_\infty |u_0|_0^2),
\end{aligned} \tag{7.31}$$

and

$$\begin{aligned}
\tilde{\chi}_1 &= \frac{3c_2}{\gamma} + c\gamma^{-10/3} |f|_0^{10/3} + cL^{-1/2} \gamma^{-3} |f|_0^3 \\
&\leq c[\gamma^{-2} |f|_1^2 + \gamma^{-3} \|f\|_\infty |f|_0^2 + \gamma^{-10/3} |f|_0^{10/3} + \gamma^{-3} L^{-1/2} |f|_0^3],
\end{aligned} \tag{7.32}$$

and c in (7.31) and (7.32) is a numerical constant.

We finally deduce from (7.30) that

$$|u(t)|_1 \le \rho_1 = \rho_1(\gamma, L, |f|_0, |f|_1, \|f\|_\infty), \qquad \forall t \ge T_1, \tag{7.33}$$

where $T_1 = T_1(\gamma, L, |f|_0, |f|_1, \|f\|_\infty, |u_0|_0, |u_0|_1)$.

Step II. *Absorbing sets in $H^m_{\text{per}}(\Omega)$ for $m = 2, 3, \ldots, k$.*

We establish their existence by an inductive argument. Let m be fixed with $2 \le m \le k$. Assume that for each $j = 0, 1, \ldots, m-1$ there exist $M_j = M_j(\gamma, L, \|f\|_j)$ and $T_j = T_j(\gamma, L, \|f\|_j, R)$ such that

$$|u(t)|_j \le M_j, \qquad \forall t \ge T_j, \quad \forall u_0 \in H^j_{\text{per}}(\Omega), \quad \|u_0\|_j \le R. \tag{7.34}$$

We want to show that the same statement holds for $j = m$. Then, we will be able to deduce that it holds up to $m = k$, which proves, in particular, the existence of an absorbing ball in $H^k_{\text{per}}(\Omega)$ for $\{S^{(k)}(t)\}_{t \in \mathbb{R}}$.

In what follows, we denote by c a generic constant which may depend on γ, L, and f, but is independent of u_0.

Consider the energy equation $(7.11)_m$, i.e.,

$$\frac{d}{dt} I_m(u) + 2\gamma I_m(u) = K_m(u), \tag{7.35}$$

where $I_m(u)$ and $K_m(u)$ are given by (7.7) and (7.12), respectively.

Using the Agmon inequality and the induction hypothesis (7.34), we deduce that

$$\begin{aligned}\|u(t)\|_\infty &\le |u(t)|_0^{1/2}(2|u(t)|_1 + L^{-1}|u(t)|_0)^{1/2} \\ &\le M_0^{1/2}(2M_1 + L^{-1}M_0)^{1/2}, \qquad \forall t \ge \max\{T_0, T_1\},\end{aligned} \tag{7.36}$$

and also, since $D^j u$ has zero average for $j \ge 1$,

$$\begin{aligned}\|D^j u(t)\|_\infty &\le c|u(t)|_j^{1/2}|u(t)|_{j+1}^{1/2} \\ &\le cM_j^{1/2}M_{j+1}^{1/2}, \qquad \forall t \ge \max\{T_j, T_{j+1}\}, \quad j = 1, 2, \ldots, m-2.\end{aligned} \tag{7.37}$$

We then have the following estimate:

$$\begin{aligned}&\left|\int_\Omega (-\alpha_m(D^{m-1}u)^2 + Q_m(u, Du, \ldots, D^{m-2}u))\,dx\right| \\ &\quad \le c|u|_{m-1}^2 + L\|Q_m(u, Du, \ldots, D^{m-2}u)\|_\infty \\ &\quad \le \text{(using (7.34), (7.36), and (7.37))} \\ &\quad \le c, \qquad \forall t \ge \max\{T_0, T_1, \ldots, T_{m-1}\} = T'.\end{aligned} \tag{7.38}$$

Therefore, we deduce from (7.7) that

$$|u(t)|_m^2 - c \le I_m(u(t)) \le |u(t)|_m^2 + c, \qquad \forall t \ge T'. \tag{7.39}$$

Now, from (7.12), we find that

$$\begin{aligned} K_m(u) &\le 2|f|_m|u|_m + c\|f\|_\infty |u|_{m-1}^2 + c\|u\|_\infty |u|_{m-1}|f|_{m-1} \\ &\quad + c\|u\|_\infty |u|_{m-1}^2 + cL\|Q_m(u, Du, \ldots, D^{m-2}u)\|_\infty \\ &\quad + \sum_{j=0}^{m-2} (|f|_j + \gamma|u|_j)L\left\|\frac{\partial Q_m}{\partial y_j}(u, Du, \ldots, D^{m-2}u)\right\|_\infty \\ &\le \text{(using again (7.34), (7.36), and (7.37))} \\ &\le \gamma|u|_m^2 + \frac{1}{\gamma}|f|_m^2 + c, \qquad \forall t \ge T'. \end{aligned} \tag{7.40}$$

From (7.35), (7.39), and (7.40) we obtain

$$\frac{d}{dt} I_m(u) + \gamma I_m(u) \le \frac{1}{\gamma}|f|_m^2 + c, \qquad \forall t \ge T'.$$

Hence, by the Gronwall lemma it follows that

$$I_m(u(t)) \le I_m(u(T'))e^{-\gamma(t-T')} + \frac{1}{\gamma}|f|_m^2 + c, \qquad \forall t \ge T', \tag{7.41}$$

which together with (7.39) gives

$$|u(t)|_m^2 \le |u(T')|_m^2 e^{-\gamma(t-T')} + \frac{1}{\gamma}|f|_m^2 + c, \qquad \forall t \ge T'. \tag{7.42}$$

By Theorem 7.2 we know that there exists $c_1 = c(R, T')$ such that

$$\|u(t)\|_m \le c_1, \qquad \forall u_0 \in H_{\text{per}}^m(\Omega), \quad \|u_0\|_m \le R, \quad \forall t \in [-T', T']. \tag{7.43}$$

From (7.42) and (7.43) we then obtain

$$|u(t)|_m^2 \le c_1 e^{-\gamma(t-T')} + \frac{1}{\gamma}|f|_m^2 + c, \qquad \forall t \ge T'. \tag{7.44}$$

Recall that c depends only on $M_0, M_1, \ldots, M_{m-1}$.

Define now

$$M_m^2 = 2\left(\frac{1}{\gamma}|f|_m^2 + c\right),$$

c as in (7.44), and set

$$T_m = \frac{1}{\gamma}\log\frac{1}{c_1}\left(\frac{1}{\gamma}|f|_m^2 + c\right) + \max\{T_0, T_1, \ldots, T_{m-1}\},$$

and note that $M_m = M_m(\gamma, L, \|f\|_m)$ and $T_m = T_m(\gamma, L, \|f\|_m, R)$. We then infer from (7.44) that

$$|u(t)|_m \leq M_m, \qquad \forall t \geq T_m. \tag{7.45}$$

Remark 7.1. A more careful computation using the particular structure of the polynomials Q_m leads to the following time-uniform a priori estimates: for each $j = 0, 1, \ldots, k$ there exist two constants

$$\kappa_j = \kappa_j(\gamma, L, |f|_0, |f|_1, \ldots, |f|_{j-1}, |u_0|_0, |u_0|_1, \ldots, |u_0|_j)$$

and

$$\tilde{\kappa}_j = \tilde{\kappa}_j(\gamma, L, |f|_0, |f|_1, \ldots, |f|_j),$$

such that

$$|u(t)|_j^2 \leq \kappa_j e^{-(\gamma/2^j)t} + \tilde{\kappa}_j, \qquad \forall t \geq 0;$$

this gives a more precise expression for the time needed for the orbits to reach the absorbing set.

We are now in position to prove the existence of the global attractor which will follow from Theorem I.1.1 modified by Remark I.1.4; it is based on the *asymptotic compactness* of the semigroup $\{S^{(k)}(t)\}_{t\geq 0}$. We recall that the semigroup $\{S^{(k)}(t)\}_{t\geq 0}$ is said to be *asymptotically compact* in $H^k_{\text{per}}(\Omega)$ if

$$\{S^{(k)}(t_n)u_n\}_n \text{ is relatively compact in } H^k_{\text{per}}(\Omega) \tag{7.46}$$

whenever

$$\{u_n\}_n \text{ is a sequence bounded in } H^k_{\text{ker}}(\Omega) \text{ and } t_n \to \infty. \tag{7.47}$$

In order to show that $\{S^{(k)}(t)\}_{t\geq 0}$ is asymptotically compact in $H^k_{\text{per}}(\Omega)$, we use the energy equations (7.11). More precisely, we consider $H^k_{\text{per}}(\Omega)$ temporarily endowed with the equivalent norm

$$\|u\| = (|u|_0^2 + |u|_k^2)^{1/2},$$

and add the energy equations (7.11) for $m = 0$ and $m = k$ to find

$$\frac{d}{dt}(\|S^{(k)}(t)u_0\|^2 + J_k(S^{(k)}(t)u_0)) + 2\gamma(\|S^{(k)}(t)u_0\|^2 + J_k(S^{(k)}(t)u_0))$$
$$= \tilde{K}_k(S^{(k)}(t)u_0), \qquad \forall u_0 \in H^k_{\text{per}}(\Omega), \tag{7.48}$$

where

$$J_k(S^{(k)}(t)u_0) = J_k(u(t)) = \int_\Omega [-\alpha_k u(D^{k-1}u)^2 + Q_k(u, Du, \ldots, D^{k-2}u)]\,dx, \tag{7.49}$$

and

$$\tilde{K}_k(S^{(k)}(t)u_0) = \tilde{K}_k(u(t)) = K_k(u(t)) + K_0(u(t))$$

$$= \int_\Omega \Big[2D^k f D^k u - \alpha_k f (D^{k-1}u)^2 + 2\alpha_k u D^{k-1} u D^{k-1} f$$

$$+ \gamma \alpha_k u (D^{k-1}u)^2 + 2\gamma Q_k(u, Du, \ldots, D^{k-2}u) + 2fu$$

$$+ \sum_{j=0}^{k-2} (D^j f - \gamma D^j u) \frac{\partial Q_k}{\partial y_j}(u, Du, \ldots, D^{k-2}u) \Big] \, dx. \tag{7.50}$$

Note that J_k and $\tilde{K}_k$ are weakly continuous on $H^k_{\text{per}}(\Omega)$ since the embedding of $H^k_{\text{per}}(\Omega)$ into $H^{k-1}_{\text{per}}(\Omega)$ is compact. Then we integrate (7.48) in time to find

$$\|S^{(k)}(t)u_0\|^2 + J_k(S^{(k)}(t)u_0) = (\|u_0\|^2 + J_k(u_0))e^{-2\gamma t}$$

$$+ \int_0^t e^{-2\gamma(t-s)} \tilde{K}_k(S^{(k)}(s)u_0) \, ds, \tag{7.51}$$

for all $u_0 \in H^k_{\text{per}}(\Omega)$ and $t \geq 0$.

To show the asymptotic compactness of the semigroup, let $B \subset H^k_{\text{per}}(\Omega)$ be bounded and consider $\{u_n\}_n \subset B$ and $\{t_n\}_n$, $t_n \geq 0$, $t_n \to \infty$.

By Proposition 7.1, the ball in $H^k_{\text{per}}(\Omega)$ of radius ρ_k and centered at the origin, which we now denote by $\mathscr{B}_k$, is absorbing for $S^{(k)}(t)$. Hence, there exists a time $T(B) > 0$ such that

$$S^{(k)}(t)B \subset \mathscr{B}_k, \qquad \forall t \geq T(B),$$

so that for n large enough ($t_n \geq T(B)$),

$$S^{(k)}(t_n)u_n \in \mathscr{B}_k. \tag{7.52}$$

Thus $\{S^{(k)}(t_n)u_n\}_n$ is weakly relatively compact in $H^k_{\text{per}}(\Omega)$ and, hence,

$$S^{(k)}(t_{n'})u_{n'} \rightharpoonup w \quad \text{weakly in } H^k_{\text{per}}(\Omega), \tag{7.53}$$

for some subsequence n' and some $w \in \mathscr{B}_k$.

Similarly, for each $T > 0$, we have

$$S^{(k)}(t_n - T)u_n \in \mathscr{B}_k, \tag{7.54}$$

for $t_n \geq T + T(B)$. Thus, $\{S^{(k)}(t_n - T)u_n\}_n$ is weakly relatively compact in $H^k_{\text{per}}(\Omega)$, and by using a diagonal process, and passing to a further subsequence if necessary, we can assume that

$$S^{(k)}(t_{n'} - T)u_{n'} \rightharpoonup w_T \quad \text{weakly in } H^k_{\text{per}}(\Omega), \qquad \forall T \in \mathbb{N}, \tag{7.55}$$

with $w_T \in \mathscr{B}_k$.

Note then by the weak continuity of $S^{(k)}(t)$ established in Theorem 7.2 that

$$w = \lim_{n'} {}_{H^k_w} S^{(k)}(t_{n'})u_{n'} = \lim_{n'} {}_{H^k_w} S^{(k)}(T)S^{(k)}(t_{n'} - T)u_{n'}$$

$$= S^{(k)}(T) \lim_{n'} {}_{H^k_w} S^{(k)}(t_{n'} - T)u_{n'} = S^{(k)}(T)w_T,$$

where $\lim_{H^k_w}$ denotes the limit taken in the weak topology of $H^k_{\text{per}}(\Omega)$. Thus,

$$w = S^{(k)}(T)w_T, \qquad \forall T \in \mathbb{N}. \tag{7.56}$$

We shall now show that

$$\limsup_{n'} \|S^{(k)}(t_{n'})u_{n'}\| \le \|w\|.$$

For $T \in \mathbb{N}$ and $t_n > T$ we have by (7.51)

$$\begin{aligned}\|S^{(k)}(t_n)u_n\|^2 + J_k(S^{(k)}(t_n)u_n) &= \|S^{(k)}(T)S^{(k)}(t_n - T)u_n\|^2 \\ &\quad + J_k(S^{(k)}(T)S^{(k)}(t_n - T)u_n) \\ &= (\|S^{(k)}(t_n - T)u_n\|^2 + J_k(S^{(k)}(t_n - T)u_n))e^{-2\gamma T} \\ &\quad + \int_0^T e^{-2\gamma(T-s)}\tilde{K}_k(S^{(k)}(s)S^{(k)}(t_n - T)u_n)\,ds.\end{aligned} \tag{7.57}$$

From (7.54) we find that

$$\limsup_{n'} e^{-2\gamma T}(\|S^{(k)}(t_{n'} - T)u_{n'}\|^2 + J_k(S^{(k)}(t_{n'} - T)u_{n'})) \le ce^{-2\gamma T}, \tag{7.58}$$

for some constant c independent of T and n'. Then, by the weak continuities (7.53) and (7.55), and the weak continuity of the operators J_k and $\tilde{K}_k$, we can pass to the lim sup in (7.57) as n' goes to infinity; we obtain

$$\limsup_{n'} \{\|S^{(k)}(t_n)u_n\|^2\} + J_k(w) \le ce^{-2\gamma T} + \int_0^T e^{-2\gamma(T-s)}\tilde{K}_k(S^{(k)}(s)w_T)\,ds. \tag{7.59}$$

On the other hand, we find from (7.51) applied to $w = S^{(k)}(T)w_T$ that

$$\begin{aligned}\|w\|^2 + J_k(w) &= \|S^{(k)}(T)w_T\|^2 + J_k(S^{(k)}(T)w_T) \\ &= (\|w_T\|^2 + J_k(w_T))e^{-2\gamma T} + \int_0^T e^{-2\gamma(T-s)}\tilde{K}_k(S^{(k)}(s)w_T)\,ds.\end{aligned} \tag{7.60}$$

From (7.59) and (7.60) we then find

$$\begin{aligned}\limsup_{n'} \|S^{(k)}(t_{n'})u_{n'}\|^2 &\le \|w\|^2 + (c - \|w_T\|^2 - J_k(w_T))e^{-2\gamma T} \\ &\le \|w\|^2 + c'e^{-2\gamma T},\end{aligned} \tag{7.61}$$

for all $T \in \mathbb{N}$ and for another constant c' independent of T. We let T go to infinity in (7.61) and obtain as claimed

$$\limsup_{n'} \|S^{(k)}(t_{n'})u_{n'}\|^2 \le \|w\|^2. \tag{7.62}$$

Since $H^k_{\text{per}}(\Omega)$ is a Hilbert space, and (7.62) and (7.53) imply that

$$S^{(k)}(t_{n'})u_{n'} \to w \quad \text{strongly in } H^k_{\text{per}}(\Omega). \tag{7.63}$$

This shows that $\{S^{(k)}(t_n)u_n\}_n$ is relatively compact in $H^k_{\text{per}}(\Omega)$, and hence, that $\{S^{(k)}(t)\}_{t\geq 0}$ is asymptotically compact in $H^k_{\text{per}}(\Omega)$. Since $\{S^{(k)}(t)\}_{t\geq 0}$ has also a bounded absorbing set $\mathscr{B}_k$ in $H^k_{\text{per}}(\Omega)$, the existence of the global attractor follows from Theorem I.1.1 and Remark I.1.4.

We have thus proved the following result.

Theorem 7.3. *The semigroup $\{S^{(k)}(t)\}_{t\geq 0}$ possesses a global attractor $\mathscr{A}_k$ in $H^k_{\text{per}}(\Omega)$, i.e., there exists a set $\mathscr{A}_k$ compact in $H^k_{\text{per}}(\Omega)$, invariant for $\{S^{(k)}(t)\}_{t\geq 0}$ and which attracts all bounded subsets of $H^k_{\text{per}}(\Omega)$ in the sense of the semidistance in $H^k_{\text{per}}(\Omega)$. Moreover, $\mathscr{A}_k$ is also connected in $H^k_{\text{per}}(\Omega)$.*

7.3. Regularity of the Attractor

Let $f \in H^k_{\text{per}}(\Omega)$, $k \geq 2$, and $\gamma > 0$ be given. For each $j = 2, 3, \ldots, k$ we can consider the group $\{S^{(j)}(t)\}_{t\in\mathbb{R}}$ on $H^j_{\text{per}}(\Omega)$ defined by

$$S^{(j)}(t)\colon H^j_{\text{per}}(\Omega) \to H^j_{\text{per}}(\Omega), \qquad S^{(j)}(t)u_0 = u(t) \qquad \text{for} \quad t \in \mathbb{R},$$

where $u(t)$ is the solution of (2.1)–(2.3) at time t. Clearly $S^{(j)}(t)|_{H^{j+1}_{\text{per}}(\Omega)} = S^{(j+1)}(t)$ for $j = 2, 3, \ldots, k-1$. By Theorem 7.3, there exists a global attractor $\mathscr{A}_j$ in $H^j_{\text{per}}(\Omega)$ for the group $\{S^{(j)}(t)\}_{t\in\mathbb{R}}$. Obviously, $\mathscr{A}_2 \supset \mathscr{A}_3 \supset \cdots \supset \mathscr{A}_k$. Since $\{S^{(j)}(t)\}_{t\in\mathbb{R}}$ is a group, there is no smoothing effect. Thus, a natural, non-obvious question is whether these attractors are equal. We will prove below that if $k \geq 4$, then $\mathscr{A}_3 = \mathscr{A}_4 = \cdots = \mathscr{A}_k$. The idea of the proof is to decompose $S^{(3)}(t)$ as $S^{(3)}_1(t) + S^{(3)}_2(t)$, somehow as in Theorem I.1.1; here $S^{(3)}_1(t)$ and $S^{(3)}_2(t)$ are not necessarily semigroups, $S^{(3)}_1(t)u_0$ is more regular than the solution $S^{(3)}(t)u_0$ (more exactly as regular as f), and $\|S^{(3)}_2(t)u_0\|_1$ tends to zero as t goes to infinity, uniformly for u_0 bounded in $H^3_{\text{per}}(\Omega)$.

Set

$$v = \sum_{l\in\mathbb{Z}} \hat{v}_l e^{2\pi i l x/L}.$$

For $N \in \mathbb{N}$, $N \geq 1$, consider

$$P_N v = \sum_{|l|\leq N} \hat{v}_l e^{2\pi i l x/L},$$

and

$$Q_N v = v - P_N v = \sum_{|l|>N} \hat{v}_l e^{2\pi i l x/L};$$

P_N and Q_N are orthogonal projectors in any H^s, $s \geq 0$.

From Lemma 7.2 (with $\varepsilon^{-p} = N$) we have the following estimates:

$$\|P_N v\|_{m+l} \le c_l N^l \|P_N v\|_m, \qquad \forall l > 0, \tag{7.64}$$

$$\|Q_N v\|_{m-l} \le c_l N^{-l} \|Q_N v\|_m, \qquad \text{for} \quad l = 0, 1, \ldots, m, \tag{7.65}$$

and also

$$|Q_N v|_{m-l} \le c_l N^{-l} |Q_N v|_m, \qquad \text{for} \quad l = 0, 1, \ldots, m. \tag{7.66}$$

Consider now $u_0 \in H^3_{\text{per}}(\Omega)$, $u(t) = S^{(3)}(t)u_0$, and set

$$y(t) = P_N u(t). \tag{7.67}$$

We split the high-frequency part of u, $Q_N u$, as

$$Q_N u = q + z,$$

where q and z satisfy the following conditions:

$$\begin{cases} \dfrac{dq}{dt} + D^3 q + \gamma q + Q_N(qDq) + Q_N D(yq) = Q_N f - Q_N(yDy), \\ q(0) = 0, \end{cases} \tag{7.68}$$

and

$$\begin{cases} \dfrac{dz}{dt} + D^3 z + \gamma z + Q_N(zDz) + Q_N D(vz) = 0, \\ z(0) = Q_N u_0, \end{cases} \tag{7.68}$$

with $v = y + q$.

We now show that equations (7.68) and (7.69) define indeed uniquely q and z.

Proposition 7.2. *Given $N \ge 1$ and $u_0 \in H^3_{\text{per}}(\Omega)$, there exist a unique solution q of* (7.68) *and a unique solution z of* (7.69) *satisfying*

$$q, z \in L^\infty(0, T; Q_N H^3_{\text{per}}(\Omega)) \cap \mathscr{C}([0, T], Q_N L^2(\Omega)), \qquad \forall T > 0. \tag{7.70}$$

Moreover, there exist $N \in \mathbb{N}$ large enough and a constant $c > 0$, which depend on γ, L, and f, such that the following estimates hold:

$$|q(t)|_k \le c N^{k-2}, \qquad \forall t \ge 0, \tag{7.71}$$

and

$$|z(t)|_1 \le c e^{-(\gamma/2)t}, \qquad \forall t \ge 0, \tag{7.72}$$

uniformly for u_0 in $\mathscr{B}_3$, the absorbing ball in $H^3_{\text{per}}(\Omega)$ for $\{S^{(3)}(t)\}_{t \in \mathbb{R}}$.

The proof of Proposition 7.2 is given in Section 7.5.

Thanks to Proposition 7.2, we can define the families $\{S_1^{(3)}(t)\}_{t \ge 0}$ and

$\{S_2^{(3)}(t)\}_{t\ge 0}$ of maps in $H^3_{\text{per}}(\Omega)$; we set

$$S_1^{(3)}(t)u_0 = y(t) + q(t) \quad \text{and} \quad S_2^{(3)}(t)u_0 = z(t), \quad \forall t \ge 0, \tag{7.73}$$

where $y(t) = P_N S^{(3)}(t)u_0 = P_N u(t)$, and q and z are, respectively, the solutions of problems (7.68) and (7.69) for a given u_0 in $H^3_{\text{per}}(\Omega)$.

Note then that $w = q + z$ is a solution of the following problem:

$$\begin{cases} \dfrac{dw}{dt} + D^3 w + \gamma w + Q_N((y + w)D(y + w)) = Q_N f, \\ w(0) = Q_N u_0. \end{cases} \tag{7.74}$$

Since $Q_N u$ is also a solution of this problem, and uniqueness is easily verified, it follows that $Q_N u = q + z$. In other words, we have that

$$S^{(3)}(t) = S_1^{(3)}(t) + S_2^{(3)}(t), \qquad \forall t \ge 0, \tag{7.75}$$

as operators defined in $H^3_{\text{per}}(\Omega)$.

We are now in position to show that $\mathscr{A}_3 = \mathscr{A}_k$. For this, consider $u \in \mathscr{A}_3$. From a well-known characterization of ω-limit sets (see Section I.1.1), there exist a sequence of elements $u_n \in \mathscr{B}_3$ and a sequence of positive real numbers t_n which tends to infinity as n goes to infinity such that

$$S^{(3)}(t_n)u_n \to u \quad \text{in } H^3_{\text{per}}(\Omega), \qquad \text{as} \quad n \to \infty. \tag{7.76}$$

We also have from (7.75) that

$$S^{(3)}(t_n)u_n = S_1^{(3)}(t_n)u_n + S_2^{(3)}(t_n)u_n, \qquad \forall n \in \mathbb{N}. \tag{7.77}$$

From the definition (7.73) and using (7.71), (7.72), and (7.156), we deduce that if N is large enough (as given by Proposition 7.2), then

$$\|S_1^{(3)}(t_n)u_n\|_k \le cN^{k-2}, \qquad \forall n \in \mathbb{N}, \tag{7.78}$$

and

$$\|S_2^{(3)}(t_n)u_n\|_1 \le ce^{-(\gamma/2)t_n}, \qquad \forall n \in \mathbb{N}, \tag{7.79}$$

for some constant c.

From (7.78) we infer that there exist subsequences $t_{n'}$ and $u_{n'}$, and $w \in H^k_{\text{per}}(\Omega)$ such that

$$S_1^{(3)}(t_{n'})u_{n'} \rightharpoonup w \quad \text{weakly in } H^k_{\text{per}}(\Omega). \tag{7.80}$$

We also have

$$\|w\|_k \le \liminf_{n' \to \infty} \|S_1^{(3)}(t_{n'})u_{n'}\|_k \le cN^{k-2}. \tag{7.81}$$

Consider $\varphi \in L^2(\Omega)$. From (7.77) we can write

$$(S^{(3)}(t_{n'})u_{n'}, \varphi)_{L^2} = (S_1^{(3)}(t_{n'})u_{n'}, \varphi)_{L^2} + (S_2^{(3)}(t_{n'})u_{n'}, \varphi)_{L^2}.$$

Passing to the limit $n' \to \infty$ in the expression above and using (7.76), (7.79), and (7.80) we obtain

$$(u, \varphi)_{L^2} = (w, \varphi)_{L^2}, \qquad \forall \varphi \in L^2(\Omega).$$

Therefore, $u = w$ in $L^2(\Omega)$ and, consequently, $u \in H^k_{\mathrm{per}}(\Omega)$. From (7.81), we also have

$$\|u\|_k \leq cN^{k-2}, \qquad \forall u \in \mathscr{A}_3.$$

In other words, $\mathscr{A}_3$ is contained and bounded in $H^k_{\mathrm{per}}(\Omega)$. Since $\mathscr{A}_k$ attracts all bounded sets in $H^k_{\mathrm{per}}(\Omega)$, we have, moreover,

$$\mathrm{dist}_{H^k}(\mathscr{A}_3, \mathscr{A}_k) = \mathrm{dist}_{H^k}(S^{(k)}(t)\mathscr{A}_3, \mathscr{A}_k) \to 0, \qquad \text{as} \quad t \to \infty.$$

Hence, since $\mathscr{A}_k$ is closed in $H^k_{\mathrm{per}}(\Omega)$,

$$\mathscr{A}_3 \subset \mathscr{A}_k.$$

Thus, we have proved the following result:

Theorem 7.4. *Let $f \in H^k_{\mathrm{per}}(\Omega)$, $k \geq 4$, and $\gamma > 0$ be given. Then $\mathscr{A}_3 = \mathscr{A}_4 = \cdots = \mathscr{A}_k$, where $\mathscr{A}_3, \ldots, \mathscr{A}_k$ are the global attractors obtained in Theorem* 7.3.

Remark 7.2. Note that Theorem 7.4 is a regularity result; it says that the global attractors $\mathscr{A}_j$ for $\{S^{(j)}(t)\}_{t \in \mathbb{R}}$ and $j \geq 3$ are as smooth[1] as the forcing term.

Remark 7.3. For regularity reasons, we were not able to prove that $\mathscr{A}_2 = \mathscr{A}_3$ using the techniques above.

7.4. Proof of the Results in Section 7.1

In this section we prove the results stated in Section 7.1.

Proof of Theorem 7.1. The proof is by parabolic regularization as in R. Temam [6]. We consider for $\varepsilon \in (0, 1]$ fixed the following regularized form of (7.1):

$$\frac{du_\varepsilon}{dt} + u_\varepsilon Du_\varepsilon + D^3 u_\varepsilon + \gamma u_\varepsilon + \varepsilon D^4 u_\varepsilon = f_\varepsilon, \tag{7.82}$$

$$u_\varepsilon(0) = u_{0\varepsilon}; \tag{7.83}$$

here f_ε and u_ε are, respectively, approximations in $H^k_{\mathrm{per}}(\Omega)$ of f and u_0 by smooth (C^∞ and L-periodic) functions.

It is well known that there exists $u_\varepsilon = u_\varepsilon(t)$, $t \in [0, \infty)$, satisfying (7.82) and (7.83) and also

$$u_\varepsilon \in L^\infty(0, T; L^2(\Omega)) \cap L^2(0, T; H^2_{\mathrm{per}}(\Omega)), \qquad \forall T > 0. \tag{7.84}$$

[1] Here $\mathscr{A}_j$ smooth means that its elements are smooth functions of x.

Then $Du_\varepsilon \in L^2(0, T; H^1_{\text{per}}(\Omega)) \subset L^2(0, T; L^\infty(\Omega))$, so that $u_\varepsilon Du_\varepsilon \in L^2(0, T; L^2(\Omega))$. Hence,

$$\frac{du_\varepsilon}{dt} + D^3 u_\varepsilon + \gamma u_\varepsilon + \varepsilon D^4 u_\varepsilon = f_\varepsilon - u_\varepsilon D u_\varepsilon \in L^2(0, T; L^2(\Omega)). \tag{7.85}$$

By a classical regularity result for linear parabolic equations we obtain

$$u_\varepsilon \in L^2(0, T; H^4_{\text{per}}(\Omega)), \qquad \frac{du_\varepsilon}{dt} \in L^2(0, T; L^2(\Omega)). \tag{7.86}$$

We remark that the right-hand side in (7.85) is actually more regular and we deduce that u_ε is also more regular. We can continue this bootstrapping reasoning indefinitely to conclude that u_ε is in $\mathscr{C}^\infty$ in $(0, \infty) \times \Omega$. Hence, all the computations below will be justified.

We then want to pass to the limit $\varepsilon \to 0$. For that purpose, we need estimates on u_ε independent of ε, more precisely we need to prove that u_ε remains bounded in $L^\infty(0, T; H^k_{\text{per}}(\Omega))$ as ε goes to 0. First, we find estimates for u_ε independent of ε in $L^\infty(0, T; H^2_{\text{per}}(\Omega))$. Using these estimates it will be easy to obtain estimates for u_ε independent of ε in $L^\infty(0, T; H^k_{\text{per}}(\Omega))$. Since we are now interested in finite-time estimates, we fix $T > 0$ and let c denote a generic constant independent of ε, but which might depend on T and on the other data (including k). We will use the fact that $u_{0\varepsilon}$ and f_ε converge, respectively, to u_0 and f in $H^k_{\text{per}}(\Omega)$, so that we can bound $\|u_{0\varepsilon}\|_k$ and $\|f_\varepsilon\|_k$ independently of ε. We also note that all the estimates obtained will be uniformly valid for $\|u_0\|_k$ bounded.

Step I. *Estimates for u_ε in $L^\infty(0, T; H^2_{\text{per}}(\Omega))$.*

First, multiply equation (7.82) by $L_0(u_\varepsilon) = 2u_\varepsilon$ and integrate the resulting equation on Ω. This gives

$$\begin{aligned}
\frac{d}{dt}|u_\varepsilon|_0^2 + 2\gamma |u_\varepsilon|_0^2 + 2\varepsilon |D^2 u_\varepsilon|_0^2 &= 2\int_\Omega f_\varepsilon u_\varepsilon \, dx \\
&\leq 2|f_\varepsilon|_0 |u_\varepsilon|_0 \\
&\leq |\gamma|\,|u_\varepsilon|_0^2 + \frac{1}{|\gamma|}|f_\varepsilon|_0^2.
\end{aligned} \tag{7.87}$$

Hence,

$$\frac{d}{dt}|u_\varepsilon|_0^2 + 2\varepsilon |D^2 u_\varepsilon|_0^2 \leq 3|\gamma|\,|u_\varepsilon|_0^2 + \frac{1}{|\gamma|}|f_\varepsilon|_0^2.$$

By multiplying the above expression by $\exp(-3|\gamma| t)$ and integrating it on t, we obtain

$$|u_\varepsilon(t)|_0^2 + 2\varepsilon \int_0^t |u_\varepsilon(s)|_2^2 \, ds \leq |u_{0\varepsilon}|_0^2 e^{3|\gamma| t} + \frac{|f_\varepsilon|_0^2}{3\gamma^2} e^{3|\gamma| t},$$

so that for $0 \le t \le T$,

$$|u_\varepsilon(t)|_0^2 + \varepsilon \int_0^t |u_\varepsilon(s)|_2^2\, ds \le c, \tag{7.88}$$

for some constant $c > 0$.

Multiplying now equation (7.82) by $L_1(u_\varepsilon) = -2D^2u_\varepsilon - u_\varepsilon^2$, we find that

$$\begin{aligned}\frac{d}{dt}\int_\Omega ((Du_\varepsilon)^2 - \tfrac{1}{3}u_\varepsilon^3)\,dx &+ 2\gamma \int_\Omega ((Du_\varepsilon)^2 - \tfrac{1}{2}u_\varepsilon^3)\,dx \\ &+ 2\varepsilon \int_\Omega ((D^3u_\varepsilon)^2 + u_\varepsilon Du_\varepsilon D^3u_\varepsilon)\,dx \\ = \int_\Omega (2Df_\varepsilon Du_\varepsilon - f_\varepsilon u_\varepsilon^2)\,dx.\end{aligned} \tag{7.89}$$

Using (7.23), (7.24), and (7.88), we can write for $0 \le t \le T$

$$\frac{\gamma}{3}\int_\Omega u_\varepsilon^3\,dx \le \frac{|\gamma|}{2}\int_\Omega ((Du_\varepsilon)^2 - \tfrac{1}{3}u_\varepsilon^3)\,dx + c. \tag{7.90}$$

Then we have

$$\begin{aligned}\left|2\int_\Omega u_\varepsilon Du_\varepsilon D^3u_\varepsilon\,dx\right| &\le 2\|u_\varepsilon\|_{L^4}\|Du_\varepsilon\|_{L^4}|D^3u_\varepsilon|_0 \\ &\le c\|u_\varepsilon\|_{H^{1/4}}\|Du_\varepsilon\|_{H^{1/4}}|D^3u_\varepsilon|_0 \\ &\le \text{(by interpolation inequalities)} \\ &\le c|u_\varepsilon|_0^{3/2}\|u_\varepsilon\|_3^{1/2}|u_\varepsilon|_3 \\ &\le c|u_\varepsilon|_0^2|u_\varepsilon|_3 + c|u_\varepsilon|_0^{3/2}|u_\varepsilon|_3^{3/2} \\ &\le |u_\varepsilon|_3^2 + c|u_\varepsilon|_0^4 + c|u_\varepsilon|_0^6 \\ &\le \text{(using (7.88))} \\ &\le |u_\varepsilon|_3^2 + c, \qquad \text{for} \quad 0 \le t \le T.\end{aligned} \tag{7.91}$$

Also

$$\begin{aligned}\left|\int_\Omega (2Df_\varepsilon\, Du_\varepsilon - f_\varepsilon u_\varepsilon^2)\,dx\right| &\le 2|f_\varepsilon|_1|u_\varepsilon|_1 + \|f_\varepsilon\|_\infty |u_\varepsilon|_0^2 \\ &\le \frac{|\gamma|}{3}|u_\varepsilon|_1^2 + \frac{3}{|\gamma|}|f_\varepsilon|_1^2 + \|f_\varepsilon\|_\infty|u_\varepsilon|_0^2 \\ &\le \text{(using (7.88) and (7.24))} \\ &\le \frac{|\gamma|}{2}I_1(u_\varepsilon) + c, \qquad \text{for} \quad 0 \le t \le T,\end{aligned} \tag{7.92}$$

where

$$I_1(u_\varepsilon) = \int_\Omega ((Du_\varepsilon)^2 - \tfrac{1}{3}u_\varepsilon^3)\,dx.$$

Inserting (7.90), (7.91), and (7.92) in (7.89), we obtain

$$\frac{d}{dt}I_1(u_\varepsilon) + 2\gamma I_1(u_\varepsilon) + \varepsilon|u_\varepsilon|_3^2 \le |\gamma|\,I_1(u_\varepsilon) + c, \qquad \text{for} \quad 0 \le t \le T.$$

But from (7.24), (7.25), and (7.88)

$$\tfrac{2}{3}|u_\varepsilon|_1^2 - c \le I_1(u_\varepsilon) \le \tfrac{4}{3}|u_\varepsilon|_1^2 + c, \qquad \text{for} \quad 0 \le t \le T,$$

from which we can deduce that

$$-2\gamma I_1(u_\varepsilon) \le 2|\gamma|\,I_1(u_\varepsilon) + c, \qquad \text{for} \quad 0 \le t \le T.$$

Therefore,

$$\frac{d}{dt}I_1(u_\varepsilon) + \varepsilon|u_\varepsilon|_3^2 \le 3|\gamma|\,I_1(u_\varepsilon) + c, \qquad \text{for} \quad 0 \le t \le T, \tag{7.93}$$

which implies

$$I_1(u_\varepsilon(t)) + \varepsilon\int_0^t |u_\varepsilon(s)|_3^2\,ds \le I_1(u_{0\varepsilon})e^{3|\gamma|t} + \frac{c}{3|\gamma|}e^{3|\gamma|t}, \qquad \text{for} \quad 0 \le t \le T. \tag{7.94}$$

Using (7.24) and (7.25) we deduce from (7.94) that

$$|u_\varepsilon(t)|_1^2 + \varepsilon\int_0^t |u_\varepsilon(s)|_3^2\,ds \le c, \qquad \text{for} \quad 0 \le t \le T. \tag{7.95}$$

We proceed now with H^2 estimates. A simple computation using the method of undetermined coefficients described in Section 7.1 yields

$$L_2(u_\varepsilon) = 2D^4u_\varepsilon + \tfrac{5}{3}(Du_\varepsilon)^2 + \tfrac{10}{3}u_\varepsilon D^2u_\varepsilon + \tfrac{5}{9}u_\varepsilon^3, \tag{7.96}$$

and

$$I_2(u_\varepsilon) = \int_\Omega ((D^2u_\varepsilon)^2 - \tfrac{5}{3}(Du_\varepsilon)^2 + \tfrac{5}{36}u_\varepsilon^4)\,dx. \tag{7.97}$$

We multiply the regularized equation (7.82) by $L_2(u_\varepsilon)$ and find

$$\frac{d}{dt}I_2(u_\varepsilon) + 2\gamma I_2(u_\varepsilon)$$
$$+ \varepsilon\int_\Omega (2(D^4u_\varepsilon)^2 + \tfrac{5}{3}(Du_\varepsilon)^2D^4u_\varepsilon + \tfrac{10}{3}u_\varepsilon D^2u_\varepsilon D^4u_\varepsilon + \tfrac{5}{9}u_\varepsilon^3D^4u_\varepsilon)\,dx$$

$$= \int_\Omega (2D^2 f_\varepsilon D^2 u_\varepsilon + \tfrac{5}{3}(Du_\varepsilon)^2 f_\varepsilon + \tfrac{10}{3} u_\varepsilon D^2 u_\varepsilon f_\varepsilon + \tfrac{5}{9} u_\varepsilon^3 f_\varepsilon)\, dx$$

$$+ \gamma \int_\Omega (\tfrac{5}{3} u_\varepsilon (Du_\varepsilon)^2 - \tfrac{5}{18} u_\varepsilon^4)\, dx. \tag{7.98}$$

We need to estimate the terms in (7.98). First,

$$\left| \gamma \int_\Omega (\tfrac{5}{3} u_\varepsilon (Du_\varepsilon)^2 - \tfrac{5}{18} u_\varepsilon^4)\, dx \right| \le c\, \|u_\varepsilon\|_\infty |u_\varepsilon|_1^2 + c\, \|u_\varepsilon\|_\infty^2 |u_\varepsilon|_0^2$$

$$\le \text{(using (7.6), (7.88), and (7.95))}$$

$$\le c, \qquad \text{for} \quad 0 \le t \le T. \tag{7.99}$$

Using the continuous embeddings $H^1_{\text{per}}(\Omega) \subset L^\infty(\Omega)$ and $H^{1/4}(\Omega) \subset L^4(\Omega)$, (7.88), and (7.95), we write similarly

$$\left| \int_\Omega (\tfrac{5}{3}(Du_\varepsilon)^2 + \tfrac{10}{3} u_\varepsilon D^2 u_\varepsilon + \tfrac{5}{9} u_\varepsilon^3) D^4 u_\varepsilon\, dx \right|$$

$$\le c(\|Du_\varepsilon\|_{H^{1/4}}^2 + \|u_\varepsilon\|_\infty |u_\varepsilon|_2 + \|u_\varepsilon\|_\infty^2 |u_\varepsilon|_0) |u_\varepsilon|_4$$

$$\le c(\|u_\varepsilon\|_{H^{5/4}}^2 + |u_\varepsilon|_2 + 1) |u_\varepsilon|_4$$

$$\le \text{(by interpolation)}$$

$$\le c\, |u_\varepsilon|_0^{11/8} \|u_\varepsilon\|_4^{5/8} |u_\varepsilon|_4 + c\, |u_\varepsilon|_0^{1/2} \|u_\varepsilon\|_4^{1/2} |u_\varepsilon|_4 + c\, |u_\varepsilon|_4$$

$$\le \text{(by the Young inequality)}$$

$$\le |u_\varepsilon|_4^2 + c, \qquad \text{for} \quad 0 \le t \le T. \tag{7.100}$$

We also have

$$\left| \int_\Omega (2D^2 f_\varepsilon D^2 u_\varepsilon + \tfrac{5}{3}(Du_\varepsilon)^2 f_\varepsilon + \tfrac{10}{3} u_\varepsilon D^2 u_\varepsilon f_\varepsilon + \tfrac{5}{9} u_\varepsilon^3 f_\varepsilon)\, dx \right|$$

$$\le 2|f_\varepsilon|_2 |u_\varepsilon|_2 + c\, \|f_\varepsilon\|_\infty |u_\varepsilon|_1^2 + c\, |f_\varepsilon|_1 \|u_\varepsilon\|_\infty |u_\varepsilon|_1 + c\, |f_\varepsilon|_0 |u_\varepsilon|_0 \|u_\varepsilon\|_\infty^2$$

$$\le \text{(using (7.95))}$$

$$\le |\gamma| |u_\varepsilon|_2^2 + \frac{1}{|\gamma|} |f_\varepsilon|_2^2 + c, \qquad \text{for} \quad 0 \le t \le T. \tag{7.101}$$

We deduce from (7.97), (7.88), and (7.95) that

$$|u_\varepsilon|_2^2 \le I_2(u_\varepsilon) + c, \qquad \text{for} \quad 0 \le t \le T, \tag{7.102}$$

and also

$$|I_2(u_\varepsilon)| \le |u_\varepsilon|_2^2 + c \le \text{(by (7.102))} \le I_2(u_\varepsilon) + c, \qquad \text{for} \quad 0 \le t \le T. \tag{7.103}$$

Using (7.99)–(7.103), we infer from (7.98) that

$$\frac{d}{dt} I_2(u_\varepsilon) + \varepsilon |u_\varepsilon|_4^2 \leq 3|\gamma| I_2(u_\varepsilon) + c, \qquad \text{for} \quad 0 \leq t \leq T. \tag{7.104}$$

Therefore,

$$I_2(u_\varepsilon(t)) + \varepsilon \int_0^t |u_\varepsilon(s)|_4^2 \, ds \leq I_2(u_{0\varepsilon}) e^{3|\gamma|t} + \frac{c}{3|\gamma|} e^{3|\gamma|t}, \qquad \text{for} \quad 0 \leq t \leq T,$$

which, together with (7.102) and (7.103), gives

$$|u_\varepsilon(t)|_2^2 + \varepsilon \int_0^t |u_\varepsilon(s)|_4^2 \, ds \leq c, \qquad \text{for} \quad 0 \leq t \leq T. \tag{7.105}$$

Step II. *Estimates for u_ε in $L^\infty(0, T; H^k_{\text{per}}(\Omega))$, $k \geq 2$.*

Since we are now only interested in finite-time estimates, we can use Lemma 7.1 and the estimates (7.88), (7.95), and (7.105) in order to obtain estimates for u_ε independent of ε in $H^k_{\text{per}}(\Omega)$, for $k \geq 2$, avoiding the use of the invariants I_k. We then multiply equation (7.82) by $2(-1)^k D^{2k} u_\varepsilon$ to find

$$\frac{d}{dt} |u_\varepsilon|_k^2 + 2 \int_\Omega D^k u_\varepsilon D^k (u_\varepsilon D u_\varepsilon) \, dx + 2\gamma |u_\varepsilon|_k^2 + 2\varepsilon |u_\varepsilon|_{k+2}^2 = 2 \int_\Omega D^k f_\varepsilon D^k u_\varepsilon \, dx. \tag{7.106}$$

We estimate

$$\begin{aligned}
&\left| 2 \int_\Omega D^k u_\varepsilon D^k (u_\varepsilon D u_\varepsilon) \, dx \right| \\
&\quad = \left| 2 \int_\Omega D^k u_\varepsilon D^{k+1} u_\varepsilon u_\varepsilon \, dx + 2 \int_\Omega D^k u_\varepsilon (D^k (u_\varepsilon D u_\varepsilon) - u_\varepsilon D^{k+1} u_\varepsilon) \, dx \right| \\
&\quad = \left| - \int_\Omega (D^k u_\varepsilon)^2 D u_\varepsilon \, dx + 2 \int_\Omega D^k u_\varepsilon (D^k (u_\varepsilon D u_\varepsilon) - u_\varepsilon D^{k+1} u_\varepsilon) \, dx \right| \\
&\quad \leq \|D u_\varepsilon\|_\infty |u_\varepsilon|_k^2 + 2 |u_\varepsilon|_k |D^k (u_\varepsilon D u_\varepsilon) - u_\varepsilon D^{k+1} u_\varepsilon|_0 \\
&\quad \leq \text{(using Lemma 7.1 with } s = k \text{ and } \beta = 1) \\
&\quad \leq \|D u_\varepsilon\|_\infty |u_\varepsilon|_k^2 + c |u_\varepsilon|_k (\|u_\varepsilon\|_k \|D u_\varepsilon\|_1 + \|u\|_2 \|D u_\varepsilon\|_{k-1}) \\
&\quad \leq \text{(using (7.88) and (7.105))} \\
&\quad \leq c |u_\varepsilon|_k^2 + c, \qquad \text{for} \quad 0 \leq t \leq T.
\end{aligned} \tag{7.107}$$

Hence, we infer from (7.106) that

$$\frac{d}{dt} |u_\varepsilon|_k^2 + \varepsilon |u_\varepsilon|_{k+2}^2 \leq c |u_\varepsilon|_k^2 + c |f_\varepsilon|_k^2 + c, \qquad \text{for} \quad 0 \leq t \leq T. \tag{7.108}$$

Therefore,

$$|u_\varepsilon(t)|_k^2 + \varepsilon \int_0^t |u_\varepsilon(s)|_{k+2}^2\, ds \le c, \qquad \text{for} \quad 0 \le t \le T. \tag{7.109}$$

Now, thanks to (7.88), (7.95), and (7.105), and since

$$\frac{du_\varepsilon}{dt} = -u_\varepsilon Du_\varepsilon - D^3 u_\varepsilon - \gamma u_\varepsilon - \varepsilon D^4 u_\varepsilon + f_\varepsilon,$$

we deduce that, for any $T > 0$,

$$\frac{du_\varepsilon}{dt} \text{ remains bounded in } L^2(0, T; H^{-1}(\Omega)), \tag{7.110}$$

where $H^{-1}(\Omega)$ denotes the dual space of $H^1_{\text{per}}(\Omega)$ when we identify $L^2(\Omega)$ with its dual.

From (7.109) and (7.110), there exist u and a subsequence of u_ε (still denoted by u_ε) such that, for every $T > 0$,

$$u_\varepsilon \to u \quad \text{in } L^\infty(0, T; H^k_{\text{per}}(\Omega)) \text{ weak-star}, \tag{7.111}$$

and

$$\frac{du_\varepsilon}{dt} \to \frac{du}{dt} \quad \text{in } L^2(0, T; H^{-1}(\Omega)) \text{ weakly}, \tag{7.112}$$

as $\varepsilon \to 0$. Since the embedding $H^k_{\text{per}}(\Omega) \subset H^{k-1}_{\text{per}}(\Omega)$ is compact, it follows from (7.111) and (7.112) that

$$u_\varepsilon \to u \quad \text{in } L^2(0, T; H^{k-1}_{\text{per}}(\Omega)) \text{ strongly}. \tag{7.113}$$

Hence,

$$u_\varepsilon Du_\varepsilon \to uDu \quad \text{in } L^2(0, T; H^{k-1}_{\text{per}}(\Omega)) \text{ weakly}. \tag{7.114}$$

Then, we can pass to the limit in (7.82) as ε goes to zero to find

$$\frac{du}{dt} + uDu + D^3 u + \gamma u = f \tag{7.115}$$

in $H^{-1}(\Omega)$ and, in particular, in the distribution sense in $\Omega \times (0, \infty)$, with u satisfying

$$u \in L^\infty(0, T; H^k_{\text{per}}(\Omega)), \qquad \forall T > 0, \tag{7.116}$$

and

$$\frac{du}{dt} \in L^\infty(0, T; H^{-1}(\Omega)), \qquad \forall T > 0. \tag{7.117}$$

The relation (7.117) follows from (7.116) and (7.115). From (7.116) and (7.117) it follows by Lemma II.3.2 that u is almost everywhere equal to a continuous function from $[0, T]$ into $L^2(\Omega)$,

$$u \in C([0, T], L^2(\Omega)), \qquad \forall T > 0. \tag{7.118}$$

Similarly, it follows from (7.111) and (7.112) that $u_\varepsilon(0)$ converges to $u(0)$ in $L^2(\Omega)$, and since $u_{0\varepsilon}$ converges to u_0 in $H^k_{\text{per}}(\Omega)$, we conclude that

$$u(0) = u_0. \tag{7.119}$$

For the uniqueness, let u and v be two solutions satisfying (7.115), (7.116), and (7.119). Then, u and v satisfy also (7.117) and (7.118), and $w = u - v$ satisfies (7.116), (7.117), (7.118), and

$$\frac{dw}{dt} + D^3 w + \gamma w = -uDu + vDv, \tag{7.120}$$

$$w(0) = 0. \tag{7.121}$$

Since w satisfies (7.116) and (7.117), it follows that

$$\frac{1}{2}\frac{d}{dt}|w|_0^2 = \left\langle \frac{dw}{dt}, w \right\rangle_{H^{-1}, H^1_{\text{per}}}.$$

Then (7.120) implies

$$\frac{1}{2}\frac{d}{dt}|w|_0^2 + \gamma|w|_0^2 = -\int_\Omega (uDu - vDv)w\,dx.$$

But note that

$$\int_\Omega (uDu - vDv)w\,dx = \int_\Omega (wDw + D(vw))w\,dx = \int_\Omega \tfrac{1}{2}w^2 Dv\,dx.$$

Hence,

$$\begin{aligned}
\frac{d}{dt}|w|_0^2 &= -2\gamma|w|_0^2 - \tfrac{1}{2}\int_\Omega w^2 Dv\,dx \\
&\le \tfrac{1}{2}\|Dv\|_\infty |w|_0^2 + 2|\gamma||w|_0^2 \\
&\le \text{(using (7.116))} \\
&\le c|w|_0^2.
\end{aligned}$$

By the Gronwall lemma and (7.121) we deduce that $w = 0$, i.e., $u = v$, which proves the uniqueness. Thus, the proof of Theorem 7.1 is complete. □

Proof of Theorem 7.2. We consider again the regularized problem (7.82) and (7.83), but we now take particular regularizations of u_0 and f. We consider the Fourier series of u_0

$$u_0 = \sum_{j \in \mathbb{Z}} \hat{u}_0^j e^{2\pi i j x/L}.$$

and, for $\varepsilon \in (0, 1]$, we set

$$u_{0\varepsilon} = \sum_{|j| \le \varepsilon^{-p}} \hat{u}_0^j e^{2\pi jx/L},$$

where $0 < p < \frac{1}{4}$ is fixed. It is clear, since the sum is finite that $u_{0\varepsilon} \in \mathscr{C}^\infty_{\text{per}}(\Omega)$. We define f_ε similarly.

We need the following technical result.

Lemma 7.2. *Let*

$$v = \sum_{j \in \mathbb{Z}} \hat{v}_j e^{2\pi ijx/L}.$$

For $p > 0$ and $\varepsilon \in (0, 1]$, set

$$v_\varepsilon = \sum_{|j| \le \varepsilon^{-p}} \hat{v}_j e^{2\pi ijx/L}.$$

Then the following estimates hold:

$$\|v_\varepsilon\|^2_{m+l} \le (1 + (2\pi)^{2l})\varepsilon^{-2pl}\|v_\varepsilon\|^2_m, \qquad \forall l \ge 0, \tag{7.122}$$

and

$$\|v - v_\varepsilon\|^2_{m-l} \le \left(\frac{1}{2\pi}\right)^{2l} \varepsilon^{2pl}\|v - v_\varepsilon\|^2_m, \qquad \textit{for} \quad 0 \le l \le m. \tag{7.123}$$

Proof. We have

$$\|v\|^2_m = \sum_{n=0}^{m} L^{2n}|v|^2_n,$$

where

$$|v|^2_n = \int_\Omega |D^n v|^2\, dx = L \sum_{j \in \mathbb{Z}} \left(\frac{2\pi j}{L}\right)^{2n} |\hat{v}_j|^2.$$

We first obtain

$$\begin{aligned}
|v_\varepsilon|^2_{m+l} &= L \sum_{|j| \le \varepsilon^{-p}} \left(\frac{2\pi j}{L}\right)^{2(m+l)} |\hat{v}_j|^2 \\
&\le L\left(\frac{2\pi\varepsilon^{-p}}{L}\right)^{2l} \sum_{|j| \le \varepsilon^{-p}} \left(\frac{2\pi j}{L}\right)^{2m} |\hat{v}_j|^2 \\
&= \left(\frac{2\pi}{L}\right)^{2l} \varepsilon^{-2pl}|v_\varepsilon|^2_m.
\end{aligned}$$

Hence, since $0 < \varepsilon \le 1$,

$$\begin{aligned}
\|v_\varepsilon\|^2_{m+l} &= \|v_\varepsilon\|^2_m + \sum_{n=m+1}^{m+l} L^{2n}|v_\varepsilon|^2_n \\
&= \|v_\varepsilon\|^2_m + \sum_{n=m-l+1}^{m} L^{2(n+l)}|v_\varepsilon|^2_{n+l}
\end{aligned}$$

$$\begin{aligned}
&\leq \|v_\varepsilon\|_m^2 + \sum_{n=m-l+1}^{m} L^{2(n+l)}\left(\frac{2\pi}{L}\right)^{2l} \varepsilon^{-2pl}|v_\varepsilon|_n^2 \\
&= \|v_\varepsilon\|_m^2 + (2\pi)^{2l}\varepsilon^{-2pl} \sum_{n=m-l+1}^{m} L^{2n}|v_\varepsilon|_n^2 \\
&\leq (1 + (2\pi)^{2l}\varepsilon^{-2pl})\,\|v_\varepsilon\|_m^2 \\
&\leq (1 + (2\pi)^{2l})\varepsilon^{-2pl}\,\|v_\varepsilon\|_m^2.
\end{aligned}$$

Also,

$$\begin{aligned}
|v - v_\varepsilon|_{m-l}^2 &= L \sum_{|j|>\varepsilon^{-p}} \left(\frac{2\pi j}{L}\right)^{2(m-l)} |\hat{v}_j|^2 \\
&\leq L\left(\frac{2\pi\varepsilon^{-p}}{L}\right)^{-2l} \sum_{|j|>\varepsilon^{-p}} \left(\frac{2\pi j}{L}\right)^{2m} |\hat{v}_j|^2 \\
&= \left(\frac{L}{2\pi}\right)^{2l} \varepsilon^{2pl}|v - v_\varepsilon|_m^2,
\end{aligned}$$

and thus

$$\begin{aligned}
\|v - v_\varepsilon\|_{m-l}^2 &= \sum_{n=0}^{m-l} L^{2n}|v - v_\varepsilon|_n^2 \\
&= \sum_{n=l}^{m} L^{2(n-l)}|v - v_\varepsilon|_{n-l}^2 \\
&\leq \sum_{n=l}^{m} L^{2(n-l)}\left(\frac{L}{2\pi}\right)^{2l} \varepsilon^{2pl}|v - v_\varepsilon|_n^2 \\
&= \varepsilon^{2pl}\left(\frac{1}{2\pi}\right)^{2l} \sum_{n=l}^{m} L^{2n}|v - v_\varepsilon|_n^2 \\
&\leq \left(\frac{1}{2\pi}\right)^{2l} \varepsilon^{2pl}\|v - v_\varepsilon\|_m^2.
\end{aligned}$$

□

Going back to the proof of Theorem 7.2, we choose, as indicated before, $0 < p < \frac{1}{4}$ and we infer from (7.122) and (7.123)

$$\begin{aligned}
&\|u_{0\varepsilon}\|_{k+1}, \|f_\varepsilon\|_{k+1} \leq c\varepsilon^{-p}, \\
&\|u_{0\varepsilon}\|_{k+2}, \|f_\varepsilon\|_{k+2} \leq c\varepsilon^{-2p}, \\
&|u_{0\varepsilon} - u_0|_0, |f_\varepsilon - f|_0 \leq c\varepsilon^{kp} \leq c\varepsilon^{2p}, \\
&\|u_{0\varepsilon} - u_0\|_1, \|f_\varepsilon - f\|_1 \leq c\varepsilon^{(k-1)p} \leq c\varepsilon^{p},
\end{aligned} \tag{7.124}$$

for some constant $c > 0$, uniformly for u_0 and f bounded in $H_{\text{per}}^k(\Omega)$. It is also clear that

$$u_{0\varepsilon} \to u_0, \qquad f_\varepsilon \to f \quad \text{in } H_{\text{per}}^k(\Omega), \qquad \text{as} \quad \varepsilon \to 0. \tag{7.125}$$

Our purpose is to prove that the solutions of the regularized problems form a Cauchy sequence in $C([0, T], H^k_{\text{per}}(\Omega))$, so that they converge strongly in this space to the solution of the original problem. As before, since we are interested in finite-time estimates, we fix $T > 0$ and let c denote a generic constant independent of ε, but which might depend on T and all the data of the problem. Again all the estimates obtained will be uniformly valid for u_0 and f bounded in $H^k_{\text{per}}(\Omega)$.

Consider $0 < \delta < \varepsilon$ and set $w = u_\varepsilon - u_\delta$. We prove that $\{u_\varepsilon\}$ is a Cauchy sequence in $C([0, T], H^k_{\text{per}}(\Omega))$ in several steps. First, we need some technical results concerning estimates for the norm of u_ε in the spaces $H^{k+1}_{\text{per}}(\Omega)$ and $H^{k+2}_{\text{per}}(\Omega)$. Then, we deduce estimates for w in $L^2(\Omega)$ and $H^1_{\text{per}}(\Omega)$ norms. In the last step we obtain estimates for w in $H^k_{\text{per}}(\Omega)$, which allow us to conclude that $\{u_\varepsilon\}$ is a Cauchy sequence in $L^\infty(0, T; H^k_{\text{per}}(\Omega))$.

Step I. *Estimates for the norm of u_ε in the spaces $H^{k+1}_{\text{per}}(\Omega)$ and $H^{k+2}_{\text{per}}(\Omega)$.*

We multiply the regularized equation (7.82) by $2(-1)^{k+1}D^{2(k+1)}u_\varepsilon$ and find

$$\frac{d}{dt}|u_\varepsilon|^2_{k+1} + 2\int_\Omega D^{k+1}u_\varepsilon D^{k+1}(u_\varepsilon Du_\varepsilon)\,dx + 2\gamma|u_\varepsilon|^2_{k+1} + 2\varepsilon|u_\varepsilon|^2_{k+3}$$
$$= 2\int_\Omega D^{k+1}f_\varepsilon D^{k+1}u_\varepsilon\,dx. \tag{7.126}$$

We write

$$\begin{aligned}
&2\int_\Omega D^{k+1}u_\varepsilon D^{k+1}(u_\varepsilon Du_\varepsilon)\,dx\\
&\quad= 2\int_\Omega D^{k+1}u_\varepsilon D^{k+2}u_\varepsilon u_\varepsilon\,dx\\
&\qquad+ 2\int_\Omega D^{k+1}u_\varepsilon(D^{k+1}(u_\varepsilon Du_\varepsilon) - u_\varepsilon D^{k+2}u_\varepsilon)\,dx\\
&\quad= -\int_\Omega (D^{k+1}u_\varepsilon)^2 Du_\varepsilon\,dx\\
&\qquad+ 2\int_\Omega D^{k+1}u_\varepsilon(D^{k+1}(u_\varepsilon Du_\varepsilon) - u_\varepsilon D^{k+2}u_\varepsilon)\,dx\\
&\quad\le \text{(using Lemma 7.1 with } s = k+1 \text{ and } \beta = 1)\\
&\quad\le |u_\varepsilon|^2_{k+1}\|Du_\varepsilon\|_\infty + c|u_\varepsilon|_{k+1}(\|u_\varepsilon\|_{k+1}\|Du_\varepsilon\|_1 + \|u_\varepsilon\|_2\|Du_\varepsilon\|_k)\\
&\quad\le c\|u_\varepsilon\|^2_{k+1}\|u_\varepsilon\|_2\\
&\quad\le \text{(using (7.88), (7.95), and (7.105))}\\
&\quad\le c|u_\varepsilon|^2_{k+1} + c, \qquad \text{for} \quad 0 \le t \le T.
\end{aligned} \tag{7.127}$$

We infer from (7.126) and (7.127) that

$$\frac{d}{dt}|u_\varepsilon|_{k+1}^2 + 2\varepsilon|u_\varepsilon|_{k+3}^2 \le c|u_\varepsilon|_{k+1}^2 + |f_\varepsilon|_{k+1}^2 + c, \qquad \text{for} \quad 0 \le t \le T. \tag{7.128}$$

We find then, using Gronwall's lemma

$$\begin{aligned}|u_\varepsilon(t)|_{k+1}^2 + \varepsilon\int_0^t |u_\varepsilon(s)|_{k+3}^2\,ds &\le c|u_{0\varepsilon}|_{k+1}^2 + c|f_\varepsilon|_{k+1}^2 + c\\ &\le (\text{using (7.124)}) \le c\varepsilon^{-2p}, \qquad \text{for} \quad 0 \le t \le T.\end{aligned} \tag{7.129}$$

Similarly, multiplying the regularized equation (7.82) by $2(-1)^{k+2}D^{2(k+2)}u_\varepsilon$ we find

$$\frac{d}{dt}|u_\varepsilon|_{k+2}^2 + 2\varepsilon|u_\varepsilon|_{k+4}^2 \le c|u_\varepsilon|_{k+2}^2 + |f_\varepsilon|_{k+2}^2 + c, \qquad \text{for} \quad 0 \le t \le T. \tag{7.130}$$

Hence,

$$\begin{aligned}|u_\varepsilon(t)|_{k+2}^2 + \varepsilon\int_0^t |u_\varepsilon(s)|_{k+4}^2\,ds &\le c|u_{0\varepsilon}|_{k+2}^2 + c|f_\varepsilon|_{k+2}^2 + c\\ &\le (\text{using (7.124)})\\ &\le c\varepsilon^{-4p}, \qquad \text{for} \quad 0 \le t \le T.\end{aligned} \tag{7.131}$$

Step II. *Estimates for the norm of* $w = u_\varepsilon - u_\delta$ *in the spaces* $L^2(\Omega)$ *and* $H^1_{\text{per}}(\Omega)$.

For $w = u_\varepsilon - u_\delta$, $0 < \delta < \varepsilon$, it follows from (7.124) that

$$\begin{aligned}|w(0)|_0 = |u_{0\varepsilon} - u_{0\delta}|_0 &\le |u_{0\varepsilon} - u_0|_0 + |u_{0\delta} - u_0|_0\\ &\le c\varepsilon^{2p} + c\delta^{2p} \le c\varepsilon^{2p},\end{aligned} \tag{7.132}$$

and

$$\|w(0)\|_1 = \|u_{0\varepsilon} - u_{0\delta}\|_1 \le c\varepsilon^p + c\delta^p \le c\varepsilon^p. \tag{7.133}$$

Thus w satisfies the following equation:

$$\frac{dw}{dt} + D(u_\varepsilon w) - wDw + D^3w + \gamma w + \delta D^4 w + (\varepsilon - \delta)D^4 u_\varepsilon = f_\varepsilon - f_\delta. \tag{7.134}$$

We multiply (7.134) by $2w$ and obtain

$$\begin{aligned}&\frac{d}{dt}|w|_0^2 - 2\int_\Omega u_\varepsilon w Dw\,dx + 2\gamma|w|_0^2 + 2\delta|w|_2^2 + 2(\varepsilon - \delta)\int_\Omega D^2u_\varepsilon D^2 w\,dx\\ &\quad = 2\int_\Omega (f_\varepsilon - f_\delta)w\,dx.\end{aligned} \tag{7.135}$$

Therefore,

$$\begin{aligned}\frac{d}{dt}|w|_0^2 + 2\delta|w|_2^2 &\le |w|_0^2 + |f_\varepsilon - f_\delta|_0^2 + \|Du_\varepsilon\|_\infty |w|_0^2 + c\varepsilon|u_\varepsilon|_2|w|_2 \\ &\le \text{(using (7.105))} \\ &\le c|w|_0^2 + |f_\varepsilon - f_\delta|_0^2 + c\varepsilon, \qquad \text{for} \quad 0 \le t \le T,\end{aligned}$$

so that by Gronwall's lemma

$$\begin{aligned}|w(t)|_0^2 &\le c|w(0)|_0^2 + c|f_\varepsilon - f_\delta|_0^2 + c\varepsilon \\ &\le \text{(using (7.124) and since } 4p < 1) \\ &\le c\varepsilon^{4p} + c\varepsilon \le c\varepsilon^{4p}, \qquad \text{for} \quad 0 \le t \le T. \qquad (7.136)\end{aligned}$$

We multiply now (7.134) by $-2D^2w$ and integrate the result in space; we find

$$\begin{aligned}&\frac{d}{dt}|w|_1^2 - 2\int_\Omega D(u_\varepsilon w)D^2w\,dx + 2\int_\Omega wDwD^2w\,dx \\ &\quad + 2\gamma|w|_1^2 + 2\delta|w|_3^2 - 2(\varepsilon - \delta)\int_\Omega D^4u_\varepsilon D^2w\,dx \\ &= 2\int_\Omega (Df_\varepsilon - Df_\delta)Dw\,dx. \qquad (7.137)\end{aligned}$$

We estimate the terms in (7.137) as follows:

$$\begin{aligned}2(\varepsilon - \delta)\int_\Omega D^4u_\varepsilon D^2w\,dx &\le 2\varepsilon|u_\varepsilon|_4|w|_2 \\ &\le \text{(using (7.105))} \le c\varepsilon|u_\varepsilon|_4; \\ -2\int_\Omega wDwD^2w\,dx &= \int_\Omega (Dw)^3\,dx \\ &\le |w|_1^2\|Dw\|_\infty \\ &\le \text{(from (7.105))} \le c|w|_1^2;\end{aligned}$$

and

$$\begin{aligned}2\int_\Omega D(u_\varepsilon w)D^2w\,dx &= -2\int_\Omega D^2u_\varepsilon wDw\,dx - 3\int_\Omega Du_\varepsilon(Dw)^2\,dx \\ &\le c|u_\varepsilon|_2|w|_1\|w\|_\infty + c\|Du_\varepsilon\|_\infty|w|_1^2 \\ &\le \text{(using (7.105))} \le c|w|_1\|w\|_1 \le c|w|_1^2 + c|w|_0^2.\end{aligned}$$

Using the above estimates, we infer from (7.137) that

$$\frac{d}{dt}|w|_1^2 + 2\delta|w|_3^2 \le c|w|_1^2 + c|w|_0^2 + |f_\varepsilon - f_\delta|_1^2 + c\varepsilon|u_\varepsilon|_4, \qquad \text{for} \quad 0 \le t \le T. \tag{7.138}$$

By Gronwall's lemma we see that

$$\begin{aligned}|w(t)|_1^2 &\le c|w(0)|_1^2 + c \sup_{0\le t\le T} |w(t)|_0^2 + c|f_\varepsilon - f_\delta|_1^2 + c\varepsilon \int_0^t |u_\varepsilon(s)|_4\, ds \\ &\le \text{(using (7.136) and (7.124))} \\ &\le c\varepsilon^{2p} + c\varepsilon^{4p} + c\varepsilon^{1/2}\left(\varepsilon \int_0^t |u_\varepsilon(s)|_4^2\, ds\right)^{1/2} \\ &\le \text{(using (7.105))} \le c\varepsilon^{2p} + c\varepsilon^{4p} + c\varepsilon^{1/2} \\ &\le c\varepsilon^{2p}, \qquad \text{for} \quad 0 \le t \le T.\end{aligned}$$

Thus,

$$|w(t)|_1^2 \le c\varepsilon^{2p}, \qquad \text{for} \quad 0 \le t \le T. \tag{7.139}$$

Step III. *Estimate for the norm of w in $H^k_{\text{per}}(\Omega)$.*

We multiply equation (7.134) by $2(-1)^k D^{2k} w$ and integrate the result in space; this yields

$$\begin{aligned}&\frac{d}{dt}|w|_k^2 + 2\int_\Omega D^k w D^{k+1}(u_\varepsilon w)\, dx - 2\int_\Omega D^k w D^k(wDw)\, dx \\ &\qquad + 2\gamma|w|_k^2 + 2\delta|w|_{k+2}^2 + 2(\varepsilon - \delta)\int_\Omega D^{k+4}u_\varepsilon D^k w\, dx \\ &= 2\int_\Omega (D^k f_\varepsilon - D^k f_\delta) D^k w\, dx.\end{aligned} \tag{7.140}$$

We bound the terms in (7.140) as follows:

$$\begin{aligned}&2\int_D D^k w D^{k+1}(u_\varepsilon w)\, dx \\ &= 2\int_\Omega D^k w D^{k+1} w u_\varepsilon\, dx + 2\int_\Omega D^k w(D^{k+1}(u_\varepsilon w) - u_\varepsilon D^{k+1}))\, dx \\ &= -\int_\Omega (D^k w)^2 D u_\varepsilon\, dx + 2\int_\Omega D^k w(D^{k+1}(u_\varepsilon w) - u_\varepsilon D^{k+1} w)\, dx \\ &\le \|Du_\varepsilon\|_\infty |w|_k^2 + 2|w|_k |D^{k+1}(u_\varepsilon w) - u_\varepsilon D^{k+1} w|_0\end{aligned}$$

$$
\begin{aligned}
&\leq (\text{by Lemma 7.1 with } s = k+1 \text{ and } \beta = \tfrac{3}{4}) \\
&\leq \|Du_\varepsilon\|_\infty |w|_k^2 + c|w|_k(\|u_\varepsilon\|_{k+1}\|w\|_{H^{3/4}} + \|u_\varepsilon\|_{H^{7/4}}\|w\|_k) \\
&\leq c\|u_\varepsilon\|_2 |w|_k^2 + c|w|_k\|w\|_k\|u_\varepsilon\|_2 + c|w|_k\|w\|_{H^{3/4}}\|u_\varepsilon\|_{k+1} \\
&\leq c|w|_k^2 + c|w|_0^2 + c\|w\|_{H^{3/4}}^2\|u_\varepsilon\|_{k+1}^2 \\
&\leq c|w|_k^2 + c|w|_0^2 + c|w|_0^{1/2}\|w\|_1^{3/2}\|u_\varepsilon\|_{k+1}^2,
\end{aligned}
$$

so that

$$
\left|2\int_\Omega D^k w D^{k+1}(u_\varepsilon w)\,dx\right| \leq c|w|_k^2 + c|w|_0^2 + c\|u_\varepsilon\|_{k+1}^2 |w|_0^{1/2}\|w\|_1^{3/2},
$$

$$
\text{for} \quad 0 \leq t \leq T; \quad (7.141)
$$

$$
\begin{aligned}
2\int_\Omega D^k w D^k(wDw)\,dx &= 2\int_\Omega D^k w D^{k+1} w w\,dx \\
&\quad + 2\int_\Omega D^k w(D^k(wDw) - wD^{k+1}w)\,dx \\
&= -\int_\Omega (D^k w)^2 Dw\,dx \\
&\quad + 2\int_\Omega D^k w(D^k(wDw) - wD^{k+1}w)\,dx \\
&\leq \|Dw\|_\infty |w|_k^2 + 2|w|_k|D^k(wDw) - wD^{k+1}w|_0 \\
&\leq (\text{using Lemma 7.1 with } s = k \text{ and } \beta = 1) \\
&\leq \|Dw\|_\infty |w|_k^2 + c|w|_k(\|w\|_k\|Dw\|_1 + \|w\|_2\|Dw\|_{k-1}) \\
&\leq c\|w\|_2\|w\|_k^2 \leq c\|w\|_k^2 \leq c|w|_k^2 + c|w|_0^2,
\end{aligned}
$$

and, hence,

$$
\left|2\int_\Omega D^k w D^k(wDw)\,dx\right| \leq c|w|_k^2 + c|w|_0^2, \qquad \text{for} \quad 0 \leq t \leq T; \quad (7.142)
$$

finally,

$$
2(\varepsilon - \delta)\int_\Omega D^{k+4}u_\varepsilon D^k w\,dx \leq 2\varepsilon|u_\varepsilon|_{k+4}|w|_k \leq c|w|_k^2 + c\varepsilon^2|u_\varepsilon|_{k+4}^2. \qquad (7.143)
$$

Using (7.141), (7.142), and (7.143), we deduce from (7.140) that

$$
\begin{aligned}
\frac{d}{dt}|w|_k^2 + 2\delta|w|_{k+2}^2 &\leq c|w|_k^2 + c|w|_0^2 + |f_\varepsilon - f_\delta|_k^2 + c\varepsilon^2|u_\varepsilon|_{k+4}^2 \\
&\quad + c\|u_\varepsilon\|_{k+1}^2|w|_0^{1/2}\|w\|_1^{3/2}, \qquad \text{for} \quad 0 \leq t \leq T. \qquad (7.144)
\end{aligned}
$$

Taking (7.136), (7.139), and (7.129) into account, we infer from (7.144) that

$$\frac{d}{dt}|w|_k^2 \le c|w|_k^2 + c\varepsilon^{\min(1,4p)} + |f_\varepsilon - f_\delta|_k^2 + c\varepsilon^2|u_\varepsilon|_{k+4}^2 + c\varepsilon^{-2p}\varepsilon^p\varepsilon^{3p/2}$$

$$\le c|w|_k^2 + |f_\varepsilon - f_\delta|_k^2 + c\varepsilon^2|u_\varepsilon|_{k+4}^2 + c\varepsilon^{p/2}, \qquad \text{for} \quad 0 \le t \le T. \quad (7.145)$$

Therefore, by Gronwall's lemma, we deduce that

$$|w(t)|_k^2 \le c|w(0)|_k^2 + c|f_\varepsilon - f_\delta|_k^2 + c\varepsilon^2 \int_0^t |u_\varepsilon(s)|_{k+4}^2\, ds + c\varepsilon^{p/2}$$

$$\le \text{(using (7.131))}$$

$$\le c|w(0)|_k^2 + c|f_\varepsilon - f_\delta|_k^2 + c\varepsilon^{1-4p} + c\varepsilon^{p/2}, \qquad \text{for} \quad 0 \le t \le T.$$

Using (7.136) we find then

$$\|u_\varepsilon(t) - u_\delta(t)\|_k \le c\|u_{0\varepsilon} - u_{0\delta}\|_k + c\|f_\varepsilon - f_\delta\|_k + c\varepsilon^{(1/2)(1-4p)} + c\varepsilon^{p/4}, \quad (7.146)$$

for $t \in [0, T]$ and for all $0 < \delta < \varepsilon \le 1$. Now we choose a particular $0 < p < \frac{1}{4}$, say $p = \frac{1}{6}$. We then deduce from (7.146) that

$$\|u_\varepsilon(t) - u_\delta(t)\|_k \le c\|u_{0\varepsilon} - u_{0\delta}\|_k + \|f_\varepsilon - f_\delta\|_k + c\varepsilon^{1/24}, \quad (7.147)$$

for $t \in [0, T]$ and for all $0 < \delta < \varepsilon \le 1$. Using (7.125) it follows from (7.147) that $\{u_\varepsilon\}$ is a Cauchy sequence in $C([0, T], H^k_{\text{per}}(\Omega))$, and, consequently, u_ε converges strongly to u in $C([0, T], H^k_{\text{per}}(\Omega))$. From the time reversibility of the equation we deduce also that

$$u \in C([-T, T], H^k_{\text{per}}(\Omega)), \qquad \forall T > 0.$$

Hence, we can define on $H^k_{\text{per}}(\Omega)$ the group $\{S^{(k)}(t)\}_{t \in \mathbb{R}}$ given by

$$S^{(k)}(t)u_0 = u(t), \qquad \forall t \in \mathbb{R}, \quad (7.148)$$

for all $u_0 \in H^k_{\text{per}}(\Omega)$, where $u = u(t)$ is the solution of (7.1)–(7.3) at time t. Estimate (7.15) follows from (7.88), (7.95), (7.105), and (7.109).

For the strong continuity of $S^{(k)}(t)$ as an operator from $H^k_{\text{per}}(\Omega)$ into itself, note first that the corresponding solution operator of the regularized problem (7.82) and (7.83) is continuous from $H^k_{\text{per}}(\Omega)$ into itself since it is a parabolic problem. In fact, it is easy to show that given $R > 0$ and $T > 0$, there exists a constant $c > 0$ such that for any $u_0, v_0 \in H^k_{\text{per}}(\Omega)$, $\|u_0\|_k, \|v_0\|_k \le R$, we have

$$\|u_\varepsilon(t) - v_\varepsilon(t)\|_k \le c\varepsilon^{-1/4}\|u_{0\varepsilon} - v_{0\varepsilon}\|_k, \qquad 0 \le t \le T, \quad (7.149)$$

where u_ε and v_ε are the solutions of the regularized problem with initial conditions $u_{0\varepsilon}$ and $v_{0\varepsilon}$, respectively ($u_{0\varepsilon}$ and $v_{0\varepsilon}$ being the regularizations of u_0 and v_0). Then if $u = u(t) = S^{(k)}(t)u_0$ and $v = v(t) = S^{(k)}(t)v_0$, we have

$$\|u(t) - v(t)\|_k \le \|u(t) - u_\varepsilon(t)\|_k + \|v(t) - v_\varepsilon(t)\|_k + \|u_\varepsilon(t) - v_\varepsilon(t)\|_k.$$

From (7.149) we obtain, at the limit, as δ goes to zero,

$$\|u_\varepsilon(t) - u(t)\|_k \leq c\|u_{0\varepsilon} - u_0\|_k + c\varepsilon^{1/24},$$

with a similar result for v_ε. Thus

$$\|u(t) - v(t)\|_k \leq c\varepsilon^{1/24} + c\|u_{0\varepsilon} - u_0\|_k + c\|v_{0\varepsilon} - v_0\|_k + c\varepsilon^{-1/4}\|u_{0\varepsilon} - v_{0\varepsilon}\|_k.$$

Assume now that v_0 converges to u_0 in $H^k_{\text{per}}(\Omega)$. Then, clearly, $v_{0\varepsilon}$ converges to $u_{0\varepsilon}$ in $H^k_{\text{per}}(\Omega)$, for each ε, and, hence,

$$\limsup_{\substack{v_0 \to u_0 \\ \text{in } H^k_{\text{per}}(\Omega)}} \sup_{t \in [0,T]} \|u(t) - v(t)\|_k \leq c\varepsilon^{1/24} + 2c\|u_{0\varepsilon} - u_0\|_k. \tag{7.150}$$

Observing that the left-hand side of (7.149) is independent of ε and that the right-hand side can be made arbitrarily small, for sufficiently small ε, we conclude that

$$\limsup_{\substack{v_0 \to u_0 \\ \text{in } H^k_{\text{per}}(\Omega)}} \sup_{t \in [0,T]} \|u(t) - v(t)\|_k = 0; \tag{7.151}$$

in particular, for each t in $[0, T]$, $S^{(k)}(t)v_0 = v(t)$ converges to $S^{(k)}(t)u_0 = u(t)$; hence the strong continuity of $S^{(k)}(t)$ in $H^k_{\text{per}}(\Omega)$.

For the weak continuity of $S^{(k)}(t)$ in $H^k_{\text{per}}(\Omega)$ at fixed t, first note from (7.120) that $S^{(k)}(t)$ is strongly continuous in the $L^2(\Omega)$ topology on bounded subsets of $H^k_{\text{per}}(\Omega)$. Now, let $\{u_n\}_n$ be a sequence in $H^k_{\text{per}}(\Omega)$ converging weakly in $H^k_{\text{per}}(\Omega)$ to some limit $u \in H^k_{\text{per}}(\Omega)$. Then, $\{u_n\}_n$ is bounded in $H^k_{\text{per}}(\Omega)$, so that, thanks to (7.15), $\{S^{(k)}(t)u_n\}_n$ is also bounded in $H^k_{\text{per}}(\Omega)$. Therefore, there exists a subsequence $\{S^{(k)}(t)u_{n'}\}_{n'}$ converging weakly in $H^k_{\text{per}}(\Omega)$ to some limit $\varphi \in H^k_{\text{per}}(\Omega)$. Since the embedding $H^k_{\text{per}}(\Omega) \subset L^2(\Omega)$ is compact, $\{S^{(k)}(t)u_{n'}\}_{n'}$ converges strongly in $L^2(\Omega)$ to φ. Then, from the strong continuity of $S^{(k)}(t)$ in $L^2(\Omega)$ on bounded subsets of $H^k_{\text{per}}(\Omega)$, $\{S^{(k)}(t)u_n\}_n$ converges strongly to $S^{(k)}(t)u$ in $L^2(\Omega)$. Therefore, $\varphi = S^{(k)}(t)u$, and $\{S^{(k)}(t)u_{n'}\}_{n'}$ converges weakly to $S^{(k)}(t)u$ in $H^k_{\text{per}}(\Omega)$. Necessarily, the whole sequence $\{S^{(k)}(t)u_n\}_n$ converges weakly to $S^{(k)}(t)u$ in $H^k_{\text{per}}(\Omega)$, which proves the weak continuity of $S^{(k)}(t)$ in $H^k_{\text{per}}(\Omega)$. This completes the proof of Theorem 7.2. □

Proof of Corollary 7.1. As a consequence of the proof of Theorem 7.2, we prove that, for $u_0 \in H^k_{\text{per}}(\Omega)$, $u = u(t) = S^{(k)}(t)u_0$ satisfies the energy equations

$$\frac{d}{dt}I_m(u) + 2\gamma I_m(u) = K_m(u), \qquad \text{for} \quad m = 0, 1, \ldots, k, \tag{7.151$_m$}$$

where $I_m(u)$ and $K_m(u)$ are as in (7.7) and (7.12), respectively.

Consider the regularized problem (7.82) and (7.83) with $u_{0\varepsilon}$, f_ε as in the proof of Theorem 7.2. We multiply equation (7.82) by $L_m(u_\varepsilon)$, where L_m is given by (7.8), and find

$$\frac{d}{dt}I_m(u_\varepsilon) + 2\gamma I_m(u_\varepsilon) + \varepsilon \int_\Omega D^4 u_\varepsilon L_m(u_\varepsilon)\, dx = K_m(u_\varepsilon). \tag{7.152$_m$}$$

Notice that

$$\int_\Omega D^4 u_\varepsilon L_m(u_\varepsilon)\,dx$$

$$= \int_\Omega D^4 u_\varepsilon \Bigg[2(-1)^m D^{2m} u_\varepsilon - \alpha_m (D^{m-1}u_\varepsilon)^2 + 2\alpha_m(-1)^m D^{m-1}(u_\varepsilon D^{m-1}u_\varepsilon)$$

$$+ \sum_{j=0}^{m-2} (-1)^j D^j\left(\frac{\partial Q_m}{\partial y_j}(u_\varepsilon, Du_\varepsilon, \ldots, D^{m-2}u_\varepsilon)\right)\Bigg]\,dx$$

$$= 2|u_\varepsilon|_{m+2}^2 + \int_\Omega D^4 u_\varepsilon[-\alpha_m(D^{m-1}u_\varepsilon)^2 + 2\alpha_m(-1)^m D^{m-1}(u_\varepsilon D^{m-1}u_\varepsilon)]\,dx$$

$$+ \int_\Omega \sum_{j=0}^{m-2} D^{j+2}u_\varepsilon D^2\left(\frac{\partial Q_m}{\partial y_j}(u_\varepsilon, Du_\varepsilon, \ldots, D^{m-2}u_\varepsilon)\right)dx.$$

We also have

$$\int_\Omega D^4 u_\varepsilon[-(D^{m-1}u_\varepsilon)^2 + 2(-1)^m D^{m-1}(u_\varepsilon D^{m-1}u_\varepsilon)]\,dx$$

$$= \int_\Omega [Du_\varepsilon D^3((D^{m-1}u_\varepsilon)^2) + 2D^{m+2}u_\varepsilon D(u_\varepsilon D^{m-1}u_\varepsilon)]\,dx$$

$$= \int_\Omega [Du_\varepsilon(2D^{m+2}u_\varepsilon D^{m-1}u_\varepsilon + 6D^{m+1}u_\varepsilon D^m u_\varepsilon)$$

$$+ 2D^{m+2}u_\varepsilon(D^m u_\varepsilon u_\varepsilon + D^{m-1}u_\varepsilon Du_\varepsilon)]\,dx$$

$$= \int_\Omega (4D^{m+2}u_\varepsilon D^{m-1}u_\varepsilon Du_\varepsilon + 2D^{m+2}u_\varepsilon D^m u_\varepsilon u_\varepsilon + 6D^{m+1}u_\varepsilon D^m u_\varepsilon Du_\varepsilon)\,dx.$$

Thus,

$$\frac{d}{dt}I_m(u_\varepsilon) + 2\gamma I_m(u_\varepsilon) + 2\varepsilon|u_\varepsilon|_{m+2}^2$$

$$+ \varepsilon\int_\Omega \sum_{j=0}^{m-2} D^{j+2}u_\varepsilon D^2\left(\frac{\partial Q_m}{\partial y_j}(u_\varepsilon, Du_\varepsilon, \ldots, D^{m-2}u_\varepsilon)\right)dx$$

$$+ \varepsilon\alpha_m \int_\Omega (4D^{m+2}u_\varepsilon D^{m-1}u_\varepsilon Du_\varepsilon + 2D^{m+2}u_\varepsilon D^m u_\varepsilon u_\varepsilon$$

$$+ 6D^{m+1}u_\varepsilon D^m u_\varepsilon Du_\varepsilon)\,dx = K_m(u_\varepsilon), \qquad (7.153)_m$$

for $m = 0, 1, \ldots, k$.

Since u_ε converges to u in $C([0, T], H^k_{\text{per}}(\Omega))$ and $|u_\varepsilon|_k$ is bounded in $C([0, T])$, it is easy to see that $(7.153)_m$ yields $(7.151)_m$ at the limit as ε goes to zero, for $m = 0, 1, \ldots, k - 2$.

Consider now $(7.153)_{k-1}$. After integration by parts, it reads

$$\frac{d}{dt} I_{k-1}(u_\varepsilon) + 2\gamma I_{k-1}(u_\varepsilon) + 2\varepsilon |u_\varepsilon|_{k+1}^2 - 2\varepsilon\alpha_{k-1} \int_\Omega ((D^k u_\varepsilon)^2 u_\varepsilon + 2D^k u_\varepsilon D^{k-2} u_\varepsilon D^2 u_\varepsilon)\, dx = K_{k-1}(u_\varepsilon). \tag{7.154}$$

Then, thanks to the estimate (7.129), it is also easy to see that (7.154) yields $(7.151)_{k-1}$ at the limit as ε goes to zero.

Finally, consider $(7.153)_k$. Using again that u_ε converges to u in $C([0, T], H^k_{\text{per}}(\Omega))$ and that $\|u_\varepsilon\|_k$ is bounded in $C(0, T)$, and using also the estimates (7.129) and (7.131), we can pass to the limit in $(7.153)_k$, as ε goes to zero, to obtain $(7.151)_k$. Note that for the energy equations $(7.151)_m$ with $m = 0, 1, \ldots, k-1$, we only need the weak convergence (7.111) and the estimate (7.129), but for the energy equation $(7.151)_k$, we do need the strong convergence in $C([0, T], H^k_{\text{per}}(\Omega))$, otherwise we would only obtain an inequality. □

7.5. Proof of Proposition 7.2

We now prove Proposition 7.2. For the existence and uniqueness results, we can use the Galerkin approximation. The a priori estimates are obtained as in the proof of Theorem 2.1 using slight appropriate modifications of the invariants I_j.

In what follows we proceed formally to obtain the a priori estimates that lead to (7.71) and (7.72). A rigorous proof follows from similar estimates derived on the Galerkin approximation.

Using (7.15) and the fact that $\mathscr{B}_3$ is an absorbing ball in $H^3_{\text{per}}(\Omega)$ for $\{S^{(3)}(t)\}_{t\geq 0}$, we deduce that there exists $\rho > 0$ such that

$$\|S^{(3)}(t)u_0\|_3 \leq \rho, \qquad \forall t \geq 0, \quad \forall u_0 \in \mathscr{B}_3. \tag{7.155}$$

Hence

$$\|y(t)\|_3 \leq \rho, \qquad \forall t \geq 0, \tag{7.156}$$

and (see (7.64))

$$\|y(t)\|_{3+l} \leq cN^l, \qquad \forall t \geq 0, \quad \forall l \in \mathbb{N}. \tag{7.157}$$

In what follows, we denote by c a generic constant which may depend on γ, L, f, and ρ, but is independent of N. We also write, for simplicity, $P = P_N$ and $Q = Q_N$.

Step I. *Estimates for q in $L^2(\Omega)$ and $H^1_{\text{per}}(\Omega)$.*

Multiplying the q equation (7.68) by $2q$ and since $q = Qq$, we find

$$\frac{d}{dt}|q|_0^2 + 2\gamma|q|_0^2 + 2(D(yq), q)_{L^2} = 2(Qf, q)_{L^2} - 2(yDy, q)_{L^2}. \quad (7.158)$$

Note that

$$2(D(yq), q)_{L^2} = -2(yq, Dq)_{L^2} = -(y, D(q^2))_{L^2} = (Dy, q^2)_{L^2}.$$

Thus,

$$\begin{aligned} 2(Qf - yDy - D(yq), q)_{L^2} &\le 2|Qf|_0|q|_0 + 2\|y\|_\infty|y|_1|q|_0 + \|Dy\|_\infty|q|_0^2 \\ &\le \gamma|q|_0^2 + \frac{2}{\gamma}|Qf|_0^2 + \frac{2}{\gamma}\|y\|_\infty^2|y|_1^2 + c\|y\|_2|q|_0^2. \end{aligned}$$

We deduce from (7.158) that

$$\frac{d}{dt}|q|_0^2 + \gamma|q|_0^2 \le \frac{2}{\gamma}|Qf|_0^2 + \frac{c}{\gamma}\rho^4 + c\rho|q|_0^2.$$

Then, we write,

$$\frac{d}{dt}|q|_0^2 + \gamma|q|_0^2 \le c_1 + c_2|q|_0^2, \quad (7.159)$$

where $c_1 = (2/\gamma)|Qf|_0^2 + c\rho^4/\gamma$ and $c_2 = c\rho$.
Since $q(0) = 0$, we obtain

$$|q(t)|_0^2 \le \frac{c_1}{c_2}e^{c_2 t}, \qquad \forall t \ge 0. \quad (7.160)$$

By replacing c_2 by $\max\{c_2, \gamma\}$, if necessary, we can assume that $c_2 \ge \gamma$.
Multiply now the q-equation (7.68) by $-2D^2q - Q(q^2) - 2Q(yq)$. We observe that

$$(D^3q + Q(qDq) + QD(yq), -2D^2q - Q(q^2) - 2Q(yq))_{L^2} = 0.$$

Also,

$$\begin{aligned} (q_t, -2D^2q - Q(q^2) - 2Q(yq))_{L^2} &= (q_t, -2D^2q - q^2 - 2yq)_{L^2} \\ &= \int_\Omega \left(\partial_t(Dq)^2 - \partial_t\left(\frac{q^3}{3}\right) - y\partial_t(q^2)\right) dx \\ &= \frac{d}{dt}\int_\Omega \left((Dq)^2 - \frac{q^3}{3} - yq^2\right) dx \\ &\quad + (y_t, q^2)_{L^2}, \end{aligned}$$

and

$$\gamma(q, -2D^2q - Q(q^2) - 2Q(yq))_{L^2} = \gamma(q, -2D^2q - q^2 - 2yq)_{L^2}$$
$$= 2\gamma \int_\Omega \left((Dq)^2 - \frac{q^3}{2} - yq^2\right) dx.$$

Therefore,

$$\frac{d}{dt}\int_\Omega \left((Dq)^2 - \frac{q^3}{3} - yq^2\right) dx + 2\gamma \int_\Omega \left((Dq)^2 - \frac{q^3}{2} - yq^2\right) dx + (y_t, q^2)_{L^2}$$
$$= (Qf - Q(yDy), -2D^2q - Q(q^2) - 2Q(yq))_{L^2}. \tag{7.161}$$

Now, we estimate

$$\begin{aligned}
&(Qf - Q(yDy), -2D^2q - Q(q^2) - 2Q(yq))_{L^2}\\
&\quad = 2(QDf, Dq)_{L^2} - (Qf, q^2) - 2(Qf, yq)_{L^2} - 2(D(yDy), Dq)_{L^2}\\
&\quad\quad + (Q(yDy), q^2)_{L^2} + 2(Q(yDy), yq)_{L^2}\\
&\quad \le 2|Qf|_1|q|_1 + \|Qf\|_\infty |q|_0^2 + 2|yQf|_0|q|_0 + |y^2|_2|q|_1\\
&\quad\quad + \tfrac{1}{2}\|QD(y^2)\|_\infty |q|_0^2 + |y^2|_1\|y\|_\infty |q|_0\\
&\quad \le (2|Qf|_1 + |y^2|_2)|q|_1 + (\|Qf\|_\infty + \tfrac{1}{2}\|QD(y^2)\|_\infty)|q|_0^2\\
&\quad\quad + (2|yQf|_0 + |y^2|_1\|y\|_\infty)|q|_0\\
&\quad \le (2|Qf|_1 + c\|y\|_2^2)|q|_1 + (c\|Qf\|_1 + c\|y\|_2^2)|q|_0^2\\
&\quad\quad + (2\|y\|_\infty |Qf|_0 + c\|y\|_1^3)|q|_0\\
&\quad \le \text{(using (7.156))}\\
&\quad \le c|q|_0 + c|q|_0^2 + c|q|_1.
\end{aligned}$$

From (7.66) we see that

$$|q|_0 \le c_0 N^{-1}|q|_1. \tag{7.162}$$

Therefore,

$$(Qf - Q(yDy), -2D^2q - Q(q^2) - 2Q(yq))_{L^2} \le cN^{-1}|q|_1 + cN^{-2}|q|_1^2 + c|q|_1$$
$$\le c|q|_1 + cN^{-2}|q|_1^2.$$

Hence, from (7.161) and the previous estimates, we obtain

$$\frac{d}{dt}\int_\Omega \left((Dq)^2 - \frac{q^3}{3} - yq^2\right) dx + (y_t, q^2)_{L^2} + 2\gamma \int_\Omega \left((Dq)^2 - \frac{q^3}{2} - yq^2\right) dx$$
$$\le c|q|_1 + cN^{-2}|q|_1^2. \tag{7.163}$$

Setting

$$\tilde{I}_1(q) = \int_\Omega \left((Dq)^2 - \frac{q^3}{3} - yq^2\right) dx,$$

we can write (7.163) as

$$\frac{d}{dt}\tilde{I}_1(q) + 2\gamma\tilde{I}_1(q) \leq -(y_t, q^2)_{L^2} + \frac{\gamma}{3}\int_\Omega q^3\, dx + c|q|_1 + cN^{-2}|q|_1^2. \tag{7.164}$$

From the equation for $y = Pu$, we have the following estimate for $|y_t|_0$:

$$|y_t|_0 = |D^3 y + \gamma y + P(uDu) - Pf|_0 \leq \text{(using (7.155) and (7.156))} \leq c, \tag{7.165}$$

so that

$$\begin{aligned}(y_t, q^2)_{L^2} &\leq |y_t|_0 |q|_0 \|q\|_\infty \\ &\leq c|q|_0 \|q\|_\infty \\ &\leq \text{(using the Agmon inequality)} \\ &\leq c|q|_0^{3/2}|q|_1^{1/2} \\ &\leq \text{(using (7.162))} \\ &\leq cN^{-3/2}|q|_1^2. \end{aligned} \tag{7.166}$$

Using again the Agmon inequality and (7.162), we estimate

$$\left|\int_\Omega \frac{q^3}{3}\, dx\right| \leq \tfrac{1}{3}|q|_0^2\|q\|_\infty \leq c|q|_0^{5/2}|q|_1^{1/2} \leq c_3 N^{-3/2}|q|_0|q|_1^2, \tag{7.167}$$

and

$$\left|\int_\Omega yq^2\, dx\right| \leq \|y\|_\infty |q|_0^2 \leq c|q|_0^2 \leq c_4 N^{-2}|q|_1^2. \tag{7.168}$$

From the definition of $\tilde{I}_1(q)$ and using (7.167) and (7.168), we deduce that

$$(1 - c_3 N^{-3/2}|q|_0 - c_4 N^{-2})|q|_1^2 \leq \tilde{I}_1(q) \leq (1 + c_3 N^{-3/2}|q|_0 + c_4 N^{-2})|q|_1^2. \tag{7.169}$$

Using now (7.166) and (7.167) in (7.164), we obtain

$$\begin{aligned}\frac{d}{dt}\tilde{I}_1(q) + 2\gamma\tilde{I}_1(q) &\leq (cN^{-2} + \gamma c_3 N^{-3/2}|q|_0 + cN^{-3/2})|q|_1^2 + c|q|_1 \\ &< \left(c_5 N^{-3/2} + \gamma c_3 N^{-3/2}|q|_0 + \frac{\gamma}{4}\right)|q|_1^2 + \frac{c_6}{\gamma}. \end{aligned} \tag{7.170}$$

We claim that for N large enough we have

$$|q(t)|_0^2 \leq \frac{3c_1}{\gamma}, \qquad \forall t \geq 0, \tag{7.171}$$

and

$$|q(t)|_1^2 \leq \frac{2c_6}{\gamma^2}, \qquad \forall t \geq 0. \tag{7.172}$$

Indeed, let N be large enough so that

$$c_5 N^{-3/2} + \sqrt{3c_1\gamma c_3} N^{-3/2} \leq \frac{\gamma}{4}, \tag{7.173}$$

$$c_4 N^{-2} + \sqrt{\frac{3c_1}{\gamma}} c_3 N^{-3/2} \leq \frac{1}{2}, \tag{7.174}$$

and

$$\frac{c_0^2 c_6}{\gamma} N^{-2} \leq c_1. \tag{7.175}$$

From (7.160) we see that

$$|q(t)|_0^2 \leq \frac{c_1}{c_2} e^{c_2 t} < \frac{3c_1}{\gamma}, \qquad \text{for} \quad 0 \leq t < \tilde{t}, \tag{7.176}$$

where $\tilde{t} = (1/c_2)\log(3c_2/\gamma)$ (recall that $c_2 \geq \gamma$, so that $3c_2/\gamma \geq 3 > 1$). Set now

$$T = \sup\left\{t \geq 0; |q(t)|_0^2 < \frac{3c_1}{\gamma}\right\}.$$

Clearly $T \geq \tilde{t} > 0$. We want to show that $T = \infty$.

Consider t such that $0 \leq t < T$; for such t, we have

$$|q(t)|_0^2 < \frac{3c_1}{\gamma}, \tag{7.177}$$

so that from (7.169) and (7.174) we deduce that

$$\tfrac{1}{2}|q|_1^2 \leq \tilde{I}_1(q) \leq \tfrac{3}{2}|q|_1^2, \tag{7.178}$$

while from (7.170) and (7.173) we find

$$\frac{d}{dt}\tilde{I}_1(q) + 2\gamma\tilde{I}_1(q) \leq \frac{\gamma}{2}|q|_1^2 + \frac{c_6}{\gamma}. \tag{7.179}$$

Using now (7.178) and (7.179), we obtain

$$\frac{d}{dt}\tilde{I}_1(q) + \gamma\tilde{I}_1(q) \leq \frac{c_6}{\gamma}, \qquad \text{for} \quad 0 \leq t < T. \tag{7.180}$$

Hence, by the Gronwall lemma and since $q(0) = 0$,

$$\tilde{I}_1(q(t)) \leq \tilde{I}_1(q(0))e^{-\gamma t} + \frac{c_6}{\gamma^2} = \frac{c_6}{\gamma^2}, \qquad \text{for} \quad 0 \leq t < T. \tag{7.181}$$

Thus, (7.181) and (7.178), we deduce that

$$|q(t)|_1^2 \leq 2\tilde{I}_1(q(t)) \leq \frac{2c_6}{\gamma^2}, \qquad \text{for} \quad 0 \leq t < T. \tag{7.182}$$

If T were finite, we would integrate (7.159) on $[T - \varepsilon, T + \varepsilon]$, for some $\varepsilon > 0$, to find

$$|q(T + \varepsilon)|_0^2 \leq |q(T - \varepsilon)|_0^2 e^{2c_2\varepsilon} + \frac{c_1}{c_2}(e^{2c_2\varepsilon} - 1). \tag{7.183}$$

But from (7.62) and (7.182) we have

$$\begin{aligned} |q(T - \varepsilon)|_0^2 &\leq c_0^2 N^{-2} |q(T - \varepsilon)|_1^2 \\ &\leq \frac{2c_0^2 c_6}{\gamma^2} N^{-2} \\ &\leq \text{(using (7.175))} \\ &\leq \frac{2c_1}{\gamma}. \end{aligned} \tag{7.184}$$

Then, from (7.183) and (7.184), we would infer that

$$|q(T + \varepsilon)|_0^2 \leq \left(\frac{2c_1}{\gamma} + \frac{c_1}{c_2}\right) e^{2c_2\varepsilon} - \frac{c_1}{c_2} < \frac{3c_1}{\gamma}, \qquad \text{for} \quad \varepsilon \in \left(0, \frac{1}{2c_2} \log \frac{3c_2 + \gamma}{2c_2 + \gamma}\right),$$

contradicting the maximality of T. Therefore, $T = \infty$ and

$$|q(t)|_0^2 < \frac{3c_1}{\gamma}, \qquad \forall t \geq 0,$$

and then from (7.182),

$$|q(t)|_1^2 \leq \frac{2c_6}{\gamma^2}, \qquad \forall t \geq 0,$$

which proves the claim and gives a priori estimates for q in $L^2(\Omega)$ and $H^1_{\text{per}}(\Omega)$.

Step II. *Estimates for q in $H^2_{\text{per}}(\Omega)$.*

Multiply the q-equation (7.68) by a special multiplier, namely, $L_2(q) - a_2 D(yDq)$, where a_2 is a constant which will be specified in the sequel.

For $m = 2$ one can find that (7.8) and (7.7) can be written, respectively, as

$$L_2(q) = 2D^4 q - \alpha_2 (Dq)^2 + 2\alpha_2 q D^2 q + 4\delta q^3, \tag{7.185}$$

and

$$I_2(q) = \int_\Omega ((D^2 q)^2 - \alpha_2 q (Dq)^2 + \delta q^4)\, dx, \tag{7.186}$$

for appropriate real numbers α_2 and δ (see J.M. Ghidaglia [4]). From (7.9) and (7.10),

$$(q_t, L_2(q))_{L^2} = \frac{d}{dt} I_2(q),$$

and

$$(D^3 q + qDq, L_2(q))_{L^2} = 0.$$

Therefore, we obtain

$$\begin{aligned}
&\frac{d}{dt} I_2(q) + (q_t, -a_2 D(yDq))_{L^2} + 2\gamma I_2(q) \\
&\qquad + \gamma(q, -a_2 D(yDq))_{L^2} - (P(qDq), L_2(q))_{L^2} \\
&\qquad + (D^3 q + Q(qQq), -a_2 D(yDq))_{L^2} + (QD(yq), L_2(q) - a_2 D(yDq))_{L^2} \\
&\quad = K_2(q) + (Qf, -a_2 D(yDq))_{L^2} - (Q(yDy), L_2(q) - a_2 D(yDq))_{L^2}, \quad (7.187)
\end{aligned}$$

where

$$\begin{aligned}
K_2(q) = \int_\Omega &(2QD^2 f D^2 q - \alpha_2 f (Dq)^2 - 2\alpha_2 qDqQDf \\
&+ 4\delta Q f q^3 + \gamma \alpha_2 q (Dq)^2 - 2\gamma\delta q^4)\, dx. \quad (7.188)
\end{aligned}$$

Note that

$$(q_t, -a_2 D(yDq))_{L^2} = a_2 (Dq_t, yDq) = \frac{a_2}{2} \frac{d}{dt} \int_\Omega (Dq)^2 y\, dx - \frac{a_2}{2} (y_t, (Dq)^2)_{L^2}, \quad (7.189)$$

and

$$\begin{aligned}
(D^3 q, -a_2 D(yDq))_{L^2} &= a_2 (D^2 q, D^2 (yDq))_{L^2} \\
&= a_2 (D^2 q, yD^3 q + 2DyD^2 q + D^2 y Dq)_{L^2} \\
&= \frac{3a_2}{2} \int_\Omega (D^2 q)^2 Dy\, dx + a_2 (D^2 q, D^2 y Dq)_{L^2}. \quad (7.190)
\end{aligned}$$

Also, we have from $(QD(yq), L_2(q))_{L^2}$ the term

$$\begin{aligned}
(QD(yq), 2D^4 q)_{L^2} &= 2(D^2 q, D^3 (yq))_{L^2} \\
&= 2(D^2 q, yD^3 q + 3DyD^2 q + 3D^2 y Dq + qD^3 y)_{L^2} \\
&= 5 \int_\Omega (D^2 q)^2\, Dy\, dx + 6(D^2 q, D^2 y Dq)_{L^2} + 2(D^2 q, qD^3 y)_{L^2}. \quad (7.191)
\end{aligned}$$

Now, we choose $a_2 = -\frac{10}{3}$ in order to cancel the term $((D^2q)^2, Dy)_{L^2}$, which we cannot control. Using (7.189)–(7.191) in (7.187), we then obtain

$$\frac{d}{dt}\left(I_2(q) - \frac{5}{3}\int_\Omega (Dq)^2 y\,dx\right) + 2\gamma\left(I_2(q) - \frac{5}{3}\int_\Omega (Dq)^2 y\,dx\right)$$

$$+ \frac{5}{3}(y_t, (Dq)^2)_{L^2} + \left(-\frac{10}{3} + 6\right)\int_\Omega D^2qDqD^2y\,dx$$

$$+ 2\int_\Omega qD^2qD^3y\,dx - (P(qDq), L_2(q))_{L^2} + \left(Q(qDq), \frac{10}{3}D(yDq)\right)_{L^2}$$

$$+ \left(QD(yq), \alpha_2(Dq)^2 + 2\alpha_2 qD^2q + 4\delta q^3 + \frac{10}{3}D(yDq)\right)_{L^2}$$

$$= K_2(q) + \left(Qf, \frac{10}{3}D(yDq)\right)_{L^2} - \left(Q(yDy), L_2(q) + \frac{10}{3}D(yDq)\right)_{L^2}. \tag{7.192}$$

We estimate below the terms in (7.192) using extensively (7.171) and (7.172) and the fact that q has zero average, in which case $|q|_0, |q|_1 \le c|q|_2$. We proceed as follows:

$$(y_t, (Dq)^2)_{L^2} \le |y_t|_0|q|_1\|Dq\|_\infty \le \text{(using also (7.165))} \le c|q|_1|q|_2 \le c|q|_2; \tag{7.193}$$

$$\left|\int_\Omega D^2qDqD^2y\,dx\right| \le |q|_2|q|_1\|D^2y\|_\infty \le c|q|_2|q|_1\|y\|_3$$

$$\le \text{(using (7.172))} \le c|q|_2; \tag{7.194}$$

$$\left|\int_\Omega qD^2qD^3y\,dx\right| \le |q|_2\|q\|_\infty|y|_3 \le c|q|_2|q|_1^{1/2}|q|_0^{1/2}$$

$$\le \text{(using (7.171) and (7.172))} \le c|q|_2; \tag{7.195}$$

$$\begin{aligned}(P(qDq), L_2(q))_{L^2} &= (P(qDq), 2D^4q + \alpha_2(Dq)^2 + 2\alpha_2 qD^2q + 4\delta q^3)_{L^2}\\ &= \text{(because } P \text{ and } Q \text{ are orthogonal)}\\ &= (P(qDq), \alpha_2(Dq)^2 + 2\alpha_2 qD^2q + 4\delta q^3)_{L^2}\\ &\le c|qDq|_0(|(Dq)^2|_0 + |qD^2q|_0 + |q^3|_0)\\ &\le c\|q\|_\infty|q|_1(|q|_1\|Dq\|_\infty + \|q\|_\infty|q|_2 + \|q\|_\infty^2|q|_0)\\ &\le c|q|_2;\end{aligned} \tag{7.196}$$

$$\begin{aligned}\left(Q(qDq), \frac{10}{3}D(yDq)\right)_{L^2} &\le c|qDq|_0|D(yDq)|_0\\ &\le c\|q\|_\infty|q|_1\|y\|_1|q|_2 \le c|q|_2;\end{aligned} \tag{7.197}$$

and

$$
\begin{aligned}
&(QD(yq), \alpha_2(Dq)^2 + 2\alpha_2 qD^2q + 4\delta q^3 + \tfrac{10}{3}D(yDq))_{L^2} \\
&\quad \leq c|D(yq)|_0(|(Dq)^2|_0 + |qD^2q|_0 + |q|_0^3 + |D(yDq)|_0) \\
&\quad \leq c\|y\|_1|q|_1(|q|_1\|Dq\|_\infty + \|q\|_\infty|q|_2 + \|q\|_\infty^2|q|_0 + \|y\|_1|q|_2) \\
&\quad \leq c|q|_2.
\end{aligned}
\tag{7.198}
$$

For the right-hand side terms in (7.192) we have the following estimates:

$$
\begin{aligned}
K_2(q) &\leq 2|Qf|_2|q|_2 + c\|f\|_\infty|q|_1 + c|q|_0|q|_1\|QDf\|_\infty \\
&\qquad + c\|Qf\|_\infty\|q\|_\infty|q|_0^2 + c\|q\|_\infty|q|_1^2 + c\|q\|_\infty^2|q|_0^2 \\
&\leq c|q|_2;
\end{aligned}
\tag{7.199}
$$

$$
\begin{aligned}
(Qf, \tfrac{10}{3}D(yDq))_{L^2} &\leq c|Qf|_0|D(yDq)|_0 \leq c|Qf|_0\|y\|_1|q|_2 \\
&\leq c|q|_2;
\end{aligned}
\tag{7.200}
$$

and

$$
\begin{aligned}
&(Q(yDy), L_2(q) + \tfrac{10}{3}D(yDq))_{L^2} \\
&\quad \leq 2(D^2(yDy), D^2q)_{L^2} \\
&\qquad + (Q(yDy), \alpha_2(Dq)^2 + 2\alpha_2 qD^2q + 4\delta q^3 + \tfrac{10}{3}D(yDq))_{L^2} \\
&\quad \leq c|D^2(yDy)|_0|q|_2 + c|q|_2 \\
&\quad \leq c\|y\|_3\|y\|_2|q|_2 + c|q|_2 \\
&\quad \leq c|q|_2.
\end{aligned}
\tag{7.201}
$$

Using now (7.193)–(7.201) in (7.192), we obtain

$$
\begin{aligned}
&\frac{d}{dt}\left(I_2(q) - \frac{5}{3}\int_\Omega (Dq)^2 y\,dx\right) + 2\gamma\left(I_2(q) - \frac{5}{3}\int_\Omega (Dq)^2 y\,dx\right) \\
&\quad \leq c|q|_2 \\
&\quad \leq \gamma|q|_2^2 + c.
\end{aligned}
\tag{7.202}
$$

Consider now

$$
\begin{aligned}
&\left|\int_\Omega (-\alpha_2(Dq)^2 + \delta q^4 - \tfrac{5}{3}(Dq)^2 y)\,dx\right| \\
&\quad \leq c(|q|_1^2 + \|q\|_\infty^2|q|_0^2 + |q|_1\|y\|_\infty) \\
&\quad \leq (\text{using again (7.171) and (7.172)}) \leq c,
\end{aligned}
\tag{7.203}
$$

so that

$$|q|_2^2 - c \leq I_2(q) - \frac{5}{3}\int_\Omega (Dq)^2 y \, dx \leq |q|_2^2 + c. \tag{7.204}$$

Then, from (7.201) and (7.203), we obtain

$$\frac{d}{dt}(I_2(q) - \tfrac{5}{3}(Dq)^2 y \, dx) + \gamma\left(I_2(q) - \frac{5}{3}\int_\Omega (Dq)^2 y \, dx\right) \leq c, \qquad \forall t \geq 0.$$

Therefore, since $q(0) = 0$, we find that

$$I_2(q(t)) - \frac{5}{3}\int_\Omega (Dq(t))^2 y(t) \, dx \leq c, \qquad \forall t \geq 0,$$

which, together with (7.204), gives

$$|q(t)|_2^2 \leq c, \qquad \forall t \geq 0. \tag{7.205}$$

Step III. *Estimates for q in $H^k_{\text{per}}(\Omega)$.*

Multiply the q-equation (7.68) by

$$2(-1)^k D^{2k} q + a_k(-1)^{k-1} D^{k-1}(y D^{k-1} q),$$

where a_k will be specified later on. We obtain

$$\begin{aligned}
&\frac{d}{dt}\left(|q|_k^2 + \frac{a_k}{2}\int_\Omega (D^{k-1}q)^2 y \, dx\right) - \frac{a_k}{2}\int_\Omega (D^{k-1}q)^2 y_t \, dx \\
&\qquad + (D^3 q, a_k(-1)^{k-1} D^{k-1}(y D^{k-1} q))_{L^2} + 2\gamma\left(|q|_k^2 + \frac{a_k}{2}\int_\Omega (D^{k-1}q)^2 y \, dx\right) \\
&\qquad + (Q(qDq), 2(-1)^k D^{2k} q + a_k(-1)^{k-1} D^{k-1}(y D^{k-1} q))_{L^2} \\
&\qquad + (Q(D(yq)), 2(-1)^k D^{2k} q + a_k(-1)^{k-1} D^{k-1}(y D^{k-1} q))_{L^2} \\
&= (Qf - Q(yDy), 2(-1)^k D^{2k} q + a_k(-1)^{k-1} D^{k-1}(y D^{k-1} q))_{L^2}.
\end{aligned} \tag{7.206}$$

Note that

$$\begin{aligned}
&(D^3 q, a_k(-1)^{k-1} D^{k-1}(y D^{k-1} q))_{L^2} \\
&\quad = (D^k q, a_k D^2(y D^{k-1} q))_{L^2} \\
&\quad = \int_\Omega a_k D^k q (y D^{k+1} q + 2 D^k q D y + D^{k-1} q D^2 y) \, dx \\
&\quad = \frac{3a_k}{2}\int_\Omega (D^k q)^2 D y \, dx + a_k \int_\Omega D^k q D^{k-1} q D^2 y \, dx,
\end{aligned} \tag{7.207}$$

and

$$\begin{aligned}(Q(D(yq), 2(-1)^k D^{2k}q)_{L^2} &= (D(yq), 2(-1)^k D^{2k}q)_{L^2} \\ &= 2(D^{k+1}(yq), D^k q)_{L^2} \\ &= 2\left(D^k q, \sum_{j=0}^{k+1}\binom{k+1}{j} D^{k+1-j}qD^j y\right)_{L^2} \\ &= (2k+1)\int_\Omega (D^k q)^2 Dy\,dx + 2\sum_{j=2}^{k+1}\binom{k+1}{j}(D^k q, D^{k+1-j}qD^j y)_{L^2}. \qquad (7.208)\end{aligned}$$

We pass now to the estimates in (7.206), taking (7.207) and (7.208) into account. We first have

$$\begin{aligned}\left|\int_\Omega D^k q D^{k-1} q D^2 y\,dx\right| &\le |q|_k |q|_{k-1} \|D^2 y\|_\infty \\ &\le c|q|_k|q|_{k-1}\|y\|_3 \le c|q|_k|q|_{k-1} \\ &\le (\text{using } (7.66)) \le cN^{-1}|q|_k^2. \qquad (7.209)\end{aligned}$$

Also,

$$\begin{aligned}(Q(qDq), 2(-1)^k D^{2k}q)_{L^2} &= (qDq, 2(-1)^k D^{2k}q)_{L^2} \\ &= (D^k(qDq), 2D^k q)_{L^2} \\ &= 2(qD^{k+1}q, D^k q)_{L^2} + 2(D^k(qDq) - qD^{k+1}q, D^k q)_{L^2} \\ &= -(Dq, (D^k q)^2)_{L^2} + 2(D^k(qDq) - qD^{k+1}q, D^k q)_{L^2} \\ &\le |q|_k^2 \|Dq\|_\infty + 2|D^k(qDq) - qD^{k+1}q|_0|q|_k \\ &\le (\text{using Lemma 7.1 with } s = k \text{ and } \beta = \tfrac{3}{4}) \\ &\le |q|_k^2\|Dq\|_\infty + c|q|_k[|q|_k|Dq|_{3/4} + |q|_{7/4}|Dq|_{k-1}] \\ &\le c|q|_k^2|q|_{7/4} \le c|q|_k^2|q|_1^{1/4}|q|_2^{3/4} \\ &\le (\text{using } (7.66)) \le cN^{-1/4}|q|_k^2|q|_2 \\ &\le (\text{using } (7.207)) \le cN^{-1/4}|q|_k^2, \qquad (7.210)\end{aligned}$$

and

$$\begin{aligned}(Q(qDq), (-1)^{k-1}D^{k-1}(yD^{k-1}q))_{L^2} &= (QD^{k-1}(qDq), yD^{k-1}q)_{L^2} \\ &\le |QD^{k-1}(qDq)|_0|q|_{k-1}\|y\|_\infty.\end{aligned}$$

Since

$$\begin{aligned}|QD^{k-1}(qDq)|_0 &\le |D^{k-1}(qDq)| \\ &\le |qD^k q|_0 + |D^{k-1}(qDq) - qD^k q|_0\end{aligned}$$

$$\begin{aligned}
&\leq \text{(using Lemma 7.1 with } s = k - 1 \text{ and } \beta = 1)\\
&\leq \|q\|_\infty |q|_k + c(|q|_{k-1}|Dq|_1 + |q|_2|Dq|_{k-2})\\
&\leq c|q|_1|q|_k + c|q|_{k-1}|q|_2\\
&\leq \text{(using (7.205))}\\
&\leq c|q|_k + c|q|_{k-1} \leq c|q|_k,
\end{aligned}$$

we also have

$$\begin{aligned}
(Q(qDq), (-1)^{k-1}D^{k-1}(yD^{k-1}q))_{L^2} &\leq c|q|_k|q|_{k-1}\|y\|_\infty\\
&\leq c|q|_k|q|_{k-1} \leq \text{(using (7.66))}\\
&\leq cN^{-1}|q|_k^2. \qquad (7.211)
\end{aligned}$$

Now, consider the remaining terms from $(QD(yq), 2(-1)^k D^{2k}q)_{L^2}$ in (7.208):

$$\begin{aligned}
\left|\sum_{j=2}^{k+1}\binom{k+1}{j}\int_\Omega D^{k+1-j}qD^kqD^jy\,dx\right| &\leq c|q|_k\left(\sum_{j=2}^{k+1}|q|_{k+1-j}\|D^jy\|_\infty\right)\\
&\leq c|q|_k\left(\sum_{j=2}^{k+1}|q|_{k+1-j}\|y\|_{j+1}\right)\\
&\leq \text{(using (7.64) and (7.66))}\\
&\leq c|q|_k\left(\sum_{j=2}^{k+1}N^{-j+1}|q|_kN^{j-2}\|y\|_3\right)\\
&\leq \text{(using (7.156))}\\
&\leq cN^{-1}|q|_k^2. \qquad (7.212)
\end{aligned}$$

We also have

$$\begin{aligned}
(QD(yq), (-1)^{k-1}D^{k-1}(yD^{k-1}q)) &= (QD^k(yq), yD^{k-1}q)_{L^2}\\
&\leq |QD^k(yq)|_0|q|_{k-1}\|y\|_\infty.
\end{aligned}$$

But as in (7.212), we can estimate

$$\begin{aligned}
|QD^k(yq)|_0 &\leq |D^k(yq)|_0 \leq c\sum_{j=0}^{k}|D^{k-j}qD^jy|_0\\
&\leq c\sum_{j=0}^{k}|q|_{k-j}\|D^jy\|_\infty\\
&\leq c\left(|q|_k\|y\|_\infty + |q|_{k-1}\|Dy\|_\infty + \sum_{j=2}^{k}N^{-j}|q|_kN^{j-2}\|y\|_3\right)\\
&\leq c|q|_k.
\end{aligned}$$

Hence,

$$\begin{aligned}(QD(yq), (-1)^{k-1}D^{k-1}(yD^{k-1}q))_{L^2} &\leq c|q|_k|q|_{k-1}\|y\|_\infty \\ &\leq c|q|_k|q|_{k-1} \\ &\leq cN^{-1}|q|_k^2.\end{aligned} \tag{7.213}$$

Moreover, using the Agmon inequality, (7.165), and (7.66),

$$\begin{aligned}(y_t, (D^{k-1}q)^2)_{L^2} &\leq |y_t|_0|q|_{k-1}\|D^{k-1}q\|_\infty \\ &\leq c|y_t|_0|q|_{k-1}^{3/2}|q|_k^{1/2} \\ &\leq cN^{-3/2}|q|_k^2.\end{aligned} \tag{7.214}$$

For the right-hand side of (7.206) we proceed as follows:

$$\begin{aligned}&(Qf, 2(-1)^kD^{2k}q + a_k(-1)^{k-1}D^{k-1}(yD^{k-1}q))_{L^2} \\ &\leq (2QD^kf, D^kq)_{L^2} + (QD^{k-1}f, a_kyD^{k-1}q)_{L^2} \\ &\leq 2|Qf|_k|q|_k + c|Qf|_{k-1}|q|_{k-1}\|y\|_\infty \\ &\leq c|q|_k + c|q|_{k-1} \leq c|q|_k,\end{aligned} \tag{7.215}$$

and

$$\begin{aligned}&(Q(yDy), 2(-1)^kD^{2k}q + a_k(-1)^{k-1}D^{k-1}(yD^{k-1}q))_{L^2} \\ &= 2(QD^k(yDy), D^kq)_{L^2} + a_k(QD^{k-1}(yDy), yD^{k-1}q)_{L^2} \\ &\leq 2|QD^k(yDy)|_0|q|_k + c|QD^{k-1}(yDy)|_0|q|_{k-1}\|y\|_\infty;\end{aligned}$$

but

$$\begin{aligned}|QD^k(yDy)|_0 &\leq |D^k(yDy)|_0 \leq |yD^{k+1}y|_0 + |D^k(yDy) - yD^{k+1}y|_0 \\ &\leq \text{(using again Lemma 7.1 with } s = k \text{ and } \beta = 1) \\ &\leq |y|_{k+1}\|y\|_\infty + c(\|y\|_k\|Dy\|_1 + \|y\|_2\|Dy\|_{k-1}) \\ &\leq |y|_{k+1}\|y\|_\infty + c\|y\|_k\|y\|_2 \\ &\leq c\|y\|_{k+1} + c\|y\|_k \leq \text{(using (7.157) and (7.156))} \\ &\leq cN^{k-2}\|y\|_3 + cN^{k-3}\|y\|_3 \leq cN^{k-2},\end{aligned}$$

and, similarly,

$$|QD^{k-1}(yDy)|_0 \leq cN^{k-3},$$

so that

$$\begin{aligned}&(Q(yDy), 2(-1)^kD^{2k}q + a_k(-1)^{k-1}D^{k-1}(yD^{k-1}q))_{L^2} \\ &\leq cN^{k-2}|q|_k + cN^{k-3}|q|_{k-1}\|y\|_\infty \\ &\leq \text{(using (7.66))} \leq cN^{k-2}|q|_k.\end{aligned} \tag{7.216}$$

Choose now $a_k = -2(2k+1)/3$ in order to cancel the term $((D^k q)^2, Dy)_{L^2}$. Then, using the estimates (7.209)–(7.216) in (7.206), we obtain

$$\frac{d}{dt}\left(|q|_k^2 - \frac{2k+1}{3}\int_\Omega (D^{k-1}q)^2 y\,dx\right) + 2\gamma\left(|q|_k^2 - \frac{2k+1}{3}\int_\Omega (D^{k-1}q)^2 y\,dx\right)$$
$$\le cN^{k-2}|q|_k + c_7 N^{-1/4}|q|_k^2$$
$$\le \frac{\gamma}{2}|q|_k^2 + cN^{2(k-2)} + c_7 N^{-1/4}|q|_k^2. \tag{7.217}$$

Note now that for some constant c_8 depending only on γ, L, and $\|f\|_k$,

$$\left|\frac{2k+1}{3}\int_\Omega (D^{k-1}q)^2 y\,dx\right| \le c|q|_{k-1}^2\|y\|_\infty \le c_8 N^{-2}|q|_k^2.$$

Assume that N is large enough so that

$$c_8 N^{-2} \le \tfrac{1}{2} \tag{7.218}$$

and

$$c_7 N^{-1/4} \le \frac{\gamma}{4}. \tag{7.219}$$

Then, we find

$$\tfrac{1}{2}|q|_k^2 \le |q|_k^2 - \frac{2k+1}{3}\int_\Omega (D^{k-1}q)^2 y\,dx \le \tfrac{3}{2}|q|_k^2, \tag{7.220}$$

which gives with (7.217)

$$\frac{d}{dt}\left(|q|_k^2 - \frac{2k+1}{3}\int_\Omega (D^{k-1}q)^2 y\,dx\right) + \gamma\left(|q|_k^2 - \frac{2k+1}{3}\int_\Omega (D^{k-1}q)^2 y\,dx\right)$$
$$\le cN^{2(k-2)}, \qquad \forall t \ge 0. \tag{7.221}$$

Therefore, by the Gronwall lemma,

$$|q(t)|_k^2 - \frac{2k+1}{3}\int_\Omega (D^{k-1}q)^2 y\,dx$$
$$\le \left(|q(0)|_k^2 - \frac{2k+1}{3}\int_\Omega (D^{k-1}q(0))^2 y(0)\,dx\right)e^{-\gamma t} + cN^{2(k-2)}$$
$$= cN^{2(k-2)}, \qquad \forall t \ge 0.$$

From this relation and (7.220), we deduce that

$$|q(t)|_k \le cN^{k-2}, \qquad \forall t \ge 0, \tag{7.222}$$

for N satisfying (7.173)–(7.175), (7.218), and (7.219).

Thus, we have prove (7.71).

Step IV. *Decay of z in $H^1_{\text{per}}(\Omega)$.*

Using (7.222) for $k = 3$ and the q-equation (7.68) we deduce that

$$|q_t|_0 \le cN, \qquad \forall t \ge 0,$$

which, together with (7.165), implies the following estimates for $v = y + q$:

$$|v_t|_0 \le cN, \qquad \forall t \ge 0, \tag{7.223}$$

and

$$\|v(t)\|_2 \le c, \qquad \forall t \ge 0, \tag{7.224}$$

Also, note that $q + z$ solves the same equation as $Q_N u$, so that by uniqueness we have $q + z = Q_N u$. Therefore, from (7.155) and the previous Step I, we find that

$$\begin{aligned} |z(t)|_0 &= |Q_N u(t) - q(t)|_0 \\ &\le |u(t)|_0 + |q(t)|_0 \le c_9, \qquad \forall t \ge 0. \end{aligned} \tag{7.225}$$

Multiply now the z-equation (7.69) by $-2D^2 z - Q(z^2) - 2Q(vz)$. Similar computations as those for q yield

$$\begin{aligned} &\frac{d}{dt}\int_\Omega \left((Dz)^2 - \frac{z^3}{3} - vz^2\right) dx + 2\gamma \int_\Omega \left((Dz)^2 - \frac{z^3}{3} - vz^2\right) dx \\ &\quad + (v_t, z^2)_{L^2} - \frac{\gamma}{3}\int_\Omega z^3\, dx = 0. \end{aligned} \tag{7.226}$$

We estimate

$$\begin{aligned} (v_t, z^2)_{L^2} &\le |v_t|_0 |z|_0 \|z\|_\infty \le c|v_t|_0 |z|_0^{3/2} |z|_1^{1/2} \\ &\le \text{(using (7.66) and (7.223))} \\ &\le cNN^{-3/2}|z|_1^2 = c_{10}N^{-1/2}|z|_1^2, \end{aligned} \tag{7.227}$$

and

$$\begin{aligned} \left|\int_\Omega \frac{z^3}{3}\, dx\right| &\le \tfrac{1}{3}|z|_0^2 \|z\|_\infty \le c|z|_0^{5/2}|z|_1^{1/2} \\ &\le \text{(using (7.66) and (7.225))} \\ &\le c_{11}N^{-3/2}|z|_1^2; \end{aligned} \tag{7.228}$$

also

$$\begin{aligned} \left|\int_\Omega vz^2\, dx\right| &\le \|v\|_\infty |z|_0^2 \le c\|v\|_1 |z|_0^2 \\ &\le \text{(using (7.66) and (7.224))} \le c_{12}N^{-2}|z|_1^2. \end{aligned} \tag{7.229}$$

For N sufficiently large we have

$$|z(t)|_1^2 \leq c e^{-\gamma t}, \qquad \forall t \geq 0, \tag{7.230}$$

for some constant c.

Indeed, let us assume that N satisfies

$$c_{11} N^{-3/2} + c_{12} N^{-2} \leq \tfrac{1}{2}, \tag{7.231}$$

$$c_{10} N^{-1/2} + \gamma c_{11} N^{-3/2} \leq \frac{\gamma}{2}. \tag{7.232}$$

Using (7.228), (7.229), and (7.231), we deduce that

$$\tfrac{1}{2}|z|_1^2 \leq |z|_1^2 - \int_\Omega \frac{z^3}{3}\, dx + \int z^2 v\, dx \leq \tfrac{3}{2}|z|_1^2. \tag{7.233}$$

Now, using (7.226), (7.227), (7.232), and (7.233), we obtain

$$\begin{aligned}
&\frac{d}{dt}\int_\Omega \left((Dz)^2 - \frac{z^3}{3} + z^2 v\right) dx + 2\gamma \int_\Omega \left((Dz)^2 - \frac{z^3}{3} + z^2 v\right) dx \\
&\leq \frac{\gamma}{2}|z|_1^2 \\
&\leq \gamma \int_\Omega \left((Dz)^2 - \frac{z^3}{3} + z^2 v\right) dx;
\end{aligned}$$

hence

$$\frac{d}{dt}\left(\int_\Omega \left((Dz)^2 - \frac{z^3}{3} + z^2 v\right) dx\right) + \gamma \int_\Omega \left((Dz)^2 - \frac{z^3}{3} + z^2 v\right) dx \leq 0. \tag{7.234}$$

Therefore, by Gronwall's lemma,

$$\begin{aligned}
&|z(t)|_1^2 - \int_\Omega \frac{(z(t))^3}{3}\, dx + \int_\Omega (z(t))^2 v(t)\, dx \\
&\leq \left(|z(0)|_1^2 - \int_\Omega \frac{(z(0))^3}{3}\, dx + \int_\Omega (z(0))^2 v(0)\, dx\right) e^{-\gamma t}, \qquad \forall t \geq 0.
\end{aligned} \tag{7.235}$$

From (7.235) and (7.233), we deduce that

$$|z(t)|_1^2 \leq 3|z(0)|_1^2 e^{-\gamma t}, \qquad \forall t \geq 0. \tag{7.236}$$

Now, since z has zero average, the norm $\|\cdot\|_1$ is equivalent to $|\cdot|_1$. Therefore, (7.236) implies

$$\|z(t)\|_1^2 \leq c e^{-\gamma t}, \qquad \forall t \geq 0. \tag{7.237}$$

This completes the proof of Proposition 7.2.

8. Unbounded Case: The Lack of Compactness

In all the examples considered in Chapters III and IV we have assumed that the physical domain Ω was bounded, so that we could obtain compactness of the injection between two function spaces V and H by using the compactness theorems between Sobolev spaces recalled in Chapter II (see II.(1.15)–II.(1.17)).

When Ω is not bounded, compactness is lost and it might be difficult to obtain the compactness properties needed in Theorem I.1.1 (either I.(1.12) or I.(1.13) or I.(1.17)). Several remedies have been found to this difficulty. One of them consists in working in weighted Sobolev spaces such as

$$L_\alpha^2(\mathbb{R}^n) = \left\{ u\colon \mathbb{R}^n \to \mathbb{R}, \int_{\mathbb{R}^n} (1 + |x|^2)^\alpha |u(x)|^2\, dx < \infty \right\},$$

$$H_{\alpha,\beta}^1(\mathbb{R}^n) = \left\{ u \in L_\alpha^2(\mathbb{R}^n), \frac{\partial u}{\partial x_i} \in L_\beta^2(\mathbb{R}^n), i = 1, \dots, n \right\},$$

for suitable α and β. For appropriate values of α and β, the embedding of $H_{\alpha,\beta}^1(\mathbb{R}^n)$ into $L_\alpha^2(\mathbb{R}^n)$ is compact and one can derive I.(1.13). The drawback here is that the theory of partial differential equations in weighted Sobolev spaces is more involved than that in regular Sobolev spaces. Furthermore, depending on the technicalities of the proofs, the data (driving force, initial value) might belong as well to a weighted space, and the attractor might be included in such a space. Extensive work in this direction leading to a number of existence results of attractors appears in, e.g., F. Abergel [1], [2], A.V. Babin [1], A.V. Babin and M.I. Vishik [4], E. Feireisl, P. Laurençot, F. Simondon, and H. Touré [1].

Other approaches developed for unbounded domains yield attractors in the usual (non-weighted) Sobolev spaces. One of them consists in checking I.(1.13) by using a suitable decomposition $S(t) = S_1(t) + S_2(t)$ of $S(t)$ (and using also weighted spaces, but only as an intermediate tool; see P. Laurençot [1]). Another one, which we will present hereafter, consists in checking I.(1.17) then applying Theorem I.1.1 as in Remark I.1.4; the derivation of I.(1.17) is based on the utilization of the energy equation and the idea of J. Ball [1] already used in Section 7.

Our aim in this section is to present the latest approach, as an example, following R. Rosa [3]. We consider the incompressible Navier–Stokes equations in an unbounded domain $\Omega \subset \mathbb{R}^2$, Ω bounded in one direction[1] so that

[1] That is, Ω is included in a region of the form $0 < x_1 < L$. In some of the other works, Ω is a more general (fully) unbounded domain, depending on the equation considered. Here, we could consider a general unbounded domain if, say, we add the term αu, $\alpha > 0$, in the left-hand side of the first equation (8.1): this allows to obtain an absorbing set in $\mathbb{L}^2(\Omega)$ without using the Poincaré inequality (8.2) for which we assumed that Ω is bounded in one direction.

the Poincaré inequality is valid. In Section 8.1 we show how to extend the general results of Section III.2.1 from the bounded to the unbounded case. In Section 8.2 we derive the existence of the attractor.

8.1. Preliminaries

As in Section III.2.1, we consider the flow of an incompressible viscous fluid of constant density enclosed in a region $\Omega \subset \mathbb{R}^2$ with rigid boundary Γ and governed by the Navier–Stokes equations. We denote, respectively, by $u = u(x, t) \in \mathbb{R}^2$ and $p = p(x, t) \in \mathbb{R}$, the velocity and the pressure of the fluid at the point $x \in \Omega$ and at time $t \geq 0$; they are determined by the following initial-boundary value problem (similar to III.(2.1), III.(2.2), III.(2.4), III.(2.7)):

$$\begin{cases} \dfrac{\partial u}{\partial t} - \nu\Delta u + (u\cdot\nabla)u + \nabla p = f \quad \text{in } \Omega, \\ \nabla\cdot u = 0 \quad \text{in } \Omega, \\ u = 0 \quad \text{on } \Gamma, \\ u(\cdot, 0) = u_0 \quad \text{in } \Omega, \end{cases} \tag{8.1}$$

where $\nu > 0$ is the kinematic viscosity of the fluid and $f = f(x) \in \mathbb{R}^2$ is the external body force (assumed to be time-independent).

The domain Ω can be an arbitrary bounded or unbounded set in $\mathbb{R}^2$ without any regularity assumption on its boundary Γ, and with the only assumption that the Poincaré inequality holds on it. More precisely, we assume only that

$$\begin{cases} \text{There exists } \lambda_1 > 0 \text{ such that} \\ \displaystyle\int_\Omega \phi^2\,dx \leq \frac{1}{\lambda_1}\int_\Omega |\nabla\phi|^2\,dx, \qquad \forall\phi \in H_0^1(\Omega). \end{cases} \tag{8.2}$$

The mathematical framework of (8.1) is now essentially as in the bounded case: first, let $\mathbb{L}^2(\Omega) = (L^2(\Omega))^2$ and $\mathbb{H}_0^1(\Omega) = (H_0^1(\Omega))^2$ be, respectively, endowed with the inner products

$$(u, v) = \int_\Omega u\cdot v\,dx, \qquad u, v \in \mathbb{L}^2(\Omega),$$

and

$$((u, v)) = \int_\Omega \sum_{j=1}^2 \nabla u_j\cdot\nabla v_j\,dx, \qquad u = (u_1, u_2), \quad v = (v_1, v_2) \in \mathbb{H}_0^1(\Omega),$$

and norms $|\cdot| = (\cdot, \cdot)^{1/2}$, $\|\cdot\| = ((\cdot, \cdot))^{1/2}$. Note that thanks to (8.2) the norm

$\|\cdot\|$ is equivalent to the usual one in $\mathbb{H}_0^1(\Omega)$. Set now

$$\mathscr{V} = \{v \in (\mathscr{D}(\Omega))^2; \nabla \cdot v = 0 \text{ in } \Omega\},$$

$$V = \text{closure of } \mathscr{V} \text{ in } \mathbb{H}_0^1(\Omega),$$

$$H = \text{closure of } \mathscr{V} \text{ in } \mathbb{L}^2(\Omega),$$

with H and V endowed with the inner product and norm of, respectively, $\mathbb{L}^2(\Omega)$ and $\mathbb{H}_0^1(\Omega)$. It follows from (8.2) that

$$|u|^2 \leq \frac{1}{\lambda_1}\|u\|^2, \qquad \forall u \in V. \tag{8.3}$$

We then consider the following weak formulation of (8.1): to find

$$u \in L^\infty(0, T; H) \cap L^2(0, T; V), \qquad \forall T > 0, \tag{8.4}$$

such that

$$\frac{d}{dt}(u, v) + \nu((u, v)) + b(u, u, v) = \langle f, v \rangle, \qquad \forall v \in V, \quad \forall t > 0, \tag{8.5}$$

and

$$u(0) = u_0, \tag{8.6}$$

where b: $V \times V \times V \to \mathbb{R}$ is given by

$$b(u, v, w) = \sum_{i,j=1}^{2} \int_\Omega u_i \frac{\partial v_j}{\partial x_i} w_j \, dx, \tag{8.7}$$

and $\langle \cdot, \cdot \rangle$ is the duality product between V' and V when we identify H with its dual, and we assumed for simplicity that $f \in V'$. The weak formulation (8.5) is equivalent to the functional equation

$$u' + \nu A u + B(u) = f, \quad \text{in } V', \quad \text{for } t > 0, \tag{8.8}$$

where $u' = du/dt$, A: $V \to V'$ is the Stokes operator defined by

$$\langle Au, v \rangle = ((u, v)), \qquad \forall u, v \in V, \tag{8.9}$$

and $B(u) = B(u, u)$ is the bilinear operator B: $V \times V \to V'$ defined by

$$\langle B(u, v), w \rangle = b(u, v, w), \qquad \forall u, v, w \in V.$$

The Stokes operator is still an isomorphism from V into V', while B satisfies the following inequality following from the last inequality III.(2.16) and III.(2.21) (see, e.g., R. Temam [2], Lemma III.3.4):

$$\|B(u)\|_{V'} \leq 2^{1/2}|u|\,\|u\|, \qquad \forall u \in V. \tag{8.10}$$

We have the following result:

Theorem 8.1. *Given $f \in V'$ and $u_0 \in H$, there exists a unique $u \in L^\infty(R^+; H) \cap L^2(0, T; V)$, $\forall T > 0$, such that (8.5) (hence (8.8)) and (8.6) hold. Moreover, $u' \in L^2(0, T; V')$, $\forall T > 0$, and $u \in \mathscr{C}(\mathbb{R}^+; H)$.*

Now, let $u = u(t)$, $t \geq 0$, be the solution of (8.5) given by Theorem 8.1. Since $u \in L^2(0, T; V)$ and $u' \in L^2(0, T; V')$, we have

$$\frac{1}{2}\frac{d}{dt}|u|^2 = \langle u', u \rangle, \tag{8.11}$$

so that from (8.8),

$$\begin{aligned}\frac{1}{2}\frac{d}{dt}|u|^2 &= \langle f - \nu A u - B(u), u \rangle \\ &= \langle f, u \rangle - \nu \|u\|^2 - b(u, u, u).\end{aligned}$$

Hence, from the orthogonality property (see III.(2.21))

$$b(u, v, v) = 0, \qquad \forall u, v \in V, \tag{8.12}$$

we deduce that

$$\frac{d}{dt}|u|^2 + 2\nu\|u\|^2 = 2\langle f, u \rangle, \tag{8.13}$$

in the distribution sense on $\mathbb{R}^+$.

From (8.13) and using (8.3) one can easily deduce the classical estimates

$$|u(t)|^2 \leq |u_0|^2 e^{-\nu\lambda_1 t} + \frac{1}{\nu^2\lambda_1}\|f\|_{V'}^2, \qquad \forall t \geq 0, \tag{8.14}$$

and

$$\frac{1}{t}\int_0^t \|u(s)\|^2\, ds \leq \frac{1}{t\nu}|u_0|^2 + \frac{1}{\nu^2}\|f\|_{V'}^2, \qquad \forall t > 0. \tag{8.15}$$

The energy equation (8.13) will be further used in Section 8.2. For the moment, note that thanks to Theorem 8.1, we can define a continuous semigroup $\{S(t)\}_{t\geq 0}$ in H by setting

$$S(t)u_0 = u(t), \qquad t \geq 0,$$

where u is the solution of (8.5) with $u(0) = u_0 \in H$. It is not difficult to see that, for $t \geq 0$, the map $S(t)$: $H \to H$ is Lipschitz continuous on bounded subsets of H. Moreover, from (8.14) it follows that the set

$$\mathscr{B} = \left\{ v \in H; |v| \leq \rho_0 \equiv \frac{1}{\nu}\sqrt{\frac{2}{\lambda_1}}\|f\|_{V'} \right\} \tag{8.16}$$

is absorbing in H for the semigroup.

We will also need in Section 8.2 the following weak continuity of the semigroup $\{S(t)\}_{t\geq 0}$:

Lemma 8.1. *Let $\{u_{0_n}\}_n$ be a sequence in H converging weakly in H to an element $u_0 \in H$. Then*

$$S(t)u_{0_n} \rightharpoonup S(t)u_0 \quad \text{weakly in } H, \qquad \forall t \geq 0, \tag{8.17}$$

and

$$S(\cdot)u_{0_n} \rightharpoonup S(\cdot)u_0 \quad \text{weakly in } L^2(0, T; V), \qquad \forall T > 0. \tag{8.18}$$

PROOF. Let $u_n(t) = S(t)u_{0_n}$ and $u(t) = S(t)u_0$, for $t \geq 0$. From (8.14) and (8.15) we find that

$$\{u_n\}_n \text{ is bounded in } L^\infty(\mathbb{R}^+, H) \cap L^2(0, T; V), \qquad \forall T > 0. \tag{8.19}$$

Hence, since

$$u_n' = f - \nu A u_n - B(u_n),$$

and since A is a bounded linear operator from V into V' and B satisfies (8.10), it follows that

$$\{u_n'\}_n \text{ is bounded in } L^2(0, T; V'), \qquad \forall T > 0. \tag{8.20}$$

Then, for all $v \in V$ and $0 \leq t \leq t + a \leq T$, with $T > 0$,

$$\begin{aligned}(u_n(t+a) - u_n(t), v) &= \int_t^{t+a} \langle u_n'(s), v\rangle \, ds \\ &\leq \|v\| a^{1/2} \|u_n'\|_{L^2(0,T;V')} \\ &\leq c_T \|v\| a^{1/2},\end{aligned} \tag{8.21}$$

where c_T is positive, independent of n. Thus, setting $v = u_n(t+a) - u_n(t)$, which belongs to V for almost every t, we find from (8.21) that

$$|u_n(t+a) - u_n(t)|^2 \leq c_T a^{1/2} \|u_n(t+a) - u_n(t)\|.$$

Hence,

$$\int_0^{T-a} |u_n(t+a) - u_n(t)|^2 \, dt \leq c_T a^{1/2} \int_0^{T-a} \|u_n(t+a) - u_n(t)\| \, dt. \tag{8.22}$$

Using the Cauchy–Schwarz inequality and (8.19) we find from (8.22) that

$$\int_0^{T-a} |u_n(t+a) - u_n(t)|^2 \, dt \leq \tilde{c}_T a^{1/2},$$

for another positive constant $\tilde{c}_T$ independent of n. Therefore,

$$\limsup_{a\to 0}\ \sup_n \int_0^{T-a} \|u_n(t+a) - u_n(t)\|^2_{\mathbb{L}^2(\Omega_r)}\, dt = 0, \tag{8.23}$$

for all $r > 0$, where $\Omega_r = \{x \in \Omega; |x| < r\}$. Moreover, from (8.19), for all $r > 0$,

$$\begin{aligned}&\{u_n|_{\Omega_r}\}_n \text{ is bounded independently of } n\\ &\text{in } L^2(0, T; \mathbb{H}^1(\Omega_r)) \cap L^\infty(0, T; \mathbb{L}^2(\Omega_r)).\end{aligned} \tag{8.24}$$

Now, assume for simplicity[1] that the domain Ω is regular enough so that $\mathbb{H}^1(\Omega_r)$ is compactly embedded in $\mathbb{L}^2(\Omega_r)$. Then, by the same compactness theorem used in Section III.2 (see III.(2.47)), we find

$$\{u_n|_{\Omega_r}\} \text{ is relatively compact in } L^2(0, T; \mathbb{L}^2(\Omega_r)), \qquad \forall T > 0, \quad \forall r > 0. \tag{8.25}$$

Then, from (8.19) and (8.25) and by a diagonal process, we can extract a subsequence $\{u_{n'}\}_{n'}$ such that

$$\begin{aligned} u_{n'} \to \tilde{u} \quad & \text{in } L^\infty(\mathbb{R}^+; H) \text{ weak-star},\\ & \text{in } L^2_{\text{loc}}(\mathbb{R}^+; V) \text{ weakly},\\ & \text{in } L^2_{\text{loc}}(\mathbb{R}^+; \mathbb{L}^2(\Omega_r)) \text{ strongly}, \qquad \forall r > 0, \end{aligned} \tag{8.26}$$

for some

$$\tilde{u} \in L^\infty(\mathbb{R}^+; H) \cap L^2_{\text{loc}}(\mathbb{R}^+; V). \tag{8.27}$$

The convergence (8.26) allows us to pass to the limit in the equation for $u_{n'}$ to find that $\tilde{u}$ is a solution of (8.5) with $\tilde{u}(0) = u_0$. By the uniqueness of the solutions we must have $\tilde{u} = u$. Then by a contradiction argument we deduce that the whole sequence $\{u_n\}_n$ converges to u in the sense of (8.26). This proves (8.18).

Now, from the strong convergence in (8.26) we also infer that $u_n(t)$ converges strongly in $\mathbb{L}^2(\Omega_r)$ to $u(t)$ for almost every $t \geq 0$ and all $r > 0$. Hence for all $v \in \mathscr{V}$,

$$(u_n(t), v) \to (u(t), v), \qquad \text{a.e.} \quad t \in \mathbb{R}^+.$$

Moreover, from (8.19) and (8.21), we see that $\{(u_n(t), v)\}_n$ is equibounded and equicontinuous on $[0, T]$, for all $T > 0$. Therefore,

$$(u_n(t), v) \to (u(t), v), \qquad \forall t \in \mathbb{R}^+, \quad \forall v \in \mathscr{V}. \tag{8.28}$$

[1] If $\mathbb{H}^1(\Omega_r)$ is not compactly embedded in $\mathbb{L}^2(\Omega_r)$, then we multiply u_n by a smooth function ψ compactly supported in Ω_r and equal to one for $|x| \leq r - 1$. Since $\mathbb{H}^1_0(\Omega_r)$ is compactly embedded in $\mathbb{L}^2(\Omega_r)$, even if Ω_r is not regular, we conclude that $\psi u_n|_{\Omega_r}$ is relatively compact in $L^2(0, T; \mathbb{L}^2(\Omega_r))$. Hence, $u_n|_{\Omega_{r-1}}$ is relatively compact in $L^2(0, T; \mathbb{L}^2(\Omega_{r-1}))$, $\forall T > 0, \forall r > 1$, which is the same as (8.25).

Finally, (8.17) follows from (8.28) by taking into account (8.19) and the fact that $\mathscr{V}$ is dense in H. □

8.2. The Global Attractor

The existence of the global attractor will follow from Theorem I.1.1 modified by Remark I.1.4 and is based on the *asymptotic compactness* of the semigroup $\{S(t)\}_{t\geq 0}$. We recall that the semigroup $\{S(t)\}_{t\geq 0}$ is said to be *asymptotically compact* in H if

$$\{S(t_n)u_n\}_n \text{ is relatively compact in } H \tag{8.29}$$

whenever

$$\{u_n\}_n \text{ is bounded in } H \text{ and } t_n \to \infty. \tag{8.30}$$

In order to show that $\{S(t)\}_{t\geq 0}$ is asymptotically compact in H, we use the energy equation (8.13).

First define $[\cdot,\cdot]\colon V \times V \to \mathbb{R}$ by

$$[u, v] = \nu((u, v)) - \nu\frac{\lambda_1}{2}(u, v), \qquad \forall u, v \in V. \tag{8.31}$$

Clearly, $[\cdot,\cdot]$ is bilinear and symmetric. Moreover, from (8.3),

$$\begin{aligned}[u]^2 \equiv [u, u] &= \nu\|u\|^2 - \nu\frac{\lambda_1}{2}|u|^2 \\ &\geq \nu\|u\|^2 - \frac{\nu}{2}\|u\|^2 = \frac{\nu}{2}\|u\|^2.\end{aligned}$$

Hence,

$$\frac{\nu}{2}\|u\|^2 \leq [u]^2 \leq \nu\|u\|^2, \qquad \forall u \in V, \tag{8.32}$$

and $[\cdot,\cdot]$ defines an inner product in V with norm $[\cdot] = [\cdot,\cdot]^{1/2}$ equivalent to $\|\cdot\|$.

Now, add and subtract $\nu\lambda_1|u|^2$ from the energy equation (8.13) to find

$$\frac{d}{dt}|u|^2 + \nu\lambda_1|u|^2 + 2[u]^2 = 2\langle f, u\rangle, \tag{8.33}$$

for any solution $u = u(t) = S(t)u_0$, $u_0 \in H$. Then, by the variation of constants formula,

$$|u(t)|^2 = |u_0|^2 e^{-\nu\lambda_1 t} + 2\int_0^t e^{-\nu\lambda_1(t-s)}(\langle f, u(s)\rangle - [u(s)]^2)\, ds,$$

which can be written as

$$|S(t)u_0|^2 = |u_0|^2 e^{-\nu\lambda_1 t} + 2\int_0^t e^{-\nu\lambda_1(t-s)}(\langle f, S(s)u_0\rangle - [S(s)u_0]^2)\,ds, \tag{8.34}$$

for all $u_0 \in H$, and $t \geq 0$.

We are now in position to show the asymptotic compactness of the semigroup. For that purpose, let $B \subset H$ be bounded and consider $\{u_n\}_n \subset B$ and $\{t_n\}_n$, $t_n \geq 0$, $t_n \to \infty$, as $n \to \infty$.

Since the set $\mathscr{B}$ defined in (8.16) is absorbing, there exists a time $T(B) > 0$ such that

$$S(t)B \subset \mathscr{B}, \qquad \forall t \geq T(B),$$

so that for t_n large enough ($t_n \geq T(B)$),

$$S(t_n)u_n \in \mathscr{B}. \tag{8.35}$$

Thus $\{S(t_n)u_n\}_n$ is weakly relatively compact in H and hence

$$S(t_{n'})u_{n'} \rightharpoonup w \quad \text{weakly in } H, \tag{8.36}$$

for some subsequence n' and some $w \in \mathscr{B}$ (since $\mathscr{B}$ is closed and convex).

Similarly, for each $T > 0$, we also have

$$S(t_n - T)u_n \in \mathscr{B}, \tag{8.37}$$

for $t_n \geq T + T(B)$. Thus, $\{S(t_n - T)u_n\}_n$ is weakly relatively compact in H, and by using a diagonal process and passing to a further subsequence if necessary we can assume that

$$S(t_{n'} - T)u_{n'} \rightharpoonup w_T \quad \text{weakly in } H, \qquad \forall T \in \mathbb{N}, \tag{8.38}$$

with $w_T \in \mathscr{B}$.

Note then by the weak continuity of $S(t)$ established in Lemma 8.1 that

$$\begin{aligned} w &= \lim_{n'} {}_{H_w} S(t_{n'})u_{n'} = \lim_{n'} {}_{H_w} S(T)S(t_{n'} - T)u_{n'} \\ &= S(T) \lim_{n'} {}_{H_w} S(t_{n'} - T)u_{n'} = S(T)w_T, \end{aligned}$$

where $\lim_{H_w}$ denotes the limit taken in the weak topology of H. Thus,

$$w = S(T)w_T, \qquad \forall T \in \mathbb{N}. \tag{8.39}$$

Now, from (8.36), we find

$$|w| \leq \liminf_{n'} |S(t_{n'})u_{n'}|, \tag{8.40}$$

and we shall now show that

$$\limsup_{n'} |S(t_{n'})u_{n'}| \leq |w|.$$

For $T \in \mathbb{N}$ and $t_n > T$ we have by (8.34)

$$\begin{aligned}
&|S(t_n)u_n|^2 \\
&\quad = |S(T)S(t_n - T)u_n|^2 \\
&\quad = |S(t_n - T)u_n|^2 e^{-\nu\lambda_1 T} \\
&\qquad + 2\int_0^T e^{-\nu\lambda_1(T-s)}\{\langle f, S(s)S(t_n - T)u_n\rangle - [S(s)S(t_n - T)u_n]^2\}\, ds. \quad (8.41)
\end{aligned}$$

From (8.37) we find

$$\limsup_{n'} (e^{-\nu\lambda_1 T}|S(t_{n'} - T)u_{n'}|^2) \leq \rho_0^2 e^{-\nu\lambda_1 T}. \quad (8.42)$$

Also, by the weak continuity (8.18) we deduce from (8.38) that

$$S(\cdot)S(t_{n'} - T)u_{n'} \rightharpoonup S(\cdot)w_T \quad \text{weakly in } L^2(0, T; V). \quad (8.43)$$

Then, since

$$s \mapsto e^{-\nu\lambda_1(T-s)}f \in L^2(0, T; V'),$$

we find

$$\lim_{n'} \int_0^T e^{-\nu\lambda_1(T-s)}\langle f, S(s)S(t_{n'} - T)u_{n'}\rangle\, ds = \int_0^T e^{-\nu\lambda_1(T-s)}\langle f, S(s)w_T\rangle\, ds. \quad (8.44)$$

Moreover, since $[\cdot]$ is a norm on V equivalent to $\|\cdot\|$ and

$$0 < e^{-\nu\lambda_1 T} \leq e^{-\nu\lambda_1(T-s)} \leq 1, \qquad \forall s \in [0, T],$$

we see that

$$\left(\int_0^T e^{-\nu\lambda_1(T-s)}[\cdot]^2\, ds\right)^{1/2}$$

is a norm on $L^2(0, T; V)$ equivalent to the usual norm. Hence, from (8.43), we deduce that

$$\int_0^T e^{-\nu\lambda_1(T-s)}[S(s)w_T]^2\, ds \leq \liminf_{n} \int_0^T e^{-\nu\lambda_1(T-s)}[S(s)S(t_n - T)u_n]^2\, ds. \quad (8.45)$$

Thus,

$$\begin{aligned}
&\limsup_{n'} \left\{-2\int_0^T e^{-\nu\lambda_1(T-s)}[S(s)S(t_{n'} - T)u_{n'}]^2\, ds\right\} \\
&\quad = -2\liminf_{n'} \int_0^T e^{-\nu\lambda_1(T-s)}[S(s)S(t_{n'} - T)u_{n'}]^2\, ds \\
&\quad \leq -2\int_0^T e^{-\nu\lambda_1(T-s)}[S(s)w_T]^2\, ds. \quad (8.46)
\end{aligned}$$

We can now pass to the lim sup as n' goes to infinity in (8.41), taking (8.42), (8.44), and (8.46) into account to obtain

$$\limsup_{n'} |S(t_{n'})u_{n'}|^2 \leq \rho_0^2 e^{-\nu\lambda_1 T} + 2\int_0^T e^{-\nu\lambda_1(T-s)}\{\langle f, S(s)w_T\rangle - [S(s)w_T]^2\}\, ds. \quad (8.47)$$

On the other hand, we obtain from (8.34) applied to $w = S(T)w_T$ that

$$|w|^2 = |S(T)w_T|^2 = e^{-\nu\lambda_1 T}|w_T|^2 + 2\int_0^T e^{-\nu\lambda_1(T-s)}\{\langle f, S(s)w_T\rangle - [S(s)w_T]^2\}\, ds. \quad (8.48)$$

From (8.47) and (8.48) we then find

$$\limsup_{n'} |S(t_{n'})u_{n'}|^2 \leq |w|^2 + (\rho_0^2 - |w_T|^2)e^{-\nu\lambda_1 T} \leq |w|^2 + \rho_0^2 e^{-\nu\lambda_1 T}, \qquad \forall T \in \mathbb{N}. \quad (8.49)$$

We let T goes to infinity in (8.49) to obtain

$$\limsup_{n'} |S(t_{n'})u_{n'}|^2 \leq |w|^2, \quad (8.50)$$

as claimed. Since H is a Hilbert space, (8.50) together with (8.36) imply

$$S(t_{n'})u_{n'} \to w \quad \text{strongly in } H. \quad (8.51)$$

This shows that $\{S(t_n)u_n\}_n$ is relatively compact in H, and hence that $\{S(t)\}_{t\geq 0}$ is asymptotically compact in H. Since $\{S(t)\}_{t\geq 0}$ has also a bounded absorbing set $\mathscr{B}$ in H, the existence of the global attractor follows from Theorem I.1.1 and Remark I.1.4.

We have proved our main result.

Theorem 8.2. *Let Ω be an open set satisfying* (8.2). *Assume $\nu > 0$ and $f \in V'$. Then, the dynamical system $\{S(t)\}_{t\geq 0}$ associated with the evolution equation* (8.5), *possesses a global attractor in H, i.e., a compact invariant set $\mathscr{A}$ in $\mathscr{H}$ which attracts all bounded sets in H. Moreover, $\mathscr{A}$ is connected in H and is maximal for the inclusion relation among all the functional invariant sets bounded in H.*

Remark 8.1. The global attractor $\mathscr{A}$ obtained in Theorem 8.2 is actually included and bounded in V. This is a particular case of more general results proven (for the bounded case) in the following Section 9. Here $\mathscr{A} \subset V$ follows from a classical a priori estimate obtained by taking the scalar product in H

of (8.8) with $tAu(t)$ in the case of smooth domains. In the case of nonsmooth domains we take the scalar product with $2tu_t$ and add this to the scalar product with $2t^2u_t$ of the differentiated equation (see O.A. Ladyzhenskaya ([9]). This yields

$$t\|S(t)u_0\|^2 \leq e^{c(1+|u_0|^4+t^2\|f\|_{V'}^4)}, \qquad \forall u_0 \in H, \quad \forall t \geq 0, \tag{8.52}$$

where c is a constant depending only on ν. Estimate (8.52) does not rely on either the Poincaré inequality or the smoothness of the domain and can be shown to hold for arbitrary unbounded domains in $\mathbb{R}^2$.

9. Regularity of Attractors

Our aim in this section is to prove some regularity results for attractors. A regularity result is to be understood here in the sense of the theory of partial differential equations, i.e., if the data are sufficiently regular, then the attractor lies in a set of more (spatially) regular functions, a subset of H or even V. Related results in the theory of partial differential equations are the results of (spatial) regularity of solutions of elliptic nonlinear equations corresponding to the stationary solutions of a dynamical system; also, the spatial regularity of a time-periodic solution of one of the dynamical systems studied in Chapters III and IV. As indicated in Chapter I, a stationary solution or the orbit of a time-periodic solution are by themselves functional-invariant sets. Therefore the regularity problem that we study in this section includes these classical regularity problems of the theory of partial differential equations.

In simplified terms, the results that we will prove here for some of the systems studied in Chapters III and IV are of the following type: if the data are sufficiently regular then the maximal attractor $\mathscr{A}$ lies in a Sobolev space $H^m(\Omega)$, for an appropriate m; or if the data are $\mathscr{C}^\infty$, the maximal attractor is included in $\mathscr{C}^\infty(\bar{\Omega})$. The data mentioned here are the different functions appearing in the partial differential equation, and the boundary Γ of Ω which must be sufficiently regular.

We do not attempt to prove systematically regularity results for the equations considered in Chapters III and IV; this would represent a lengthy and very technical task. Instead, we restrict ourselves to two typical results using an argument which can be, in principle, extended to many other situations.

This section is organized as follows. Section 9.1 contains a preliminary result. Section 9.2 gives a partial regularity result for a reaction–diffusion equation (i.e., a regularity result in H^m). Section 9.3 contains a result of $\mathscr{C}^\infty$ regularity for the maximal attractor of the two-dimensional Navier–Stokes equation.

9.1. A Preliminary Result

We consider, as in Section II.3, an evolution equation of the form

$$\frac{du}{dt} + Au = f. \tag{9.1}$$

The abstract framework and the hypotheses are exactly the same as in Section II.3.1. Instead of II.(3.3), we assume that f is given in $L^\infty(\mathbb{R}; H)$

$$f \in L^\infty(\mathbb{R}; H). \tag{9.2}$$

We prove below that, in such a case, there exists a unique solution u of (6.1) which remains bounded as $t \to -\infty$. Beside this existence and uniqueness result, the main object of Proposition 9.1 below is to prove some useful a priori estimates of this solution.

Proposition 9.1. *Under the above assumptions there exists a unique solution u of* (9.1) *defined on the whole line* $\mathbb{R}$ *satisfying*

$$u \in L^\infty(\mathbb{R}; H) \cap L^\infty_{\text{loc}}(\mathbb{R}; V). \tag{9.3}$$

If, furthermore, A is symmetric, u is in $L^\infty(\mathbb{R}; V)$ *and for any* $r > 0$ *fixed, the integrals*

$$\int_t^{t+r} |Au|^2\, ds \quad \text{and} \quad \int_t^{t+r} |u'|^2\, ds$$

are uniformly bounded.

Proof. (i) We set

$$\lambda_1 = \operatorname*{Inf}_{v \in V} \frac{\|v\|^2}{|v|^2}; \tag{9.4}$$

$\lambda_1 > 0$, since the injection of V in H is continuous.

The uniqueness amounts to proving that if u, denoted by φ, satisfies (9.1) with $f = 0$ and (9.3), then $\varphi = 0$. We take the scalar product of (9.1) with φ in H. Then using II.(3.2), II.(3.18), and (9.4) we obtain

$$\begin{gathered} \frac{d}{dt}|\varphi|^2 + 2a(\varphi, \varphi) = 0, \\ \frac{d}{dt}|\varphi|^2 + 2\lambda_1 \alpha |\varphi|^2 \le 0. \end{gathered} \tag{9.5}$$

We integrate between t_0 and t, $-\infty < t_0 < t < \infty$, and find

$$|\varphi(t)|^2 \le |\varphi(t_0)|^2 \exp(-2\lambda_1 \alpha (t - t_0)). \tag{9.6}$$

Since $\varphi \in L^\infty(\mathbb{R}; H)$, by letting $t_0 \to -\infty$ in this inequality, we find $\varphi(t) = 0$.

By application of Theorem II.3.1 we see that the initial-value problem

$$\frac{d\varphi}{dt} + A\varphi = 0, \qquad \varphi(0) = \varphi_0 \, (\in H), \tag{9.7}$$

possesses a unique solution φ which belongs to $\mathscr{C}([0, T]; H) \cap L^2(0, T; V)$, $\forall T > 0$. Let $\Sigma(t)$, $t > 0$, denote the operator

$$\Sigma(t)\colon \varphi(0) \to \varphi(t).$$

Clearly, $\Sigma(t)$ is a linear continuous operator in H and (9.6) implies that

$$|\Sigma(t)|_{\mathscr{L}(H)} \leq \exp(-\lambda_1 \alpha t). \tag{9.8}$$

We claim that the function

$$t \to u(t) = \int_{-\infty}^{t} \Sigma(t - \tau) f(\tau) \, d\tau \tag{9.9}$$

satisfies (9.1) and (9.3). Indeed, thanks to (9.8),

$$|u(t)| \leq \int_{-\infty}^{t} \exp(-\lambda_1 \alpha (t - \tau)) |f|_\infty \, d\tau,$$

where $|f|_\infty$ is the norm of f in $L^\infty(\mathbb{R}; H)$; thus

$$|u(t)| \leq \frac{1}{\alpha \lambda_1} |f|_\infty, \qquad \forall t \in \mathbb{R}. \tag{9.10}$$

Similarly, we can show that $u \in L^\infty_{\text{loc}}(\mathbb{R}; V)$; finally, by definition of Σ, $(\Sigma(t)\varphi_0)' = -A\Sigma(t)\varphi_0$, and it is then clear by differentiation of (9.9) that u satisfies (9.1). This completes the proof of existence and uniqueness.

(ii) We now derive the a priori estimates on u; (9.9) already provides an estimate of u in $L^\infty(\mathbb{R}; H)$

$$|u|_{L^\infty(\mathbb{R}; H)} \leq \frac{1}{\alpha \lambda_1} |f|_{L^\infty(\mathbb{R}; H)}. \tag{9.11}$$

We take the scalar product of (9.1) with u in H and by standard computations we find

$$\frac{1}{2} \frac{d}{dt} |u|^2 + a(u, u) = (f, u) \leq |f|_\infty |u|_\infty.$$

Hence, using II.(3.2) and (9.11),

$$\frac{d}{dt} |u|^2 + 2\alpha \|u\|^2 \leq \frac{2}{\alpha \lambda_1} |f|_\infty^2;$$

then by integration between t and $t + r$,

$$
\begin{aligned}
2\alpha \int_t^{t+r} \|u\|^2\, ds &\le |u(t)|^2 + \frac{2r}{\alpha\lambda_1} |f|_\infty^2, \\
\int_t^{t+r} \|u\|^2\, ds &\le \frac{1}{\alpha^3 \lambda_1^2} (\tfrac{1}{2} + r\alpha\lambda_1) |f|_\infty^2, \qquad \forall t \in H.
\end{aligned}
\tag{9.12}
$$

We assume from now on that A is symmetric. We infer from Theorem II.3.3 that u belongs to $L^\infty_{\text{loc}}(\mathbb{R}; V) \cap L^2_{\text{loc}}(\mathbb{R}; D(A))$. Then we use relation II.(3.29) which is valid a.e. on $\mathbb{R}$

$$
\frac{d}{dt} a(u, u) + 2|Au|^2 = 2(f, Au) \le 2|f||Au| \le |f|^2 + |Au|^2.
$$

Therefore

$$
\frac{d}{dt} a(u, u) + |Au|^2 \le |f|_\infty^2. \tag{9.13}
$$

For every $\varphi \in D(A)$, $\varphi \neq 0$, we have

$$
\alpha\lambda_1 \le \frac{a(\varphi, \varphi)}{|\varphi|^2} \le \frac{|A\varphi|^2}{a(\varphi, \varphi)}; \tag{9.14}
$$

this is due to the conjunction of II.(3.2), (9.4), and the fact that

$$
a(\varphi, \varphi) = (\varphi, A\varphi) \le |\varphi||A\varphi|, \qquad \forall \varphi \in D(A).
$$

Hence we infer from (9.13) that

$$
\frac{d}{dt} a(u, u) + \alpha\lambda_1 a(u, u) \le |f|_\infty^2. \tag{9.15}
$$

In the case where $f = 0$, i.e., in the case of equation (9.6), the analog of (9.15) reads

$$
\frac{d}{dt} a(\varphi, \varphi) + \alpha\lambda_1 a(\varphi, \varphi) \le 0. \tag{9.16}
$$

After multiplication by t and some elementary computations, this yields

$$
\begin{aligned}
&\frac{d}{dt} t a(\varphi, \varphi) + \alpha\lambda_1 t a(\varphi, \varphi) \le a(\varphi, \varphi), \\
&\qquad t a(\varphi(t), \varphi(t)) \le \int_0^t a(\varphi(s), \varphi(s)) \exp(\alpha\lambda_1(s - t))\, ds.
\end{aligned}
\tag{9.17}
$$

Similarly, starting from (9.5), we obtain after some elementary computations

$$
\int_0^t a(\varphi(s), \varphi(s)) \exp(\alpha\lambda_1(t - s))\, ds \le |\varphi(0)|^2 \exp(-\alpha\lambda_1 t),
$$

and finally

$$a(\varphi(t), \varphi(t)) \leq \frac{1}{t} |\varphi(0)|^2 \exp(-\alpha\lambda_1 t). \tag{9.18}$$

It is easy to check that (9.18) can be rephrased in the following way: the operator $\Sigma(t)$ in (9.7), (9.8) maps H into V, $\forall t > 0$, and

$$|\Sigma(t)|_{\mathscr{L}(H, V_a)} \leq \frac{1}{\sqrt{t}} \exp\left(-\frac{\alpha\lambda_1 t}{2}\right), \qquad \forall t > 0, \tag{9.19}$$

where $V_a =$ the space V endowed with the norm $\{a(\varphi, \varphi)\}^{1/2}$ equivalent to $\|\varphi\|$.

At this point we can prove that the solution u of (9.1), (9.3) belongs to $L^\infty(\mathbb{R}; V)$, exactly as in (9.10),

$$\begin{aligned} \{a(u(t), u(t)\}^{1/2} &\leq |f|_\infty \int_{-\infty}^{t} \frac{1}{\sqrt{t-\tau}} \exp\left(-\frac{\alpha\lambda_1}{2}(t-\tau)\right) d\tau, \\ a(u(t), u(t)) &\leq \frac{c_1}{\alpha\lambda_1} |f|_\infty^2, \qquad \forall t \in \mathbb{R}, \end{aligned} \tag{9.20}$$

where

$$c_1 = 2 \int_0^\infty \frac{\exp(-\sigma)}{\sqrt{\sigma}} \, d\sigma = 4 \int_0^\infty e^{-\sigma^2} \, d\sigma = 2\sqrt{\pi}.$$

(iii) In order to obtain the last estimates we return to (9.13) which we integrate between t and $t + r$, $t \in \mathbb{R}$, $r > 0$,

$$\begin{aligned} \int_t^{t+r} |Au|^2 \, ds &\leq a(u(t), u(t)) + r|f|_\infty^2 \\ &\leq \text{(with (9.20))} \\ &\leq \left(r + \frac{c_1}{\alpha\lambda_1}\right) |f|_\infty^2. \end{aligned} \tag{9.21}$$

Equation (9.1) then gives

$$\begin{aligned} u' &= f - Au, \\ |u'|^2 &\leq 2|f|_\infty^2 + 2|Au|^2, \\ \int_t^{t+r} |u'|^2 \, ds &\leq 2r|f|_\infty^2 + 2 \int_t^{t+r} |Au|^2 \, ds \\ &\leq \left(3r + \frac{c_1}{\alpha\lambda_1}\right) |f|_\infty^2. \end{aligned} \tag{9.22}$$

Proposition 9.1 is proved. □

9.2. Example of Partial Regularity

For the sake of simplicity we consider a special case of equation III.(1.2), where the polynomial g reads

$$g(s) = s^3 - s,$$

and we take $d = 1$. Equation III.(1.2) becomes

$$\frac{\partial u}{\partial t} - \Delta u + u^3 - u = 0.$$

We also restrict ourselves, for the moment, to the boundary condition III.(1.3) although the boundary conditions III.(1.3′), (1.3″) in Remark III.1.5 could be considered as well.

Let $\mathscr{A}$ denote the maximal attractor and let $r > 0$ be given. We introduce the following notations: for any function θ defined on H we set

$$\begin{cases} \overline{\theta(u)} = \operatorname{Sup}_{u_0 \in \mathscr{A}} \operatorname{Sup}_{t \in \mathbb{R}} \dfrac{1}{t} \displaystyle\int_t^{t+r} \theta(S(\tau)u_0)\, d\tau, \\ \overline{\overline{\theta(u)}} = \operatorname{Sup}_{u_0 \in \mathscr{A}} \operatorname{Sup}_{t \in \mathbb{R}} \theta(S(t)u_0). \end{cases} \tag{9.23}$$

The first regularity result consists of showing that the universal attractor $\mathscr{A}$ is included in $L^\infty(\Omega)$.[1]

Proposition 9.2. *The universal attractor* $\mathscr{A}$ *of equation* (9.23) *is included in* $L^\infty(\Omega)$.

Proof. If $\delta \in L^2(\Omega)$ we denote by φ_+ the function $\varphi_+ = \max(\varphi, 0)$. We recall that if φ is in $H^1(\Omega)$ then φ_+ and $\varphi_- = \varphi_+ - \varphi = \max(-\varphi, 0)$ belong to $H^1(\Omega)$.[2] We multiply equation (6.23) by $(u - M)_+$ with M chosen below, and integrate over Ω. Using the properties of the truncation operator, recalled in Section II.1.4, we find

$$\frac{1}{2}\frac{d}{dt}|(u - M)_+|^2 + |\nabla(u - M)_+|^2 + \int_\Omega g(u)(u - M)_+\, dx = 0. \tag{9.24}$$

As in III.(1.8), there exists a constant c_2 such that

$$\tfrac{1}{2}s^4 - c_2 \le g(s)s \le \tfrac{3}{2}s^4 + c_2, \qquad \forall s \in \mathbb{R}. \tag{9.25}$$

[1] This result, related to the maximum principle, is valid only for reaction–diffusion equations, but the following results and methods are much more general.

[2] More results on the truncation operators are recalled in Section II.1.4. See also, in Section III.3.5, a similar utilization of these operators for the thermohydraulic equations.

If $u \le M$, then $g(u)(u - M)_+ = 0$, and if $u > M$, we write

$$\begin{aligned} g(u)(u - M)_+ &= g(u)u\frac{(u - M)_+}{4} = g(u)u\left(1 - \frac{M}{u}\right) \\ &\ge (\text{by } (9.25)) \\ &\ge \left(\tfrac{1}{2}u^4 - c_2\right)\left(1 - \frac{M}{u}\right) \\ &\ge \left(\tfrac{1}{2}M^4 - c_2\right)\left(1 - \frac{M}{u}\right). \end{aligned}$$

We choose

$$M = (2c_2)^{1/4}, \tag{9.26}$$

and conclude that $g(u)(u - M)_+ \ge 0$ a.e. Hence (9.24) yields

$$\frac{d}{dt}|(u - M)_+|^2 + 2|\nabla(u - M)_+|^2 \le 0,$$

and using the constant λ_1 in (9.4)

$$\frac{d}{dt}|(u - M)_+|^2 + 2\lambda_1|(u - M)_+|^2 \le 0.$$

By integration between t_0 and t, $t_0 < t$, we find

$$|(u - M)_+(t)|^2 \le |(u - M)_+(t_0)|^2 \exp(-2\lambda_1(t - t_0)). \tag{9.27}$$

Let $u = u(t)$ be an orbit on the attractor $\mathscr{A}$. Since $\mathscr{A}$ is bounded in $H = L^2(\Omega)$, $|(u - M)_+(t_0)|$ is bounded independently of t_0; letting $t_0 \to -\infty$ in (9.27) we find $|(u - M)_+(t)|^2 = 0$. But $u(t)$ can be any point of $\mathscr{A}$ and we conclude that $\varphi(x) \le M$ a.e., $\forall \varphi \in \mathscr{A}$.

In a similar manner, using the operator $(u - M)_-$, we can prove that $\varphi(x) \ge -M$ a.e., $\forall \varphi \in \mathscr{A}$, M given by (9.25), (9.26). The proof is complete. □

An easy consequence of Proposition 9.2 is that $|\varphi| \le M|\varphi|^{1/2}$, $\forall \varphi \in \mathscr{A}$, or using the notations (9.23)

$$\overline{\overline{|u|}} \le M|\Omega|^{1/2} = \kappa_1. \tag{9.28}$$

We now combine Propositions 9.1 and 9.2 to obtain regularity results and a priori bounds for the point of $\mathscr{A}$.

Let $u = u(t)$ be an orbit on $\mathscr{A}$. We rewrite (9.23) in the form

$$u' + Au = u - u^3, \tag{9.29}$$

with $A = -\Delta$, $D(A) = H_0^1(\Omega) \cap H^2(\Omega)$, as usual. Then Proposition 9.2 shows that $u - u^3 \in L^\infty(\mathbb{R}; H)$ and, more precisely,

$$|u^3 - u|_{L^\infty(\mathbb{R}; H)} \le (M^3 + M)|\Omega|^{1/2}. \tag{9.30}$$

Then Proposition 9.1 implies that u belongs to $L^\infty(\mathbb{R}; V)$ with a bound of u in $L^\infty(\mathbb{R}; V)$ depending only on the right-hand side of (9.30). Since u is an arbitrary orbit of $\mathscr{A}$, we conclude that $\mathscr{A}$ is bounded in V. Proposition 6.1 also gives uniform bounds for the quantities

$$\int_t^{t+r} |Au|^2\, ds, \qquad \int_t^{t+r} |u'|^2\, ds,$$

and we conclude as above that

$$\overline{|Au|^2} \quad \text{and} \quad \overline{|u'|^2} \quad \text{are finite,} \tag{9.31}$$

and can be estimated explicitly in terms of the data.

The proof continues by successive differentiation of equation (9.29) with respect to time. We differentiate once and, setting $v = u'$, we find

$$v' + Av = v - 3u^2 v. \tag{9.32}$$

The L^2-norm of (9.32) is majorized by $(1 + 3M^2)|v|$. It can be shown that $u' = v$ belongs to $L^\infty_{\text{loc}}(\mathbb{R}; H) \cap L^2_{\text{loc}}(\mathbb{R}; V)$. Taking the scalar product of (9.32) with v in H $(= L^2(\Omega))$, we obtain

$$\begin{aligned} &\frac{1}{2}\frac{d}{dt}|v|^2 + |\nabla v|^2 = (v - 3u^2 v, v), \\ &\frac{1}{2}\frac{d}{dt}|v|^2 + |\nabla v|^2 \le (1 + 3M^2)|v|^2. \end{aligned} \tag{9.33}$$

We apply the uniform Gronwall lemma (Lemma III.1.1) with g replaced by $2(1 + 3M^2)$, $h = 0$, $y = |v|^2$, and we conclude, thanks to (9.31), that $|u'| = |v|$ is uniformly bounded with a bound depending only on the quantity κ_1 in (9.28). Hence

$$\overline{\overline{|u'|^2}} \quad \text{is finite.} \tag{9.34}$$

Returning to (9.33) we find that

$$\overline{|\nabla u'|^2} \quad \text{is finite.} \tag{9.35}$$

Now we rewrite (9.29) as

$$Au = u - u^3 - u', \tag{9.36}$$

$u = u(t)$ still denoting an arbitrary orbit on $\mathscr{A}$. Thanks to (9.30), (9.34), the right-hand side of (9.36) is bounded in $L^\infty(\mathbb{R}; H)$ and therefore u is bounded in $L^\infty(\mathbb{R}; D(A))$. In the present case, $D(A) = H^1_0(\Omega) \cap H^2(\Omega)$ and we can conclude that $\mathscr{A}$ is included and bounded in $H^1_0(\Omega) \cap H^2(\Omega)$.

In fact, with very minor modifications, the above proof applies to the general equation III.(1.2) associated with any of the boundary conditions III.(1.3), (1.3′), (1.3″).

In conclusion, we have

Theorem 9.1. *The maximal attractor* $\mathscr{A}$ *given by Theorem* III.1.2 *is included and bounded in* $L^\infty(\Omega) \cap H_0^1(\Omega) \cap H^2(\Omega)$.

Remark 9.1.

(i) As indicated above we can prove several other regularity results for this attractor $\mathscr{A}$.

(ii) Once we know that $\mathscr{A}$ is included in a Banach space $W \subset H$ (here $W = L^\infty(\Omega) \cap H_0^1(\Omega) \cap H^2(\Omega)$), a natural question is whether $\mathscr{A}$ is an attractor in W, i.e., does $\mathscr{A}$ attract, in the norm of W, the points or the bounded sets of W. We will not develop this point here; results of this type for different equations can be found in J.M. Ghidaglia and R. Temam [4].

9.3. Example of $\mathscr{C}^\infty$ Regularity

We now develop an example of $\mathscr{C}^\infty$ regularity for which the iteration procedure mentioned above is fully accomplished.

We consider, as in Section III.2, the two-dimensional Navier–Stokes equations III.(2.1), III.(2.2) supplemented by one of the boundary conditions III.(2.4) or III.(2.5), (2.6) or III.(2.39). For boundary conditions other than space-periodicity, we assume that

$$\Omega \text{ is of class } \mathscr{C}^\infty. \tag{9.37}$$

As already stated in Theorem III.2.2, the maximal attractor is bounded in H and the existence of an absorbing set in V also implies that $\mathscr{A}$ is included and bounded in V. Our aim is to prove

Theorem 9.2. *The attractor* $\mathscr{A}$ *is included in* $\mathscr{C}^\infty(\overline{\Omega})^2$ *and bounded in all the spaces* $H^l(\Omega)^2$, $l \geq 0$.

The proof consists of showing that $\mathscr{A}$ is included and bounded in all the spaces $H^l(\Omega)^2$. This is the object of Propositions 9.3 and 9.4.

We start from the functional form of equation III.(2.18),

$$u' + \nu A u + B(u) = f, \tag{9.38}$$

and denote by $u = u(t)$ an arbitrary orbit on $\mathscr{A}$. We will write

$$u^m = \frac{d^m u}{dt^m}, \qquad u^0 = u. \tag{9.39}$$

By successive time differentiation of equation (9.38) we will obtain some estimates on the norms of the derivatives u^m in H, V, and $D(A)$ (Proposition 9.3); then, in Proposition 9.4, we will obtain the estimates on u in the norm of $H^l(\Omega)^2$.

Proposition 9.3. *For every $m \geq 1$, the following expressions are finite:*

$$\overline{\overline{|u^m|^2}}, \quad \overline{\overline{|Au^{m-1}|^2}}, \tag{9.40}$$

$$\overline{\|u^m\|^2}, \quad \overline{\|Au^m\|^2}, \quad \overline{\|u^{m+1}\|^2}. \tag{9.41}$$

The proof relies on the following lemma whose proof, similar to that of Proposition 9.1, will be given below:

Lemma 9.1. *The hypotheses are those of Proposition 9.1. We assume that A is symmetric and that $u = u(t)$ satisfies*

$$u \in L^\infty_{\text{loc}}(\mathbb{R}; H) \cap L^2_{\text{loc}}(\mathbb{R}; V), \tag{9.42}$$

$$\frac{du}{dt}(t) + Au(t) = g(t, u(t)), \qquad t \in \mathbb{R}, \tag{9.43}$$

where

$$(g(t, \varphi), \varphi) \leq \frac{\alpha}{2}\|\varphi\|^2 + h_1(t), \tag{9.44}$$

$$|g(t, \varphi)| \leq h_2(t)|A\varphi|^{1-\varepsilon} + h_3(t)\|\varphi\| + h_4(t), \quad 0 \leq \varepsilon < 1, \forall t, \forall \varphi \in D(A), \tag{9.45}$$

and for some $r > 0$,

$$\begin{aligned} &\sup_{t \in \mathbb{R}} \int_t^{t+r} h_1(s)\, ds < \infty, \\ &\sup_{t \in \mathbb{R}} \int_t^{t+r} (h_2(s)^{2/\varepsilon} + h_3(s)^2 + h_4(s)^2)\, ds < \infty. \end{aligned} \tag{9.46}$$

Then $u \in L^\infty(\mathbb{R}; V) \cap L^2_{\text{loc}}(\mathbb{R}; D(A))$, $u' \in L^2_{\text{loc}}(\mathbb{R}; H)$, and the quantities

$$\sup_{t \in \mathbb{R}} \int_t^{t+r} |Au|^2\, ds, \qquad \sup_{t \in \mathbb{R}} \int_t^{t+r} |u'|^2\, ds, \tag{9.47}$$

are finite.

Proof of Proposition 9.3. The result is proved by induction; c denotes a constant depending on $\mathscr{A}$ and on the data which may be different at different places.

(i) *Initialization of the induction* ($m = 1$)
We know already that $\mathscr{A}$ is bounded in V; hence

$$\overline{\|u\|^2} \leq \overline{\overline{\|u\|^2}} < \infty. \tag{9.48}$$

Equation (9.38) is then written in the form (9.43) with $g = f - B(u)$. Owing to III.(2.16)

$$|B(u)| \leq c\,|u|^{1/2}\,|Au|^{1/2}\,\|u\|. \tag{9.49}$$

Hence, thanks to the previous estimate, $|B(u)| \leq k\,|Au|^2$, $|g(u)| \leq k(|Au|^2 + 1)$; (9.45) follows readily and (9.44) is satisfied since $(B(u), u) = 0$ (see III.(2.21)). We conclude from Lemma 9.1 that the expressions (9.47) are finite, and since $u(\cdot)$ is an arbitrary trajectory on $\mathscr{A}$ and the bounds depend only on $\mathscr{A}$

$$\overline{\|Au\|^2} < \infty, \qquad \overline{\overline{\|u'\|^2}} < \infty. \tag{9.50}$$

We now differentiate (9.38) with respect to t and find

$$\frac{du'}{dt} + \nu A u' + B(u', u) + B(u, u') = 0. \tag{9.51}$$

This equation is of the form (9.43) with u replaced by u' and $g(t, u') = -B(u', u(t)) - B(u(t), u')$. We have, because of III.(2.16), III.(2.21),

$$\begin{aligned}
(g(t, u'), u') &= -(B(u', u(t)), u') \\
&\leq c\,|u'|\,\|u'\|\,\|u(t)\| \\
&\leq \frac{\nu}{2}\|u'\|^2 + \frac{c^2}{2\nu}|u'|^2\,\|u(t)\|^2.
\end{aligned}$$

Therefore (9.44) is satisfied with $h_1(t) = c\,|u'(t)|^2\,\|u(t)\|^2$, and the first condition (9.46) holds true because of (9.50).

We also infer from III.(2.16) that

$$|g(t, u')| \leq c\{|u|^{1/2}\,|Au|^{1/2}\,\|u'\| + |u'|^{1/2}\,|Au'|^{1/2}\,\|u\|\}.$$

Therefore (9.45) and the second inequality (9.46) are satisfied thanks to the previous estimates. Lemma 9.1 allows us to conclude that $u' \in L^\infty(\mathbb{R}; V) \cap L^2_{\mathrm{loc}}(\mathbb{R}; D(A))$ and that

$$\overline{\overline{\|u'\|^2}} < \infty, \qquad \overline{|Au'|^2} < \infty, \qquad \overline{|u''|^2} < \infty.$$

This shows that all the conditions (9.40), (9.41) with $m = 1$ are satisfied, except the condition

$$\overline{\overline{|Au|^2}} < \infty. \tag{9.52}$$

In order to prove this property we infer from (9.38) and III.(2.16) that

$$\nu A u = -u' - Bu + f,$$

$$|Au|^2 \leq \frac{3}{\nu^2}(|u'|^2 + |Bu|^2 + |f|^2),$$

$$|Au|^2 \leq \frac{3}{\nu^2}|u'|^2 + \frac{3}{\nu^2}|f|^2 + c\,|u|\,|Au|\,\|u\|^2,$$

$$|Au|^2 \leq c(|u'|^2 + |f|^2 + |u|^2\,\|u\|^4);$$

(9.52) follows readily.

(ii) *The induction argument*

We now assume that the expressions (9.40), (9.41) are finite for m replaced by $1, \dots, m-1$ and we prove that the same is true for m.

We differentiate equation (9.38) m times with respect to t

$$\frac{du^m}{dt} + \nu A u^m + \sum_{j=1}^{m} \binom{j}{m} B(u^j, u^{m-j}) = 0. \tag{9.53}$$

This equation is of the form (9.43) with u replaced by u^m and

$$g(t, u^m) = -\sum_{j=1}^{m-1} \binom{j}{m} B(u^j(t), u^{m-j}(t)) - B(u(t), u^m) - B(u^m, u(t)).$$

We have

$$(g(t, \varphi), \varphi) = -\sum_{j=1}^{m-1} \binom{j}{m} (B(u^j, u^{m-j}), \varphi) - (B(\varphi, u), \varphi),$$

the term $(B(u, \varphi), \varphi)$ vanishing because of III.(2.21). Thanks to III.(2.16)

$$|(g(t, u^m), u^m)| \le c \sum_{j=1}^{m-1} |u^j|^{1/2} \|u^j\|^{1/2} \|u^{m-j}\| \, |u^m|^{1/2} \|u^m\|^{1/2} + c|u|\,|u^m|\,\|u^m\|.$$

Thanks to the induction assumption this is majorized by

$$\frac{\nu}{2}\|u^m\|^2 + c(|u^m|^2 + \|u^{m-1}\|^2).$$

Hence (9.44) is satisfied with

$$h_1(t) = c(|u^m(t)|^2 + \|u^{m-1}(t)\|^2,$$

and the first inequality (9.46) is satisfied because of the induction assumption (9.41) at the order $m - 1$.

We then prove (9.45); due to III.(2.16)

$$|g(t, u^m)| \le c \sum_{j=1}^{m-1} |u^j|^{1/2} \|Au^j\|^{1/2} \|u^{m-j}\| + c|u|^{1/2} \|Au\|^{1/2} \|u^m\|$$
$$+ c|u^m|^{1/2} |Au^m|^{1/2} \|u\|.$$

Thanks to the induction assumption this is majorized by an expression similar to (9.45) with $\varepsilon = 1/2$ and

$$h_2(t) = c|u^m(t)|^{1/2}, \qquad h_3(t) = c,$$
$$h_4(t) = c(1 + \|u^{m-1}(t)\| + |Au^{m-1}(t)|).$$

We infer from the induction assumption that the second inequality (9.46) is satisfied. Lemma 9.1 can then be applied to (9.53) and it shows that the expressions

$$\overline{\|u^m\|^2}, \quad \overline{|Au^m|^2}, \quad \overline{|u^{m+1}|^2} \tag{9.54}$$

are finite. In order to complete the induction argument there remains to show

that

$$\overline{\overline{|Au^{m-1}|^2}} < \infty. \tag{9.55}$$

For that purpose we write equation (9.53) with m replaced by $m - 1$

$$\nu A u^{m-1} = -u^m - \sum_{j=0}^{m-1} \binom{j}{m-1} B(u^j, u^{m-1-j}).$$

Then

$$|Au^{m-1}| \leq \frac{1}{\nu}|u^m| + c \sum_{j=0}^{m-1} |B(u^j, u^{m-1-j})|,$$

and (9.55) follows readily from the estimates III.(2.16) on B, from the induction assumptions, and (9.54).

Proposition 9.3 is proved. □

We now state our last result which implies Theorem 9.1.

Proposition 9.4. *The maximal attractor $\mathscr{A}$ is included and bounded in $H^l(\Omega)^2$, $\forall l \geq 1$.*

The proof of Proposition 9.4 consists of writing (9.53) in the form

$$\nu A u^m = -\frac{du^m}{dt} - \sum_{j=0}^{m} \binom{j}{m} B(u^j, u^{m-j}). \tag{9.56}$$

Using Proposition 9.3, and assuming that $u = u(t)$ is an arbitrary orbit on $\mathscr{A}$, we notice that the right-hand side of (9.56) belongs to $L^\infty(\mathbb{R}; H)$. This shows that u^m is bounded in $D(A)$ whose norm is equivalent to that of $H^2(\Omega)^2$. This is valid for any $m \geq 0$. In particular, for $m = 0$, we find that $\mathscr{A}$ is included and bounded in $H^2(\Omega)^2$.

The proof then continues by induction on l, $l \geq 2$. The details of the proof are technical; they pertain to the theory of partial differential equations and have no implications for the dynamical system point of view. We omit them.

Finally, it remains to prove Lemma 9.1.

Proof of Lemma 9.1. The proof is similar to that of Proposition 9.1.

We take the scalar product of (9.43) with $u(t)$ in H and use the coercivity of A (see II.(3.2)) and (9.44)

$$\frac{1}{2}\frac{d}{dt}|u|^2 + a(u, u) = (g(u), u),$$

$$\frac{1}{2}\frac{d}{dt}|u|^2 + \alpha\|u\|^2 \leq \frac{\alpha}{2}\|u\|^2 + h_1, \tag{9.57}$$

$$\frac{d}{dt}|u|^2 + \alpha\|u\|^2 \leq 2h_1.$$

We apply the uniform Gronwall lemma with $y = |u|^2$, $g = 0$, $h = 2h_1$, and we infer from (9.46) that u belongs to $L^\infty(\mathbb{R}; H)$.[1] Then (9.57) shows that

$$\operatorname*{Sup}_{t \in \mathbb{R}} \int_t^{t+r} \|u\|^2 \, ds < \infty. \tag{9.58}$$

We then take the scalar product of (9.43) with $Au(t)$ in H; we obtain

$$\begin{aligned}
\frac{1}{2}\frac{d}{dt} a(u, u) + |Au|^2 &= (g(u), Au) \\
&\leq |g(u)||Au| \\
&\leq h_2 |Au|^{2-\varepsilon} + h_3 \|u\| \, |Au| + h_4 |Au| \\
&\leq \tfrac{1}{2}|Au|^2 + c_\varepsilon(h_2^{2/\varepsilon} + h_3^2 \|u\|^2 + h_4^2),
\end{aligned}$$

$$\frac{d}{dt} a(u, u) + |Au|^2 \leq 2c_\varepsilon(h_2^{2/\varepsilon} + h_4^2 + h_3^2 \|u\|^2). \tag{9.59}$$

Similarly, by utilization of (9.46), (9.58) and the uniform Gronwall lemma, we find that u belongs to $L^\infty(\mathbb{R}; V)$. Then (9.59) yields

$$\operatorname*{Sup}_{t \in \mathbb{R}} \int_t^{t+r} |Au|^2 \, ds < \infty. \tag{9.60}$$

Finally, we return to equation (9.43) and thanks to (9.45), we find

$$\begin{aligned}
u' &= -Au + g(u), \\
|u'|^2 &\leq (|Au| + |g(u)|)^2 \\
&\leq 4|Au|^2 + 4h_2^2 |Au|^{2(1-\varepsilon)} + 4h_3^2 \|u\|^2 + 4h_4^2 \\
&\leq 5|Au|^2 + c_\varepsilon h_2^{2/\varepsilon} + 4h_3^2 \|u\|^2 + 4h_4^2.
\end{aligned}$$

Thanks to (9.60), (9.46) and the estimate of u in $L^\infty(\mathbb{R}; V)$ we conclude easily that

$$\operatorname*{Sup}_{t \in \mathbb{R}} \int_t^{t+r} |u'|^2 \, ds < \infty. \qquad \square$$

10. Stability of Attractors

Our aim in this section is to illustrate with an example the result for the stability of attractors presented in Chapter I, Section 1.4. This result (Theorem I.1.2) is quite general and applies to various types of approximations of a semigroup $S(t)$ by a family of semigroups $S_\eta(t)$. In particular, with a view to numerical computations, the semigroups $S_\eta(t)$ may arise from the finite-dimensional

[1] We apply the uniform Gronwall lemma on (t_0, ∞) for some $t_0 \in \mathbb{R}$. This lemma provides a uniform bound of $y = |u|^2$ for $t \geq t_0 + r$. Since the explicit expression of the bound is independent of t_0, the majorization is valid on the whole line.

approximation to an infinite-dimensional dynamical system. It would be beyond the scope of this book to study the discretization of the equations considered in Chapters III and IV and its effect on the dynamics of these equations; actually, much remains to be done in that direction.

Instead of a systematic study of the approximation of attractors, we will describe here one typical example, namely the Galerkin approximation of an evolution equation of the first order in time. We have chosen to consider the equation studied in Section 3 of Chapter III, i.e., equation III.(3.10). As indicated above, numerous other types of equations and perturbations can be considered; the reader is referred, for instance, to B. Bréfort, J.M. Ghidaglia, and R. Temam [1], J.M. Ghidaglia and R. Temam [5], for some different applications of Theorem I.1.2 to fluid mechanics equations. It is also possible to treat in the same manner the perturbations corresponding to the modification of the coefficients in (most of) the equations; for example, one can consider suitable perturbations of ν and f in equation III.(2.1), (2.2), or of α and f in equation (2.1). □

The notations and hypotheses are those of Section III.3.1, in particular, III.(3.3)–(3.9). We consider the semigroup $S(t)$ defined by the initial-value problem III.(3.10), (3.11)

$$\frac{du}{dt} + Au + B(u) + Ru = 0, \tag{10.1}$$

$$u(0) = u_0. \tag{10.2}$$

We recall that for u_0 given in H, (10.1), (10.2) possesses a unique solution u which belongs to $\mathscr{C}([0, T]; H)$ and $L^2(0, T; V)$, $\forall T > 0$ (see Theorem III.3.1), and $S(t)$ is the nonlinear mapping in H

$$S(t)\colon u_0 \to u(t). \tag{10.3}$$

Let w_j, λ_j, $j \in \mathbb{N}$, be the family of eigenvectors and eigenvalues of A

$$Aw_j = \lambda_j w_j, \qquad \forall j \in \mathbb{N}. \tag{10.4}$$

The family w_j constitutes an orthonormal basis of H and the sequence λ_j converges to $+\infty$ as $j \to \infty$

$$0 < \lambda_1 \le \lambda_2 \le \cdots. \tag{10.5}$$

For every $m \in \mathbb{N}$, we denote by $V_m = H_m$ the space spanned by $w_1, \ldots, w_m$ and we denote by P_m the orthogonal projector in H onto H_m. We recall that the spectral projector P_m is also an orthogonal projector in V, V', $D(A), \ldots$; we also recall that A and P_m commutes.

A Galerkin approximation of (10.1), (10.2) using this basis $\{w_j\}_{j \in \mathbb{N}}$ of H consists of solving for every integer $m \in \mathbb{N}$ the following problem

$$\frac{du_m}{dt} + Au_m + P_m B(u_m) + P_m R u_m = P_m f, \tag{10.6}$$

$$u_m(0) = u_{0m} = P_m u_0. \tag{10.7}$$

This is a finite-dimensional problem, the function u_m is searched as a function from $\mathbb{R}_+$ into H_m. There exists a unique solution u_m to (10.6), (10.7) $\forall m$, and as m converges to infinity, u_m converges to the solution u of (10.1), (10.2). Actually, equations (10.6), (10.7) are precisely those used in the proof of existence in Theorem III.3.1 (see a related situation in the proof of Theorems III.1.1 and III.2.1). We denote by $S_m(t)$, $t \geq 0$, the nonlinear operator in H_m defined by

$$S_m(t)\colon u_m(0) \to u_m(t). \tag{10.8}$$

Theorem III.3.2 provides the existence of a maximal attractor $\mathscr{A}$ for equation (10.1) which attracts the bounded sets of H. The proof of Theorem III.3.2 shows the existence of a set $\mathscr{B}_0$ bounded in H which absorbs the bounded sets of H. Let us denote by $\mathscr{U}$ an open neighborhood of $\mathscr{B}_0$, say an ε-neighborhood with $\varepsilon = 1$; it is clear that $\mathscr{A}$ absorbs $\mathscr{U}$.

In a totally similar manner we can show that S_m possesses a maximal attractor $\mathscr{A}_m$ which is compact and connected in H_m, and which attracts the bounded sets of H_m. In fact, Theorem III.3.2 is directly applicable to equation (10.6) which is just a particular case of III.(3.10) with H, V, replaced by H_m, V_m, A replaced by its restriction to H_m, B, R, f replaced by $P_m B$, $P_m R$, and $P_m f$: it is easy to see that the analogs of III.(3.3)–(3.9) are satisfied; the hypotheses of Theorem III.3.2 are satisfied and therefore its conclusions are valid. Furthermore, by repeating explicitly the computations showing the existence of an absorbing set in H, we find that $\mathscr{B}_0 \cap H_m$ is also an absorbing set for the semigroup S_m. This implies that $\mathscr{A}_m \subset \mathscr{B}_0 \cap H_m \subset \mathscr{U} \cap H_m$ and, obviously, $\mathscr{A}_m$ attracts $\mathscr{U} \cap H_m$.

Our task is now to let $m \to \infty$ and to show that $\mathscr{A}_m$ approximates $\mathscr{A}$. For that purpose we will apply Theorem I.1.2 with $\eta = 1/m$.

We recall the definition of the semidistance d appearing in I.(1.20)

$$d(\mathscr{B}, \mathscr{C}) = \operatorname*{Sup}_{x \in \mathscr{B}} \operatorname*{Inf}_{y \in \mathscr{C}} |x - y|. \tag{10.9}$$

Theorem 10.1. *When $m \to \infty$, $\mathscr{A}_m$ converges to $\mathscr{A}$ in the sense of the semidistance d*

$$d(\mathscr{A}_m, \mathscr{A}) \to 0 \quad \text{as } m \to \infty. \tag{10.10}$$

Equivalently, this means that for every neighborhood $\mathscr{V}$ of $\mathscr{A}$ in H, there exists $m_0 = m_0(\mathscr{V})$ such that

$$\mathscr{A}_m \subset \mathscr{V}, \qquad \forall m \geq m_0(\mathscr{V}). \tag{10.11}$$

Proof. The theorem is a direct consequence of Theorem I.1.2 applied in the conditions indicated above ($\mathscr{U}' = \mathscr{U}$). We only need to verify the hypothesis I.(1.18). □

In the present case I.(1.18) is just a result of convergence of the Galerkin method. The convergence of u_m to u is known and is part of the proof of the existence of u; I.(1.18) is a refinement of the result of convergence of u_m to u: it amounts to proving that $u_m(t)$ converges to $u(t)$ in H, uniformly for t in a compact set of $]0, +\infty[$ and u_0 $(= u(0))$ in a bounded set of H. This is precisely the object of

Lemma 10.1. *When $m \to \infty$, u_m converges to u in $\mathscr{C}([t_0, T]; H)$, uniformly for u_0 in a bounded set of H, $|u_0| \leq r_1$, $\forall 0 < t_0 < T < \infty$.*

PROOF. (i) The initial data u_0 are given in H, not necessarily in V. A slight modification of the point III.(3.13) in Theorem III.3.1 shows that

$$u \in \mathscr{C}(]0, T]; V) \cap L^2(\delta, T; D(A)), \qquad \forall \delta > 0. \tag{10.12}$$

This follows from III.(3.2), i.e.,

$$u \in \mathscr{C}([0, T]; H) \cap L^2(0, T; V), \tag{10.13}$$

and from the a priori estimate resulting from III.(3.16) after multiplication by t

$$\frac{d}{dt} a(u, u) + |Au|^2 \leq 2|f|^2 + 2c_1' \|u\|^2,$$

$$\frac{d}{dt} ta(u, u) + t|Au|^2 \leq 2t|f|^2 + a(u, u) + 2tc_1' \|u\|^2. \tag{10.14}$$

With standard methods we easily obtain that

$$\sqrt{t}u \in L^\infty(0, T; V) \cap L^2(0, T; D(A)) \tag{10.15}$$

with an explicit bound on the norm of u in that space which depends on u_0 only through r_1.

In a totally parallel way we can derive a priori bounds for the norm of u_m in the spaces appearing in (10.13), (10.16). Furthermore, these bounds are uniform when u_0 belongs to the ball $B(0, r_1)$ of H centered at 0 of radius r_1

> u_m is bounded in $L^\infty(0; T; H) \cap L^2(0, T; V)$ and $\sqrt{t}u_m$ is bounded in $L^\infty(0, T; V) \cap L^2(0, T; D(A))$, uniformly with respect to m and $u_0 \in B(0, r_1)$. (10.16)

(ii) At this point we can show that for $t_0 > 0$, $u_m(t_0)$ converges to $u(t_0)$ in H, uniformly with respect to u_0, i.e.,

$$\sup_{u_0 \in B(0, r_1)} |S_m(t_0)P_m u_0 - S(t_0)u_0| \to 0 \quad \text{as } m \to \infty. \tag{10.17}$$

If this were not true, we could find $\delta > 0$ and a sequence $m_j \to \infty$ such that

$$\sup_{u_0 \in B(0, r_1)} |S_{m_j}(t_0)P_{m_j} u_0 - S(t_0)u_0| \geq \delta > 0, \qquad \forall j.$$

Then for each j there exists $u_{0_j} \in B(0, r_1)$ such that

$$|S_{m_j}(t_0)P_{m_j} u_{0_j} - S(t_0)u_{0_j}| \geq \frac{\delta}{2} > 0, \qquad \forall j. \tag{10.18}$$

By extracting a subsequence, if necessary, we can assume that, as $j \to \infty$,

$$u_{0_j} \to u_0, \quad \text{weakly in } H.$$

The a priori estimates provided by (10.16) concerning $S_{m_j}(\cdot)P_{m_j}u_{0_j}$ and the analog estimates for $S(\cdot)u_{0_j}$ allow us to prove, in a standard manner, various convergence results; in particular,

$$S(t)u_{0_j} \to S(t)u_0 \quad \text{weakly in } V, \qquad \forall t > 0,$$

$$S_{m_j}(t)P_{m_j}u_{0_j} \to S(t)u_0 \quad \text{weakly in } V, \qquad \forall t > 0.$$

Hence $S_{m_j}(\cdot)P_{m_j}u_{0_j} - S(t_0)u_{0_j}$ converges to 0 weakly in V; since the embedding of V in H is compact, this convergence holds strongly in H, in contradiction with (10.18).

The proof of (10.17) is complete.

(iii) We set $v_m = u - u_m$ and substract (10.6), (10.7) from (10.1), (10.2)

$$v_m' + Av_m + P_m B(u, v_m) + P_m B(v_m, u) + P_m B(v_m, v_m) + P_m R v_m = Q_m g, \quad (10.19)$$

$$v_m(0) = Q_m u_0, \quad (10.20)$$

with $Q_m = I - P_m$ and

$$g = f - B(u) - Ru. \quad (10.21)$$

Thanks to III.(3.3), III.(3.8) we write

$$|Ru| \le c_1 \|u\|^{1-\theta_1} |Au|^{\theta_1},$$

$$|B(u)| \le c_4 \|u\|^{2-\theta_4} |Au|^{\theta_4},$$

$$|g| \le c_2'\{1 + (1 + \|u\|)|Au|\}. \quad (10.22)$$

We take the scalar product of (10.19) with v_m in H and we find

$$\frac{1}{2}\frac{d}{dt}|v_m|^2 + ((A + R)v_m, v_m) = -(B(v_m, u), v_m) + (B(u, v_m) + B(v_m, u)$$

$$+ B(v_m, v_m) + Rv_m, Q_m u) + (g, Q_m u). \quad (10.23)$$

We majorize the right-hand side of (10.23) using III.(3.3) and III.(3.7) repeatedly. After some lengthy but easy calculations we find that this quantity is bounded by

$$c_3'|v_m|^2 + h_m,$$

with

$$\int_{t_0}^T h_m(t)\, dt \le c_4' \int_{t_0}^T \{\|Q_m u\|^2 + \|Q_m u\|^{2\theta_3/(1+\theta_3)}\}\, dt$$

$$+ c_4'\left\{\int_{t_0}^T |Q_m u|^2\, dt\right\}^{1/(1+\theta_1)} + c_4'\left\{\int_{t_0}^T |Q_m u|^2\, dt\right\}^{1/2}. \quad (10.24)$$

If we can show that

$$\int_{t_0}^T \|Q_m u\|^2\, dt \to 0 \quad \text{as } m \to \infty, \quad \text{uniformly for } u_0 \in B(0, r_1), \quad (10.25)$$

then it will readily follow that

$$\int_{t_0}^{T} h_m(t)\,dt \to 0 \quad \text{as } m \to \infty, \quad \text{uniformly for } u_0 \in B(0, r_1). \tag{10.26}$$

To prove (10.25) we observe that

$$\begin{aligned}\int_{t_0}^{T} \|Q_m u\|^2\,dt &\le \lambda_{m+1} \int_{t_0}^{T} |Q_m A u|^2\,dt \\ &\le \lambda_{m+1} \int_{t_0}^{T} |Au|^2\,dt.\end{aligned} \tag{10.27}$$

Due to (10.15) the norm of $\sqrt{t}Au$ in $L^2(0, T; H)$ is bounded independently of u_0; therefore, the right-hand side of (10.27) is majorized by $c_5'\lambda_{m+1}$ and (10.25) follows.

Then we infer from (10.23) and III.(3.5) that

$$\frac{1}{2}\frac{d}{dt}|v_m|^2 + \alpha'\|v_m\|^2 \le c_3'|v_m|^2 + h_m,$$

$$\frac{d}{dt}|v_m|^2 \le c_3'|v_m|^2 + h_m'. \tag{10.28}$$

The Gronwall lemma yields

$$|v_m(t)|^2 \le |v_m(t_0)|^2 \exp(c_3'(t - t_0)) + \exp(c_3' t) \int_{t_0}^{T} h_m(s)\,ds$$

$$\operatorname*{Sup}_{t \in [t_0, T]} |v_m(t)|^2 \le \exp(c_3' T) \cdot \left\{ |v_m(t_0)|^2 + \int_{t_0}^{T} h_m(s)\,ds \right\}.$$

Thanks to (10.17) and (10.26),

$$\operatorname*{Sup}_{t \in [t_0, T]} |v_m(t)|^2 \to 0 \quad \text{as } m \to \infty,$$

uniformly for $u_0 \in B(0, r_1)$.

Lemma 10.1 is proved. □

CHAPTER V
Lyapunov Exponents and Dimension of Attractors

Introduction

This chapter contains the essential definitions and results which concern the study of the geometry of the attractors and functional sets, viz. the concept of Lyapunov exponents and Lyapunov numbers and general abstract results concerning the dimensions of attractors and functional sets. The Lyapunov numbers have a geometrical interpretation. They indicate how volumes are distorted in dimension m by the semigroup: the semigroup on the attractor is contracting in some directions and expanding in other directions, leading to a dynamics which can be complicated. Because the Lyapunov numbers (or the Lyapunov exponents which are the logarithms of the Lyapunov numbers) indicate the exponential rates of variation of lengths, surfaces, volumes in dimension $1, 2, 3, \ldots$, they provide valuable information about the dynamics.

The attractor and functional sets can be very complicated and, at the moment, very few tools are available for their description and the study of their geometrical properties. Essentially, the concept of dimension is one of the few pieces of information which is related to general sets. There are several definitions of dimension for complicated sets; two of them will be studied here, namely, the Hausdorff dimension and the fractal dimension, also called the capacity. The abstract results given in this chapter are general results, expressing a bound on these dimensions of the attractors and functional sets in terms of the Lyapunov exponents. In Chapter VI, we will apply these results to the concrete examples of dynamical systems introduced in Chapters III and IV and, finally, on estimating the Lyapunov exponents in terms of the physical parameters, we obtain some majorizations of the dimension in terms of the physical data.

We now describe how this chapter is organized. Section 1 contains some preliminary material on linear and multilinear algebra. We recall the concept of the exterior product of Hilbert spaces and review a few important properties of these spaces (Section 1.1). Then we study some linear and multilinear operators acting on exterior products (Section 1.2). Finally, in Section 1.3, we study the image of a ball by a linear operator acting in a Hilbert space; this is easy in the compact case but more delicate in the noncompact case.

Section 2 contains the definition of the Lyapunov exponents and numbers. We start in Section 2.1 with some geometrical remarks on the distortion of volumes, then in Section 2.2 we give the definitions and properties of the Lyapunov exponents. Section 2.3 contains a useful remark concerning the exponential decay of the volume element (in high dimension) in the function space.

Section 3 deals with the Hausdorff and fractal dimensions of attractors. The definitions of Hausdorff and fractal dimensions are recalled in Section 3.1. Section 3.2 contains some technical covering lemmas. Finally, Section 3.3 provides the main results on the Hausdorff and fractal dimensions; first, we state the results without utilization of the Lyapunov exponents (Theorems 3.1 and 3.2), and then with the utilization of these exponents.

1. Linear and Multilinear Algebra

Our aim in this section is to recall a few results of linear and multilinear algebra and to provide some complements needed in the sequel. In Section 1.1 we briefly recall the definitions of the exterior product of spaces and vectors in the case of Hilbert spaces. For further details the reader is referred to R. Abraham and J.E. Marsden [1], N. Bourbaki [1], [2], or L. Schwartz [3]; the Hilbert space case is explicitly considered in N. Bourbaki [2]. Then, in Section 1.2, we give the properties of some linear and multilinear operators acting in exterior products of spaces. Finally, in Section 1.3, we are concerned with the question of the image of a ball by a linear operator in a Hilbert space. We first consider the simpler case of a compact operator (Section 1.3.1) and then, after, the more delicate case of a noncompact operator.

1.1. Exterior Product of Hilbert Spaces

Let E be a Hilbert space endowed with the scalar product and the norm $(\cdot, \cdot)_E$, $|\cdot|_E$; the index E will be omitted when there is no possible confusion. If $\varphi_1, \ldots, \varphi_m$ belong to E, we denote by $\varphi_1 \otimes \cdots \otimes \varphi_m$ the m-linear form on E defined by

$$(\varphi_1 \otimes \cdots \otimes \varphi_m)(\psi_1, \ldots, \psi_m) = \prod_{i=1}^{m} (\varphi_i, \psi_i)_E, \qquad \forall \psi_1, \ldots, \psi_m \in E.$$

This m-form $\varphi_1 \otimes \cdots \otimes \varphi_m$ is the tensor product of $\varphi_1, \ldots, \varphi_m$. We denote by

$\bigotimes^m E$ the space of m-linear forms spanned by such tensor products, and we endow it with an inner product which we define by linearity and its expression for tensor products

$$(\varphi_1 \otimes \cdots \otimes \varphi_m, \psi_1 \otimes \cdots \otimes \psi_m)_{\bigotimes^m E}$$
$$= (\varphi_1, \psi_1) \ldots (\varphi_m, \psi_m)_E, \qquad \forall \varphi_1, \ldots, \varphi_m, \psi_1, \ldots, \psi_m \in E. \tag{1.1}$$

It is easy to see that this product is well defined (i.e., $(\xi, \eta)_{\bigotimes^m E}$ does not depend on the representation of $\xi, \eta \in \bigotimes^m E$ as sums of tensor products), and that it is positive definite, so that $\{(\xi, \xi)_{\bigotimes^m E}\}^{1/2}$ is a norm on $\bigotimes^m E$. The Hilbert m-tensor product of E denoted by $\widehat{\bigotimes}^m E$ is the completion of $\bigotimes^m E$ under this norm; we denote by $(\cdot, \cdot)_{\widehat{\bigotimes}^m E}$ and $|\cdot|_{\widehat{\bigotimes}^m E}$ the scalar product and the norm on $\widehat{\bigotimes}^m E$ induced by (1.1), and we omit the indices $\widehat{\bigotimes}^m E$ when no confusion can occur.

It is clear that the mapping

$$\{\varphi_1, \ldots, \varphi_m\} \to \varphi_1 \otimes \cdots \otimes \varphi_m$$

is an m-linear mapping from E into $\bigotimes^m E$ (or $\widehat{\bigotimes}^m E$), i.e, we have

$$(\alpha\varphi_1 + \alpha'\varphi_1') \otimes \varphi_2 \otimes \cdots \otimes \varphi_m = \alpha\varphi_1 \otimes \varphi_2 \otimes \cdots \otimes \varphi_m + \alpha'\varphi_1' \otimes \varphi_2 \otimes \varphi_m, \tag{1.2}$$

$\forall \varphi_1, \varphi_1', \varphi_2, \ldots, \varphi_m \in E$, $\alpha, \alpha' \in \mathbb{R}$, and similar relations for the other indices.

If $\{e_i\}_{i \in I}$ is an orthonormal basis of E, then the products $e_{i_1} \otimes \cdots \otimes e_{i_m}$, $i_1, \ldots, i_m \in I$, constitute an orthornormal basis of $\bigotimes^m E$, and if $\{e_i\}_{i \in I}$ is a Hilbert orthonormal basis of E, then the products $e_{i_1} \otimes \cdots \otimes e_{i_m}$, $i_\alpha \in I$, constitute a Hilbert orthonormal basis of $\widehat{\bigotimes}^m E$.

If $E_1, \ldots, E_m$ are m Hilbert spaces we define, in a similar manner, their tensor products $E_1 \otimes \cdots \otimes E_m$ and $E_1 \hat{\otimes} \cdots \hat{\otimes} E_m$. These tensor products (and the tensor product $\varphi_1 \otimes \cdots \otimes \varphi_m$ of elements φ_i of E_i) are associative.

We then consider the Hilbert m-exterior product of E, $\widehat{\bigwedge}^m E$. In order to define this space we first consider the m-exterior product $\bigwedge^m E$ of E

$$\bigwedge^m E = \underbrace{E \wedge \cdots \wedge E}_{m \text{ times}}.$$

This is the subspace of $\bigotimes^m E$ spanned by all the sums

$$\sum_\sigma (-1)^\sigma \varphi_{\sigma(1)} \otimes \cdots \otimes \varphi_{\sigma(m)}, \tag{1.3}$$

where $\varphi_1, \ldots, \varphi_m \in E$, σ is a permutation of $\{1, \ldots, m\}$, and the sum in (1.3) is extended to all such permutations ($(-1)^\sigma$ is the sign of σ). The expression in (1.3) is denoted by $\varphi_1 \wedge \cdots \wedge \varphi_m$ and is called the exterior product of $\varphi_1, \ldots, \varphi_m$. The usual multilinear rules follow readily from (1.2)

$$(\alpha\varphi_1 + \alpha'\varphi_1') \wedge \varphi_2 \wedge \cdots \wedge \varphi_m$$
$$= \alpha\varphi_1 \wedge \varphi_2 \wedge \cdots \wedge \varphi_m + \alpha'\varphi_1' \wedge \varphi_2 \wedge \cdots \wedge \varphi_m, \tag{1.4}$$

$\forall \varphi_1, \varphi_1', \varphi_2, \ldots, \varphi_m \in E$, $\alpha, \alpha' \in \mathbb{R}$, and similar relations for the other indices. The usual exterior multilinear rules also follow from (1.3).

$$\varphi_1 \wedge \cdots \wedge \varphi_m = 0 \quad \text{if } \varphi_i = \varphi_j \quad \text{for } i \neq j, \tag{1.5}$$

which implies

$$\varphi_1 \wedge \cdots \wedge \varphi_m \text{ changes to its opposite if we permute } \varphi_i \text{ and } \varphi_j. \tag{1.6}$$

We define on $\bigwedge^m E$ an inner product $(\cdot, \cdot)_{\bigwedge^m E}$ by first defining the inner product of two exterior products $\varphi_1 \wedge \cdots \wedge \varphi_m, \psi_1 \wedge \cdots \wedge \psi_m$,

$$\begin{aligned}(\varphi_1 \wedge \cdots \wedge \varphi_m, \psi_1 \wedge \cdots \wedge \psi_m)_{\bigwedge^m E} \\ = \det\{(\varphi_i, \psi_j)_E\}_{1 \leq i,j \leq m}, \qquad \forall \varphi_1, \ldots, \varphi_m, \psi_1, \ldots, \psi_m \in E,\end{aligned} \tag{1.7}$$

and then extending it to $\bigwedge^m E$ by linearity. It can be shown that the inner product $(\xi, \eta)_{\bigwedge^m E}$ is well defined, i.e., $(\xi, \eta)_{\bigwedge^m E}$ does not depend on the representation of ξ and η as sums of exterior products, and that it is positive definite. The space $\hat{\bigwedge}^m E$ is the completion of $\bigwedge^m E$ under the associated norm $\{(\xi, \eta)_{\bigwedge^m E}\}^{1/2}$.

We denote by $(\cdot, \cdot)_{\hat{\bigwedge}^m E}$ and $|\cdot|_{\hat{\bigwedge}^m E}$ the scalar product and the norm on $\hat{\bigwedge}^m E$ induced by (1.7).[1]

Let $\{e_i\}_{i \in I}$ be an orthonormal basis of E for which we assume that the family of indices I is totally ordered.[2] We recall that the family of exterior products

$$e_{i_1} \wedge \cdots \wedge e_{i_m}, \qquad i_1 < \cdots < i_m, \qquad i_\alpha \in I, \tag{1.8}$$

constitutes an orthonormal basis of $\bigwedge^m E$. Similarly, if $\{e_i\}_{i \in I}$ is a Hilbert orthonormal basis of E, then the family of exterior products (1.8) constitutes a Hilbert orthonormal basis of $\hat{\bigwedge}^m E$. For instance, if $\varphi_1, \ldots, \varphi_m \in E$, $\varphi_i = \sum_{j_i \in I} a_{ij_i} e_{j_i}$, $i = 1, \ldots, m$, $I \subset \mathbb{N}$, then by (1.4)–(1.6)

$$\begin{aligned}\varphi_1 \wedge \cdots \wedge \varphi_m &= \left(\sum_{j_1 \in I} a_{ij_1} e_{j_1}\right) \wedge \cdots \wedge \left(\sum_{j_m \in I} a_{ij_m} e_{j_m}\right) \\ &= \sum_{i_1 < \cdots < i_m} \kappa_{i_1 \ldots i_m} e_{i_1} \wedge \cdots \wedge e_{i_m},\end{aligned} \tag{1.9}$$

where the sum is extended to all families $i_1 < \cdots < i_m$, $i_\alpha \in I$, and $\kappa_{i_1 \ldots i_m}$ is the determinant of the $m \times m$ matrix of elements

$$a_{\alpha\beta}, \qquad \alpha = 1, \ldots, m, \quad \beta = i_i, \ldots, i_m. \tag{1.10}$$

We recall that the exterior product $\varphi_1 \wedge \cdots \wedge \varphi_m$ is different from 0 if and

[1] Of course, (1.7) also reads as

$$\begin{aligned}(\varphi_1 \wedge \cdots \wedge \varphi_m, \psi_1 \wedge \cdots \wedge \psi_m)_{\hat{\bigwedge}^m E} &= (\varphi_1 \wedge \cdots \wedge \varphi_m, \psi_1 \wedge \cdots \wedge \psi_m)_{\bigwedge^m E} \\ &= \det\{(\varphi_i, \psi_j)_E\}_{1 \leq i,j \leq n}, \qquad \forall \varphi_1, \ldots, \varphi_m, \psi_1, \ldots, \psi_m \in E.\end{aligned}$$

[2] We are especially interested in the case $I = \mathbb{N}$ (or $I \subset \mathbb{N}$ finite when E is finite dimensional).

only if the family $\varphi_1, \ldots, \varphi_m$ is linearly independent in E. If $\varphi_1, \ldots, \varphi_m$ and $\psi_1, \ldots, \psi_m$ are two linearly independent families of elements of E, then the m-dimensional linear space spanned by $\varphi_1, \ldots, \varphi_m$ and by $\psi_1, \ldots, \psi_m$ are the same if and only if the exterior products $\varphi_1 \wedge \cdots \wedge \varphi_m$ and $\psi_1 \wedge \cdots \wedge \psi_m$ are proportional.

Remark 1.1. More generally, we can define the exterior product $\varphi \wedge \psi \in \bigwedge^{m+n} E$ when $\varphi \in \bigwedge^m E$ and $\psi \in \bigwedge^n E$; its expression as an $(m+n)$-linear form is given by

$$(\varphi \wedge \psi)(\theta_1, \ldots, \theta_{m+n})$$
$$= \sum{}' (-1)^\sigma \varphi(\theta_{\sigma(1)}, \ldots, \theta_{\sigma(m)})\psi(\theta_{\sigma(m+1)}, \ldots, \theta_{\sigma(m+n)}), \qquad \forall \theta_1, \ldots, \theta_{m+n} \in E, \tag{1.11}$$

where the sum $\sum'$ extends to all permutations σ of $\{1, \ldots, m+n\}$ such that $\sigma(1) < \cdots < \sigma(m)$ and $\sigma(m+1) < \cdots < \sigma(m+n)$. The basic properties (1.4)–(1.6) of the product $\bigwedge$ extends as follows:

$$\bigwedge \text{ is bilinear continuous on } \bigwedge^m E \times \bigwedge^n E, \tag{1.12}$$

$$\alpha \wedge \beta = (-1)^{mn} \beta \wedge \alpha, \tag{1.13}$$

$$\alpha \wedge (\beta \wedge \gamma) = (\alpha \wedge \beta) \wedge \gamma. \tag{1.14}$$

Due to (1.12), $\bigwedge$ can be extended by continuity as a bilinear continuous operator on $\hat{\bigwedge}^m E \times \hat{\bigwedge}^n E$ which enjoys the same properties (1.13), (1.14). The direct sum of the spaces $\bigwedge^m E$, $m \in \mathbb{N}$, i.e.,

$$\bigwedge E = (\bigwedge^0 E = \mathbb{R}) \oplus (\bigwedge^1 E = E) \oplus \bigwedge^2 E \oplus \cdots,$$

endowed with its structure as a real vector space and with the multiplication induced by $\bigwedge$, is the exterior algebra of E or the Grassman algebra of E. Similarly, we can consider the Hilbert direct sum

$$\hat{\bigwedge} E = \hat{\bigwedge}^0 E \oplus \hat{\bigwedge}^1 E \oplus \hat{\bigwedge}^2 E \oplus \cdots,$$

which is both an algebra and a Hilbert space.

For more details and the proof of the results recalled above, the reader is referred to R. Abraham and J.E. Marsden [1], N. Bourbaki [1], [2], or L. Schwartz [3].

Of particular interest to us will be the relation between exterior products and volumes. If $\varphi_1, \ldots, \varphi_m \subset E$, the norm of $\varphi_1 \wedge \cdots \wedge \varphi_m$ in $\bigwedge^m E$ is the m-dimensional volume of the parallelepiped generated by $\varphi_1, \ldots, \varphi_m$, i.e, the set of points $\lambda_1\varphi_1 + \cdots + \lambda_m\varphi_m$, $0 \le \lambda_i \le 1$, $\forall i$. Indeed, according to (1.9), (1.10), if $e_1, \ldots, e_m$ is an orthonormal basis in the m-space spanned by $\varphi_1, \ldots, \varphi_m$ (assuming $\varphi_1 \wedge \cdots \wedge \varphi_m \neq 0$), then $\varphi_i = \sum_{j=1}^m a_{ij} e_j$, $1 \le i \le m$, and

$$\varphi_1 \wedge \cdots \wedge \varphi_m = \det(a_{ij})_{1 \le i,j \le m}(e_1 \wedge \cdots \wedge e_m)$$

and since $|e_1 \wedge \cdots \wedge e_m|_{\bigwedge^m E} = 1$

$$|\varphi_1 \wedge \cdots \wedge \varphi_m|_{\bigwedge^m E} = |\det(a_{ij})_{1 \le i,j \le m}|. \tag{1.15}$$

The norm $|\varphi_1 \wedge \cdots \wedge \varphi_m|_{\bigwedge^m E}$ is majorized by the products of the norms of the φ_i

$$|\varphi_1 \wedge \cdots \wedge \varphi_m|_{\bigwedge^m E} \le |\varphi_1|_E \cdots |\varphi_m|_E. \tag{1.16}$$

For example, if $m = 2$

$$\begin{aligned} |\varphi_1 \wedge \varphi_2|^2_{\bigwedge^m E} &= \begin{vmatrix} (\varphi_1, \varphi_1)_E & (\varphi_1, \varphi_2)_E \\ (\varphi_2, \varphi_1)_E & (\varphi_2, \varphi_2)_E \end{vmatrix} \\ &= \sin^2 \theta |\varphi_1|^2_E |\varphi_2|^2_E, \end{aligned}$$

where θ is the angle of the vectors φ_1, φ_2

$$\cos \theta = \frac{(\varphi_1, \varphi_2)_E}{|\varphi_1|_E |\varphi_2|_E}.$$

Similar (and more complicated) relations hold for $m \ge 3$; they are proved by induction on m.

1.2. Multilinear Operators and Exterior Products

Again let E be a Hilbert space which, for simplicity, we assume to be separable and let L be a linear continuous operator on E. We can associate with L several multilinear continuous operators on E; two of them will be of particular interest to us, namely, $\bigwedge^m L$ and another operator denoted L_m. They are studied respectively in Sections 1.2.1 and 1.2.2.

1.2.1. The Operator $\bigwedge^m L$

We first define the operator $\bigwedge^m L$ as an m-linear continuous operator from E^m into $\bigwedge^m E$ by setting

$$\left(\bigwedge^m L\right)(\varphi_1, \ldots, \varphi_m) = L\varphi_1 \wedge \cdots \wedge L\varphi_m, \qquad \forall \varphi_1 \wedge \cdots \wedge \varphi_m \in E. \tag{1.17}$$

This operator is clearly m-linear and its continuity follows from (1.15)

$$\begin{aligned} |L\varphi_1 \wedge \cdots \wedge L\varphi_m|_{\bigwedge^m E} &\le |L\varphi_1|_E \cdots |L\varphi_m|_E \\ &\le \|L\|^m_{\mathscr{L}(E)} |\varphi_1|_E \cdots |\varphi_m|_E, \qquad \forall \varphi_1, \ldots, \varphi_m \in E. \end{aligned} \tag{1.18}$$

We define the norm of $\bigwedge^m L$ as an m-linear operator from E^m into $\bigwedge^m E$ by

$$\left\|\bigwedge^m L\right\|_{\mathscr{L}_m(E^m, \bigwedge^m E)} = \sup_{\substack{\varphi_1, \ldots, \varphi_m \in E \\ |\varphi_i|_E \le 1, \forall i}} \left|\left(\bigwedge^m L\right)(\varphi_1, \ldots, \varphi_m)\right|_{\bigwedge^m E}, \tag{1.19}$$

and according to (1.18)

$$\left\|\bigwedge^m L\right\|_{\mathscr{L}_m(E^m, \bigwedge^m E)} \le \|L\|^m_{\mathscr{L}(E)}. \tag{1.20}$$

It is interesting to observe that $L\varphi_1 \wedge \cdots \wedge L\varphi_m$ depends only on $\varphi_1 \wedge \cdots \wedge \varphi_m$ and not on $\varphi_1, \ldots, \varphi_m$. Indeed, if $\varphi_1 \wedge \cdots \wedge \varphi_m = 0$, then as indicated above, the family $\varphi_1, \ldots, \varphi_m$ is not free in E, one of the vectors, say φ_m, is a linear combination of the others $\varphi_1, \ldots, \varphi_{m-1}$; then $L\varphi_m$ is also a linear combination of $L\varphi_1, \ldots, L\varphi_{m-1}$ and $L\varphi_1 \wedge \cdots \wedge L\varphi_m = 0$. Now if $\varphi_1 \wedge \cdots \wedge \varphi_m \neq 0$ and if $\psi_1, \ldots, \psi_m$ are m elements of E such that $\varphi_1 \wedge \cdots \wedge \varphi_m = \psi_1 \wedge \cdots \wedge \psi_m$, then as indicated above the m-space spanned by $\varphi_1, \ldots, \varphi_m$ is the same as that spanned by $\psi_1, \ldots, \psi_m$, and we can write

$$\psi_i = \sum_{j=1}^{m} a_{ij}\varphi_j, \qquad i = 1, \ldots, m. \tag{1.21}$$

Hence by the properties of exterior products

$$\begin{aligned}\psi_1 \wedge \cdots \wedge \psi_m &= \left(\sum_{j_1=1}^{m} a_{1j_1}\varphi_{j_1}\right) \wedge \cdots \wedge \left(\sum_{j_m=1}^{m} a_{mj_m}\varphi_{j_m}\right)\\ &= \det(a_{ij})(\varphi_1 \wedge \cdots \wedge \varphi_m),\end{aligned}$$

and, furthermore, the determinant of the matrix $(a_{ij})_{1 \le i,j \le m}$ is one. In such a case it follows readily from (1.17), (1.19) and relation $\det(a_{ij}) = 1$ that

$$\left(\textstyle\bigwedge^m L\right)(\psi_1, \ldots, \psi_m) = \left(\textstyle\bigwedge^m L\right)(\varphi_1, \ldots, \varphi_m). \tag{1.22}$$

Thus $\bigwedge^m L$ induces a linear operator on the space $\bigwedge^m E$; this linear operator on $\bigwedge^m E$ is still denoted by $\bigwedge^m L$. The next lemma shows that $\bigwedge^m L$ is continuous on the normed space $\bigwedge^m E$ and can thus be extended by continuity as a linear continuous operator on $\bigwedge^m E$. Although the continuity of $\bigwedge^m L$ on $\bigwedge^m E$ can be proved more simply than in Lemma 1.1 (see N. Bourbaki [1]), we point out that the proof of Lemma 1.1 provides some additional information which will be useful in the sequel.

Before stating Lemma 1.1 we introduce the following notation: for every $L \in \mathscr{L}(E)$ and for every integer m we write

$$\alpha_m(L) = \sup_{\substack{F \subset E\\ \dim F = m}} \ \operatorname*{Inf}_{\substack{\varphi \in F\\ |\varphi|_E = 1}} |L\varphi|_E, \tag{1.23}$$

and

$$\omega_m(L) = \alpha_1(L) \ldots \alpha_m(L). \tag{1.24}$$

It is easy to see that the sequence $\alpha_m(L)$ is nonincreasing and that $\alpha_1(L) = \|L\|_{\mathscr{L}(E)}$ is the norm of L in $\mathscr{L}^m(E)$; further properties of the numbers $\alpha_m(L)$ will appear in Section 1.3.

Lemma 1.1. *If $L \in \mathscr{L}(E)$, the linear operator $\bigwedge^m L$ from $\bigwedge^m E$ into itself defined by*

$$\left(\textstyle\bigwedge^m L\right)(\varphi_1 \wedge \cdots \wedge \varphi_m) = L\varphi_1 \wedge \cdots \wedge L\varphi_m, \qquad \forall \varphi_1, \ldots, \varphi_m \in E,$$

is continuous on $\bigwedge^m E$ for the norm $|\cdot|_{\bigwedge^m E}$. It can be extended by continuity as *a linear continuous operator in* $\hat{\bigwedge}^m E$ *and*

$$\|\bigwedge^m L\|_{\mathscr{L}(\bigwedge^m E)} \le \omega_m(L). \tag{1.25}$$

PROOF. Let $\{e_i\}_{i \in I}$ be an orthonormal basis of E and let φ be an element of $\bigwedge^m E$

$$\varphi = \sum_{i_1 < \cdots < i_m} a_{i_1 \ldots i_m} e_{i_1} \wedge \cdots \wedge e_{i_m}. \tag{1.26}$$

The sum in (1.26) is finite and we denote by M the largest index i_α appearing in this sum, and by F_M the space spanned by $e_1, \ldots, e_M$.

The quadratic form $\psi \in F_m \to |L\psi|_E^2$ is well defined, continuous, and non-negative on F_M. There exists therefore an orthonormal basis $\psi_1, \ldots, \psi_M$ of F_M consisting of the eigenvectors of this quadratic form

$$(L\psi_i, L\psi_j)_E = r_i \delta_{ij}, \tag{1.27}$$

where $r_1 \ge \cdots \ge r_M$ are the eigenvalues associated with $\psi_1, \ldots, \psi_M$. The formula of change of basis are standard

$$e_i = \sum_{j=1}^{M} a_{ij} \psi_j,$$

$$e_{i_1} \wedge \cdots \wedge e_{i_m} = \sum_{j_1 < \cdots < j_m} \kappa_{j_1 \ldots j_m}^{i_1 \ldots i_m} \psi_{j_1} \wedge \cdots \wedge \psi_{j_m}, \qquad \forall i_1 < \cdots < i_m \, (\le M),$$

where, according to (1.10), $\kappa_{j_1 \ldots j_m}^{i_1 \ldots i_m}$ is the determinant of the matrix $(a_{\alpha\beta})$, $\alpha = i_1, \ldots, i_m$, $\beta = j_1, \ldots, j_m$. As mentioned above, the family $e_{i_1} \wedge \cdots \wedge e_{i_m}$, $i_1 < \cdots < i_m \le M$, is orthonormal in $\bigwedge^m E$ and the same is true for the family $\psi_{j_1} \wedge \cdots \wedge \psi_{j_m}, j_1 < \cdots < j_m \le M$. Thus

$$|\varphi|^2_{\bigwedge^m E} = \sum_{i_1 < \cdots < i_m} |a_{i_1 \ldots i_m}|^2$$

$$|e_{i_1} \wedge \cdots \wedge e_{i_m}|^2_{\bigwedge^m E} = 1 = \sum_{j_1 < \cdots < j_m} |\kappa_{j_1 \ldots j_m}^{i_1 \ldots i_m}|^2.$$

Now

$$\begin{aligned}(\bigwedge^m L)(\varphi) &= \sum_{i_1 < \cdots < i_m} a_{i_1 \ldots i_m} L e_{i_1} \wedge \cdots \wedge L e_{i_m} \\ &= \sum_{i_1 < \cdots < i_m} \sum_{j_1 < \cdots < j_m} a_{i_1 \ldots i_m} \kappa_{j_1 \ldots j_m}^{i_1 \ldots i_m} L\psi_{j_1} \wedge \cdots \wedge L\psi_{j_m}.\end{aligned}$$

Due to (1.7) and (1.27), the family $L\psi_{j_1} \wedge \cdots \wedge L\psi_{j_m}$, $j_1 < \cdots < j_m$, is orthogonal in $\bigwedge^m E$ and

$$|L\psi_{j_1} \wedge \cdots \wedge L\psi_{j_m}|^2_{\bigwedge^m E} = |L\psi_{j_1}|_E^2 \ldots |L\psi_{j_m}|_E^2 = r_{j_1} \ldots r_{j_m}, \qquad \forall j_1 < \cdots < j_m.$$

Hence

$$\begin{aligned}|(\bigwedge^m L)(\varphi)|^2_{\bigwedge^m E} &= \sum_{i_1 < \cdots < i_m} \sum_{j_1 < \cdots < j_m} |a_{i_1 \ldots i_m} \kappa_{j_1 \ldots j_m}^{i_1 \ldots i_m}| r_{j_1} \ldots r_{j_m} \\ &\le r_1 \ldots r_m \sum_{i_1 < \cdots < i_m} \sum_{j_1 < \cdots < j_m} |a_{i_1 \ldots i_m} \kappa_{j_1 \ldots j_m}^{i_1 \ldots i_m}|^2 \\ &\le r_1 \ldots r_m |\varphi|^2_{\bigwedge^m E}.\end{aligned}$$

By the classical max–min properties of the eigenvalues of a symmetric operator in finite dimension we have

$$r_j = \operatorname*{Max}_{\substack{G \subset F_m \\ \dim G = j}} \ \operatorname*{Min}_{\substack{\theta \in G \\ |\theta|_{F_M} = 1}} |L\theta|^2, \qquad j = 1, \ldots, M \tag{1.28}$$

(see, for instance, R. Courant and D. Hilbert [1]). Thus

$$r_j = \operatorname*{Sup}_{\substack{G \subset F_m \\ \dim G = j}} \ \operatorname*{Inf}_{\substack{\theta \in G \\ |\theta|_E = 1}} |L\theta|^2 = \alpha_j(L)^2, \tag{1.29}$$

and

$$|(\textstyle\bigwedge^m L)(\varphi)|_{\bigwedge^m E} \le \alpha_1(L) \ldots \alpha_m(L) |\varphi|_{\bigwedge^m E}, \qquad \forall \varphi \in \textstyle\bigwedge^m E. \tag{1.30}$$

All the statements in Lemma 1.1 follow immediately from (1.30) and (1.24).

□

Remark 1.2. If $\varphi_1, \ldots, \varphi_m \in E$ and $|\varphi_j|_E \le 1$, $\forall j$, we can write

$$\begin{aligned} |(\textstyle\bigwedge^m L)(\varphi_1, \ldots, \varphi_m)|_{\bigwedge^m E} &= |L\varphi_1 \wedge \cdots \wedge L\varphi_m|_{\bigwedge^m E} \\ &\le \|\textstyle\bigwedge^m L\|_{\mathscr{L}(\bigwedge^m E)} |\varphi_1 \wedge \cdots \wedge \varphi_m|_{\bigwedge^m E} \\ &\le (\text{with } (1.16) \le \|\textstyle\bigwedge^m L\|_{\mathscr{L}(\bigwedge^m E)}. \end{aligned}$$

Therefore using (1.25) we can majorize the norm of $\bigwedge^m L$ considered as an m-linear operator from E^m into $\bigwedge^m E$ by

$$\|\textstyle\bigwedge^m L|_{\mathscr{L}_m(E^m, \bigwedge^m E)} \le \|\textstyle\bigwedge^m L\|_{\mathscr{L}(\bigwedge^m E)} \le \omega_m(L). \tag{1.31}$$

This statement will be improved in Section 1.3 where it will be shown that the three quantities in (1.31) are actually equal.

1.2.2. The Operator L_m

We now define the operator L_m as an m-linear continuous operator from E^m into $\bigwedge^m E$. Given $\varphi_1, \ldots, \varphi_m \in E$ we set

$$\begin{aligned} L_m(\varphi_1, \ldots, \varphi_m) = L\varphi_1 \wedge \varphi_2 \wedge \cdots \wedge \varphi_m + \varphi_1 \wedge L\varphi_2 \wedge \cdots \wedge \varphi_m + \cdots \\ + \varphi_1 \wedge \cdots \wedge \varphi_{m-1} \wedge L\varphi_m. \end{aligned} \tag{1.32}$$

Using (1.16) we see that

$$|L_m(\varphi_1, \ldots, \varphi_m)|_{\bigwedge^m E} \le m \|L\|_{\mathscr{L}(E)} |\varphi_1|_E \ldots |\varphi_m|_E, \qquad \forall \varphi_1, \ldots, \varphi_m \in E,$$

hence L_m is continuous from E^m into $\bigwedge^m E$, and its norm $\|L_m\|$ defined as in (1.19) satisfies

$$\|L_m\| \le m \|L\|_{\mathscr{L}(E)}. \tag{1.33}$$

As in the previous case we can show that $L_m(\varphi_1, \ldots, \varphi_m)$ depends only on $\varphi_1 \wedge \cdots \wedge \varphi_m$ and not on $\varphi_1, \ldots, \varphi_m$. For that purpose, we first observe that

$$L_m(\varphi_1, \ldots, \varphi_m) = 0 \quad \text{if } \varphi_i = \varphi_j \quad \text{for } i \ne j. \tag{1.34}$$

Indeed, in this case, all the exterior products on the right-hand side of (1.32) vanish except (assuming $i < j$)

$$\varphi_1 \wedge \cdots \wedge L\varphi_i \wedge \cdots \wedge \varphi_j \wedge \cdots \wedge \varphi_m \quad \text{and} \quad \varphi_1 \wedge \cdots \wedge \varphi_i \wedge \cdots \wedge L\varphi_j \wedge \cdots \wedge \varphi_m,$$

which are opposite in direction and whose sum vanishes. Now if $\varphi_1 \wedge \cdots \wedge \varphi_m = 0$, then one of the vectors, say φ_m, is a linear combination of the others, $\varphi_1, \ldots, \varphi_{m-1}$, and it follows immediately from (1.34) that $L_m(\varphi_1, \ldots, \varphi_m) = 0$. More generally, if $\psi_1, \ldots, \psi_m$ are m other elements of E such that $\psi_1 \wedge \cdots \wedge \psi_m = \varphi_1 \wedge \cdots \wedge \varphi_m$, then the above relation (1.21) holds again with $\det(a_{ij}) = 1$; using (1.21), the multilinearity of L_m and the relation $\det(a_{ij}) = 1$, we find that

$$L_m(\psi_1, \ldots, \psi_m) = L_m(\varphi_1, \ldots, \varphi_m). \tag{1.35}$$

Thus L_m induces a linear operator (still denoted by L_m) on the exterior product $\bigwedge^m E$. Except for the case (that we will consider in Lemma 3.1) where L is self-adjoint, we do not know if this operator is continuous for the norm of $\bigwedge^m E$ and if it can be extended to the space $\widehat{\bigwedge}^m E$. This is not important for our purposes, and for the moment we notice the following lemma of algebraic nature which will be particularly useful in the sequel.

Lemma 1.2. *For every family of elements $\varphi_1, \ldots, \varphi_m$ of E,*

$$(L_m(\varphi_1, \ldots, \varphi_m), \varphi_1 \wedge \cdots \wedge \varphi_m)_{\bigwedge^m E} = |\varphi_1 \wedge \cdots \wedge \varphi_m|^2_{\bigwedge^m E} \operatorname{Tr}(L \circ Q), \tag{1.36}$$

where Q is the orthogonal projector in E onto the space spanned by $\varphi_1, \ldots, \varphi_m$, and $\operatorname{Tr}(L \circ Q)$ *is the trace of the linear operator (of finite rank) $L \circ Q$.*

Proof. It suffices to consider the case where $\varphi_1, \ldots, \varphi_m$ are linearly independent, $\varphi_1 \wedge \cdots \wedge \varphi_m \neq 0$ and, without loss of generality, we can assume that $\varphi_1, \ldots, \varphi_m$ is an orthogonal family of E. If $\varphi_1, \ldots, \varphi_m$ is not orthogonal we can consider the orthogonal family $\psi_1, \ldots, \psi_m$ associated with $\varphi_1, \ldots, \varphi_m$, by the usual Gram–Schmidt process. Since

$$\psi_j \in \varphi_j + \operatorname{Span}\{\varphi_1, \ldots, \varphi_{j-1}\}, \qquad j = 1, \ldots, m,$$

we have clearly $\psi_1 \wedge \cdots \wedge \psi_m = \varphi_1 \wedge \cdots \wedge \varphi_m$ and by the preceding observation $L_m(\psi_1, \ldots, \psi_m) = L_m(\varphi_1, \ldots, \varphi_m)$. Hence the relation (1.6) is equivalent to the same relation for the ψ_j's.

Assuming now that the family j is orthogonal in E, we find with (1.7) and (1.17) that

$$|\varphi_1 \wedge \cdots \wedge \varphi_m|^2_{\bigwedge^m E} = |\varphi_1|^2_E \cdots |\varphi_m|^2_E,$$

$$\begin{aligned}
&(L_m(\varphi_1, \ldots, \varphi_m), \varphi_1 \wedge \cdots \wedge \varphi_m)_{\bigwedge^m E} \\
&\quad = (L\varphi_1 \wedge \varphi_2 \wedge \cdots \wedge \varphi_m, \varphi_1 \wedge \cdots \wedge \varphi_m) \cdots \\
&\qquad + (\varphi_1 \wedge \cdots \wedge \varphi_{m-1} \wedge L\varphi_m, \varphi_1 \wedge \cdots \wedge \varphi_m) \\
&\quad = (L\varphi_1, \varphi_1)_E |\varphi_2|^2_E \cdots |\varphi_m|^2_E + \cdots \\
&\qquad + |\varphi_1|^2_E \cdots |\varphi_{m-1}|^2_E (L\varphi_m, \varphi_m)_E
\end{aligned}$$

$$= |\varphi_1|_E^2 \dots |\varphi_m|_E^2 \cdot \sum_{j=1}^{m} \left(L \frac{\varphi_j}{|\varphi_j|_E}, \frac{\varphi_j}{|\varphi_j|_E} \right)_E$$
$$= |\varphi_1 \wedge \cdots \wedge \varphi_m|^2_{\bigwedge^m E} \cdot \mathrm{Tr}(L \circ Q).$$

The proof is complete. □

As indicated above we can show that if $L \in \mathscr{L}(E)$ is self-adjoint, then L_m is linear continuous on $\bigwedge^m E$ and can be extended to $\hat{\bigwedge}^m E$.

Lemma 1.3. *If $L \in \mathscr{L}(E)$ is self-adjoint, the linear operator L_m from $\bigwedge^m E$ into itself defined by*

$$L_m(\varphi_1 \wedge \cdots \wedge \varphi_m) = L\varphi_1 \wedge \varphi_2 \wedge \cdots \wedge \varphi_m + \cdots + \varphi_1 \wedge \cdots \wedge \varphi_{m-1} \wedge L\varphi_m,$$
$$\forall \varphi_1, \dots, \varphi_m \in E,$$

is self-adjoint and continuous on $\bigwedge^m E$ for the norm $|\cdot|_{\bigwedge^m E}$. It can be extended by continuity as a linear continuous operator in $\hat{\bigwedge}^m E$ and

$$\|L_m\|_{\mathscr{L}(\hat{\bigwedge}^m E)} \leq \alpha_1(L) + \cdots + \alpha_m(L). \tag{1.37}$$

Proof. In order to prove that L_m is self-adjoint in $\bigwedge^m E$, it suffices to show that

$$(L_m(\varphi_1 \wedge \cdots \wedge \varphi_m), \psi_1 \wedge \cdots \wedge \psi_m)_{\bigwedge^m E}$$
$$= (\varphi_1 \wedge \cdots \wedge \varphi_m), L_m(\psi_1 \wedge \cdots \wedge \psi_m))_{\bigwedge^m E}$$

for any two families of vectors $\varphi_1, \dots, \varphi_m, \psi_1, \dots, \psi_m$ of E and this is shown as follows:

$$(L_m(\varphi_1 \wedge \cdots \wedge \varphi_m), \psi_1 \wedge \cdots, \psi_m)_{\bigwedge^m E}$$
$$= (L\varphi_1 \wedge \varphi_2 \wedge \cdots \wedge \varphi_m + \varphi_1 \wedge L\varphi_2 \wedge \cdots \wedge \varphi_m + \cdots$$
$$+ \cdots + \varphi_1 \wedge \cdots \wedge L\varphi_m, \psi_1 \wedge \cdots \wedge \psi_m)_{\bigwedge^m E}$$
$$= \begin{bmatrix} (L\varphi_1, \psi_1) & (L\varphi_1, \psi_2) & \cdots & (L\varphi_1, \psi_m) \\ (\varphi_2, \psi_1) & (\varphi_2, \psi_2) & \cdots & (\varphi_2, \psi_m) \\ \vdots & \vdots & & \vdots \\ (\varphi_m, \psi_1) & (\varphi_m, \psi_2) & \cdots & (\varphi_m, \psi_m) \end{bmatrix} + \cdots$$
$$= (\text{L self-adjoint})$$
$$= \begin{bmatrix} (\varphi_1, L\psi_1) & (\varphi_1, L\psi_2) & \cdots & (\varphi_1, L\psi_m) \\ (\varphi_2, \psi_1) & (\varphi_2, \psi_2) & \cdots & (\varphi_2, \psi_m) \\ \vdots & \vdots & & \vdots \\ (\varphi_m, \psi_1) & (\varphi_m, \psi_2) & \cdots & (\varphi_m, \psi_m) \end{bmatrix} + \cdots$$
$$= (\varphi_1 \wedge \cdots \wedge \varphi_m, L\psi_1 \wedge \psi_2 \wedge \cdots \wedge \psi_m) + \cdots$$
$$= (\varphi_1 \wedge \cdots \wedge \varphi_m, L_m(\psi_1 \wedge \cdots \wedge \psi_m))_{\bigwedge^m E}.$$

For the continuity of L_m in $\bigwedge^m E$, we proceed as in Lemma 1.1: we consider an element φ of the form (1.26), the corresponding space F_M, and the eigenvectors φ_j, $j = 1, \ldots, M$, of the quadratic form $|L\psi|_E^2$. Now since L is self-adjoint we have, besides (1.27),

$$(L\psi_i, \psi_j)_E = r_i^{1/2}\delta_{ij}, \qquad i, j = 1, \ldots, M. \tag{1.38}$$

Then

$$\begin{aligned} L_m(\varphi) &= \sum_{i_1 < \cdots < i_m} a_{i_1 \ldots i_m} L_m(e_{i_1} \wedge \cdots \wedge e_{i_m}) \\ &= \sum_{i_1 < \cdots < i_m} \sum_{j_1 < \cdots < j_m} a_{i_1 \ldots i_m} \kappa_{j_1 \ldots j_m}^{i_1 \ldots i_m} L_m(\psi_{j_1} \wedge \cdots \wedge \psi_{j_m}). \end{aligned}$$

It is easy to see with (1.7) and (1.32) that two vectors $L_m(\psi_{i_1} \wedge \cdots \wedge \psi_{i_m})$ and $L_m(\psi_{j_1} \wedge \cdots \wedge \psi_{j_m})$ are orthogonal in $\bigwedge^m E$ unless $i_1 = j_1, \ldots, i_m = j_m$; also, thanks to (1.38),

$$\begin{aligned} |L_m(\psi_{i_1} \wedge \cdots \wedge \psi_{i_m})|_{\bigwedge^m E}^2 &= (r_{i_1}^{1/2} + \cdots + r_{i_m}^{1/2})^2 \\ &\leq (r_1^{1/2} + \cdots + r_m^{1/2})^2. \end{aligned} \tag{1.39}$$

Therefore

$$\begin{aligned} |L_m(\varphi)|_{\bigwedge^m E}^2 &= \sum_{i_1 < \cdots < i_m} \sum_{j_1 < \cdots < j_m} (a_{i_1 \ldots i_m} \kappa_{j_1 \ldots j_m}^{i_1 \ldots i_m})(r_{i_1}^{1/2} + \cdots + r_{i_m}^{1/2})^2 \\ &\leq (r_1^{1/2} + \cdots + r_m^{1/2})^2 |\varphi|_{\bigwedge^m E}^2 \\ &\leq \text{(with (1.29))} \\ &\leq (\alpha_1(L) + \cdots + \alpha_m(L))^2 |\varphi|_{\bigwedge^m E}^2, \qquad \forall \varphi \in \textstyle\bigwedge^m E. \end{aligned}$$

The lemma follows easily from this inequality. □

Remark 1.3. Under the assumptions of Lemma 1.3, if $\varphi_1, \ldots, \varphi_m$ is an orthonormal family in E, then $|\varphi_1 \wedge \cdots \wedge \varphi_m|_{\bigwedge^m E} = 1$, and using (1.36) we find

$$\begin{aligned} \operatorname{Tr}(L \circ Q) &= (L_m(\varphi_1 \wedge \cdots \wedge \varphi_m), \varphi_1 \wedge \cdots \wedge \varphi_m)_{\bigwedge^m E} \\ &\leq |L_m(\varphi_1 \wedge \cdots \wedge \varphi_m)|_{\bigwedge^m E} \\ &\leq \text{(with Lemma 1.3)} \\ &\leq \alpha_1(L) + \cdots + \alpha_m(L). \end{aligned}$$

Of course, since L is self-adjoint, $\alpha_1, \ldots, \alpha_m$ are the largest m eigenvalues of L (see Section 1.3 below).

Remark 1.4. If L is a linear unbounded operator in E with domain $D(L)$, we can define exactly as above the operators $\bigwedge^m L$ and L_m operating in $\bigwedge^m D(L) \subset \bigwedge^m E$. Formula (1.36) is still valid when $\varphi_1, \ldots, \varphi_m \in D(L)$ but, of course, Lemmas 1.1 and 1.3 do not apply as stated. However, the methods of Lemmas 1.1 and 1.3 can be used to show that $\bigwedge^m L$ (and L_m when L is self-adjoint) can be extended as a linear continuous operator from $\bigwedge^m D(L)$ into $\bigwedge^m E$.

1.3. Image of a Ball by a Linear Operator

Let L be a linear continuous operator in the Hilbert space E. Our aim in Section 1.3 is to describe the image by L of the unit ball B of E.

1.3.1. The Compact Case

We start with the simple case where the operator L is compact. In this case, L^*L is a compact self-adjoint nonnegative operator in E; we can define its square root $(L^*L)^{1/2}$ which enjoys the same properties. Therefore there exists an orthonormal basis of E, $\{e_i\}_{i\in I}$, which consists of eigenvectors of $(L^*L)^{1/2}$ and we denote by $\alpha_i = \alpha_i(L)$ the corresponding eigenvalues

$$\begin{cases}\alpha_1(L) \geq \alpha_2(L) \geq \cdots \geq 0, \\ (L^*L)^{1/2}e_i = \alpha_i e_i, \qquad \forall i.\end{cases} \tag{1.40}$$

This definition of the numbers $\alpha_i(L)$ agrees totally with (1.23) because of the classical min–max characterization of the eigenvalues of a compact self-adjoint operator, namely $(L^*L)^{1/2}$ (see, for instance, R. Courant and D. Hilbert [1]).

We observe that

$$(Le_i, Le_j)_E = (L^*Le_i, e_j)_E = \alpha_i^2(e_i, e_j)_E = \alpha_i^2\delta_{ij}, \tag{1.41}$$

so that the vectors Le_i are orthogonal and $|Le_i|_E = \alpha_i$, $Le_i \neq 0$, if and only if $\alpha_i \neq 0$. In this way we obtain an orthogonal decomposition of E into the space E_0 equal to the kernel of L and the space E_1 spanned by the vectors $Le_i \neq 0$ (i.e., $\alpha_i \neq 0$). Now, if φ is any vector of E, $\varphi = \sum_i a_i e_i$, then

$$L\varphi = \sum_i a_i Le_i = \sum_{\alpha_i>0} a_i\alpha_i \frac{Le_i}{\alpha_i},$$

and setting $\xi_i = a_i\alpha_i$ we see that φ belongs to the unit ball B of E if and only if

$$\sum_i a_i^2 \leq 1 \Leftrightarrow \sum_{\alpha_i>0}\left(\frac{\xi_i}{\alpha_i}\right)^2 \leq 1 \quad \text{and} \quad \xi_i = 0 \quad \text{when } \alpha_i = 0.$$

Thus

When L is compact $L(B)$ is the ellipsoid of the space E_1, whose axes are the vectors Le_i for the values of i for which $\alpha_i(L) > 0$, the length of the axes being the numbers $\alpha_i(L)$. (1.42)

1.3.2. The Noncompact Case

We now continue and consider the case where L is not compact. The operator $T = L^*L$ is positive, self-adjoint, and continuous but no longer compact, and in general we cannot find an orthonormal basis of E consisting of eigenvectors of T, We will use instead a spectral result proved in M. Reed and B. Simon

[1]. For that purpose, we introduce the sequence of numbers $\mu_n(T)$, $n \geq 1$, defined for any self-adjoint linear continuous operator T

$$\mu_n(T) = \operatorname*{Inf}_{\substack{F \subset E \\ \dim F \leq n-1}} \operatorname*{Sup}_{\substack{\varphi \in F^\perp \\ |\varphi|_E = 1}} (T\varphi, \varphi)_E. \tag{1.43}$$

The sequence $\mu_n(T)$ is nonincreasing and we can easily see that the definition of $\mu_n(T)$ is unchanged if we replace the infimum in (1.43) by the infimum for $F \subset E$, $\dim F = n - 1$. If T were compact then, according to the well known min–max principle (see R. Courant and D. Hilbert [1]), the $\mu_n(T)$ would be the eigenvalues of T. In the present case, some partial results are still valid and they are recalled in Proposition 1.1 hereafter (see Theorem XIII.1 of M. Reed and B. Simon [1] or N. Dunford and J.T. Schwartz [1], p. 1544). We set

$$\mu_\infty(T) = \lim_{n \to \infty} \mu_n(T) = \operatorname*{Inf}_{n \geq 1} \mu_n(T), \tag{1.44}$$

and we have

Proposition 1.1. *Let T be a linear bounded self-adjoint operator in a Hilbert space E. Then:*

(i) *If $\mu_n(T) > \mu_\infty(T)$ for some n, $\mu_1(T), \ldots, \mu_n(T)$ are n eigenvalues of T counting multiplicities.*
(ii) *The space E is the direct sum of two orthogonal spaces $E = E_v \oplus E_v^\perp$ stable for T;[1] E_v is spanned by an orthonormal family of eigenvectors e_i corresponding to the eigenvalues $\mu_i(T) > \mu_\infty(T)$, if any*

$$Te_i = \mu_i(T) e_i, \qquad \forall i.$$

Furthermore,

$$(T\varphi, \varphi)_E \leq \mu_\infty(T) |\varphi|_E^2, \qquad \forall \varphi \in E_v^\perp. \tag{1.45}$$

PROOF. (i) We first prove point (i) of the proposition and we proceed by induction on n.

We start the induction by assuming that $\mu_1 = \mu_1(T) > \mu_\infty(T)$, and we want to prove that μ_1 is an eigenvalue of T. We first show that μ_1 belongs to the spectrum of T, $\sigma(T)$; if this were not true, $\mu_1 - T$ would be invertible and $(\mu_1 - T)^{-1}$ would be bounded. By definition of μ_1, $\mu_1 - T$ is nonnegative and with the above properties we would have necessarily

$$\operatorname*{Inf}_{\substack{\varphi \in E \\ |\varphi|_E = 1}} ((\mu_1 - T)\varphi, \varphi) > 0$$

in contradiction with the definition of μ_1.

Let us now assume that μ_1 belongs to the spectrum of T but that $\mu_1 - T$

[1] One of these spaces can be the whole space E and the other one is then reduced to $\{0\}$.

is injective. For any subspace F of E of finite dimension, we set

$$\mu(F) = \operatorname*{Sup}_{\substack{\varphi \in F^\perp \\ |\varphi|_E = 1}} (T\varphi, \varphi)_E.$$

We claim that if $\mu_1 - T$ is injective then $\mu(F) = \mu_1$, $\forall F$. Indeed $\delta = \mu_1 - \mu(F) \geq 0$ and if $\delta > 0$ then

$$\operatorname*{Inf}_{\substack{\varphi \in F^\perp \\ |\varphi|_E = 1}} (A\varphi, \varphi)_E = \delta > 0,$$

where $A = \mu_1 - T$. Since A is invertible, the positiveness of δ implies that A is an isomorphism from $F^\perp$ onto $A(F^\perp)$. However, we know that A^{-1} is not continuous on $E(\mu_1 \in \sigma(T))$; hence there exists a sequence φ_n of elements of E such that

$$|\varphi_n|_E = 1, \quad \forall n, \quad \text{and} \quad |A\varphi_n|_E \to 0 \quad \text{as } n \to \infty.$$

We write $\varphi_n = \varphi_n^1 + \varphi_n^2$, with $\varphi_n^1 \in F^\perp$, $\varphi_n^2 \in F$. The sequences φ_n^1, φ_n^2, are bounded in $F^\perp$ and F; thus there exists a subsequence still denoted by n such that

$$\varphi_n^1 \to \varphi^1 \in F^\perp, \quad \text{weakly in } E,$$

$$\varphi_n^2 \to \varphi^2 \in F \quad \text{for the norm of } E.$$

Since $|A\varphi_n|_E \to 0$, we see that $A\varphi = A\varphi^1 + A\varphi^2 = 0$, and since A is injective, $\varphi^1 + \varphi^2 = 0$. Furthermore, we conclude from the preceding that $|A\varphi_n^1 - A\varphi^1|_E \to 0$ as $n \to \infty$, and since A is an isomorphism from $F^\perp$ onto its image, we find that $|\varphi_n^1 - \varphi^1|_E \to 0$ as $n \to \infty$; finally, φ_n converges to $\varphi = 0$ in the norm of E and this contradicts the fact that $|\varphi_n|_E = 1, \forall n$. We have thus proved that $A = \mu_1 - T$ is not injective and μ_1 is indeed an eigenvalue of T.

(ii) We complete the induction argument which proves point (i) of Proposition 1.1. Let us assume that (i) is valid at the order n and let us prove it at the order $n + 1$. Hence we assume that $\mu_{n+1} > \mu_\infty$; since the sequence μ_j is non-increasing, $\mu_j \geq \mu_{n+1} > \mu_\infty$ for $j = 1, \ldots, n$, and by the induction hypotheses $\mu_1, \ldots, \mu_n$ are eigenvalues of T with associated eigenvectors e_j

$$Te_j = \mu_j e_j, \qquad j = 1, \ldots, n.$$

$$(e_i, e_j)_E = \delta_{ij}, \qquad \forall i, j.$$

Let G denote the space spanned by $e_1, \ldots, e_n$; both G and $G^\perp$ are stable for T and it is easy to see that

$$\mu_{n+1} = \operatorname*{Sup}_{\substack{\varphi \in F^\perp \\ |\varphi|_E = 1}} (T\psi, \psi)_E.$$

Since $\mu_{n+1} > \mu_\infty$, we can apply the proposition at the order one (as in (i)) to the restriction of T to $G^\perp$; we obtain a new eigenvector of T which belongs to $G^\perp$ and corresponds to the eigenvalue μ_{n+1}.

Point (i) of the proposition is proved.

(iii) Concerning the sequence $\mu_n = \mu_n(T)$, two situations can occur. Either the sequence is stationary at some stage, i.e.,

$$\mu_1(T) \geq \cdots \geq \mu_{n_0}(T) > \mu_{n_0+1}(T) = \mu_m(T) = \mu_\infty(T), \qquad \forall m \geq n_0 + 1, \quad (1.46)$$

or

$$\mu_m(T) > \mu_\infty(T), \qquad \forall m. \quad (1.47)$$

In the first case it follows from the above result that $\mu_1, \ldots, \mu_{n_0}$ are eigenvalues of T, while in case (1.47) each μ_m is an eigenvalue. In both cases we decompose E into the direct sum $E_v \oplus E_v^\perp$ where E_v is the space spanned by the eigenvectors of T, e_i, $i \in J$, which we suppose orthonormalized ($J = (1, \ldots, n_0)$ when (1.46) occurs, $J = \mathbb{N}$ when (1.47) holds). Of course, it may happen that $E_v = \{0\}$ or $E_v = E$. The proof of the proposition will be complete after we show (1.45). In both cases (i.e., whether the sequence μ_j is stationary or not), it suffices to notice that, whenever $\mu_{n+1} < \mu_n$,

$$\mu(G_n) = \mu_{n+1} < \mu_n = \operatorname*{Inf}_{\substack{\varphi \in E \\ |\varphi|_E = 1}} (T\varphi, \varphi)_E,$$

where G_n is the space spanned by the eigenvectors $e_1, \ldots, e_n$ corresponding to $\mu_1, \ldots, \mu_n$, and

$$\mu(F) = \operatorname*{Sup}_{\substack{\varphi \in F^\perp \\ |\varphi|_E = 1}} (T\varphi, \varphi)_E.$$

Indeed, $\mu(G_n) \geq \mu_{n+1}$ and if the inequality were strict we could find F with $\dim F = n$ and

$$\mu_{n+1} \leq \mu(F) < \min(\mu(G_n), \mu_n).$$

Then $F \neq G_n$; there exists $\theta \in G_n \cap F^\perp$, $\theta \neq 0$, and hence $|\theta|_E = 1$ and we would have the contradicting inequalities

$$\mu_n \leq (T\theta, \theta)_E \leq \mu(F) < \mu_n. \qquad \square$$

In the compact case we have an alternative expression for the eigenvalues with the sup–inf formula, i.e., $\mu_n(T) = \lambda_n(T)$, $\forall n$, where

$$\lambda_n(T) = \operatorname*{Sup}_{\substack{F \subset E \\ \dim F = n}} \operatorname*{Inf}_{\substack{\varphi \in F \\ |\varphi|_E = 1}} (T\varphi, \varphi)_E. \quad (1.48)$$

The same result holds here and is proved in a similar manner. We recall the proof for the sake of completeness.

Proposition 1.2. *For a self-adjoint linear continuous operator $T \in \mathscr{L}(E)$ we have*

$$\lambda_n(T) = \mu_n(T), \qquad \forall n. \quad (1.49)$$

PROOF. Let F and G be two subspaces of E, with $\dim F = n - 1$, $\dim G = n$. The space $G \cap F^\perp$ is not reduced to $\{0\}$ and it contains at least one vector φ_0, $|\varphi_0|_E = 1$. We have

$$\operatorname*{Inf}_{\substack{\varphi \in F \\ |\varphi|_E = 1}} (T\varphi, \varphi)_E \leq (T\varphi_0, \varphi_0)_E \leq \operatorname*{Sup}_{\substack{\varphi \in F^\perp \\ |\varphi|_E = 1}} (T\varphi, \varphi)_E,$$

and therefore

$$\lambda_n(T) \leq \mu_n(T).$$

In order to prove the opposite inequality we consider a subspace G of E such that $\dim G = n$, and let $\varphi_0 \in G$, $|\varphi_0|_E = 1$, such that

$$(T\varphi_0, \varphi_0)_E = \operatorname*{Inf}_{\substack{\varphi \in G \\ |\varphi|_E = 1}} (T\varphi, \varphi)_E.$$

The space $F_0 = \{\psi \in G, (\psi, \psi_0)_E = 0\}$ has dimension $n - 1$ and, since $F_0^\perp \cap G = \mathbb{R}\varphi_0$, we can write

$$\operatorname*{Sup}_{\substack{\varphi \in F_0^\perp \cap G \\ |\varphi|_E = 1}} (T\varphi, \varphi)_E = (T\varphi_0, \varphi_0)_E = \operatorname*{Inf}_{\substack{\varphi \in G \\ |\varphi|_E = 1}} (T\varphi, \varphi)_E \leq \lambda_n(T).$$

Therefore

$$\operatorname*{Inf}_{\substack{F \subset E \\ \dim F = n-1}} \operatorname*{Sup}_{\substack{\varphi \in F^\perp \cap G \\ |\varphi|_E = 1}} (T\varphi, \varphi)_E \leq \lambda_n(T),$$

and, since $G \subset E$ satisfying $\dim G = n - 1$ is arbitrary, we obtain

$$\mu_n(T) \leq \lambda_n(T).$$

Proposition 1.2 is proved. □

Now we consider again a linear continuous operator L in E, $L \in \mathscr{L}(E)$, which is not necessarily self-adjoint. We associate with L the positive self-adjoint continuous operator $T = L^*L$. Let $E = E_v \oplus E_v^\perp$ denote the decomposition of E given by Proposition 1.1 and let e_i, $i \in J$, denote the orthonormal family of eigenvectors of T which constitute a basis of E_v. As in the compact case, we observe that the vectors Le_i, $i \in J$, are orthogonal

$$(Le_i, Le_j)_E = (L^*Le_i, e_j)_E = \mu_i(e_i, e_j)_E = \mu_i\delta_{ij}, \qquad \forall i, j \in J, \tag{1.50}$$

and that $L(E_v)$ is orthogonal to $L(E_v^\perp)$: if $\varphi \in E_v^\perp$, $L\varphi \in L(E_v^\perp)$, then

$$(Le_i, L\varphi)_E = (L^*Le_i, \varphi)_E = \mu_i(e_i, \varphi)_E = 0, \qquad \forall i \in J. \tag{1.51}$$

We also have by (1.45)

$$(T\varphi, \varphi)_E = |L\varphi|_E \leq \mu_\infty(T)|\varphi|_E^2, \qquad \forall \varphi \in E_v^\perp. \quad \square \tag{1.52}$$

In accordance with (1.23), (1.24) we now set, for every $m \in \mathbb{N}$,

$$\alpha_m(L) = \mu_m((L^*L)^{1/2}) = \underset{\substack{F \subset E \\ \dim F = m}}{\text{Sup}} \; \underset{\substack{\varphi \in F \\ |\varphi|_E = 1}}{\text{Inf}} |L\varphi|_E, \tag{1.53}$$

$$\omega_m(L) = \alpha_1(L) \dots \alpha_m(L), \tag{1.54}$$

and we are able to show the following result, similar to (1.42),

The set $L(B)$ is included in the sum of the ellipsoid of E_v

$$\sum_{i \in J} \frac{1}{\alpha_i^2} \left(\xi, \frac{Le_i}{\alpha_i} \right)^2 \leq 1, \qquad \xi \in E_v,$$

and of the ball of $E_v^\perp$ centered at 0 of radius

$$\alpha_\infty(L) = \{\mu_\infty(L^*L)\}^{1/2}. \tag{1.55}$$

Let us prove (1.55). If $\varphi \in B$ we write $\varphi = \sum_{i \in J} a_i e_i + \varphi'$, $\varphi' \in E_v^\perp$. Then

$$L\varphi = \sum_{i \in J} a_i \alpha_i \frac{Le_i}{\alpha_i} + L\varphi' = \sum_{i \in J} \xi_i \frac{Le_i}{\alpha_i} + L\varphi',$$

$\xi_i = a_i \alpha_i$. Since $|\varphi|_E^2 \leq 1$, we have

$$\sum_{i \in J} a_i^2 = \sum_{i \in J} \frac{\xi_i^2}{\alpha_i^2} \leq 1.$$

Similarly, $|\varphi'|_E \leq 1$ and we infer from (1.45) that

$$|L\varphi'|_E \leq \{\mu_\infty(L^*L)\}^{1/2} |\varphi'|_E \leq \alpha_\infty(L). \qquad \square$$

We collect the statements (1.42), (1.55) into

Proposition 1.3. *Let H be a Hilbert space and B its unit ball. Let L be a linear continuous operator in H and, if L is not compact, let E_v be defined as in* (1.46)–(1.50). *Then $L(B)$ is included in an ellipsoid $\mathscr{E}$:*

(i) *If L is compact or L is not compact, but $E_v = E$, the axes of $\mathscr{E}$ are directed along the vectors Le_i, and their length is $\alpha_i(L)$, the e_i being the eigenvectors of L^*L.*
(ii) *If L is not compact and $E_v \neq E$, $\mathscr{E}$ is the product of the ball centered at 0 of radius $\alpha_\infty(L)$ in $E_v^\perp$, and of the ellipsoid of E_v whose axes are directed along the vectors Le_i, with lengths $\alpha_i(L)$, the e_i being the eigenvectors of L spanning E_v.*[1]

We conclude this section with a result announced in Remark 1.1 which will be particularly useful especially for its consequence in Corollary 1.1.

[1] We recall that their number may be finite or infinite.

Proposition 1.4. *Let L be a linear continuous operator in E. Then for every integer m, $\omega_m(L)$ is equal to the norm of $\bigwedge^m L$ in $\bigwedge^m E$, and to the norm of $\bigwedge^m L$ as an m-linear operator from E^m into $\bigwedge^m E$*

$$\begin{aligned}\alpha_1(L)\dots\alpha_m(L) &= \sup_{\substack{\varphi\in\bigwedge^m E\\ |\varphi|_{\bigwedge^m E}=1}} |\textstyle\bigwedge^m L(\varphi)|_{\bigwedge^m E}\\ &= \sup_{\substack{\varphi_1,\dots,\varphi_m\in E\\ |\varphi_i|_E\le 1,\forall i}} |L\varphi_1\wedge\cdots\wedge L\varphi_m|_E. \end{aligned}\tag{1.56}$$

PROOF. The result is obvious when $m = 1$ since $\alpha_1(L) = \|L\|_{\mathscr{L}(E)}$. We will then proceed by induction and we assume that (1.56) has been proved for m replaced by $1, \dots, m-1$. In order to prove (1.56) at the order m we notice the partial result provided by (1.31) and Remark 1.1

$$\|\textstyle\bigwedge^m L\|_{\mathscr{L}_m} \le \omega_m(L) = \omega_{m-1}(L)\alpha_m(L), \tag{1.57}$$

where we have written $\mathscr{L}_m$ instead of $\mathscr{L}_m(E^m, \bigwedge^m E)$. By the induction assumption

$$\|\textstyle\bigwedge^{m-1} L\|_{\mathscr{L}_{m-1}} = \alpha_1(L)\dots\alpha_{m-1}(L) = \omega_{m-1}(L)$$

and (1.56) is obvious if $\omega_{m-1}(L) = 0$. We can thus assume that $\|\bigwedge^{m-1} L\|_{\mathscr{L}_m} = \omega_{m-1}(L) > 0$, and we consider $0 < \varepsilon < \omega_{m-1}(L)^2$. By the definition of $\|\bigwedge^{m-1}\|_{\mathscr{L}_{m-1}}$ (see (1.19)), there exists $\varphi_1, \dots, \varphi_{m-1}$ in E such that $|\varphi_i|_E \le 1$, $\forall i$, and

$$|L\varphi_1 \wedge \cdots \wedge L\varphi_{m-1}|^2_{\bigwedge^m E} \ge \omega_{m-1}(L)^2 - \varepsilon > 0. \tag{1.58}$$

Let $\psi_1, \dots, \psi_{m-1}$ be the orthonormal sequence of eigenvectors for the quadratic form $\psi \to |L\psi|_E^2$ restricted to the space spanned by $\varphi_1, \dots, \varphi_{m-1}$. By expressing the vectors φ_j in terms of the vectors ψ_j we easily see that

$$|L\psi_1 \wedge \cdots \wedge L\psi_{m-1}|_E^2 \ge |L\varphi_1 \wedge \cdots \wedge L\varphi_{m-1}|_E^2,$$

and thus

$$\omega_{m-1}(L)^2 = \|\textstyle\bigwedge^{m-1} L\|^2_{\mathscr{L}_{m-1}} \ge |L\psi_1 \wedge \cdots \wedge L\psi_{m-1}|_E^2 \ge \omega_{m-1}(L)^2 - \varepsilon.$$

Furthermore, since the vectors $L\psi_i$ are orthogonal,

$$|L\psi_1 \wedge \cdots \wedge L\psi_{m-1}|_E^2 = |L\psi_1|_E^2 \dots |L\psi_{m-1}|_E^2. \tag{1.59}$$

Let now $F_1 = \{\psi \in E, (L\psi, \psi_i)_E = 0, i = 1, \dots, m-1\}$. Its orthogonal (in E), $G = F_1^\perp$ has dimension $\le m - 1$ and hence

$$\alpha_m(L) = \inf_{\substack{F\subset E\\ \dim F=m-1}} \sup_{\substack{\varphi\in F^\perp\\ |\varphi|_E=1}} |L\varphi|_E \le \sup_{\substack{\varphi\in G^\perp\\ |\varphi|_E=1}} |L\varphi|_E.$$

This shows that there exists $\psi \in G^\perp = F_1$ such that $|\psi|_E = 1$ and $|L\psi|_E^2 \ge$

$\alpha_m(L)^2 - \varepsilon$ and then

$$|L\psi_1 \wedge \cdots \wedge L\psi_{m-1} \wedge L\psi|^2 = |L\psi_1|_E^2 \ldots |L\psi_{m-1}|_E^2 |L\psi|_E^2$$
$$\geq (\omega_{m-1}(L)^2 - \varepsilon)(\alpha_m(L)^2 - \varepsilon).$$

Since $|\psi_j|_E \leq 1$, $|\psi|_E \leq 1$, the left-hand side of this inequality is less than or equal to $\|\bigwedge^m L\|_{\mathscr{L}_m}^2$; thus

$$\|\bigwedge^m L\|_{\mathscr{L}_m}^2 \geq (\omega_{m-1}(L)^2 - \varepsilon)(\alpha_m(L)^2 - \varepsilon),$$

and by letting $\varepsilon \to 0$,

$$\|\bigwedge^m L\|_{\mathscr{L}_m} \geq \omega_m(L).$$

This inequality, in conjunction with (1.57), completes the proof. □

We deduce from Proposition 1.4

Corollary 1.1. *If L, L' are two linear continuous operators in E then, for every integer m,*

$$\omega_m(LL') \leq \omega_m(L)\omega_m(L'). \tag{1.60}$$

PROOF. If $\varphi_1, \ldots, \varphi_m$ belong to E,

$$\bigwedge^m(LL')(\varphi_1 \wedge \cdots \wedge \varphi_m) = LL'\varphi_1 \wedge \cdots \wedge LL'\varphi_m$$
$$= \bigwedge^m(L)(L'\varphi_1 \wedge \cdots \wedge L'\varphi_m)$$
$$= \bigwedge^m(L)\bigwedge^m(L')(\varphi_1 \wedge \cdots \wedge \varphi_m).$$

Therefore we have the following equality in $\mathscr{L}(\bigwedge^m E)$:

$$\bigwedge^m(LL') = (\bigwedge^m L)\cdot(\bigwedge^m L'). \tag{1.61}$$

It follows readily from (1.61) that

$$\|\bigwedge^m L\|_{\mathscr{L}(\bigwedge^m E)} \leq \|\bigwedge^m L\|_{\mathscr{L}(\bigwedge^m E)}\|\bigwedge^m L'\|_{\mathscr{L}(\bigwedge^m E)} \tag{1.62}$$

and (1.60) follows then, thanks to Proposition 1.4. We also infer from (1.56) that

$$\|\bigwedge^m LL'\|_{\mathscr{L}_m(E^m,\bigwedge^m E)} \leq \|\bigwedge^m L\|_{\mathscr{L}_m(E^m,\bigwedge^m E)}\cdot\|\bigwedge^m L'\|_{\mathscr{L}_m(E^m,\bigwedge^m E)}. \quad \square \tag{1.63}$$

Remark 1.5. If L is a linear continuous operator in E and $d \in \mathbb{R}_+$, $d = n + s$, n integer ≥ 1, $0 < s < 1$, we define $\omega_d(L)$ as

$$\omega_d(L) = \omega_n(L)^{1-s}\omega_{n+1}(L)^s. \tag{1.64}$$

It is easy to see that (1.60) is also valid for $m = d \in [1, \infty[$ and that

$$d \to \omega_d(L) \text{ is a nonincreasing function from } [1, \infty[\text{ into } \mathbb{R}_+. \tag{1.65}$$

2. Lyapunov Exponents and Lyapunov Numbers

We consider a dynamical system described by a semigroup $\{S(t)\}_{t\geq 0}$, with $t \in \mathbb{R}_+$ in the continuous case and $t \in \mathbb{N}$ in the discrete case. We are interested in studying the distortion of distances, areas, and, more generally, m-dimensional volumes which are produced locally by the semigroup, i.e., we want to see how an infinitesimal m-dimensional volume element around a point u_0 evolves as t increases. In Section 2.1 we show how this is related to certain properties of linear operators, namely, the differential of the mapping $S(t)$: $u_0 \to S(t)u_0$, $t \geq 0$. In Section 2.2 we introduce the concept of Lyapunov numbers and Lyapunov exponents. These quantities are appropriate and convenient tools for the study of the evolution of the m-volume elements for large time; we successively define the pointwise Lyapunov numbers and the *global* (*or uniform*) *Lyapunov numbers* on an invariant set of the semigroup. Then, in Section 2.3, we return to the study of the evolution of the volume element, and, in some cases, its exponential decay. We consider here the continuous case ($t \in \mathbb{R}_+$), the semigroup $S(t)$ being associated with an evolution equation; we provide an abstract result which will then be applied to specific equations in the next chapter.

2.1. Distortion of Volumes Produced by the Semigroup

We are given a Hilbert space H and a semigroup of continuous operators $\{S(t)\}_{t\geq 0}$ acting on H (assumptions I.(1.1), I.(1.4)); in the continuous case $t \in \mathbb{R}_+$, and in the discrete case $t = n \in \mathbb{N}$,

$$S(n) = S^n, \qquad S = S(1). \tag{2.1}$$

We assume for the moment that

$$u_0 \to S(t)\cdot u_0 \quad \text{is Fréchet differentiable in } H. \tag{2.2}$$

although this assumption will be weakened later on. The differential at point u_0 is denoted by $L(t, u_0)$ ($\in \mathscr{L}(H)$). In the discrete case, (2.2) amounts to assuming that

$$S\ (= S(1)) \quad \text{is differentiable in } H, \tag{2.3}$$

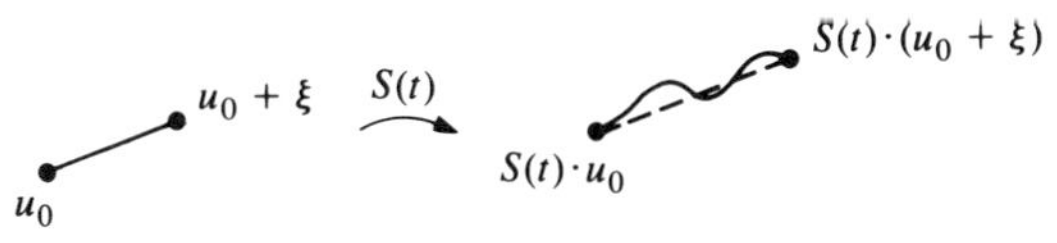

Figure 2.1

and, in this case, the differential of S is denoted by L ($\in \mathscr{L}(H)$).

Now let u_0 and $u_0 + \xi$ be two points in H ($\xi \in H$). At time t ($\in \mathbb{R}_+$ or $\mathbb{N}$) their image by $S(t)$ are the points $S(t)\cdot u_0$ and $S(t)\cdot(u_0 + \xi)$ (see Figure 2.1), and the distance of these points is $|S(t)\cdot(u_0 + \xi) - S(t)\cdot u_0|$ where $|\cdot|$ denotes the norm in H. Instead of u_0 and $u_0 + \xi$ we may consider u_0 and $u_0 + \varepsilon\xi$, $\varepsilon > 0$. When ε is small we use (2.2) (or (2.3)) and write

$$S(t)\cdot(u_0 + \varepsilon\xi) = S(t)\cdot u_0 + \varepsilon L(t, u_0)\cdot\xi + o(\varepsilon), \tag{2.4}$$

where, using the Landau notation, $o(\varepsilon) \to 0$ in H as $\varepsilon \to 0$. Since ε is the distance of $u_0 + \varepsilon\xi$ to u_0, we have

$$\frac{\text{distance of } S(t)\cdot(u_0 + \varepsilon\xi) \text{ to } S(t)u_0}{\text{distance of } u_0 + \varepsilon\xi \text{ to } u_0} = |L(t, u_0)\cdot\xi + o(1)|, \tag{2.5}$$

and for ε small this ratio is close to $|L(t, u_0)\cdot\xi|$. Assuming that ξ has norm one, we see that the largest possible value of $|L(t, u_0)\cdot\xi|$ is the norm of $L(t, u_0)$ in $\mathscr{L}(H)$ or, alternatively, the number $\alpha_1(L(t, u_0))$ defined in (1.53).

More generally, let us consider the points $u_0, u_0 + \varepsilon\xi_1, \ldots, u_0 + \varepsilon\xi_m$ of H where $\xi_1, \ldots, \xi_m \in H$ and $\varepsilon > 0$. They define an m-parallelepiped consisting of the points $u_0 + \varepsilon\rho_1\xi_1 + \cdots + \varepsilon\rho_m\xi_m$, $0 \le \rho_i \le 1$, $i = 1, \ldots, m$. The image of this parallelepiped generated by $S(t)$ is a curvilinear parallelepiped-like domain with edges $S(t)u_0, S(t)(u_0 + \varepsilon\xi_1), \ldots, S(t)\cdot(u_0 + \varepsilon\xi_m), \ldots$, and we are interested in comparing the m-volume of the parallelepiped with edges u_0, $u_0 + \varepsilon\xi_1, \ldots, u_0 + \varepsilon\xi_m$, to the m-volume of the parallelepiped with edges $S(t)u_0, S(t)\cdot(u_0 + \varepsilon\xi_1), \ldots, S(t)\cdot(u_0 + \varepsilon\xi_m)$ (see Figure 2.2). As indicated in Section 1, the m-volume of the first parallelepiped is the norm of $(\varepsilon\xi_1) \wedge \cdots \wedge (\varepsilon\xi_m)$ in $\bigwedge^m H$, while the m-volume of the second parallelepiped is

$$|(S(t)\cdot(u_0 + \varepsilon\xi_1) - S(t)\cdot u_0) \wedge \cdots \wedge (S(t)\cdot(u_0 + \varepsilon\xi_m) - S(t)\cdot u_0)|_{\bigwedge^m H}.$$

Using relation (2.4) with ξ replaced by $\xi_1, \ldots, \xi_m$, we find that

$$\text{ratio of } m\text{-volumes of the } m\text{-parallelepipeds} = \frac{|L(t, u_0)\xi_1 \wedge \cdots \wedge L(t, u_0)\xi_m + o(1)|_{\bigwedge^m H}}{|\xi_1 \wedge \cdots \wedge \xi_m|_{\bigwedge^m H}}. \tag{2.6}$$

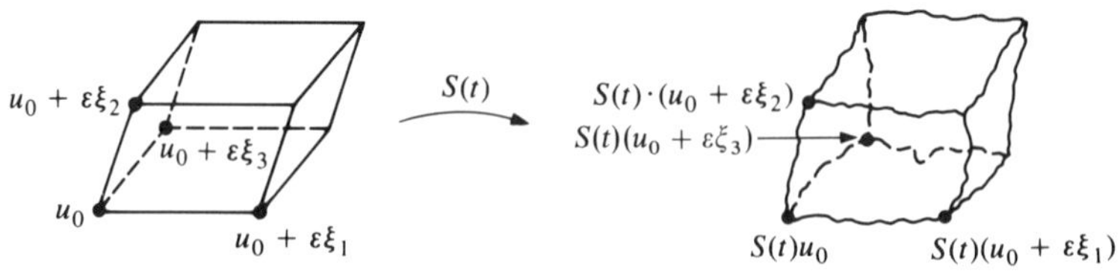

Figure 2.2

Therefore, when ε is small, this number is close to

$$\frac{|L(t,u_0)\xi_1 \wedge \cdots \wedge L(t,u_0)\xi_m|_{\wedge^m H}}{|\xi_1 \wedge \cdots \wedge \xi_m|_{\wedge^m H}}. \tag{2.7}$$

The largest value of the ratio (2.7) for $\xi_1, \ldots, \xi_m$ in H (and u_0 and t fixed), is nothing other than the number $\omega_m(L(t, u_0))$ defined in Section 1 (see, in particular, Proposition 1.2). The statement concerning the distortion of the volume of m-dimensional parallelepipeds with edge u_0 extends easily to any infinitesimal volume around u_0 included in an affine subspace of dimension m, and to its image. In conclusion, we can say that

> *The numbers $\omega_m(L(t, u_0))$ indicate the largest distortion of an infinitesimal m-dimensional volume produced by $S(t)$ around the point $u_0 \in H$.*[1,2] (2.8)

2.2. Definition of the Lyapunov Exponents and Lyapunov Numbers

The Lyapunov numbers are a good indication of the distortion of infinitesimal m-dimensional volumes produced by $S(t)$ *for large t*. They are not, however, necessarily defined at every point $u_0 \in H$.

Definition 2.1. When they exist, the Lyapunov numbers of the semigroup $S(t)$, at point u_0, are the numbers

$$\lambda_j(u_0) = \lim_{t\to\infty} \{\alpha_j(L(t, u_0))\}^{1/t}, \qquad j \in \mathbb{N}, \tag{2.9}$$

when the limit exists. The corresponding Lyapunov exponents are the numbers

$$\mu_j(u_0) = \log \lambda_j(u_0). \tag{2.10}$$

In the discrete case we write, of course,

$$\lambda_j(u_0) = \lim_{n\to\infty} \{\alpha_j(L^n)\}^{1/n}, \mu_j(u_0) = \log \lambda_j(u_0). \tag{2.11}$$

When the limit does not exist it may sometimes be useful to consider the numbers

$$\begin{cases} \Lambda_j(u_0) = \limsup\limits_{t\to\infty} \{\alpha_j(L(t, u_0))\}^{1/t}, \qquad j \in \mathbb{N}, \\ \mu_j(u_0) = \log \Lambda_j(u_0). \end{cases} \tag{2.12}$$

[1] Implicit in what we said before is the easy-to-check fact that for ε small the image by $S(t)$ of the parallelepiped with edges u_0, $u_0 + \varepsilon\xi_1, \ldots, u_0 + \varepsilon\xi_m$ has an m-dimensional volume of the same order as that of the parallelepiped with edges $S(t)u_0$, $S(t)(u_0 + \varepsilon\xi_1), \ldots, S(t)(u_0 + \varepsilon\xi_m)$.

[2] Of course, in the discrete case $t = n \in \mathbb{N}$, we replace $\omega_m(L(n, u_0))$ by $\omega_m(L^n) \leq$ (by Corollary 1.1) $\leq [\omega_m(L)]^n$.

In the ergodic theory of dynamical systems an invariant set X of the semigroup $S(t)$ is also the support of a measure ν which is invariant for $S(t)$. Then, according to a theorem of V.I. Oseledec [1], the Lyapunov numbers $\lambda_j(u_0)$ exist $d\nu$-almost everywhere.

Instead of the pointwise Lyapunov numbers which may not exist we will use, in the following, the concept of *uniform Lyapunov numbers* which are always defined on an invariant set. These numbers were introduced in the context of the Navier–Stokes equations in P. Constantin and C. Foias [1] for dimension 2, and in P. Constantin, C. Foias, and R. Temam [1] for dimension 3. The latter article also contains an abstract general presentation which has been used systematically in subsequent works.

As indicated above, the differentiability assumption (2.2), (2.3) can be weakened and the assumptions are then the following. We are still given a Hilbert space H and a semigroup of continuous operators $\{S(t)\}_{t\geq 0}$ ($t \in \mathbb{R}_+$ or $\mathbb{N}$) acting in H. We are also given a functional invariant set X for the semigroup

$$S(t)X = X, \qquad \forall t \geq 0. \tag{2.13}$$

In the discrete case ($SX = X$) we assume, instead of (2.3), that S is "uniformly differentiable on X". By this we mean the following:

For every $u \in X$, there exists a linear operator $L(u) \in \mathscr{L}(H)$ and

$$\operatorname*{Sup}_{\substack{u,v\in X\\ 0<|u-v|\leq\varepsilon}} \frac{|Sv - Su - L(u)\cdot(v-u)|}{|v-u|} \to 0 \quad \text{as } \varepsilon \to 0. \tag{2.14}$$

Note that the operators $L(u)$ satisfying (2.14) may not be unique. It is also easy to see that if S is "uniformly differentiable on X" (and $SX = X$), then the same is true for S^p for any $p \in \mathbb{N}$, the operators $L(u)$ being replaced by

$$L_p(u) = L(S^{p-1}(u)) \circ \cdots \circ L(S(u)) \circ L(u), \qquad p \in \mathbb{N}. \tag{2.15}$$

We also assume that the norm of the operators $L(u)$ in $\mathscr{L}(H)$ is uniformly bounded for $u \in X$

$$\operatorname*{Sup}_{u\in X} |L(u)|_{\mathscr{L}(H)} \leq m < +\infty, \tag{2.16}$$

which easily implies that

$$\operatorname*{Sup}_{u\in X} |L_p(u)|_{\mathscr{L}(H)} \leq m^p < +\infty, \qquad \forall p \in \mathbb{N}. \tag{2.17}$$

Let us then consider the numbers $\omega_j(L_p(u))$, j, $p \in \mathbb{N}$, $u \in X$, and

$$\begin{cases} \bar{\omega}_j = \operatorname*{Sup}_{u\in X} \omega_j(L(u)), \\ \bar{\omega}_j(p) = \operatorname*{Sup}_{u\in X} \omega_j(L_p(u)), \qquad j, p \in \mathbb{N}. \end{cases} \tag{2.18}$$

All these numbers are finite since, by Proposition 1.3, Corollary 1.1, and (2.16),

(2.17)

$$\omega_j(L(u)) \leq |L(u)|_{\mathscr{L}(H)}, \qquad \bar{\omega}_j \leq m,$$
$$\omega_j(L_p(u)) \leq |L_p(u)|_{\mathscr{L}(H)}, \qquad \bar{\omega}_j(p) \leq m^p.$$

Concerning the numbers $\bar{\omega}_j(p)$ we note the following: for $j, p, q \in \mathbb{N}$ and $u \in X$ we can write

$$\begin{aligned}
\omega_j(L_{p+q}(u)) &= \omega_j(L(S^{p+q-1}(u)) \circ \cdots \circ L(u)) \\
&\leq \text{(by Corollary 1.1)} \\
&\leq \omega_j(L(S^{p+q-1}(u)) \circ \cdots \circ L(S^q(u))) \cdot \omega_j(L(S^{q-1}(u)) \circ \cdots \circ L(u)) \\
&\leq (\text{setting } v = S^q(u) \in X) \\
&\leq \omega_j(L(S^{p-1}(v)) \circ \cdots \circ L(v)) \cdot \omega_j(L(S^{q-1}(u)) \circ \cdots \circ L(u)) \\
&\leq \bar{\omega}_j(p) \cdot \bar{\omega}_j(q),
\end{aligned}$$

from which we infer that

$$\bar{\omega}_j(p + q) \leq \bar{\omega}_j(p) \cdot \bar{\omega}_j(q), \qquad \forall j, p, q \in \mathbb{N}. \tag{2.19}$$

The inequality (2.19) is similar to the equality which characterizes the exponential function, and a function satisfying this inequality is called a *subexponential function*. Thus $\bar{\omega}_j$ is subexponential on $\mathbb{N}$, $\forall j \in \mathbb{N}$, and, due to Lemma 2.1 below, we see that the following limit exists, $\forall j \in \mathbb{N}$:

$$\lim_{p \to \infty} \{\bar{\omega}_j(p)\}^{1/p} = \inf_{p \in \mathbb{N}} \{\bar{\omega}_j(p)\}^{1/p}. \tag{2.20}$$

This limit is denoted Π_j and we denote by $\Lambda_1, \ldots, \Lambda_m, \ldots$, the sequence of numbers recursively defined by

$$\Lambda_1 = \Pi_1, \qquad \Lambda_1 \Lambda_2 = \Pi_2, \ldots, \qquad \Lambda_1 \ldots \Lambda_m = \Pi_m, \ldots,$$

or, equivalently,

$$\begin{aligned}
&\Lambda_1 = \Pi_1, \qquad \Lambda_m = \Pi_m / \Pi_{m-1}, \qquad m \geq 2, \\
&\Lambda_m = \lim_{p \to \infty} \left(\frac{\bar{\omega}_m(p)}{\bar{\omega}_{m-1}(p)} \right)^{1/p}, \qquad m \geq 2.
\end{aligned} \tag{2.21}$$

The numbers Λ_m are the *global* (*or uniform*) *Lyapunov numbers on* X and the *global* (*uniform*) *Lyapunov exponents* are defined by

$$\mu_m = \log \Lambda_m, \qquad m \geq 1. \tag{2.22}$$

Let us mention some related quantities which will be useful. We set

$$\bar{\alpha}_j(p) = \operatorname*{Sup}_{u \in X} \alpha_j(L_p(u)), \tag{2.23}$$

and

$$\bar{\Lambda}_j = \limsup_{p \to \infty} \{\bar{\alpha}_j(p)\}^{1/p}, \qquad \bar{\mu}_j = \log \bar{\Lambda}_j, \qquad \forall j \geq 1. \tag{2.24}$$

Since the sequence of numbers $\alpha_j(L)$ is decreasing for any $L \in \mathscr{L}(H)$, it is obvious that the sequences $\lambda_j(u)$, $\mu_j(u)$ (when they exist), and $\bar{\Lambda}_j$, $\bar{\mu}_j$ are decreasing. It is also easy to verify that

$$\begin{cases} \bar{\omega}_j(p) \le \bar{\omega}_{j-1}(p)\bar{\alpha}_j(p), & \forall j \ge 2, \quad \forall p \in \mathbb{N}, \\ \bar{\alpha}_j(p) \le \{\bar{\omega}_j(p)\}^{1/j}, & \forall j \ge 1, \quad \forall p \in \mathbb{N}. \end{cases} \tag{2.25}$$

and these inequalities imply at the limit

$$\Lambda_j \le \bar{\Lambda}_j \le (\Lambda_1 \dots \Lambda_j)^{1/j}, \tag{2.26}$$

$$\mu_j \le \bar{\mu}_j \le \frac{1}{j}(\mu_1 + \cdots + \mu_j), \qquad \forall j \ge 1. \tag{2.27}$$

In the continuous case the assumptions and definitions are similar. We assume that $S(t)$ is "uniformly differentiable on X", for every $t \ge 0$, and we denote by $L(t, u)$ the corresponding linear operator. Furthermore, we assume that

$$\operatorname*{Sup}_{t \in [0,1]} \operatorname*{Sup}_{u \in X} |L(t, u)|_{\mathscr{L}(H)} \le m < +\infty, \tag{2.28}$$

which implies that

$$\operatorname*{Sup}_{u \in X} |L(t, u)|_{\mathscr{L}(H)} \le m^{[t]+1}, \qquad \forall t \ge 0, \tag{2.29}$$

where $[t]$ is the integer part of t.[1] In order to show (2.29) it suffices to write

$$S(t) = S(t - [t]) \cdot S(1)^{[t]},$$

and

$$L(t, u) = L(t - [t], S([t])u) \circ L(1, S([t] - 1)u) \circ \cdots \circ (L(1, u).^2 \tag{2.30}$$

We then consider the numbers $\omega_j(L(t, u))$, defined for $j \in \mathbb{N}$, $t \ge 0$, $u \in X$, and

$$\bar{\omega}_j(t) = \operatorname*{Sup}_{u \in X} \omega_j(L(t, u)), \qquad j \in \mathbb{N}, \quad t \in \mathbb{R}_+. \tag{2.31}$$

As in the discrete case we check that the functions $t \to \bar{\omega}_j(t)$ are *subexponential*

$$\bar{\omega}_j(t + s) \le \bar{\omega}_j(t) \cdot \bar{\omega}_j(s), \qquad \forall s, t \in \mathbb{R}_+, \tag{2.32}$$

so that, by Lemma 2.1, $\lim_{t \to \infty} \{\bar{\omega}_j(t)\}^{1/t}$ exists and is equal to

$$\operatorname*{Inf}_{t > 0} \{\bar{\omega}_j(t)\}^{1/t}. \tag{2.33}$$

[1] $[t] \le t < [t] + 1$.

[2] Since $L(t, u)$ may not be unique we can decide to use (2.25) to define $L(t, u)$, $t \ge 1$, in terms of the $L(\tau, u)$, $0 \le \tau \le 1$; of course, (2.25) is nothing other than the chain differentiation rule if the operators $S(t)$ are Fréchet differentiable in the classical sense.

This number is denoted Π_j and we then define exactly as above the numbers Λ_m and μ_m, $m \geq 1$, which are the *global* (*or uniform*) *Lyapunov numbers and the global* (*or uniform*) *Lyapunov exponents on* X.

We also define the related quantities

$$\bar{\alpha}_j(t) = \operatorname*{Sup}_{u \in X} \alpha_j(L(t, u)), \qquad j \geq 1, \quad t \in \mathbb{R}_+, \tag{2.34}$$

$$\bar{\Lambda}_j = \limsup_{t \to \infty} \{\bar{\alpha}_j(t)\}^{1/t}, \qquad \bar{\mu}_j = \log \bar{\Lambda}_j, \tag{2.35}$$

and with the analog of (2.25) we see that the inequalities (2.26), (2.27) are satisfied.

Remark 2.1. We have used, for the definition of the uniform Lyapunov exponents, assumptions (2.14), (2.16) in the discrete case and assumptions (2.28) and $S(t)$ "uniformly differentiable on X" for the continuous case. Obviously, we can replace these differentiability assumptions by the stronger assumptions (2.2), (2.3) expressing the differentiability of $S(t)$ (or S) in the classical Fréchet sense.

Remark 2.2. In both cases, the discrete and the continuous one, we can define uniform Lyapunov numbers Λ_d, μ_d on an invariant set X with $d \in [1, \infty[$. The definition is exactly the same as above, using Remark 1.1 and the definition (1.64) of $\omega_d(L)$ for $L \in \mathscr{L}(H)$.

We conclude Section 2.2 with the following elementary result used above.

Lemma 2.1. *Let* φ *be a subexponential function from* $\mathbb{R}_+$ *or* $\mathbb{N}$ *into* $\mathbb{R}_+$

$$\varphi(t + s) \leq \varphi(t)\varphi(s), \qquad \forall s, t \in \mathbb{R}_+ \text{ or } \mathbb{N}, \tag{2.36}$$

such that

$$\operatorname*{Sup}_{t \in [a, b]} \varphi(t) < \infty \quad \text{for some } a, b, \quad 0 < a < b < \infty.[1] \tag{2.37}$$

Then $\lim_{t \to \infty} \{\varphi(t)\}^{1/t}$ *exists as* $t \to \infty$ ($t \in \mathbb{R}_+$ *or* $\mathbb{N}$) *and this limit is equal to*

$$l = \operatorname*{Inf}_{t > 0} \{\varphi(t)\}^{1/t}.$$

Proof. Let p, q belong to $\mathbb{R}_+$ or $\mathbb{N}$, $p \geq a$, $q \geq p$, and let k be the integer such that

$$(k + 1)p \leq q < (k + 2)p.$$

[1] Due to (2.36), (2.37), φ is bounded on any interval $[a, c]$, $\forall c$, $a < c < \infty$.

Due to (2.36)

$$\varphi(q) \leq \varphi(kp)\varphi(q-kp) \leq \varphi(p)^k \varphi(q-kp),$$

$$\varphi(q)^{1/q} \leq (\varphi(p)^{1/p})^{kp/q} \cdot \left(\operatorname*{Sup}_{p \leq t \leq 2p} \varphi(t) \right)^{1/q}.$$

We let $k, q \to \infty$, p remaining fixed and we obtain

$$\limsup_{q \to \infty} \varphi(q)^{1/q} \leq \varphi(p)^{1/p}.$$

Then, taking the lower limit as $p \to \infty$, we find

$$\limsup_{q \to \infty} \varphi(q)^{1/q} \leq \liminf_{p \to \infty} \varphi(p)^{1/p},$$

which shows that $\lim_{p \to \infty} \varphi(p)^{1/p}$ exists. In order to show that this limit is equal to l, we consider for $\varepsilon > 0$ fixed p_ε such that $\varphi(p_\varepsilon)^{1/p_\varepsilon} \leq l + \varepsilon$. Clearly,

$$\lim_{p \to \infty} \varphi(p)^{1/p} = \lim_{k \to \infty} \varphi(kp_\varepsilon)^{1/kp_\varepsilon} \leq l + \varepsilon,$$

and the equality follows since $\varepsilon > 0$ can be chosen arbitrarily small. □

2.3. Evolution of the Volume Element and Its Exponential Decay: The Abstract Framework

We provide a framework which is adapted to the study of the evolution of the volume element and the estimate of the Lyapunov numbers for dynamical systems associated to evolution equations.

Let there be given a Banach subspace W of H, the injection of W in H being continuous, and let there be given a function F from W into H. We assume that the initial-value problem

$$\begin{cases} \dfrac{du}{dt}(t) = F(u(t)), \qquad t > 0, \\ u(0) = u_0, \end{cases} \tag{2.38}$$

is well posed for every $u_0 \in H$, with $u(t) \in W$, $\forall t \geq 0$, and the mapping $S(t)$: $u_0 \in H \to u(t) \in H$ enjoys the properties I.(1.1) and I.(1.4).

It is also assumed that F is Fréchet differentiable from W into H with differential F' and that the (linear) initial-value problem

$$\begin{cases} \dfrac{dU}{dt}(t) = F'(S(t)u_0) \cdot U(t), \\ U(0) = \xi, \end{cases} \tag{2.39}$$

is well posed for every $u_0, \xi \in H$. Finally, we assume that $S(t)$ is differentiable in H with the differential $L(t, u_0)$ defined by

$$L(t, u_0)\cdot \xi = U(t), \qquad \forall \xi \in H,$$

and U the solution of (2.39). Since (2.39) is the first variation equation for (2.38), all these assumptions are very natural and very easily verified, for instance, for ordinary differential equations, when H is finite dimensional. Of course, in the infinite-dimensional case, the verification of these assumptions may lead to technical difficulties.

Now, for u_0 fixed in H, let $\xi_1, \dots, \xi_m$ be m elements of H and let $U_1, \dots, U_m$ be the corresponding solutions of (2.39). Then, in view of studying the ratio (2.7), we seek an evolution equation satisfied by

$$|L(t, u_0)\xi_1 \wedge \cdots \wedge L(t, u_0)\xi_m|_{\bigwedge^m H} = |U_1(t) \wedge \cdots \wedge U_m(t)|_{\bigwedge^m H}.$$

By time differentiation we write

$$\begin{aligned}
&\frac{1}{2}\frac{d}{dt}|U_1(t) \wedge \cdots \wedge U_m(t)|^2_{\bigwedge^m H}\\
&\quad = \left(\frac{d}{dt}(U_1(t) \wedge \cdots \wedge U_m(t)), U_1(t) \wedge \cdots \wedge U_m(t)\right)_{\bigwedge^m H}\\
&\quad = (U_1'(t) \wedge U_2(t) \wedge \cdots \wedge U_m(t), U_1(t) \wedge \cdots \wedge U_m(t))_{\bigwedge^m H}\\
&\qquad + \cdots + (U_1(t) \wedge \cdots \wedge U_{m-1}(t) \wedge U_m'(t), U_1(t) \wedge \cdots \wedge U_m(t))_{\bigwedge^m H}\\
&\quad = (F'(u(t))U_1(t) \wedge U_2(t) \wedge \cdots \wedge U_m(t), U_1(t) \wedge \cdots \wedge U_m(t))_{\bigwedge^m H}\\
&\qquad + \cdots + (U_1(t) \wedge \cdots \wedge U_{m-1}(t) \wedge F'(u(t))U_m(t), U_1(t) \wedge \cdots \wedge U_m(t))_{\bigwedge^m H}\\
&\quad = (\text{see the definition of } F'(u(t))_m \text{ in } (1.32))\\
&\quad = (F'(u(t))_m(U_1(t), \dots, U_m(t)), U_1(t) \wedge \cdots \wedge U_m(t))_{\bigwedge^m H}\\
&\quad = (\text{thanks to Lemma 1.2})\\
&\quad = |U_1(t) \wedge \cdots \wedge U_m(t)|^2_{\bigwedge^m H} \operatorname{Tr}(F'(u(t)) \circ Q_m),
\end{aligned}$$

where $Q_m = Q_m(t, u_0; \xi_1, \dots, \xi_m)$ is the projector in H onto the space spanned by $U_1(t), \dots, U_m(t)$. Therefore

$$\frac{1}{2}\frac{d}{dt}|U_1 \wedge \cdots \wedge U_m|^2_{\bigwedge^m H} = |U_1 \wedge \cdots \wedge U_m|^2_{\bigwedge^m H} \operatorname{Tr}(F'(u) \circ Q_m),$$

$$\frac{d}{dt}|U_1 \wedge \cdots \wedge U_m|_{\bigwedge^m H} = |U_1 \wedge \cdots \wedge U_m|_{\bigwedge^m H} \operatorname{Tr}(F'(u) \circ Q_m), \tag{2.40}$$

$$\begin{aligned}
&|U_1(t) \wedge \cdots \wedge U_m(t)|_{\bigwedge^m H}\\
&\qquad = |\xi_1 \wedge \cdots \wedge \xi_m|_{\bigwedge^m H} \exp\left(\int_0^t \operatorname{Tr} F'(S(\tau)u_0) \circ Q_m(\tau)\, d\tau\right).
\end{aligned} \tag{2.41}$$

We can then estimate

$$\omega_m(L(t, u_0)) = \underset{\substack{\xi_i \in H \\ |\xi_i| \leq 1 \\ i=1,\ldots,m}}{\text{Sup}} \ |U_1(t) \wedge \cdots \wedge U_m(t)|_{\wedge^m H},$$

$$\omega_m(L(t, u_0)) \leq \underset{\substack{\xi_i \in H \\ |\xi_i| \leq 1 \\ i=1,\ldots,m}}{\text{Sup}} \ \exp\left(\int_0^t \text{Tr } F'(S(\tau)u_0) \circ Q_m(\tau)\, d\tau\right). \tag{2.42}$$

At this point it is useful to introduce the quantities

$$q_m(t) = \underset{u_0 \in X}{\text{Sup}} \ \underset{\substack{\xi_i \in H \\ |\xi_i| \leq 1 \\ i=1,\ldots,m}}{\text{Sup}} \left(\frac{1}{t}\int_0^t \text{Tr } F'(S(\tau)u_0) \circ Q_m(\tau)\, d\tau\right), \tag{2.43}$$

$$q_m = \limsup_{t \to \infty} q_m(t),^1 \tag{2.44}$$

where, as above, $Q_m(\tau) = Q_m(\tau, u_0; \xi_1, \ldots, \xi_m)$, and where it has been assumed that $u_0 \in X = a$ functional-invariant set.

We infer from (2.42), (2.43) that

$$\bar{\omega}_m(t) = \underset{u_0 \in X}{\text{Sup}} \ \omega_m(L(t, u_0)) \leq \exp(tq_m(t)), \tag{2.45}$$

or, alternatively,

$$\bar{\omega}_m(t)^{1/t} \leq \exp q_m(t), \qquad \frac{1}{t}\log \bar{\omega}_m(t) \leq q_m(t). \tag{2.46}$$

Then, at the limit $m \to \infty$, we bound the uniform Lyapunov numbers and Lyapunov exponents for X

$$\Lambda_1 \ldots \Lambda_m \leq \exp(q_m), \tag{2.47}$$

$$\mu_1 + \cdots + \mu_m \leq q_m. \tag{2.48}$$

At this point it is easy to conclude with

Proposition 2.1. *Under the above assumptions, if for some m and some $t_0 > 0$,*

$$q_m(t) \leq -\delta < 0, \qquad \forall t \geq t_0, \tag{2.49}$$

then the volume element $|U_1(t) \wedge \cdots \wedge U_m(t)|_{\wedge^m H}$ decays exponentially as $t \to \infty$, uniformly for $u_0 \in X, \xi_1, \ldots, \xi_m \in H$,

$$|U_1(t) \wedge \cdots \wedge U_m(t)|_{\wedge^m H} \leq c \exp(-\delta t).^2 \tag{2.50}$$

If X is a functional-invariant set for the semigroup $S(t)$ and

$$q_m < 0 \tag{2.51}$$

[1] In principle, we can strengthen the results below by defining q_m as the lim inf of $q_m(t)$ as $t \to \infty$, instead of the lim sup, but in practical applications we cannot distinguish these two limits.

[2] $c = |U_1(t_0) \wedge \cdots \wedge U_m(t_0)|_{\wedge^m H} \exp(\delta t_0)$.

for some m, then $\Pi_m = \Lambda_1 \ldots \Lambda_m < 1, \mu_1 + \cdots + \mu_m < 0$, *which implies (at least) that* $\Lambda_m < 1$, *i.e.*, $\mu_m < 0$.

In Chapter VI the method and the proposition above will be explicitly applied to the equations considered in Chapters III and IV, and will provide conditions for the exponential decay of the volume element and estimates on the Lyapunov exponents. The exponential decay of the volume element was proved in P. Constantin and C. Foias [2] for the two-dimensional Navier–Stokes equation.

3. Hausdorff and Fractal Dimensions of Attractors

In this section our aim is to derive the two fundamental results which provide an estimate of the Hausdorff and fractal dimensions of an attractor, or a functional-invariant set, using the uniform Lyapunov exponents. The general results were proved in P. Constantin, C. Foias, and T. Temam [1] in the compact case; it was then shown in J.M. Ghidaglia and R. Temam [1] that the results can be extended to the noncompact case; we essentially follow the presentation in these articles.

In Section 3.1 we first recall the definition and the most elementary properties of the Hausdorff and fractal dimensions of a set X, the latter sometimes also being called the capacity of X. In Section 3.2 we present some technical lemmas concerning the covering of ellipsoids and ellipsoid-like sets. Section 3.3 then provides the main results concerning the Hausdorff and fractal dimensions of an invariant set; they are first expressed without utilization of the Lyapunov numbers (Theorems 3.1 and 3.2), Theorem 3.3 then provides an alternative form of the results using the Lyapunov numbers.

Let us mention that although Theorem 3.1–3.3 are stated in the discrete case, they are also applicable in the continuous case (see Remark 3.2).

3.1. Hausdorff and Fractal Dimensions

Let E be a metric space and let $Y \subset E$ be a subset of E. Given $d \in \mathbb{R}_+$ and $\varepsilon > 0$, we denote by $\mu_H(Y, d, \varepsilon)$ the quantity

$$\operatorname{Inf} \sum_{i \in I} r_i^d,$$

where the infimum is for all coverings of Y by a family $(B_i)_{i \in I}$ of balls of E of radii $r_i \leq \varepsilon$; $\mu_H(Y, d, \varepsilon)$ is clearly a nonincreasing function of ε and the number $\mu_H(Y, d) \in [0, \infty]$ defined by

$$\mu_H(Y, d) = \lim_{\varepsilon \to 0} \mu_H(Y, d, \varepsilon) = \operatorname{Sup}_{\varepsilon > 0} \mu_H(Y, d, \varepsilon) \tag{3.1}$$

is the d-dimensional Hausdorff measure of Y. It is easy to see that if $\mu_H(Y, d') < \infty$ for some d', then $\mu_H(Y, d) = 0$ for every $d > d'$; thus there exists $d_0 \in [0, \infty]$ such that $\mu_H(Y, d) = 0$ for $d > d_0$ and $\mu_H(Y, d) = +\infty$ for $d < d_0$, while $\mu_H(Y, d_0)$ can be any number in $[0, \infty]$. This number d_0 is called the *Hausdorff dimension of* Y and is denoted $d_H(Y)$. The reader is referred to H. Federer [1] for the properties of the Hausdorff measure and dimension. See also K.J. Falconer [1].

Now let $n_Y(\varepsilon)$, $\varepsilon > 0$, denote the minimum number of balls of H of radius ε which is necessary to cover Y. The *fractal dimension of* Y, which is also called the *capacity of* Y, is the number

$$d_F(Y) = \limsup_{\varepsilon \to 0} \frac{\log n_Y(\varepsilon)}{\log 1/\varepsilon}. \tag{3.2}$$

The following alternative expression for $d_F(Y)$ is given in B. Mandelbrot [1]:

$$d_F(Y) = \operatorname{Inf}\{d > 0,\ \mu_F(Y, d) = 0\}, \tag{3.3}$$

where

$$\mu_F(Y, d) = \limsup_{\varepsilon \to 0} \varepsilon^d n_Y(\varepsilon).$$

The difference between the Hausdorff and the fractal dimensions lies in the fact that we consider, in one case, the covering of Y by balls of radius $\leq \varepsilon$, and in the other case the covering of Y by balls of radius ε. The number $\mu_F(Y, d, \varepsilon) = \varepsilon^d n_Y(\varepsilon)$ is the analog of $\mu_H(Y, d, \varepsilon)$ for the fractal dimension; it is clear that

$$\mu_H(Y, d, \varepsilon) \leq \mu_F(Y, d, \varepsilon),$$

so that

$$\mu_H(Y, d) \leq \mu_F(Y, d)$$

and

$$d_H(Y) \leq d_F(Y). \tag{3.4}$$

It is worth mentioning that inequality (3.4) can indeed be a strict inequality. The simplest example of strict inequality is given by $Y = \{1/p, p \in \mathbb{N} \setminus \{0\}\}$, $E = \mathbb{R}$: indeed, if

$$\varepsilon = \varepsilon_p = \frac{1}{4}\left(\frac{1}{p} - \frac{1}{p+1}\right) = \frac{1}{4p(p+1)},$$

then the points $1, \ldots, 1/p$ necessarily belong to p different balls of radius ε_p. Thus $n_Y(\varepsilon_p) \geq p$ and we conclude that the fractal dimension of Y is $\geq \frac{1}{2}$, while its Hausdorff dimension is 0, like that of any denumberable set. If $E = \mathbb{R}$ again and Y consists of the points 2^{-p}, $p \in \mathbb{N}$, and of $2^{(p+1)^2} - 1$ equidistributed points between $2^{-(p+1)}$ and 2^{-p}, $\forall p \in \mathbb{N}$, we obtain a set of Hausdorff dimension

0 and of fractal dimension 1, i.e., that of $\mathbb{R}$ itself. In infinite dimension we can even construct a set of Hausdorff dimension 0 and fractal dimension ∞ (see, for example, A. Eden, C. Foias, and R. Temam [1]).

3.2. Covering Lemmas

We give two lemmas concerning the covering by balls of an ellipsoid and of an ellipsoid-like set. If $\mathscr{E}$ is an ellipsoid in H we denote by $\alpha_j(\mathscr{E})$, $j \geq 1$, its axes, $\alpha_1(\xi) \geq \alpha_2(\xi) \geq \cdots$, and we set $\omega_n(\mathscr{E}) = \alpha_1(\mathscr{E}) \ldots \alpha_n(\mathscr{E})$ when n is an integer and $\omega_d(\mathscr{E}) = \omega_n(\mathscr{E})^{1-s}\omega_{n+1}(\mathscr{E})^s = \omega_n(\mathscr{E})\alpha_{n+1}(\mathscr{E})^s$ when $d = n + s$, $n \in \mathbb{N}$, and $0 < s < 1$. It is clear that the function $d \in [1, \infty] \rightarrow \{\omega_d(\mathscr{E})\}^{1/d}$ is nonincreasing.

If L is a linear-continuous operator in H, it follows from (1.42) and (1.55) that the image by L of the unit ball $\mathscr{B}$ of H is included in an ellipsoid $\mathscr{E}$ whose axes $\alpha_j(\mathscr{E})$ are precisely equal to $\alpha_j(L)$, $\forall j \in \mathbb{N}$. This remark will be used in Sections 3.3 and 3.4 and for the moment we prove the following:

Lemma 3.1. *Let there be given $d > 0$, $d = n + s$, $n \in \mathbb{N}$, $n \geq 1$, and $0 < s \leq 1$, and an ellipsoid $\mathscr{E} \subset H$. For any r, $\alpha_{n+1}(\mathscr{E}) \leq r \leq \alpha_1(\mathscr{E})$, the minimum number of balls of radii $\sqrt{n+1}\, r$ which is necessary to cover $\mathscr{E}$ is*

$$n_{\mathscr{E}}(\sqrt{n+1}\,r) \leq 2^n \frac{\omega_l(\mathscr{E})}{r^l},^1 \tag{3.5}$$

where l is the largest integer $\leq n$ such that $r \leq \alpha_l(\mathscr{E})$.

Consequently, if $\varepsilon \geq (\omega_d(\mathscr{E}))^{1/d}$, then

$$\mu_H(\mathscr{E}, d, \sqrt{n+1}\,\varepsilon) \leq \beta_d \omega_d(\mathscr{E}), \qquad \beta_d = 2^n(n+1)^{d/2}. \tag{3.6}$$

Proof. Let $\alpha_i = \alpha_i(\mathscr{E})$, $\rho = \alpha_{n+1}$, and consider an orthonormal basis φ_j, $j \geq 1$, of H corresponding to the ordered axes of $\mathscr{E}$. The ellipsoid is included in the product of the set $\prod_{i=1}^n [-\alpha_i, \alpha_i]$ of QH and the ball of $(I - Q)H$, centered at 0 of radius ρ, where Q is the orthonormal projector onto the space spanned by $\varphi_1, \ldots, \varphi_n$.

The set $\prod_{i=1}^n [-\alpha_i, \alpha_i]$ is covered by N cubes of QH of edge $2r$, with

$$N \leq \prod_{i=1}^n \left(\left[\frac{\alpha_i}{r}\right] + 1\right) \leq \prod_{i=1}^l \left(\frac{2\alpha_i}{r}\right) \cdot 2^{n-l} \leq 2^n \frac{\omega_l(\mathscr{E})}{r^l}.$$

Hence $\mathscr{E}$ is covered by the product of these N cubes with the ball $B(0, \rho)$ of

[1] Of course, $n_{\mathscr{E}}(\sqrt{n+1}\,r) = n_{\mathscr{E}}(r) = 1$ for $r > \alpha_1(\mathscr{E})$. Hence, in the form

$$n_{\mathscr{E}}(\sqrt{n+1}\,r) \leq \max\left\{1, 2^n \frac{\omega_l(\mathscr{E})}{r^l}\right\}, \tag{3.5'}$$

formula (3.5) is valid for any $r \geq \alpha_{n+1}(\mathscr{E})$.

$(I - Q)H$ and each of these sets is included in a ball of H of radius $\sqrt{n+1}r$; hence (3.5).

For (3.6) we apply (3.5) with $r = \rho$ (in which case $l = n$), and we observe that $\rho \leq (\omega_n(E))^{1/n} \leq (\omega_d(E))^{1/d} \leq \varepsilon$; thus

$$\mu_H(\mathscr{E}, d, \sqrt{n+1}\varepsilon) \leq \mu_H(\mathscr{E}, d, \sqrt{n+1}\rho)$$

$$\leq 2^n \frac{\omega_n(\mathscr{E})}{\rho^n} \cdot (n+1)^{d/2} \rho^d = 2^n (n+1)^{d/2} \omega_d(\mathscr{E}). \qquad \square$$

Lemma 3.2. *Let $\mathscr{E}$ be an ellipsoid such that $\alpha_1(\mathscr{E}) \leq m$, $\omega_d(\mathscr{E}) \leq k$ with $k \leq m^d$, $d = n + s$, n integer ≥ 1, $0 < s \leq 1$. Then, for any $\eta > 0$, the sum $\mathscr{E} + B(0, \eta)$ is included in an ellipsoid $\mathscr{E}'$ such that*

$$\omega_d(\mathscr{E}') \leq (1 + K\eta)^d k, \tag{3.7}$$

$$K = \left(\frac{m^n}{k}\right)^{1/s}. \tag{3.8}$$

Proof. Since $k \leq m^d$, we can, by increasing the α_j, embed $\mathscr{E}$ into an ellipsoid $\bar{\mathscr{E}}$ such that $\omega_d(\bar{\mathscr{E}}) = k$ and $\bar{\alpha}_j = \alpha_j(\bar{\mathscr{E}}) = \alpha_{n+1}(\bar{\mathscr{E}})$ (hereafter denoted ρ), $\forall j \geq n+1$, with $\bar{\alpha}_1 = \alpha_1(\bar{\mathscr{E}}) \leq m$.

Then $k = \bar{\alpha}_1 \ldots \bar{\alpha}_n (\bar{\alpha}_{n+1})^s \leq m^n \rho^s$, so that $\rho \geq K^{-1} = (k/m^n)^{1/s}$. Since the ball $B(0, \rho)$ is included in $\bar{\mathscr{E}}$, we can write

$$\bar{\mathscr{E}} + B(0, \eta) \subset \left(1 + \frac{\eta}{\rho}\right) \bar{\mathscr{E}} \subset (1 + K\eta)\bar{\mathscr{E}}.$$

We conclude that $\mathscr{E}' = (1 + K\eta)\bar{\mathscr{E}}$ satisfies the desired properties. $\square$

Remark 3.1. Although they become obvious in this case, it is worth noticing that Lemmas 3.1 and 3.2 remain valid if $n = 0$ and $0 < s \leq 1/d$, i.e., $0 < d \leq 1$, if we agree that $\omega_0(\mathscr{E}) = 1$, $\omega_d(\mathscr{E}) = (\omega_1(\mathscr{E}))^d$. In Lemma 3.1 we simply have $r = \omega_d(\mathscr{E}) = \alpha_1(\mathscr{E})$; (3.5) completed with (3.5′) says that $n_{\mathscr{E}}(r) \leq 1$, which is obvious since $n_{\mathscr{E}}(r) = 1$, and (3.6) amounts to saying that $\mu_H(\mathscr{E}, d, \varepsilon) \leq \omega_d(\mathscr{E})$ ($\beta_1 = 1$) when $\varepsilon \geq \omega_1(\mathscr{E})$ and this follows readily from the definition. In Lemma 3.2, $k = m = \alpha_1(\mathscr{E}) = \omega_1(\mathscr{E}) = \rho$, $\bar{\mathscr{E}}$ reduces to the ball centered at 0 of radius ρ, $\mathscr{E}'$ is the ball $B(0, \eta + \rho)$, $K = 1/k = 1/\rho$, and (3.7) is obvious.

3.3. The Main Results

We now state and prove the main results concerning the Hausdorff and fractal dimensions of invariant sets.

Let H be a Hilbert space (norm $|\cdot|$), $X \subset H$ a compact set, and S a (nonlinear) continuous mapping from X into H such that

$$SX = X.^1 \tag{3.9}$$

[1] For Theorems 3.1–3.3 it suffices to assume that $SX \supset X$, provided S is defined on all of H.

As in (2.14) we assume that L is "uniformly differentiable on X", i.e.,

For every $u \in X$, there exists a linear operator $L(u) \in \mathscr{L}(H)$ and

$$\operatorname*{Sup}_{\substack{u, v \in X \\ 0 < |u-v| \leq \varepsilon}} \frac{|Su - Sv - L(u)\cdot(v-u)|}{|v-u|} \to 0 \quad \text{as } \varepsilon \to 0. \tag{3.10}$$

As indicated above, these operators $L(u)$ may not be unique. Concerning these operators, we assume the following:[1]

$$\operatorname*{Sup}_{u \in X} |L(u)|_{\mathscr{L}(H)} < +\infty, \tag{3.11}$$

$$\operatorname*{Sup}_{u \in X} \omega_d(L(u)) < 1 \quad \text{for some} \quad d > 0, \tag{3.12}$$

and then we have

Theorem 3.1. *Under the above assumptions, and in particular* (3.9)–(3.12), *the Hausdorff dimension of X is finite and is less than or equal to d.*

Proof. (i) Let k and m be such that

$$\operatorname*{Sup}_{u \in X} |L(u)|_{\mathscr{L}(H)} \leq m < \infty, \tag{3.13}$$

$$\operatorname*{Sup}_{u \in X} \omega_d(L(u)) \leq k < 1. \tag{3.14}$$

For any integer p, the mapping S^p is well defined from X into H and it enjoys similar properties: this is obvious for (3.9) and it is elementary to check that (3.10) is satisfied with $L(u)$ replaced by $L_p(u)$, whose expression was given in (2.15)

$$L_p(U) = L(S^{p-1}(u)) \circ \cdots \circ L(S(u)) \circ L(u).$$

Then (3.11), (3.12) are also satisfied and due to (3.13), (3.14), (1.60)

$$\operatorname*{Sup}_{u \in X} |L_p(u)|_{\mathscr{L}(H)} \leq m^p,$$

$$\operatorname*{Sup}_{u \in X} \omega_d(L_p(u)) \leq k^p.$$

We infer from this remark that, by replacing if necessary S by S^p, we can assume that the number k in (3.14) is arbitrarily small. In particular, it will be convenient for our later purposes to assume that (3.13), (3.14) hold with

$$\sqrt{d+1}k^{1/d} \leq \tfrac{1}{4} \quad \text{and} \quad \beta_d k \leq (\tfrac{1}{2})^{d+1}, \tag{3.15}$$

$\beta_d = 2^n(n+1)^{d/2}$ as in Lemma 3.1.

[1] We recall from (1.64) the definition of $\omega_d(L)$ for $d = n + s$, $n \in \mathbb{N}$, $0 < s < 1$: $\omega_d(L) = \omega_d(L)^{1-s}\omega_{n+1}(L)^s$. For $n = 0$ we set, as in Remark 1.5, $\omega_0(L) = 1$ and then $\omega_d(L)^{1/d}$ is a non-increasing function for $d > 0$.

(ii) *Iterated Coverings of X*

By increasing m if necessary, we can assume that $k \le m^d$ as in Lemma 3.2 (see also Remark 3.1). We are then given $\eta > 0$, $\eta < 1/K$, K as in (3.8), and we choose $\varepsilon > 0$ such that the supremum in (3.10) is less than or equal to η.

Since X is compact we can cover this set by a finite number of balls $B(u_i, r_i)$ centered at u_i of radius $r_i \le \varepsilon$, $i = 1, \ldots, N$,

$$X \subset \bigcup_{i=1}^{N} B(u_i, r_i) \cap X.$$

Then by (3.9)

$$SX = X \subset \bigcup_{i=1}^{N} S(B(u_i, r_i) \cap X).$$

Due to (3.10) and the choice of ε,

$$|Sv - Su_i - L(u_i)(v - u_i)| \le \eta |v - u_i|, \tag{3.16}$$

for every $v \in B(u_i, r_i) \cap X$, $i = 1, \ldots, N$. We set $B(u_i, r_i) = u_i + B_i$, $B_i = B(0, r_i)$, $\forall i$, and we infer from Proposition 1.3 that $L(u_i)(B_i)$ is included in an ellipsoid $\mathscr{E}_i$ described in Proposition 1.3, in particular, the length of its axes is $r_i \alpha_j(L(u_i))$, $\forall j \in \mathbb{N}$, $i = 1, \ldots, N$. It then follows from (3.16) that $S(u_i + B_i \cap X)$ is included in the sum

$$Su_i + \mathscr{E}_i + B(0, \eta r_i).$$

Since $\omega_d((1/r_i)\mathscr{E}_i)$ is exactly equal to $\omega_d(L(u_i))$ we can write

$$\omega_d(\mathscr{E}_i) = r_i^d \omega_d\left(\frac{\mathscr{E}_i}{r_i}\right) = r_i^d \omega_d(L(u_i)) \le k r_i^d.$$

and we deduce from Lemma 3.2 that $\mathscr{E}_i + B(0, \eta r_i)$ is included in an ellipsoid $\mathscr{E}_i'$ such that

$$\omega_d(\mathscr{E}_i') \le (1 + K\eta)^d k r_i^d \le 2^d k r_i^d, \qquad i = 1, \ldots, N. \tag{3.17}$$

At this point we conclude that we have covered $X = SX$ by the sets $Su_i + \mathscr{E}_i'$, $i = 1, \ldots, N$, $\mathscr{E}_i'$ satisfying (3.17).

(iii) We now apply inequality (3.6) in Lemma 3.1 to each ellipsoid $\mathscr{E}_i'$

$$\begin{aligned} \mu_H\left(\mathscr{E}_i', d, \frac{\varepsilon}{2}\right) &\le \text{(due to (3.15))} \\ &\le \mu_H(\mathscr{E}_i', d, \sqrt{d+1}2k^{1/d}\varepsilon) \\ &\le \mu_H(\mathscr{E}_i', d, \sqrt{n+1}2k^{1/d}r_i) \\ &\le \beta_d \omega_d(\mathscr{E}_i') \le 2^d k \beta_d r_i^d \\ &\le \text{(due to (3.15))} \le \tfrac{1}{2} r_i^d. \end{aligned}$$

Then we can write

$$\mu_H\left(X, d, \frac{\varepsilon}{2}\right) \leq \sum_{i=1}^{N} \mu_H\left(\mathscr{E}_i', d, \frac{\varepsilon}{2}\right)$$
$$\leq \frac{1}{2}\sum_{i=1}^{N} r_i^d,$$

and by taking the infimum for all the coverings of X by balls $B(u_i, r_i)$, $i = 1, \dots, N$, with $r_i \leq \varepsilon$, we find

$$\mu_H\left(X, d, \frac{\varepsilon}{2}\right) \leq \tfrac{1}{2}\mu_H(X, d, \varepsilon). \tag{3.18}$$

By reiteration we obtain

$$\mu_H\left(X, d, \frac{\varepsilon}{2^j}\right) \leq \frac{1}{2^j}\mu_H(X, d, \varepsilon), \qquad \forall j \in \mathbb{N}, \tag{3.19}$$

and letting $j \to \infty$, we obtain $\mu_H(X, d) = 0$.

The proof is complete. □

Remark 3.2. Under the assumptions of Theorem 3.1, the Hausdorff dimension of X is 0 if $d \leq 1$. Indeed, for $d \leq 1$, $\bar{\omega}_d < 1$ is equivalent to $\bar{\omega}_1 < 1$. Therefore if $\bar{\omega}_{d_0} < 1$ for some $d_0 < 1$, we also have $\bar{\omega}_d < 1, \forall d, 0 < d < 1$, thus $d_H(X) \leq d$, $\forall d > 0$ and $d_H(X) = 0$.

We now proceed and establish the analog of Theorem 3.1 for the fractal dimension. The assumptions are essentially the same except that we replace (3.12) by a stronger hypothesis, which is motivated by the fact that a larger number of balls of a given radius may be necessary to cover ellipsoids $\mathscr{E}_i$ if they are very thin. We assume that, for some $d = n + s$, n integer, $0 < s \leq 1$, we have

$$\bar{\omega}_j \bar{\omega}_{n+1}^{(d-j)/(n+1)} < 1 \quad \text{for} \quad j = 1, \dots, n, \tag{3.20}$$

where as in (2.18)

$$\bar{\omega}_j = \operatorname*{Sup}_{u \in X} \omega_j(L(u)). \tag{3.21}$$

Setting $i = n$ in (3.20) and using (1.64), (1.65) we see easily that the left-hand side of (3.20) is larger than or equal to $\bar{\omega}_d$ and (3.20) is indeed stronger than (3.12).

Theorem 3.2. *The hypotheses are those of Theorem* 3.1, (3.12) *being replaced by* (3.20) *and* $d > 1$. *Then the fractal dimension of* X *is finite and less than or equal to* d.

Proof. The proof is similar to that of Theorem 3.1. We use the same notations in this proof and points (i) and (ii) are exactly the same: as observed above,

$\bar{\omega}_d < 1$ since $\bar{\omega}_d$ is majorized by the left-hand side of (3.20) with $i = n$. In point (iii) the utilization of (3.6) is not appropriate and instead we use (3.5).

We set $r = 2\varepsilon\bar{\alpha}_{n+1}$ where (compare with (2.23))

$$\bar{\alpha}_j = \operatorname*{Sup}_{u \in X} \alpha_j(L(u)), \qquad \forall j \geq 1; \tag{3.22}$$

we have

$$\begin{aligned}
\alpha_{n+1}(\mathscr{E}_i') &= (1 + K\eta)\alpha_{n+1}(\mathscr{E}_i) \\
&= (1 + K\eta) r_i \alpha_{n+1}(L(u_i)) \\
&\leq (1 + K\eta) r_i \bar{\alpha}_{n+1} \\
&\leq 2 r_i \bar{\alpha}_{n+1} \leq 2\varepsilon\bar{\alpha}_{n+1}.
\end{aligned}$$

With this value of r we can then apply (3.5) to $\mathscr{E}_i'$. Taking into account the footnote after (3.5), and the fact that the value of l in (3.5) is not known ($1 \leq l \leq n$), we can write that the minimum number of balls of radius $2\sqrt{n+1}\,\varepsilon\bar{\alpha}_{n+1}$, which is necessary to cover $\mathscr{E}_i'$, is

$$\begin{aligned}
n_{\mathscr{E}_i'}(2\sqrt{n+1}\,\varepsilon\bar{\alpha}_{n+1}) &\leq \max\left\{1, \max_{1 \leq j \leq n} 2^n \frac{\omega_j(\mathscr{E}_i')}{(2\varepsilon\bar{\alpha}_{n+1})^j}\right\} \\
&\leq \max\left\{1, \max_{1 \leq j \leq n} 2^n \frac{r_i^j \omega_j(L(u_i))}{(\varepsilon\bar{\alpha}_{n+1})^j}\right\} \\
&\leq \max\left\{1, 2^n \max_{1 \leq j \leq n} \frac{\bar{\omega}_j}{(\bar{\alpha}_{n+1})^j}\right\}.
\end{aligned}$$

Since $\bar{\omega}_1 = \bar{\alpha}_1 \geq \bar{\alpha}_{n+1}$, we obtain

$$n_{\mathscr{E}_i'}(2\sqrt{n+1}\,\varepsilon\bar{\alpha}_{n+1}) \leq 2n \max_{1 \leq j \leq n} \frac{\bar{\omega}_j}{(\bar{\alpha}_{n+1})^j}. \tag{3.23}$$

Now once we have chosen $\eta < 1/K$ and ε as in the proof of Theorem 3.1, we can assume that the initial covering of X by the balls $B(u_i, r_i)$, $i = 1, \ldots, N$, considered in point (ii) corresponds to a minimal covering of X by balls of radius ε, so that $N = n_X(\varepsilon)$ and $r_i = \varepsilon$, $\forall i$. The minimum number of balls of radius $2\sqrt{d+1}\,\varepsilon\bar{\alpha}_{n+1}$ which is necessary to cover X can then be majorized according to (3.23) by

$$\begin{gathered}
n_X(2\sqrt{d+1}\,\varepsilon\bar{\alpha}_{n+1}) \leq n_X(\varepsilon) \max_{1 \leq i \leq n} n_{\mathscr{E}_i'}(2\sqrt{d+1}\,\varepsilon\bar{\alpha}_{n+1}), \\
n_X(2\sqrt{d+1}\,\varepsilon\bar{\alpha}_{n+1}) \leq 2^d n_X(\varepsilon) \max_{1 \leq j \leq n} \frac{\bar{\omega}_j}{(\bar{\alpha}_{n+1})^j}.
\end{gathered} \tag{3.24}$$

It is clear that $\bar{\alpha}_{n+1} \leq (\bar{\omega}_{n+1})^{1/(n+1)}$ and thus

$$\begin{aligned}
&(2\sqrt{d+1}\,\varepsilon\bar{\alpha}_{n+1})^d n_X(2\sqrt{d+1}\,\varepsilon\bar{\alpha}_{n+1}) \\
&\qquad \leq 2^{2d}(d+1)^{d/2} \max_{1 \leq j \leq n} [\bar{\omega}_j (\bar{\omega}_{n+1})^{(d-j)/(n+1)}] \varepsilon^d n_X(\varepsilon).
\end{aligned} \tag{3.25}$$

As in point (i) of Theorem 3.1 we can replace S by S^p and $L(u)$ by $L_p(u)$

$$L_p(u) = L(S^{p-1}(u)) \circ \cdots \circ L(S(u)) \circ L(u),$$

and this allows us to render centain quantities small by choosing p large. For instance, replacing $\overline{\alpha}_{n+1}$ and $\overline{\omega}_j$ by $\overline{\alpha}_{n+1}(p)$ and $\overline{\omega}_j(p)$,

$$\overline{\alpha}_{n+1}(p) = \operatorname*{Sup}_{u \in X} \alpha_{n+1}(L_p(u)), \tag{3.26}$$

$$\overline{\omega}_j(p) = \operatorname*{Sup}_{u \in X} \omega_j(L_p(u)) \qquad (\leq (\overline{\omega}_j)^p), \tag{3.27}$$

inequality (3.25) becomes

$$(\alpha\varepsilon)^d n_X(\alpha\varepsilon) \leq \theta\varepsilon^d n_X(\varepsilon), \tag{3.28}$$

where

$$\begin{aligned}
\alpha &= 2\sqrt{d+1}\,\overline{\alpha}_{n+1}(p),\\
\theta &= 2^{2d}\{d+1\}^{d/2} \max_{1 \leq j \leq n} [\overline{\omega}_j(p)(\overline{\omega}_{n+1}(p))^{(d-j)/(n+1)}]\\
&\leq 2^{2d}\{d+1\}^{d/2} \left\{ \max_{1 \leq j \leq n} [\overline{\omega}_j(\overline{\omega}_{n+1})^{(d-j)/(n+1)}] \right\}^p.
\end{aligned}$$

It is easy to see that

$$\overline{\alpha}_{n+1} \leq \overline{\omega}_j(\overline{\omega}_{n+1})^{(d-j)/(n+1)} \quad \text{for} \quad j = 1, \ldots, n,$$

and similarly,

$$\overline{\alpha}_{n+1}(p) \leq \overline{\omega}_j(p)(\overline{\omega}_{n+1}(p))^{(d-j)/(n+1)} \leq \{\overline{\omega}_j(\overline{\omega}_{n+1}(p))^{(d-j)/(n+1)}\}^p \quad \text{for} \quad j = 1, \ldots, n.$$

Hence, thanks to (3.20), we have $\alpha \leq \frac{1}{2}$ and $\theta \leq \frac{1}{2}$ if p is large enough. With this choice of α and θ, (3.28) can be written in the form

$$\varphi(\alpha\varepsilon) \leq \tfrac{1}{2}\varphi(\varepsilon) \quad \text{for} \quad 0 < \varepsilon \leq \varepsilon_0, \tag{3.29}$$

where $\varphi(\varepsilon) = \varepsilon^d n_X(\varepsilon)$, $0 < \alpha < 1$, and ε_0 is sufficiently small.[1]

As we show below, inequality (3.29) implies that

$$\varphi(\varepsilon) \to 0 \quad \text{as } \varepsilon \to 0. \tag{3.30}$$

For $\varepsilon \leq \varepsilon_0$, we define $j = j(\varepsilon)$ by the condition $\alpha^{j+1}\varepsilon_0 \leq \varepsilon < \alpha^j\varepsilon_0$, and by reiteration of (3.29) we can write

$$\varphi(\varepsilon) = \varphi(\alpha^j\alpha^{-j}\varepsilon) \leq 2^{-j}\varphi(\alpha^{-j}\varepsilon) \leq 2^{-j}M, \tag{3.31}$$

$$M = \operatorname*{Sup}_{\alpha\varepsilon_0 \leq \varepsilon' \leq \varepsilon_0} \varphi(\varepsilon') = \operatorname*{Sup}_{\alpha\varepsilon_0 \leq \varepsilon' \leq \varepsilon_0} (\varepsilon')^d n_X(\varepsilon') \leq \varepsilon_0^d n_X(\alpha\varepsilon_0) < \infty.$$

We easily deduce (3.30) from (3.31). It follows from (3.30) that there exists

[1] We first choose p such that $\alpha \leq \frac{1}{2}, \theta \leq \frac{1}{2}$; then ε_0 is the largest number making the left-hand side of (3.10) less than or equal to $\eta_0 = 1/2K$, L being replaced by L_p.

$\varepsilon_1 > 0$ such that

$$\varphi(\varepsilon) = \varepsilon^d n_X(\varepsilon) \leq 1 \quad \text{for} \quad \varepsilon \leq \varepsilon_1.$$

Hence

$$\frac{\log n_X(\varepsilon)}{\log 1/\varepsilon} \leq d \quad \text{for} \quad \varepsilon \leq \varepsilon_1,$$

$$\limsup_{\varepsilon \to 0} \frac{\log n_X(\varepsilon)}{\log 1/\varepsilon} \leq d,$$

and in view of (3.2) this completes the proof of Theorem 3.2.

Remark 3.3. As in Remark 3.2 we notice that if $\bar{\omega}_1 < 1$, in Theorem 3.2, then the fractal dimension of X vanishes. Indeed, $\bar{\omega}_1 < 1$ is equivalent to $\bar{\omega}_d < 1$ for any $d \leq 1$, and the proof of Theorem 3.1 and the proof of Theorem 3.2 can be extended to this case, setting $n = 0$, $\bar{\alpha}_{n+1} = \bar{\alpha}_1 = \bar{\omega}_1$.

Finally, using the Lyapunov exponents defined above, we give an alternative form of Theorems 3.1 and 3.2.

Theorem 3.3. *Under assumptions* (3.9)–(3.11) *and, if for some* $n \geq 1$,

$$\mu_1 + \cdots + \mu_{n+1} < 0, \tag{3.32}$$

then

$$\mu_{n+1} < 0, \qquad \frac{\mu_1 + \cdots + \mu_n}{|\mu_{n+1}|} < 1, \tag{3.33}$$

and

(i) *the Hausdorff dimension of* X *is less than or equal to*

$$n + \frac{(\mu_1 + \cdots + \mu_n)_+}{|\mu_{n+1}|}; \tag{3.34}$$

(ii) *the fractal dimension of* X *is less than or equal to*

$$(n+1)\left\{\operatorname*{Max}_{1 \leq j \leq n} 1 + \frac{(\mu_1 + \cdots + \mu_j)_+}{|\mu_1 + \cdots + \mu_{n+1}|}\right\}. \tag{3.35}$$

Proof. Because of (2.28), (3.32), μ_{n+1} cannot be ≥ 0; thus $\mu_{n+1} < 0$ and then $\mu_1 + \cdots + \mu_n < -\mu_{n+1} = |\mu_{n+1}|$.

The proof of the result concerning the Hausdorff dimension is essentially that of Theorem 3.1. We replace S by S^p and $L(u)$ by $L_p(u)$ as in point (i) of the proof of Theorem 3.1, and we only need to verify that

$$\sup_{u \in X} \omega_d(L_p(u)) < 1$$

for p sufficiently large and d larger than the number in (3.34). Due to (2.20), (2.22),

$$\mu_1 + \cdots + \mu_j = \lim_{p\to\infty} \frac{1}{p} \log \bar{\omega}_j(p), \qquad \forall j \geq 1, \tag{3.36}$$

and it is already clear that $\bar{\omega}_{n+1}(p_0) < 1$ for some p_0 sufficiently large. In order that $\bar{\omega}_{d'}(p_0) < 1$ for p_0 sufficiently large and $d' = n + s', 0 < s' < 1$, we should have

$$\lim_{p\to\infty} \frac{1}{p} \log \bar{\omega}_{d'}(p) < 0. \tag{3.37}$$

But with (1.64)

$$\bar{\omega}_{d'}(p) \leq (\bar{\omega}_n(p))^{1-s'}(\bar{\omega}_{n+1}(p))^{s'},$$

$$\begin{aligned} \lim_{p\to\infty} \frac{1}{p} \log \bar{\omega}_{d'}(p) &\leq \lim_{p\to\infty} \left(\frac{1-s'}{p} \log \bar{\omega}_n(p)\right) + \lim_{p\to\infty} \left(\frac{s'}{p} \log \bar{\omega}_{n+1}(p)\right) \\ &\leq (1-s')(\mu_1 + \cdots + \mu_n) + s'(\mu_1 + \cdots + \mu_{n+1}) \\ &\leq \mu_1 + \cdots + \mu_n + s'\mu_{n+1}. \end{aligned} \tag{3.38}$$

A sufficient condition for (3.37) is then

$$s' > \frac{\mu_1 + \cdots + \mu_n}{(-\mu_{n+1})},$$

and since we also require $s' > 0$, we find

$$s' > \frac{(\mu_1 + \cdots + \mu_1)_+}{|\mu_{n+1}|}.$$

In conclusion, (3.37) is valid for any d' strictly larger than the number d_0 in (3.34); for such a d', $\bar{\omega}_{d'}(p_0) < 1$ for p_0 sufficiently large and, by Theorem 3.1, the Hausdorff dimension of X is $\leq d'$ for every $d' > d_0$. The Hausdorff dimension of X is thus $\leq d_0$.

For the fractal dimension it suffices to prove that this dimension is less than or equal to d_1 where

$$d_1 = \max_{1\leq j\leq n} \left\{n + \frac{(\mu_1 + \cdots + \mu_j)_+}{|\bar{\mu}_{n+1}|}\right\} \tag{3.39}$$

and $\bar{\mu}_{n+1}$ is given by (2.24)

$$\bar{\mu}_{n+1} = \limsup_{p\to\infty} \frac{1}{p} \log(\bar{\alpha}_{n+1}(p)),$$

since, due to (2.27), d_1 is less than or equal to the number (3.35).

We proceed as for Theorem 3.2 again replacing S by $L(u)$ and S^p by $L_p(u)$. A perusal of the proof of Theorem 3.2 shows that (3.24) does not necessitate

(3.20), and we can then write this inequality with S, $L(u)$ replaced by S^p, $L_p(u)$, as mentioned before, and $d > d_1$

$$n_X(2\sqrt{d+1}\,\varepsilon\bar{\alpha}_{n+1}(p)) \le 2^d n_X(\varepsilon) \max_{1\le j\le n} \frac{\bar{\omega}_j(p)}{(\bar{\alpha}_{n+1}(p))^j}, \qquad \forall p \ge 1.^1 \quad (3.40)$$

Thus

$$(2\sqrt{d+1}\,\varepsilon\bar{\alpha}_{n+1}(p))^d n_X(2\sqrt{d+1}\,\varepsilon\bar{\alpha}_{n+1}(p)) \le \theta n_X(\varepsilon), \quad (3.41)$$

with

$$\theta = \theta(p) 2^{2d}(d+1)^{d/2} \max_{1\le j\le n} [\bar{\omega}_j(p)(\bar{\alpha}_{n+1}(p))^{d-j}].$$

We have

$$\begin{aligned}
&\limsup_{p\to\infty} \frac{1}{p} \log\{\bar{\omega}_j(p)(\bar{\alpha}_{n+1}(p))^{d-j}\}\\
&\quad = \mu_1 + \cdots + \mu_j + (d-j)\bar{\mu}_{n+1}\\
&\quad \le (\text{since } \bar{\mu}_{n+1} < 0 \text{ by (2.28), (3.32)})\\
&\quad \le \mu_1 + \cdots + \mu_j - (d-j)|\bar{\mu}_{n+1}|\\
&\quad \le (\mu_1 + \cdots + \mu_j)_+ - (d-n)|\bar{\mu}_{n+1}|, \qquad j = 1, \ldots, n.
\end{aligned}$$

This last term is < 0 due to (3.39) and since $d > d_1$. Consequently, $\limsup_{p\to\infty} (1/p) \log \theta(p) < 0$ and we can find arbitrarily large p's such that $\theta = \theta(p) \le \frac{1}{2}$. Similarly,

$$\limsup_{p\to\infty} \{\bar{\alpha}_{n+1}(p)\}^{1/p} = \exp(\bar{\mu}_{n+1}) < 1,$$

and $\alpha = 2\sqrt{d+1}\,\varepsilon\bar{\alpha}_{n+1}(p) \le \frac{1}{2}\varepsilon^d n_X(\varepsilon)$.

We conclude, as in Theorem 3.2, that the fractal dimension of X is $\le d$, and since $d > d_1$ is arbitrarily close to d_1, this dimension is $\le d_1$.

Theorem 3.3 is proved. □

Remark 3.4. If, under the assumptions of Theorem 3.3, $\mu_1 < 0$, then the Hausdorff and fractal dimensions of X vanish. The proof of Theorem 3.3 can easily be extended to this case using Remarks 3.2 and 3.3.

Remark 3.5.

(i) The bound (3.34) for the Hausdorff dimension of X is particularly interesting when $n = n_0$ is the smallest integer for which (3.32) occurs, in which case,

$$d_H(X) \le n_0 + \frac{\mu_1 + \cdots + \mu_{n_0}}{|\mu_{n_0+1}|}. \quad (3.42)$$

[1] Note that n is not the same here as in Theorem 3.2 and Lemma 3.1, i.e., the relation $d = n + s$, n integer, $0 < s \le 1$, does not apply here.

(ii) The number on the right-hand side of (3.46) (which belongs to the interval $]n_0, n_0 + 1[$) is called the *Lyapunov dimension* of X (see J. Kaplan and J. Yorke [1], D. Farmer [1]).

Remark 3.6. In the compact case, i.e., assuming that the operators $L(u)$ are compact, Theorem 3.1 is due to A. Douady and J. Oesterlé [1], Theorems 3.2 and 3.3 are due to P. Constantin, C. Foias, and R. Temam [1]. The extension to the noncompact case was made in J.M. Ghidaglia and R. Temam [1].

Remark 3.7. An improved form of Theorem 3.3 is proved in A. Eden, C. Foias, and R. Temam [1] and in A. Eden [1]. The Lyapunov exponents at a point $u_0 \in X$ are recursively defined by

$$\mu_1(u_0) = \limsup_{p\to\infty} \frac{1}{p} \log \omega_1(L_p(u_0)),$$

$$(\mu_1 + \cdots + \mu_m)(u_0) = \limsup_{p\to\infty} \frac{1}{p} \log \omega_m(L_p(u_0)), \qquad \forall m \geq 2.$$

Then (compare to Theorem 3.3), if n is such that

$$(\mu_1 + \cdots + \mu_{n+1})(u_0) < 0, \qquad \forall u_0 \in X,$$

we can assert that the Hausdorff dimension of X is less than or equal to

$$n + \sup_{u_0 \in X} \frac{(\mu_1 + \cdots + \mu_n)_+(u_0)}{|\mu_{n+1}(u_0)|}.$$

Furthermore, there exists $u_0 \in X$ such that the Hausdorff dimension of X is less than or equal to

$$n + \frac{(\mu_1 + \cdots + \mu_n)_+(u_0)}{|\mu_{n+1}(u_0)|}.$$

The orbit of such a u_0 is called a *critical* orbit. The proof uses some methods developed in G. Choquet and C. Foias [1].

3.4. Application to Evolution Equations

Although Theorems 3.1–3.3 are stated in the discrete case, they are also applicable in the continuous case. Indeed, if $X \subset H$ is invariant for the semigroup $\{S(t)\}_{t\geq 0}$, it is also invariant for $S(t_0)$ for an arbitrary $t_0 > 0$. Thus Theorem 3.1 is applicable if $S(t)X = X, \forall t$, if the operators $S(t)$ are "uniformly differentiable on X" and for some $t_0 > 0$ and $d \geq 1$

$$\operatorname{Sup}_{u\in X} |L(t_0, u)|_{\mathscr{L}(H)} < \infty, \tag{3.43}$$

$$\operatorname{Sup}_{u\in X} \omega_d(L(t_0, u)) < 1. \tag{3.44}$$

For Theorem 3.2 we replace (3.43) by the hypothesis that for some $t_0 > 0$ and some $d > 1$, $d = n + s$, n integer, $0 < s \leq 1$,

$$\bar{\omega}_j(t_0)(\bar{\omega}_{n+1}(t_0))^{(d-j)/(n+1)} < 1 \quad \text{for} \quad i = 1, \ldots, n. \tag{3.45}$$

Finally, Theorem 3.3 is applicable without modification; we notice that the Lyapunov exponents that we can associate with X (by considering either X as an invariant set for the continuous semigroup $\{S(t)\}_{t \geq 0}$ or an invariant set for an operator $S(t_0)$ ($t_0 > 0$)) are proportional

$$\lim_{t \to \infty} \frac{1}{t} \log \bar{\omega}_j(t) = \frac{1}{t_0} \lim_{t \to \infty} \frac{1}{n} \log \bar{\omega}_j(nt_0). \tag{3.46}$$

Therefore condition (3.32) and the numbers in (3.34), (3.35) are the same in both cases.

Practical Applications

In practical applications to evolution equations, Theorem 3.3 will be used most often. We proceed as follows: we consider the first variation equation (2.39) and we compute (estimate) the numbers $q_m(t)$ and q_m as in (2.43), (2.44)

$$q_m(t) = \operatorname*{Sup}_{u_0 \in X} \operatorname*{Sup}_{\substack{\xi_i \in H \\ |\xi_i| \leq 1 \\ i=1,\ldots,m}} \left\{ \frac{1}{t} \int_0^t \operatorname{Tr} F'(S(\tau)u_0) \circ Q_m(\tau)\, d\tau \right\}, \tag{3.47}$$

$$q_m = \limsup_{t \to \infty} q_m(t) \tag{3.48}$$

(see before (2.43) the explanation of the different quantities). By (2.48)

$$\mu_1 + \cdots + \mu_m \leq q_m, \tag{3.49}$$

and we look for an m such that

$$q_m < 0. \tag{3.50}$$

For such an m, the main assumptions of Proposition 2.1 and Theorem 3.3 are fulfilled, namely, (2.49) and (3.32) (with $n + 1 = m$). Proposition 2.1 is applicable and produces the exponential decay of the m-volume element. Theorem 3.3 implies that the Hausdorff dimension of X is less than or equal to m, and the fractal dimension of X is less than or equal to

$$m \max_{1 \leq j \leq m-1} \left(1 + \frac{(q_j)_+}{|q_m|}\right). \tag{3.51}$$

The following simple case will appear repeatedly, namely

$$q_j \leq -\alpha j^\theta + \beta, \qquad \forall j \in \mathbb{N}, \tag{3.52}$$

for some positive numbers α, β, θ. Then we choose m such that

$$m - 1 < \left(\frac{2\beta}{\alpha}\right)^{1/\theta} \leq m, \tag{3.53}$$

in which case $q_m \leq -\alpha m^\theta + \beta \leq -\beta$. Hence (2.49) is satisfied with $\beta = \delta$ and (3.32) is satisfied with $n + 1 = m$, from which we conclude that

$$\text{The Hausdorff dimension of } X \text{ is less than or equal to } m. \tag{3.54}$$

Then concerning the fractal dimension of X, we observe that $(q_j)_+ \leq \beta, \forall j$ and $|q_m| = -q_m \geq \beta$, so that

$$1 + \frac{(\mu_1 + \cdots + \mu_j)_+}{|\mu_1 + \cdots + \mu_{n+1}|} \leq 1 + \frac{(q_j)_+}{|q_m|} \leq 2,$$

and

$$\text{The fractal dimension of } X \text{ is less than or equal to } 2m. \tag{3.55}$$

CHAPTER VI

Explicit Bounds on the Number of Degrees of Freedom and the Dimension of Attractors of Some Physical Systems

Introduction

This chapter is aimed at applying the general results of Chapter V to the attractors of all the physical equations that we have considered in Chapters III and IV. It appears as one of the culminating points of the theory of attractors for dissipative partial differential equations presented in this book.

As indicated on several occasions in Chapters III and IV and before, the ω-limit set of a point and, more generally, the global attractor of a dissipative equation are the mathematical objects which represent the permanent regime that can be observed when the excitation starts from any point in the function (phase) space or when the excitation starts from that particular point. All of the information of a quantitative or qualitative nature concerning the attractor then yields valuable information concerning the flows that this physical system can generate.

One of the desirable pieces of information concerning a given flow is to try to measure or to predict its level of complexity. Although chaotic behavior can already be observed with a low number of degrees of freedom, it is expected that for many phenomena of physical interest exhibiting complex behaviors the number of degrees of freedom is high. For example, in wind-tunnel fluid-mechanics experiments this number of degrees of freedom is of the order of 10^8 to 10^9; for the geophysical flows of meteorology, it is of the order of 10^{18} and comparable numbers can be attained for plasmas. Although these numbers are very high, intermediate values of experimental parameters lead to numbers of degrees of freedom which are within the present computational capacity of computers, and of course we can expect still higher capacities in the future.

It is our understanding here that the number of degrees of freedom of a turbulent phenomenon is the dimension of the attractor which represents it.

Of course, from the strict mathematical point of view, an attractor can be a very complicated set and if its Hausdorff dimension is N we cannot expect to parametrize it in $\mathbb{R}^N$, as if it were a smooth manifold of dimension N. Nevertheless, it is known that a set of finite Hausdorff dimension is homeomorphic to a subset of $\mathbb{R}^p$ for a suitable p; a much stronger result is that almost all the projectors of dimension $2N + 1$ are injective on a compact set of Hausdorff dimension N (or $\leq N$) (see R. Mañé [1]).

As indicated above, our aim in this chapter is to apply to various equations the results of Chapter V for the estimate of the Lyapunov exponents, and of the Hausdorff and fractal dimensions of the attractors; we also give some sufficient conditions on N for the exponential decay of the N-volume element in the phase space.

The chapter is organized as follows. We start in Section 1 with a finite-dimensional example and we provide a bound on the Hausdorff dimension of the Lorenz attractor. This bound is not of course very sharp, but it shows how powerful are the theorems of Chapter V since they already give a nontrivial result in the lowest dimensional case. The other interest of this introductory section is to provide in a simple example a very explicit form of the details of the computations which will be repeated constantly in the following sections.

In Sections 2–7 we consider most of the equations considered in Chapters III and IV. Each of these sections should be considered as a continuation of the corresponding section of Chapter III or IV, where the same equation was studied. Thus we automatically adopt here the notations of the corresponding section of Chapter III or IV and we refer to it constantly. Section 2 deals with the reaction–diffusion equations considered in Section III.1; Section 3 and then Section 4 concern the two-dimensional Navier–Stokes equations of Section III.2, and the other fluid mechanics equations of Section III.3. In Section 5 we continue the study of the pattern formation equations of Section III.4. The evolution equations of Section III.5 are not studied because of technical difficulties encountered in the differentiation of the semigroup $S(t)$. Then in Section 6 we consider the nonlinear damped-wave equations of Section IV.3, which include as a particular case those of Sections IV.1 and IV.2. Finally, Section 7 deals with the nonlinear Schrödinger equation of Section IV.4. This chapter concludes with a technical section, Section 8, which sketches the proof of differentiability properties of the semigroups $S(t)$.

1. The Lorenz Attractor

We return to the Lorenz equations studied in Section I.2.3. We consider the equations in the form I.(2.16), namely,

$$\begin{cases} x' + \sigma x - \sigma y = 0, \\ y' + \sigma x + y + xz = 0, \\ z' + bz - xy = -b(r + \sigma), \end{cases} \tag{1.1}$$

where σ, r, and b are either the numbers considered in the original work of E.N. Lorenz [1],

$$\sigma = 10, \qquad r = 28, \qquad b = \tfrac{8}{3}, \tag{1.2}$$

or more general numbers satisfying

$$\sigma > 0, \qquad r > 0, \qquad b > 1. \tag{1.3}$$

In Section I.2.3 we have set $u = (x, y, z)$, $H = \mathbb{R}^3$, $u(t) \in H$, and we have shown that

$$\limsup_{t\to\infty} |u(t)| \leq \rho_0, \qquad \rho_0 = \frac{b(r+\sigma)}{2\sqrt{l(b-1)}}, \tag{1.4}$$

where $|\cdot|$ is the usual Euclidean norm on $\mathbb{R}^3$. This shows, of course, that the Lorenz attractor is included in the ball $B(0, \rho_0)$ of $\mathbb{R}^3$ centered at 0 of radius ρ_0.

Our aim is now to study the evolution of the two- and three-dimensional volume elements in the phase space $\mathbb{R}^3$ following the general approach discussed in Section V.2 (and especially Section V.2.3), then to deduce some information concerning the Lyapunov exponents, and finally, to estimate the Hausdorff dimension of the Lorenz attractor $\mathscr{A}$.

We rewrite (1.1) in the form

$$u' = F(u), \tag{1.5}$$

$$F(u) = F(x, y, z) = -\begin{pmatrix} \sigma x - \sigma y \\ \sigma x + y + xz \\ bz - xy + b(r+\sigma) \end{pmatrix}, \tag{1.6}$$

and the first variation equation V.(2.39) is

$$\frac{dU}{dt} = F'(u)\cdot U, \tag{1.7}$$

where $F'(u)$ is the following 3×3 matrix:

$$-F'(u)\cdot U = A_1 U + A_2 U + B(u)U,$$

$$A_1 = \begin{pmatrix} \sigma & 0 & 0 \\ 0 & 1 & 0 \\ 0 & 0 & b \end{pmatrix}, \qquad A_2 = \begin{pmatrix} 0 & -\sigma & 0 \\ 0 & 0 & 0 \\ 0 & 0 & 0 \end{pmatrix}, \tag{1.8}$$

$$B(u) = \begin{pmatrix} 0 & 0 & 0 \\ z & 0 & x \\ -y & -x & 0 \end{pmatrix}, \qquad \forall u = (x, y, z) \in \mathbb{R}^3.$$

We consider the initial-value problem for (1.7) with initial data

$$U(0) = \xi, \qquad \xi \in H \quad (= \mathbb{R}^3). \tag{1.9}$$

More precisely, we consider three initial-value problems (1.7), (1.9) with initial data $\xi = \xi_1, \xi_2, \xi_3$ ($\in H$), and solutions $U = U_1, U_2, U_3$. At each time $t > 0$,

$U_i(t) = L(t, u_0)\xi_i$, where u_0 is the initial data for (1.1), (1.5) (producing the solution $u = u(t) = S(t)u_0$) and $L(t, u_0)$ is simply the linear operator in $\mathbb{R}^3$

$$L(t, u_0)\colon U(0) = \xi \in \mathbb{R}^3 \to U(t) \in \mathbb{R}^3.$$

The evolution of the two- and three-dimensional volume elements obey equation V.(2.40) in $\bigwedge^m H$, $m = 2, 3$,

$$\frac{d}{dt}|U_1 \wedge U_2 \wedge U_3| = |U_1 \wedge U_2 \wedge U_3| \operatorname{Tr} F'(u), \tag{1.10}$$

$$\frac{d}{dt}|U_1 \wedge U_2| = |U_1 \wedge U_2| \operatorname{Tr}(F'(u) \circ Q), \tag{1.11}$$

where $Q = Q_2(t, u_0; \xi_1, \xi_2)$ is the orthogonal projector in $\mathbb{R}^3$ on the space spanned by $U_1(t)$, $U_2(t)$. We have omitted the indices $\bigwedge^m H$ and written the norm $|\cdot|_{\bigwedge^m H}$ as $|\cdot|$ in order to simplify notation.

By (1.8) the trace of $F'(u)$ is $-(\sigma + b + 1)$ and from (1.10) (and V.(2.41)) we recover the well-known result that

For the Lorenz equation, the infinitesimal three-dimensional volume element is exponentially decaying

$$|U_1(t) \wedge U_2(t) \wedge U_3(t)| = |\xi_1 \wedge \xi_2 \wedge \xi_3| \exp(-(\sigma + b + 1)t). \tag{1.12}$$

Then, as shown in V.(2.43)–(2.48),

$$\omega_3(L(t, u_0)) = \sup_{\substack{\xi_i \in H \\ |\xi_i| \le 1 \\ i=1,2,3}} |U_1(t) \wedge U_2(t) \wedge U_3(t)| \le \exp(-(\sigma + b + 1)t), \tag{1.13}$$

$$\bar{\omega}_3(t) = \exp(-(\sigma + b + 1)t), \tag{1.14}$$

$$\Lambda_1\Lambda_2\Lambda_3 = \lim_{t\to\infty} \bar{\omega}_3(t)^{1/t} = \exp(-(\sigma + b + 1)), \tag{1.15}$$

$$\mu_1 + \mu_2 + \mu_3 = -(\sigma + b + 1), \tag{1.16}$$

where Λ_1, Λ_2, Λ_3 and μ_1, μ_2, μ_3 are the uniform Lyapunov numbers and Lyapunov exponents of the Lorenz attractor.

Similarly, we deduce from (1.11) that

$$|U_1(t) \wedge U_2(t)| = |\xi_1 \wedge \xi_2| \exp \int_0^t (\operatorname{Tr} F'(u(\tau)) \circ Q(\tau))\, d\tau.$$

Thus $|U_1(t) \wedge U_2(t)| \neq 0$, $\forall t > 0$, if and only if $\xi_1 \wedge \xi_2 \neq 0$. In this case, let φ_1, φ_2, φ_3 be an orthonormal basis of $\mathbb{R}^3$ such that φ_1, φ_2 constitute a basis of $\operatorname{Span}[U_1(t), U_2(t)]$. Then

$$\operatorname{Tr}(A_1 + A_2) \circ Q = \operatorname{Tr} A_1 \circ Q \ge 1 + b + \sigma - m,$$

where $m = \max(1, b, \sigma)$, and

$$\operatorname{Tr}(B(u) \circ Q) = \sum_{i=1}^{2} (B(u)\varphi_i) \cdot \varphi_i.$$

Setting $\varphi_i = (x_i, y_i, z_i)$, we find

$$\operatorname{Tr}(B(u) \circ Q) = \sum_{i=1}^{2} (zx_i y_i - x_i z_i y) = -zx_3 y_3 + x_3 z_3 y,$$

$$\begin{aligned} |\operatorname{Tr}(B(u) \circ Q)| &\le |x_3| \sqrt{y_3^2 + z_3^2} \sqrt{y^2 + z^2} \\ &\le \tfrac{1}{2} \sqrt{|x_3|^2 + |y_3|^2 + |z_3|^2} \sqrt{y^2 + z^2} \\ &\le \tfrac{1}{2} \sqrt{y^2 + z^2} \le \tfrac{1}{2}|u| = \tfrac{1}{2}|u(t)|. \end{aligned}$$

Using (1.4) we find that for t large, $t \ge t_1(\delta)$,

$$\operatorname{Tr}(B(u) \circ Q) \ge -\frac{b(r+\sigma)}{4\sqrt{l(b-1)}} - \delta,$$

$\delta > 0$ arbitrarily small. Thus

$$\begin{aligned} |U_1(t) \wedge U_2(t)| &\le |\xi_1 \wedge \xi_2| \exp((k_2 + \delta)t), \\ k_2 &= -(\sigma + b + 1) + m + \frac{b(r+\sigma)}{4\sqrt{l(b-1)}}, \end{aligned} \tag{1.17}$$

$$\begin{aligned} \omega_2(L(t, u_0)) &\le \sup_{\substack{\xi_i \in H \\ |\xi_i| \le 1 \\ i=1,2}} |U_1(t) \wedge U_2(t)| \\ &\le \exp((k_2 + \delta)t) \quad \text{for} \quad t \ge t_1(\delta). \end{aligned} \tag{1.18}$$

$$\bar{\omega}_2(t) \le \exp((k_2 + \delta)t), \qquad t \ge t_1(\delta). \tag{1.19}$$

Since $\delta > 0$ is arbitrarily small we obtain, by letting $t \to \infty$,

$$\Lambda_1 \Lambda_2 = \lim_{t \to \infty} \bar{\omega}_2(t)^{1/t} \le \exp k_2, \tag{1.20}$$

$$\mu_1 + \mu_2 \le k_2. \tag{1.21}$$

Now if $d = 2 + s$, $0 < s < 1$, $t \ge t_1(\delta)$, and

$$k(\delta) = -s(\sigma + b + 1) + (1 - s)(k_2 + \delta) < 0, \tag{1.22}$$

we infer from (1.13), (1.18) that

$$\begin{aligned} \omega_d(L(t, u_0)) &\le \omega_2(L(t, u_0))^{1-s} \omega_3(L(t, u_0))^s \\ &\le \exp(k(\delta)t), \\ \bar{\omega}_d(t) &= \sup_{u_0 \in X} \omega_d(L(t, u_0)) \le \exp(k(\delta)t) < 1. \end{aligned}$$

Then, by application of Theorem V.3.1, we conclude that the Hausdorff dimension of the Lorenz attractor $\mathscr{A}$ is less than or equal to d, when (1.22) holds, namely, when

$$s > \frac{k_2 + \delta}{\sigma + b + 1 + k_2 + \delta}. \tag{1.23}$$

This shows that

$$d_H(\mathscr{A}) \le 2 + \frac{k_2 + \delta}{\sigma + b + 1 + k_2 + \delta}, \tag{1.24}$$

and since $\delta > 0$ is arbitrarily small we conclude that

$$d_H(\mathscr{A}) \le 2 + \frac{k_2}{\sigma + b + 1 + k_2}. \tag{1.25}$$

In conclusion, we have proved

Theorem 1.1. *Under the assumption that σ, r, b satisfy* (1.3), *the Hausdorff dimension of the Lorenz attractor $\mathscr{A}$ is bounded by the expression in* (1.24), *k_2 as in* (1.17), $m = \max(1, b, \sigma)$.

If $\sigma = 10$, $r = 28$, $b = \frac{8}{3}$ (see (1.12)), *then*

$$d_H(\mathscr{A}) \le 2.538 \ldots .$$

Remark 1.1.

(i) The bound on $d_H(\mathscr{A})$ has been derived from Theorem V.3.1 and not from Theorem V.3.3. Thus the estimates (1.16), (1.21) on the Lyapunov exponents μ_j were only given for the sake of completeness. However, in most of the subsequent applications, the estimates will be derived directly from Theorem V.3.3.

(ii) A more precise computation leads to a bound of $d_H(\mathscr{A})$ in case (1.2), $d_H(\mathscr{A}) \le 2.468$. The value of the dimension of the Lorenz attractor, reported in the literature and computed numerically, is approximately 2.05, see I. Procacia, P. Grassberger, and H.G.E. Hentschel [1]. This is, however, the probabilistic dimension which is always less than or equal to the Hausdorff dimension.

Remark 1.2. Theorems V.3.2 and V.3.3 do not provide a useful bound (i.e., < 3) on the fractal dimension of the Lorenz attractor. This is not surprising since the analysis in the proof of these theorems is not sufficiently refined for low dimensions but is rather intended for high dimensions.

2. Reaction–Diffusion Equations

This section contains the first application of the results of Chapter V to partial differential equations. We consider here the reaction–diffusion equations treated in Section III.1. For the convenience of the reader we will treat the equations in the same sequential order as they appear in Section III.1: hence we will successively consider the equations with a polynomial nonlinearity associated with the Dirichlet boundary condition and then with other boundary conditions (Section 2.1). In Section 2.2 we consider the equations leaving a region invariant. The Dirichlet and the Neumann boundary conditions are treated successively. The expression of the bounds on the dimension of the attractors in terms of the data are here very explicit and the application to Examples 1.1–1.5 of Chapter III is very easy.

As indicated in the Introduction to this chapter the main tools that we use here are those of Chapter V and the collective Sobolev estimates provided in the Appendix.

2.1. Equations with a Polynomial Nonlinearity

2.1.1. The Linearized Equations

We consider the dynamical system associated with the boundary-value problem III.(1.2), (1.3). Writing this problem in the abstract form

$$u' = F(u) \tag{2.1}$$

as in V.(2.38), we see that the first variation equation V.(2.39)

$$U' = F'(u)U, \tag{2.2}$$

$$U(0) = \xi, \tag{2.3}$$

can be written formally as

$$\frac{\partial U}{\partial t} - d\Delta U + g'(u)U = 0 \quad \text{in} \quad \Omega \times \mathbb{R}_+, \tag{2.4}$$

$$U = 0 \quad \text{on} \quad \Gamma \times \mathbb{R}_+, \tag{2.5}$$

$$U(x, 0) = \xi(x) \quad \text{in } \Omega, \tag{2.6}$$

where ξ is given in $H = L^2(\Omega)$. In a more rigorous manner, we can say the following:

(i) If u is the solution of III.(1.2)–(1.4) given by Theorem III.1.1, then the linear initial- and boundary-value problem (2.4)–(2.6) possesses a unique solution

$$U \in L^2(0, T; V) \cap \mathscr{C}([0, T]; H), \qquad \forall T > 0, \tag{2.7}$$

$V = H_0^1(\Omega)$, $H = L^2(\Omega)$. This follows from Theorem II.3.4, and the verification of this point (and of the assumptions of Theorem II.3.4) is left as an exercise for the reader.

(ii) The semigroup $S(t)$ is Fréchet differentiable in H

$$\begin{aligned}&\text{For every } t > 0, \text{the function } u_0 \to S(t)u_0 \text{ is Fréchet differentiable}\\&\text{in } H \text{ at } u_0 \text{ with differential } L(t, u_0)\colon \xi \in H \to U(t) \in H, \text{ where}\\&U(\cdot) \text{ is the solution of (2.4)–(2.7).}\end{aligned} \tag{2.8}$$

This technical point, like all the Fréchet differentiability results for the semigroup $S(t)$ that we will state and use in this chapter, is proved in Section 8.

Orientation

Once these results are available we can define the uniform Lyapunov exponents of the global attractor $\mathscr{A}$ given by Theorem II.1.2 and estimate them using the methods of Chapter V.

2.1.2. Dimension of the Attractor

We now estimate the dimension of $\mathscr{A}$.

Following the procedure in Section V.2.3 we consider, for $m \in \mathbb{N}$,

$$
\begin{aligned}
&|U_1(t) \wedge \cdots \wedge U_m(t)|_{\bigwedge^m H} \\
&\quad = |\xi_1 \wedge \cdots \wedge \xi_m|_{\bigwedge^m H} \exp \int_0^t \operatorname{Tr} F'(S(\tau)u_0 \circ Q_m(\tau)\, d\tau,
\end{aligned}
$$

where $S(\tau)u_0 = u(\tau)$, u is the solution of III.(1.2), (1.3), $U_1, \ldots, U_m$ are m solutions of (2.2)–(2.6) corresponding to $\xi = \xi_1, \ldots, \xi_m$, and $Q_m(\tau) = Q_m(\tau, u_0; \xi_1, \ldots, \xi_m)$ is the orthogonal projector in H onto the space spanned by $U_1(\tau), \ldots, U_m(\tau)$.

At a given time τ, let $\varphi_j(\tau), j \in \mathbb{N}$, be an orthonormal basis of H, with $\varphi_1(\tau), \ldots, \varphi_m(\tau)$ spanning $Q_m(\tau)H = \operatorname{Span}[U_1(\tau), \ldots, U_m(\tau)]$; $\varphi_j(\tau) \in V\,(= H_0^1(\Omega))$ for $j = 1, \ldots, m$ (and if we wish $\forall j \in \mathbb{N}$); since $U_1(\tau), \ldots, U_m(\tau) \in V$ (a.e. $\tau \in \mathbb{R}_+$, see (2.7)).

We have

$$
\begin{aligned}
\operatorname{Tr} F'(u(\tau)) \circ Q_m(\tau) &= \sum_{j=1}^{\infty} (F'(u(\tau)) \circ Q_m(\tau)\varphi_j(\tau), \varphi_j(\tau)) \\
&= \sum_{j=1}^{m} (F'(u(\tau))\varphi_j(\tau), \varphi_j(\tau)), \qquad (2.9)
\end{aligned}
$$

$(\cdot, \cdot)$ denoting, as in Section III.1.1, the scalar product in H. Then, omitting temporarily the dependence on τ

$$
\begin{aligned}
(F'(u)\varphi_j, \varphi_j) &= +d(\Delta\varphi_j, \varphi_j) - (g'(u)\varphi_j, \varphi_j) \\
&= -d\|\varphi_j\|^2 - \int_\Omega g'(u)(\varphi_j)^2\, dx, \qquad (2.10)
\end{aligned}
$$

$$
\sum_{j=1}^{m} (F'(u)\varphi_j, \varphi_j) = -d \sum_{j=1}^{m} \|\varphi_j\|^2 - \int_\Omega g'(u)\rho\, dx,
$$

where

$$
\rho = \rho(x, \tau) = \sum_{j=1}^{m} |\varphi_j(x, \tau)|^2. \qquad (2.11)
$$

We infer from II.(1.1), (1.18) that g' is bounded from below by a constant $-c_3'$ that we rename here $-\kappa_1$ for consistency of notation

$$g'(s) \geq -\kappa_1 \qquad (\kappa_1 \geq 0), \qquad \forall s \in \mathbb{R}. \tag{2.12}$$

Therefore (2.9)–(2.11) imply

$$\text{Tr } F'(u(\tau)) \circ Q_m(\tau) \leq -d \sum_{j=1}^{m} \|\varphi_j\|^2 + \kappa_1 \int_\Omega \rho \, dx.$$

Recalling that the family φ_j, $j = 1, \ldots, m$, is orthonormal in H, we see that $\int_\Omega \rho(x)\, dx = m$ and

$$\text{Tr } F'(u(\tau)) \circ Q_m(\tau) \leq -d \sum_{j=1}^{m} \|\varphi_j\|^2 + m\kappa_1. \tag{2.13}$$

We now apply Lemma 2.1 below (see the end of Section 2.1.3). We recall that the sequence λ_j of eigenvalues of the Dirichlet operator ($A = -\Delta$ in $L^2(\Omega)$ with domain $H^2(\Omega) \cap H_0^1(\Omega)$) satisfies

$$\lambda_j \sim c\lambda_1 j^{2/n},$$

where c depends only on the shape of Ω and not of its size and n is the space dimension (see R. Courant and D. Hilbert [1]). By changing the constant c we can also write

$$\lambda_j \sim c|\Omega|^{-2/n} j^{2/n},$$

$|\cdot|$ denoting the measure (volume) of Ω.

Then Lemma 2.1 and (2.24) show that there exists a dimensionless constant c_1'' which depends only on the shape of Ω (and n), such that

$$\sum_{j=1}^{m} \|\varphi_j\|^2 \geq c_1'' \frac{m^{1+(2/n)}}{|\Omega|^{2/n}}.[1] \tag{2.14}$$

Hence

$$\text{Tr } F'(u(\tau)) \circ Q_m(\tau) \leq -c_1'' d \frac{m^{1+(2/n)}}{|\Omega|^{2/n}} + m\kappa_1.$$

We easily deduce from Young's inequality that

$$m\kappa_1 \leq \frac{1}{2} c_1'' d \frac{m^{1+(2/n)}}{|\Omega|^{2/n}} + c_2'' \frac{\kappa_1^{1+(n/2)}}{d^{n/2}} |\Omega|,$$

where the constant c_2'' depends only on n. Thus

$$\text{Tr } F'(u(\tau)) \circ Q_m(\tau) \leq -\frac{1}{2} c_1'' d \frac{m^{1+(2/n)}}{|\Omega|^{2/n}} + c_2'' \frac{\kappa_1^{1+(n/2)}}{d^{n/2}} |\Omega|. \tag{2.15}$$

[1] This also follows from Corollary A.5.1.

We can now majorize the quantities $q_m(t)$, q_m defined in V.(3.47), (3.48),

$$q_m(t) = \operatorname*{Sup}_{u_0 \in X} \operatorname*{Sup}_{\substack{\xi_i \in H \\ |\xi_i| \le 1 \\ i=1,\dots,m}} \left(\frac{1}{t} \int_0^t \operatorname{Tr} F'(S(\tau)u_0) \circ Q_m(\tau)\, d\tau \right),$$

$$q_m = \limsup_{t \to \infty} q_m(t).$$

We have

$$q_m(t) \le -\kappa_3 m^{1+(2/n)} + \kappa_4, \tag{2.16}$$

$$q_m \le -\kappa_3 m^{1+(2/n)} + \kappa_4, \tag{2.17}$$

$$\kappa_3 = \frac{1}{2} c_1'' \frac{d}{|\Omega|^{2/n}}, \qquad \kappa_4 = c_2'' \kappa_1^{1+(n/2)} \frac{|\Omega|}{d^{n/2}}.$$

This inequality already provides a bound for the Lyapunov exponents μ_j, $j \in \mathbb{N}$ (see V.(3.49))

$$\mu_1 + \cdots + \mu_j \le q_j \le -\kappa_3 j^{1+2/n} + \kappa_4, \qquad \forall j \in \mathbb{N}. \tag{2.18}$$

If m is sufficiently large so that

$$q_m \le -\kappa_3 m^{1+(2/n)} + \kappa_4 < 0, \tag{2.19}$$

Proposition V.2.1 implies that the m-volume element is exponentially decaying in H and Theorem V.3.3 implies that the Hausdorff dimension of $\mathscr{A}$ is $\le m$. In order to evaluate the fractal dimension of $\mathscr{A}$ we use the elementary Lemma 2.2 below: it implies that if m is the first integer for which a slightly stronger form of (2.19) is valid

$$m - 1 < \left(\frac{2\kappa_4}{\kappa_3} \right)^{n/(2+n)} \le m, \tag{2.20}$$

then

$$\frac{(\mu_1 + \cdots + \mu_j)_+}{|\mu_1 + \cdots + \mu_j|} \le 1, \qquad \forall j = 1, \dots, m-1, \tag{2.21}$$

the expression V.(3.35) is less than or equal to $2m$, and by Theorem V.3.3 the fractal dimension of $\mathscr{A}$ is less than or equal to $2m$.

From the expression (2.17) of κ_3, κ_4, we have

$$\frac{\kappa_4}{\kappa_3} = \frac{2c_2''}{c_1''} \kappa_1^{1+(n/2)} \frac{|\Omega|^{1+(2/n)}}{d^{1+(n/2)}},$$ [1]

and this gives a very explicit form of the dependence of the number m in (2.20)

[1] Note that from the physical point of view this formula is dimensionally correct. Assuming that u (= a concentration) has no dimension then $\kappa_1 \sim T^{-1}$, $d \sim L^2/T$, $|\Omega| \sim L^n$, where T and L are the units of time and space variables and the ratio κ_4/κ_3 is indeed dimensionless as it should be for (2.20) (c_1'', c_2'' = dimensionless constants).

on d, $|\Omega|$, and κ_1 (i.e., g)

$$m - 1 < c_3''\left(\frac{\kappa_1}{d}\right)^{n/2} |\Omega| \leq m, \tag{2.22}$$

c_3'' a dimensionless constant depending only on n and the shape of Ω.

We recapitulate the results that we have proved in the following

Theorem 2.1. *The hypotheses are those of Theorem* III.1.2.

We consider the dynamical system defined by III.(1.2), (1.3) *and let m satisfy* (2.22). *Then*:

(i) *the m-dimensional volume element is exponentially decaying in the phase space as* $t \to \infty$;
(ii) *the corresponding global attractor* $\mathscr{A}$ *defined by Theorem* III.1.2 *has a Hausdorff dimension less than or equal to m and a fractal dimension less than or equal to* $2m$.

We conclude Section 2.13 with two simple lemmas used above and which will be used repeatedly in this chapter.

Lemma 2.1. *Let A be a linear positive unbounded self-adjoint operator in a Hilbert space H with domain* $D(A) \subset H$, *and assume that* A^{-1} *is compact.*

Then, for any family of elements $\varphi_1, \ldots, \varphi_m$ *of V which is orthonormal in H*,

$$\sum_{j=1}^{m} (A\varphi_j, \varphi_j) \geq \lambda_1 + \cdots + \lambda_m, \tag{2.23}$$

where $(\lambda_j)_{j \in \mathbb{N}}$ *is the complete sequence of eigenvalues of A* $(0 < \lambda_1 \leq \lambda_2 \leq \cdots)$.

If, furthermore, $\lambda_j \sim c\lambda_1 j^\alpha$, $\alpha > 0$, *as* $j \to \infty$, *c depending on A, then*

$$\sum_{j=1}^{m} (A\varphi_j, \varphi_j) \geq \lambda_1 + \cdots + \lambda_m \geq c'\lambda_1 m^{\alpha+1}, \tag{2.24}$$

with another constant c' *depending on A and* α.

Proof. We consider the operator A_m defined in Section V.1.2.2 (see, in particular, Remark V.1.4). Since the family φ_j is orthonormal in H, we have

$$\begin{aligned}(A_m(\varphi_1 \wedge \cdots \wedge \varphi_m), \varphi_1 \wedge \cdots \wedge \varphi_m)_{\wedge^m H} \\ &= (A\varphi_1, \varphi_1)(\varphi_2, \varphi_2)\ldots(\varphi_m, \varphi_m) + \cdots + (\varphi_1, \varphi_1)\ldots(\varphi_{m-1}, \varphi_{m-1})(A\varphi_m, \varphi_m) \\ &= \sum_{j=1}^{m} (A\varphi_j, \varphi_j),\end{aligned}$$

and since $|\varphi_1 \wedge \cdots \wedge \varphi_m|_{\wedge^m H} = 1$,

$$\sum_{j=1}^{m} (A\varphi_j, \varphi_j) \geq \operatorname*{Inf}_{\substack{\psi \in \wedge^m D(A^{1/2}) \\ |\psi|_{\wedge^m H - 1}}} (A_m\psi, \psi)_{\wedge^m H}.$$

Let $(w_j)_{j \in \mathbb{N}}$ denote the complete orthonormal sequence of eigenvectors of A ($Aw_j = \lambda_j w_j$, $\forall j$). Clearly, the vectors $w_{i_1} \wedge \cdots \wedge w_{i_m}$ are eigenvectors of A_m with corresponding eigenvalues $\lambda_{i_1} + \cdots + \lambda_{i_m}$. Since the $(w_j)_{j \in \mathbb{N}}$ form an orthonormal basis of H, the vectors $w_{i_1} \wedge \cdots \wedge w_{i_m}$ constitute, for $i_1 < \cdots < i_m$, $i_\alpha \in \mathbb{N}$, a complete orthonomral basis of $\bigwedge^m H$. Thus these vectors are all the eigenvectors of A_m (which is self-adjoint with A_m^{-1} compact in $\bigwedge^m H$), and the numbers $\lambda_{i_1} + \cdots + \lambda_{i_m}$ are all the eigenvalues of A_m. In particular, $\lambda_1 + \cdots + \lambda_m$ is the smallest eigenvalue of A_m and

$$(A_m \psi, \psi)_{\bigwedge^m H} \geq \lambda_1 + \cdots + \lambda_m, \qquad \forall \psi \in \textstyle\bigwedge^m D(A^{1/2}), \quad |\psi|_{\bigwedge^m H} = 1.$$

Hence (2.23).

The proof of (2.24) is totally elementary; there exists c'' such that $\lambda_j \geq c'' \lambda_1 j^\alpha$, $\forall j$, and:

$$\lambda_1 + \cdots + \lambda_m \geq c'' \lambda_1 (1^\alpha + \cdots + m^\alpha) \geq c' \lambda_1 m^{\alpha+1}. \qquad \square$$

Lemma 2.2. *We assume that the sequence of numbers μ_j, $j \geq 1$, satisfies the following inequalities*

$$\mu_1 + \cdots + \mu_j \leq -\alpha j^\theta + \beta, \qquad \forall j \geq 1,$$

where $\alpha, \beta, \theta > 0$. Let $m \in \mathbb{N}$ be defined by

$$m - 1 < \left(\frac{2\beta}{\alpha}\right)^{1/\theta} \leq m.$$

Then $\mu_1 + \cdots + \mu_m < 0$ and

$$\frac{(\mu_1 + \cdots + \mu_j)_+}{|\mu_1 + \cdots + \mu_m|} \leq 1, \qquad j = 1, \ldots, m. \tag{2.25}$$

Proof. We have

$$\mu_1 + \cdots + \mu_m \leq -\alpha m^\theta + \beta \leq \text{(by (2.23))} \leq -\beta,$$
$$(\mu_1 + \cdots + \mu_j)_+ \leq (-\alpha j^\theta + \beta)_+ \leq \beta, \qquad \forall j,$$

and (2.25) follows. $\square$

2.1.3. Other Boundary Conditions

The treatment above extends with minor modifications to the case where the boundary condition III.(1.3) is replaced by one of the boundary conditions (1.3′), (1.3″) mentioned in Remark 1.5.

For the first variation equation we replace (2.5) by the analog of (1.3′) or (1.3″), i.e.,

$$\frac{\partial U}{\partial \mathbf{v}} = 0 \quad \text{on} \quad \Gamma \times \mathbb{R}_+, \tag{2.5′}$$

or

$$\Omega = (0, L)^n \quad \text{and } U \text{ is } \Omega\text{-periodic.} \tag{2.5''}$$

The statements (2.7), (2.8) hold true and are proved in a similar manner. The estimate of $\operatorname{Tr} F'(u(\tau)) \circ Q_m(\tau)$ starts as in Section 2.1.2. The first difference is in (2.14); it is simpler here to apply Corollary A.5.1 instead of Lemma 2.1. We obtain, instead of (2.14),

$$\sum_{j=1}^{m} \|\varphi_j\|^2 \geq c_4'' \frac{m^{1+(2/n)}}{|\Omega|^{2/n}} - c_5'' \frac{m}{|\Omega|^{2/n}}, \tag{2.26}$$

where the constants c_4'', c_5'' (the same as κ_1'', κ_2'' in Corollary A.5.1), are dimensionless constants depending only on n and the shape of Ω. Thus

$$\begin{aligned} \operatorname{Tr} F'(u(\tau)) \circ Q_m(\tau) &\leq -c_4'' d \frac{m^{1+(2/n)}}{|\Omega|^{2/n}} + c_5'' \frac{dm}{|\Omega|^{2/n}} + m\kappa_1 \\ &\leq \text{(with the Young inequality)} \\ &\leq -\frac{1}{2} c_4'' \frac{dm^{1+(2/n)}}{|\Omega|^{2/n}} \\ &\quad + c_6'' \frac{d}{|\Omega|^{2/n}} \left(1 + \frac{\kappa_1 |\Omega|^{2/n}}{d}\right)^{1+(n/2)}. \end{aligned} \tag{2.27}$$

The formulas (2.16)–(2.21) are all valid with now

$$\kappa_3 = \frac{1}{2} c_4'' \frac{d}{|\Omega|^{2/n}}, \qquad \kappa_4 = c_6'' \frac{d}{|\Omega|^{2/n}} \left(1 + \frac{\kappa_1 |\Omega|^{2/n}}{d}\right)^{1+(n/2)}. \tag{2.28}$$

The expression of m in (2.22) is replaced by

$$m - 1 < 2 \frac{c_6''}{c_4''} \left(1 + \frac{\kappa_1 |\Omega|^{2/n}}{d}\right)^{n/2} \leq m. \tag{2.29}$$

We derive from these inequalities exactly the same conclusions as in Theorem 2.1, condition (2.22) defining m being replaced by (2.29).

2.2. Equations with an Invariant Region

2.2.1. The Linearized Equations

We now consider the evolution equation

$$u' = F(u) \tag{2.30}$$

corresponding to the boundary-value problem III.(1.41)–(1.43). The hypotheses are the same as in Chapter III, in particular, the assumptions III.(1.44)–(1.46); for the study of the differentiability of the corresponding semigroup $S(t)$ we will also need the property III.(6.22) used for backward uniqueness.

The first variation equation V.(2.39) is formally written

$$U' = F'(u)U, \tag{2.31}$$

$$U(0) = \xi, \tag{2.32}$$

and this corresponds to the initial- and boundary-value problem

$$\frac{\partial U}{\partial t} - D\Delta U + g'_u(u, x)U = 0 \quad \text{in} \quad \Omega \times \mathbb{R}_+, \tag{2.33}$$

$$U = 0 \quad \text{on} \quad \Gamma \times \mathbb{R}_+, \tag{2.34}$$

$$U(x, 0) = \xi(x) \quad \text{in } \Omega. \tag{2.35}$$

The notation is the same as in Section III.1.2; like u, U is a vector function from $\Omega \times \mathbb{R}_+$ into $\mathbb{R}^l$, D is the diagonal matrix in III.(1.41), $g'_u = g'_u(u(x, t), x)$ is an $l \times l$ matrix, namely, the differential of g with respect to u at point $u = u(x, t), x$; ξ is given in $H = L^2(\Omega)^m$.

The precise relations between (2.33)–(2.35) and III.(1.41)–(1.43) are expressed by the following statements:

(i) If u is the solution to III.(1.41)–(1.43) whose existence and uniqueness is assumed in III.(1.45), then the linear initial- and boundary-value problem (2.33)–(2.35) possesses a unique solution

$$U \in L^2(0, T; V) \cap \mathscr{C}([0, T]; H), \qquad \forall T > 0, \tag{2.36}$$

$V = H_0^1(\Omega)^l$, $H = L^2(\Omega)$.

(ii) The semigroup $S(t)$ is differentiable in H

For every $t > 0$, for every $u_0 \in H$, the function $t \to S(t)u_0$ is Fréchet differentiable in H with differential $L(t, u_0)$: $\xi \in H \to U(t) \in H$, where U is the solution of (2.33)–(2.35). (2.37)

Point (i) follows from Theorem III.3.4; the verification of this point and of the necessary assumptions of Theorem III.3.4 is easy (note that, thanks to III.(1.46), the function $\{x, t\} \to g(u(x, t), x)$ is in $L^\infty(\Omega \times (0, T))$. Point (ii) (the differentiability property) is proved in Section 8.

2.2.2. *Dimension of the Attractor*

Our aim now is to apply the results of Chapter V to the global attractor $\mathscr{A}$ whose existence is asserted by Theorem III.1.3, in particular, we want to estimate the dimension of $\mathscr{A}$.

We follow the procedure in Section V.2.3 and we consider, for $m \in \mathbb{N}$,

$$|U_1(t) \wedge \cdots \wedge U_m(t)|_{\wedge^m H}$$

$$= |\xi_1 \wedge \cdots \wedge \xi_m|_{\wedge^m H} \exp \int_0^t \operatorname{Tr} F'(S(\tau)u_0) \circ Q_m(\tau)\, d\tau,$$

where $S(\tau)u_0 = u(\tau)$, u is the solution of III.(1.41)–(1.43), $U_1, \ldots, U_m$ are m solutions of (2.57) and (2.58) corresponding to initial data $\xi = \xi_1, \ldots, \xi_m$, and

$Q_m(\tau) = Q_m(\tau, u_0; \xi_1, \ldots, \xi_m)$ is the orthogonal projector in H onto the space spanned by $U_1(\tau), \ldots, U_m(\tau)$.

At a given time τ, let $\varphi_j(\tau)$, $j \in \mathbb{N}$, be an orthonormal basis of $Q_m(\tau)H = \text{Span}[U_1(\tau), \ldots, U_m(\tau)]$; $\varphi_j(\tau) \in V$ $(= H_0^1(\Omega))$ for $j = 1, \ldots, m$ (and if we wish, for every $j \in \mathbb{N}$) since $U_1(\tau), \ldots, U_m(\tau) \in V$ for a.e. $\tau \in \mathbb{R}_+$ (see (2.60)). Although this is not essential hereafter, it is noteworthy that the $\varphi_j(\tau)$ can be chosen so that they depend measurably on τ.

Now we write

$$\begin{aligned}\text{Tr } F'(u(\tau)) \circ Q_m(\tau) &= \sum_{j=1}^{\infty} (F'(u(\tau) \circ Q_m(\tau)\varphi_j(\tau), \varphi_j(\tau)) \\ &= \sum_{j=1}^{m} (F'(u(\tau))\varphi_j(\tau), \varphi_j(\tau)),\end{aligned} \tag{2.38}$$

$(\cdot, \cdot)$ denoting as before the scalar product in H. For the sake of simplicity we omit the dependence on τ and we write

$$\begin{aligned}(F'(u)\varphi_j, \varphi_j) &= (D\Delta\varphi_j, \varphi_j) - (g_u'(u)\varphi_j, \varphi_j) \\ &= -a(\varphi_j, \varphi_j) - (g_u'(u)\, \varphi_j, \varphi_j),\end{aligned}$$

where a is the quadratic form defined in III.(1.48). Then using III.(1.51), (6.22) we obtain[1]

$$(F'(u)\varphi_j, \varphi_j) \leq -d_0 \|\varphi_j\|^2 + \kappa_5 |\varphi_j|^2,$$

provided

$$(g_u'(u, x) \cdot \xi) \cdot \xi \geq -\kappa_5 |\xi|^2, \qquad \forall u \in \mathscr{D}, \quad \forall x \in \Omega, \quad \forall \xi \in \mathbb{R}^n. \tag{2.39}$$

Since the family φ_j is orthonormal in H, we find

$$\text{Tr } F'(u(\tau)) \circ Q_m(\tau) \leq -d_0 \sum_{j=1}^{m} \|\varphi_j\|^2 + \kappa_5 m. \tag{2.40}$$

We then apply Corollary A.5.1. which implies, in the present case ($\varphi_j \in H_0^1(\Omega)^l$),

$$\sum_{j=1}^{m} \|\varphi_j\|^2 \geq c_7'' \frac{m^{1+(2/n)}}{|\Omega|^{2/n}}, \tag{2.41}$$

where n is the space dimension, $|\Omega|$ is the measure of Ω, and c_7'' is a dimensionless constant depending only on n and the shape of Ω.[2] Hence

$$\text{Tr } F'(u(\tau)) \circ Q_m(\tau) \leq -c_7'' \frac{d_0 m^{1+(2/n)}}{|\Omega|^{2/n}} + \kappa_5 m. \tag{2.42}$$

We apply Young's inequality, as in the computations preceding (2.15), to

[1] The existence of κ_5, $0 \leq \kappa_5 < \infty$, follows from III.(6.22)

$$\kappa_5 \geq \sup_{\substack{u \in \mathscr{D} \\ x \in \Omega}} |g_u'(u, x)|,$$

where the norm $|\cdot|$ is here the norm of $g_u'(u, x)$ in $\mathscr{L}(\mathbb{R}^n)$.

[2] c_7'' is κ_3'' in Corollary A.5.1, while $\kappa_4'' = 0$.

obtain

$$\kappa_5 m \leq \frac{1}{2} c_7'' \frac{d_0 m^{1+(2/n)}}{|\Omega|^{2/n}} + c_8'' \frac{|\Omega|}{d_0^{n/2}} \kappa_5^{1+(n/2)},$$

and (2.42) becomes

$$\operatorname{Tr} F'(u(\tau)) \circ Q_m(\tau) \leq -\frac{1}{2} c_7'' \frac{d_0 m^{1+(2/n)}}{|\Omega|^{2/n}} + c_8'' \frac{|\Omega|}{d_0^{n/2}} \kappa_5^{1+(n/2)}, \tag{2.43}$$

c_8'' also denoting a nondimensional constant depending on n and the shape of Ω, and (2.42) becomes (2.43).

Hence

$$\frac{1}{t} \int_0^t \operatorname{Tr} F'(u(\tau)) \circ Q_m(\tau)\, d\tau \leq -\kappa_6 m^{1+(2/n)} + \kappa_7,$$

$$\kappa_6 = \frac{1}{2} c_7'' \frac{d_0}{|\Omega|^{2/n}}, \qquad \kappa_7 = c_8'' \frac{|\Omega|}{d_0^{n/2}} \kappa_5^{1+(n/2)}, \tag{2.44}$$

and we can majorize the quantities $q_m(t)$, q_m defined in V.(3.47), (3.48)

$$q_m(t) = \operatorname*{Sup}_{u_0 \in X} \operatorname*{Sup}_{\substack{\xi_i \in H \\ |\xi_i| \leq 1 \\ i=1,\dots,m}} \left(\frac{1}{t} \int_0^t \operatorname{Tr} F'(u(\tau)) \circ Q_m(\tau)\, d\tau \right\}, \tag{2.45}$$

$$q_m(t) \leq -\kappa_6 m^{1+(2/n)} + \kappa_7,$$

$$q_m = \limsup_{t \to \infty} q_m(t) \leq -\kappa_6 m^{1+(2/n)} + \kappa_7. \tag{2.46}$$

We infer from (2.46) and V.(2.49) the following bound on the Lyapunov exponents μ_j, $j \in \mathbb{N}$,

$$\mu_1 + \cdots + \mu_j \leq q_j \leq -\kappa_6 j^{1+(2/n)} + \kappa_7, \qquad \forall j \in \mathbb{N}. \tag{2.47}$$

Using Lemma 2.2 we see that if m is defined by

$$m - 1 < \left(\frac{2\kappa_7}{\kappa_6} \right)^{n/(2+n)} \leq m, \tag{2.48}$$

then $\mu_1 + \cdots + \mu_m < 0$ and

$$\frac{(\mu_1 + \cdots + \mu_j)_+}{|\mu_1 + \cdots + \mu_m|} \leq 1, \qquad \forall j = 1, \dots, m-1. \tag{2.49}$$

We note that

$$\left(\frac{2\kappa_7}{\kappa_6} \right)^{n/(2+n)} = c_9'' |\Omega| \left(\frac{\kappa_5}{d_0} \right)^{n/2}, \tag{2.50}$$

$c_9'' = 4c_8''/c_7''$.[1]

We conclude with

[1] As expected from (2.48), the ratio in (2.50) is nondimensional.

Theorem 2.2. *We consider the dynamical system defined by* III.(1.41), (1.42) *and we assume that* III.(6.22) *and the hypotheses of Theorem* III.1.3 *are fulfilled.*

Let m be defined by (2.48), (2.50), *i.e.,*

$$(m-1) < c_9''|\Omega|\left(\frac{\kappa_5}{d_0}\right)^{n/2} \le m, \tag{2.51}$$

κ_5 *given in* (2.39). *Then*:

(i) *the m-dimensional volume element is exponentially decaying in the phase space as $t \to \infty$;*

(ii) *the corresponding global attractor $\mathscr{A}$ defined by Theorem* III.1.3 *has a Hausdorff dimension less than or equal to m and a fractal dimension less than or equal to 2m.*

2.2.3. Other Boundary Conditions

The treatment above extends with minor modifications to the case where the boundary condition III.(1.42) is replaced by III.(1.83).

For the first variation equation we replace (2.34) by

$$\frac{\partial U_i}{\partial \mathbf{v}} = 0 \quad \text{on } \Gamma, \qquad i = 1, \ldots, l, \tag{2.52}$$

where $U = (U_1, \ldots, U_l)$. The statements (2.36), (2.37) remain valid and are proved in a similar manner. Then the estimate of Tr $F'(u(\tau)) \circ Q_m(\tau)$ starts as in Section 2.2.2, but instead of (2.41) we infer from Corollary A.5.1 that

$$\sum_{j=1}^{m} \|\varphi_j\|^2 \ge c_{10}'' \frac{m^{1+(2/n)}}{|\Omega|^{2/n}} - c_{11}'' \frac{m}{|\Omega|^{2/n}}, \tag{2.53}$$

where c_{10}'', c_{11}'' are dimensionless constants depending only on n and the shape of Ω. Then (2.42)–(2.44) are replaced by

$$\begin{aligned} \operatorname{Tr} F'(u(\tau)) \circ Q_m(\tau) &\le -c_{10}'' \frac{d_0 m^{1+(2/n)}}{|\Omega|^{2/n}} + c_1'' \frac{d_0 m}{|\Omega|^{2/n}} \kappa_5 m \\ &\le \text{(with the Young inequality)} \\ &\le -\frac{1}{2} c_{10}'' \frac{d_0 m^{1+(2/n)}}{|\Omega|^{2/n}} \\ &\quad + c_{12}'' \frac{d_0}{|\Omega|^{2/n}} \left(1 + \frac{\kappa_5 |\Omega|^{2/n}}{d_0}\right)^{1+(2/n)}, \end{aligned} \tag{2.54}$$

$$\frac{1}{t}\int_0^t \operatorname{Tr} F'(u(\tau)) \circ Q_m(\tau)\, d\tau \le -\kappa_8 m^{1+(2/n)} + \kappa_9,$$

$$\kappa_8 = \frac{1}{2} c_{10}'' \frac{d_0}{|\Omega|^{2/n}}, \qquad \kappa_9 = c_{12}'' \frac{d_0}{|\Omega|^{2/n}} \left(1 + \frac{\kappa_5 |\Omega|^{2/n}}{d_0}\right)^{1+(2/n)}. \tag{2.55}$$

The formulas (2.45), (2.46) are valid with κ_6, κ_7 replaced by κ_8, κ_9. We derive from these inequalities exactly the same conclusions as in Theorem 2.2, condition (2.51) defining m being replaced by

$$m - 1 < c_{13}''\left(1 + \frac{\kappa_5|\Omega|^{2/n}}{d_0}\right)^{n/2} \leq m. \tag{2.56}$$

Remark 2.1. The bounds on m that we obtain in (2.51) and (2.56) are remarkably simple and their explicit dependence on the data is transparent. If we want to apply these results to Examples 1.1–1.5 of Chapter III,[1] the only computations which remain to be done are those related to the determination of κ_5 (see (2.39)): this involves a tedious but wholly elementary computation of the minimum of $g_4'(u, x)$ on $\mathscr{D} \times \Omega$. We do not develop this point here.

Remark 2.2. It is natural to interpret the dimension of the attractor as the number of degrees of freedom of the corresponding phenomena when the "permanent regime" is established. From that point of view it is noteworthy to observe that the bounds (2.22) and (2.56) are proportional to the volume $|\Omega|$ of Ω. This property of the number of degrees of freedom which seems physically reasonable has not been proved for all the equations. In particular, this is an open problem in the case of the Kuramoto–Sivashinsky equation (see Remark 5.1).

3. Navier–Stokes Equations ($n = 2$)

In Section III.2 we have shown the existence of a maximal attractor for the Navier–Stokes equations in space dimension 2. This attractor is the mathematical object describing the permanent regime in two-dimensional turbulent flows and, in particular, its dimension expresses the number of degrees of freedom of the flow. In conventional turbulence theory a heuristical estimate of the number of degrees of freedom of a turbulent flow is given in the two-dimensional case by

$$N \sim \left(\frac{L_0}{L_d}\right)^2, \tag{3.1}$$

where L_0 denotes a characteristic length of the region occupied by the fluid, as the diameter of Ω or $|\Omega|^{1/2}$. The length L_d is the diffusion length below which the viscosity effects determine entirely the motion, the eddies of size smaller than L_d being rapidly damped. In two- and three-dimensional spaces, $L_d = (\nu^3/\varepsilon)^{1/4}$ where ε is the energy dissipation (see L. Landau and I.M. Lifschitz [1]); in two dimensions, it is also useful to define $L_\chi = (\nu^3/\eta)^{1/6}$, where η is

[1] $\mathscr{D}$ is unbounded in Example III.1.4 and the Neumann boundary condition cannot be considered in this case.

the enstrophy dissipation (see G.K. Batchelor [1], R.H. Kraichnan [1] and below). Our aim here is to interpret the dimension of the attractor as the number of degrees of freedom of the flow, and to provide an estimate of its dimension in terms of the nondimensional numbers defined in Chapter III (G, Re, see III.(2.37), (2.38)). The ratio (L_0/L_d) is also expressed in terms of these numbers and we obtain an estimate of the type (3.1) as well.

In Section 3.1 we consider the flow in a bounded region Ω with one of the three boundary conditions studied in Section III.2, i.e.,

the Dirichlet boundary condition III.(2.4), (3.2a)

the space-periodicity boundary condition III.(2.5), (2.6), (3.2b)

the free boundary condition III.(2.39). (3.2c)

In these cases we use the mechanism of energy flux to define L_d and we are able to estimate N and the ratio $(L_0/L_d)^2$ in the form

$$N \sim (L_0/L_d)^2 \sim G. \tag{3.3}$$

We achieve this result by using the general framework of Chapter V, the tools developed in Section II.2, and collective Sobolev inequalities from the Appendix.

Then in Section 3.2 we improve these estimates for the space-periodic boundary condition (ii); L_d is replaced by the length L_χ and we obtain for N and $(L_0/L_\chi)^2$ an estimate of the form

$$N \sim (L_0/L_\chi)^2(1 + \log(L_0/L_\chi))^{1/3} \sim G^{2/3}(1 + \log G)^{1/3}. \tag{3.4}$$

This result is obtained by utilization of the same tools as above together with some specific properties of the equation in the space-periodic case and an alternative form of the collective Sobolev inequalities.

3.1. General Boundary Conditions

3.1.1. *Preliminary Properties*

We consider the dynamical system associated with equation III.(2.18); as explained in Section III.2, this equation is the abstract form of the boundary-value problem corresponding to equations III.(2.1), (2.2) associated with one of the boundary conditions (3.2a, b, c); the main difference between these cases lies essentially in the definition of the spaces V, H, and of the operator A. We will treat the three cases simultaneously.

We write this problem in the abstract form

$$u' = F(u), \tag{3.5}$$

as in V.(2.38), and we see that the first variation equation V.(2.39)

$$U' = F'(u)U \tag{3.6}$$

is also equivalent to

$$\frac{dU}{dt} + \nu AU + B(u, U) + B(U, u) = 0, \tag{3.7}$$

or to

$$\frac{\partial U}{\partial t} - \nu\Delta U + (u \cdot \nabla)U + (U \cdot \nabla)u + \nabla\Pi = 0 \quad \text{in} \quad \Omega \times \mathbb{R}_+, \tag{3.8}$$

$$\operatorname{div} U = 0 \quad \text{in} \quad \Omega \times \mathbb{R}_+, \tag{3.9}$$

together with (3.2a, b, or c). Equation (3.6) is supplemented as usual by the initial condition

$$U(0) = \xi, \qquad \xi \in H \quad \text{given}. \tag{3.10}$$

We can prove the following properties rigorously:

If u is the solution of III.(2.18), (2.19) given by Theorem III.2.1, then the initial- and boundary-value problem (3.6)–(3.10) possesses a unique solution

$$U \in L^2(0, T; V) \cap \mathscr{C}([0, T]; H), \qquad \forall T > 0. \tag{3.11}$$

For every $t > 0$, the function $u_0 \to S(t)u_0$ is Fréchet differentiable in H at u_0 with differential $L(t, u_0)$: $\xi \in H \to U(t) \in H$, where U is the solution of (3.6)–(3.10). (3.12)

Property (3.11) follows readily from Theorem II.3.4, while the proof of (3.12) will be sketched in Section 8.

Estimate of the Energy Dissipation Flux

Before proceeding with the estimate of the Lyapunov exponents, we want to compute here a bound for the energy dissipation flux. This number ε is defined in turbulence theory as

$$\varepsilon = \nu\langle \text{grad } u\rangle^2, \tag{3.13}$$

where u is the velocity and $\langle\ \rangle$ denotes ensemble averaging. We reinterpret this averaging as

$$\varepsilon = \nu\lambda_1 \limsup_{t\to\infty} \frac{1}{t}\int_0^t \|u(s)\|^2\, ds,^1 \tag{3.14}$$

and estimate this quantity in terms of the data, and more specifically in terms of the generalized Grashof number G (see III.(2.37))

$$G = \frac{|f|}{\nu^2\lambda_1}, \tag{3.15}$$

[1] $\lambda_1 \|u(\cdot\, t)\|^2$ is proportional to the L^2-space average of $|\text{grad } u(x, t)|^2$ and then we take the lim sup of the time average of this quantity.

λ_1 representing the first eigenvalue of A (which is not the same in the three cases (3.2a, b, c)). We start from III.(2.33), i.e.,

$$\frac{d}{dt}|u|^2 + \nu\|u\|^2 \le \frac{1}{\nu\lambda_1}|f|^2. \tag{3.16}$$

By integration

$$\frac{1}{t}|u(t)|^2 + \frac{\nu}{t}\int_0^t \|u(s)\|^2\, ds \le \frac{1}{t}|u_0|^2 + \frac{1}{\nu\lambda_1}|f|^2. \tag{3.17}$$

Therefore

$$\varepsilon \le \nu^3\lambda_1^2 G^2. \tag{3.18}$$

More generally, we can consider, instead of a specific trajectory $S(t)u_0$, all the trajectories for u_0 in a bounded functional-invariant set $X \subset H$. In this case ε is defined as

$$\varepsilon = \nu\lambda_1 \limsup_{t\to\infty} \operatorname*{Sup}_{u_0\in X} \frac{1}{t}\int_0^t \|u(s)\|^2\, ds. \tag{3.19}$$

In this case we deduce from (3.17) that

$$\frac{\nu}{t}\operatorname*{Sup}_{u_0\in X}\int_0^t \|u(s)\|^2\, ds \le \frac{1}{t}\operatorname*{Sup}_{u_0\in X}|u_0|^2 + \frac{1}{\nu\lambda_1}|f|^2$$

and (3.18) remains valid.

Remark 3.1. The length L_d associated with the energy dissipation flux is the only length which is deduced by dimensional analysis from ε and ν. Hence

$$L_d = \nu^{3/4}\varepsilon^{-1/4}, \tag{3.20}$$

and setting $L_0 = \lambda_1^{-1/2}$ we find

$$\left(\frac{L_0}{L_d}\right)^2 = \frac{\varepsilon^{1/2}}{\nu^{3/2}\lambda_1} \le G. \tag{3.21}$$

3.1.2. Dimension of the Attractor

We estimate the dimension of $\mathscr{A}$ by following the procedure described in Section V.2.3.

We consider, for $m \in \mathbb{N}$,

$$|U_1(t) \wedge \cdots \wedge U_m(t)|_{\wedge^m H}$$
$$= |\xi_1 \wedge \cdots \wedge \xi_m|_{\wedge^m H} \exp \int_0^t \operatorname{Tr} F'(S(\tau)u_0) \circ Q_m(\tau)\, d\tau,$$

where $S(\tau)u_0 = u(\tau)$ is the solution of III.(2.18), (2.19); $U_1, \ldots, U_m$ are m solutions of (3.6)–(3.10) corresponding to $\xi = \xi_1, \ldots, \xi_m$; $Q_m(\tau) = Q_m(\tau, u_0; \xi_1, \ldots, \xi_m)$ is the orthogonal projector in H onto the space spanned by $U_1(\tau), \ldots, U_m(\tau)$.

At a given time τ, let $\varphi_j(\tau)$, $j = 1, \ldots, m$, be an orthonormal basis of $Q_m(\tau)H = \text{Span}[U_1(\tau), \ldots, U_m(\tau)]$: $\varphi_j(\tau) \in V$ for $j = 1, \ldots, m$, since $U_1(\tau), \ldots, U_m(\tau) \in V$ (a.e. $\tau \in \mathbb{R}_+$, see (3.11)), and we have

$$\begin{aligned}\text{Tr } F'(S(\tau)u_0) \circ Q_m(\tau) &= \sum_{j=1}^{\infty} (\text{Tr } F'(u(\tau)) \circ Q_m(\tau)\varphi_j(\tau), \varphi_j(\tau)) \\ &= \sum_{j=1}^{m} (F'(u(\tau))\varphi_j(\tau), \varphi_j(\tau)),\end{aligned} \tag{3.22}$$

$(\cdot, \cdot)$ denoting the scalar product in H. Omitting temporarily the dependence on τ, we write, using III.(2.21), (2.17)

$$\begin{aligned}(F'(u)\varphi_j, \varphi_j) &= -\nu(A\varphi_j, \varphi_j) - (B(\varphi_j, \varphi_j) \\ &= -\nu\|\varphi_j\|^2 - b(\varphi_j, u, \varphi_j), \\ \sum_{j=1}^{m} (F'(u)\varphi_j, \varphi_j) &= -\nu \sum_{j=1}^{m} \|\varphi_j\|^2 - \sum_{j=1}^{m} b(\varphi_j, u, \varphi_j).\end{aligned} \tag{3.23}$$

Using the explicit expression of b provided by III.(2.15), we obtain

$$\sum_{j=1}^{m} b(\varphi_j, u, \varphi_j) = \int_\Omega \sum_{j=1}^{m} \sum_{i,k=1}^{2} \varphi_{ji}(x) D_i u_k(x) \varphi_{jk}(x)\, dx.$$

Pointwise (i.e., for a.e. $x \in \Omega$), we can write

$$\left| \sum_{j=1}^{m} \sum_{i,k=1}^{2} \varphi_{ji}(x) D_i u_k(x) \varphi_{jk}(x) \right| \leq |\text{grad } u(x)| \rho(x), \tag{3.24}$$

where

$$|\text{grad } u(x)| = \left\{ \sum_{i,k=1}^{2} |D_i u_k(x)|^2 \right\}^{1/2}, \tag{3.25}$$

$$\rho(x) = \sum_{i=1}^{2} \sum_{j=1}^{m} (\varphi_{ji}(x))^2. \tag{3.26}$$

Therefore

$$\begin{aligned}\left| \sum_{j=1}^{m} b(\varphi_j, u, \varphi_j) \right| &\leq \int_\Omega |\text{grad } u(x)| \rho(x)\, dx, \\ &\leq \text{(with the Schwarz inequality)}, \\ &\leq \|u\| |\rho|_{L^2},\end{aligned} \tag{3.27}$$

where $|\rho|_{L^2}$ is the norm of ρ in $L^2(\Omega)$.

We recall that the dependence on τ has been omitted and in fact $u = u(x, \tau)$, $\rho = \rho(x, \tau)$, etc.... At this point we have established the following inequality

$$\text{Tr } F'(u(\tau)) \circ Q_m(\tau) \leq -\nu \sum_{j=1}^{m} \|\varphi_j(\tau)\|^2 + \|u(\tau)\| |\rho(\tau)|_{L^2}. \tag{3.28}$$

We infer from Theorem A.3.1. (see, in particular, Example A.4.1), the existence of a dimensionless constant c_2' depending only on the shape of Ω

such that

$$|\rho(\tau)|_{L^2}^2 = \int_\Omega |\rho(x, \tau)|^2 \, dx \le c_2' \sum_{j=1}^{m} \|\varphi_j(\tau)\|^2. \tag{3.29}$$

Hence

$$\begin{aligned} \operatorname{Tr} F'(u(\tau)) \circ Q_m(\tau) &\le -\nu \sum_{j=1}^{m} \|\varphi_j(\tau)\|^2 + \|u(\tau)\| \left(c_2' \sum_{j=1}^{m} \|\varphi_j(\tau)\|^2 \right)^{1/2} \\ &\le \text{(with the Schwarz inequality)} \\ &\le -\frac{\nu}{2} \sum_{j=1}^{m} \|\varphi_j(\tau)\|^2 + \frac{c_2'}{2\nu} \|u(\tau)\|^2. \end{aligned} \tag{3.30}$$

We recall from G. Métivier [1] that the eigenvalues λ_j of A satisfy

$$\lambda_j \sim c\lambda_1 j^{2/n} \quad \text{as} \quad j \to \infty \tag{3.31}$$

(n, the space dimension equals 2 here). Then Lemma 2.1 shows the existence of a dimensionless constant c_1' depending only on the shape of Ω such that

$$\sum_{j=1}^{m} \|\varphi_j(\tau)\|^2 \ge \lambda_1 + \cdots + \lambda_m \ge c_1' \lambda_1 m^2. \tag{3.32}$$

This allows us to majorize $\operatorname{Tr} F'(u) \circ Q_m$ as follows

$$\operatorname{Tr} F'(u(\tau)) \circ Q_m(\tau) \le -c_1' \frac{\nu\lambda_1}{2} m^2 + \frac{c_2'}{2\nu} \|u(\tau)\|^2. \tag{3.33}$$

Hence

$$\frac{1}{t} \int_0^t \operatorname{Tr} F'(u(\tau)) \circ Q_m(\tau) \, d\tau \le -c_1' \frac{\nu\lambda_1}{2} m^2 + \frac{c_2'}{2\nu} \frac{1}{t} \int_0^t \|u(\tau)\|^2 \, d\tau,$$

and using (3.18), (3.19) we can majorize the quantities $q_m(t)$, q_m defined in V.(3.47), (3.48)

$$q_m(t) = \operatorname*{Sup}_{u_0 \in X} \operatorname*{Sup}_{\substack{\xi_i \in H \\ |\xi_i| \le 1 \\ i=1,\dots,m}} \left(\frac{1}{t} \int_0^t \operatorname{Tr} F'(u(\tau)) \circ Q_m(\tau) \, d\tau \right),$$

$$q_m(t) \le -c_1' \frac{\nu\lambda_1}{2} m^2 + \frac{c_2'}{2\nu} \operatorname*{Sup}_{u_0 \in X} \frac{1}{t} \int_0^t \|S(\tau)u_0\|^2 \, d\tau, \tag{3.34}$$

$$q_m = \limsup_{t \to \infty} q_m(t) \le -\kappa_1 m^2 + \kappa_2,$$

$$\kappa_1 = \frac{c_1'}{2} \nu\lambda_1, \qquad \kappa_2 = \frac{c_2'}{2} \frac{\varepsilon}{\nu^2 \lambda_1}, \tag{3.35}$$

ε as in (3.19) with $X = \mathscr{A}$.

We infer from (3.35) and V.(3.49) the following bound on the Lyapunov exponents $\mu_j, j \in \mathbb{N}$,

$$\mu_1 + \cdots + \mu_j \le q_j \le -\kappa_1 m^2 + \kappa_2, \qquad \forall j \in \mathbb{N}. \tag{3.36}$$

Using Lemma 2.2 we see that if m is defined by

$$m - 1 < \left(\frac{2\kappa_2}{\kappa_1}\right)^{1/2} = \left(\frac{2c_2'}{c_1'}\right)^{1/2} \frac{\varepsilon^{1/2}}{\nu^{3/2}\lambda_1} \le m, \tag{3.37}$$

then $\mu_1 + \cdots + \mu_m \le 0$ and

$$\frac{(\mu_1 + \cdots + \mu_j)_+}{|\mu_1 + \cdots + \mu_m|} \le 1, \qquad \forall j = 1, \dots, m-1. \tag{3.38}$$

Setting $c_3' = (2c_2'/c_1')^{1/2}$, $L_0 = \lambda_1^{-1/2}$ (the macroscopical length), and $L_d = \varepsilon^{1/4}/\nu^{3/4}$ (the dissipation length), we can rewrite (3.37) in the form

$$m - 1 < c_3' \left(\frac{L_0}{L_d}\right)^2 \le m. \tag{3.39}$$

By application of the results of Chapter V (in particular, Proposition V.2.1 and Theorem V.3.3), we have now proved the following

Theorem 3.1. *We consider the dynamical system associated with the two-dimensional Navier–Stokes equations* III.(2.1), (2.2), *with boundary conditions* (3.2a, b, c). *We define m by*

$$m - 1 < c_3' \left(\frac{L_0}{L_d}\right)^2 = c_3' \left(\frac{\varepsilon}{\nu^3}\right)^{1/2} \frac{1}{\lambda_1} \le m, \tag{3.40}$$

where c_3' is a dimensionless constant depending only on the shape of Ω. Then:

(i) *the m-dimensional volume element in H is exponentially decaying in the phase space as $t \to \infty$;*
(ii) *the corresponding global attractor $\mathcal{A}$ defined by Theorem* III.2.2 *has a Hausdorff dimension less than or equal to m and a fractal dimension less than or equal to 2m.*

Remark 3.2. Thanks to (3.18),

$$\left(\frac{L_0}{L_d}\right)^2 = \frac{\varepsilon^{1/2}}{\nu^{3/2}} \frac{1}{\lambda_1} \le G, \tag{3.41}$$

and we can replace, in the statement of Theorem 3.1, m by another larger number m_1 given by

$$m_1 - 1 < c_3' G \le m_1. \tag{3.42}$$

Remark 3.3. The estimates on the dimension of the attractor for the unbounded case considered in Section IV.7 are derived essentially as in the bounded case. For the details see R. Rosa [3].

3.2. Improvements for the Space-Periodic Case

We now consider the case of a space-periodicity boundary condition, i.e., (3.2b) or III.(2.5), (2.6). Of course, all the previous results in Section III.2 and above in Section 3.1 remain valid.

We will be led in this section to consider the dimension of $\mathscr{A}$ in the space $V(\subset \dot{H}^1_{\text{per}}(\Omega)^2)$ instead of considering as usual its dimension in H: the analysis is different but the dimensions are in fact the same, as this follows from Propositions 3.1 and 3.2 below. Section 3.2.1 contains some preliminary properties, including a property of the operator B which is specific to the space-periodic case. Section 3.2.2 contains a comparison of the dimension of $\mathscr{A}$ in V and H via a method of general interest. The main result is given in Section 3.2.3.

3.2.1. Preliminary Results

In the two-dimensional space-periodic case the operator B enjoys the following important orthogonality property (besides III.(2.21)).

Lemma 3.1.

$$(B(v, v), Av) = b(v, v, Av) = 0, \qquad \forall v \in D(A). \tag{3.43}$$

PROOF. Let P denote the orthogonal projector in $L^2(\Omega)^2$ onto H. Using Fourier series expansions it is easy to see that $P\Delta v = \Delta v$, $\forall v \in D(A)$, i.e., $Av = -\Delta v$, $\forall v \in D(A)$. Therefore

$$\begin{aligned}(B(v, v), Av) &= b(v, v, Av) = -b(v, v, \Delta v)\\ &= -\sum_{i,j,k=1}^{2} \int_\Omega v_i D_i v_j D_k^2 v_j \, dx\\ &= \text{(after integration by parts)}\\ &= -\sum_{i,j,k=1}^{2} \left(\int_\Omega v_i D_i D_k v_j D_k v_j \, dx + \int_\Omega D_k v_i D_i v_j D_k v_j \, dx \right).\end{aligned}$$

Now the sum of the first integrals vanishes since $\operatorname{div} v = 0$,

$$-\sum_{i,j,k=1}^{2} \frac{1}{2} \int_\Omega v_i D_i (D_k v_j)^2 \, dx = \frac{1}{2} \sum_{j,k=1}^{2} \int_\Omega \operatorname{div} v (D_k v_j)^2 \, dx = 0,$$

while the second sum of integrals vanishes too since

$$\sum_{i,j,k=1}^{2} D_k v_i(x) D_i v_j(x) D_k v_j(x) = 0, \quad \text{a.e. } x; \tag{3.44}$$

(3.44) can be verified by elementary algebraic computations, remembering that $D_1 v_1(x) = -D_2 v_2(x)$ a.e.

The lemma is proved. □

By Fréchet differentiation of the function

$$v \in D(A) \to (B(v, v), Av),$$

we deduce immediately from Lemma 3.1 another identity

Lemma 3.2.

$$(B(v, v), Aw) + (B(v, w), Av) + (B(w, v), Av) = 0, \qquad \forall v, w \in D(A). \quad (3.45)$$

Estimate of the Enstrophy Dissipation Flux

The enstrophy dissipation flux χ is defined in two-dimensional conventional theory of turbulence by

$$\chi = \nu \langle \Delta u \rangle^2, \quad (3.46)$$

where u is the velocity and $\langle \ \rangle$ denotes ensemble averaging. We reinterpret this averaging as

$$\chi = \nu \lambda_1 \limsup_{t \to \infty} \frac{1}{t} \int_0^t |Au(s)|^2 \, ds \quad (3.47)$$

in the case of a single trajectory ($Au = -\Delta u$). If we consider all the trajectories $u(t) = S(t)u_0$ for u_0 in a bounded functional-invariant set $X \subset H$, then χ is defined as

$$\chi = \nu \lambda_1 \limsup_{t \to \infty} \operatorname*{Sup}_{u_0 \in X} \frac{1}{t} \int_0^t |Au(s)|^2 \, ds. \quad (3.48)$$

We start from III.(2.31), where $(B(u), Au) = 0$ thanks to (3.43). There remains

$$\frac{1}{2} \frac{d}{dt} \|u\|^2 + \nu |Au|^2 = (f, Au). \quad (3.49)$$

We write

$$(f, Au) \leq |f||Au| \leq \frac{\nu}{2}|Au|^2 + \frac{1}{2\nu}|f|^2,$$

$$\frac{d}{dt} \|u\|^2 + \nu |Au|^2 \leq \frac{1}{\nu}|f|^2, \quad (3.50)$$

and by integration

$$\frac{1}{t} \|u(t)\|^2 + \frac{\nu}{t} \int_0^t |Au(s)|^2 \, ds \leq \frac{1}{t} \|u_0\|^2 + \frac{1}{\nu}|f|^2$$

$$\leq \frac{1}{t} \operatorname*{Sup}_{u_0 \in X} \|u_0\|^2 + \frac{1}{\nu}|f|^2.$$

A functional-invariant set bounded in H is also bounded in V, and we then obtain

$$\chi \le \frac{\lambda_1}{\nu}|f|^2 = \nu^3\lambda_1^3 G^2. \tag{3.51}$$

3.2.2. *Comparison of Dimensions in V and H*

The following simple result is of general interest.

Proposition 3.1. *Let X and Y be two metric spaces and let Φ be a Lipschitz application from X into Y*

$$d_Y(\Phi(u), \Phi(v)) \le k d_X(u, v), \qquad \forall u, v \in X, \tag{3.52}$$

d_X, d_Y representing the distances in X and Y.

For any subset $\mathscr{C}$ of X we have

$$d^Y_{\mathscr{H}}(\Phi(\mathscr{C})) \le d^X_{\mathscr{H}}(\mathscr{C}), \tag{3.53}$$

$$d^Y_F(\Phi(\mathscr{C})) \le d^Y_F(\mathscr{C}), \tag{3.54}$$

$d^X_{\mathscr{H}}$, $d^Y_{\mathscr{H}}$ representing the Hausdorff dimensions in X and Y, d^X_F, d^Y_F the fractal dimensions.

Proof. Due to the Lipschitz property, a ball of X centered at u_i of radius r_i, $B^X(u_i, r_i)$, is mapped by Φ into the ball of Y centered at $\Phi(u_i)$ of radius kr_i. Thus if $\mathscr{C}$ is covered by the set of balls $B^X(u_i, r_i)$, $i \in I$, then $\Phi(\mathscr{C})$ is covered by the set of balls $B^Y(\Phi(u_i), kr_i)$, $i \in I$. The proposition then follows readily from this observation and the definitions. □

From this we deduce the following

Proposition 3.2. *The attractor $\mathscr{A}$ given by Theorem* III.2.2 *has the same Hausdorff (resp. fractal) dimension in V and H.*

Proof. This results from two applications of Proposition 3.1 to $\mathscr{C} = \mathscr{A}$. We first apply this proposition with $X = V$, $Y = H$, $\Phi = I$, the identity

$$d^H_{\mathscr{H}}(\mathscr{A}) \le d^V_{\mathscr{H}}(\mathscr{A}), \qquad d^H_F(\mathscr{A}) \le d^V_F(\mathscr{A}).$$

Then we observe that $S(t)$ is locally Lipschitz from H into V, $\forall t > 0$, and since $S(t)\mathscr{A} = \mathscr{A}$, we obtain the reverse inequalities.

The Lipschitz property of $S(t)$ which reads

$$\|S(t)u - S(t)v\| \le k(t)|u - v|, \qquad \forall u, v \in V, \quad \forall t > 0, \tag{3.55}$$

is a particular case of more general results proved in Section 8. □

3.2.3. *Main Result*

Our aim now is to show that the result of Theorem 3.1 is valid with the ratio $(L_0/L_d)^2$ replaced by $(L_0/L_\chi)^2$ (up to a logarithmic order correction), i.e., that the dimension of the attractor is of the order of $G^{2/3}(\log G)^{1/3}$ for large G.

The proof consists of applying the general method described in Chapter V in the space V instead of H, remembering from Proposition 3.2 that the dimension of $\mathscr{A}$ is the same in V and H.

We first notice that (3.11), (3.12) can be extended to V: the mapping $u_0 \to S(t)u_0$ is differentiable in V, $\forall t > 0$, and its differential is the linear mapping $\xi \in V \to U(t)$, where U is the solution of (3.6)–(3.10); for $\xi \in V$ (and $u_0 \in V$) the solution U of (3.6)–(3.10) satisfies

$$U \in L^2(0, T; D(A)) \cap \mathscr{C}([0, T]; V). \tag{3.56}$$

Then we consider m solutions $U = U_1, \ldots, U_m$, of (3.6)–(3.10) corresponding to initial data $\xi = \xi_1, \ldots, \xi_m$ ($\in V$), with $u = u(\tau) = S(\tau)u_0$, $u_0 \in V$. We have

$$\begin{aligned}&\|U_1(t) \wedge \cdots \wedge U_m(t)\|_{\wedge^m V}\\ &\quad = \|\xi_1 \wedge \cdots \wedge \xi_m\|_{\wedge^m V} \exp \int_0^t \operatorname{Tr} F'(S(\tau)u_0 \cdot \tilde{Q}_m(\tau)\, d\tau,\end{aligned}$$

where $\tilde{Q}_m(\tau) = \tilde{Q}_m(\tau, u_0; \xi_1, \ldots, \xi_m)$ is the *orthogonal projector in* V onto the space spanned by $U_1(\tau), \ldots, U_m(\tau)$. At a given time τ, let $\varphi_j(\tau)$, $j = 1, \ldots, m$, be a basis, *orthonormal in* V, of $\tilde{Q}_m(\tau)V$. We have

$$\begin{aligned}\operatorname{Tr} F'(u(\tau) \circ \tilde{Q}_m(\tau) &= \sum_{j=1}^m ((F'(u(\tau)) \circ \tilde{Q}_m \varphi_j(\tau), \varphi_j(\tau)))\\ &= \sum_{j=1}^m (F'(u(\tau))\varphi_j(\tau), A\varphi_j(\tau)),\end{aligned} \tag{3.57}$$

$((\cdot, \cdot))$, $(\cdot, \cdot)$ denoting as usual the scalar products in V and H; we also recall that $((\varphi, \psi)) = (A\varphi, \psi)$, $\forall \varphi \in D(A)$, $\forall \psi \in V$. We omit temporarily the dependence on τ and write

$$\begin{aligned}(F'(u)\varphi_j, A\varphi_j) &= -\nu|A\varphi_j|^2 - (B(\varphi_j, u), A\varphi_j) - (B(u, \varphi_j), A\varphi_j)\\ &= (\text{with } (3.45))\\ &= -\nu|A\varphi_j|^2 + (B(\varphi_j, \varphi_j), Au).\end{aligned}$$

But

$$\begin{aligned}(B(\varphi_j, \varphi_j), Au) &= b(\varphi_j, \varphi_j, Au) = -b(\varphi_j, \varphi_j, \Delta u),\\ \sum_{j=1}^m (B(\varphi_j, \varphi_j), Au) &= -\sum_{j=1}^m \sum_{i,k=1}^2 \int_\Omega \varphi_{ji} D_i \varphi_{jk} \Delta u_k \, dx\\ &\le \sum_{j=1}^m \int_\Omega |\varphi_j(x)|\,|\operatorname{grad} \varphi_j(x)|\,|\Delta u(x)| \, dx\\ &\le \int_\Omega \rho(x)^{1/2} \sigma(x)^{1/2} |\Delta u(x)| \, dx\\ &\le |\rho|_{L^\infty(\Omega)}^{1/2} |\sigma|_{L^2(\Omega)}^{1/2} |\Delta u|_{L^{4/3}(\Omega)},\end{aligned} \tag{3.58}$$

where

$$\rho(x) = \sum_{j=1}^{m} |\varphi_j(x)|^2, \tag{3.59}$$

$$\sigma(x) = \sum_{j=1}^{m} |\text{grad } \varphi_j(x)|^2 = \sum_{j=1}^{m} \sum_{i,k=1}^{2} |D_i \varphi_{jk}(x)|^2. \tag{3.60}$$

The functions grad φ_j are orthonormal in $L^2(\Omega)^4$ since by assumption $((\varphi_i, \varphi_j)) = \delta_{ij}$. Therefore we can apply Theorem A.3.1 to $\mathbb{R}^4$ vector-valued functions which are space-periodic and have 0 average, $V = \dot{H}^2_{\text{per}}(\Omega)$, $H = \dot{H}^1_{\text{per}}(\Omega)$, $a(u, v) = (\Delta u, \Delta v)$. The verification of the hypotheses is made exactly as in Example A.4.3. Theorem A.3.1 implies the existence of a constant c_4' depending only on the shape of Ω (i.e., the ratio L_1/L_2) such that

$$|\sigma|^2_{L^2(\Omega)} \leq c_4' \sum_{j=1}^{m} |A\varphi_j|^2. \tag{3.61}$$

We also prove, in Lemma 3.3 hereafter, the existence of a constant c_5' such that

$$|\rho|_{L^\infty(\Omega)} \leq c_5' \left(1 + \log\left(\lambda_1^{-1} \sum_{j=1}^{m} |A\varphi_j|^2\right)\right). \tag{3.62}$$

Finally, thanks to the Hölder inequality

$$|\Delta u|_{L^{4/3}(\Omega)} \leq |\Omega|^{1/4} |\Delta u| \leq c_6' \lambda_1^{-1/4} |Au|.$$

We can majorize Tr $F'(u(\tau)) \circ \tilde{Q}_m(\tau)$ as follows:

$$\begin{aligned} \text{Tr } F'(u) \circ \tilde{Q}_m \leq -\nu \sum_{j=1}^{m} |A\varphi_j|^2 &+ c_7' \left(\lambda_1^{-1} \sum_{j=1}^{m} |A\varphi_j|^2\right)^{1/4} \\ &\times \left(1 + \log\left(\lambda_1^{-1} \sum_{j=1}^{m} |A\varphi_j|^2\right)\right)^{1/2} |Au|, \end{aligned} \tag{3.63}$$

$c_7' = (c_4')^{1/4}(c_5')^{1/2} c_6'$. We set $x_m(\tau) = \lambda_1^{-1} \sum_{j=1}^{m} |A\varphi_j(\tau)|^2$. By integration of (3.63) and utilization of the Schwarz inequality we obtain

$$\begin{aligned} &\frac{1}{t} \int_0^t \text{Tr } F'(u(\tau)) \circ \tilde{Q}_m(\tau)\, d\tau \\ &\leq -\frac{\nu\lambda_1}{t} \int_0^t x_m(\tau)\, d\tau + c_7' \frac{1}{t} \int_0^t x_m(\tau)^{1/4} (1 + \log x_m(\tau))^{1/2} |Au(\tau)|\, d\tau \\ &\leq -\frac{\nu\lambda_1}{t} \int_0^t x_m(\tau)\, d\tau + c_7' \left(\frac{1}{t} \int_0^t x_m(\tau)^{1/2} (1 + \log x_m(\tau))\, d\tau\right)^{1/2} \\ &\quad \cdot \left(\frac{1}{t} \int_0^t |Au(\tau)|^2\, d\tau\right)^{1/2}. \end{aligned} \tag{3.64}$$

We then apply Lemma 2.1 with H replaced by V and taking into account (3.31), we obtain the existence of a constant c_8' depending only on the shape of Ω (i.e., the ratio L_1/L_2) such that

$$\sum_{j=1}^{m} |A\psi_j|^2 \geq \lambda_1 + \cdots + \lambda_m \geq c_8' \lambda_1 m^2 \tag{3.65}$$

for any family ψ_j orthonormal in V and in particular $\psi_j = \varphi_j(\tau)$, $j = 1, \ldots, m$. Hence $x_m(\tau) \geq c_8' m^2$ and also $x_m(\tau) \geq 1 + \lambda_2/\lambda_1 + \cdots + \lambda_m/\lambda_1 \geq 1$. We observe that the function $x \to x^{1/2}(1 + \log x)$ is concave for $x \geq 1/e$. It is then possible to apply the Jensen inequality to the integral in (3.64) and we find

$$\frac{1}{t}\int_0^t \operatorname{Tr} F'(u(\tau)) \circ \tilde{Q}_m(\tau)\, d\tau$$
$$\leq -\nu\lambda_1 y_m(t) + c_7' \nu\lambda_1 y_m(t)^{1/4}(1 + \log y_m(t))^{1/2}\theta(t),$$

where

$$y_m(t) = \frac{1}{t}\int_0^t x_m(\tau)\, d\tau \geq 1, \quad \text{and} \quad \theta(t) = \left(\frac{1}{\nu^2\lambda_1^2 t}\int_0^t |Au(\tau)|^2\, d\tau\right)^{1/2}.$$

It follows from the definition (3.48) of χ that for t sufficiently large, $t \geq t_1$ (t_1 depending only on $\mathscr{A}$)

$$\operatorname*{Sup}_{u_0 \in \mathscr{A}} \frac{1}{\nu^2\lambda_1^2 t}\int_0^t |Au(\tau)|^2\, d\tau \leq \frac{2\chi}{\nu^3\lambda_1^3}, \quad \text{i.e.,} \quad \theta(t) \leq \left(\frac{2\chi}{\nu^3\lambda_1^3}\right)^{1/2}.$$

We set $\rho = c_7'(2\chi/\nu^3\lambda_1^3)^{1/2}$ and observing that the function $y \to -y/2 + \rho y^{1/4}(1 + \log y)^{1/2}$ is bounded from above, we denote by $h(\rho)$ an upper bound of its supremum for $y \geq 1$. An explicit expression of h is given below; for the moment, we write

$$\frac{1}{t}\int_0^t \operatorname{Tr} F'(S(\tau)u_0) \circ \tilde{Q}_m(\tau)\, d\tau \leq -\frac{\nu}{2}\lambda_1 c_8' m^2 + \nu\lambda_1 h(\rho), \qquad t \geq t_1,$$

($y_m \geq c_8' m^2$ since $x_m \geq c_8' m^2$).

Assuming that u_0 belongs to $\mathscr{A}$, we can estimate the quantities $\tilde{q}_m(t)$, $\tilde{q}_m$

$$\tilde{q}_m(t) = \operatorname*{Sup}_{u_0 \in \mathscr{A}} \operatorname*{Sup}_{\substack{\xi_i \in V \\ \|\xi_i\| \leq 1 \\ i=1,\ldots,m}} \frac{1}{t}\int_0^t \operatorname{Tr} F'(S(\tau)u_0) \circ \tilde{Q}_m(\tau)\, d\tau,$$

$$\tilde{q}_m(t) \leq -\frac{\nu\lambda_1}{2} c_8' m^2 + \nu\lambda_1 h(\rho), \qquad \forall m \in \mathbb{N}, \quad \forall t \geq 1,$$

$$\tilde{q}_m = \limsup_{t\to\infty} \tilde{q}_m(t) \leq -\kappa_3 m^2 + \kappa_4, \tag{3.66}$$

$$\kappa_3 = \frac{\nu\lambda_1}{2} c_8', \qquad \kappa_4 = \nu\lambda_1 h(\rho).$$

We infer from (3.66) and V.(3.49) the following bound for the Lyapunov exponents $\tilde{\mu}_j$ (in V)

$$\tilde{\mu}_1 + \cdots + \tilde{\mu}_j \leq \tilde{q}_j \leq -\kappa_3 m^2 + \kappa_4, \qquad \forall j \in \mathbb{N}. \tag{3.67}$$

Using Lemma 2.2 we see that if $m = m_2$ is defined by

$$m_2 - 1 < \left(\frac{2\kappa_4}{\kappa_3}\right)^{1/2} \leq m_2, \tag{3.68}$$

then $\tilde{\mu}_1 + \cdots + \tilde{\mu}_m < 0$ and

$$\frac{(\tilde{\mu}_1 + \cdots + \tilde{\mu}_j)_+}{|\tilde{\mu}_1 + \cdots + \tilde{\mu}_{m_2}|} \leq 1, \qquad \forall j = 1, \ldots, m_2 - 1. \tag{3.69}$$

Thus the Hausdorff dimension of $\mathscr{A}$ in V (and in H) is less than or equal to m, its fractal dimension less than or equal to $2m$.

In order to determine m_2 more precisely we need a more explicit expression of $h(\rho)$. This is elementary: by determining the supremum of $\rho^2(1 + \log y) - \varepsilon y^{3/2}$ (we choose $\varepsilon = \frac{1}{16}$ below), we find

$$\rho^2(1 + \log y) \leq \varepsilon y^{3/2} + \rho^2\left(1 + \tfrac{2}{3}\log\left(\frac{2\rho^2}{3\varepsilon}\right)\right),$$

$$\begin{aligned} \rho y^{1/4}(1 + \log y)^{1/2} &\leq \varepsilon^{1/2} y + \rho y^{1/4}\left(1 + \tfrac{2}{3}\log\left(\frac{2\rho^2}{3\varepsilon}\right)\right)^{1/2} \\ &\leq 2\varepsilon^{1/2} y + c\rho^{4/3}\left(1 + \tfrac{2}{3}\log\left(\frac{2\rho^2}{3\varepsilon}\right)\right)^{2/3}. \end{aligned}$$

Thus with $\varepsilon = \frac{1}{16}$ and c_9' an absolute constant

$$h(\rho) = c_9' \rho^{4/3}(1 + \log \rho)^{2/3}, \tag{3.70}$$

that we can replace by

$$c_{10}'\left(\frac{\chi}{\nu^3\lambda_1^3}\right)^{2/3}\left(1 + \log\left(\frac{\chi}{\nu^3\lambda_1^3}\right)\right)^{2/3}.$$

This gives for (3.68)

$$m_2 - 1 < (2c_{10}')^{1/2}\left(\frac{\chi}{\nu^3\lambda_1^3}\right)^{1/3}\left(1 + \log\left(\frac{\chi}{\nu^3\lambda_1^3}\right)\right)^{1/3} \leq m_2. \tag{3.71}$$

Alternatively, we define the macroscopical length $L_0 = \lambda_1^{-1/2}$ and microscopical length $L_\chi = (\nu^3/\chi)^{1/6}$. Then $\chi/\nu^3\lambda_1^3 = (L_0/L_\chi)^6$ and (3.71) becomes

$$m_2 - 1 < c_{11}'\left(\frac{L_0}{L_\chi}\right)^2\left(1 + \log\left(\frac{L_0}{L_\chi}\right)\right)^{1/3} \leq m_2. \tag{3.72}$$

In conclusion, we have

Theorem 3.2. *We consider the dynamical system associated with the two-dimensional Navier–Stokes equations* III.(2.1), (2.2) *with space-periodicity boundary condition* III.(2.5), (2.6). *We define* m_2 *by* (3.72). *Then:*

(i) *the* m_2*-dimensional volume element in* V *is exponentially decaying in the phase space as* $t \to \infty$;

(ii) *the corresponding global attractor defined by Theorem* III.2.2 *has a Hausdorff dimension less than or equal to* m_2 *and a fractal dimension less than or equal to* $2m_2$.

Remark 3.4. Thanks to (3.51)

$$\left(\frac{L_0}{L_\chi}\right)^2 = \left(\frac{\chi}{\nu^3\lambda_1^3}\right)^{1/3} \le G^{2/3}$$

and we can replace (in Theorem 3.2) m_2 by another larger number $m = m_3$ given by

$$m_3 - 1 < c''_{12} G^{2/3}(1 + \log G)^{1/3} \le m_3 \tag{3.73}$$

(compare with (3.42)). For large G we obtain the bound on the dimension of the attractor, $m \sim cG^{2/3}(\log G)^{1/3}$ instead of cG (see (3.42)). This improvement which is consistent with the predictions of the conventional theory of turbulence is due to P. Constantin, C. Foias, and R. Temam [2].

We conclude this section with the lemma promised above

Lemma 3.3. *There is a constant c depending only on the shape of* $\Omega = (0, L_1) \times (0, L_2)$ (i.e., on L_1/L_2) *such that for every family of functions* $\varphi_1, \dots, \varphi_m$ *in* $\dot{H}^2_{\text{per}}(\Omega)^l$, *which is orthonormal in* $\dot{H}^1_{\text{per}}(\Omega)^l$, *we have*

$$\sum_{i=1}^m |\varphi_i(x)|^2 \le c\left(1 + \log\left(\lambda_1^{-1} \sum_{i=1}^m |\Delta\varphi_i|^2_{L^2(\Omega)}\right)\right), \tag{3.74}$$

where λ_1 *is the first eigenvalue of* $-\Delta$ *in* $\dot{H}^2_{\text{per}}(\Omega)$.

Proof. We infer from H. Brézis and T. Gallouet [1] the existence of c' such that

$$|\varphi|_{L^\infty(\Omega)} \le c'\|\varphi\|\left(1 + \log\frac{|\Delta\varphi|^2_{L^2(\Omega)}}{\lambda_1\|\varphi\|^2}\right)^{1/2}.$$

We write this inequality with $\varphi = \sum_{j=1}^m \alpha_j\varphi_j$, $\alpha_j \in \mathbb{R}$, $\sum_{j=1}^m \alpha_j^2 \le 1$. Since the family φ_j is orthonormal in $\dot{H}^1_{\text{per}}(\Omega)^l$,

$$\|\varphi\| = \left(\sum_{j=1}^m \alpha_j^2\right)^{1/2} \le 1,$$

and therefore

$$\left|\sum_{j=1}^m \alpha_j\varphi_j(x)\right| \le |\varphi|_{L^\infty(\Omega)} \le c'\left(1 + \log\frac{\left(\sum_{j=1}^m \alpha_j^2\right)\left(\sum_{j=1}^m |\Delta\varphi_j|^2_{L^2}\right)}{\lambda_1}\right)^{1/2}$$

$$\le c'\left(1 + \log \lambda_1^{-1}\sum_{j=1}^m |\Delta\varphi_j|^2_{L^2}\right)^{1/2}, \qquad \forall x \in \Omega.$$

Taking the supremum with respect to the α_j we obtain (3.74). □

Remark 3.5. An alternative proof of (3.74) is given in P. Constantin [1]; see also E. Lieb [1].

4. Other Equations in Fluid Mechanics

Various equations of fluid mechanics have been studied in Chapter III, Section 3. They correspond to the Navier–Stokes equation in particular situations different from that studied in Section III.2 (and Section 3 in this chapter), or to systems representing the coupling of these equations with other ones (the Maxwell equations in the case of magnetohydrodynamics, and the heat equation for thermohydraulics).

We will now pursue our task in this chapter and apply the results of Chapter V to all these equations. In Section 4.1 we consider the abstract equation of Section III.3.1, i.e., III.(3.10), (3.14) and derive its first variation equation and the differentiability of the semigroup $S(t)$. Then, the estimates on the dimension of the attractors necessitate more specific information on the equations and they will be derived for each equation separately. In Section 4.2 we consider the fluid driven by its boundary; Section 4.3 is related to magnetohydrodynamics, Section 4.4 is related to flows on a manifold. Finally, in Section 4.5, we study thermohydraulics equations.

4.1. The Linearized Equations (The Abstract Framework)

This section is the continuation of Section III.3.1 of which we retain all the notations and hypotheses, in particular III.(3.3)–(3.9).[1] By virtue of Theorem III.3.1, the initial-value problem III.(3.10), (3.11) possesses a unique solution u for any $u_0 \in H$, $u(t) = S(t)u_0$, $\forall t \geq 0$. Writing equation III.(3.10) in the general form

$$u' = F(u), \tag{4.1}$$

as in V.(2.38), we see that the first variation equation V.(2.39)

$$U' = F'(u)U, \tag{4.2}$$

$$U(0) = \xi, \tag{4.3}$$

can be written heuristically

$$\frac{dU}{dt} + AU + B(u, U) + B(U, u) + RU = 0, \qquad t > 0, \tag{4.4}$$

$$U(0) = \xi, \tag{4.5}$$

[1] Hypothesis III.(3.5) is not necessary at this point but will be needed subsequently.

where ξ is given in H. More precisely, it is elementary to check that the conditions of Theorem II.3.4 are fulfilled and then we infer from this theorem the following:

(i) If $u_0 \in H$ and u is the solution of III.(3.10), (3.11) given by Theorem III.3.1, the linear initial-value problem (4.2)–(4.5) possesses a unique solution U

$$U \in L^2(0, T; V) \cap \mathscr{C}([0, T]; H), \qquad \forall T > 0. \tag{4.6}$$

Furthermore,

(ii) The semigroup $S(t)$: $u_0 \to u(t)$ is Fréchet differentiable in H,

For every $t > 0$, the function $u_0 \to S(t)u_0$ is Fréchet differentiable in H at u_0 with differential $L(t, u_0)$: $\xi \in H \to L(t, u_0) \cdot \xi = U(t) \in H$, where U is the solution of (4.2)–(4.5). (4.7)

This technical point will be proved in Section 8.

Orientation

We will now give the explicit form of problem (4.4), (4.5) for the examples studied in Section III.3 and we will derive estimates on the uniform Lyapunov exponents and on the dimension of the corresponding attractor.

4.2. Fluid Driven by Its Boundary

4.2.1. The General Case

The linearized equations (4.2)–(4.5) corresponding to III.(3.19)–(3.22) are the abstract form of the following boundary-value problem

$$\frac{\partial U}{\partial t}(t) + ((u + \Phi)\cdot\nabla)U + (u\cdot\nabla)(U + \Phi) - \nu\Delta U + \nabla\Pi = 0 \quad \text{in} \quad \Omega \times \mathbb{R}_+, \tag{4.8}$$

$$\operatorname{div} U = 0 \quad \text{in} \quad \Omega \times \mathbb{R}_+, \tag{4.9}$$

$$U = 0 \quad \text{on} \quad \Gamma \times \mathbb{R}_+. \tag{4.10}$$

Alternatively, the analog of III.(3.22) is

$$\frac{d}{dt}(U, v) + \nu((U, v)) + b(u + \Phi, U, v) + b(U, u + \Phi, v) = 0, \qquad \forall v \in V. \tag{4.11}$$

The spaces V, H, the form b, and Φ are the same as in Section III.3.2; Π is the pressure function implicitly contained in (4.4). Finally, u is the solution of the (initial) nonlinear problem III.(3.19)–(3.22).

We follow the steps of the procedure summarized in Section V.3.4 and consider, for every $m \in \mathbb{N}$,

$$|U_1(t) \wedge \cdots \wedge U_m(t)|_{\wedge^m H}$$
$$= |\xi_1 \wedge \cdots \wedge \xi_m|_{\wedge^m H} \exp \int_0^t \operatorname{Tr} F'(S(\tau)u_0) \circ Q_m(\tau)\, d\tau,$$

where $S(\tau)u_0 = u(\tau)$, u as above; $U_1, \ldots, U_m$ are m solutions of (4.2)–(4.7) corresponding to $\xi = \xi_1, \ldots, \xi_m$, and $Q_m(\tau) = Q_m(\tau, u_0; \xi_1, \ldots, \xi_m)$ is the orthogonal projector in H onto the space spanned by $U_1(\tau), \ldots, U_m(\tau)$.

At a given time τ, let $\varphi_j(\tau)$, $j \in \mathbb{N}$,[1] be an orthonormal basis of H, with $\varphi_1(\tau), \ldots, \varphi_m(\tau)$ spanning $Q_m(\tau)H = \operatorname{Span}[U_1(\tau), \ldots, U_m(\tau)]$; hence $\varphi_j(\tau) \in V$ for $j = 1, \ldots, m$ and for a.e. $\tau \in \mathbb{R}_+$ (see (4.6)). The trace of $F'(u(\tau)) \circ Q_m(\tau)$ is given by

$$F'(u(\tau)) \circ Q_m(\tau) = \sum_{j=1}^{\infty} (F'(u(\tau)) \circ Q_m(\tau)\varphi_j(\tau), \varphi_j(\tau))$$
$$= \sum_{j=1}^{m} (F'(u(\tau))\varphi_j(\tau), \varphi_j(\tau)),$$

$(\cdot, \cdot)$ denoting as before the scalar product in H. Omitting for the moment the variable τ we write

$$(F'(u)\varphi_j, \varphi_j) = -\nu \|\varphi_j\|^2 - b(\varphi_j, u + \Phi, \varphi_j);$$

the terms $b(u + \Phi, \varphi_j, \varphi_j)$ disappeared owing to III.(3.6). Recalling the definition of b in III.(2.15) we notice that, pointwise (i.e., for a.e. $x \in \Omega$),

$$\left| \sum_{j=1}^{m} \sum_{i,k=1}^{2} \varphi_{ji}(x)(D_i u_k(x) + D_i \Phi_k(x))\varphi_{jk}(x) \right| \le \sum_{j=1}^{m} |\varphi_j(x)|^2 |\operatorname{grad}(u + \Phi)(x)|,$$
$$\le \rho(x)|\operatorname{grad}(u + \Phi)(x)|,$$

with[2]

$$\rho(x) = \sum_{j=1}^{m} |\varphi_j(x)|^2, \tag{4.12}$$

$$|\operatorname{grad}(u + \Phi)(x)| = \left\{ \sum_{i,k=1}^{2} |D_i u_k(x) + D_i \Phi_k(x)|^2 \right\}^{1/2}.$$

Hence, after integration with respect to x,

$$\left| \sum_{j=1}^{m} b(\varphi_j, u + \Phi, \varphi_j) \right| \le \|u + \Phi\| \, |\rho|_{L^2}, \tag{4.13}$$

where $|\rho|_{L^2}$ is the norm of ρ in $L^2(\Omega)$. Returning to the quantity we want to

[1] These $\varphi_j(\tau)$ are totally unrelated to the function Φ associated with the boundary data in Section III.3.2.

[2] Recall that these quantities depend on τ although this does not appear explicitly for the moment.

estimate, we see that

$$\operatorname{Tr} F'(u(\tau)) \circ Q_m(\tau) \leq -\nu \sum_{j=1}^{m} \|\varphi_j(\tau)\|^2 + \|u(\tau) + \Phi(\tau)\| \, |\rho(\tau)|_{L^2}. \tag{4.14}$$

At this point we make use of Corollary A.5.1 and infer from this corollary the existence of a constant c_1' depending only on the shape of Ω such that

$$|\rho(\tau)|^2_{L^2(\Omega)} \leq c_1' \sum_{j=1}^{m} \|\varphi_j(\tau)\|^2 \quad \text{for a.e. } \tau. \tag{4.15}$$

Together with (4.14) and the Schwarz inequality this implies

$$\operatorname{Tr} F'(u(\tau)) \circ Q_m(\tau) \leq -\frac{\nu}{2} \sum_{j=1}^{m} \|\varphi_j(\tau)\|^2 + \frac{(c_1')^2}{2\nu} \|u(\tau) + \Phi\|^2. \tag{4.16}$$

The underlying linear operator A is nothing other than the Stokes operator with (homogeneous) Dirichlet boundary condition on Γ. We have recalled in (3.31) that, according to G. Métivier [1], the asymptotic distribution of its eigenvalues λ_j is given by

$$\lambda_j \sim c\lambda_1 j^{2/n} \quad \text{as} \quad j \to \infty \quad (n = 2 \text{ here}). \tag{4.17}$$

We then deduce from Lemma 2.1 the existence of a constant c_2' depending only on the shape of Ω such that

$$\sum_{j=1}^{m} \|\varphi_j\|^2 \geq c_2' \lambda_1 m^2.$$

Thus (4.16) yields

$$\operatorname{Tr} F'(u(\tau)) \circ Q_m(\tau) \leq -\frac{c_2'}{2} \nu \lambda_1 m^2 + \frac{(c_1')^2}{2\nu} \|u(\tau) + \Phi\|. \tag{4.18}$$

The quantities $q_m(t)$, q_m defined in V.(3.47), (3.48) can now be majorized; for q_m we have

$$\begin{gathered} q_m \leq -\kappa_1 m^2 + \kappa_2, \\ \kappa_1 = \frac{c_2'}{2} \nu \lambda_1, \qquad \kappa_2 = \frac{(c_1')^2}{\nu} (\gamma + \|\Phi\|^2), \\ \gamma = \limsup_{t \to \infty} \operatorname*{Sup}_{u_0 \in X} \frac{1}{t} \int_0^t \|S(\tau) u_0\|^2 \, d\tau. \end{gathered} \tag{4.19}$$

Here X is a bounded functional-invariant set for the semigroup $S(t)$ which could be, in particular, the maximal attractor $\mathscr{A}$ given by Theorem III.3.2.

We infer from (4.19) and V.(3.49) the following bound on the Lyapunov exponents μ_j, $j \in \mathbb{N}$,

$$\mu_1 + \cdots + \mu_j \leq q_j \leq -\kappa_1 j^2 + \kappa_2, \qquad \forall j \in \mathbb{N}. \tag{4.20}$$

Using Lemma 2.2 we see that if m is defined by

$$m - 1 < \left(\frac{2\kappa_2}{\kappa_1}\right)^{1/2} \leq m, \tag{4.21}$$

then $\mu_1 + \cdots + \mu_m < 0$ and

$$\frac{(\mu_1 + \cdots + \mu_j)_+}{|\mu_1 + \cdots + \mu_m|} \leq 1, \qquad \forall j = 1, \ldots, m-1. \tag{4.22}$$

By application of the results of Chapter V (see Section V.3.4), we then have proved the following

Theorem 4.1. *We consider the dynamical system associated with the two-dimensional Navier–Stokes equations* III.(2.1), (2.2) *with boundary condition* III.(3.17) (*see also* III.(3.19)–(3.21)). *We denote by* $\mathscr{A}$ *the corresponding maximal attractor given by Theorem* III.3.2. *Then the uniform Lyapunov exponents of* $\mathscr{A}$ *are majorized according to* (4.20) *and if m is defined by* (4.21):

(i) *the m-dimensional volume element in H is exponentially decaying in the phase space as $t \to \infty$;*
(ii) *the maximal attractor $\mathscr{A}$ of the system has a Hausdorff dimension less than or equal to m and a fractal dimension less than or equal to $2m$.*

Estimate of m in terms of the data.

According to (4.19)

$$\begin{aligned}\left(\frac{2\kappa_2}{\kappa_1}\right)^{1/2} &= \frac{(c_2')^{1/2}}{c_1'} \frac{1}{\nu\lambda_1^{1/2}} (\gamma + \|\Phi\|^2)^{1/2} \\ &= c_3' \frac{1}{\nu\lambda_1^{1/2}} (\gamma + \|\Phi\|^2)^{1/2},\end{aligned} \tag{4.23}$$

and Theorem 4.1 will be complete and totally relevant once we have expressed m (i.e., γ and $\|\Phi\|$) in terms of the physical entities ν, Ω, f, and φ. This is now our aim.

Setting $v = u$ in III.(3.22) and using III.(2.21) and the expression of $\bar{f}$ in III.(3.19), we find

$$\frac{1}{2}\frac{d}{dt}|u|^2 + \nu\|u\|^2 + b(u, \Phi, u) = (\bar{f}, u) = (f, u) - \nu((u, \Phi)) - b(\Phi, \Phi, \Phi).$$

We write

$$(f, u) \leq |f||u| \leq \lambda_1^{-1/2}|f|\,\|u\| \leq \frac{\nu}{4}\|u\|^2 + \frac{1}{\nu\lambda_1}|f|^2$$

$$-\nu((u, \Phi)) \leq \nu\|u\|\,\|\Phi\| \leq \frac{\nu}{4}\|u\|^2 + \nu\|\Phi\|^2;$$

III.(2.21) does not apply to $b(\Phi, \Phi, \Phi)$ and this quantity does not vanish; returning to the explicit expression of b and using the last inequality III.(3.23) we see that

$$|b(\Phi, \Phi, \Phi)| = \left| \sum_{i,j=1}^{2} \int_\Omega \Phi_i D_i \Phi_j \Phi_j \, dx \right|$$
$$\leq \text{const } |\Phi| \, \|\Phi\|_1^2, \tag{4.24}$$

$\|\Phi\|_1$ is the norm of Φ in $H^1(\Omega)^2$. It is convenient to write the constant above in the form $c_4' \lambda_1$. Finally,

$$\frac{d}{dt}|u|^2 + \nu \|u\|^2 + 2b(u, \Phi, u) \leq \frac{2}{\nu \lambda_1}|f|^2 + 2\nu \|\Phi\|^2 + 2c_4' \lambda_1 |\Phi| \, \|\Phi\|_1^2. \tag{4.25}$$

The function Φ is an extension inside Ω of the given (imposed) boundary value φ of u on Γ (see III.(3.17)); Φ is arbitrary provided III.(3.18) is satisfied. We indicated in III.(3.29) that the extension $\Phi = \Phi_\nu$ can be chosen so that

$$|b(v, \Phi_\nu, v)| \leq \frac{\nu}{4}\|v\|^2, \qquad \forall v \in V. \tag{4.26}$$

We then infer from (4.25) and (4.26) that

$$\frac{d}{dt}|u^2| + \frac{\nu}{2}\|u\|^2 \leq \frac{2}{\nu \lambda_1}|f|^2 + 2\nu \|\Phi_\nu\|^2 + 2c_4' \lambda_1 |\Phi_\nu| |\Phi_\nu|_1^2.$$

By integration,

$$\frac{1}{t}\int_0^t \|u(s)\|^2 \, ds \leq \frac{2}{\nu t} \operatorname*{Sup}_{u_0 \in X} |u_0|^2 + \frac{4}{\nu^2 \lambda_1}|f|^2 + 4\|\Phi_\nu\|^2 + \frac{4c_4' \lambda_1}{\nu} |\Phi_\nu| |\Phi_\nu|_1^2,$$

and the expression (4.19) of γ is bounded by

$$\gamma \leq \frac{4}{\nu^2 \lambda_1}|f|^2 + 4\|\Phi_\nu\|^2 + \frac{4c_4' \lambda_1}{\nu} |\Phi_\nu| |\Phi_\nu|_1^2.$$

It remains to insert these values (these bounds) of γ and Φ_ν into (4.23). We obtain

$$m \simeq \left(\frac{2\kappa_2}{\kappa_1}\right)^{1/2} \simeq c_5' \left(\frac{|f|}{\nu^2 \lambda_1} + \frac{1}{\nu \lambda_1^{1/2}} \|\Phi_\nu\| + \frac{1}{\nu^3} |\Phi_\nu|_1^3 \right). \tag{4.27}$$

A bound of the norm of Φ_ν was given in Remark III.3.1. A more precise form of the bound explained in Remark 4.2 below shows that

$$|\Phi_\nu|_1 \leq c_6' \nu \left(\frac{|\varphi|_{H^{1/2}(\Gamma)^2}}{\nu \lambda_1^{1/2}} \right) \exp\left(c_7' \frac{|\varphi|_{H^{1/2}(\Gamma)^2}}{\nu \lambda_1^{1/2}} \right), \tag{4.28}$$

with two suitable constants c_6', c_7'.

Remark 4.1 (Physical Comments). As explained below, the constants c_i' are nondimensional (and depend only on Ω). Therefore the data f, ν, φ (and Ω) occur in (4.29) and (4.30) through two nondimensional numbers which characterize the flow, namely,

$$\frac{|f|}{\nu^2 \lambda_1} \quad \text{and} \quad \frac{|\varphi|_{H^{1/2}(\Gamma)^2}}{\nu \lambda_1^{1/2}}. \tag{4.29}$$

We can show through a dimension analysis that these numbers are indeed nondimensional, $|\varphi|_{H^{1/2}(\Gamma)^2}$ having the dimension of a velocity.[1] The first number in (4.29) is nothing other than the Grashof number introduced in Remark III.2.3 in the case of the homogeneous boundary condition. Since $|\varphi|_{H^{1/2}(\Gamma)^2}$ is of the order of the given boundary velocities of the fluid, the second number in (4.29) naturally plays the role of a Reynolds number attached to the data.

Then (4.27), (4.28) express the fact that

$$m \simeq c_8'(G + \mathrm{Re}^3 \exp(3c_7' \,\mathrm{Re})). \tag{4.30}$$

The linear dependence of m on G is consistent with the results of the homogeneous case (Section 3). The appearance of terms of the order of exp(Re) does not seem natural and is probably due to the technical choice of the extension $\Phi = \Phi_\nu$, of φ inside Ω. It is likely that the diameter of $\mathscr{A}$ in the $|\cdot|_1$ norm is much smaller that $|\Phi_\nu|_1$, and that the exp(Re) term is not necessary.

Remark 4.2. The passage of III.(3.30), (3.31) to (4.30) is due to the introduction of dimensionalized quantities and occurs as follows.

First, for dimensional reasons, we set $\|\Phi\|_1 = (|\varphi|^2 + \lambda_1^{-1}\|\Phi\|^2)^{1/2}$. Then a perusal of the proof of Theorem 1.6 in Section II.1.4 of R. Temam [2] shows that III.(3.30) is of the form

$$\max\{|\Phi_\varepsilon|, \lambda_1^{-1/2}\|\Phi_\varepsilon\|\} \le |\Phi_\varepsilon|_1 \le c'' \frac{\nu}{\varepsilon} \exp\left(\frac{c'''}{\varepsilon}\right) |\varphi|_{H^{1/2}(\Gamma)^2},$$

Re as in Remark 4.1. The appropriate choice of ε is $\varepsilon = c^{\mathrm{IV}}\,\mathrm{Re}^{-1}$, with an absolute nondimensional constant c^{IV}; this yields (4.30).

4.2.2. An Improved Result

In this section, we assume that Ω is smooth (at least $\mathscr{C}^3$) and that φ satisfies III.(3.32). The bound on the dimension of the global attractor, derived in the previous section for the general case, contains an exponential of the Reynolds number, and thus is not physically satisfactory. Our aim in this section is to show that when III.(3.32) is satisfied, we can obtain an upper bound that is, in fact, polynomial with respect to the Reynolds number.

We proceed exactly as in Section 4.2.1, except that now $\boldsymbol{\Phi}$ is the function $\boldsymbol{\Phi}_\varepsilon$ constructed in Lemma III.3.1 corresponding to $\varepsilon = \nu/cUL$, where $U =$

[1] φ is a velocity and $|\varphi|_{H^{1/2}(\Gamma)^2}$ has the same dimension as $|\varphi|_{L^2(\Gamma)}^{1/2}|\nabla\varphi|_{L^2(\Gamma)}^{1/2}$.

$\operatorname{Max}(|\varphi|_{L^\infty(\Gamma)^2}, |\nabla\varphi|_{L^\infty(\Gamma)^4} \operatorname{diam}\Omega)$ and $L = |\Omega|^{1/2}$. We then find

$$q_m \le -\frac{c\nu}{|\Omega|}m^2 + \frac{c'}{\nu}(\gamma + \|\Phi\|^2),$$

where

$$\gamma = \operatorname*{Lim\,sup}_{t\to+\infty} \sup_{v_0 \in \mathscr{A}} \left(\frac{1}{t}\int_0^t \|v(\tau)\|^2\, d\tau\right).$$

Taking the scalar product of III.(3.25) with v in H, we obtain

$$\frac{1}{2}\frac{d}{dt}|v|^2 + \nu\|v\|^2 + b(v, \Phi, v) = -\nu((\Phi, v)) + (f, v) - b(\Phi, \Phi, v).$$

We have

$$|b(\Phi, \Phi, v)| \le \left|\int_\Omega (\Phi\cdot \operatorname{dist}((x, y), \Gamma)\nabla)\Phi \cdot \frac{v}{\operatorname{dist}((x, y), \Gamma)}\, dx\right|$$
$$\le |\Phi|\,|(\operatorname{dist}((x, y), \Gamma)\nabla\Phi|_{L^\infty}\left|\frac{v}{\operatorname{dist}((x, y), \Gamma)}\right|.$$

We then proceed as in Lemma III.3.1 and find

$$|b(\Phi, \Phi, v)| \le c|\varphi|_{L^\infty(\Gamma)^2}|\Phi|\,\|v\|.$$

Therefore

$$\frac{1}{2}\frac{d}{dt}|v|^2 + \frac{\nu}{2}\|v\|^2 \le \frac{\nu}{4}\|v\|^2 + \frac{c}{\nu\lambda_1}|f|^2$$
$$+ c'\nu\|\Phi\|^2 + \frac{c''}{\nu}|\varphi|^2_{L^\infty(\Gamma)^2}|\Phi|^2,$$

that is to say

$$\frac{d}{dt}|v|^2 + \frac{\nu}{2}\|v\|^2 \le \frac{c}{\nu\lambda_1}|f|^2 + c'\nu\|\Phi\|^2 + \frac{c''}{\nu}|\varphi|^2_{L^\infty(\Gamma)^2}|\Phi|^2,$$

which yields

$$\gamma \le c\|\Phi\|^2 + \frac{c'|f|^2}{\lambda_1\nu^2} + \frac{c''}{\nu^2}|\varphi|^2_{L^\infty(\Gamma)^2}|\Phi|^2,$$

and

$$q_m \le -\frac{c\nu}{|\Omega|}m^2 + \frac{c'}{\nu}\|\Phi\|^2 + \frac{c''}{\lambda_1\nu^3}|f|^2 + \frac{c'''}{\nu^3}|\Phi|^2|\varphi|^2_{L^\infty(\Gamma)^2}.$$

We then obtain

$$q_m \le -\frac{c\nu}{|\Omega|}m^2 + \frac{c'U^2}{\nu\varepsilon} + \frac{c''|f|^2}{\lambda_1\nu^3} + c'''\varepsilon\frac{U^4L^2}{\nu^3}. \tag{4.31}$$

We infer from (4.36) the following bound on the Lyapunov exponents μ_j, $j \in \mathbb{N}$,

$$\mu_1 + \cdots + \mu_j \leq -\kappa_1 j^2 + \kappa_2, \qquad \forall j \in \mathbb{N},$$

where

$$\kappa_1 = \frac{c\nu}{|\Omega|}, \qquad \kappa_2 = \frac{c' U^2}{\nu\varepsilon} + \frac{c''|f|^2}{\lambda_1 \nu^3} + c'''\varepsilon \frac{U^4 L^2}{\nu^3}.$$

By application of the results of Chapter V, we prove the analog of Theorem 4.1 for our case. Moreover,

$$\left(\frac{2\kappa_2}{\kappa_1}\right)^{1/2} = c\left(\frac{|\Omega|^{1/2}|f|}{\lambda_1^{1/2}\nu^2} + \left(\frac{U^3 L^3}{\nu^3}\right)^{1/2}\right).$$

We set

$$G = \frac{|\Omega|^{1/2}|f|}{\nu^2 \lambda_1^{1/2}} \quad \text{(Grashof number)},$$

$$\mathrm{Re} = \frac{UL}{\nu} \quad \text{(Reynolds number)},$$

and we finally deduce that the Hausdorff and fractal dimensions of $\mathscr{A}$ are bounded by

$$c(G + \mathrm{Re}^{3/2}), \tag{4.32}$$

where c is a nondimensional constant. From the physical viewpoint, this bound is much more satisfactory than (4.30).

4.3. Magnetohydrodynamics

We continue the study of the magnetohydrodynamic equations initiated in Chapter III. In this section the notations are the same as in Section III.3.3.

The linearized equations (4.2)–(4.5) corresponding to III.(3.35)–(3.41) are the abstract form of a boundary-value problem that we now make explicit. Here U will denote a pair (w, C) and u represents, as in Section III.3.3, a pair (v, B). We have

$$\frac{\partial w}{\partial t} + (w \cdot \nabla)v + (v \cdot \nabla)w - \frac{1}{R_e}\Delta w - S(C \cdot \nabla)B - S(B \cdot \nabla)C + \nabla(\Pi + SBC) = 0, \tag{4.33}$$

$$\frac{\partial C}{\partial t} + (w \cdot \nabla)B + (v \cdot \nabla)C + \frac{1}{R_m}\operatorname{curl}(\operatorname{curl} C) - (C \cdot \nabla)v - (B \cdot \nabla)w = 0, \tag{4.34}$$

$$\operatorname{div} w = 0, \tag{4.35}$$

$$\operatorname{div} C = 0, \tag{4.36}$$

$$w = 0 \quad \text{on } \Gamma, \tag{4.37}$$

$$C \cdot \mathbf{v} = 0 \quad \text{on } \Gamma, \tag{4.38}$$

$$\operatorname{curl} C \times \mathbf{v} = 0 \quad \text{on } \Gamma. \tag{4.39}$$

Alternatively, the weak form of this problem reads (compare with III.(3.54))

$$\frac{d}{dt}[U, \varphi] + a(U, \varphi) + b(u, U, \varphi) + b(U, u, \varphi) = 0, \qquad \forall \varphi \in V. \tag{4.40}$$

The spaces V, H, the forms $[\cdot, \cdot]$, and a, b, are the same as in Section III.3.3 and $u = u(t) = S(t)u_0$ is the solution of III.(3.54), (3.55) satisfying $u(0) = u_0$; (4.6) asserts the existence and uniqueness of a solution $U \in L^2(0, T; V) \cap L^\infty(0, T; H)$ ($\forall T > 0$) of (4.40) which satisfies $U(0) = \xi \in H$.

We now consider, for every $m \in \mathbb{N}$, the expression

$$\begin{aligned} &|U_1(t) \wedge \cdots \wedge U_m(t)|_{\wedge^m H} \\ &\quad = |\xi_1 \wedge \cdots \wedge \xi_m|_{\wedge^m H} \exp \int_0^t \operatorname{Tr} F'(S(\tau)u_0) \circ Q_m(\tau)\, d\tau, \end{aligned}$$

where $u(\tau) = S(\tau)u_0$ is as above; $U_1, \ldots, U_m$ are m solutions of (4.2)–(4.7) (or (4.40)) corresponding to $\xi = \xi_1, \ldots, \xi_m$ and $Q_m(\tau) = Q_m(\tau, u_0; \xi_1, \ldots, \xi_m)$ is the orthogonal projector in H onto the space spanned by $U_1(\tau), \ldots, U_m(\tau)$.

At a given time τ, we consider an orthonormal basis of H, $\varphi_j(\tau), j \in \mathbb{N}$, such that $\varphi_1(\tau), \ldots, \varphi_m(\tau)$ span $Q_m(\tau)H = \operatorname{Span}[U_1(\tau), \ldots, U_m(\tau)]$; hence $\varphi_j(\tau) \in V$ for $j = 1, \ldots, m$, and for a.e. $\tau \in \mathbb{R}_+$. Using these functions φ_j, we express the trace of $F'(u(\tau)) \circ Q_m(\tau)$ as

$$\begin{aligned} \operatorname{Tr} F'(u(\tau)) \circ Q_m(\tau) &= \sum_{j=1}^{\infty} [F'(u(\tau)) \circ Q_m(\tau)\varphi_j(\tau), \varphi_j(\tau)] \\ &= \sum_{j=1}^{m} [F'(u(\tau)) \circ Q_m(\tau)\varphi_j(\tau), \varphi_j(\tau)], \end{aligned}$$

where $[\cdot, \cdot]$ is the scalar product on H (see Section III.3.3). Omitting temporarily the variable τ we observe that

$$[F'(u)\varphi_j, \varphi_j] = -a(\varphi_j, \varphi_j) - b(\varphi_j, u, \varphi_j);$$

the term $b(u, \varphi_j, \varphi_j)$ vanishes because of III.(3.52). We then write $\varphi_j(\tau)$ as a pair $\{w_j(\tau), C_j(\tau)\}$ and return to the explicit expression III.(3.49) of b

$$b(\varphi_j, u, \varphi_j) = b_1(w_j, v, w_j) - Sb_1(C_j, B_j, w_j) + Sb_1(w_j, B_j, C_j) - Sb_1(C_j, v, C_j),$$

$$b_1(\varphi, \psi, \theta) = \sum_{i,k=1}^{2} \int_\Omega \varphi_i \frac{\partial \psi_k}{\partial x_i} \theta_k \, dx.$$

We apply several times the Schwarz inequality pointwise, i.e., for a.e. $x \in \Omega$,

$$\left|\sum_{j=1}^{m} \sum_{i,k=1}^{2} w_{ji}(x) D_i v_k(x) w_{jk}(x)\right| \le \rho_1(x) |\text{grad } v(x)|,$$

$$\left|\sum_{j=1}^{m} \sum_{i,k=1}^{2} C_{ji}(x) D_i B_{jk}(x) w_{jk}(x)\right| \le \rho_1(x)^{1/2} \rho_2(x)^{1/2} |\text{grad } B(x)|,$$

$$\left|\sum_{j=1}^{m} \sum_{i,k=1}^{2} w_{ji}(x) D_i B_{jk}(x) C_{jk}(x)\right| \le \rho_1(x)^{1/2} \rho_2(x)^{1/2} |\text{grad } B(x)|,$$

$$\left|\sum_{j=1}^{m} \sum_{i,k=1}^{2} C_{ji}(x) D_i v_k(x) C_{jk}(x)\right| \le \rho_2(x) |\text{grad } v(x)|.$$

We have set

$$\rho_1(x) = \sum_{j=1}^{m} |w_j(x)|^2, \qquad \rho_2(x) = \sum_{j=1}^{m} |C_j(x)|^2, \tag{4.41}$$

$$\begin{aligned} |\text{grad } v(x)| &= \left\{ \sum_{i,k=1}^{2} |D_i v_k(x)|^2 \right\}^{1/2}, \\ |\text{grad } B(x)| &= \left\{ \sum_{i,k=1}^{2} |D_i B_k(x)|^2 \right\}^{1/2}, \end{aligned} \tag{4.42}$$

and we will also consider

$$\rho(x) = \rho_1(x) + S\rho_2(x). \tag{4.43}$$

We infer from the previous inequalities that

$$\begin{aligned} \left|\sum_{j=1}^{m} b(\varphi_k, u, \varphi_j)\right| &\le \int_\Omega (\rho(x) |\text{grad } v(x)| + 2S\rho_1(x)^{1/2} \rho_2(x)^{1/2} |\text{grad } B(x)|)\, dx \\ &\le \int_\Omega \rho(x) (|\text{grad } v(x)|^2 + S|\text{grad } B(x)|^2)^{1/2}\, dx \\ &\le |\rho|_{L^2(\Omega)} ([\![v]\!]_1^2 + S[\![B]\!]_1^2)^{1/2}. \end{aligned} \tag{4.44}$$

Since $[\![\cdot]\!]_2$ is a norm on V_2 equivalent to the norm of $H^1(\Omega)^2$, there exists a constant c_1' depending only on Ω satisfying

$$[\![\theta]\!]_1 \le c_1' [\![\theta]\!]_2, \qquad \forall \theta \in V_2, \tag{4.45}$$

and we can write with another constant c_2'

$$\begin{aligned} [\![v]\!]_1^2 + S[\![B]\!]_1^2 &\le [\![v]\!]_1^2 + (c_1')^2 S [\![B]\!]_2^2 \\ &\le (c_2')^2 ([\![v]\!]_1^2 + S[\![B]\!]_2^2) = (c_2')^2 [\![u]\!]^2. \end{aligned} \tag{4.46}$$

Similarly,

$$\begin{aligned}a(\varphi_j, \varphi_j) &= \frac{1}{R_e}[\![w_j]\!]_1^2 + \frac{S}{R_m}[\![C_j]\!]_2^2 \\ &\geq \frac{1}{R_e + R_m}([\![w_j]\!]_1^2 + S[\![C_k]\!]_2^2) \\ &\geq \frac{1}{R_e + R_m}[\![\varphi_j]\!]^2, \end{aligned} \tag{4.47}$$

and we arrive at

$$\operatorname{Tr} F'(u(\tau)) \circ Q_m(\tau) \leq -\frac{1}{R_e + R_m}\sum_{j=1}^m [\![\varphi_j(\tau)]\!]^2 + c_2'|\rho(\tau)|_{L^2(\Omega)}[\![u(\tau)]\!]. \tag{4.48}$$

At this point we apply an inequality following from Theorem A.5.1. Since the family $\varphi_j(\tau)$ is orthonormal in H for the scalar product $[\cdot, \cdot]$, the family of pairs $\{w_j(\tau), S^{1/2}C_j(\tau)\}$ is orthonormal in the space $L^2(\Omega)^{2n}$ ($n = 2$) endowed with the usual scalar product. Corollary A.5.1. then implies the existence of a constant c_3' depending only on the shape of Ω such that

$$\begin{aligned}|\rho(\tau)|_{L^2(\Omega)}^2 &\leq c_3' \sum_{j=1}^m \{[\![w_j(\tau)]\!]_1^2 + S[\![C_j(\tau)]\!]_1^2\} \\ &\leq c_4' \sum_{j=1}^m [\![\varphi_j(\tau)]\!]^2\end{aligned} \tag{4.49}$$

($c_4' = c_2'c_3'$, c_2' appearing in the inequality (4.46)). Setting $c_5' = \frac{1}{2}c_4'(c_2')^2$ we obtain

$$\operatorname{Tr} F'(u(\tau)) \circ Q_m(\tau) \leq -\frac{1}{2(R_e + R_m)}\sum_{j=1}^m [\![\varphi_j(\tau)]\!]^2 + c_5'(R_e + R_m)[\![u(\tau)]\!]^2.$$

We then deduce from (4.49), using the Hölder inequality and $\int_\Omega \rho(x, \tau)\, dx = m$,

$$[\![\varphi_j(\tau)]\!]^2 \geq \frac{m^2}{c_4'|\Omega|}.$$

Thus

$$\operatorname{Tr} F'(u(\tau)) \circ Q_m(\tau) \leq -\frac{1}{c_4'(R_e + R_m)|\Omega|}m^2 + c_5'(R_e + R_m)[\![u(\tau)]\!]^2. \tag{4.50}$$

Now we assume that u_0 belongs to a bounded functional-invariant set X which may be the maximal attractor $\mathscr{A}$ and we write (see V.(3.47), (3.48))

$$q_m(t) = \operatorname*{Sup}_{u_0 \in X} \operatorname*{Sup}_{\substack{\xi_i \in H \\ |\xi_i| \leq 1 \\ i=1,\dots,m}} \left(\frac{1}{t}\int_0^t \operatorname{Tr} F'(u(\tau)) \circ Q_m(\tau)\, d\tau\right),$$

$$q_m = \limsup_{t\to\infty} q_m(t),$$

$$q_m \le -\frac{m^2}{c_4'(R_e + R_m)|\Omega|} + c_5'(R_e + R_m)\gamma,$$

with

$$\gamma = \limsup_{t\to\infty} \operatorname*{Sup}_{u_0 \in X} \frac{1}{t}\int_0^t [\![S(\tau)u_0]\!]^2 \, d\tau.$$

In order to be able to conclude, we need an estimate of γ in terms of the data.

Bound on γ

Thanks to III.(3.46) and (4.47)

$$[\![u(\tau)]\!]^2 = [\![v(\tau)]\!]_1^2 + S[\![B(\tau)]\!]_2^2 \le (R_e + R_m)a(u(\tau), u(\tau)).$$

We set $\varphi = u$ in III.(3.54) and in view of III.(3.52), (3.53) and the preceding

$$\frac{1}{2}\frac{d}{dt}[u]^2 + a(u, u) = [f, u],$$

$$\frac{1}{2}\frac{d}{dt}[u]^2 + \frac{1}{R_e + R_m}[\![u]\!]^2 \le [f][u].$$

The injection of V in H is continuous and there exists a constant depending only on Ω such that $[\theta] \le \text{const}[\![u]\!]$, $\forall \theta \in V$. We write this constant in the form $c_8'|\Omega|^{1/2}$ and we obtain

$$\frac{1}{t}\frac{d}{dt}[u]^2 + \frac{1}{R_e + R_m}[\![u]\!]^2 \le c_8'|\Omega|^{1/2}[f][\![u]\!]$$

$$\le \frac{1}{2(R_e + R_m)}[\![u]\!]^2 + \frac{(c_8')^2}{2}(R_e + R_m)|\Omega|[f]^2,$$

$$\frac{d}{dt}[u]^2 + \frac{1}{R_e + R_m}[\![u]\!]^2 \le (c_8')^2(R_e + R_m)|\Omega|[f]^2,$$

and by integration

$$[u(t)]^2 + \frac{1}{R_e + R_m}\int_0^t [\![u(t)]\!]^2 \, ds \le [u_0]^2 + (c_8')^2(R_e + R_m)|\Omega|[f]^2 t,$$

$$\frac{1}{t}\int_0^t [\![u(s)]\!]^2 \, ds \le \frac{1}{t(R_e + R_m)} \operatorname*{Sup}_{u_0 \in X} [u_0]^2 + (c_8')^2(R_e + R_m)^2|\Omega|[f]^2.$$

At the limit $t \to \infty$,

$$\gamma \le (c_8')^2(R_e + R_m)^2|\Omega|[f]^2. \tag{4.51}$$

Conclusions

We now have a bound of q_m

$$q_m \leq \frac{1}{2(c_2')^2(R_e + R_m)|\Omega|}\left(-\frac{c_6'}{2}m^2 + \frac{c_7'}{c_6'}\right) + c_5'(c_8')^2(R_e + R_m)^3|\Omega|[f]^2,$$

$$q_m \leq -\kappa_1 m^2 + \kappa_2, \tag{4.52}$$

$$\kappa_1 = \frac{c_9'}{(R_e + R_m)|\Omega|}, \qquad \kappa_2 = \frac{c_{10}'}{(R_e + R_m)|\Omega|}(1 + (R_e + R_m)^4|\Omega|^2[f]^2),$$

$$c_9' = \frac{c_6'}{4(c_2')^2}, \qquad c_{10}' = \frac{c_7'}{2(c_2')^2 c_6'} + c_5'(c_8')^2.$$

We can apply the procedure in Chapter V (see Section V.3.4) and Lemma 2.2 to obtain the main conclusions which are exactly the same as in Theorem 4.1, the present dynamical system replacing that of Section 4.2:

(i) The uniform Lyapunov exponents μ_j associated with a bounded functional-invariant set X (and, in particular, with the maximal attractor $\mathscr{A}$) satisfy

$$\mu_1 + \cdots + \mu_j \leq -\kappa_1 j^2 + \kappa_2, \qquad \forall j \in \mathbb{N}.$$

(ii) If m is given by

$$m - 1 < \left(\frac{2\kappa_2}{\kappa_1}\right)^{1/2} \leq m,$$

i.e.,

$$m - 1 < 2\frac{c_{10}'}{c_9'}(1 + (R_e + R_m)^4|\Omega|^2[f]^2)^{1/2} \leq m, \tag{4.53}$$

then the m-dimensional volume element in the phase space H is exponentially decaying, the Hausdorff dimension of X or $\mathscr{A}$ is less than or equal to m, and its fractal dimension less than or equal to $2m$.

4.4. Flows on a Manifold

We pursue the study of the Navier–Stokes equations describing the flow on a manifold. This study continues that initiated in Section III.3.4 of which we keep all the assumptions and notations.

The linearized equations (4.2)–(4.5) corresponding to III.(3.65), (3.66) are the abstract form of the equations

$$\frac{\partial U}{\partial t} - \nu\Delta U + u^l D_l U + U^l D_l u + \nabla\Pi = 0, \tag{4.54}$$

$$\operatorname{div} U = 0, \tag{4.55}$$

where Π is the pressure function naturally associated with U. Alternatively, the weak form of this problem reads (compare with III.(3.74))

$$\frac{d}{dt}(U, v) + \nu((U, v)) + b(u, U, v) + b(U, u, v) = 0. \qquad \forall v \in V. \tag{4.56}$$

Here $u = u(t) = S(t)u_0$ is the solution of III.(3.74), (3.76) satisfying $u(0) = u_0$. We know that for every ξ given in H, equations (4.54)–(4.56) possess a unique solution U belonging to $L^2(0, T; V) \cap \mathscr{C}([0, T]; H)$, $\forall T > 0$.

We then consider, for every $m \in \mathbb{N}$, the expression

$$|U_1(t) \wedge \cdots \wedge U_m(t)|_{\wedge^m H} = |\xi_1 \wedge \cdots \wedge \xi_m|_{\wedge^m H} \exp \int_0^t \operatorname{Tr} F'(S(\tau)u_0) \circ Q_m(\tau)\, d\tau,$$

with u as above; $U_1, \ldots, U_m$ are m solutions of (4.2)–(4.7) (or (4.54)–(4.56)) corresponding to $\xi = \xi_1, \ldots, \xi_m$, and $Q_m(\tau) = Q_m(\tau, u_0; \xi_1, \ldots, \xi_m)$ is the orthogonal projector in H onto the space spanned by $U_1(\tau), \ldots, U_m(\tau)$.

For $\tau > 0$ fixed we choose an orthonormal basis $\varphi_j(\tau)$ of H such that $\varphi_1(\tau), \ldots, \varphi_m(\tau)$ constitute an orthonormal basis of $Q_m(\tau)H$ (made of elements of V since $Q_m(\tau)H = \operatorname{Span}[U_1(\tau), \ldots, U_m(\tau)] \subset V$ for a.e. τ). We express the trace of $F'(u(\tau)) \circ Q_m(\tau)$ as

$$\begin{aligned} \operatorname{Tr} F'(u(\tau)) \circ Q_m(\tau) &= \sum_{j=1}^{\infty} (F'(u(\tau)) \circ Q_m(\tau)\varphi_j(\tau), \varphi_j(\tau)) \\ &= \sum_{j=1}^{m} (F'(u(\tau))\varphi_j(\tau), \varphi_j(\tau)). \end{aligned}$$

Omitting the variable τ we write

$$(F'(u)\varphi_j, \varphi_j) = -\nu\|\varphi_j\|^2 - b(\varphi_j, u, \varphi_j),$$

since $b(u, \varphi_j, \varphi_j) = 0$ owing to III.(3.72). By the expression III.(3.71) of b, we have

$$b(\varphi_j, u, \varphi_j) = \int g_{ik}(\varphi_j)^l D_l u^i (\varphi_j)^k \, dM. \tag{4.57}$$

Since $g_{\alpha\beta}g^{\beta\gamma} = \delta_\alpha^\gamma$, we have $(\varphi_j)^l D_l u^i = g_{\alpha\beta}g^{\beta\gamma}(\varphi_j)^\alpha D_\gamma u^i$. Hence the integrand for b is

$$g_{ik}g^{\beta\gamma}D_\gamma u^i d_{\alpha\beta}(\varphi_j)^\alpha(\varphi_j)^k.$$

By application of the Schwarz inequality pointwise (for dM — a.e. $x \in M$), we majorize the absolute value of this expression by

$$(g_{ik}g^{\beta\gamma}D_\gamma u^i D_\beta u^k)^{1/2}(g_{ik}g^{\beta\gamma}g_{\alpha\beta}(\varphi_j)^\alpha(\varphi_j)^k g_{\mu\gamma}(\varphi_j)^\mu(\varphi_j)^i)^{1/2}.$$

We set $[\![Du(x)]\!] = (g_{ik}(x)g^{\beta\gamma}(x)D_\gamma u^i(x)D_\beta u^k(x))^{1/2}$, and we observe that $\int [\![Du(x)]\!]^2 \, dM = \|u\|^2$. Also, $g_{\alpha\beta}g^{\beta\gamma} = \delta_\alpha^\gamma$ and hence

$$(g_{ik}g^{\beta\gamma}g_{\alpha\beta}(\varphi_j)^\alpha(\varphi_j)^k g_{\mu\gamma}(\varphi_j)^\mu(\varphi_j)^i)^{1/2} = g_{ik}(\varphi_j)^i(\varphi_j)^k = g(x; \varphi_j(x), \varphi_j(x)).$$

Finally, the integrand in (4.57) is majorized in absolute value by

$$[\![Du(x)]\!]g(x; \varphi_j(x), \varphi_j(x))$$

and

$$\left|\sum_{j=1}^{m} b(\varphi_j, u, \varphi_j)\right| \leq \int [\![Du(x)]\!]\rho(x)\, dM$$

$$\leq \|u\| \left(\int \rho(x)^2\, dM\right)^{1/2}, \tag{4.58}$$

where

$$\rho = \rho(x, \tau) = \sum_{j=1}^{m} g(x; \varphi_j(x, \tau), \varphi_j(x, \tau)). \tag{4.59}$$

Finally,

$$\operatorname{Tr} F'(u(\tau)) \circ Q_m(\tau) \leq -\nu \sum_{j=1}^{m} \|\varphi_j(\tau)\|^2 + |\rho(\tau)|_{L^2} \|u(\tau)\|. \tag{4.60}$$

The collective Sobolev inequalities proved in the Appendix do not apply to the present situation and we simply use the Sobolev inequalities which are valid as in the Euclidean (no curvature) case (see Section III.3.4).[1] Also taking into account the Poincaré-type inequality III.(3.64), we find the existence of a constant c_1' depending only on M such that

$$\int \varphi^4\, dM \leq c_1' \left(\int \varphi^2\, dM\right) \cdot \left(\int |\nabla\varphi|^2\, dM\right), \qquad \forall \varphi \in \mathscr{C}^\infty(M). \tag{4.61}$$

Applying this to each function $|\varphi_j(x)|$, and summing for $j = 1, \ldots, m$, we obtain

$$\int \rho^2\, dM \leq c_2' m \int \sum_{j=1}^{m} |\varphi_j(x)|^4\, dM,$$

$$\leq \text{(thanks to (4.61) and since } \varphi_j \text{ is orthonormal)},$$

$$\leq c_1' c_2' m \sum_{j=1}^{m} \int |\nabla\varphi_j|^2\, dM = c_1' c_2' m \sum_{j=1}^{m} \|\varphi_j\|^2.$$

Thus

$$\operatorname{Tr} F'(u) \circ Q_m \leq -\nu \sum_{j=1}^{m} \|\varphi_j\|^2 + \left(c_1' c_2' m \sum_{j=1}^{m} \|\varphi_j\|^2\right)^{1/2} \|u\|$$

$$\leq -\frac{\nu}{2} \sum_{j=1}^{m} \|\varphi_j\|^2 + \frac{c_1' c_2' m}{2\nu} \|u\|^2.$$

The eigenvalues of the Laplace operator Δ satisfy $\lambda_j \sim c\lambda_1 j^2$, as $j \to \infty$.

[1] See Remark 4.1.

Lemma 2.1 implies the existence of c_3' depending only on M such that

$$\sum_{j=1}^{m} \|\varphi_j\|^2 \geq c_3' \lambda_1 m^2, \qquad \forall m \in \mathbb{N}.$$

Hence

$$\operatorname{Tr} F'(u(\tau)) \circ Q_m(\tau) \leq -\frac{\nu}{2} c_3' \lambda_1 m^2 + \frac{c_1' c_2' m}{2\nu} \|u\|^2. \tag{4.62}$$

Assuming that $u_0 \in \mathscr{A}$, the maximal attractor, we can estimate the quantities $q_m(t)$, q_m, and then the uniform Lyapunov exponents μ_j (see Section V.3.4)

$$q_m(t) = \operatorname*{Sup}_{u_0 \in \mathscr{A}} \operatorname*{Sup}_{\substack{\xi_i \in H \\ |\xi_i| \leq 1 \\ i=1,\dots,m}} \left(\frac{1}{t} \int_0^t \operatorname{Tr} F'(S(\tau)u_0) \circ Q_m(\tau)\, d\tau \right),$$

$$q_m = \limsup_{t \to \infty} q_m(t).$$

We have

$$\begin{aligned} q_m &\leq -\frac{\nu}{2} c_3' \lambda_1 m^2 + \frac{c_1' c_2'}{2\nu} m\gamma \\ &\leq -\frac{\nu}{4} c_3' \lambda_1 m^2 + \frac{c_1' c_2'}{2 c_3' \nu^3 \lambda_1} \gamma^2, \end{aligned}$$

where

$$\gamma = \limsup_{t \to \infty} \operatorname*{Sup}_{u_0 \in \mathscr{A}} \frac{1}{t} \int_0^t \|S(\tau)u_0\|^2\, d\tau. \tag{4.63}$$

It is easy to derive from Section III.3.4 a bound on γ. Setting $v = u$ in III.(3.74) and taking into account III.(3.71), we find

$$\begin{aligned} \frac{1}{2}\frac{d}{dt}|u|^2 + \nu\|u\|^2 = (f, u) &\leq |f||u| \leq \lambda_1^{1/2}|f|\|u\| \\ &\leq \frac{\nu}{2}\|u\|^2 + \frac{1}{2\nu\lambda_1}|f|^2, \end{aligned} \tag{4.64}$$

$$\frac{d}{dt}|u|^2 + \nu\|u\|^2 \leq \frac{1}{\nu\lambda_1}|f|^2,$$

where λ_1 the first eigenvalue of A, which is positive thanks to III.(3.64), satisfies

$$\lambda_1 = \operatorname*{Inf}_{\substack{u \in V \\ u \neq 0}} \frac{\|u\|^2}{|u|^2}.$$

By integration (4.64) yields

$$\frac{\nu}{t} \int_0^t \|u(s)\|^2\, ds \leq \frac{1}{t} \operatorname*{Sup}_{u_0 \in \mathscr{A}} |u_0|^2 + \frac{1}{\nu\lambda_1}|f|^2,$$

and then

$$\gamma \leq \frac{1}{\nu^2 \lambda_1} |f|^2, \tag{4.65}$$

$$q_m \leq -\kappa_1 m^2 + \kappa_2, \tag{4.66}$$

$$\kappa_1 = \tfrac{1}{4} c_3' \nu \lambda_1, \qquad \kappa_2 = \frac{c_1' c_2'}{2c_3'} \frac{|f|^4}{\nu^7 \lambda_1^3} = c_4' \nu \lambda_1 G^4,$$

where $c_4' = c_1' c_2' / 2c_3'$ and

$$G = \frac{|f|}{\nu^2 \lambda_1}. \tag{4.67}$$

Conclusions

By application of the procedure in Section V.3 and of Lemma 2.2 we arrive at exactly the same conclusion as in Theorem 4.1, with the only difference that the present dynamical system replaces that of Section 4.2:

(i) The uniform Lyapunov exponents μ_j associated with the maximal attractor $\mathscr{A}$ satisfy

$$\mu_1 + \cdots + \mu_j \leq -\kappa_1 j^2 + \kappa_2, \qquad \forall j \in \mathbb{N}.$$

(ii) If m is given by

$$m - 1 < \left(\frac{2\kappa_2}{\kappa_1}\right)^{1/2} = \left(\frac{8c_4'}{c_3'}\right)^{1/2} G^2 \leq m, \tag{4.68}$$

then the m-dimensional volume element in the phase space H is exponentially decaying, the Hausdorff dimension of $\mathscr{A}$ is less than or equal to m, and its fractal dimension less than or equal to $2m$.

Remark 4.3. The number G is similar to that introduced in Section 3 for the two-dimensional Navier–Stokes equations in a flat domain (compare (4.67) and (3.15)); however, in Section 3 we had $m \sim G$ (see (3.42)), while in (4.68), $m \sim G^2$. This difference (deterioration) is due to the fact that we have used the Sobolev embeddings instead of the Lieb–Thirring inequalities. However, we give in J.M. Ghidaglia, M. Marion, and R. Temam [1], [2] an extension of the Sobolev–Lieb–Thirring inequalities to functions defined on a Riemannian manifold and, as usual, this leads to an improvement of the value of m in (4.68). In fact, we recover $m \sim G$ as in the linear case.

Remark 4.4. Sharp estimates on the dimension of the attractor for flows around the sphere, with or without the Coriolis force, appear in A.A. Ilyin [2–5]. See also J.L. Lions, R. Temam and S. Wang [1, 2], for the study of the attractors of the Primitive Equations for the atmosphere, the ocean and the coupled ocean-atmosphere.

4.5. Thermohydraulics

We complete our study of the examples of Section III.3 and consider now the equations of thermohydraulics. This section is a continuation of the study initiated in Section III.3.5 and we use the same assumptions and the same notations as in that section.

The linearized equations (4.2)–(4.5) for the present problem III.(3.84)–(3.88) are the abstract form of a boundary problem that we now make explicit. Here U will denote a pair (w, ψ) and u represents, as in Section III.3.5, the pair (v, θ), $v =$ the velocity vector, $\theta =$ the reduced temperature (see the relations before III.(3.84)).

The equations of the linearized problem read

$$\frac{\partial w}{\partial t} + (v \cdot \nabla)w + (w \cdot \nabla)v - \nu\Delta w + \nabla\Pi = e_n\psi, \tag{4.69}$$

$$\frac{\partial \psi}{\partial t} + (v \cdot \nabla)\psi + (w \cdot \nabla)\theta - w_n - \kappa\Delta\psi = 0, \tag{4.70}$$

$$\operatorname{div} w = 0, \tag{4.71}$$

and the boundary conditions are

$$w \text{ and } \psi = 0 \quad \text{at} \quad x_n = 0 \quad \text{and} \quad x_n = 1, \tag{4.72}$$

Π, w, ψ and the first derivatives of w and ψ are periodic of period 1 in the direction x_1; Π is the "pressure" naturally associated with w. (4.73)

The equivalent weak form of the problem is (compare with III.(3.95))

$$\frac{d}{dt}(U, \varphi) + a(U, \varphi) + b(u, U, \varphi) + b(U, u, \varphi) + (RU, \varphi) = 0, \qquad \forall \varphi \in V. \tag{4.74}$$

The spaces V, H, the forms a, b, and the operator R are the same as in Section III.3.5: $(\cdot, \cdot)$ is the scalar product in H, $u = u(t) = S(t)u_0$ is the solution of III.(3.95) satisfying $u(0) = u_0$; (4.6) provides the existence and uniqueness of a solution $U \in L^2(0, T; V) \cap L^\infty(0, T; H)$, $\forall T > 0$, of (4.74), such that $U(0) = \xi \in H$.

For $\xi = \xi_1, \dots, \xi_m \in H$, let $U_1, \dots, U_m$ denote the corresponding solutions of (4.2)–(4.7) (or (4.74)); following the steps of the procedure summarized in Section V.3.4 we consider, for every $m \in \mathbb{N}$,

$$|U_1(t) \wedge \cdots \wedge U_m(t)|_{\wedge^m H} = |\xi_1 \wedge \cdots \wedge \xi_m|_{\wedge^m H} \exp \int_0^t \operatorname{Tr} F'(S(\tau)u_0) \circ Q_m(\tau)\, d\tau,$$

where $S(\tau)u_0 = u(\tau)$, u as above and $Q_m(\tau) = Q_m(\tau, u_0; \xi_1, \dots, \xi_m)$ is the orthogonal projector in H onto the space spanned by $U_1(\tau), \dots, U_m(\tau)$.

At a given time τ, let $\varphi_j(\tau)$, $j \in \mathbb{N}$, be an orthonormal basis of H, such that $\varphi_1(\tau), \ldots, \varphi_m(\tau)$, span $Q_m(\tau)H = \mathrm{Span}[U_1(\tau), \ldots, U_m(\tau)]$; hence $\varphi_j(\tau) \in V$ for $j = 1, \ldots, m$, and for a.e. $\tau \in \mathbb{R}_+$. The trace of $F'(u(\tau)) \circ Q_m(\tau)$ is given by

$$\begin{aligned}\operatorname{Tr} F'(u(\tau)) \circ Q_m(\tau) &= \sum_{j=1}^{\infty} (F'(u(\tau)) \circ Q_m(\tau)\varphi_j(\tau), \varphi_j(\tau)) \\ &= \sum_{j=1}^{m} (F'(u(\tau)\varphi_j(\tau), \varphi_j(\tau)).\end{aligned}$$

Omitting for the moment the variable τ we notice that

$$(F'(u)\varphi_j, \varphi_j) = -a(\varphi_j, \varphi_j) - b(\varphi_j, U, \varphi_j) - (R\varphi_j, \varphi_j),$$

since $b(u, \varphi_j, \varphi_j) = 0$ thanks to III.(3.6). The vector φ_j consists of a pair $\{w_j, \psi_j\}$; considering then the explicit expressions of a, b, R we write

$$\begin{aligned}(F'(u)\varphi_j, \varphi_j) &= -\nu\|w_j\|^2 - \kappa\|\psi_j\|^2 + 2\int_\Omega \psi_j(w_j)_n \, dx \\ &= \int_\Omega [((w_j \cdot \nabla)v)w_j + (w_j \cdot \nabla)\theta\psi_j] \, dx.\end{aligned}$$

The last integral is integrated by parts and majorized as follows

$$\begin{aligned}\left|\int_\Omega [((w_j \cdot \nabla)\theta)\psi_j] \, dx\right| &= \left|\int_\Omega \theta w_j(\nabla\psi_j) \, dx\right| \\ &\le \text{(since } |\theta(x, t)| \le 1 \text{ a.e., see Remark III.3.5)} \\ &\le |w_j||\nabla\psi_j| \le \|w_j\|;\end{aligned}$$

the last inequality, $|w_j| \le 1$, holds since $|w_j| \le (|w_j|^2 + |\psi_j|^2)^{1/2} = |\varphi_j|$, and $|\varphi_j| = 1$, the family $\{\varphi_j\}_j$ being orthonormal in H. Thus

$$\begin{aligned}\left|\sum_{j=1}^m \int_\Omega [((w_j \cdot \nabla)\theta)\psi_j] \, dx\right| &\le \sum_{j=1}^m \|\psi_j\| \\ &\le \frac{\kappa}{4}\sum_{j=1}^m \|\psi_j\|^2 + \frac{m}{\kappa}.\end{aligned} \tag{4.75}$$

Thanks to the Schwarz inequality we then write the pointwise inequality

$$|[((w_j \cdot \nabla)v)w_j](x)| \le |\operatorname{grad} v(x)||w_j(x)|^2$$

and summing for $j = 1, \ldots, m$ we find

$$\begin{aligned}\left|\sum_{j=1}^m \int_\Omega [((w_j \cdot \nabla)v)w_j] \, dx\right| &\le \int_\Omega |\operatorname{grad} v(x)|\rho(x) \, dx \\ &\le \|v\||\rho|_{L^2}^{1/2},\end{aligned} \tag{4.76}$$

where $|\rho|_{L^2}$ is the L^2-norm of ρ and this function depending in fact on x and

τ is

$$\rho(x) = \sum_{j=1}^{m} \{|w_j(x)|^2 + |\psi_j(x)|^2\}. \tag{4.77}$$

We now use Theorem A.3.1. (see also Theorem A.5.1) which provides the existence of a constant c_1' depending only on Ω such that

$$\begin{aligned}\int_\Omega \rho(x)^2\,dx &\le c_1' \sum_{j=1}^{m} \int_\Omega (|\operatorname{grad} w_j(x)|^2 + |\operatorname{grad} \psi_j(x)|^2)\,dx \\ &\le c_1' \sum_{j=1}^{m} (\|w_j\|^2 + \|\psi_j\|^2).\end{aligned} \tag{4.78}$$

Hence, the right-hand side of (4.76) is majorized by

$$\begin{aligned}&(c_1')^{1/2}\|v\| \left\{\sum_{j=1}^{m} (\|w_j\|^2 + \|\psi_j\|^2)\right\}^{1/2} \\ &\le (c_1')^{1/2}\|v\| \left\{\left(\sum_{j=1}^{m} \|w_j\|^2\right)^{1/2} + \left(\sum_{j=1}^{m} \|\psi_j\|^2\right)^{1/2}\right\} \\ &\le \frac{\nu}{4}\sum_{j=1}^{m} \|w_j\|^2 + \frac{\kappa}{4}\sum_{j=1}^{m} \|\psi_j\|^2 + c_1'\left(\frac{1}{\nu} + \frac{1}{\kappa}\right)\|v\|^2.\end{aligned}$$

The sum

$$2\sum_{j=1}^{m} \int_\Omega \psi_j (w_j)_n\,dx$$

is bounded in absolute value by

$$2\sum_{j=1}^{m} |w_j||\psi_j|$$

(where $|\cdot|$ denotes, as usual, the L^2-norms) and this expression is majorized by

$$\sum_{j=1}^{m} (|w_j|^2 + |\psi_j|^2) = \sum_{j=1}^{m} |\varphi_j|^2 = m,$$

since the family φ_j is orthonormal in H.

Collecting all these inequalities we obtain the following

$$\begin{aligned}\operatorname{Tr} F'(u(\tau)) \circ Q_m(\tau) \le &-\sum_{j=1}^{m} \left(\frac{\nu}{2}\|w_j(\tau)\|^2 + \frac{\kappa}{2}\|\psi_j(\tau)\|^2\right) \\ &+ m\left(1 + \frac{1}{\kappa}\right) + c_1'\frac{\nu + \kappa}{\nu\kappa}\|v(\tau)\|^2.\end{aligned} \tag{4.79}$$

We now want to apply Lemma 2.1 to the operator $\nu^{-1}A_0 \times \kappa^{-1}A_1$ (see Section III.3.5). The operator $\nu^{-1}A_0$ is the (two-dimensional) Stokes operator with the appropriate boundary conditions; according to G. Métivier [1], its eigenvalues satisfy $\lambda_j' \sim j/|\Omega|$ when $j \to \infty$. Similarly, $\kappa^{-1}A_1$ is the two-dimensional Laplace operator with the present boundary conditions; according to R. Courant and D. Hilbert [1], its eigenvalues λ_j'' satisfy, $\lambda_j'' \sim c'j/|\Omega|$ as $j \to \infty$. The eigenvalues of $\nu^{-1}A_0 \times \kappa^{-1}A_1$ are the union of the numbers λ_j' and λ_j''; therefore, they are bounded from below as $j \to \infty$ by $c[\nu\kappa/(\nu + \kappa)]j/|\Omega|$. Lemma 2.1 implies the existence of a constant c_2' depending only on Ω (i.e., on l) such that

$$\sum_{j=1}^{m} (\|w_j(\tau)\|^2 + \|\psi_j(\tau)\|^2) \geq c_2' \frac{m^2}{|\Omega|}, \quad \text{a.e.} \quad \tau > 0. \tag{4.80}$$

Consequently,

$$\sum_{j=1}^{m} (\nu\|w_j(\tau)\|^2 + \kappa\|\psi_j(\tau)\|^2) \geq \frac{\nu\kappa}{\nu+\kappa} \sum_{j=1}^{m} (\|w_j(\tau)\|^2 + \|\psi_j(\tau)\|^2) \geq c_2' \frac{\nu\kappa}{\nu+\kappa} \frac{m^2}{|\Omega|},$$

and (4.79) becomes

$$\operatorname{Tr} F'(u(\tau)) \circ Q_m(\tau) \leq -c_2' \frac{\nu\kappa}{\nu+\kappa} \frac{m^2}{|\Omega|} + m\left(1 + \frac{1}{\kappa}\right) + c_1' \frac{\nu+\kappa}{\nu\kappa} \|v(\tau)\|^2. \tag{4.81}$$

Now we assume that u_0 belongs to the maximal attractor $\mathscr{A}$ and introduce the quantities $q_m(t)$, q_m as indicated in V.(3.47), (3.48)

$$q_m(t) = \operatorname*{Sup}_{u_0 \in \mathscr{A}} \operatorname*{Sup}_{\substack{\xi_i \in H \\ |\xi_i| \leq 1 \\ i=1,\ldots,m}} \left(\frac{1}{t} \int_0^t \operatorname{Tr} F'(u(\tau)) \circ Q_m(\tau)\, d\tau \right)$$

$$q_m = \limsup_{t \to \infty} q_m(t)$$

$$q_m \leq -c_2' \frac{\nu\kappa}{\nu+\kappa} \frac{m^2}{|\Omega|} + m\left(1 + \frac{1}{\kappa}\right) + c_1' \frac{\nu+\kappa}{\nu\kappa} \gamma,$$

with

$$\gamma = \limsup_{t \to \infty} \operatorname*{Sup}_{u_0 \in \mathscr{A}} \frac{1}{t} \int_0^t \|v(\tau)\|^2\, d\tau.$$

In order to conclude, we need an estimate of γ in terms of the data. This bound is easily derived from III.(3.112) where we can bound $|\theta|^2$ by $|\Omega|$ owing to the fact that $|\theta(x, t)| \leq 1$ a.e. on the attractor $\mathscr{A}$ (see Remark 3.5). Therefore

$$\frac{d}{dt}|v|^2 + \nu\|v\|^2 \leq \frac{|\Omega|}{\nu}.$$

By integration

$$\frac{\nu}{t} \int_0^t \|v(s)\|^2\, ds \leq \frac{1}{t} \operatorname*{Sup}_{\substack{u_0 = \{v_0, \theta_0\} \\ \in \mathscr{A}}} |v_0|^2 + \frac{|\Omega|}{\nu},$$

and as $t \to \infty$,

$$\gamma \leq \frac{|\Omega|}{\nu^2}. \tag{4.82}$$

Returning to q_m we find

$$q_m \leq -\frac{1}{2}\kappa_1 m^2 + \kappa_4,$$

$$\kappa_1 = \frac{c_2'}{|\Omega|}\frac{\nu\kappa}{\nu + \kappa}, \qquad \kappa_2 = \frac{\kappa + 1}{\kappa}, \qquad \kappa_3 = c_1'|\Omega|\frac{\nu + \kappa}{\nu^3\kappa}, \tag{4.83}$$

$$\kappa_4 = c_3'|\Omega|\frac{\nu + \kappa}{\nu\kappa}\left(\frac{1}{\nu^2} + \frac{(\kappa + 1)^2}{\kappa^2}\right).$$

Conclusions

We can apply, at this point, the results summarized in Section V.3.4 and Lemma 2.2 to obtain the main conclusions which can be stated exactly as in Theorem 4.1, the present dynamical system replacing that of Section 4.2:

(i) The uniform Lyapunov exponents μ_j associated with the global attractor $\mathscr{A}$ satisfy

$$\mu_1 + \cdots + \mu_j \leq -\frac{\kappa_1}{2}j^2 + \kappa_4, \qquad \forall j \in \mathbb{N}.$$

(ii) If m is given by

$$m - 1 < 2\left(\frac{\kappa_4}{\kappa_1}\right)^{1/2} \leq m, \tag{4.84}$$

then the m-dimensional volume element in the phase space H is exponentially decaying as $t \to \infty$; the Hausdorff dimension of the global attractor $\mathscr{A}$ is less than or equal to m and its fractal dimension is less than or equal to $2m$.

In terms of the nondimensional numbers introduced in III.(3.80), i.e., the Grashoff number $G_r = 1/\nu^2$, the Prandtl number $P_r = \nu/\kappa$, and the Rayleigh number $R_a = 1/\nu\kappa$, we express the number m determined by (4.84) in the form

$$\begin{aligned} m &\sim c|\Omega|\frac{\nu + \kappa}{\nu\kappa}\left(\frac{1}{\nu} + \frac{\kappa + 1}{\kappa}\right) \\ &\sim c|\Omega|(R_a + G_r^{1/2} + G_r)(1 + P_r) \\ &\sim c|\Omega|\{G_r^{1/2}(1 + P_r) + G_r(1 + P_r)^2\}. \end{aligned}$$

5. Pattern Formation Equations

In this section we consider the pattern formation equations studied in Section III.4. We successively consider the Kuramoto–Sivashinsky equation (Section 5.1) and the Cahn–Hilliard equation (Section 5.2).

5.1. The Kuramoto–Sivashinsky Equation

We continue the study of the Kuramoto–Sivashinsky equation; this section is a continuation of Section III.4.1 of which we retain the notations and all the hypotheses.

The dynamical system is defined by equation III.(4.12)–(4.14) which is the abstract setting of the initial- and boundary-value problem III.(4.4)–(4.7). This initial- and boundary-value problem is well posed and Theorem III.4.1 provides the existence of a compact global attractor $\mathscr{A}$ for the corresponding semigroup of H, $S(t)$: $v_0 = v(0) \to v(t)$. We aim at applying to this system (and its attractor $\mathscr{A}$) the results and methods developed in Chapter V. We begin with

5.1.1. The Linearized Equations

Writing equation III.(4.13) in the form

$$v' = F(v), \tag{5.1}$$

as in V.(2.38), we see that the first variation equation V.(2.39)

$$U' = F'(v)U \tag{5.2}$$

is equivalent to

$$\frac{dU}{dt} + \nu AU + B(v, U) + B(U, v) + RU = 0. \tag{5.3}$$

This is also an abstract form of the boundary-value problem

$$\frac{\partial U}{\partial t} + \nu\frac{\partial^4 U}{\partial x^4} + v\frac{\partial U}{\partial x} + U\frac{\partial v}{\partial x} + \frac{\partial^2 U}{\partial x^2} = 0, \tag{5.4}$$

$$\frac{\partial^j U}{\partial x^j}\left(-\frac{L}{2}, t\right) = \frac{\partial^j U}{\partial x^j}\left(\frac{L}{2}, t\right), \qquad j = 0, \ldots, 3, \tag{5.5}$$

$$U(x, t) = -U(-x, t), \qquad \int_\Omega U(x, t)\, dx = 0, \qquad \forall t. \tag{5.6}$$

The associated initial condition is

$$U(0) = \xi. \tag{5.7}$$

The precise relations of the linear problem with the dynamical system are the following:

(i) If v is the solution of III.(4.13), (4.14), $v \in L^2(0, T; V) \cap L^\infty(0, T; H)$, $\forall T > 0$, then by application of Theorem II.3.4, it is easy to see that for ξ given in H, (5.3)–(5.7) has a unique solution U satisfying

$$U \in L^2(0, T; V) \cap L^\infty(0, T; H), \qquad \forall T > 0. \tag{5.8}$$

(ii) The semigroup $S(t)$ is Fréchet differentiable in H

For every $t > 0$, the function $v_0 \to S(t)v_0$ is Fréchet differentiable in H at v_0 with differential $L(t, v_0)$: $\xi \in H \to U(t) \in H$, where $U(\cdot)$ is the solution of (5.3)–(5.7). (5.9)

The differentiation property is established in Section 8.

5.1.2. Dimension of the Attractor

We now estimate the dimension of $\mathscr{A}$ as indicated in Section V.3.4.

For $m \in \mathbb{N}$, we consider $\xi = \xi_1, \ldots, \xi_m$, m elements of H, and the corresponding solutions $U_1, \ldots, U_m$ of (5.3)–(5.7); $v = v(\tau) = S(\tau)v_0$ is a fixed orbit. According to Section V.2.3 we have

$$|U_1(t) \wedge \cdots \wedge U_m(t)|_{\wedge^m H}$$
$$= |\xi_1 \wedge \cdots \wedge \xi_m|_{\wedge^m H} \exp \int_0^t \operatorname{Tr} F'(S(\tau)v_0) \circ Q_m(\tau)\, d\tau,$$

where $Q_m(\tau) = Q_m(\tau, v_0; \xi_1, \ldots, \xi_m)$ is the orthogonal projector in H onto the space spanned by $U_1(\tau), \ldots, U_m(\tau)$.

At a given time τ, let $\varphi_j(\tau)$, $j \in \mathbb{N}$, be an orthonormal basis of H, such that $\varphi_1(\tau), \ldots, \varphi_m(\tau)$ span $Q_m(\tau)H = \operatorname{Span}[U_1(\tau), \ldots, U_m(\tau)]$; since $U_j(\tau) \in V$ for a.e. τ, $\varphi_1(\tau), \ldots, \varphi_m(\tau)$ also belong to V for a.e. τ. We have

$$\begin{aligned} \operatorname{Tr} F'(v(\tau)) \circ Q_m(\tau) &= \sum_{j=1}^{\infty} (F'(v(\tau)) \circ Q_m(\tau)\varphi_j(\tau), \varphi_j(\tau)) \\ &= \sum_{j=1}^{m} (F'(v(\tau))\varphi_j(\tau), \varphi_j(\tau)). \end{aligned} \tag{5.10}$$

Omitting temporarily the variable τ, we see that

$$\begin{aligned} (F'(v)\varphi_j, \varphi_j) &= -\nu\left|\frac{\partial^2 \varphi_j}{\partial x^2}\right|^2 + \left|\frac{\partial \varphi_j}{\partial x}\right|^2 \\ &\quad - \int_{-L/2}^{L/2} v\frac{\partial \varphi_j}{\partial x}\varphi_j\, dx - \int_{-L/2}^{L/2} \varphi_j^2 \frac{\partial v}{\partial x}\, dx \\ &= -\nu\left|\frac{\partial^2 \varphi_j}{\partial x^2}\right|^2 + \left|\frac{\partial \varphi_j}{\partial x}\right|^2 - \frac{1}{2}\int_{-L/2}^{L/2} \varphi_j^2 \frac{\partial v}{\partial x}\, dx; \end{aligned}$$

$|\cdot|$ denotes as usual the norm in H (i.e., in $\dot{L}^2(\Omega)$) and $(\cdot, \cdot)$ is the scalar product in H. Hence

$$\sum_{j=1}^{m} (F'(v)\varphi_j, \varphi_j) = -\nu \sum_{j=1}^{m} \left|\frac{\partial^2 \varphi_j}{\partial x^2}\right|^2 + \sum_{j=1}^{m} \left|\frac{\partial \varphi_j}{\partial x}\right|^2 - \frac{1}{2}\int_{-L/2}^{L/2} \rho \frac{\partial v}{\partial x}\, dx, \tag{5.11}$$

$$\rho = \rho(x, \tau) = \sum_{j=1}^{m} |\varphi_j(x, \tau)|^2. \tag{5.12}$$

By interpolation (see III.(4.20)),

$$\left|\frac{\partial \varphi_j}{\partial x}\right|^2 \leq |\varphi_j|\left|\frac{\partial^2 \varphi_j}{\partial x^2}\right|,$$

$$\sum_{j=1}^{m}\left|\frac{\partial \varphi_j}{\partial x}\right|^2 \leq \sum_{j=1}^{m}|\varphi_j|\left|\frac{\partial^2 \varphi_j}{\partial x^2}\right| \leq \left(\sum_{j=1}^{m}|\varphi_j|^2\right)^{1/2}\left(\sum_{j=1}^{m}\left|\frac{\partial^2 \varphi}{\partial x^2}\right|^2\right)^{1/2}.$$

Since the family φ_j is orthonormal in H, $|\varphi_j|^2 = 1$ and this expression is majorized by

$$m^{1/2}\left(\sum_{j=1}^{m}\left|\frac{\partial^2 \varphi_j}{\partial x^2}\right|^2\right)^{1/2} \leq \frac{\nu}{4}\sum_{j=1}^{m}\left|\frac{\partial^2 \varphi_j}{\partial x^2}\right|^2 + \frac{m}{\nu}.$$

Thanks to the Hölder inequality

$$\left|\int_{-L/2}^{L/2} \rho \frac{\partial v}{\partial x}\, dx\right| \leq |\rho|_{L^5(\Omega)}\left|\frac{\partial v}{\partial x}\right|_{L^{5/4}(\Omega)}.$$

By application of Theorem A.3.1 (see Example A.4.4), we find the existence of an absolute constant c_1' such that

$$|\rho(\tau)|_{L^5(\Omega)}^5 \leq c_1' \sum_{j=1}^{m}\left|\frac{\partial^2 \varphi_j(\tau)}{\partial x^2}\right|^2, \quad \text{a.e. } \tau. \tag{5.13}$$

Hence

$$\begin{aligned}\left|\frac{1}{2}\int_{-L/2}^{L/2} \rho \frac{\partial v}{\partial x}\, dx\right| &\leq \frac{1}{2}|\rho|_{L^5(\Omega)}\left|\frac{\partial v}{\partial x}\right|_{L^{5/4}(\Omega)} \\ &\leq \frac{(c_1')^{1/5}}{2}\left(\sum_{j=1}^{m}\left|\frac{\partial^2 \varphi_j}{\partial x^2}\right|^2\right)^{1/5}\left|\frac{\partial v}{\partial x}\right|_{L^{5/4}(\Omega)} \\ &\leq \text{(with the Young inequality)} \\ &\leq \frac{\nu}{2}\sum_{j=1}^{m}\left|\frac{\partial^2 \varphi}{\partial x^2}\right|^2 + \frac{c_2'}{\nu^{1/4}}\left|\frac{\partial v}{\partial x}\right|_{L^{5/4}(\Omega)}^{5/4},\end{aligned}$$

c_2' an absolute constant.

Collecting all these inequalities we find

$$\sum_{j=1}^{m}(F'(v)\varphi_j, \varphi_j) \leq -\frac{\nu}{2}\sum_{j=1}^{m}\left|\frac{\partial^2 \varphi_j}{\partial x^2}\right|^2 + \frac{c_2'}{\nu^{1/4}}\left|\frac{\partial v}{\partial x}\right|_{L^{5/4}(\Omega)}^{5/4} + \frac{m}{\nu}.$$

The eigenvalues λ_j of the operator $A = \partial^4/\partial x^4$ with domain $D(A)$ are given explicitly in Remark III.4.2, and $\lambda_j \sim (2\Pi j/L)^4$ as $j \to \infty$. Lemma 2.1 then implies the existence of an absolute constant c_3' such that

$$\sum_{j=1}^{m}\left|\frac{\partial^2 \varphi_j}{\partial x^2}\right|^2 \geq c_3' \frac{1}{L^4} m^5, \qquad \forall m,$$

and then

$$\operatorname{Tr} F'(v(\tau)) \circ Q_m(\tau) \leq -\frac{c_3'}{2}\frac{\nu}{L^4} m^5 + \frac{c_2'}{\nu^{1/4}}\left|\frac{\partial v}{\partial x}\right|_{L^{5/4}(\Omega)}^{5/4} + \frac{m}{\nu}. \tag{5.14}$$

Assuming that $v_0 \in \mathscr{A}$, the universal attractor, we can now majorize the quantities $q_m(t)$, q_m defined in V.(3.47), (3.48)

$$q_m(t) = \operatorname*{Sup}_{v_0 \in \mathscr{A}} \operatorname*{Sup}_{\substack{\xi_i \in H \\ |\xi_i| \leq 1 \\ i=1,\dots,m}} \left(\frac{1}{t} \int_0^t \operatorname{Tr} F'(S(\tau)v_0) \circ Q_m(\tau)\, d\tau \right),$$

$$q_m = \limsup_{t \to \infty} q_m(t).$$

We find

$$q_m \leq -\frac{c_3'}{2} \frac{\nu}{L^4} m^5 + \frac{c_2'}{\nu^{1/4}} \gamma + \frac{m}{\nu},$$

with

$$\gamma = \limsup_{t \to \infty} \operatorname*{Sup}_{v_0 \in \mathscr{A}} \frac{1}{t} \int_0^t \left| \frac{\partial v}{\partial x} \right|_{L^{5/4}(\Omega)}^{5/4} d\tau.$$

Estimate of γ

An estimate of γ can be derived from III.(4.44), (4.45) which are also valid in a slightly stronger form (see B. Nicolaenko, B. Scheurer, and R. Temam [1])

$$\begin{cases} \limsup\limits_{t \to \infty} \operatorname*{Sup}\limits_{v_0 \in \mathscr{B}} |S(\tau)v_0| \leq c_1 \nu^{-1/4} \tilde{L}^{5/2}, \\ \limsup\limits_{t \to \infty} \operatorname*{Sup}\limits_{v_0 \in \mathscr{B}} \dfrac{1}{t} \displaystyle\int_0^t \left| \frac{\partial^2 v}{\partial x^2} \right|^2 d\tau \leq c_2 \nu^{-5/2} \tilde{L}^5, \end{cases} \tag{5.15}$$

where $\tilde{L} = L/2\Pi\sqrt{\nu}$, c_1, c_2 are absolute constants and $\mathscr{B}$ is *any* bounded set of H; we take $\mathscr{B} = \mathscr{A}$ below.

Due to the Hölder inequality

$$\left| \frac{\partial v}{\partial x} \right|_{L^{5/4}(\Omega)} \leq L^{3/10} \left| \frac{\partial v}{\partial x} \right|$$

and by interpolation (see III.(4.20))

$$\left| \frac{\partial v}{\partial x} \right| \leq |v|^{1/2} \left| \frac{\partial^2 v}{\partial x^2} \right|^{1/2},$$

$$\left| \frac{\partial v}{\partial x} \right|_{L^{5/4}(\Omega)}^{5/4} \leq L^{3/8} |v|^{5/8} \left| \frac{\partial^2 v}{\partial x^2} \right|^{5/8}.$$

Then

$$\begin{aligned} \frac{1}{t} \int_0^t \left| \frac{\partial v}{\partial x} \right|_{L^{5/4}(\Omega)}^{5/4} d\tau &\leq \frac{L^{3/8}}{t} \int_0^t |v|^{5/8} \left| \frac{\partial^2 v}{\partial x^2} \right|^{5/8} d\tau, \\ &\leq \text{(with the Hölder inequality)}, \\ &\leq L^{3/8} \left(\frac{1}{t} \int_0^t |v|^{10/11}\, d\tau \right)^{11/16} \left(\frac{1}{t} \int_0^t \left| \frac{\partial^2 v}{\partial x^2} \right|^2 d\tau \right)^{5/16}, \end{aligned}$$

and thanks to (5.15)

$$\begin{aligned}\gamma &\le L^{3/8}(c_1 v^{-1/4}\tilde{L}^{5/2})^{5/8}(c_2 v^{-5/2}\tilde{L}^5)^{5/16}\\ \gamma &\le c_4' v^{-3/4}\tilde{L}^{7/2}.\end{aligned} \tag{5.16}$$

With this estimate on γ, the bound on q_m becomes

$$\begin{aligned}q_m &\le -\frac{c_3'}{2}\frac{v}{L^4}m^5 + \frac{c_2'}{v}\tilde{L}^{7/2} + \frac{m}{v}\\ &\le -\frac{c_3'}{4}\frac{v}{L^4}m^5 + c_2'\frac{\tilde{L}^{7/2}}{v} + c_5'\frac{\tilde{L}}{v},\end{aligned}$$

$$q_m \le -\kappa_1 m^5 + \kappa_2, \tag{5.17}$$

$$\kappa_1 = \frac{c_3'}{4\sqrt{2\Pi}}\frac{1}{L^2\tilde{L}^2}, \qquad \kappa_2 = c_2'\frac{\tilde{L}^{7/2}}{v} + c_5'\frac{\tilde{L}}{v}.$$

We infer from (5.17) and V. (3.49) the following bound on the Lyapunov exponents μ_j, $j \in \mathbb{N}$, associated with $\mathscr{A}$

$$\mu_1 + \cdots + \mu_j \le q_j \le -\kappa_1 j^5 + \kappa_2, \qquad \forall j \in \mathbb{N}. \tag{5.18}$$

Using Lemma 2.2 we see that if m is defined by

$$m - 1 < \left(\frac{2\kappa_2}{\kappa_1}\right)^{1/5} \le m, \tag{5.19}$$

then $\mu_1 + \cdots + \mu_m < 0$ and

$$\frac{(\mu_1 + \cdots + \mu_j)_+}{|\mu_1 + \cdots + \mu_m|} \le 1, \qquad \forall j = 1, \ldots, m-1.$$

By application of the results of Chapter V (in particular, Proposition V.2.1 and Theorem V.3.3) we have then proved the following

Theorem 5.1. *We consider the dynamical system associated with the Kuramoto–Sivashinsky equation restricted to odd functions with space-periodic boundary conditions, see* III.(4.4)–(4.7). *We denote by $\mathscr{A}$ its maximal attractor introduced in Section* III.4.1, *and we let m be defined by* (5.19),

$$m - 1 < c(\tilde{L}^{3/2} + \tilde{L}^{4/5}) \le m, \tag{5.20}$$

i.e., $m \sim c\tilde{L}^{3/2}$ for $\tilde{L}$ large, $\tilde{L} = L/2\Pi\sqrt{v}$. Then:

(i) *the m-dimensional volume element in the phase space is exponentially decaying as $t \to \infty$;*
(ii) *the Hausdorff dimension of $\mathscr{A}$ is less than or equal to m, and its fractal dimension is less than or equal to 2m.*

Remark 5.1. It was conjectured by Y. Pomeau and Ph. Manneville [1] that the number of degrees of freedom of the equation is of the order of $\tilde{L}$ for large

$\tilde{L}$. The improvement from $\tilde{L}^{3/2}$ to $\tilde{L}$ remains an open question. The extension of Theorem 5.1 to all (not necessarily odd) solutions is also unresolved.

5.1.3. A Collective Sobolev Estimate

We give in this section a simple direct proof of the collective Sobolev estimate used for (5.13) (C. Foias).

We first notice the Agmon-type inequality (see II.(1.40)) which is very easy to prove here:

$$|u|_{L^\infty(\Omega)} \leq |u|^{1/2}|Du|^{1/2} \leq |u|^{3/4}|D^2u|^{1/4}, \qquad \forall u \in \dot{H}^2_{\text{per}}(\Omega).$$

Now let $\varphi_1, \dots, \varphi_m$ be a finite family of $D(A^{1/2})$ which is orthonormal in H; we denote by $P_m H$ the space that they span in H, and P_m is the orthogonal projector from H onto $P_m H$. We observe that the quantity

$$\rho(x) = \sum_{j=1}^m (\varphi_j(x))^2; \qquad x \in \Omega,$$

depends only on $P_m H$ and not on the actual family $\varphi_1, \dots, \varphi_m$: this is due to the fact that any change of orthonormal basis in $P_m H$ introduces an orthogonal matrix of order m, whose coefficients are obviously independent of x.

Hence we can choose for $\varphi_1, \dots, \varphi_m$ a particular family. We choose the eigenvalues of the linear self-adjoint operator $\bar{A}_m$ that is the restriction of $P_m A$ to $P_m H$; we also denote by $\alpha_1, \dots, \alpha_m$ the corresponding eigenvalues. We then have, for a.e. $x \in \Omega$, and for any vector $\xi = (\xi_1, \dots, \xi_m) \in \mathbb{R}^m$:

$$\begin{aligned}
\left|\sum_{j=1}^m \xi_j \varphi_j(x)\right| &\leq \left|\sum_{j=1}^m \xi_j \varphi_j\right|_{L^\infty(\Omega)} \\
&\leq \left|\sum_{j=1}^m \xi_j \varphi_j\right|^{3/4} \left|\sum_{j=1}^m \xi_j D^2\varphi_j\right|^{1/4} \\
&\leq \left(\sum_{j=1}^m \xi_j^2\right)^{3/8} \left|\sum_{i,j=1}^m \xi_i\xi_j (D^2\varphi_i, D^2\varphi_j)\right|^{1/8} \\
&\leq \left(\sum_{j=1}^m \xi_j^2\right)^{3/8} \left|\sum_{i,j=1}^m \xi_i\xi_j (\bar{A}_m\varphi_i, \varphi_j)\right|^{1/8} \\
&\leq \left(\sum_{j=1}^m \xi_j^2\right)^{3/8} \left(\sum_{j=1}^m \alpha_j\xi_j^2\right)^{1/8}.
\end{aligned}$$

We now choose $\xi_j = \varphi_j(x)$ and obtain

$$\rho(x) \leq \rho(x)^{3/8}\left(\sum_{j=1}^m \alpha_j(\varphi_j(x))^2\right)^{1/8},$$

$$\rho(x)^5 \leq \sum_{j=1}^m \alpha_j(\varphi_j(x))^2,$$

and by integration in x

$$\int_{-(L/2)}^{L/2} \rho(x)^5\,dx \le \sum_{j=1}^{m} \alpha_j = \operatorname{Tr} A_m = \sum_{j=1}^{m} \int_{-(L/2)}^{L/2} |D^2\varphi_j(x)|^2\,dx.$$

5.2. The Cahn–Hilliard Equations

In this section we continue the study of the Cahn–Hilliard equations initiated in Section III.4.2. Our aim is to apply to these equations the general results and methods of Chapter V. We restrict ourselves to the estimate of the dimension of the attractors $X_\beta = \mathscr{A} \cap H_\beta$, i.e., we will work in the affine spaces H_β corresponding to a fixed average of u,

$$m(u_0) = m(u(t)) = \beta, \qquad \forall t.$$

Also, to avoid unnecessary technicalities, we will assume that $p = 2$ when $n = 3$.

In Section 5.2.1 we present the linearized equations; in Section 5.2.2 we study the Lyapunov exponents and establish the results on dimension. Section 5.2.3 is devoted to technical computations leading to a more explicit form of the estimates in terms of the data.

5.2.1. The Linearized Equation

The exact problem consists of the equation III.(4.47) supplemented by one of the boundary conditions III.(4.52), (4.53), and the initial condition III.(4.55). Since $m(u(t)) = m(u_0)$, $\forall t$, and we restrict ourselves to the affine subspace H_β of H,

$$H_\beta = \{\varphi \in H, m(\varphi) = \beta\}, \tag{5.21}$$

we can, alternatively, write equations III.(4.47) as

$$\begin{gathered} \frac{\partial \bar{u}}{\partial t} - \Delta K(\bar{u} + \beta) = 0, \\ K(\bar{u} + \beta) = -\nu\Delta\bar{u} + f(\bar{u} + \beta), \end{gathered} \tag{5.22}$$

where $\bar{u} = u - m(u) = u - \beta$. The boundary conditions for $\bar{u}$ are the same as for u and the initial condition III.(4.55) is the same as

$$\bar{u}(x, 0) - \bar{u}_0(x), \qquad x \subset \Omega. \tag{5.23}$$

The point of view in this section amounts to considering this dynamical system in $\bar{u}$, with which we associate the semigroup S_β acting in H_0

$$S_\beta(t)\colon \bar{u} \in H_0 \to S(t)(\bar{u}_0 + \beta) - \beta$$

and the maximal attractor $\tilde{X}_\beta = X_\beta - \beta \subset H_0$.

In the sequel we will consider either u or $\bar{u}$ without particular preference.

The exact problem, being written in the abstract form

$$u' = F(u), \tag{5.24}$$

as in V.(2.38), the linearized equation V.(2.39)

$$U' = F'(u)U, \tag{5.25}$$

is equivalent to

$$\frac{d}{dt}(U, v) + \nu(\Delta U, \Delta v) - (f'(u)U, \Delta v) = 0, \qquad \forall v \in V. \tag{5.26}$$

This is the functional form of a boundary-value problem consisting of the evolution equation

$$\frac{\partial U}{\partial t} + \nu\Delta^2 U - \Delta(f'(u)U) = 0, \tag{5.27}$$

supplemented by one of the boundary conditions III.(4.52), (4.53), the same as for u.

The initial condition is

$$U(0) = \xi. \tag{5.28}$$

The semigroup $S(t)$ mapping H_β into itself, $\forall \beta \in \mathbb{R}$, U will take its values in H_0. It is easy to derive from Theorem II.3.4 the following:

> If $\xi \in H_0$, $u_0 \in H_\beta$, $\beta \in \mathbb{R}$, and $u = u(t) = S(t)u_0$ is the corresponding solution of III.(4.47), (4.48), (4.52)–(4.55) given by Theorem III.4.2, then the initial- and boundary-value problem (5.25)–(5.28) possesses a unique solution
>
> $$U \in L^2(0, T; V) \cap \mathscr{C}([0, T]; H_0), \qquad \forall T > 0. \tag{5.29}$$

Concerning the differentiability of the mapping $S(t)$: $u_0 \to S(t)u_0$, the following property holds, whose proof will be sketched in Section 8:

> For every $t > 0$, the function $u_0 \to S(t)u_0$ is differentiable in H and its differential at a point u_0 is the linear operator $L(t, u_0) \in \mathscr{L}(H_0)$: $\xi \in H_0 \to U(t) \in H_0$, where U is the solution of (5.25)–(5.28). (5.30)

5.2.2. *Dimension of the Attractor*

Our aim is now to estimate the dimension of X_β (or $\tilde{X}_\beta$) by following the procedure described in Section V.2.3.

We consider for $m \in \mathbb{N}$, m elements $\xi_1, \ldots, \xi_m$ of H_0 and the corresponding solutions $U = U_1, \ldots, U_m$ of (5.25)–(5.28). We denote by $Q_m(\tau) = Q_m(\tau, u_0; \xi_1, \ldots, \xi_m)$ the orthogonal projector in H_0 onto the space spanned by $U_1(\tau), \ldots, U_m(\tau)$; $u_0 \in H$, $\beta = m(u_0)$, and $u(t) = S(t)u_0$ is the solution of III.(4.47), (4.48), (4.52)–(4.55). We recall that

$$
\begin{aligned}
&|U_1(t) \wedge \cdots \wedge U_m(t)|_{\bigwedge^m H} \\
&\quad = |\xi_1 \wedge \cdots \wedge \xi_m|_{\bigwedge^m H} \exp \int_0^t \operatorname{Tr} F'(S(\tau)u_0) \circ Q_m(\tau)\, d\tau.
\end{aligned}
$$

At a given time τ, let $\varphi_j(\tau)$, $j = 1, \ldots, m$, be an orthonormal basis of $Q_m(\tau)H = \operatorname{Span}[U_1(\tau), \ldots, U_m(\tau)]$; $\varphi_j(\tau) \in V$ for $j = 1, \ldots, m$, since $U_1(\tau), \ldots, U_m(\tau) \in V$ (for a.e. $\tau \in \mathbb{R}_+$), and we have

$$
\begin{aligned}
\operatorname{Tr} F'(u(\tau)) \circ Q_m(\tau) &= \sum_{j=1}^{\infty} (F'(u(\tau)) \circ Q_m(\tau)\varphi_j(\tau), \varphi_j(\tau)) \\
&= \sum_{j=1}^{m} (F'(u(\tau))\varphi_j(\tau), \varphi_j(\tau)).
\end{aligned} \tag{5.31}
$$

Omitting temporarily the dependence on τ, we write

$$
(F'(u)\varphi_j, \varphi_j) = -\nu|\Delta\varphi_j|^2 + (f'(u)\varphi_j, \Delta\varphi_j),
$$

$$
\sum_{j=1}^{m} (F'(u)\varphi_j, \varphi_j) \le -\nu \sum_{j=1}^{m} |\Delta\varphi_j|^2 + \int_\Omega |f'(u(x))|\rho(x)^{1/2}\left(\sum_{j=1}^{m} |\Delta\varphi_j(x)|^2\right)^{1/2} dx
$$

with

$$
\rho = \rho(x, \tau) = \sum_{j=1}^{m} |\varphi_j(x, \tau)|^2 \quad \text{a.e.} \tag{5.32}
$$

Setting $\alpha_n = 1 + 4/n$, $\beta_n = 2 + n/2$, we observe that $1/(2\alpha_n) + 1/\beta_n = \frac{1}{2}$, and we apply the Hölder inequality with exponents 2, $2\alpha_n$, β_n. This yields

$$
\sum_{j=1}^{m} (F'(u)\varphi_j, \varphi_j) \le -\nu \sum_{j=1}^{m} |\Delta\varphi_j|^2 + |f'(u)|_{L^{\beta_n}}|\rho|_{L^{\alpha_n}(\Omega)}^{1/2}\left(\sum_{j=1}^{m} |\Delta\varphi_j|^2\right). \tag{5.33}
$$

Thanks to Theorem 4.1., we know that there exists a nondimensional constant c_1'' depending only on the shape of Ω, such that

$$
|\rho|_{L^{\alpha_n}(\Omega)}^{\alpha_n} = \int_\Omega \rho(x, \tau)^{\alpha_n}\, dx \le c_1'' \sum_{j=1}^{m} |\Delta\varphi_j(\tau)|^2, \quad \text{for a.e. } \tau. \tag{5.34}
$$

Also, using Corollary A.4.1, we see that the sum $\sum_{j=1}^m |\Delta\varphi_j|^2$ can be bounded from below

$$
\sum_{j=1}^{m} |\Delta\varphi_j(\tau)|^2 \ge c_2'' \frac{m^{1+4/m}}{|\Omega|^{4/n}} - c_3'' \frac{m}{|\Omega|^{4/n}}, \tag{5.35}
$$

where c_2'', c_3'' depend only on the shape of Ω and n.

Using (5.34), (5.35), we deduce from (5.33) that

$$\sum_{j=1}^{m} (F'(u)\varphi_j, \varphi_j) \leq -\nu \sum_{j=1}^{m} |\Delta\varphi_j|^2 + (c_1'')^{1/\alpha_n} |f'(u)|_{L^{\beta_n}(\Omega)} \left(\sum_{j=1}^{m} |\Delta\varphi_j|^2 \right)^{1/2+1/(2\alpha_n)}$$

$$\leq \text{(with the Young inequality)}$$

$$\leq -\frac{\nu}{2} \sum_{j=1}^{m} |\Delta\varphi_j|^2 + c_4'' \nu^{1-\beta_n} |f'(u)|^{\beta_n}_{L^{\beta_n}(\Omega)}$$

$$\leq -\frac{\nu c_2''}{2} \frac{m^{1+4/n}}{|\Omega|^{4/n}} + \frac{\nu}{2} c_3'' \frac{m}{|\Omega|^{4/n}} + c_4'' \nu^{-1-n/2} |f'(u)|^{\beta_n}_{L^{\beta_n}(\Omega)}. \tag{5.36}$$

We now assume that $m(u_0) = \beta$ is fixed, and that $u_0 \in X_\beta$. The inequalities above allow us to majorize the expressions $q_m(t)$, q_m appearing in V.(3.47), (3.48)

$$q_m(t) = \operatorname*{Sup}_{u_0 \in X_\beta} \operatorname*{Sup}_{\substack{\xi_i \in H_0 \\ |\xi_i| \leq 1 \\ i=1,\dots,m}} \left(\frac{1}{t} \int_0^t \operatorname{Tr} F'(u(\tau)) \circ Q_m(\tau)\, d\tau \right),$$

$$q_m(t) \leq -\frac{\nu c_2''}{2} \frac{m^{1+4/n}}{|\Omega|^{4/n}} + \frac{\nu}{2} c_3'' \frac{m}{|\Omega|^{4/n}}$$

$$+ c_4'' \nu^{-1-n/2} \operatorname*{Sup}_{u_0 \in X_\beta} \frac{1}{t} \int_0^t |f'(S(\tau)u_0)|^{\beta_n}_{L^{\beta_n}(\Omega)}\, d\tau, \tag{5.37}$$

$$q_m = \limsup_{t \to \infty} q_m(t) \leq -\frac{\nu c_2''}{2} \frac{m^{1+4/n}}{|\Omega|^{4/n}} + \frac{\nu}{2} \frac{c_3'' m}{|\Omega|^{4/n}} + c_4'' \nu^{-1-n/2} \gamma_\beta,$$

where

$$\gamma_\beta = \limsup_{t \to \infty} \operatorname*{Sup}_{u_0 \in X_\beta} \frac{1}{t} \int_0^t |f'(S(\tau)u_0)|^{\beta_n}_{L^{\beta_n}(\Omega)}\, d\tau. \tag{5.38}$$

Thanks to the Young inequality

$$c_3'' m \leq \frac{c_2'' m^{1+4/n}}{2} + c_5'',$$

and, finally,

$$q_m \leq -\kappa_1 m^{1+4/n} + \kappa_2, \tag{5.39}$$

$$\kappa_1 = \frac{c_2''}{4} \frac{\nu}{|\Omega|^{4/n}}, \qquad \kappa_2 = \frac{c_5''}{4} \frac{\nu}{|\Omega|^{4/n}} + c_4'' \nu^{-1-n/2} \gamma_\beta,$$

the nondimensional constants c_2'', c_4'', c_5'' depending only on the shape of Ω and n.

By application of the results of Chapter V (in particular, Proposition V.2.1 and Theorem V.3.3), we have now proved the following

Theorem 5.2. *We consider the dynamical system associated with the Cahn–Hilliard equations* III.(4.47), (4.48), (4.52), (4.53). *We assume that* $n = 1, 2, 3$, $p \geq 2$ *arbitrary if* $n = 1, 2$ *and* $p = 2$ *if* $n = 3$. *For* $\beta \in \mathbb{R}$ *fixed we denote by* $X_\beta \subset \beta + H_0$ *the corresponding maximal attractor given by Theorem* III.4.3 *and Remark* III.4.6(i). *Then*:

(i) *the corresponding uniform Lyapunov exponents* μ_j *are majorized as follows*

$$\mu_1 + \cdots + \mu_j \leq -\kappa_1 j^{1+4/n} + \kappa_2, \qquad \forall j \in \mathbb{N}, \tag{5.40}$$

κ_1, κ_2 *as in* (5.39) *and if m is defined by*

$$m - 1 < \left(\frac{2\kappa_2}{\kappa_1}\right)^{n/(n+4)} \leq m; \tag{5.41}$$

then:

(ii) *the m-dimensional volume element in* H_0 *decays exponentially as* $t \to \infty$;

(iii) *the Hausdorff dimension of* X_β *is less than or equal to m, and its fractal dimension is less than or equal to 2m.*

5.2.3. An Explicit Estimate

As indicated in the introduction of Section 5.2 we now show how we can complete Theorem 5.2 and give an explicit bound of γ_β (and thus a more explicit expression for m) in terms of the data. This relies on a careful inspection of the inequalities established in Chapter III, especially those at the end of Section III.4.2.3. In particular, we need to make more explicit the dependence of some constants in terms of the data ν, f, Ω. In this respect, in the following, the c_j'' denote as above various nondimensional constants depending on the shape of Ω but not on its size.

First, we rewrite III.(4.82), (4.83) as

$$\|\varphi - m(\varphi)\|_{-1} \leq c_6''|\Omega|^{1/n}|\varphi|, \qquad \forall \varphi \in L^2(\Omega),$$

$$|\varphi - m(\varphi)| \leq c_7''|\Omega|^{1/n}|\nabla\varphi|, \qquad \forall \varphi \in H^1(\Omega),$$

i.e., $c_1' = c_6''|\Omega|^{1/n}$, $c_2' = c_7''|\Omega|^{1/n}$.

Then, due to III.(4.95), we have

$$|\nabla\varphi| \geq \sqrt{k_2(\alpha)}, \qquad \forall \varphi \in X_\beta, \tag{5.42}$$

where $\alpha \geq |\beta|$, $\beta = m(\varphi)$. The explicit expression of $k_2(\alpha)$ appears in III.(4.94), r is arbitrary, c_1, c_2, c_3 depend on f, and c_1', c_2' are replaced by the expressions above.

The next step is to obtain a uniform estimate of $|\Delta\varphi|$ on X_β and this can be derived from (5.42) and the calculations following III.(4.104). Inequality III.(4.109) is left unchanged.[1] The constants c in III.(4.110), (4.111) are dimen-

[1] u is dimensionless, k_4' like f has dimension L^2/T, ν has dimension L^4/T, $L =$ the reference length, and $T =$ the reference time.

sionless and then in (4.112),

$$k_3(\alpha) = c_8''(k_2(\alpha))^{1/3}, \qquad k_4(\alpha) = c_9''(k_2(\alpha))^{1/2-n/24}.$$

In III.(4.116) we write

$$\begin{aligned} \beta_1 &= 0, & k_{\varepsilon,1} &= c_{10}''|\Omega|\sqrt{k_2(\alpha)}, \\ \beta_2 &= \varepsilon, & k_{\varepsilon,2} &= c_{11}''(\varepsilon)|\Omega|^{(1-\varepsilon)/2}(k_2(\alpha))^{(1-\varepsilon)/2}, \\ \beta_3 &= \tfrac{1}{6}, & k_{\varepsilon,3} &= c_{12}''(\varepsilon)|\Omega|^{5/18}(k_2(\alpha))^{5/12}. \end{aligned}$$

Inequality III.(4.105) being suitably modified, we then deduce from III.(4.106) that

$$\frac{d}{dt}|\Delta u|^2 + \nu c_{13}''|\Omega|^{-4/n}|\Delta u|^2 \le k_1'. \tag{5.43}$$

This implies

$$\limsup_{t\to\infty} |\Delta u(t)| \le \frac{k_1'|\Omega|^{4/n}}{\nu c_{13}''},$$

i.e.,

$$|\Delta\varphi| \le \frac{k_1'|\Omega|^{4/n}}{\nu c_{13}''}, \qquad \forall \varphi \in X_\beta. \tag{5.44}$$

Finally, we estimate $\{\mathrm{Sup}|\varphi|_{L^\infty(\Omega)}, \varphi \in X_\beta\}$ by using (5.42), (5.44) and the following analog of III.(4.113)–(4.115):

$$|\varphi - m(\varphi)|_{L^\infty(\Omega)} \le \begin{cases} c_{14}''|\Omega|^{1/2}|\nabla\varphi|, & \forall\varphi \in H^1(\Omega) \quad \text{if } n = 1, \\ c_{15}''(\varepsilon)|\Omega|^{\varepsilon/n}|\nabla\varphi|^{1-\varepsilon}|\Delta\varphi|^{\varepsilon}, & \forall\varphi \in H^2(\Omega) \quad \text{if } n = 2, \\ c_{16}''|\nabla\varphi|^{1/2}|\Delta\varphi|^{1/2}, & \forall\varphi \in H^2(\Omega) \quad \text{if } n = 3. \end{cases} \tag{5.45}$$

Since $|\varphi|_{L^\infty(\Omega)} \le |\varphi - m(\varphi)|_{L^\infty(\Omega)} + |\beta|$, $\forall\varphi \in X_\beta$, we can estimate

$$\sup_{u_0 \in X_\beta} \ \sup_{x\in\Omega, \tau\in\mathbb{R}} f'((S(\tau)u_0)(x,\tau))$$

and we easily obtain a bound for γ_β.

6. Dissipative Wave Equations

In this section our object is the application of the results and methods of Chapter V to the nonlinear dissipative wave equations considered in Chapter IV, Section 2.3.4. In order to avoid the repetition of tedious computations we will restrict ourselves to the abstract wave equation of Section IV.4; the transcription of the results to the equations of Sections IV.2 and IV.3 is left as an exercise for the reader (see, however, the end of Section 6.2 for the sine–Gordon equations).

As mentioned in Chapter IV, the equations treated here are slightly different from the first-order evolution equations studied in Sections 2–5, since the semigroup $S(t)$ and its differential $L(t, u_0)$ are not any more compact, and they even appear to be *invertible operators* in the appropriate function spaces. Some of the difficulties which appeared in Chapter V, Section 1 (see 1.3, in particular), which were due to the lack of compactness of the operators, were motivated by the examples considered here.

This section is a continuation of Section IV.4 of which we keep all the hypotheses and notations. In Section 6.1 we present the linearized equation and then in Section 6.2 we derive the bound on the dimensions.

6.1. The Linearized Equation

The abstract evolution equation considered here is IV.(4.1), i.e.,

$$u'' + \alpha u' + Au + g(u) = f, \tag{6.1}$$

supplemented by the initial conditions IV.(4.18), i.e.,

$$u(0) = u_0, \qquad u'(0) = u_1. \tag{6.2}$$

The hypotheses are those of Theorem IV.4.2, in particular, the hypotheses IV.(4.8) and IV.(4.10)–(4.15) on g.

The operators $S(t)$ are defined for all $t \in \mathbb{R}$ and constitute a group of operators in $E_0 = V \times H$

$$S(t)\{u_0, u_1\} = \{u(t), u'(t)\}.$$

It will also be convenient to consider the operators $S_\varepsilon(t) = R_\varepsilon(S(t)R_{-\varepsilon}$ for the values of ε allowed by IV. (1.19)

$$0 < \varepsilon \leq \varepsilon_0, \qquad \varepsilon_0 = \min\left(\frac{\alpha}{4}, \frac{\lambda_1}{2\alpha}\right). \tag{6.3}$$

We refer the reader to IV.(1.29) for the definition of R_ε: $\{a, b\} \to \{a, b + \varepsilon a\}$, so that $S_\varepsilon(t)$ is the mapping

$$S_\varepsilon(t)\colon \{u_0, v_1 = u_1 + \varepsilon u_0\} \to \{u(t), v(t) = u'(t) + \varepsilon u(t)\}.$$

The operators $S_\varepsilon(t)$, $t \in \mathbb{R}$, also form a group and any result concerning $S(t)$ is easily transcribed for $S_\varepsilon(t)$; for example, if $\mathscr{A}$ is the maximal attractor for $S(t)$, then $R_\varepsilon\mathscr{A}$ is the maximal attractor for $S_\varepsilon(t)$ and it attracts all the orbits $\{u(t), u'(t) + \varepsilon u(t)\}$ starting from any point $\{u_0, u_1 + \varepsilon u_0\} \in E_0 = V \times H$.

The first step will be to study the differentiability of $S(t)$. The formal linearization of equation (6.1) leads to

$$U'' + \alpha U' + AU + g'(u)U = 0, \tag{6.4}$$

which we supplement with the initial conditions

$$U(0) = \xi, \qquad U'(0) = \zeta. \tag{6.5}$$

It follows from Theorem II.4.3 that

> If u is the solution of (6.1), (6.2) given by Theorem IV.4.1, then (6.4) (6.5) possesses a unique solution U such that
>
> $$U \in \mathscr{C}(\mathbb{R}; V), \qquad U' \in \mathscr{C}(\mathbb{R}; H). \tag{6.6}$$

Indeed, we set $A_1(t) = g'(u(t))$ which belongs to $\mathscr{L}(V, H)$ for every $t \in \mathbb{R}$, thanks to IV.(4.11). Also, if $t \in [-T, T]$, since $u \in \mathscr{C}([-T, T]; V)$, the norm of $g'(u(t))$ in $\mathscr{L}(V, H)$ is bounded for $t \in [-T, T]$: indeed, $\|u(t)\|$ is bounded and g' maps bounded sets of V into bounded sets of $\mathscr{L}(V, H)$; hence assumption II.(4.26) of Theorem II.4.3. We obtain the existence and uniqueness of U, $U \in \mathscr{C}([-T, T]; V)$, $U' \in \mathscr{C}([-T, T]; H)$, $\forall T > 0$, and (6.6) follows.

We then show

Lemma 6.1. *For any $t > 0$, the mapping $S(t)$ is Fréchet differentiable on E_0. Its differential at $\varphi_0 = \{u_0, u_1\}$ is the linear operator on E_0,*

$$L(t, \varphi_0)\colon \{\xi, \zeta\} \to \{U(t), U'(t)\},$$

where U is the solution of (6.4), (6.5).

PROOF. We first prove a Lipschitz property of $S(t)$ on the bounded sets of E_0.

Let $\varphi_0 = \{u_0, u_1\}$, $\tilde{\varphi}_0 = \varphi_0 + \{\xi, \zeta\} = \{u_0 + \xi, u_1 + \xi\}$ with $\|\varphi_0\|_{E_0} \leq R$, $\|\tilde{\varphi}_0\|_{E_0} \leq R$. We write $S(t)\varphi_0 = \varphi(t) = \{u(t), u'(t)\}$, $S(t)\tilde{\varphi}_0 = \{\tilde{\varphi}(t), \tilde{\varphi}'(t)\}$. It follows from Theorem IV.4.1 and Lemma IV.4.1 that $S(t)\varphi_0$ is uniformly bounded in E_0 for $t \geq 0$ and $\|\tilde{\varphi}_0\|_{E_0} \leq R$. Let R' denote the corresponding supremum of $\|S(t)\varphi_0\|_{E_0}$.

The difference $\psi = \tilde{u} - u$ satisfies

$$\psi'' + \alpha\psi' + A\psi + g(\tilde{u}) - g(u) = 0, \tag{6.7}$$

$$\psi(0) = \xi, \qquad \psi'(0) = \xi. \tag{6.8}$$

Due to IV.(4.9)(i) and the previous remark, there exists a constant $c_1' = c_1'(R')$ such that

$$|g(\tilde{u}(\tau)) - g(u(\tau))| \leq c_1'(R')\|\tilde{u}(\tau) - u(\tau)\|, \qquad \forall \tau \geq 0.$$

Taking the scalar product of (6.7) with ψ in H, we then find

$$\frac{1}{2}\frac{d}{dt}\{|\psi'|^2 + \|\psi\|^2\} + \alpha|\psi'|^2 = -(g(\tilde{u}) - g(u), \psi') \leq c_1'\|\psi\|\,|\psi'|,$$

$$\frac{d}{dt}\{|\psi'|^2 + \|\psi\|^2\} \leq 2c_1'\{|\psi'|^2 + \|\psi\|^2\}.$$

Hence the Lipschitz property

$$\|\tilde{\varphi}(t) - \varphi(t)\|_{E_0}^2 = |\tilde{u}'(t) - u'(t)|^2 + \|\tilde{u}(t) - u(t)\|^2$$
$$\le \exp(2c_1' t)\{|\zeta|^2 + \|\xi\|^2\}, \qquad \forall t \ge 0. \tag{6.9}$$

We now consider the difference $\theta = \tilde{u} - u - U$, with U the solution of (6.4), (6.5), ξ, ζ in (6.5) being the same as above. Clearly,

$$\theta(0) = 0, \qquad \theta'(0) = 0, \tag{6.10}$$

and θ satisfies

$$\theta'' + \alpha\theta' + A\theta + g'(u)\theta = h, \tag{6.11}$$

with

$$h = g(u) - g(\tilde{u}) - g'(u)(u - \tilde{u}). \tag{6.12}$$

We have

$$h(\tau) = \int_0^1 \{g'(s\tilde{u}(\tau) + (1 - s)u(\tau)) - g'(u(\tau))\} \cdot (u(\tau) - \tilde{u}(\tau))\, ds.$$

We observed before that $\|u(\tau)\|$ and $\|\tilde{u}(\tau)\|$ are uniformly bounded for $\tau \ge 0$ by R'. Therefore, thanks to IV.(4.11)(ii),

$$|g'(s\tilde{u}(\tau) + (1 - s)u(\tau)) - g'(u(\tau))|_{\mathscr{L}(V,H)}$$
$$\le c_0'(R')s^\delta \|\tilde{u}(\tau) - u(\tau)\|^\delta, \qquad \forall \rho \ge 0, \quad \forall s \in [0, 1], \tag{6.13}$$
$$|h(\tau)| \le c_1'(R')\|\tilde{u}(\tau) - u(\tau)\|^{1+\delta}, \qquad \tau \ge 0.$$

Similarly, since by IV.(4.11)(i), g' maps bounded sets of V into bounded sets of $\mathscr{L}(V, H)$, there exists a constant $c_2' = c_2'(R')$ such that

$$|g'(u(\tau))|_{\mathscr{L}(V,H)} \le c_2', \qquad \tau \ge 0. \tag{6.14}$$

We take the scalar product of each side of (6.11) with θ' in H. Taking into account (6.12)–(6.14), we find

$$\frac{1}{2}\frac{d}{dt}\{|\theta'|^2 + \|\theta\|^2\} + \alpha|\theta'|^2 = (h - g'(u)\theta, \theta')$$
$$\le c_1'|\theta'|\,\|\tilde{u}(\tau) - u(\tau)\|^{1+\delta} + c_2'|\theta'|\,\|\theta\|, \tag{6.15}$$
$$\frac{d}{dt}\{|\theta'|^2 + \|\theta\|^2\} \le c_2'\{|\theta'|^2 + \|\theta\|^2\} + c_3'\|\tilde{u}(\tau) - u(\tau)\|^{2+2\delta}.$$

Using the usual Gronwall lemma and (6.10) we deduce from (6.15) that

$$\{|\theta'(t)|^2 + \|\theta(t)\|^2\} \le \frac{c_3'}{c_2'}\exp(c_2' t)\int_0^t \|\tilde{u}(\tau) - u(\tau)\|^{2+2\delta}\, d\tau$$
$$\le \text{(with (6.9))}$$
$$\le c_4' \exp(c_5' t)\{|\zeta|^2 + \|\xi\|^2\}^{1+\delta}. \tag{6.16}$$

This is equivalent to

$$\|\tilde{\varphi}(t) - \varphi(\tau) - U(t)\|_{E_0}^2 \le c_4' \exp(c_5' t)\|\{\xi, \zeta\}\|_{E_0}^{2+2\delta}, \tag{6.17}$$

and consequently

$$\frac{\|\tilde{\varphi}(t) - \varphi(\tau) - U(t)\|_{E_0}^2}{\|\{\xi, \zeta\}\|_{E_0}^2} \to 0 \quad \text{as} \quad \{\xi, \zeta\} \to 0 \quad \text{in } E_0.$$

The differentiability of $S(t)$ is proved. □

6.2. Dimension of the Attractor

It is convenient for the moment to rewrite the initial-value problem (6.1), (6.2) as a first-order evolution equation for the variable $\psi = R_\varepsilon \varphi = \{u, v = u' + \varepsilon u\}$ as we did in IV.(1.25) for the linear equation. Hence we choose ε as in (6.3) and set

$$\psi' + \Lambda_\varepsilon \psi + \tilde{g}(\psi) = \tilde{f}, \tag{6.18}$$

where $\tilde{f} = \{0, f\}$, $\tilde{g}(\psi) = \{0, g(u)\}$, $\psi = \{u, v\}$, and

$$\Lambda_\varepsilon = \begin{pmatrix} \varepsilon I & -I \\ A - \varepsilon(\alpha - \varepsilon)I & (\alpha - \varepsilon)I \end{pmatrix}. \tag{6.19}$$

With this setting the abstract equation V.(2.38) is written

$$\psi' = F(\psi) = \tilde{f} - \Lambda_\varepsilon \psi - \tilde{g}(\psi), \tag{6.20}$$

and the first variation equation V.(2.39)

$$\Psi' = F'(\psi)\Psi \tag{6.21}$$

corresponds to the similar form of (6.4), i.e.,

$$\Psi' + \Lambda_\varepsilon \Psi + \tilde{g}'(\psi)\Psi = 0, \tag{6.22}$$

where $\Psi = \{U, U' + \varepsilon U\}$ and $\tilde{g}'(\psi)\Psi = \{0, g'(u)U\}$. The initial condition (6.5) is written

$$\Psi(0) = \eta, \qquad \eta = \{\xi, \zeta\} \in E_0. \tag{6.23}$$

It was shown in Section 6.1 that (6.22), (6.23) is a well-posed problem in E_0 and the relation of (6.21) to (6.20) was made explicit in Lemma 6.1.

We now follow the procedure described in Section V.2.3 with some slight change of notations due to the replacement of u, U by ψ, Ψ. Hence, for $m \in \mathbb{N}$, we consider m solutions $\Psi = \Psi_1, \ldots, \Psi_m$ of (6.21)–(6.23) corresponding to initial data $\eta = \eta_1, \ldots, \eta_m$, $\eta_j \in E_0$, and we recall that

$$|\Psi_1(t) \wedge \cdots \wedge \Psi_m(t)|_{\bigwedge^m E_0}$$
$$= |\eta_1 \wedge \cdots \wedge \eta_m|_{\bigwedge^m E_0} \exp \int_0^t \operatorname{Tr} F'(S_\varepsilon(\tau)\psi_0) \circ Q_m(\tau)\, d\tau;$$

$\psi(\tau) = S_\varepsilon(\tau)\psi_0$ and with the definition of S_ε recalled above we see that $\psi(\tau) =$

$\{u(\tau), v(\tau) = u'(\tau) + \varepsilon u(\tau)\}$, where u is the solution of (6.1), (6.2); $Q_m(\tau) = Q_m(\tau, \psi_0; \eta_1, \ldots, \eta_m)$ is the orthogonal projector in $E_0 = V \times H$ onto the space spanned by $\Psi_1(\tau), \ldots, \Psi_m(\tau)$.

At a given time τ, let $\Phi_j(\tau) = \{\xi_j(\tau), \zeta_j(\tau)\}$, $j = 1, \ldots, m$, denote an orthonormal basis of $Q_m(\tau)E_0 = \mathrm{Span}[\Psi_1(\tau), \ldots, \Psi_m(\tau)]$. We can write

$$\begin{aligned}\operatorname{Tr} F'(\psi(\tau)) \circ Q_m(\tau) &= \sum_{j=1}^{\infty} (F'(\psi(\tau)) \circ Q_m(\tau)\Phi_j(\tau), \Phi_j(\tau))_{E_0} \\ &= \sum_{j=1}^{m} (F'(\psi(t))\Phi_j(\tau), \Phi_j(\tau))_{E_0},\end{aligned} \tag{6.24}$$

where $(\cdot, \cdot)_{E_0}$ is the scalar product in E_0,

$$(\{\xi, \zeta\}, \{\tilde{\xi}, \tilde{\zeta}\})_{E_0} = ((\xi, \tilde{\xi})) + (\zeta, \tilde{\zeta}).$$

We omit for the moment the variable τ and we write

$$\begin{aligned}(F'(\psi)\Phi_j, \Phi_j)_{E_0} &= -(\Lambda_\varepsilon \Phi_j, \Phi_j)_{E_0} - (g'(u)\xi_j, \zeta_j), \\ (\Lambda_\varepsilon \Phi_j, \Phi_j) &= \varepsilon\|\xi_j\|^2 + (\alpha - \varepsilon)|\zeta_j|^2 - \varepsilon(\alpha - \varepsilon)(\xi_j, \zeta_j) \\ &\geq (\text{by IV.(1.22)}) \\ &\geq \alpha_1(\|\xi_j\|^2 + |\zeta_j|^2), \qquad \alpha_1 = \varepsilon/2.\end{aligned}$$

The term $(g'(u)\xi_j, \zeta_j)$ is majorized in absolute value by $|g'(u)\xi_j||\zeta_j|$. Let us assume from now on that $\{u_0, u_1\}$ belongs to $\mathscr{A}$ which, according to Theorem IV.4.3, is a bounded set of E_1; then $\psi(t) = \{u(t), u'(t) + \varepsilon u(t)\}$ belongs to a bounded set of E_1 (namely $R_\varepsilon \mathscr{A}$) and $u(t)$ belongs to a bounded set of $D(A)$. According to IV.(4.11)(i) there exists $s \in [0, 1[$, such that g' maps this bounded set of $D(A)$ into a bounded set of $\mathscr{L}(V_s, H)$

$$R'' = \sup_{\{\xi, \zeta\} \in \mathscr{A}} |A\xi| < \infty, \tag{6.25}$$

$$\sup_{\substack{w \in D(A) \\ |Aw| \leq R''}} |g'(w)|_{\mathscr{L}(V_s, H)} \leq \gamma < \infty. \tag{6.26}$$

This allows us to majorize $|(g'(u)\xi_j, \zeta_j)|$ by

$$|g'(u)\xi_j||\zeta_j| \leq \gamma|\xi_j|_s|\zeta_j|.$$

At this point we have proved that

$$\begin{aligned}(F'(\psi)\Phi_j, \Phi_j)_{E_0} &\leq -\alpha_1(\|\xi_j\|^2 + |\zeta_j|^2) + \gamma|\xi_j|_s|\zeta_j| \\ &\leq (\text{with the Schwarz inequality}) \\ &\leq -\frac{\alpha_1}{2}(\|\xi_j\|^2 + |\zeta_j|^2) + \frac{\gamma^2}{2\alpha_1}|\xi_j|_s^2.\end{aligned} \tag{6.27}$$

We observe that $\|\xi_j\|^2 + |\zeta_j|^2 = \|\Phi_j\|_{E_0}^2 = 1$ since the family Φ_j is orthonormal in E_0 and hence

$$\sum_{j=1}^{m} (F'(\psi)\Phi_j, \Phi_j)_{E_0} \leq -\frac{m\alpha_1}{2} + \frac{\gamma^2}{2\alpha_1}\sum_{j=1}^{m} |\xi_j|_s^2. \tag{6.28}$$

For almost every τ, the sum $\sum_{j=1}^{m} |\xi_j(\tau)|_s^2$ can be majorized as follows using Lemma 6.3 below (see Section 6.4):

$$\sum_{j=1}^{m} |\xi_j(\tau)|_s^2 \leq \sum_{j=1}^{m-1} \lambda_j^{s-1}. \tag{6.29}$$

Hence

$$\operatorname{Tr} F'(\psi(\tau)) \circ Q_m(\tau) \leq -\frac{m\alpha_1}{2} + \frac{\gamma^2}{2\alpha_1} \sum_{j=1}^{m} \lambda_j^{s-1}. \tag{6.30}$$

Assuming as we did before that $\{u_0, u_1\} \in \mathscr{A}$, or equivalently $\psi_0 = \{u_0, u_1 + \varepsilon u_0\} \in R_\varepsilon \mathscr{A}$, we can introduce the expressions $q_m(t)$, q_m associated with $R_\varepsilon \mathscr{A}$,

$$q_m(t) = \sup_{\psi_0 \in R_\varepsilon \mathscr{A}} \operatorname*{Sup}_{\substack{\eta_i \in E_0 \\ \|\eta_i\|_{E_0} \leq 1 \\ i=1,\dots,m}} \left(\frac{1}{t} \int_0^t \operatorname{Tr} F'(S_\varepsilon(\tau)\psi_0) \circ Q_m(\tau)\, d\tau \right),$$

$$q_m = \limsup_{t\to\infty} q_m(t).$$

Due to (6.30),

$$q_m(t) \leq -\frac{m\alpha_1}{2} + \frac{\gamma^2}{2\alpha_1} \sum_{i=1}^{m} \lambda_i^{s-1},$$

$$q_m \leq -\frac{m\alpha_1}{2} + \frac{\gamma^2}{2\alpha_1} \sum_{i=1}^{m} \lambda_i^{s-1}. \tag{6.31}$$

We infer from (6.31) and V.(2.49) the following bound on the uniform Lyapunov exponents $\mu_j, j \in \mathbb{N}$ of $\mathscr{A}$ (or $R_\varepsilon \mathscr{A}$):

$$\mu_1 + \cdots + \mu_j \leq -\frac{m\alpha_1}{2} + \frac{\gamma^2}{2\alpha_1} \sum_{i=1}^{j} \lambda_i^{s-1}. \tag{6.32}$$

Now $q_m \to 0$ as $m \to \infty$. More precisely, from the compactness of A^{-1}, we know that $\lambda_i \to \infty$ as $i \to \infty$. Hence the Cesaro mean $(1/m)\sum_{i=1}^{m} \lambda_i^{s-1}$ tends to 0 as $m \to \infty$ and there exists $m \geq 1$ such that

$$\frac{1}{m} \sum_{i=1}^{m} \lambda_i^{s-1} \leq \frac{\alpha_1^2}{2\gamma^2}. \tag{6.33}$$

For this value of m,

$$q_m \leq -\frac{m\alpha_1}{2}\left(1 - \frac{\gamma^2}{\alpha_1^2 m} \sum_{i=1}^{m} \lambda_i^{s-1}\right) \leq -\frac{3m\alpha_1}{4},$$

and for $j = 1, \dots, m$,

$$(q_j)_+ \leq \frac{\gamma^2}{2\alpha_1} \sum_{i=1}^{j} \lambda_i^{s-1} \leq \frac{\gamma^2}{2\alpha_1} \sum_{i=1}^{m} \lambda_i^{s-1} \leq \frac{m\alpha_1}{4},$$

$$\operatorname*{Max}_{1 \leq j \leq m-1} \frac{(q_j)_+}{|q_m|} \leq \tfrac{1}{3}.$$

Thanks to Theorem V.3.3 and V.(3.51) we conclude that the Hausdorff dimension of $\mathscr{A}$ (or $R_\varepsilon \mathscr{A}$) is less than or equal to m and the fractal dimension is majorized by $4m/3$.

In conclusion, we have proved the

Theorem 6.1. *We consider the dynamical system defined in Section* IV.4 *and the hypotheses are the same as for Theorems* V.4.1–V.4.3. *Then*:

(i) *the uniform Lyapunov exponents μ_j associated with the global attractor $\mathscr{A}$ defined by Theorem* V.4.2 *are majorized according to* (6.32).
Furthermore, if m is defined by (6.33),
(ii) *the m-dimensional volume element is exponentially decaying in the phase space E_0*;
(iii) *the Hausdorff dimension of $\mathscr{A}$ is less than or equal to m, and its fractal dimension is less than or equal to $4m/3$.*

Remark 6.1. The derivation of an estimate of m more explicit than (6.33) depends on a more explicit estimate of the number γ in (6.26). For the examples in Sections IV.2 and IV.3, A is the Laplace operator with Dirichlet boundary conditions, hence $\lambda_j \sim c\lambda_1 j^{2/n}$ as $j \to \infty$, $\lambda_j \geq c'\lambda_1 j^{2/n}$, $\forall j$,

$$\sum_{i=1}^{m} \lambda_i^{s-1} \leq (c'\lambda_1)^{s-1} \sum_{i=1}^{m} i^{2(s-1)/n}.$$

If $2s + n \neq 2$ this is majorized by

$$c''\lambda_1 m^{(2s-2+n)/n},$$

and a sufficient condition for (6.33) is

$$m^{2(s-1)/n} \leq \frac{\alpha_1^2}{2c''\gamma^2},$$

$$m \geq \left(\frac{2c''\gamma^2}{\alpha_1^2}\right)^{n/2(1-s)}. \tag{6.34}$$

For the sine–Gordon equation see Section 6.3 hereafter.

6.3. Sine–Gordon Equations

In the case of the sine–Gordon equation of Section IV.2 we are able to give an expression of m more explicit than (6.33).

We have $g'(u) = \beta \cos u$, and $g'(u)$ belongs to $\mathscr{L}(H)$ for any u in $D(A)$ (or for any measurable function $u\colon \Omega \to \mathbb{R}$). Thus IV.(4.11)(i) holds with $s = 0$. Furthermore, $|g'(u)|_{\mathscr{L}(H)} = |\beta|$ for any u and the number γ in (6.26) is simply equal to $|\beta|$.

It is permitted by IV.(2.18) to take $\varepsilon = \varepsilon_0$; the definition IV.(2.20) of α_1 shows then that $\alpha_1 = \min(\alpha/8, \lambda_1/4\alpha)$; thus

$$\frac{1}{\alpha_1} = \max\left(\frac{8}{\alpha}, \frac{4\alpha}{\lambda_1}\right).$$

We can make m larger by replacing $1/\alpha_1$ in (6.33), (6.34) by

$$\frac{8}{\alpha}\left(1 + \frac{\alpha^2}{\lambda_1}\right).$$

Condition (6.33) then reads

$$\frac{2^7\beta^2}{\alpha^2}\left(1 + \frac{\alpha^2}{\lambda_1}\right)^2 \frac{1}{m}\sum_{i=1}^{m} \lambda_i^{-1} \leq 1.$$

As observed in Remark 6.1

$$\sum_{i=1}^{m} \lambda_i^{-1} \leq c''\lambda_1 m^{1-2/n} \tag{6.35}$$

if $n \neq 2$; with the same reasoning we see that for $n = 2$ this sum is majorized by $c''\lambda_1 \log m$. Hence we require that

$$m \log m \geq c''' \frac{\beta^2}{\lambda_1\alpha^2}\left(1 + \frac{\alpha^2}{\lambda_1}\right)^2 \quad \text{if} \quad n = 2,$$

$$m^{2/n} \geq c''' \frac{\beta^2}{\lambda_1\alpha^2}\left(1 + \frac{\alpha^2}{\lambda_1}\right)^2 \quad \text{if} \quad n \neq 2.$$

Let J and D denote the nondimensional numbers

$$J = \frac{\beta^2}{\lambda_1\alpha^2}, \qquad D = \frac{\alpha^2}{\lambda_1}, \tag{6.36}$$

which "measure", respectively, the electrical current ($\beta \sin u$) and the damping ($\alpha u'$) in the physical problem. The number m in Theorem 6.1 is then the first integer such that

$$\begin{cases} m \log m \geq cJ(1 + D^2) & \text{if} \quad n = 2, \\ \qquad m \geq c^{n/2} J^{n/2}(1 + D^2)^{n/2} & \text{if} \quad n \neq 2, \end{cases} \tag{6.37}$$

$n =$ the space dimension, $c = 8c''$, c'' given by (6.35) (and its analog if $n = 2$).

6.4. Some Lemmas

We conclude Section 6 with two lemmas, viz. Lemma 6.3 justifying (6.29) and a preliminary result in Lemma 6.2.

The following lemma is similar to Lemma 2.1.

Lemma 6.2. *Let K be a positive linear self-adjoint compact operator in a Hilbert space H, with eigenvalues μ_j, $j \in \mathbb{N}$,*

$$\mu_1 \geq \mu_2 \geq \cdots, \mu_j \geq 0, \qquad \mu_j \to 0 \quad \text{as } j \to \infty.$$

Then for any orthonormal family of elements of H, $\varphi_1, \ldots, \varphi_m$,

$$\sum_{j=1}^{m} (K\varphi_j, \varphi_j) \leq \mu_1 + \cdots + \mu_m. \tag{6.38}$$

PROOF. We consider the operator $K_m \in \mathscr{L}(\bigwedge^m H)$ defined as in Section V.1.2.2, and we observe that since the family φ_j is orthonormal in H

$$\begin{aligned}(K_m(\varphi_1 \wedge \cdots \wedge \varphi_m), \varphi_1 \wedge \cdots \wedge \varphi_m)_{\bigwedge^m H} \\ &= (K\varphi_1, \varphi_1)(\varphi_2, \varphi_2)\ldots(\varphi_m, \varphi_m) + \cdots + (\varphi_1, \varphi_1)\ldots(\varphi_{m-1}, \varphi_{m-1})(K\varphi_m, \varphi_m) \\ &= \sum_{j=1}^{m} (K\varphi_j, \varphi_j).\end{aligned}$$

Hence, since $|\varphi_1 \wedge \cdots \wedge \varphi_m|_{\bigwedge^m H} = 1$,

$$\sum_{j=1}^{m} (K\varphi_j, \varphi_j) \leq \sup_{\substack{\psi \in \bigwedge^m H \\ |\psi|_{\bigwedge^m H = 1}}} (K_m\psi, \psi)_{\bigwedge^m H}. \tag{6.39}$$

We can prove, exactly as in Lemma 2.1, that the eigenvectors of K_m are the numbers $\mu_{i_1} + \cdots + \mu_{i_m}$; the largest eigenvalue is $\mu_1 + \cdots + \mu_m$ and this number is obviously larger than the right-hand side of (6.39). □

Lemma 6.3. *Let there be given V, H, and A as above. Then for any s, $0 \leq s < 1$, and for any orthonormal family of elements of $V \times H$, $\{\xi_i, \zeta_i\}$, $i = 1, \ldots, m$, we have*

$$\sum_{i=1}^{m} |A^{s/2}\xi_i|^2 \leq \sum_{i=1}^{m} \lambda_i^{s-1}. \tag{6.40}$$

PROOF. The orthonormality property is expressed as

$$((\xi_i, \xi_j)) + (\zeta_i, \zeta_j) = \delta_{ij}, \qquad \forall i, j = 1, \ldots, m, \tag{6.41}$$

δ_{ij} denoting the Kronecker symbol.

We consider the following subspace of V:

$$X = \{v \in V, ((v, \xi_i)) = (A^{s/2}v, A^{s/2}\xi_i) = 0, j = 1, \ldots, m\}.$$

This space has finite codimension ($\leq 2m$), and it contains a family of vectors v_j, $j = 1, \ldots, m$, which are orthonormal in V. Consider then the vectors

$$\theta_i = \xi_i + \sum_{k=1}^{i} \alpha_{ik} v_k, \tag{6.42}$$

where the $\alpha_{ik} \in \mathbb{R}$ are searched such that

$$((\theta_i, \theta_j)) = (1 + \varepsilon)\delta_{ij}, \qquad \forall i, j, \tag{6.43}$$

$\varepsilon > 0$ given. Due to (6.41) and the properties of the v_j's, (6.43) is equivalent to

$$\sum_{k=1}^{i} \alpha_{ik}\alpha_{jk} = (1 + \varepsilon)\delta_{ij} - ((\xi_i, \xi_j)), \qquad \forall i, j. \tag{6.44}$$

We have set for convenience $\alpha_{ik} = 0$ for $k > i$; such a matrix $\alpha = \{\alpha_{ij}\}$ exists if and only if the matrix in the right-hand side of (6.44) is positive definite, i.e.,

$$q(x) = (1 + \varepsilon)|x|^2 - \sum_{i,j=1}^{m} ((\xi_i, \xi_j))x_i x_j > 0, \qquad \forall x = \{x_1, \ldots, x_m\} \in \mathbb{R}^m, \quad x \neq 0.$$

Thanks to (6.41)

$$q(x) = \varepsilon|x|^2 + \left|\sum_{i=1}^{m} x_i \zeta_i\right| \geq \varepsilon|x|^2.$$

The existence of α and of the vectors θ_i in (6.42) follows; the vectors $\tilde{\theta}_i = (1 + \varepsilon)^{-1/2}\theta_i$ are obviously orthonormal in V.

We apply Lemma 6.2 with H replaced by V, φ_i by $\tilde{\theta}_i$, and K by A^{s-1}. The eigenvalues of A are the numbers $\mu_j = \lambda_j^{s-1}$ and hence

$$\sum_{i=1}^{m} ((A^{s-1}\tilde{\theta}_i, \tilde{\theta}_i)) = \frac{1}{1 + \varepsilon} \sum_{i=1}^{m} |A^{s/2}\theta_i|^2 \leq \sum_{i=1}^{m} \lambda_i^{s-1}.$$

Using again (6.42) and the fact that $v_j \in X$ we write

$$|A^{s/2}\theta_i|^2 = |A^{s/2}\xi_i|^2 + \left|\sum_{k=1}^{i} \alpha_{ik} A^{s/2} v_k\right|^2 \geq |A^{s/2}\xi_i|^2,$$

$$\sum_{i=1}^{m} |A^{s/2}\xi_i|^2 \leq \sum_{i=1}^{m} |A^{s/2}\theta_i|^2 \leq (1 + \varepsilon) \sum_{i=1}^{m} \lambda_i^{s-1}.$$

On letting $\varepsilon \to 0$, we obtain the desired result. □

7. The Ginzburg–Landau Equation

We continue the analysis of the Ginzburg–Landau equation studied in Section IV.5; hereafter the notations and hypotheses are the same as in that section.

7.1. The Linearized Equation

Writing equation IV.(5.6) in the form

$$u' = F(u), \tag{7.1}$$

as in V.(2.38), we see that the first variation equation V.(2.39)

$$U' = F'(u)U \tag{7.2}$$

is equivalent to

$$U' + (\lambda + i\alpha)AU + (\kappa + i\beta)\{|u|^2 U + 2\,\mathrm{Re}(\bar{u}U)\} - \gamma U = 0. \tag{7.3}$$

This is the abstract form of the boundary-value problem consisting of the evolution equation (similar to IV.(5.1))

$$\frac{\partial U}{\partial t} - (\lambda + i\alpha)\Delta U + (\kappa + i\beta)\{|u|^2 U + 2u \operatorname{Re}(\bar{u}U)\} - \gamma U = 0, \tag{7.4}$$

supplemented by one of the boundary conditions IV.(5.3a)–(5.3c), of course, the same as that satisfied by u. The associated initial condition is

$$U(0) = \xi. \tag{7.5}$$

The precise relations of the linear problem with the dynamical system are the following:

(i) If u is the solution of IV.(5.6), (5.7) given by Theorem IV.5.1, then by application of Theorem II.3.4 it is easy to see that for every ξ given in H, (7.3)–(7.5) possesses a unique solution U satisfying

$$U \in L^2(0, T; V) \cap L^\infty(0, T; H), \qquad \forall T > 0. \tag{7.6}$$

(ii) The semigroup $S(t)$ is Fréchet differentiable in H

> For every $t > 0$, the function $u_0 \to S(t)u_0$ is Fréchet differentiable in H; its differential $L(t, u_0)$ at u_0 is the linear operator: $\xi \in H \to U(t) \in H$, where $U(\cdot)$ is the solution of (7.3)–(7.5). (7.7)

The differentiability property is established in Section 8.

7.2. Dimension of the Attractor

We now estimate the dimension of the maximal attractor $\mathscr{A}$ provided by Theorem IV.5.2. We follow the method summarized in Section V.3.4, with some slight differences due to the fact that here the function u is complex valued; as usual, we can rewrite the complex equation as a system of real equations and adapt the results accordingly.

For $m \in \mathbb{N}$, we consider $\xi = \xi_1, \ldots, \xi_m$, m elements of H, and the corresponding solutions $U = U_1, \ldots, U_m$ of (7.3)–(7.5); $u = u(\tau) = S(\tau)u_0$ is a fixed orbit. According to Section V.2.3 we have

$$|U_1(t) \wedge \cdots \wedge U_m(t)|_{\bigwedge^m H}$$
$$= |\xi_1 \wedge \cdots \wedge \xi_m|_{\bigwedge^m H} \exp \int_0^t \operatorname{Re} \operatorname{Tr} F'(S(\tau)v_0) \circ Q_m(\tau)\, d\tau,$$

where $Q_m(\tau) = Q_m(\tau, u_0; \xi_1, \ldots, \xi_m)$ is the orthogonal projector in H onto the space spanned by $U_1(\tau), \ldots, U_m(\tau)$.

At a given time τ, let $\varphi_j(\tau)$, $j \in \mathbb{N}$, be an orthonormal basis of H such that $Q_m(\tau)H = \operatorname{Span}[U_1(\tau), \ldots U_m(\tau)]$; since $U_j(\tau) \in V$ for a.e. τ, $\varphi_1(\tau), \ldots, \varphi_m(\tau)$

belong to V for a.e. τ. Furthermore, since $H = \mathbb{L}^2(\Omega) = L^2(\Omega)^2$, the $\varphi_j(\tau)$ considered as vectors of $L^2(\Omega)^2$ are orthogonal too.

We have

$$\begin{aligned}\operatorname{Re}\operatorname{Tr} F'(u(\tau)) \circ Q_m(\tau) &= \sum_{j=1}^{\infty} \operatorname{Re}(F'(u(\tau)) \circ Q_m(\tau), \varphi_j(\tau), \varphi_j(\tau)) \\ &= \sum_{j=1}^{m} \operatorname{Re}(F'(u(\tau))\varphi_j(\tau), \varphi_j(\tau)),\end{aligned} \tag{7.8}$$

where $(\cdot, \cdot)$ is the scalar product in $H = \mathbb{L}^2(\Omega)$. Omitting temporarily the variable τ, we see that

$$\begin{aligned}\operatorname{Re}(F'(u)\varphi_j, \varphi_j) = &-\operatorname{Re}(\lambda + i\alpha) \int_\Omega \nabla\varphi_j \nabla\bar{\varphi}_j \, dx \\ &-\operatorname{Re}(\kappa + i\beta) \int_\Omega \{|u|^2\varphi_j + 2u \operatorname{Re}(\bar{u}\varphi_j\}\bar{\varphi}_j \, dx + \gamma \int_\Omega \varphi_j\bar{\varphi}_j \, dx \\ = &-\lambda\|\varphi_j\|^2 - \kappa \int_\Omega |u|^2|\varphi_j|^2 \, dx \\ &+ 2\int_\Omega \operatorname{Re}(\bar{u}\varphi_j)\{\beta \operatorname{Im}(u\bar{\varphi}_j) - \kappa \operatorname{Re}(u\bar{\varphi}_j)\} \, dx + \gamma|\varphi_j|^2.\end{aligned}$$

Since $|\varphi_j| = 1$, the family φ_j being orthonormal, and

$$\int_\Omega \beta \operatorname{Re}(\bar{u}\varphi_j) \operatorname{Im}(u\bar{\varphi}_j) \, dx \leq |\beta| \int_\Omega |u|^2|\varphi_j|^2 \, dx,$$

we see that

$$(F(u)\varphi_j, \varphi_j) \leq -\lambda\|\varphi_j\|^2 - (\kappa - 2|\beta|) \int_\Omega |u|^2|\varphi_j|^2 \, dx + \gamma;$$

hence ($\kappa > 0$)

$$\sum_{j=1}^{m} (F'(u)\varphi_j, \varphi_j) \leq -\lambda \sum_{j=1}^{m} \|\varphi_j\|^2 + 2|\beta| \int_\Omega |u|^2\rho \, dx + \gamma m, \tag{7.9}$$

where we have set

$$\rho = \rho(x, \tau) = \sum_{j=1}^{m} |\varphi_j(x, \tau)|^2 \quad \text{a.e. } x, \tau. \tag{7.10}$$

We infer from Corollary A.4.1 the existence of two nondimensional constants c_4', c_5', depending only on n and the shape of Ω such that

$$\int_\Omega \rho^{1+2/n} \, dx \leq \frac{c_4' m}{|\Omega|^{2/n}} + c_5' \sum_{j=1}^{m} \|\varphi_j\|^2.^1 \tag{7.11}$$

[1] $c_4' = 0$ in the case of the Dirichlet boundary condition IV.(5.3a).

Thanks to the Hölder and Young inequalities

$$
\begin{aligned}
2|\beta| \int_\Omega |u|^2 \rho \, dx &\le 2|\beta| |u|^2_{L^{n+2}(\Omega)} |\rho|_{L^{1+2/n}(\Omega)}, \\
|\rho|_{L^1(\Omega)} = m &\le |\Omega|^{2/(n+2)} |\rho|_{L^{1+2/n}(\Omega)}
\end{aligned}
\tag{7.12}
$$

and we can majorize the right-hand side of (7.9) by

$$
\begin{aligned}
&-\frac{\lambda}{c_5'} |\rho|^{1+2/n}_{L^{1+2/n}(\Omega)} + \frac{\lambda c_4'}{c_5'} \frac{m}{|\Omega|^{2/n}} + 2|\beta| |u|^2_{L^{n+2}(\Omega)} |\rho|_{L^{1+2/n}(\Omega)} \\
&\le \text{(thanks to the Young inequality)} \\
&\le -\frac{\lambda}{2c_5'} |\rho|^{1+2/n}_{L^{1+2/n}(\Omega)} + \frac{\lambda c_4'}{c_5'} \frac{m}{|\Omega|^{2/n}} + c_6' |\beta|^{1+n/2} \lambda^{-n/2} |u|^{n+2}_{L^{n+2}(\Omega)} \\
&\le \text{(with (7.12))} \\
&\le -\frac{\lambda}{2c_5'} \frac{m^{1+2/n}}{|\Omega|^{2/n}} + \frac{\lambda c_4'}{c_5'} \frac{m}{|\Omega|^{2/n}} + c_6' |\beta|^{1+n/2} \lambda^{-n/2} |u|^{n+2}_{L^{n+2}(\Omega)} \\
&\le -\frac{\lambda}{4c_5'} \frac{m^{1+2/n}}{|\Omega|^{2/n}} + 2c_7' \frac{\lambda}{|\Omega|^{2/n}} + c_6' |\beta|^{1+n/2} \lambda^{-n/2} |u|^{n+2}_{L^{n+2}(\Omega)}
\end{aligned}
$$

Finally,

$$
\begin{aligned}
\operatorname{Re} \operatorname{Tr} F'(S(\tau)u_0) \circ Q_m(\tau) \le &- c_8' \frac{\lambda}{|\Omega|^{2/n}} m^{1+2/n} \\
&+ c_9' \frac{\lambda}{|\Omega|^{2/n}} \left(1 + \left(\frac{|\beta|}{\lambda} \right)^{1+n/2} |\Omega|^{2/n} |u(\tau)|^{n+2}_{L^{n+2}(\Omega)} \right).
\end{aligned}
\tag{7.13}
$$

Assuming now that u_0 belongs to the global attractor $\mathscr{A}$, we can majorize the quantity

$$
q_m = \limsup_{t \to \infty} \operatorname*{Sup}_{\substack{\xi_i \in H \\ |\xi_i| \le 1 \\ i=1,\dots,m}} \frac{1}{t} \int_0^t \operatorname{Re} \operatorname{Tr} F'(S(\tau)u_0) \circ Q_m(\tau) \, d\tau
$$

as follows:

$$
\begin{aligned}
&q_m \le -\kappa_1 m^{1+2/n} + \kappa_2, \\
&\kappa_1 = c_8' \frac{\lambda}{|\Omega|^{2/n}}, \qquad \kappa_2 = c_9' \frac{\lambda}{|\Omega|^{2/n}} \left(1 + \left(\frac{|\beta|}{\lambda} \right)^{1+n/2} |\Omega|^{2/n} \delta \right), \\
&\delta = \limsup_{t \to \infty} \operatorname*{Sup}_{u_0 \in \mathscr{A}} \frac{1}{t} \int_0^t |S(\tau)u_0|^{n+2}_{L^{n+2}(\Omega)} \, dx.
\end{aligned}
\tag{7.14}
$$

We infer from (7.14) and V.(3.49) the following bound on the uniform

Lyapunov exponents $\mu_j, j \in \mathbb{N}$, associated with $\mathscr{A}$:

$$\mu_1 + \cdots + \mu_j \le q_j \le -\kappa_1 j^{1+2/n} + \kappa_2, \qquad \forall j \in \mathbb{N}. \tag{7.15}$$

Using Lemma 2.2, we see that if m is defined by

$$m - 1 < \left(\frac{2\kappa_2}{\kappa_1}\right)^{n/(n+2)} \le m, \tag{7.16}$$

then $\mu_1 + \cdots + \mu_m < 0$ and

$$\frac{(\mu_1 + \cdots + \mu_j)_+}{|\mu_1 + \cdots + \mu_m|} \le 1, \qquad \forall j = 1, \ldots, m-1. \tag{7.17}$$

By application of Proposition V.2.1 and Theorem V.3.3 we have thus proved

Theorem 7.1. *We consider the dynamical system associated with the Ginzburg–Landau equation* IV.(5.1) *with boundary condition* IV.(5.3a), (5.3b), *or* (5.3c). *We denote by* $\mathscr{A}$ *the corresponding maximal attractor whose existence is stated in Theorem* IV.5.2, *and let m be defined by* (7.16). *Then*:

(i) *the uniform Lyapunov exponents* μ_j *are majorized as in* (7.15), $\forall j \in \mathbb{N}$;
(ii) *the m-dimensional volume element in the phase space H is exponentially decaying as* $t \to \infty$;
(iii) *the Hausdorff dimension of* $\mathscr{A}$ *is less than or equal to m, and its fractal dimension is less than or equal to 2m.*

Remark 7.1. The relation (7.16) defining m also reads

$$m - 1 < c'_{10}\left(1 + \left(\frac{|\beta|}{\lambda}\right)^{1+n/2} |\Omega|^{2/n}\delta\right)^{n/(n+2)} \le m. \tag{7.18}$$

A more explicit evaluation of m in terms of the data depends on a more explicit evaluation of δ. If $n = 1$, 2, or 3, then by the Sobolev embeddings, $H^1(\Omega) \subset L^{n+2}(\Omega)$, and we can write the Sobolev inequality in the form

$$|\varphi|_{L^{n+2}(\Omega)} \le c'_{11}\left(\|\varphi\| + \frac{1}{|\Omega|^{2/n}}|\varphi|\right), \qquad \forall \varphi \in H^1(\Omega),$$

where the nondimensional constant c'_{11} depends on the shape of Ω but not on its size.

According to the results in Section IV.5.2 (existence of absorbing sets),

$$\operatorname*{Sup}_{\varphi \in \mathscr{A}} |\varphi| \le \operatorname*{Sup}_{\varphi \in \mathscr{B}_0} |\varphi| \le \rho'_0,$$

$$\operatorname*{Sup}_{\varphi \in \mathscr{A}} \|\varphi\| \le \operatorname*{Sup}_{\varphi \in \mathscr{B}_1} \|\varphi\| \le \rho_1,$$

ρ'_0, ρ_1 given in IV.(5.22) and IV.(5.32). Hence

$$\delta \le c'_{11}\left(\rho_1 + \frac{1}{|\Omega|^{2/n}}\rho'_0\right).$$

8. Differentiability of the Semigroup

In this last section of Chapter VI, we show how to prove the various results of differentiability of the semigroup $S(t)$ which have been used in the previous sections. The proofs are only sketched.

Differentiability in H

Let H be a Hilbert space (scalar product $(\cdot, \cdot)$, norm $|\cdot|$), and let A be a closed linear positive self-adjoint unbounded operator in H with domain $D(A) \subset H$. We denote by V the domain $D(A^{1/2})$ of $A^{1/2}$, which is endowed with the norm

$$\begin{aligned} \|v\| &= |A^{1/2}v|, \qquad \forall v \in D(A^{1/2}), \\ &= \{(Av, v)\}^{1/2}, \qquad \forall v \in D(A). \end{aligned}$$

We are also given a nonlinear operator G which maps V into V'[1] such that

There exists $0 < \sigma_0 \leq 1$ and for every $R > 0$ there exists $k_0(R)$ such that

$$\begin{aligned} |(G(v) - G(u), v - u)| &\leq k_0(R)|v - u|^{\sigma_0}\|v - u\|^{2-\sigma_0}, \\ &\forall u, v \in V, \quad |u| \leq R, \quad |v| \leq R. \end{aligned} \tag{8.1}$$

If $u, v \in L^2(0, T; V) \cap L^\infty(0, T; H)$, then for a.e. $t \in (0, T)$, $G(v(t)) - G(u(t)) = l_0(t)\cdot(v(t) - u(t)) + l_1(t; v(t) - u(t))$, where $l_0(t) \in \mathscr{L}(V, V')$, and

(i) $\|l_0(t)\|_{\mathscr{L}(V,V')} \leq N_T$;
(ii) for some $0 < \varepsilon \leq 1$,

$$|(l_0(t)\varphi, \varphi)| \leq (1 - \varepsilon)\|\varphi\|^2 + C_\varepsilon|\varphi|^2, \qquad \forall \varphi \in V;$$

(iii) there exists $\sigma_1 > 0$ and $k_0' > 0$ such that

$$|l_1(t; v(t) - u(t))|_{-1} \leq k_0'\|v(t) - u(t)\|^{1+\sigma_1}. \tag{8.2}$$

Now let u, v belong to $L^2(0, T; V) \cap L^\infty(0, T; H)$ and satisfy

$$\frac{du}{dt} + Au + G(u) = 0, \qquad u(0) = u_0, \tag{8.3}$$

$$\frac{dv}{dt} + Av + G(v) = 0, \qquad v(0) = v_0. \tag{8.4}$$

First we show a Lipschitz property of the mapping $S(t)\colon u_0 \to u(t)$. We set $w = v - u$, and choose R such that

$$R \geq |u|_{L^\infty(0,T;H)}, \qquad R \geq |v|_{L^\infty(0,T;H)},$$

and consider the corresponding constant $k_0 = k_0(R)$ from (8.1).

[1] $V' = D(A^{-1/2})$ is endowed with the norm $|v|_{-1} = |A^{-1/2}v|$.

The difference $w = v - u$ satisfies the equation

$$\frac{dw}{dt} + Aw + G(v) - G(u) = 0. \tag{8.5}$$

Hence

$$\begin{aligned}\frac{1}{2}\frac{d}{dt}|w|^2 + \|w\|^2 &= -(G(v) - G(u), w)\\ &\leq k_0|w|^{\sigma_0}\|w\|^{2-\sigma_0}\\ &\leq \text{(with the Young inequality)}\\ &\leq \tfrac{1}{2}\|w\|^2 + \frac{c_1'}{2}k_0^{2/\sigma_0}|w|^2,\end{aligned}$$

$$\frac{d}{dt}|w|^2 + \|w\|^2 \leq c_1' k_0^{2/\sigma_0}|w|^2. \tag{8.6}$$

From (8.6) we easily deduce that

$$|u(t) - v(t)|^2 \leq |u_0 - v_0|^2 \exp(c_1' k_0^{2/\sigma_0} T), \qquad \forall t \in (0, T), \tag{8.7}$$

$$\int_0^t \|u(t) - v(t)\|^2\, dt \leq |u_0 - v_0|^2 \exp(c_1' k_0^{2/\sigma_0} T), \qquad \forall t \in (0, T). \tag{8.8}$$

Now let us consider the "linearized equation"

$$\frac{dU}{dt} + AU + l_0(t)U = 0, \qquad U(0) = \xi = v_0 - u_0. \tag{8.9}$$

Owing to the assumptions (8.2)(i), (ii), we are able to apply Theorem II.3.4 which provides the existence and uniqueness of a solution $U \in L^2(0, T; V) \cap L^\infty(0, T; H)$ for (8.9). Let

$$\varphi = v - u - U = w - U.$$

Clearly, φ satisfies

$$\frac{d\varphi}{dt} + A\varphi + l_0(t)\cdot\varphi = -l_1(t; w(t)), \qquad \varphi(0) = 0, \tag{8.10}$$

and taking the scalar product of (8.10) with φ we find

$$\frac{1}{2}\frac{d}{dt}|\varphi|^2 + \|\varphi\|^2 + (l_0(t)\varphi, \varphi) = -(l_1(t; w), \varphi),$$

$$\begin{aligned}\frac{1}{2}\frac{d}{dt}|\varphi|^2 + \varepsilon\|\varphi\|^2 &\leq C_\varepsilon|\varphi|^2 + k_0'\|w(t)\|^{1+\sigma_1}\|\varphi(t)\|\\ &\leq C_\varepsilon|\varphi|^2 + \frac{\varepsilon}{2}\|\varphi(t)\|^2 + \frac{(k_0')^2}{2\varepsilon}\|w(t)\|^{2(1+\sigma_1)},\end{aligned}$$

$$\frac{d}{dt}|\varphi|^2 + \varepsilon\|\varphi\|^2 \le C_\varepsilon|\varphi|^2 + \frac{(k_0')^2}{2\varepsilon}\|w(t)\|^{2(1+\sigma_1)}, \tag{8.11}$$

$$\frac{d}{dt}|\varphi|^2 \le C_\varepsilon|\varphi|^2 + \frac{(k_0')^2}{2\varepsilon}\|w(t)\|^{2(1+\sigma_1)}. \tag{8.12}$$

We apply the Gronwall lemma to (8.12) and find, since $\varphi(0) = 0$,

$$\begin{aligned} |\varphi(t)|^2 &\le \frac{(k_0')^2}{2\varepsilon}\int_0^T \|w(s)\|^{2(1+\sigma_1)}\,ds, \qquad \forall t,\\ |\varphi(t)|^2 &\le \frac{(k_0')^2}{2\varepsilon}\exp(c_1' k_0^{2/\sigma_0}T(1+\sigma_1))|v_0 - u_0|^{2(1+\sigma_1)}. \end{aligned} \tag{8.13}$$

This shows that

$$\frac{|v(t) - u(t) - U(t)|^2}{|v_0 - u_0|^2} \le \text{const}\,\|v_0 - u_0\|^{2\sigma_1} \to 0, \tag{8.14}$$

as $v_0 \to u_0$ and this proves that the mapping

$$S(t)\colon u_0 \to u(t)$$

is differentiable in H, its differential at a point $u_0 \in H$ being the mapping

$$U(0) = \xi \to U(t),$$

U the solution of (8.9).

Application

We briefly show how the hypotheses (8.1), (8.2) can be verified for the equations considered in Sections III.2, 3, 4.1 (= Section 3, 4, 5.1 of this chapter). The differentiability for the semigroup in Section 6 has been studied directly in that section. The other examples in Sections 2, 5.2, and 7 necessitate some slight modifications of the proof which are left as an exercise for the reader.

For the examples in Sections 3, 4, and 5.1 we have

$$G(u) = B(u, u) + Ru - f,$$

with B, R satisfying assumptions III.(3.3), (3.4), (3.6)–(3.9). Therefore

$$\begin{aligned} G(v) - G(u) &= B(v, w) + B(w, u) + Rw,\\ (G(v) - G(u), v - u) &= (B(w, u), w) + (Rw, w) \quad \text{(due to III.(3.6))},\\ |(G(v) - G(u), v - u)| &\le |(B(w, u), w)| + |(Rw, w)|\\ &\le \text{(by III. (3.4) and III.(3.7))}\\ &\le c_3\|u\|\,\|w\|^{2-\theta_3}|w|^{\theta_3} + c_2\|w\|^{1+\theta_2}|w|^{1-\theta_2}; \end{aligned}$$

hence (8.1) since $\theta_2, \theta_3 < 1$.

Then for (8.2) we set

$$l_0(t)\cdot w = B(u(t), w) + B(w, u(t)) + Rw,$$

$$l_1(t; w) = B(w, w),$$

and again (8.2)(i), (ii) follow easily from the hypotheses.

Differentiability from V into H

A stronger result is valid if we make proper assumptions on G; for instance, those corresponding to the examples of Sections 3, 4 and 5.1 are sufficient (i.e., III.(3.3), (3.4), (3.6)–(3.9)).

The principle of the proof consists in taking the scalar product of (8.10) with $A\varphi$ in H

$$\frac{1}{2}\frac{d}{dt}\|\varphi\|^2 + |A\varphi|^2 + (l_0(t)\varphi, A\varphi) = -(l_1(t; w), A\varphi).$$

After multiplication by t

$$\frac{d}{dt}(t\|\varphi\|^2) + t|A\varphi|^2 \leq \|\varphi\|^2 + -t(l_0(t)\varphi, A\varphi) - t(l_1(t; w), A\varphi).$$

Taking into account the hypotheses and, the differentiability and Lipschitz properties of $S(t)$ which are available, we arrive at a differential inequality

$$\frac{d}{dt}(t\|\varphi\|^2) \leq \|v_0 - u_0\|^{2+\sigma_4} k_3(t),$$

where k_3 is a positive function in $L^1(0, T)$ and $\sigma_4 > 0$. It follows immediately that

$$\frac{\|v(t) - u(t) - U(t)\|^2}{|v_0 - u_0|^2} \leq \frac{1}{t}|v_0 - u_0|^{\sigma_4} \int_0^T k_3(s)\, ds \to 0 \quad \text{as} \quad |v_0 - u_0| \to 0,$$

and this proves that $S(t)$ is Fréchet differentiable from H into V.

CHAPTER VII

Non-Well-Posed Problems, Unstable Manifolds, Lyapunov Functions, and Lower Bounds on Dimensions

Introduction

Three different topics are addressed in this chapter:

The first one is that of non-well-posed problems.

This corresponds to the case where the semigroup $S(t)$ is not defined everywhere, or is not defined for all times t. In that case, a concept of functional-invariant sets can be introduced by modifying slightly that given in Chapter I. Of course, we can no longer expect to prove the existence of absorbing sets and attractors, and the results of Chapter I do not apply anymore. However, the results of Chapter V concerning the dimensions of invariant sets have been proved at a level of generality which makes them applicable to such situations. Some examples are given without any attempt at generality.

The second topic is that of Lyapunov functions.

When a Lyapunov function exists, the dynamics is simple since, as $t \to \infty$, all the orbits converge to a stationary solution. It appears that in general the universal attractor has a simple structure and consists of the stationary solutions and their unstable manifolds. Here too some examples are presented without any attempt of generality.

The third subject is that of unstable manifolds.

We prove or recall without proofs a certain number of results providing a more precise description of the unstable manifold of a stationary point. In particular, it is shown that under appropriate assumptions, it is a smooth manifold; sometimes its Euclidean dimension (related to the number of unstable eigenvalues of the linearized operator) can also be estimated. When this happens we obtain a lower bound on the dimension of the attractor.

The chapter is organized as follows. Sections 1 and 2 relate to the first topic. In Section 1, we give, for some specific equations, a relation between well posedness, dissipativity, and the existence of a maximal attractor. Section 2 contains an explicit application of the results and methods of Chapter V to a non-well-posed problem. Sections 3–5 are devoted to the second topic. In Section 3 we study the structure of the unstable manifold of a hyperbolic fixed point and derive some consequences for the attractor. Section 4 gives some results on the structure of the global attractor of a semigroup possessing a Lyapunov function. Finally, Section 5 shows, for example, how the results of Sections 3 and 4 can be used to derive lower bounds on dimension of attractors.

PART A: NON-WELL-POSED PROBLEMS

1. Dissipativity and Well Posedness

It was indicated earlier (see, e.g., Remark I.1.3(ii)) that the existence of an absorbing set for an evolution equation was an aspect of the dissipative character of the equation, and in the case of differential equations, this was actually the definition of dissipativity. More complex situations arise in infinite dimensions, where sometimes the theory of existence and uniqueness of solutions is not fully understood: for some systems, considered as dissipative from the physical point of view, we are unable to prove the existence of absorbing sets and consequently of attracting sets. Our aim in Sections 1 and 2 is to present some limited results that we can nevertheless derive in such situations.

The present section contains the definition of invariant sets and attractors when the given semigroup $S(t)$ is not defined everywhere (Section 1.1) and, for some specific equations, a fundamental relation between the well posedness of the initial-value problem and the existence of an absorbing set and a maximal attractor. The result that we obtain for those equations can be stated in the following schematic way: if the initial-value problem is well posed for all initial data, then there exists an absorbing set and a maximal attractor. Section 1.2 contains the description of the class of equations considered. Section 1.3 presents the main result.

1.1. General Definitions

We consider a Hilbert space H and a family of operators $S(t)$ which map some part of H into H and are not necessarily defined for all times $t \geq 0$.

More precisely, the framework is the following: we are given another Hilbert space W, $W \subset H$, with a continuous injection and we assume that for every $t \geq 0$, $S(t)$ is defined and continuous from some part $\mathscr{D}(S(t))$ of W

into W

$$\begin{cases} \mathscr{D}(S(t)) = \bigcup_{\rho>0} \mathscr{D}_\rho(S(t)), \\ \mathscr{D}_\rho(S(t)) = \{u_0 \in W, \|S(\tau)u_0\|_W \leq \rho, 0 \leq \tau \leq t\}; \end{cases} \tag{1.1}$$

$\mathscr{D}(S(t))$ is the domain of $S(t)$ in W and when the dependence on W must appear we write $\mathscr{D}(S(t); W)$, $\mathscr{D}_\rho(S(t); W)$. It is clear that

$$\mathscr{D}(S(t)) \subset \mathscr{D}(S(t')), \qquad \mathscr{D}_\rho(S(t)) \subset \mathscr{D}(S(t')) \quad \text{for} \quad t \geq t',$$

and if $u_0 \in W$ we will say that $S(t)u_0$ is defined if and only if $u_0 \in \mathscr{D}(S(t))$. Hence $S(\tau)u_0$ is defined and $\|S(\tau)u_0\|_W$ is bounded for $0 \leq \tau \leq t$.

Obviously, we require that the semigroup properties similar to I.(1.1) be satisfied in the following sense:

$$S(0) = I, \qquad \mathscr{D}(S(0)) = W, \tag{1.2}$$

$$S(t+s) = S(t)S(s) \quad \text{on } \mathscr{D}(S(t+s)), \qquad \forall s, t \geq 0. \tag{1.3}$$

Given these general hypotheses we now extend Definition I.1.1 of a functional-invariant set as follows:

Definition 1.1. A set $X \subset W$ is a functional-invariant set for the semigroup $S(t)$ if

$$S(t)u_0 \text{ exists}, \qquad \forall u_0 \in X, \quad \forall t \geq 0, \tag{1.4}$$

$$S(t)X = X, \qquad \forall t \geq 0. \tag{1.5}$$

If X is a functional-invariant set and if the operators $S(t)$ are injective (backward uniqueness), relation (1.5) implies that $S(-t)$ is defined on X, $\forall t \geq 0$, and

$$S(t)X = X, \qquad \forall t \in \mathbb{R}. \tag{1.6}$$

Of course, an attractor (in W) can be defined in a similar manner

Definition 1.2. An attractor is a set $\mathscr{A} \subset W$ which enjoys the following properties:

(i) $\mathscr{A}$ is an invariant set (in the sense of Definition 1.1);
(ii) $\mathscr{A}$ possesses an open neighborhood $\mathscr{U} \subset W$ such that, for every $u_0 \in \mathscr{U}$, $S(t)u_0 \to \mathscr{A}$ in H as $t \to \infty$.

1.2. The Class of Problems Studied

We will develop our study for a class of problems similar to those of Section III.3. The initial-value problem is formally written as in III.(3.9), (3.10)

$$\frac{du}{dt} + Au + B(u) + Ru = f, \tag{1.7}$$

$$u(0) = u_0. \tag{1.8}$$

The general assumptions on A and H are the usual ones: we are given two Hilbert spaces V and H whith $V \subset H$, V dense in H, the injection of V in H being *compact*. The scalar product and the norm in V and H are denoted $((\cdot, \cdot))$, $\|\cdot\|$, $(\cdot, \cdot)$, $|\cdot|$. We can identify H with a subspace of V' (= the dual of V) and then $V \subset H \subset V'$ where the injections are continuous and each space is dense in the following one.

We consider a symmetric bilinear continuous form on V, $a(u, v)$, which is coercive

$$a(u, u) \geq \alpha \|u\|^2, \qquad \forall u \in V \quad (\alpha > 0). \tag{1.9}$$

We associate with a the linear operator A which is an isomorphism from V into V' and can be considered alternatively as a linear unbounded self-adjoint operator in H with the domain

$$D(A) = \{u \in V, Au \in H\}.$$

We have

$$D(A) \subset V \subset H \subset V', \tag{1.10}$$

where the injections are continuous and each space is dense in the following one.

We are given a linear continuous operator R from V into V' which maps $D(A)$ into H and such that there exists θ_1, $\theta_2 \in [0, 1[$ and two positive constants c_1, c_2 such that

$$|Ru| \leq c_1 \|u\|^{1-\theta_1} |Au|^{\theta_1}, \qquad \forall u \in D(A), \tag{1.11}$$

$$|(Ru, u)| \leq c_2 \|u\|^{1+\theta_2} |u|^{1-\theta_2}, \qquad \forall u \in V, \tag{1.12}$$

$$a(u, u) + (Ru, u) \geq \alpha' \|u\|^2, \qquad \forall u \in V \quad (\alpha' > 0). \tag{1.13}$$

These hypotheses on R are exactly the same as in Section III.3.1, but the assumptions satisfied by B are only part of those of Section III.3.1: we are given a bilinear continuous operator B mapping $V \times V$ into V' and $D(A) \times D(A)$ into H and such that

$$(B(u, v), v) = 0, \qquad \forall u, v \in V, \tag{1.14}$$

$$|(B(u, v), w)| \leq c_3 \|u\| \, \|v\| \, \|w\|^{\theta_3} |w|^{1-\theta_3}, \qquad \forall u, v, w \in V, \tag{1.15}$$

$$|B(u, v)| + |B(v, u)| \leq c_4 \|u\| \, \|v\|^{1-\theta_4} |Av|^{\theta_4}, \qquad \forall u \in V, \quad \forall v \in D(A), \tag{1.16}$$

where c_3, c_4 are appropriate constants and $\theta_i \in [0, 1[$, $i = 3, 4$.

We set $B(u) = B(u, u)$ and for f and u_0 given in H we consider the initial-value problem (1.7), (1.8).

We have two relevant results concerning the existence and uniqueness of solutions of (1.7), (1.8): one is related to the existence of weak solutions which may not be unique and thus do not lead to the existence of a semigroup in H;

however, they will be useful as an intermediate tool. The second result is related to the concept of strong solutions which are not necessarily defined for all times and lead to a semigroup $S(t)$ which is not everywhere defined as in Section 1.1.

Theorem 1.1. *Under the above assumptions, for f and u_0 given in H, there exists a (not necessarily unique) solution u of* (1.7), (1.8) *satisfying*

$$u \in L^2(0, T; V) \cap L^\infty(0, T; H), \qquad \forall T > 0. \tag{1.17}$$

If, furthermore, $u_0 \in V$, $\|u_0\| \le R$, then there exists $T_1 = T_1(R) > 0$ depending on R and the other data, and there exists on $[0, T_1]$ a unique strong solution u of (1.7), (1.8)

$$u \in L^2(0, T_1; D(A)) \cap \mathscr{C}([0, T_1]; V). \tag{1.18}$$

The proof of Theorem 1.1 follows closely that of Theorems III.2.1 and III.3.1 and we will omit it. Nevertheless, we will make explicit the a priori estimates on which the proof is based since they will be used subsequently.

Remark 1.1. Let us also observe that Theorem 1.1 allows the definition of an operator $S(t)$ as in Section 1.1 with W replaced by V: this is the mapping

$$S(t)\colon u_0 \in V \to u(t) \in V,$$

whenever this mapping is defined. More precisely, Theorem 1.1 shows that the ball of V centered at 0 of radius M is included in $\mathscr{D}(S(T_1(M)))$. A more explicit result below (see (1.29)) shows that this ball is even included in $\mathscr{D}_\rho(S(T_1(M)))$, $\rho = 2(1 + M)$.

A Priori Estiamtes

We first take the scalar product of (1.7) with u in H; thanks to (1.13), (1.14) we find

$$\begin{gathered} \frac{1}{2}\frac{d}{dt}|u|^2 + a(u, u) + (Ru, u) = (f, u), \\ \frac{1}{2}\frac{d}{dt}|u|^2 + \alpha' \|u\|^2 \le |f|\,|u|. \end{gathered} \tag{1.19}$$

Let λ_1 denote the first eigenvalue of A, so that

$$\begin{cases} |\varphi| \le \lambda_1^{-1/2} \|\varphi\|, & \forall \varphi \in V, \\ \|\varphi\| \le \lambda_1^{-1/2} |A\varphi|, & \forall \varphi \in D(A). \end{cases} \tag{1.20}$$

We can majorize the right-hand side of (1.19) by

$$\lambda_1^{-1/2} |f|\, \|u\| \le \frac{\alpha'}{2} \|u\|^2 + \frac{1}{2\alpha' \lambda_1} |f|^2,$$

and hence

$$\frac{d}{dt}|u|^2 + \alpha' \|u\|^2 \leq \frac{1}{\alpha' \lambda_1}|f|^2, \tag{1.21}$$

$$\frac{d}{dt}|u|^2 + \alpha' \lambda_1 |u|^2 \leq \frac{1}{\alpha' \lambda_1}|f|^2. \tag{1.22}$$

We infer from (1.22), using the Gronwall lemma, that

$$|u(t)|^2 \leq |u_0|^2 \exp(-\alpha' \lambda_1 t) + \frac{1}{(\alpha' \lambda_1)^2}|f|^2(1 - \exp(-\alpha' \lambda_1 t)), \quad \forall t \geq 0. \tag{1.23}$$

A priori inequality (1.23) is used in the proof of the first part of Theorem 1.1. We also infer from (1.23) that

$$\limsup_{t \to \infty} |u(t)| \leq \rho_0, \qquad \rho_0 = \frac{|f|}{\alpha' \lambda_1}, \tag{1.24}$$

and if we are given $\delta > 0$, $M_0 > 0$, $\rho_0' = (\rho_0^2 + \delta^2)^{1/2}$, then

$$|u(t)|^2 \leq \delta^2 + \rho_0^2 = (\rho_0')^2 \tag{1.25}$$

for every $u_0 \in H$, $|u_0| \leq M_0$, and for every $t \geq t_0$, where t_0 depends on ρ_0, δ, M_0, and the data f, A

$$t_0 = t_0(\rho_0', M_0) = \frac{1}{2\alpha' \lambda_1} \log \frac{M_0}{\delta}. \tag{1.26}$$

This amounts to the existence of an absorbing set in H if the semigroup $S(t) = u_0 \to u(t)$ were defined everywhere in H (which is not always the case).

We also have by integration of (1.21)

$$\begin{aligned} &\int_t^{t+r} \|u(\tau)\|^2 \, d\tau \leq \frac{1}{\alpha'}|u(t)|^2 + \frac{r}{\lambda_1(\alpha')^2}|f|^2, \qquad \forall t, r \geq 0, \\ &\int_t^{t+r} \|u(\tau)\|^2 \, d\tau \leq \frac{(\rho_0')^2}{\alpha'} + \frac{r}{\lambda_1(\alpha')^2}|f|^2, \qquad \forall t, r \geq 0, \quad \forall r \geq 0. \end{aligned} \tag{1.27}$$

A second useful estimate is obtained by taking the scalar product of (1.7) with $2Au$ in H

$$\frac{d}{dt}\|u\|^2 + 2|Au|^2 + 2(B(u), Au) + 2(Ru, Au) = 2(f, Au).$$

Using the Schwarz inequality (1.11) and (1.16) we write

$$\begin{aligned} 2|(B(u), Au)| &\leq 2|B(u)||Au| \\ &\leq 2c_4 \|u\|^{2-\theta_4} |Au|^{1+\theta_4} \\ &\leq \text{(with the Young inequality)} \\ &\leq \tfrac{1}{3}|Au|^2 + c_1' \|u\|^{2(2-\theta_4)/(1-\theta_4)}, \end{aligned}$$

the constant c_1' depending only on c_4 and c_1';

$$\begin{aligned} 2|Ru, Au)| &\leq 2|Ru||Au| \\ &\leq 2c_1\|u\|^{1-\theta_1}|Au|^{1+\theta_1} \\ &\leq \tfrac{1}{3}|Au|^2 + c_2'\|u\|^2, \end{aligned}$$

c_2' depending on c_1 and θ_1;

$$2(f, Au) \leq 2|f||Au| \leq \tfrac{1}{3}|Au|^2 + 3|f|^2.$$

Hence

$$\frac{d}{dt}\|u\|^2 + |Au|^2 \leq 3|f|^2 + c_1'\|u\|^{2(2-\theta_4)/(1-\theta_4)} + c_2'\|u\|^2. \tag{1.28}$$

If u is sufficiently regular so that (1.28) is valid, then for t sufficiently small

$$\|u(t)\|^2 \leq 2(1 + \|u_0\|^2), \tag{1.29}$$

and, for such values of t, (1.28) yields

$$\frac{d}{dt}(1 + \|u(t)\|^2) \leq \kappa_1(1 + \|u(t)\|^2),$$

$$\kappa_1 = 3|f|^2 + c_2' + c_1'\{2(1 + \|u_0\|^2)\}^{1/(1-\theta_4)}.$$

By integration

$$1 + \|u(t)\|^2 \leq (1 + \|u_0\|^2)\exp(\kappa_1 t),$$

and (1.29) remains valid at least for

$$t \leq T_1(\|u_0\|) = \frac{\log 2}{\kappa_1} = \frac{c_2'}{(1 + \|u_0\|^2)^{1/(1-\theta_4)}}. \tag{1.30}$$

The time $T_1(\|u_0\|)$ appearing in (1.30) is the same as in Theorem 1.1; indeed, (1.28) and the computations above guarantee that u belongs to an a priori bounded set of $L^2(0, T_1; D(A))$ and $L^\infty(0, T_1; V)$. More precisely, (1.29) shows that

$$\|u(t)\| \leq 2(1 + \|u_0\|) \quad \text{if} \quad 0 \leq t \leq T_1(\|u_0\|). \tag{1.31}$$

1.3. The Main Result

We can now proceed and state the main result which provides us with a striking relation between well posedness and the existence of an absorbing set in V.

According to the definitions in Section 1.1 and to Theorem 1.1, we will say that the initial-value problem (1.7), (1.8) is well posed in V on $[0, T]$ for some initial value $u_0 \in V$, if $u_0 \in \mathscr{D}(S(T)) = \mathscr{D}(S(T); V)$. This implies the existence (and of course the uniqueness) of a solution u of (1.7), (1.8) belonging to $\mathscr{C}([0, T]; V)$ and moreover, as we can easily check, $u \in L^2(0, T; D(A))$.

We have

Theorem 1.2. *The hypotheses are the same as in Theorem* 1.1. *If the initial-value problem* (1.7), (1.8) *is well posed in* V *for every* $u_0 \in V$ *and every time* $T > 0$, *then the corresponding semigroup* $\{S(t)\}_{t \geq 0}$ *possesses an absorbing set in* V *and a maximal attractor.*

Remark 1.2. A partial restatement of Theorem 1.2 is the following: if we know that the solution u of (1.7), (1.8) remains bounded in V for *all* u_0 and all finite intervals $[0, T]$, then we know that $\|u(t)\|$ remains bounded for all $t \geq 0$ and uniformly for u_0 in a bounded set of V.

PROOF OF THEOREM 1.2. The proof is divided into several steps; the main difficulty, of course, is to show the existence of an absorbing set in V. We first show that $\|u(t)\|$ is uniformly bounded for $u_0 \in V$, $\|u_0\| \leq M$, and $t \in [0, T]$

$$\operatorname*{Sup}_{\substack{t \in [0,T] \\ \|u_0\| \leq M}} \|u(t)\| = K_T < \infty, \qquad \forall T > 0. \tag{1.32}$$

Then we prove the existence of the absorbing set and eventually, that of the attractor.

(i) We argue by contradiction and assume that the supremum in (1.32) is not finite for some $T > 0$ and some $M > 0$. We obtain a sequence t_n, u_{0n}, $t_n \in [0, T]$, $\|u_{0n}\| \leq M$, such that

$$\|u_n(t_n)\| \to +\infty \quad \text{as} \quad n \to \infty, \tag{1.33}$$

where u_n is the solution of

$$\frac{du_n}{dt} + Au_n + B(u_n) + Ru_n = f, \tag{1.34}$$

$$u_n(0) = u_{0n}. \tag{1.35}$$

By extracting if necessary a subsequence we can assume that $t_n \to t_* \in [0, T]$ as $n \to \infty$ and u_{0n} converges weakly in V to some limit v_0, $\|v_0\| \leq M$.

Since $\|u_{0n}\| \leq M, \forall n$, it is easy to deduce from the relations analog to (1.21)–(1.23) for u_n, that u_n remains in a bounded set of $L^2(0, T; V) \cap L^\infty(0, T; H)$. Hence there exists $v \in L^2(0, T; V) \cap L^\infty(0, T; H)$ and a subsequence, still denoted n, such that, as $n \to \infty$,

$$u_n \to v \quad \text{in } L^2(0, T; V) \text{ weakly and in } L^\infty(0, T; H) \text{ weak-star.} \tag{1.36}$$

Using standard methods (similar to those used in the proofs of Theorems III.2.1, 3.1), we can pass to the limit in (1.34), (1.35). We find that v satisfies

$$\frac{dv}{dt} + Av + B(v) + Rv = f, \tag{1.37}$$

$$v(0) = v_0. \tag{1.38}$$

Now since $v_0 \in V$, (1.37), (1.38) possess, by assumption, a unique solution in $L^2(0, T; D(A)) \cap \mathscr{C}([0, T]; V)$. A strong form of the uniqueness property which is easy to check shows that there is no other solution of (1.37), (1.38) in $L^2(0, T; V) \cap L^\infty(0, T; H)$; therefore

$$v \in L^2(0, T; D(A)) \cap \mathscr{C}([0, T]; V). \tag{1.39}$$

(ii) We want to make the convergence of u_n to v more precise; this will allow us to reach a contradiction with (1.33) and prove (1.32).

The difference $w_n = u_n - v$ satisfies

$$\frac{dw_n}{dt} + Aw_n + B(w_n) + B(v, w_n) + B(w_n, v) + Rw_n = 0, \tag{1.40}$$

$$w_n(0) = u_{0n} - v_0. \tag{1.41}$$

We take the scalar product of (1.40) with w_n in H. Taking into account (1.13)–(1.15), we obtain

$$\begin{aligned} \frac{1}{2}\frac{d}{dt}|w_n|^2 + \alpha' \|w_n\|^2 &\leq -(B(w_n, v)w_n) \\ &\leq c_3 \|v\| \, \|w_n\|^{2-\theta_3} |w_n|^{\theta_3} \\ &\leq \text{(with the Young inequality)} \\ &\leq \frac{\alpha'}{2} \|w_n\|^2 + \tfrac{1}{2} c_3' \|v\|^{2/\theta_3} |w_n|^2, \end{aligned} \tag{1.42}$$

$$\frac{d}{dt}|w_n|^2 + \alpha' \|w_n\|^2 \leq c_3' \|v\|^{2/\theta_3} |w_n|^2.$$

We recall that $w_n(0) = u_{0n} - v_0$ converges weakly to 0 in V as $n \to \infty$; since the injection of V in H is compact, $w_n(0)$ converges to 0 strongly in H. Then using the Gronwall lemma we deduce from (1.42) that

$$|w_n(t)|^2 \leq |w_n(0)|^2 \exp\left(\int_0^t c_3' \|v(s)\|^{2/\theta_3} \, ds\right), \qquad 0 \leq t \leq T, \tag{1.43}$$

and therefore w_n converges to 0 in $\mathscr{C}([0, T]; H)$. Returning to (1.41) we then find that w_n also converges to 0 in $L^2(0, T; V)$ strongly.

Since $u_n - v \to 0$ strongly in $L^2(0, T; V)$ we conclude by extracting a subsequence that, for almost $s \in (0, T)$,

$$u_n(s) \to v(s) \quad \text{in } V. \tag{1.44}$$

Let us consider a particular $s = s_1$, for which (1.44) is valid. The sequence $\|u_n(s_1)\|$ is bounded and for n sufficiently large

$$\|u_n(s_1)\| \leq M_1 = 1 + \|v\|_{L^\infty(0, T; V)}.$$

Because of (1.31), for n sufficiently large,

$$\|u_n(s)\| \leq 2(1 + M_1) \quad \text{for} \quad s \in [s_1, s_1 + T_1(M_1)]. \tag{1.45}$$

Since $T_1(R_1)$ is actually independent of s_1, we can cover the whole interval

$(0, T)$ by a finite number of intervals $[s_k, s_k + T_1(M_1)]$, $k = 1, \ldots, N$, such that $u_n(s_k) \to v(s_k)$ for every k and (1.45) holds for $s \in [s_k, s_k + T_1(M_1)]$. It follows that the norm of u_n in $L^\infty(0, T; V)$ remains bounded as $n \to \infty$. This contradicts (1.33) and completes the proof of (1.32).

(iii) We fix $r > 0$ and rewrite (1.27) in the form

$$\int_t^{t+r} \|u(\tau)\|^2 \, d\tau \le \rho_1^2, \qquad \forall t \ge t_0(\rho_0'), \tag{1.46}$$

where ρ_1^2 is the right-hand side of (1.27) and t_0 is given by (1.25); $\delta > 0$, determining ρ_0' is also fixed.

We consider for $t \ge t_0$ the set of points τ in $[t, t + r]$ where $\|u(\tau)\|^2 \ge 2\rho_1^2/r$. Its measure M satisfies

$$\frac{2\rho_1^2}{r} M \le \int_t^{t+r} \|u(\tau)\|^2 \, d\tau \le \rho_1^2;$$

hence $M \le r/2$. This shows that

$$\|u(\tau)\|^2 \le 2\rho_1^2/r, \text{ on a set of measure at least } r/2 \text{ on any interval } [t, t + r],\ t \ge t_0. \tag{1.47}$$

Let $\rho_2 = \rho_1 \sqrt{2/r}$; we claim that

$$\operatorname*{Sup}_{\substack{t \ge 0 \\ \|u_0\| \le \rho_2}} \|u(t)\| = M_2 < \infty. \tag{1.48}$$

If this were not true, we could find two sequences $t_n \ge 0$, u_{0n}, $\|u_{0n}\| \le R$, such that

$$\|u_n(t_n)\| \to +\infty \quad \text{as} \quad n \to \infty, \tag{1.49}$$

where u_n is (again) the solution of (1.34), (1.35). If the sequence t_n is bounded, $t_n \le T, \forall n$, then we reach a contradiction exactly as above. Hence we consider now the case where t_n is not bounded and, after extracting a subsequence, $t_n \to \infty$ as $n \to \infty$, and we can assume that $t_n \ge t_0$ where t_0 is associated by (1.26) to ρ_0' and $M_0 = \rho_2/\lambda_1^{-1/2}$.[1]

Thanks to (1.47), for every n there exists $s_n \in (t_n - r, t_n)$ such that $\|u(s_n)\|^2 \le \rho_2^2$. By translation, we consider the functions

$$v_n(t) = u_n(t - s_n), \qquad 0 \le t \le r.$$

They satisfy

$$\frac{dv_n}{dt} + Av_n + B(v_n) + Rv_n = f \quad \text{on } (0, r), \tag{1.50}$$

$$v_n(0) = v_{0n}, \qquad \|v_{0n}\| \le \rho_2, \tag{1.51}$$

$$\|v_n(a_n)\| \to +\infty \quad \text{as} \quad n \to \infty, \qquad a_n = t_n - s_n \in (0, r). \tag{1.52}$$

[1] Note that $|u_0| \le (1/\lambda_1^{1/2})\|u_0\| \le \rho_2/\lambda_1^{1/2}$ for all the u_0 in (1.48).

This sequence v_n enjoys the same properties as the sequence u_n in (1.34), (1.35) with T simply replaced by r, M by ρ_2, and t_n by a_n. Exactly as before we obtain the contradiction with the initial assumption, by using the fact that the initial-value problem is well posed in V for any initial data belonging to V. This completes the proof of (1.48).

(iv) We are now able to conclude: the ball $\mathscr{B}_1$ of V centered at 0 of radius M_2 given by (1.48) is an absorbing set in V and it absorbs the bounded sets. Indeed, any bounded set $\mathscr{B}$ of V is included in a ball of H centered at 0 of radius M_0. Any orbit initiating in $\mathscr{B}$ enters the ball of V centered at 0 of radius ρ_2 at a time which does not exceed $t_0 + r$, t_0 given by (1.26). Thanks to (1.48), it remains in $\mathscr{B}_1$ for all subsequent times.

The existence of a maximal attractor in H, which consists of the ω-limit set of $\mathscr{B}_1$, is then given by Theorem I.1.1. A slight modification of this theorem is necessary since the operators $S(t)$ are defined only on V and not on H, but this is easy.[1]

Theorem 1.2 is proved. □

Remark 1.3. The hypotheses of Theorems 1.1 and 1.2 and thus their conclusions, are satisfied by all the equations of fluid mechanics presented in Sections III.3 and III.2, when the space dimension $n = 3$. In the case of thermohydraulics (Section III.3.5), assumption (1.13) is not satisfied but the conclusions remain valid with some slight modifications of the proofs using the maximum principle; see Section III.3.5 and C. Foias, O. Manley, and R. Temam [1]. For the three-dimensional Navier–Stokes equations with the boundary conditions considered in Section III.2, Theorem 1.1 is standard, while Theorem 1.2 was proved, with a slightly different presentation of the conclusions, in P. Constantin, C. Foias, and R. Temam [1].

Remark 1.4 (Open Problems). A similar result is probably valid for more general equations than those of Section 1.2, but we did not intend to make a systematic study. It is likely that these methods apply to the Kuramoto–Sivashinsky equations in dimension > 1, to the Cahn–Hilliard equations when the restrictions on the degree of the polynome are not satisfied or to the wave equations of Section IV.3 when IV.(3.6) is not satisfied.

2. Estimate of Dimension for Non-Well-Posed Problems: Examples in Fluid Dynamics

As indicated in the introduction of Section 1, the results of Chapter V were sufficiently general to apply to situations where the semigroup $S(t)$ is not defined or differentiable everywhere in H. Again, without any attempt at

[1] Alternatively, by proving that the oribits are bounded in $D(A)$, we can apply Theorem I.1.1 with H replaced by V and obtain in this way an attracting set in V, bounded in $D(A)$.

generality, we will show how the results can be applied to the equations of Section 1.2.

2.1. The Equations and Their Linearization

The equations are those of Section 1.2 and Theorem 1.1 applies. We consider a bounded set X of V that is a functional-invariant set in the sense of Definition 1.1. Note that if

$$M_1 = \operatorname*{Sup}_{\varphi \in X} \|\varphi\|, \tag{2.1}$$

then $S(t)$ is well defined on X by Theorem 1.1 for $0 \leq t \leq T_1(M_1)$.

The hypotheses of Section 1.2 are not sufficient to prove the existence of such an invariant set. However, we observe that if $\{u(t), t \geq 0\}$ is an orbit uniformly bounded in V

$$\operatorname*{Sup}_{t \geq 0} \|u(t)\| < \infty, \tag{2.2}$$

u the solution of (1.7), (1.8), then the ω-limit set of u_0 is indeed a bounded functional-invariant set. Also the set X can be the maximal attractor $\mathscr{A}$ given by Theorem 1.2 if this theorem is applicable, i.e., if we know that the initial-value problem (1.7), (1.8) is well posed, $\forall u_0 \in V$.[1]

We let u_0 be an arbitrary point of X and consider the corresponding orbit defined by (1.7), (1.8). We introduce the first variation equation corresponding to the linearization around this orbit. Writing (1.7) in the form

$$u' = F(u), \tag{2.3}$$

as in V.(2.38), we see that the first variation equation V.(2.39)

$$U' = F'(u)U \tag{2.4}$$

is

$$U' + AU + B(u, U) + B(U, u) + RU = 0, \tag{2.5}$$

which we supplement with the initial condition

$$U(0) = \xi. \tag{2.6}$$

Using the hypotheses in Section 1.2 and the fact that

$$u \in L^\infty(0, \infty; V), \tag{2.7}$$

it is easy to see that Theorem II.3.4 applies to the initial-value problem (2.5),

[1] It was emphasized before that the assumptions of Section 1.2 do not guarantee the well posedness.

(2.6) and

If u is a solution of (1.7), (1.8) satisfying (2.7), then for every ξ in H there exists a unique solution U of (2.5), (2.6) such that

$$U \in L^2(0, T; V) \cap L^\infty(0, T; H), \qquad \forall T > 0. \tag{2.8}$$

Now the semigroup $S(t)$ is not a priori everywhere differentiable on H. However, we have

Proposition 2.1. *For every $t_1 > 0$, $\rho > 0$, $\mathscr{D}_\rho(S(t_1))$ is open in V, and $S(t_1)$ is differentiable in $\mathscr{D}_\rho(S(t_1))$ equipped with the norm of H. The differential of $S(t_1)$ at a point u_0 of $\mathscr{D}_\rho(S(t_1))$ is the mapping*

$$\xi \in H \to L(t_1, u_0)\cdot \xi = U(t_1),$$

where U is the solution of (2.5), (2.6).

The proof of this proposition is very technical and will not be given here; the reader is referred to P. Constantin, C. Foias, and R. Temam [1] where the case $R = 0$ is fully treated; the case $R \neq 0$ necessitates only minor modifications.

It is then easy to show

Proposition 2.2. *Let X be a functional-invariant set bounded in V, $u_0 \in X$, $u(t) = S(t)u_0$. Then for every $t \geq 0$ the mapping $S(t)$ is "uniformly differentiable in X" in the sense of* V.(2.14), *where $L(t, u_0)$ is as above. Furthermore, the boundedness property* V.(2.28) *holds.*

Due to the lack of differentiability of $S(t)$, this "uniform differentiability property on X" is the essential technicality before we can fully apply the results of Chapter V and the program summarized in Section V.3.4. This difficulty being overcome by Propositions 2.1 and 2.2, we can now estimate the dimensions of X.

2.2. Estimate of the Dimension of X

Let X denote as above a functional-invariant set bounded in V and let $u_0 \in X$; $u(t) = S(t)u_0$ is the corresponding solution of (1.7), (1.8).

Let m denote an integer and consider the solutions $U = U_1, \ldots, U_m$ of (2.5), (2.6) associated with the initial data $\xi = \xi_1, \ldots, \xi_m$. We have

$$|U_1(t) \wedge \cdots \wedge U_m(t)|_{\wedge^m H} = |\xi_1 \wedge \cdots \wedge \xi_m|_{\wedge^m H} \exp \int_0^t \operatorname{Tr} F'(u(\tau)) \circ Q_m(\tau)\, d\tau,$$

where $Q_m(\tau) = Q_m(\tau, u_0; \xi_1, \ldots, \xi_m)$ is the orthogonal projector in H onto the space spanned by $U_1(\tau), \ldots, U_m(\tau)$.

At a given time τ, let $\varphi_j(\tau)$, $j \in \mathbb{N}$, be an orthonormal basis of H, such that $\varphi_1(\tau), \ldots, \varphi_m(\tau)$ span $Q_m(\tau)H = \text{Span}[U_1(\tau), \ldots, U_m(\tau)]$; $\varphi_j(\tau) \in V$ for $j = 1, \ldots, m$ for a.e. τ, since $U_1(\tau), \ldots, U_m(\tau) \in V$ for a.e. $\tau \in \mathbb{R}_+$.

We have

$$\begin{aligned} \operatorname{Tr} F'(u(\tau)) \circ Q_m(\tau) &= \sum_{j=1}^{\infty} (F'(u(\tau)) \circ Q_m(\tau)\varphi_j(\tau), \varphi_j(\tau)) \\ &= \sum_{j=1}^{m} (F'(u(\tau))\varphi_j(\tau), \varphi_j(\tau)). \end{aligned} \tag{2.9}$$

We omit temporarily the dependence on τ and using (1.13)–(1.15) we see that

$$\begin{aligned} (F'(u)(\varphi_j, \varphi_j) &= -a(\varphi_j, \varphi_j) - (B(\varphi_j, u), \varphi_j) - (R\varphi_j, \varphi_j) \\ &\leq -\alpha' \|\varphi_j\|^2 + c_3 \|u\| \, \|\varphi_j\|^{1+\theta_3} |\varphi_j|^{1-\theta_3}. \end{aligned}$$

We recall that $|\varphi_j| = 1$ and, using the Young inequality, majorize the right-hand side of this inequality by

$$-\frac{\alpha'}{2} \|\varphi_j\|^2 + c_1'(\alpha')^{(\theta_3+1)/(\theta_3-1)} \|u\|^{2/(1-\theta_3)}.$$

Let λ_j denote the sequence of eigenvalues of A,

$$0 < \lambda_1 \leq \lambda_2 \leq \ldots, \quad \lambda_j \to \infty \quad \text{as} \quad j \to \infty,$$

and let

$$\bar{\lambda}_m = \sum_{j=1}^{m} \lambda_j.$$

We can write, using Lemma VI.2.2,

$$\sum_{j=1}^{m} \|\varphi_j\|^2 \geq \bar{\lambda}_m,$$

and then

$$\begin{aligned} \operatorname{Tr} F'(u(\tau)) \circ Q_m(\tau) &= \sum_{j=1}^{m} (F'(u)\varphi_j, \varphi_j) \\ &\leq -\alpha' \sum_{j=1}^{m} \|\varphi_j\|^2 + c_2' \|u\|^{2/(1-\theta_3)} \\ &\leq -\alpha' \bar{\lambda}_m + c_2' \|u\|^{2/(1-\theta_3)}. \end{aligned}$$

If we know that, for some $\beta > 1$,

$$\bar{\lambda}_m \geq c_3' \lambda_1 m^{\beta}, \tag{2.10}$$

we obtain

$$\begin{aligned} \operatorname{Tr} F'(u(\tau)) \circ Q_m(\tau) &\leq -c_3' \alpha' \lambda_1 m^{\beta} + c_2' \|u\|^{2/(1-\theta_3)} \\ &\leq \text{(with the Young inequality)} \\ &\leq -\tfrac{1}{2} c_3' \alpha' \lambda_1 m^{\beta} + c_4' \|u\|^{2\beta'/(1-\theta_3)}, \end{aligned}$$

$\beta' = \beta/(\beta - 1)$. We can then majorize the quantities $q_m(t)$, q_m associated with X (see V.(3.48))

$$q_m(t) = \operatorname*{Sup}_{u_0 \in X} \operatorname*{Sup}_{\substack{\xi_i \in H \\ |\xi_i| \le 1 \\ i=1,\dots,m}} \frac{1}{t} \int_0^t \operatorname{Tr} F'(u(\tau)) \circ Q_m(\tau)\, d\tau,$$

$$q_m = \limsup_{t \to \infty} q_m(t).$$

We find

$$q_m \le -\kappa_1 m^\beta + \kappa_2, \tag{2.11}$$

$$\kappa_1 = \tfrac{1}{3} c_3' \alpha' \lambda_1, \qquad \kappa_2 = c_4' \gamma,$$

$$\gamma = \limsup_{t \to \infty} \operatorname*{Sup}_{u_0 \in X} \frac{1}{t} \int_0^t \|S(\tau) u_0\|^{2\beta'/(1-\theta_3)}\, d\tau.$$

This inequality provides a bound for the uniform Lyapunov exponents μ_j, $j \in \mathbb{N}$, associated with X

$$\mu_1 + \cdots + \mu_j \le q_j \le -\kappa_1 j^\beta + \kappa_2, \qquad \forall j \in \mathbb{N}. \tag{2.12}$$

Furthermore, if m is defined by

$$m - 1 < \left(\frac{2\kappa_2}{\kappa_1}\right)^{1/\beta} \le m, \tag{2.13}$$

then using Lemma VI.2.1 and Theorem V.3.3 we conclude that the Hausdorff dimension of X is less than or equal to m, and its fractal dimension is less than or equal to $2m$.

Theorem 2.1. *We consider the dynamical system defined by equation* (1.7). *The hypotheses are those of Theorems* 1.1 *and* (2.10), *and X is a functional-invariant set bounded in V. Then*:

(i) *the uniform Lyapunov exponents μ_j are majorized as in* (2.12), *and if m is defined by* (2.13),

(ii) *the Hausdorff dimension of X is less than or equal to m, and its fractal dimension is less than or equal to $2m$.*

Remark 2.1.

(i) Of course, X can be the maximal attractor $\mathscr{A}$ if Theorem 1.2 is applicable.

(ii) We cannot prove an exponential decay of the m-dimensional volume element as in Chapter VI.

2.3. The Three-Dimensional Navier–Stokes Equations

We conclude this section by considering the case of the three-dimensional Navier–Stokes equations to which the analysis above applies. Furthermore, the estimates can be made slightly sharper and they lead then to a result of particular interest as shown in P. Constantin, C. Foias, and R. Temam [1], P. Constantin, C. Foias, O. Manley, and R. Temam [1].

The equations are the same as in Section III.2, the space dimension being $n = 3$ instead of $n = 2$. In their explicit form (i.e., partial differential equation form), the equations are similar to III.(2.1), (2.2), with $\rho_0 = 1$ for simplification. These equations are supplemented by one of the boundary conditions III.(2.4), (2.5) (or III.(2.39)), and we then obtain the functional form of the equation which is similar to III.(2.18): this is the same as (1.7) with $R = 0$.

Of course, the analysis of Sections 1 and 2 applies to this case; our aim now is to give a slightly different form of the bound of the dimension m of the attractor, the new expression producing a physically relevant bound: namely, it is the same as the estimate of the number of degrees of freedom of a turbulent flow, following directly from the Kolmogorov point of view on turbulence (see, for instance, L. Landau and I.M. Lifschitz [1]).

When we estimate the trace of $\operatorname{Tr} F'(u(\tau)) \circ Q_m(\tau)$ after (2.9) we write

$$(F'(u)\varphi_j, \varphi_j) = -\nu \|\varphi_j\|^2 - (B(\varphi_j, u), \varphi_j).$$

Considering the explicit expression of B (see III.(2.15)) we write

$$(B(\varphi_j, u), \varphi_j) = \sum_{i,k=1}^{3} \int_\Omega \varphi_{ji} D_i u_k \varphi_{jk}\, dx.$$

Pointwise (for a.e. x and τ) we write

$$\left| \sum_{j=1}^{m} \sum_{i,k=1}^{3} \varphi_{ji}(x) D_i u_k(x) \varphi_{jk}(x) \right| \leq \rho(x) |\operatorname{grad} u(x)|,$$

where

$$\begin{aligned} |\operatorname{grad} u(x)| &= \left\{ \sum_{i,k=1}^{3} |D_i u_k(x)|^2 \right\}^{1/2}, \\ \rho(x) &= \sum_{j=1}^{m} |\varphi_j(x)|^2 = \sum_{j=1}^{m} \sum_{i=1}^{3} |\varphi_{ji}(x)|^2. \end{aligned} \tag{2.14}$$

Hence

$$\left| \sum_{j=1}^{m} (B(\varphi_j, u), \varphi_j) \right| \leq \int_\Omega \rho(x) |\operatorname{grad} u(x)|\, dx. \tag{2.15}$$

Of course, all these quantities depend on τ, but we omit the variable τ for the moment.

We infer from Theorem 4.1 of the Appendix, the existence of a constant c_1' depending only on the shape of Ω, such that (see A.(4.11))

$$\int_\Omega \rho(x)^{5/3}\, dx \leq c_1' \sum_{j=1}^{m} \|\varphi_j\|^2. \tag{2.16}$$

Then, the right-hand side of (2.15) can be majorized by

$$|\rho|_{L^{5/3}(\Omega)}|\text{grad } u|_{L^{5/2}(\Omega)} \leq \left\{c'_1 \sum_{j=1}^m \|\varphi_j\|^2\right\}^{3/5} |\text{grad } u|_{L^{5/2}(\Omega)}$$
$$\leq \text{(thanks to the Young inequality)}$$
$$\leq \frac{\nu}{2} \sum_{j=1}^m \|\varphi_j\|^2 + \frac{c'_2}{\nu^{3/2}} |\text{grad } u|^{5/2}_{L^{5/2}(\Omega)}.$$

Hence

$$\text{Tr } F'(u) \circ Q_m = \sum_{j=1}^m (F'(u)\varphi_j, \varphi_j)$$
$$\leq -\frac{\nu}{2} \sum_{j=1}^m \|\varphi_j\|^2 + \frac{c'_2}{\nu^{3/2}} |\text{grad } u|^{5/2}_{L^{5/2}(\Omega)}. \tag{2.17}$$

Thanks to the Hölder inequality and (2.16), and since the functions φ_j are orthonormal in H ($\subset L^2(\Omega)^3$),

$$\begin{gathered} m = \int_\Omega \rho(x)\, dx \leq |\Omega|^{2/5} |\rho|_{L^{5/3}}, \\ m^{5/3} \leq c'_1 |\Omega|^{2/3} \sum_{j=1}^m \|\varphi_j\|^2. \end{gathered} \tag{2.18}$$

Note that (2.18) also follows from Corollary 4.1 of the Appendix.

Taking into account (2.18), (2.17) then yields

$$\text{Tr } F'(u(\tau)) \circ Q_m(\tau) \leq -\frac{\nu m^{5/3}}{2c'_1 |\Omega|^{2/3}} + \frac{c'_2}{\nu^{3/2}} |\text{grad } u(\tau)|^{5/2}_{L^{5/2}(\Omega)}.$$

Then we consider the expressions $q_m(t)$, q_m and instead of (2.11) we find

$$\begin{gathered} q_m \leq -\kappa_1 m^{5/3} + \kappa_2, \\ \kappa_1 = \frac{\nu}{2c'_1 |\Omega|^{2/3}}, \qquad \kappa_2 = \frac{c'_2 |\Omega|}{\nu^{11/4}} \varepsilon^{5/4}, \end{gathered} \tag{2.19}$$

where

$$\varepsilon = \nu \limsup_{t\to\infty} \operatorname*{Sup}_{u_0 \in X} \left\{\frac{1}{|\Omega| t} \int_0^t \int_\Omega |\text{grad } u(x, \tau)|^{5/2}\, dx\, d\tau\right\}^{4/5}. \tag{2.20}$$

Then Theorem 2.1 holds without any modification, except that we replace the definition of m in (2.13) by

$$m - 1 < c'_3 |\Omega| \left(\frac{\varepsilon}{\nu^3}\right)^{3/4} \leq m \qquad (c'_3 = (4c'_1 c'_2)^{3/5}). \tag{2.21}$$

In three-dimensional turbulence theory we consider the local rate of dissipation of energy defined by

$$\varepsilon(x, \tau) = \nu |\text{grad } u(x, \tau)|^2.$$

The quantity ε in (2.20) is an average of $\varepsilon(x, \tau)$ on the attractor, which is very similar to the statistical (ensemble) averages considered in the conventional theory of turbulence. In this theory the Kolmogorov dissipation length is defined as

$$L_d = \left(\frac{\nu^3}{\varepsilon}\right)^{1/4}. \tag{2.22}$$

We can also define a macroscopical length by setting $L_0 = |\Omega|^{1/3}$. This allows us to rewrite (2.21) as

$$m - 1 < c_3' \left(\frac{L_0}{L_d}\right)^3 \leq m. \tag{2.23}$$

Hence

Theorem 2.2. *Theorems* 1.1 *and* 2.1 *apply to the three-dimensional Navier–Stokes equations. In Theorem* 2.1, *the definition* (2.13) *of m is to be replaced by* (2.23).

Remark 2.2. The conclusions of Theorem 2.2 could be slightly improved if we replace ε by the following smaller number $\bar{\varepsilon}$ corresponding to another average:[1]

$$\bar{\varepsilon} = \nu \limsup_{t \to \infty} \operatorname*{Sup}_{u_0 \in X} \frac{1}{|\Omega| t} \int_0^t \int_\Omega |\operatorname{grad} u(x, \tau)|^2 \, dx \, d\tau.$$

PART B: UNSTABLE MANIFOLDS, LYAPUNOV FUNCTIONS, AND LOWER BOUNDS ON DIMENSIONS

In part B of this chapter we study some properties of the stable and unstable manifolds of a fixed point (Section 3) and we study the structure of the global attractor of a semigroup with a Lyapunov function (Section 4). Finally, in Section 5, by taking advantage of the results proved in Section 3, we indicate how we can derive lower bounds on the dimension of attractors.

3. Stable and Unstable Manifolds

In Section 3.1, after recalling the necessary definitions, we state without proof a result of J.C. Wells [1] on the structure of a mapping in the neighborhood of a fixed point. We apply this result in Section 3.2 to the study of the structure

[1] The same as in space dimension 2 (see VI.(3.20)).

of an attractor near a fixed point. Finally, in Section 3.3, we define and show an elementary property of the unstable manifold of a compact invariant set.

3.1. Structure of a Mapping in the Neighborhood of a Fixed Point

Let E be a Banach space (norm $\|\cdot\|$) and let S be a continuous mapping from E into itself. We denote by z a fixed point of S,

$$S(z) = z, \tag{3.1}$$

and we assume that S is Fréchet differentiable in a neighborhood $\mathcal{O}$ of z, with differential $S'(u)$. We will also assume that S' satisfies a Hölder condition

$$\|S'(u_1) - S'(u_2)\| \leq c_1 \|u_1 - u_2\|^\alpha, \qquad 0 < \alpha \leq 1, \tag{3.2}$$

$\forall u_1, u_2 \in \mathcal{O}$, where c_1 is independent of u_1, u_2.

We recall the definition of a hyperbolic fixed point to be used here.

Definition 3.1. Let z be a fixed point of S, $S(z) = z$. We say that z is a hyperbolic fixed point of z if the following two conditions are satisfied:

(i) The spectrum $\sigma(S'(z))$ of $S'(z)$ does not intersect the circle $\{\lambda \in \mathbb{C}, |\lambda| = 1\}$;
(ii) E_+ has finite dimension, where $E_+ = E_+(z)$ and $E_- = E_-(z)$ are the linear invariant subspaces of E corresponding to the subsets of $S'(z)$ contained in $\{\lambda \in \mathbb{C}, |\lambda| > 1\}$ and $\{\lambda \in \mathbb{C}, |\lambda| < 1\}$.

When z is a hyperbolic fixed point of S, we consider E_+ and E_- as above and let $\Pi_+ = \Pi_+(z)$ and $\Pi_- = \Pi_-(z)$ denote the Riesz projectors onto these subspaces and $S'_+ = S'(z) \circ \Pi_+$, $S'_- = S'(z) \circ \Pi_-$. It is clear that $S'(z)$ coincide on E_+ with S'_+ and on E_- with S'_-. Also, since z is hyperbolic, the spectral radius of S'_-, $r(S'_-) = \sup\{|\lambda|, \lambda \in \sigma(S'_-)\}$, is less than unity, $r(S'_-) = 1 - 2\delta_-$, $\delta_- > 0$. Similarly, since $r(S'_+) > 1$ and E_+ is finite dimensional, S'_+ is invertible on E_+ and $r(S'^{-1}_+) = 1 - 2\delta_+$, $\delta_+ > 0$.

It is convenient to associate with such a point z the following norms on E_+, E_-, and E:

$$\begin{cases} \|u_+\|_0 = \operatorname*{Sup}_{n \geq 0} (\|S'^{-n}_+ u_+\|/(1 - \delta_+)^n), & \forall u_+ \in E_+, \\ \|u_-\|_0 = \operatorname*{Sup}_{n \geq 0} (\|S'^{n}_- u_-\|/(1 - \delta_-)^n), & \forall u_- \in E_-, \\ \|u_+ + u_-\|_0 = \max(\|u_+\|_0, \|u_-\|_0). \end{cases} \tag{3.3}$$

We can easily check that $\| \quad \|_0$ is a norm on E and that it is equivalent to the initial norm $\|\cdot\|$

$$c_2^{-1} \|u\| \leq \|u\|_0 \leq c_2 \|u\|, \qquad \forall u \in E. \tag{3.4}$$

The estimate from below for $\|u\|_0$ in (3.4) is obvious while the estimate from above is a consequence of the relation

$$r(L) = \lim_{n\to\infty} \|L^n\|^{1/n},$$

valid for any linear bounded operator $L \in \mathscr{L}(E)$. We also have

$$\|S_+'^{-1}u_+\|_0 \le (1-\delta_+)\|u_+\|_0, \qquad \forall u_+ \in E_+, \tag{3.5}$$

$$\|S_-'u_-\|_0 \le (1-\delta_-)\|u_-\|_0, \qquad \forall u_- \in E_-. \tag{3.6}$$

We now recall the definitions of the stable and unstable manifolds of S at z

$$\mathscr{M}_-(z) = \{u_0 \in E, \forall p \in \mathbb{N}, \exists u_p \in E, u_0 = S^p(u_p) \text{ and } S^n u_0 \to z \text{ as } n \to \infty\}$$

is the stable manifold, while the unstable manifold is

$$\mathscr{M}_+(z) = \{u_0 \in E, \forall p \in \mathbb{N}, \exists u_p \in E, u_0 = S^p(u_p) \text{ and } u_p \to z \text{ as } p \to \infty\}.$$

It follows readily from the definition of $\mathscr{M}_+(z)$ and $\mathscr{M}_-(z)$ that these sets are invariant under S

$$S\mathscr{M}_+(z) = \mathscr{M}_+(z), \qquad S\mathscr{M}_-(z) = \mathscr{M}_-(z). \tag{3.7}$$

In view of a local study of these sets, we consider for $R > 0$ the ball

$$\mathscr{O}_R(z) = \{y \in E, \|y - z\|_0 \le R\},$$

and

$$\mathscr{M}_+^R(z) = \{u_0 \in \mathscr{O}_R(z), \forall n \in \mathbb{N}, \exists u_n \in S^{-n}(u_0) \cap \mathscr{O}_R(z) \text{ and } u_n \to z \text{ as } n \to \infty\},$$

$$\mathscr{M}_-^R(z) = \{u_0 \in \mathscr{O}_R(z), S^n(u_0) \to z \text{ as } n \to \infty\}.$$

Concerning the positive invariance of $\mathscr{M}_+(z)$, $\mathscr{M}_-(z)$ for S we make the following observation: for $r > 0$ let $\mathscr{B}_r(z)$ denote the open ball of E centered at 0 of radius r (norm $\|\cdot\|$). Then due to the continuity of S at z, for $r > 0$ sufficiently small $\mathscr{B}_r(z) \subset \mathscr{O}_R(z)$ and $S(B_r(z)) \subset \mathscr{O}_R(z)$. It follows readily from the definition of $\mathscr{M}_+^R(z)$, $\mathscr{M}_-^R(z)$ that for such r

$$\begin{cases} S(\mathscr{M}_+^R(z) \cap B_r(z)) \subset \mathscr{M}_+^R(z), \\ S(\mathscr{M}_-^R(z) \cap B_r(z)) \subset \mathscr{M}_-^R(z). \end{cases} \tag{3.8}$$

Now we have the following result proved in J.C. Wells [1] (see also A.V. Babin and M.I. Vishik [2]).

Theorem 3.1. *We consider a mapping $S: E \to E$, which is Fréchet differentiable, its differential S' satisfying the Hölder condition* (3.2). *We assume that z is a hyperbolic fixed point of S and that $E_+(z)$, $E_-(z)$ are defined as above.*

Then, for R sufficiently small, there exist two mappings g_+, g_-, satisfying

$$g_+: \mathscr{O}_R(z) \cap E_+(z) \to E_-(z), \qquad g_+(z) = 0,$$

$$g_-: \mathscr{O}_R(z) \cap E_-(z) \to E_+(z), \qquad g_-(z) = 0,$$

and such that the sets $\mathcal{M}_+^R(z)$, $\mathcal{M}_-^R(z)$ are represented as

$$\mathcal{M}_+^R(z) = \{u \in E, u = u_+ + g_+(u_+), u_+ \in \mathcal{O}_R(z) \cap E_+(z)\}, \tag{3.9}$$

$$\mathcal{M}_-^R(z) = \{u \in E, u = u_- + g_-(u_-), u_- \in \mathcal{O}_R(z) \cap E_-(z)\}. \tag{3.10}$$

The mappings g_+ and g_- are Fréchet differentiable and their differentials g'_+, g'_- satisfy a Hölder inequality with the same exponent α as in (3.2). *Furthermore, $g'_-(z) = g'_+(z) = 0$.*

Remark 3.1. It follows, in particular from Theorem 3.1, that $\mathcal{M}_+^R(z)$ is in $\mathcal{O}_R(z)$ a $\mathscr{C}^{1,\alpha}$ manifold of finite dimension, equal to that of $E_+(z)$.

3.2. Application to Attractors

Our aim is now to use Theorem 3.1 to obtain some information about the structure of an attractor near a fixed point z. We consider successively the discrete and the continuous cases.

3.2.1. The Discrete Case

We are given a Banach space E and a mapping $S: E \to E$, the hypotheses being the same as in Theorem 3.1.

We consider the dynamical system associated with S (i.e., with the discrete semigroup defined by $S(n) = S^n$, $\forall n \in \mathbb{N}$), and assume that S possesses a global attractor $\mathscr{A}$ which attracts the bounded sets of E.[1] We then make the following remarks

$$\mathcal{M}_+^R(z) \subset \mathcal{M}_+(z), \tag{3.11}$$

$$\mathcal{M}_+(z) \subset \mathscr{A}. \tag{3.12}$$

The first inclusion follows immediately from the definitions. In order to prove the inclusion (3.12) we consider a point $u_0 \in \mathcal{M}_+(z)$; by definition u_0 belongs to a complete orbit $\mathcal{O} = \{u(n), n \in \mathbb{Z}, u(0) = u_0\}$ and this orbit is bounded: $u(n)$ is bounded as $n \to -\infty$ since $u(n) \to z$ and $u(n)$ is bounded as $n \to \infty$, since the distance of $u(n)$ to $\mathscr{A}$ converges to 0. Thus $\mathcal{O}$ is attracted by $\mathscr{A}$ and since $\mathcal{O}$ is invariant, $\mathcal{O} \subset \mathscr{A}$, $u_0 \in \mathscr{A}$.

One of the consequences of (3.12), Theorem 3.1, and Remark 3.1 is

> If $\mathscr{A}$ is the global attractor of S and $z \in \mathscr{A}$ is a hyperbolic fixed point then $\mathscr{A}$ contains in a neighborhood of z a manifold of finite dimension, equal to that of $E_+(z)$. (3.13)

We will now extend this result to the continuous case.

[1] See Definition I.1.3. For instance, Theorem I.1.1 gives sufficient conditions which ensure the existence of $\mathscr{A}$.

3.2.2. The Continuous Case

We consider a Banach space E (norm $\|\cdot\|$) and as in Section I.1.1 a continuous semigroup of operators $S(t)\colon E \to E$[1]

$$\begin{cases} S(t+s) = S(t)\cdot S(s), & \forall s, t \geq 0, \\ \quad S(0) = I \quad \text{(Identity in } E). \end{cases} \tag{3.14}$$

The continuity property I.(1.4) is strengthened as follows:

The mapping $\{t, u_0) \to S(t)u_0$ from $R_+ \times E$ into E is continuous. (3.15)

Let z be a fixed point of $S(t)$

$$S(t)z = z, \qquad \forall t \in \mathbb{R}_+. \tag{3.16}$$

We assume that the mapping $u \to S(t)u$ is Fréchet differentiable in a neighborhood $\mathcal{O}$ of z, $\forall t \in \mathbb{R}_+$ ($\mathcal{O}$ independent of t), with differential $S'(t)$; we also assume that S' satisfies a Hölder condition similar to (3.2)

$$\|S'(t)u_1 - S'(t)u_2\| \leq c_3(T)\|u_1 - u_2\|^\alpha, \qquad 0 < \alpha \leq 1,$$
$$\forall u_1, u_2 \in \mathcal{O}, \quad \forall t \in [0, T], \tag{3.17}$$

c_3 depending on T, but not on u_1, u_2.

We extend as follows the definition of hyperbolicity:

Definition 3.2. We say that the fixed point z is hyperbolic if the following conditions are satisfied:

(i) z is a hyperbolic fixed point for $S(t)$, $\forall t > 0$ (as in Definition 3.1);
(ii) the invariant linear subspaces E_+ and E_- corresponding to the operators $S'(t)(z)$ are independent of t.

The stable and unstable manifolds at z are defined, as in Chapter I,

$$\mathcal{M}_+(z) = \{u_0 \in E, \forall t \leq 0, \exists u(t) \in S(-t)^{-1}u_0 \text{ and } u(t) \to z \text{ as } t \to -\infty\},$$

$$\mathcal{M}_-(z) = \{u_0 \in E, \forall t \leq 0, \exists u(t) \in S(-t)^{-1}u_0 \text{ and } S(t)u_0 \to z \text{ as } t \to +\infty\}.$$

It follows immediately from the definitions that

$$S(t)\mathcal{M}_+(z) = \mathcal{M}_+(z), \qquad S(t)\mathcal{M}_-(z) = \mathcal{M}_-(z). \qquad \forall t \geq 0.\text{[2]} \tag{3.18}$$

We want to study the structure of $\mathcal{M}_+(z)$ by considering z as a fixed point of $S(t_0)$ for some $t_0 > 0$ and by using the results of Section 3.2.1.

[1] The basic space denoted by H in Chapter I is denoted here by E.

[2] We required, in the definition of $\mathcal{M}_-(z)$, that u_0 belongs to a complete orbit. This is not automatically true in infinite dimension, unlike the case of finite dimension where the operators $S(t)$ are reversible in general, or at least defined for some negative interval $(-\delta, 0)$. If we remove the condition that u_0 belongs to a complete orbit, we can only assert that $S(t)\mathcal{M}_-(z) \subset \mathcal{M}_-(z)$.

Let $S_1 = S(1)$; we denote temporarily by $\mathscr{W}_+(z)$ the unstable manifold of the hyperbolic fixed point z of the *mapping* S_1. We have

Lemma 3.1.

$$\mathscr{W}_+(z) = \mathscr{M}_+(z).$$

PROOF. The inclusion $\mathscr{M}_+(z) \subset \mathscr{W}_+(z)$ is easy: if $u_0 \in \mathscr{M}_+(z)$, then there is a complete orbit $\{u(t), t \in \mathbb{R}\}$ such that $u(0) = u_0$ and $u(t) \to z$ as $t \to -\infty$. By considering the partial orbit $\{u(n), n \in \mathbb{Z}\}$, we find that $u_0 \in \mathscr{W}_+(z)$.

Now let $u_0 \in \mathscr{W}_+(z)$ and let u_n denote the sequence provided by the definition of $\mathscr{W}_+(z)$: $S_1^n u_n = S(n)u_n = u_0, \forall n \in \mathbb{N}, u_n \to z$ as $n \to +\infty$. We define as follows a complete orbit $\{u(t), t \in \mathbb{R}\}$ of $S(t)$: $u(t) = S(t)u_0$ for $t \geq 0$, $u(-n) = u_n$ for $n \in \mathbb{N}$, and if $t = -n + \tau, 0 < \tau < 1, n \in \mathbb{N}$, we set

$$u(t) = S(\tau)u_n.$$

It is easy to see that $\{u(t), t \in \mathbb{R}\}$ is a complete orbit of $S(t)$, and there remains to show that

$$u(t) \to z \quad \text{as} \quad t \to -\infty. \tag{3.19}$$

Since we already know that $u(-n) = u_n \to z$ as $n \to \infty$, $n \in \mathbb{N}$, it suffices to show that

$$\operatorname*{Sup}_{0 \leq \tau \leq 1} \|S(\tau)u_n - z\| \to 0 \quad \text{as} \quad n \to \infty. \tag{3.20}$$

Since u_m converges to z, the set X composed of the union of z and the sequence u_n is compact; thus $[0, 1] \times X$ is compact and by (3.15) the mapping $\{t, u\} \to S(t)(u)$ is uniformly continuous on $[0, 1] \times X$. This implies that

$$\operatorname*{Sup}_{\substack{|\tau-\tau'| \leq \delta \\ \|u_n - z\| \leq \delta}} \|S(\tau)u_n - S(\tau')z\| \to 0 \quad \text{as} \quad \delta \to 0.$$

In particular,

$$\operatorname*{Sup}_{\substack{0 \leq \tau \leq 1 \\ \|u_n - z\| \leq \delta}} \|S(\tau)u_n - S(\tau)z\| \to 0 \quad \text{as} \quad \delta \to 0;$$

hence (3.20), since $S(\tau)z = z$. □

Given $R > 0$ and a norm $\|\cdot\|_0$ equivalent to $\|\cdot\|$, we define $\mathcal{O}_R(z)$, $\mathscr{M}_+^R(z)$, $\mathscr{M}_-^R(z)$ as in Section 3.1

$$\mathcal{O}_R(z) = \{y \in E, \|y - z\|_0 \leq R\},$$

$$\mathscr{M}_+^R(z) = \{u_0 \in \mathcal{O}_R(z), \forall t \leq 0, \exists u(t) \in S(-t)^{-1}(u_0) \cap \mathcal{O}_R(z) \text{ and } u(t) \to z \text{ as } t \to -\infty\},$$

$$\mathscr{M}_-^R(z) = \{u_0 \in \mathcal{O}_R(z), S(t)u_0 \to z \text{ as } t \to +\infty\}.$$

We also define the similar sets for S_1, that we denote $\mathscr{W}_+^R(z)$, $\mathscr{W}_-^R(z)$,

$$\mathscr{W}_+^R(z) = \{u_0 \in \mathcal{O}_R(z),\ \forall n \in \mathbb{N},\ \exists u_n \in \mathcal{O}_R(z),\ S_1^n u_n = S(n)u_n = u_0 \text{ and } u_n \to z \text{ as } n \to \infty\}, \tag{3.21}$$

$$\mathscr{W}_-^R(z) = \{u_0 \in \mathcal{O}_R(z),\ S_1^n u_0 \to z \text{ as } n \to \infty\}. \tag{3.22}$$

It is clear that $\mathscr{M}_+^R(z) \subset \mathscr{W}_+^R(z)$, $\mathscr{M}_-^R(z) \subset \mathscr{W}_-^R(z)$. Concerning $\mathscr{W}_+^R(z)$, we show the following:

Lemma 3.2. *Let $S(t)$ satisfy* (3.14)–(3.16). *Then for any $R > 0$ and any norm $\|\cdot\|_0$ equivalent to $\|\cdot\|$,*

$$\mathscr{M}_+(z) = \bigcup_{k=0}^{\infty} S_1^k(\mathscr{W}_+^R(z)), \tag{3.23}$$

where $\mathscr{M}_+(z)$, $\mathscr{W}_+^R(z)$ are defined above.

If r is sufficiently small so that $B_r(z) \subset \mathcal{O}_R(z)$, then

$$\mathscr{M}_+(z) \subset \mathscr{W}_+^R(z) \cup \{x \in E,\ x = S(t)y,\ y \in \Gamma_r(z),\ t > 0\}, \tag{3.24}$$

with

$$\Gamma_r(z) = \mathscr{W}_+^R(z) \cap \{y \in E,\ \|y - z\| = r\}. \tag{3.25}$$

Proof. Let $\mathscr{U}$ denote the right-hand side of (3.23). If $u_0 \in \mathscr{M}_+(z)$, then u_0 belongs to a complete orbit of $S(t)$, $\{u(t), t \in \mathbb{R}\}$, such that $u(0) = u_0$ and $u(t) \to z$ as $t \to -\infty$. Thus there exists $n_0 \in \mathbb{N}$ such that $u(-n) \in \mathcal{O}_R(z)$ for $n \geq n_0$; $u(-n_0) \in \mathscr{W}_+^R(z)$ and $u_0 = S_1^{n_0} u(-n_0) \in \mathscr{U}$. Hence $\mathscr{M}_+(z) \subset \mathscr{U}$.

We then prove that $\mathscr{U} \subset \mathscr{M}_+(z)$ by using Lemma 3.1. If $v_0 \in \mathscr{U}$, then $v_0 = S_1^k u_0$ for some u_0 in $\mathscr{W}_+^R(z)$. But $\mathscr{W}_+^R(z) \subset \mathscr{W}_+(z) = \mathscr{M}_+(z)$; thus $v_0 \in S_1^k \mathscr{M}_+(z) = S(k)\mathscr{M}_+(z) = \mathscr{M}_+(z)$.

Let us prove (3.24). If $u_0 \in \mathscr{M}_+(z)$, then $u_0 = u(0)$, where $\{u(t), t \in \mathbb{R}\}$ is a complete orbit of $S(t)$, and $u(t) \to z$ as $t \to -\infty$. Let n_0 denote the smallest integer such that $u(t) \in B_r(z)$ for $t \leq -n_0$. If $n_0 = 0$, then $u_0 \in \mathscr{W}_+^R(z)$; if not, $u(n_0 - 1) \notin B_r(z)$ and since $t \to S(t)u_0$ is continuous by (3.15) there exists a smallest $\tau \in [0, 1]$ such that $\|u(-n_0 + \tau) - z\| = r$. Then $u(-n_0 + \tau) \in \Gamma_r(z)$ and $u_0 = S(n_0 - \tau)\cdot u(-n_0 + \tau)$ indeed belongs to the set in the right-hand side of (3.23).

Lemma 3.2 is proved. □

The structure of $\mathscr{W}_+^R(z)$ near z is fully described by Theorem 3.1 applied to $S = S_1$.

Lemma 3.3. *Under the above assumptions, for R sufficiently small, there exist two mappings g_+, g_-, satisfying*

$$g_+ : \mathcal{O}_R(z) \cap E_+(z) \to E_-(z), \qquad g_+(z) = 0,$$

$$g_- : \mathcal{O}_R(z) \cap E_-(z) \to E_+(z), \qquad g_-(z) = 0,$$

and such that the sets $\mathscr{W}_+^R(z)$, $\mathscr{W}_-^R(z)$ *are represented as*

$$\mathscr{W}_+^R(z) = \{u \in E, u = u_+ + g_+(u_+), u_+ \in \mathcal{O}_R(z) \cap E_+(z)\}, \tag{3.26}$$

$$\mathscr{W}_-^R(z) = \{u \in E, u = u_- + g_-(u_-), u_- \in \mathcal{O}_R(z) \cap E_-(z)\}. \tag{3.27}$$

The mappings g_+ *and* g_- *are Fréchet differentiable and their differentials* g_+', g_-' *satisfy a Hölder condition with the same exponent* α *as in* (3.17). *Furthermore,* $g_-'(z) = g_+'(z) = 0$.

Let us now assume that the semigroup $S(t)$ possesses a global attractor $\mathscr{A}$. We have then proved the following:

Theorem 3.2. *We consider in a Banach space E, a semigroup of operators* $S(t)$, $t \in \mathbb{R}_+$, *which satisfy the hypotheses above, in particular* (3.14)–(3.17). *We also assume that* $S(\cdot)$ *possesses a global attractor* $\mathscr{A}$ *and that* $z \in \mathscr{A}$ *is a hyperbolic fixed point of* $S(\cdot)$. *Then*

$$\mathscr{A} \supset \mathscr{M}_+(z) \supset \mathscr{W}_+^R(z), \tag{3.28}$$

and for $R > 0$ *sufficiently small,* $\mathscr{W}_+^R(z)$ *can be represented in the form* (3.26), *and in particular,* $\mathscr{W}_+^R(z)$ *is a* $\mathscr{C}^{1,\alpha}$ *manifold of dimension equal to that of* $E_+(z)$.

Remark 3.2. Since $\mathcal{O}_R(z)$ is closed and $E_+(z)$ is finite dimensional, $\mathcal{O}_R(z) \cap E_+(z)$ is compact. Thus, thanks to (3.26), we see that under the assumptions of Theorem 3.2 (and Lemma 3.3), the set $\mathscr{W}_+^R(z)$ is *compact* when R is sufficiently small so that Lemma 3.3 and Theorem 3.2 apply.

3.3. Unstable Manifold of a Compact Invariant Set

We consider in a Banach space E, a semigroup $S(t)_{t \geq 0}$ which satisfies the properties (3.14), (3.15). Let $X \subset E$ be an invariant set

$$S(t)X = X, \qquad \forall t \geq 0. \tag{3.29}$$

As we did in Chapter I in the case of a fixed point, we define the stable and unstable sets of X. The *stable set* $\mathscr{M}_-(X)$ of X is the (possibly empty) set of points u_* which belong to a complete orbit $\{u(t), t \in \mathbb{R}\}$ and such that

$$d(u(t), X) \to 0 \quad \text{as} \quad t \to \infty.$$

The *unstable set* $\mathscr{M}_+(X)$ of X is the (possibly empty) set of points u_* which belong to a complete orbit $\{u(t), t \in \mathbb{R}\}$ and such that

$$d(u(t), X) \to 0 \quad \text{as} \quad t \to -\infty.$$

These definitions make sense even without the assumption that X is invariant; if X is invariant then we can easily verify that these sets are invariant sets

$$\begin{cases} S(t)\mathscr{M}_+(X) = \mathscr{M}_+(X), \\ S(t)\mathscr{M}_-(X) = \mathscr{M}_-(X), \qquad \forall t \geq 0. \end{cases} \tag{3.30}$$

Assuming now that X is compact and that $\{S(t)\}_{t\geq 0}$ possesses a global attractor we notice the following results which partly complement (3.28):

Theorem 3.3. *Let there be given in the Banach space E a semigroup of operators* $\{S(t)\}_{t\geq 0}$ *satisfying* (3.14), (3.15) *and which possesses a global attractor* $\mathscr{A}$. *Let* $X \subset F$ *be a compact set invariant for* $S(t)$. *Then*

$$\mathscr{M}_+(X) \subset \mathscr{A}. \tag{3.31}$$

For $X = \mathscr{A}$

$$\mathscr{M}_+(\mathscr{A}) = \mathscr{A}. \tag{3.32}$$

PROOF. Let $u_* \in \mathscr{M}_+(X)$, and let $Y = \{u(t), t \in \mathbb{R}\}$ be the orbit containing u_*, with say $u_* = u(0)$. The orbit Y is bounded: it is bounded at $-\infty$ since $d(u(t), X) \to 0$ as $t \to -\infty$; it is bounded at $+\infty$ since $d(u(t), \mathscr{A}) \to 0$ as $t \to +\infty$; finally, any set $\{u(t), |t| \leq T\}$ is bounded thanks to the continuity property (3.15). Hence

$$d(S(t)Y, \mathscr{A}) \to 0 \quad \text{as} \quad t \to \infty,$$

and since $S(t)Y = Y$, $Y \subset \mathscr{A}$, and (3.31) follows.

For the proof of (3.32) it remains to show that $\mathscr{A} \subset \mathscr{M}_+(\mathscr{A})$. If u_* belongs to $\mathscr{A}$, then u_* belongs to a complete trajectory $\{u(t), t \in \mathbb{R}\} \subset \mathscr{A}$, with say $u_* = u(0)$. The positive part of this orbit $\{u(t), t \geq 0\}$ is uniquely determined, while its negative part $\{u(t), t < 0\}$ is uniquely determined only when the operators $S(t)$ are injective. Now $d(u(t), \mathscr{A}) = 0$, $\forall t < 0$, and by the definition we see immediately that $u_* \in \mathscr{M}_+(\mathscr{A})$.

The theorem is proved. □

4. The Attractor of a Semigroup with a Lyapunov Function

In Section 4.1 we recall the definition of a Lyapunov function for a semigroup, and we give a general result concerning the global attractor $\mathscr{A}$ of a semigroup possessing a Lyapunov function. In Section 4.2 we improve the results on $\mathscr{A}$ under the supplementary assumptions that the equilibrium set of $S(\cdot)$ is finite and that each fixed point is hyperbolic. Finally, in Section 4.3, we return briefly to the examples encountered in Chapters III and IV.

4.1. A General Result

We consider in a Banach space E a semigroup $\{S(t)\}_{t\geq 0}$ which enjoys the properties (3.14), (3.15).

Definition 4.1. A Lyapunov function of the semigroup $S(\cdot)$ on a set $\mathscr{F} \subset E$ is a continuous function $F: \mathscr{F} \to \mathbb{R}$, such that:

(i) for any $u_0 \in \mathscr{F}$, the function $t \to F(S(t)u_0)$ is decreasing;
(ii) if $F(S(\tau)u_1) = F(u_1)$, for some $\tau > 0$, then u_1 is a fixed point of the semigroup $S(\cdot)$.

Our first result is the following:

Theorem 4.1. *Let there be given a semigroup $S(\cdot)$ which enjoys the properties (3.14), (3.15). We assume that $S(\cdot)$ possesses a Lyapunov function F defined and continuous on $\mathscr{F} \subset E$, and a global attractor $\mathscr{A} \subset \mathscr{F}$. Let $\mathscr{E}$ denote the set of fixed points of the semigroup. Then*

$$\mathscr{A} = \mathscr{M}_+(\mathscr{E}). \tag{4.1}$$

Furthermore, if $\mathscr{E}$ is discrete, $\mathscr{A}$ is the union of $\mathscr{E}$ and of the heteroclinic curves joining one point of $\mathscr{C}$ to another point of $\mathscr{E}$ and

$$\mathscr{A} = \bigcup_{z \in \mathscr{E}} \mathscr{M}_+(z). \tag{4.2}$$

Proof. Since $\mathscr{E} \subset \mathscr{A}$, it is easy to check that $\mathscr{M}_+(\mathscr{E}) \subset \mathscr{M}_+(\mathscr{A})$; by Theorem 3.3, $\mathscr{M}_+(\mathscr{A}) = \mathscr{A}$, hence $\mathscr{M}_+(\mathscr{E}) \subset \mathscr{A}$.

In order to show the opposite inclusion assume that u_0 is a point of $\mathscr{A}$. Then u_0 belongs to a complete orbit $\{u(t), t \in \mathbb{R}\}$ included in $\mathscr{A}$, with say $u_0 = u(0)$. Since $\mathscr{A}$ is compact, we infer from (3.15) that the following set γ is nonempty, compact, connected, and invariant

$$\gamma = \bigcap_{s<0} \overline{\{u(t), t < s\}}.$$

Indeed, γ is part of the α-limit set $\alpha(u_0)$ of u_0, and it is all of it if the operators $S(t)$ are injective. The sets $\overline{\{u(t), t < s\}}$ are included in $\mathscr{A}$ and thus compact; for $s < 0$ they form a decreasing sequence of compact sets and γ is then compact and nonempty. These sets are connected, thanks to (3.15),[1] and γ is connected too; finally, we show that γ is invariant

$$S(t)\gamma = \gamma, \qquad \forall t \geq 0, \tag{4.3}$$

exactly as we did in Chapter I for $\alpha(u_0)$.

We also observe that F is constant on γ,

$$F|_\gamma = \lim_{t \to -\infty} F(u(t)) = \operatorname*{Sup}_{t \in \mathbb{R}} F(u(t)). \tag{4.4}$$

Since F is decreasing along the trajectories, and F is bounded from above on the compact set $\mathscr{A}$, the limit of $F(u(t))$ as $t \to -\infty$ exists. Each point of γ is the limit of a sequence $u(t_n)$, $t_n \to -\infty$, and (4.4) follows.

[1] $\overline{\{u(t), t < s\}} = \bigcap_{n \geq 0} S([0, n])(u(s - n))$.

Thanks to (4.3), (4.4) and the definition of a Lyapunov function, γ consists only of stationary points of $S(\cdot)$, $\gamma \subset \mathscr{E}$. Since $u_0 \in \mathscr{M}_+(\gamma)$, $u_0 \in \mathscr{M}_+(\mathscr{E})$, and thus $\mathscr{A} \subset \mathscr{M}_+(\mathscr{E})$.

If we assume furthermore that $\mathscr{E}$ is discrete, then the set γ above reduces to one stationary solution z and the whole family $u(t)$ converges to z as $t \to -\infty$. In particular, $u_0 \in \mathscr{M}_+(z)$ and (4.2) is proved.

In a similar manner we can prove that $u(t)$ converges to a stationary solution z' as $t \to +\infty$ by considering the set γ' similar to γ

$$\gamma' = \bigcap_{s>0} \overline{\{u(t), t > s\}};$$

γ' is nothing other than the ω-limit set of u_0, $\omega(u_0)$, and we prove, exactly as before, that $\omega(u_0) \subset \mathscr{E}$ and $\omega(u_0)$ is reduced to a single point.

The theorem is proved. □

4.2. Additional Results

We now give a more complete description of $\mathscr{M}_+(\mathscr{E})$ and thus of $\mathscr{A}$, when $\mathscr{E}$ is finite and each equilibrium point is hyperbolic.

The hypotheses are the same as in Theorem 4.1, and we assume that the differentiability property (3.17) is satisfied so that we can use the results of Section 3.

We start with a technical lemma.

Lemma 4.1. *The assumptions are the same as above and we consider a fixed point z of the semigroup $S(\cdot)$.*

Then the restriction of F to $\mathscr{M}_+(z)$ has a strict maximum at z and, furthermore, for every $r > 0$, there exists $\varepsilon > 0$ such that

$$F(u) \leq F(z) - \varepsilon, \qquad \forall u \in \mathscr{M}_+(z)\backslash B_r(z). \tag{4.5}$$

Proof. We first prove that

$$F(u_0) < F(z), \qquad \forall u_0 \in \mathscr{M}_+(z)\backslash z. \tag{4.6}$$

For that purpose, consider the sequence $u_n \in \mathscr{M}_+(z)$ such that $u_0 = S_1^n u_n = S(n)u_n$, $\forall n \in \mathbb{N}$, and $u_n \to z$ as $n \to \infty$. It is clear that

$$F(u_n) \leq F(u_{n-1}) \leq \cdots \leq F(u_0)$$

for every n, and that $F(u_n)$ converges to $F(z)$ as $n \to \infty$. If $F(z) = F(u_0)$, the sequence $F(u_n)$ is stationary; hence each u_n is a fixed point and

$$u_0 = S_1^n u_n = u_n, \qquad \forall n,$$

converges to z as $n \to \infty$; a contradiction.

For the proof of (4.5), we choose $R > 0$ such that $B_r(z) \subset \mathscr{O}_R(z)$ and first prove that

$$F(u) \leq F(z) - \varepsilon, \qquad \forall u \in \mathscr{W}_+^R(z) \backslash B_r(z).^1 \tag{4.7}$$

Due to Remark 3.1, $\mathscr{W}_+^R(z) \backslash B_r(z)$ is compact; then thanks to (4.5) the maximum of F on $\mathscr{W}_+^R(z) \backslash B_r(z)$ is $< F(z)$. Then we use the inclusion (3.24), (3.25) in Lemma 3.2, which implies that any point x in $\mathscr{M}_+(z) \backslash B_r(z)$ either belongs to $\mathscr{W}_+^R(z) \backslash B_r(z)$ or is of the form $S(t)y$, $t > 0$, $y \in \Gamma_r(z) =$ the boundary of $B_r(z)$. In the first case (4.6) suffices, $F(x) \leq F(u) - \varepsilon$; in the second case we write

$$F(x) = F(S(t)y) \leq F(y) \leq F(z) - \varepsilon,$$

since $y \in \mathscr{W}_+^R(z) \backslash B_r(z)$. Hence (4.4). □

Proposition 4.1. *Under the hypotheses above and if z is a hyperbolic fixed point of the semigroup, then for every $R > 0$ fixed, sufficiently small, there exists $r > 0$ such that*

$$\mathscr{M}_+(z) \cap B_r(z) \subset \mathscr{W}_+^R(z). \tag{4.8}$$

Proof. We must show that

$$B_r(z) \cap (\mathscr{M}_+(z) \backslash \mathscr{W}_+^R(z)) = \varnothing \tag{4.9}$$

for r sufficiently small.

Fix R as in Theorem 3.2. Due to Lemma 4.1 the maximum of $F(u)$ on $\mathscr{M}_+(z) \backslash \mathscr{O}_R(z)$ is $< F(z)$ and we denote this supremum $F(z) - \varepsilon$, $\varepsilon = \varepsilon(R) > 0$. Now if (4.8) is not true, there exists a sequence u_j converging to z as $j \to \infty$, $u_j \in \mathscr{M}_+(z) \backslash \mathscr{W}_+^R(z)$, $\forall j$. By definition, for each j there exists a sequence $u_{jk} \in \mathscr{M}_+(z)$, $k \in \mathbb{N}$, such that $u_j = S_1^k u_{jk} = S(k) u_{jk}$, $\forall k \in \mathbb{N}$, and $u_{jk} \to z$ as $k \to \infty$. Since $u_j \notin \mathscr{W}_+^R(z)$, $u_{jk_j} \notin \mathscr{O}_R(z)$ for some k_j; thus

$$F(u_j) \leq F(u_{jk_j}) \leq F(z) - \varepsilon, \qquad \forall j \in \mathbb{N}.$$

This contradicts the fact that $F(u_j)$ converges to $F(z)$ as $j \to \infty$; (4.9) is proved. □

We now prove the following:

Theorem 4.2. *Let $S(\cdot)$ be a semigroup satisfying the hypotheses* (3.14), (3.15), (3.17). *We assume that z is a hyperbolic fixed point of $S(\cdot)$, that S possesses a Lyapunov function F defined on a set $\mathscr{F} \supset \mathscr{M}_+(z)$, and*

$$\textit{the restriction of } S(t) \textit{ to } \mathscr{M}_+(z) \textit{ is a bijection for any } t > 0; \tag{4.10}$$

$$\textit{the inverse mapping } S(t)^{-1} \textit{ is continuous on } \mathscr{M}_+(z); \tag{4.11}$$

$$\textit{for every } u_0 \in \mathscr{M}_+(z),\ S'(t) \cdot u_0 \textit{ is injective.} \tag{4.12}$$

[1] $\mathscr{W}_+^R(z)$ defined in (3.21).

Then $\mathcal{M}_+(z)$ is a finite-dimensional $\mathscr{C}^1$ submanifold of E of dimension n, $n =$ the dimension of $E_+(z)$.

PROOF. Proposition 4.1 applies and (4.8) holds for some $r > 0$. We consider $\rho > 0$ such that $\rho > R$ and $\mathcal{O}_R(z) \subset B_r(z)$. Then

$$\mathcal{M}_+(z) \cap \mathcal{O}_{2\rho}(z) \subset \mathcal{W}_+^R(z) \cap \mathcal{O}_{2\rho}(z) = \mathcal{W}_+^\rho(z),$$

and since $\mathcal{W}_+^\rho(z) \subset \mathcal{M}_+(z)$ by (3.11), $\mathcal{M}_+(z) \cap \mathcal{O}_{2\rho}(z)$ is equal to $\mathcal{W}_+^\rho(z)$ and we have in $\mathcal{O}_\rho(z)$ the representation (3.26) of $\mathcal{M}_+(z)$. Hence for $\rho > 0$ sufficiently small, $\mathcal{M}_+(z)$ is in $\mathcal{O}_\rho(z)$ a $\mathscr{C}^1$ submanifold of E of dimension n.

Now the hypotheses (4.11), (4.12) show that $S_k = S(k)$ is a homeomorphism of a neighborhood of $x \in \mathcal{M}_+(z) \cap \mathcal{O}_\rho(z) = \mathcal{W}_+^\rho(z)$ on a neighborhood of $y = S_k x$. Since by (3.23) any point $y \in \mathcal{M}_+(z)$ can be written as $y = S_k x$, $x \in \mathcal{W}_+^\rho(z)$, we see that $\mathcal{M}_+(z)$ is a $\mathscr{C}^1$ manifold.

Let us prove that $\mathcal{M}_+(z)$ is a $\mathscr{C}^1$ manifold of E of dimension n. To do that we show that in some neighborhood of any point $y \in \mathcal{M}_+(z)$, the set $\mathcal{M}_+(z)$ coincides with the graph of a $\mathscr{C}^1$ function defined in a domain of an n-dimensional subspace $E_n = E_n(y) \subset E$ and taking its values in a complementary space E'_n of E_n. Given $y \in \mathcal{M}_+(z)$, $y = S_k x$, $x \in \mathcal{W}_+^\rho(z)$, we consider the n-dimensional space $T(x)$ tangent to $\mathcal{M}_+(z)$ at x (which exists by Lemma 3.3 and Proposition 4.1). Then we set $T(y) = S'_k(T(x))$ which is also of dimension n since S'_k is injective due to (4.12). We then choose arbitrarily a supplementary subspace $T'(y)$ to $T(y)$ and denote by $\Pi(y)$ and $I - \Pi(y)$ the projectors on $T(y)$ and $T'(y)$. Use of the injectivity of $S'_k(x)$ (due to (4.12)) and the implicit function theorem shows that the mapping $\Pi(x)(S_k - y)$ is a diffeomorphism of a neighborhood of the origin in $T(y)$. This implies that in a neighborhood of y, $\mathcal{M}_+(z)$ is the graph of an operator from $T(y)$ into $y + T'(y)$ of class $\mathscr{C}^1$; thus $\mathcal{M}_+(z)$ is a $\mathscr{C}^1$ submanifold of E of dimension n.

Theorem 4.2 is proved. □

Remark 4.1. A.V. Babin and M.I. Vishik show in [2] that $\mathcal{M}_+(z)$ is diffeomorphic to $\mathbb{R}^n$ and they also give a more precise description of $\mathcal{M}_+(z)$ corresponding to an ordering of the points $z_j \in \mathscr{E}$ by the values (necessarily distinct) of the numbers $F(z_j)$

$$F(z_1) < F(z_2) < \cdots < F(z_n).$$

Remark 4.2. Concerning the verification of the hypotheses of Theorem 4.2 and its application to practical examples: the assumptions (3.14), (3.15), (3.17), (4.10), (4.11) are technical but easy. The other assumptions are more delicate and in general they may be true only in a generic sense. For instance, the Sard–Smale theorem (see S. Smale [1]) can be applied as in C. Foias and R. Temam [1] to show the generic finiteness of $\mathscr{E}$ (see also J.C. Saut and R. Temam [1]); (4.12) and the hyperbolicity of the fixed points are probably generic hypotheses too, i.e., valid for "almost all values" of certain parameters.

4.3. Examples

We exhibit briefly some semigroups possessing a Lyapunov function; we do not develop the technical details: Theorem 4.1 is automatically applicable while Theorem 4.2 can be only applied after verifying the necessary assumptions (see Remark 4.2).

The equation in Section III.1 provides a first example of a Lyapunov function. Indeed, consider III.(1.2) with the boundary condition III.(1.3) (or III.(1.3′), (1.3″), see Remark III.(1.5)):

$$\frac{\partial u}{\partial t} - d\Delta u + g(u) = 0 \quad \text{in} \quad \Omega \times \mathbb{R}_+, \tag{4.13}$$

$$u = 0 \quad \text{on} \quad \Gamma \times \mathbb{R}_+, \qquad \Gamma = \partial\Omega. \tag{4.14}$$

Let G denote the primitive of g

$$G(s) = \sum_{j=1}^{2p-1} \frac{b_j}{j+1} s^{j+1}$$

(see III.(1.1)). Then we set

$$F(u) = \frac{d}{2}\|u\|^2 + G(u)$$

and we observe that

$$\begin{aligned}\frac{d}{dt}F(u(t)) &= d((u(t), u'(t))) + \int_\Omega g(u(t))u'(t)\,dx,\\ &= -\left|\frac{du}{dt}(t)\right|^2 \le 0.\end{aligned}$$

Similarly in Section III.4.2, equation III.(4.63) shows precisely that $J(u)$ is a Lyapunov function for the corresponding semigroup. Finally, consider in Section IV.2, the sine–Gordon equation. Set

$$F(u, u') = \tfrac{1}{2}\|u\|^2 + \tfrac{1}{2}|u'|^2 + G(u) - (f, u),$$

$\forall (u, u') \in F = V \times H$. Then IV.(2.12) shows precisely that F is a Lyapunov function.

Remark 4.3. In the examples above (and this is the general situation in infinite dimension), F is not finite and continuous on all of H; hence the set $\mathscr{F}$ appearing in the definition of F must be chosen properly. For example, for equation (4.13), (4.14) we observe that F is defined and continuous on $H_0^1(\Omega) \cap L^{2p}(\Omega)$. We choose $\mathscr{F}$ as a bounded set of $H^2(\Omega) \cap L^\infty(\Omega)$ containing $\mathscr{A}$ (see the regularity results in Section IV.6); on $\mathscr{F}$ the topology of $H = L^2(\Omega)$ coincides with that of $H_0^1(\Omega) \cap L^{2p}(\Omega)$ and F is thus continuous on $\mathscr{F}$.

5. Lower Bounds on Dimensions of Attractors: An Example

We can derive from Theorem 3.2 lower bounds on dimensions of attractor. Under the assumptions of Theorem 3.2, z being a hyperbolic fixed point of the semigroup we have by (3.28)

$$\mathscr{A} \supset \mathscr{W}_+^R(z), \tag{5.1}$$

where $\mathscr{W}_+^R(z)$ is a $\mathscr{C}^{1,\alpha}$ manifold of (Euclidean) dimension equal to $n = \dim E_+(z)$; thus

$$\dim \mathscr{A} \geq n, \tag{5.2}$$

where $\dim \mathscr{A}$ is the Hausdorff or fractal dimension of $\mathscr{A}$.

For example, we consider the semigroup associated with equation III.(1.2), (1.3), i.e., (4.13), (4.14). Then $z = 0$ is a stationary solution for which

$$g'(0) = b_1.$$

The linearized equation (see VI.(2.4)–(2.6)) reads

$$\frac{\partial U}{\partial t} - d\Delta U + b_1 U = 0, \tag{5.3}$$

$$U(0) = \xi. \tag{5.4}$$

Then according to the results in Section 2.1, $L(t, 0) \cdot \xi = U(t)$, where $L(t, 0) = S'(t) \cdot (0)$. Let λ_k, $k \in \mathbb{N}$, denote the eigenvalues associated with the Dirichlet problem

$$\begin{cases} -\Delta w_k = \lambda_k w_k, & w_k \in H_0^1(\Omega), \\ 0 < \lambda_1 \leq \lambda_2, \ldots, & \lambda_k \to \infty \quad \text{as} \quad k \to \infty. \end{cases} \tag{5.5}$$

Similarly, let μ_k, $k \in \mathbb{N}$, denote the sequence of eigenvalues of the linearized operator

$$-d\Delta\varphi_k + b_1 \varphi_k = \mu_k \varphi_k, \qquad \varphi_k \in H_0^1(\Omega). \tag{5.6}$$

By comparison $\varphi_k = w_k$ and

$$\mu_k = b_1 + d\lambda_k, \qquad \forall k. \tag{5.7}$$

The spectrum of $S'(t)$ consists of its eigenvalues and the number 0. The eigenvalues of $S'(t)$ are the numbers $\exp(-\mu_k t)$, the unstable ones correspond to $\mu_k < 0$, the stable ones correspond to $\mu_k > 0$, and the system is hyperbolic if $\mu_k \neq 0$, $\forall k$, i.e.,

$$\frac{-b_1}{d} \notin \bigcup_{j \in \mathbb{N}} \lambda_j. \tag{5.8}$$

Assuming that (5.8) holds, the number n of eigenvalues of $S'(t)$ in the disk

$\{|\lambda| < 1\}$ is the same as the number of k such that

$$\lambda_k < -\frac{b_1}{d}. \tag{5.9}$$

We conclude that

$$\dim \mathscr{A} \geq n; \tag{5.10}$$

and, of course, this number n is arbitrarily large if $b_1 < 0$ is sufficiently large in absolute value.

Remark 5.1. Lower bounds on the dimension of the attractor for the Navier Stokes equations using Theorem 3.2 and other suitable techniques can be found in A.V. Babin and M.I. Vishik [1], J.M. Ghidaglia and R. Temam [6]. An optimal lower bound (i.e. a lower bound of the same order as the upper bound), can be found in M. Ziane [1] for flows in thin two and three dimensional domains; for other flows see A. Miranville and M. Ziane [1] for thermohydraulics, and C.R. Doering and X. Wang [1] for shear flows.

CHAPTER VIII

The Cone and Squeezing Properties. Inertial Manifolds

Introduction

Our aim in this final chapter is to introduce the concept of *inertial manifolds* for dissipative dynamical systems. This is a new concept which has recently emerged in nonlinear dynamics and, here, we shall restrict ourselves to describing a few typical results. Our presentation follows with some slight simplifications and generalizations C. Foias, G. Sell, and R. Temam [1], [2], C. Foias, B. Nicolaenko, G. Sell, and R. Temam [1], [2]. The reader is referred to Remark 4.3 for some bibliographical references on this rapidly expanding subject.

At our present level of understanding of dynamical systems, the attractors are expected to be very complicated objects (fractals) and their practical utilization, for instance for numerical simulations, may be difficult. As we will see later the inertial manifolds, when they exist, are more convenient objects which are able to describe the large-time behavior of dynamical systems and permit the reduction of the infinite-dimensional case to the finite-dimensional one.

Consider in a finite- or infinite-dimensional Hilbert space H a dynamical system

$$u' = F(u), \tag{0.1}$$

with which we can associate the semigroup $\{S(t)\}_{t\geq 0}$, where $S(t)$ is the mapping

$$S(t)\colon u_0 \to u(t),$$

$u(\cdot)$ denoting the solution of (0.1) satisfying $u(0) = u_0$. An *inertial manifold* (I.M.) of this system is a finite-dimensional Lipschitz manifold $\mathcal{M}$, which enjoys the following properties:

(a) $\mathcal{M}$ is positively invariant for the semigroup (i.e., $S(t)\mathcal{M} \subset \mathcal{M}$, $\forall t \geq 0$);
(b) $\mathcal{M}$ attracts exponentially all the orbits of (0.1).

Although the finite-dimensionality condition can be relaxed in some specific cases, it will be imposed in most cases. An *interial system* for (0.1) is the system obtained by restricting (0.1) to $\mathcal{M}$; when $\mathcal{M}$ is given as the graph of a Lipschitz map, the inertial system can be written very easily as will be seen later. Of course, if $\mathcal{M}$ is finite-dimensional so is the inertial system, and then the inertial system is a finite-dimensional replica of the initial system. Property (b) for inertial manifolds is a drastic difference with attractors which often attract the orbits at a slower rate; actually, we can construct attractors which attract the orbits at an arbitrarily slow speed.

This chapter is a first introduction to inertial manifolds; another approach to inertial manifolds is presented is Chapter IX while Chapter X addresses approximation and other aspects of inertial manifolds. The chapter is organized as follows. Section 1 is devoted to the *cone property*; this is a remarkable geometric property of the orbits of certain dynamical systems which has emerged from the study of I.M., and is used in this study (C. Foias, B. Nicolaenko, G. Sell, and R. Temam [1], [2]). Section 2 contains the description of the framework and the general hypotheses. The main results (existence of an inertial manifold) are stated and proved in Section 3 (see Theorems 3.1 and 3.2). Section 4 contains some examples, and finally in Section 5 we study the stability of I.M. with respect to certain perturbations of the equation. Further aspects and results on inertial manifolds are described in Remark 4.1.

1. The Cone Property

As indicated in the Introduction, the cone property is a remarkable geometric property of the orbits of certain differential equations. In Section 1.1 we present this property in a transparent but simplified form for which the hypotheses are not minimal. Then in Section 1.2 we give a more general form of this property with minimal (or nearly minimal) hypotheses. Finally, Section 1.3 gives as a complementary remark, the relation with another important property of trajectories of dissipative dynamical systems, namely the squeezing property.

1.1. The Cone Property

Let H be a Hilbert space and let F be a continuous mapping from H into itself. We consider the initial-value problem

$$\frac{du}{dt} = F(u), \tag{1.1}$$

$$u(0) = u_0, \tag{1.2}$$

and assume that, for every u_0 in H, this problem possesses a unique solution u which is defined for all $t > 0$ and continuously differentiable from $\mathbb{R}_+$ into H. As usual, we denote by $S(t)$ the mapping $u_0 \to u(t)$ from H into itself.

We assume that we are given an orthogonal projector P in H, so that H is the direct sum of its orthogonal subspaces PH and QH, $Q = I - P$. Furthermore, we assume the following:

$$\begin{cases} (F(v_1) - F(v_2), y) \geq -\lambda |y|^2 - \mu_1 |y||z|, \\ (F(v_1) - F(v_2), z) \leq -\Lambda |z|^2 - \mu_2 |y||z|, \\ \forall v_1, v_2 \in H, \qquad y = P(v_1 - v_2), \qquad z = Q(v_1 - v_2), \end{cases} \tag{1.3}$$

$$\Lambda - \lambda > \frac{\mu_1 \gamma^2 + \mu_2}{\gamma}, \tag{1.4}$$

where Λ, λ, μ_1, μ_2, γ are positive constants; $(\cdot, \cdot)$ and $|\cdot|$ denote the scalar product and the norm in H.

We now consider *two* different solutions of (1.1), u_1, u_2, corresponding to two different initial data, $u_1(0) = u_{01}$, $u_2(0) = u_{02}$, and we set $u = u_1 - u_2$, $u(0) = u_0 = u_{01} - u_{02}$ (see Figure 1.1). We also denote by $\mathscr{C}_\gamma$ the cone of H

$$\mathscr{C}_\gamma = \{v \in H, |Qv| \leq \gamma |Pv|\}. \tag{1.5}$$

The cone property is stated in

Theorem 1.1. *The hypotheses are those above, u_1, u_2 denote two solutions of* (1.1). *Then*:

(i) *if $u_1(0) \in u_2(0) + \mathscr{C}_\gamma$, then $u_1(t) \in u_2(t) + \mathscr{C}_\gamma$, $\forall t \geq 0$;*

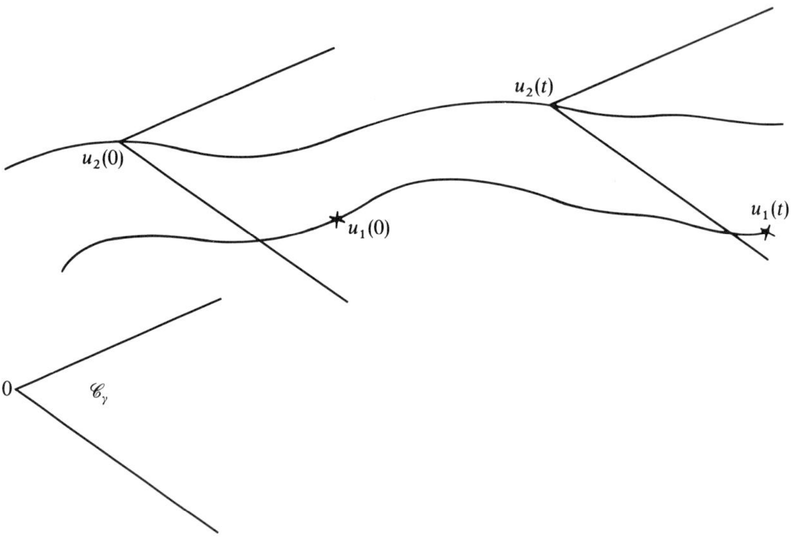

Figure 1.1. The cone property.

(ii) *if* $u_1(0) \notin u_2(0) + \mathscr{C}_\gamma$, *then*:

either $u_1(t_0) \in u_2(t_0) + \mathscr{C}_\gamma$ *for some* $t_0 > 0$, *and consequently* $u_1(t) \in u_2(t) + \mathscr{C}_\gamma$, $\forall t \geq t_0$;
or $u_1(t) \notin u_2(t) + \mathscr{C}_\gamma$, $\forall t \geq 0$, *in which case* $u_1(t) - u_2(t)$ *decays exponentially as* $t \to \infty$: *there exists* $\nu > 0$ *such that*

$$|u_1(t) - u_2(t)| \leq \frac{(1+\gamma^2)^{1/2}}{\gamma}|u_1(0) - u_2(0)| \exp(-\nu t), \qquad \forall t > 0. \quad (1.6)$$

PROOF. The proof is very simple. The difference $u = u_1 - u_2$ satisfies

$$\frac{du}{dt} = F(u_1) - F(u_2). \quad (1.7)$$

We then set $y = Pu$, $z = Qu$, take the scalar product of (1.7) with y and z successively, and use (1.3), (1.4)

$$\frac{1}{2}\frac{d}{dt}|y|^2 = (F(u_1) - F(u_2), y),$$

$$\frac{1}{2}\frac{d}{dt}|y|^2 \geq -\lambda|y|^2 - \mu_1|y||z|, \quad (1.8)$$

$$\frac{1}{2}\frac{d}{dt}|z|^2 = (F(u_1) - F(u_2), z),$$

$$\frac{1}{2}\frac{d}{dt}|z|^2 \leq -\Lambda|z|^2 + \mu_2|y||z|. \quad (1.9)$$

Combining (1.8) and (1.9) we find

$$\frac{1}{2}\frac{d}{dt}(|z|^2 - \gamma^2|y|^2) \leq -\Lambda|z|^2 + \lambda y^2|y|^2 + (\mu_1\gamma^2 + \mu_2)|y||z|. \quad (1.10)$$

For the proof of (i) we observe that if $u(t) \in \partial\mathscr{C}_\gamma$, then $|z(t)| = \gamma|y(t)|$ and

$$\frac{1}{2}\frac{d}{dt}(|z|^2 - \gamma^2|y|^2) \leq -\left((\Lambda - \lambda) - \frac{(\mu_1\gamma^2 + \mu_2)}{\gamma}\right)|z|^2,$$

and this is strictly negative thanks to (1.4), unless $z = 0$. This shows that if $u(t_1) \in \mathscr{C}_\gamma$, $u(t)$ cannot leave $\mathscr{C}_\gamma$ at $t > t_1$ through its lateral boundary $\partial\mathscr{C}_\gamma \backslash \{0\}$. It cannot leave $\mathscr{C}_\gamma$ through its vertex 0, as if $u(t_1) = 0$ then $u_1(t_1) = u_2(t_1)$ and by uniqueness, $u(t) = 0$, $\forall t > t_1$. Hence (i) and the first part of (ii).

We now prove the second part of (ii) and for that purpose assume that $u(t) \notin \mathscr{C}_\gamma$, $\forall t \geq 0$, i.e., $|z(t)| > \gamma|y(t)|$, $\forall t \geq 0$. We infer from (1.9) that

$$\frac{1}{2}\frac{d}{dt}|z|^2 \leq -\nu|z|^2, \qquad \forall t > 0,$$

with $\nu = \Lambda - \mu_2/\gamma$ (> 0 thanks to (1.4)).

Thus

$$|z(t)|^2 \le |z(0)|^2 \exp(-2\nu t), \tag{1.11}$$

$$|y(t)|^2 \le \frac{1}{\gamma^2}|z(t)|^2 \le \frac{1}{\gamma^2}|z(0)|^2 \exp(-2\nu t), \qquad \forall t \ge 0, \tag{1.12}$$

and

$$|u(t)|^2 \le \frac{1+\gamma^2}{\gamma^2}|u(0)|^2 \exp(-\nu t), \qquad \forall t \ge 0. \tag{1.13}$$

The proof of Theorem 1.1 is completed. □

Remark 1.1. We may assume that (1.3) is only valid when v_1, v_2 belong to a subset $\mathscr{B}$ of H, in which case Theorem 1.1 applies to orbits lying in $\mathscr{B}$. For instance, $\mathscr{B}$ can be a positively invariant set for the semigroup $S(\cdot)$ ($S(t)\mathscr{B} \subset \mathscr{B}$, $\forall t \ge 0$) or $\mathscr{B}$ can be an *absorbing set*, in which case Theorem 1.1 applies after the orbits $u_1(\cdot)$, $u_2(\cdot)$ enter $\mathscr{B}$.

Of course, Theorem 1.1 is related to cone invariant properties in evolution equations, but this point of view is not particularly useful here.

1.2. Generalizations

The hypotheses of Theorem 1.1 are not realistic when applications to partial differential equations are envisaged. For instance, the fact that F maps H into itself or the fact that $t \to S(t)u_0$ is $\mathscr{C}^1$ on $\mathbb{R}_+$, $\forall u_0 \in H$, is satisfied by none of the equations studied in Chapters III and IV. We will now generalize Theorem 1.1 in two directions:

by weakening the hypotheses;

by considering a function $u(\cdot)$ which is not necessarily the difference of two solutions $u_1(\cdot)$, $u_2(\cdot)$.

First, we generalize the result by considering a function $u = u(t)$, $u(t) = p(t) + q(t)$, $p(t) = Pu(t)$, $q(t) = Qu(t)$, which may not be the difference of two solutions of (1.1). Second, we observe that the proof can be completed by using only the inequalities (1.8), (1.9) (and (1.4)); and no reference to equation (1.1) is necessary.

Theorem 1.2. *We assume that H is a Hilbert space, $H = PH \oplus QH$, where P is a projector in H and $Q = I - P$. Let u be a continuous function from $\mathbb{R}_+$ into H, $y = Pu$, $z = Qu$, and assume that:*

(a) *if $u(t_0) = 0$ at some time $t_0 \ge 0$, then $u(t) = 0$, $\forall t \ge t_0$;*
(b) *the differential inequalities* (1.8), (1.9) *are satisfied when $|z(t)| \ge \gamma|y(t)|$;*
(c) (1.4) *holds.*

Finally, let $\mathscr{C}_\gamma$ be the cone (1.5). *Then we have*:

(i) *if* $u(0) \in \mathscr{C}_\gamma$, *then* $u(t) \in \mathscr{C}_\gamma$, $\forall t \geq 0$;
(ii) *if* $u(0) \notin \mathscr{C}_\gamma$, *then*:
either $u(t_0) \in \mathscr{C}_\gamma$ *for some* $t_0 > 0$, *in which case* $u(t) \in \mathscr{C}_\gamma$, $\forall t \geq t_0$,
or $u(t) \notin \mathscr{C}_\gamma$, $\forall t \geq 0$, *in which case* $u(t)$ *decays exponentially to* 0:

$$\begin{cases} |Qu(t)| \leq |Qu(0)| \exp(-\nu t), \\ |u(t)| \leq \dfrac{(1+\gamma^2)^{1/2}}{\gamma} |u(0)| \exp(-\nu t), \end{cases} \tag{1.14}$$

$\nu = \Lambda - \mu_2/\gamma > 0$.

Proof. The proof is exactly the same as that of Theorem 1.1, after (1.8), (1.9) have been established. We observe that (1.8), (1.9) are not needed in the interior of $\mathscr{C}_\gamma$ since they are used once on the boundary of $\mathscr{C}_\gamma$ and once in the exterior of $\mathscr{C}_\gamma$. The role of hypothesis (a) is to prevent the orbit $u(\cdot)$ from leaving the cone $\mathscr{C}_\gamma$ through its vertice 0; of course, this hypothesis is trivial when u is the difference of two solutions of a well-posed initial-value problem. Let us observe also that (1.14) is valid at any instant $t > 0$ such that $u(t) \notin \mathscr{C}_\gamma$, even if, subsequently, at some time $\tau > t$, $u(\tau) \in \mathscr{C}_\gamma$. □

Remark 1.2.
(i) As indicated above u may not be the difference of two solutions of an equation like (1.1). An example where u is not the difference of two solutions will appear in Section 3.3.
(ii) When applying Theorem 1.2 to the difference $u = u_1 - u_2$ of two solutions of an equation (1.1), the properties (1.8), (1.9) appear to be easy to check (and are satisfied for most of the examples in Chapters III and IV, for an appropriate P). On the contrary, the hypothesis (1.4) is a restrictive one; see the examples below.
(iii) As in Remark 1.1, inequalities (1.8), (1.9) (and thus Theorem 1.2) can be limited to orbits lying entirely in some subset $\mathscr{B}$ of H.

Remark 1.3. An "infinitesimal" version of the cone property is the following: assume that the cone property applies to the solutions of (1.1) (thanks to either (1.3) or directly (1.8), (1.9)), and assume that the solutions of (1.1) are continuously differentiable with values in H. Then

If u is a solution of (1.1), (1.2) and if at some time t_0, $u'(t_0)$ belongs to the interior of $\mathscr{C}$, then $u'(t)$ belongs to $\mathscr{C}$, $\forall t \geq t_0$ (1.15)

Indeed for small h, $0 < h \leq h_0$, $u(t_0 + h) = u(t_0) + hu'(t_0) + o(h)$ belongs to $u(t_0) + \mathscr{C}$. For any such h we can consider that we have two solutions of (1.1), namely u_1, u_2, $u_1(t) = u(t+h)$, $u_2(t) = u(t)$. By the cone property, $\forall h \leq h_0$, $\forall t \geq t_0$, $u(t+h) \in u(t) + \mathscr{C}$, and thus $u'(t) \in \mathscr{C}$.

Remark 1.4. Section 4 contains examples of evolution equations enjoying the cone property.

1.3. The Squeezing Property

The cone property is an improvement of the squeezing property first introduced for the Navier–Stokes equations in C. Foias and R. Temam [2]. For that reason the cone property is also sometimes called the strong squeezing property.

The squeezing property does not necessitate a spectral gap condition like (1.4) and is valid under more general conditions, but hereafter we will restrict ourselves to deriving it as a consequence of the cone property.

The hypotheses are those of Theorem 1.2, but, as in Theorem 1.1, $u = u_1 - u_2$, where u_1, u_2 are two solutions of an equation similar to (1.1) which lie in some set $\mathscr{B}$ where (1.8), (1.9) are valid (see Remark 1.1 and Remark 1.2). We infer from Theorem 1.2 that if $u(t_0) \notin \mathscr{C}_\gamma$, then $u(t) \notin \mathscr{C}_\gamma$ for $t \in (0, t_0)$ and (1.14) is valid at t_0.

Alternatively, we can say (and this is the *squeezing property*)

$$\begin{cases} \text{At each time } t, \text{ either} \\ \qquad |Q(u(t))| \le \gamma |Pu(t)|, \\ \text{or} \\ \qquad |u(t)| \le \dfrac{(1+\gamma^2)^{1/2}}{\gamma} |u(0)| \exp(-\nu t). \end{cases} \tag{1.15}$$

Also, the coefficient in front of $|u(0)|$ can be made arbitrarily small if we restrict t to some interval $[t_0, t_1]$, $t_0 > 0$, sufficiently large; for instance, it is clear that we have

$$|u(t)| \le \tfrac{1}{2}|u(0)| \quad \text{for} \quad t \ge t_0, \tag{1.16}$$

if we set

$$t_0 = \frac{1}{\nu} \log \frac{2(1+\gamma^2)^{1/2}}{\gamma}. \tag{1.17}$$

Thus

$$\begin{cases} \text{At each time } t \ge t_0, \text{ either} \\ \qquad |Qu(t)| \le \gamma |Pu(t)|, \\ \text{or} \\ \qquad |u(t)| \le \tfrac{1}{2}|u(0)|. \end{cases} \tag{1.18}$$

The reader is referred to C. Foias and R. Temam [2] for more details about the squeezing property; see also P. Constantin, C. Foias, and R. Temam [1]. For the cone property, see C. Foias, B. Nicolaenko, G. Sell, and R. Temam [1], [2].

2. Construction of an Inertial Manifold: Description of the Method

In Section 2.1 we describe in a fuzzy way the method employed to construct the inertial manifold. Then in Section 2.2 we describe the class of evolution equations which will be studied and give the precise assumptions; we also present a modified form of the initial equation called the prepared equation. In Section 2.3 we introduce a mapping $\mathscr{T}$ which appears in the construction process and whose fixed points define the inertial manifold. The results of existence of an inertial manifold will be stated and proved in Section 3.

2.1. Inertial Manifolds: The Method of Construction

We start by describing the method for constructing the inertial manifold in a fuzzy way, i.e., without any precise assumption on the equation and without checking that the steps of the construction can indeed be accomplished.

We consider an evolution equation

$$\frac{du}{dt} = F(u), \tag{2.1}$$

where u is a function from some interval I of $\mathbb{R}$ into H and F is a differentiable mapping from H into itself; as before H is a Hilbert space whose scalar product and norm are denoted by $(\cdot, \cdot)$, $|\cdot|$.

Let there be given a projector P in H, and let $Q = I - P$; hence H is the direct sum of the orthogonal spaces PH and QH. The elements of PH are denoted by y, y_i, those of QH are denoted by z, $z_i, \dots$.

We intend to search the inertial manifold $\mathscr{M}$ of (2.1) in the form of a graph $\{y, \Phi(y)\}$, where $y \in H$ and Φ is a Lipschitz mapping from PH into QH. If $u(t) \in \mathscr{M}$, $\forall t \in I$, then (2.1) implies

$$\frac{du}{dt} = F(y + \Phi(y)), \tag{2.2}$$

with $y = Pu$; projecting (2.2) on PH and QH we find

$$\begin{cases} \dfrac{dy}{dt} = PF(y + \Phi(y)), \\[2ex] \dfrac{d\Phi(y)}{dt} = QF(y + \Phi(y)). \end{cases} \tag{2.3}$$

Assume now that Φ (thus $\mathscr{M}$) is known and that $u_0 = \{y_0, \Phi(y_0)\}$ lies on $\mathscr{M}$. Then, with appropriate assumptions (and, in particular, that PH has finite dimension), (2.3) together with the initial condition

$$y(0) = y_0 \tag{2.4}$$

define in a unique way $y(t) = y(t; y_0, \Phi)$, *for all* $t \in \mathbb{R}$.

Then we modify the second equation (2.3) and define z ($\neq \Phi(y)$) in the following way:

$$\frac{dz}{dt} = QF_1(y + z) + QF_2(y + \Phi(y)), \tag{2.5}$$

where $F = F_1 + F_2$ is a decomposition of F such that (2.5) possesses a unique solution z defined and bounded from $\mathbb{R}$ into QH (y as defined above); hence $z(t) = z(t; y_0, \Phi)$.

Now if the graph of Φ is invariant under the semigroup S, then

$$z(t) = \Phi(y(t))$$

for all t and, in particular, at $t = 0$,

$$z(0) = z(0; y_0, \Phi) = z_0 = \Phi(y_0). \tag{2.6}$$

Thus, the construction of $\mathcal{M}$ proceeds as follows:

Given Φ in an appropriate class $\mathcal{F}$ of functions from PH into QH, we define $\mathcal{T}\Phi$ by setting

$$\mathcal{T}\Phi(y_0) = z(0; y_0, \Phi).$$

Thus, when properly defined, $\mathcal{T}\Phi$ maps PH into QH and under appropriate hypotheses $\mathcal{T}\Phi$ belongs to $\mathcal{F}$.

Equation (2.6) is then equivalent to

$$\mathcal{T}\Phi(y_0) = \Phi(y_0), \qquad \forall y_0 \in PH,$$

or

$$\mathcal{T}\Phi = \Phi, \tag{2.7}$$

i.e., Φ is a fixed point of $\mathcal{T}$.

A fixed point of $\mathcal{T}$ will be obtained by application of the strict contraction principle and it will be proved that the graph of the fixed point Φ of $\mathcal{T}$ is indeed an inertial manifold for (2.1).

2.2. The Initial and Prepared Equations

As before H denotes a Hilbert space. Let there be given a linear closed unbounded positive self-adjoint operator A in H, with domain $D(A) \subset H$; $D(A)$ is a Hilbert space when endowed with the norm $|Au|$, and A is an isomorphism from $D(A)$ onto H. We also assume that A^{-1} is compact in H.

As in Section II.2, we associate with A its powers A^s, $s \in \mathbb{R}$, defined on $D(A^s)$ and we consider the eigenvalues λ_j and the eigenvectors w_j of A, which form an orthonormal basis of H

$$\begin{cases} Aw_j = \lambda_j w_j, & j = 1, \ldots, \\ 0 < \lambda_1 \leq \lambda_2, \ldots, \lambda_j \to \infty & \text{as} \quad j \to \infty. \end{cases} \tag{2.8}$$

We are also given a (nonlinear) operator R which enjoys the following property:

$$\begin{gathered}\text{For some } \alpha \in \mathbb{R}, R \text{ is Lipschitz on the bounded sets of } D(A^\alpha) \\ \text{with values in } D(A^{\alpha-1/2}) \\ |A^{\alpha-1/2}R(u) - A^{\alpha-1/2}R(v)| \le c_M |A^\alpha(u-v)|, \\ \forall u, v \in D(A^\alpha), \quad |A^\alpha u| \le M, \quad |A^\alpha v| \le M. \end{gathered} \tag{2.9}$$

This obviously implies

$$R \text{ is a bounded mapping from } D(A^\alpha) \text{ into } D(A^{\alpha-1/2}). \tag{2.10}$$

The evolution equation that we will consider is then

$$\frac{du}{dt} + Au + R(u) = 0; \tag{2.11}$$

thus by comparison with (2.1), $F(u) = -Au - R(u)$.

We are not interested in this chapter in studying the questions which were the object of the previous ones: existence and uniqueness of solutions to (2.11), existence of an absorbing set and of a global attractor. Hence we assume that with appropriate assumptions, the following properties have been proved for (2.11):

$$\begin{gathered}\text{For every } u_0 \in D(A^\alpha), \text{(2.11) possesses a unique solution } u \\ \text{defined on } \mathbb{R}_+, \text{ satisfying } u(0) = u_0 \text{ and} \\ u \in \mathscr{C}(\mathbb{R}_+; D(A^\alpha)) \cap L^2(0, T; D(A^{\alpha+1/2})), \qquad \forall T > 0. \\ \text{Furthermore, the mapping } S(t)\colon u_0 \to u(t) \text{ is continuous from} \\ D(A^\alpha) \text{ into itself, } \forall t \ge 0. \end{gathered} \tag{2.12}$$

$$\begin{gathered}\text{The semigroup } S(\cdot) \text{ possesses an absorbing set } \mathscr{B}_0 \text{ in } D(A^\alpha), \\ \text{which is positively invariant } (S(t)\mathscr{B}_0 \subset \mathscr{B}_0, \forall t \ge 0). \\ \text{The } \omega\text{-limit set of } \mathscr{B}_0 \text{ denoted } \mathscr{A} \text{ is the maximal attractor for} \\ S(\cdot) \text{ in } D(A^\alpha). \end{gathered} \tag{2.13}$$

In order to avoid the difficulties related to the behavior of the (nonlinear) term $R(u)$ for large values of $|A^\alpha u|$, we associate with (2.11) a truncated form of the equation, called the prepared equation, and that we now define.

We choose $\rho > 0$ such that the absorbing set $\mathscr{B}_0$ (and hence $\mathscr{A}$) is included in the ball of $D(A^\alpha)$ centered at 0 of radius $\rho/2$. We also choose a $\mathscr{C}^\infty$ function θ from $\mathbb{R}_+$ into $[0, 1]$

$$\theta(s) = 1 \quad \text{for} \quad 0 \le s \le 1, \qquad \theta(s) = 0 \quad \text{for} \quad s \ge 2,$$

$$\sup_{s \ge 0} |\theta'(s)| \le 2,$$

and we set $\theta_\rho(s) = \theta(s/\rho)$. We also write

$$R_\theta(u) = \theta_\rho(|A^\alpha u|)R(u), \qquad \forall u \in D(A^\alpha), \tag{2.14}$$

and the prepared equation is that in which R_θ replaces R

$$\frac{du}{dt} + Au + R_\theta(u) = 0. \tag{2.15}$$

It is easy to see that the initial value problem for (2.15) is well posed and we denote by $S_\theta(t)$ the corresponding mapping $u(0) \to u(t)$.

Of course, if $u \in \mathscr{B}_0$, (2.11) and (2.15) coincide. Thus, for every solution u of (2.11), there exists a time $t_0 = t_0(u_0)$ after which u is (also) the solution of (2.15). Furthermore, for the converse to be true we assume the following:

The ball of $D(A^\alpha)$ centered at 0 of radius ρ is absorbing for (2.15). (2.16)

We conclude Section 2.2 by proving some technical but useful properties of R_θ.

Lemma 2.1. *R_θ is a (globally) bounded operator from $D(A^\alpha)$ into $D(A^{\alpha-1/2})$*

$$\operatorname*{Sup}_{u \in D(A^\alpha)} |A^{\alpha-1/2} R_\theta(u)| \le \operatorname*{Sup}_{|A^\alpha u| \le 2\rho} |A^{\alpha-1/2} R(u)| = M_1. \tag{2.17}$$

This lemma is obvious. We also have

Lemma 2.2. *R_θ is a (globally) Lipschitz mapping from $D(A^\alpha)$ into $D(A^{\alpha-1/2})$.*

PROOF. We consider $u_1, u_2 \in D(A^\alpha)$ and set $\theta_i = \theta_\rho(|A^\alpha u_i|)$, $i = 1, 2$, and

$$L = |\theta_1 A^{\alpha-1/2} R(u_1) - \theta_2 A^{\alpha-1/2} R(u_2)|. \tag{2.18}$$

We want to show that

$$L \le \text{const}\, |A^\alpha(u_1 - u_2)|,$$

and for that purpose we consider three different cases:

If $|A^\alpha u_1|$ and $|A^\alpha u_2| \ge 2\rho$, then $\theta_i = 0$, $L = 0$, and the inequality is obvious.

If $|A^\alpha u_1| \le 2\rho \le |A^\alpha u_2|$, let u_* be the intersection of the segment $[u_1, u_2]$ with the boundary of the ball $B(\theta, 2\rho)$ of $D(A^\alpha)$, centered at θ of radius 2ρ; then $\theta_\rho(|A^\alpha u_*|) = 0$ and

$$\begin{aligned} L &= (\theta_\rho(|A^\alpha u_1|) - \theta_\rho(|A^\alpha u_*|))|A^{\alpha-1/2}R(u_1)| \\ &\le (\text{see (2.14) and (2.17)}) \\ &\le M_1 \cdot \frac{2}{\rho}|A^\alpha(u_1 - u_*)| \le M_1 \cdot \frac{2}{\rho}|A^\alpha(u_1 - u_2)|. \end{aligned}$$

Of course, the case $|A^\alpha u_2| \le 2\rho \le |A^\alpha u_1|$ is similar. Finally,

If $|A^\alpha u_1|$ and $|A^\alpha u_2| \le 2\rho$, then

$$\begin{aligned} L &\le |\theta_1 - \theta_2||A^{\alpha-1/2}R(u_1)| + \theta_2|A^{\alpha-1/2}(R(u_1) - R(u_2))| \\ &\le (\text{see (2.9)}) \\ &\le \frac{2M_1}{\rho}|A^\alpha(u_1 - u_2)| + c_{2\rho}|A^\alpha(u_1 - u_2)|. \end{aligned}$$

Hence, in all cases,

$$L \leq M_2 |A^\alpha(u_1 - u_2)|, \qquad M_2 = \frac{2M_1}{\rho} + c_{2\rho}, \tag{2.19}$$

and Lemma 2.2 is proved. □

2.3. The Mapping $\mathscr{T}$

We consider for some $N \in \mathbb{N}$ the operators

$$P = P_N, \qquad Q = I - P = I - P_N, \tag{2.20}$$

where P_N is the projector in H onto the space spanned by $w_1, \dots, w_N$. We recall that P and Q commute with A^β, $\forall \beta \in \mathbb{R}$.

For b, $l > 0$ given we define the set $\mathscr{F} = \mathscr{F}_{b,l}^\alpha$ of Lipschitz functions from $PD(A^\alpha)$ into $QD(A^\alpha)$ satisfying

$$\begin{cases} \text{Supp } \Phi \subset \{y \in PD(A^\alpha), |A^\alpha y| \leq 2\rho\}, \\ |A^\alpha \Phi(y)| \leq b, \qquad \forall y \in PD(A^\alpha), \\ |A^\alpha \Phi(y_1) - A^\alpha \Phi(y_2)| \leq l |A^\alpha(y_1 - y_2)|, \qquad \forall y_1, y_2 \in D(A^\alpha). \end{cases} \tag{2.21}$$

Note that

$$d(\Phi_1, \Phi_2) = \operatorname*{Sup}_{y \in PD(A^\alpha)} |A^\alpha(\Phi_1(y) - \Phi_2(y))| \tag{2.22}$$

is a distance on $\mathscr{F} = \mathscr{F}_{b,l}^\alpha$ and that $\mathscr{F}$ is complete for this distance.

For Φ given in $\mathscr{F}$ and y_0 given in $PD(A^\alpha)$ we consider (see (2.3)) $y = y(t; y_0, \Phi)$ the solution of

$$\frac{dy}{dt} + Ay + PR_\theta(y + \Phi(y)) = 0, \qquad y(0) = y_0. \tag{2.23}$$

This equation is the same as the first equation (2.3) (when $F = -A - R_\theta$) since, owing to (2.20), $PAy = APy = Ay$. Due to Lemma 2.2 and (2.21), the mapping $\sigma \to PR_\theta(\sigma + \Phi(\sigma))$ is Lipschitzian. Since $PD(A^\alpha)$ is finite dimensional, we infer from standard theorems on ordinary differential equations that (2.23) possesses a unique solution $y = y(t; y_0, \Phi)$, which is defined for all $t \in \mathbb{R}$.

We consider then the analog of (2.5)

$$\frac{dz}{dt} + Az + QR_\theta(y + \Phi(y)) = 0. \tag{2.24}$$

At this point, $\sigma = -QR_\theta(y + \Phi(y))$ is known and, thanks to Lemma 2.1, belongs to $L^\infty(\mathbb{R}; D(A^{\alpha - 1/2}))$. Because of Lemma 2.3 below, (2.24) possesses a unique solution z which is continuous and bounded from $\mathbb{R}$ into $QD(A^\alpha)$. We denote this solution by $z = z(t; y_0, \Phi)$

$$z \in \mathscr{C}_b(\mathbb{R}; QD(A^\alpha)). \tag{2.25}$$

In particular, $z(0) = z(0; y_0, \Phi)$ makes sense and belongs to $QD(A^\alpha)$. The function

$$y_0 \in PD(A^\alpha) \to z(0; y_0, \Phi) \in QD(A^\alpha) \tag{2.26}$$

maps $PD(A^\alpha)$ into $QD(A^\alpha)$ and we call this function $\mathscr{T}\Phi$. Subsequently, we will show that $\mathscr{T}\Phi$ belongs to $\mathscr{F}$ and for the moment we conclude that we have defined a function $\mathscr{T}$ which maps $\mathscr{F}$ into the set $\tilde{\mathscr{F}}$ of all functions from $PD(A^\alpha)$ into $QD(A^\alpha)$

$$\mathscr{T}: \mathscr{F} = \mathscr{F}_{b,l}^{\alpha} \to \tilde{\mathscr{F}}. \tag{2.27}$$

The study of the properties of $\mathscr{T}$ is part of Theorem 3.1 given in Section 3.

For the moment we end this section with the lemma promised above. In this lemma (related to Proposition IV.6.1) H is an arbitrary Hilbert space; in particular, it was applied above with H replaced by QH. Note also that the expression of $z(0; y_0, \Phi)$ resulting from formula (2.35) below provides an integral expression for $\mathscr{T}\Phi(y_0)$ which is made explicit hereafter in Remark 2.1.

Lemma 2.3. *For any $\alpha \in \mathbb{R}$, let σ be given in $L^\infty(\mathbb{R}; D(A^{\alpha-1/2}))$. Then there exists a unique function ξ which is continuous and bounded from $\mathbb{R}$ into $D(A^\alpha)$ and satisfies*

$$\frac{d\xi}{dt} + A\xi = \sigma. \tag{2.28}$$

Proof. (i) The first part of the proof consists in studying the initial-value problem for the homogeneous analog of (2.28)

$$\frac{d\xi}{dt} + A\xi = 0, \qquad \xi(0) = \xi_0. \tag{2.29}$$

This equation is standard; for any $\xi_0 \in D(A^\alpha)$, $\alpha \in \mathbb{R}$, (2.29) possesses a unique solution belonging to $\mathscr{C}([0, \infty[; D(A^\alpha)) \cap \mathscr{C}(]0, \infty[; D(A^{\alpha+1/2}))$. In particular, the linear operator

$$e^{-tA}: \xi_0 \to \xi(t)$$

is continuous from $D(A^{\alpha-1/2})$ into $D(A^\alpha)$, $\forall t > 0$, and from $D(A^\alpha)$ into itself, $\forall t \geq 0$.

The norm of e^{-tA} in $D(A^\alpha)$ is estimated in the usual manner: (2.29) yields

$$\frac{1}{2}\frac{d}{dt}|A^\alpha \xi|^2 + |A^{\alpha+1/2}\xi|^2 = 0, \tag{2.30}$$

$$\frac{d}{dt}|A^\alpha \xi|^2 + 2\lambda_1 |A^\alpha \xi|^2 \leq 0, \tag{2.31}$$

$$|A^\alpha \xi(t)|^2 \leq |A^\alpha \xi(0)|^2 \exp(-2\lambda_1 t),$$

$$|e^{-tA}|_{\mathscr{L}(D(A^\alpha))} \leq \exp(-\lambda_1 t), \qquad \forall \alpha \in \mathbb{R}, \quad \forall t \geq 0. \tag{2.32}$$

We then estimate the norm of e^{-tA} in $\mathscr{L}(D(A^{\alpha-1/2}), D(A^\alpha))$. Replacing first α

by $\alpha - \frac{1}{2}$ in (2.30) we find

$$\frac{d}{dt}|A^{\alpha-1/2}\xi|^2 + 2|A^\alpha\xi|^2 = 0,$$

$$\frac{d}{dt}\{\exp(2\lambda_1 t)|A^{\alpha-1/2}\xi(t)|^2\} + 2\exp(2\lambda_1 t)|A^\alpha\xi(t)|^2$$
$$= 2\lambda_1 \exp(2\lambda_1 t)|A^{\alpha-1/2}\xi(t)|^2,$$

$$2\int_0^t \exp(2\lambda_1 s)|A^\alpha\xi(s)|^2\,ds$$
$$\leq |A^{\alpha-1/2}\xi(0)|^2 + 2\lambda_1\int_0^t \exp(2\lambda_1 s)|A^{\alpha-1/2}\xi(s)|^2\,ds.$$

Thus using (2.32) for $\alpha - \frac{1}{2}$ we obtain

$$\int_0^t \exp(2\lambda_1 s)|A^\alpha\xi(s)|^2\,ds \leq (\tfrac{1}{2} + t\lambda_1)|A^{\alpha-1/2}\xi(0)|^2. \tag{2.33}$$

Then we infer from (2.31) that

$$\frac{d}{dt}(t|A^\alpha\xi|^2) + 2\lambda_1 t|A^\alpha\xi(t)|^2 \leq |A^\alpha\xi(t)|^2,$$

and by utilization of Gronwall's lemma

$$t|A^\alpha\xi(t)|^2 \leq \int_0^t |A^\alpha\xi(s)|^2\exp(-2\lambda_1(t-s))\,ds$$
$$\leq (\text{with } (2.33))$$
$$\leq (\tfrac{1}{2} + t\lambda_1)\exp(-2\lambda_1 t)|A^{\alpha-1/2}\xi(0)|^2.$$

Hence

$$|e^{-tA}|_{\mathscr{L}(D(A^{\alpha-1/2}),D(A^\alpha))} \leq \left(\frac{1}{2t} + \lambda_1\right)^{1/2}\exp(-\lambda_1 t), \qquad \forall\alpha\in\mathbb{R},\quad \forall t > 0. \tag{2.34}$$

(ii) We prove the uniqueness of a solution remaining bounded as $t \to -\infty$; if ξ_1, ξ_2 are two such solutions of (2.28), then $\xi = \xi_1 - \xi_2$ satisfies (2.28) with $\sigma = 0$ and thus (2.29)–(2.31). By integration of (2.31) between t_0 and t, $t_0 < t$, we obtain

$$|A^\alpha\xi(t)|^2 \leq |A^\alpha\xi(t_0)|^2\exp(-2\lambda_1(t - t_0)).$$

When $t_0 \to -\infty$, the right-hand side of this inequality tends to 0 since $|A^\alpha\xi(t_0)|$ remains bounded. Thus $|A^\alpha\xi(t)| = 0$.

(iii) Integrating (2.28) between t_0 and t, $t_0 < t$, we can write

$$\xi(t) = e^{-(t-t_0)A}\xi(t_0) + \int_{t_0}^t e^{-(t-\tau)A}\sigma(\tau)\,d\tau.$$

If ξ is a bounded solution of (2.28) then, as $t_0 \to -\infty$, $e^{-(t-t_0)A}\xi(t_0)$ tends to 0 thanks to (2.32) and we obtain at the limit

$$\xi(t) = \int_{-\infty}^{t} e^{-(t-\tau)A}\sigma(\tau)\, d\tau. \tag{2.35}$$

Thus if ξ exists, then it is necessarily equal to the right-hand side of (2.35) and it just remains to show that the integral in (2.35) makes sense and enjoys the required properties.

Due to (2.34)

$$\int_{-\infty}^{t} |A^\alpha e^{-(t-\tau)A}\sigma(\tau)|\, d\tau \le |\sigma|_\infty \int_{-\infty}^{t} \|e^{-(t-\tau)A}\|\, d\tau,$$

where $|\sigma|_\infty$ is the norm of σ in $L^\infty(\mathbb{R}; D(A^{\alpha-1/2}))$ and $\|e^{-(t-\tau)A}\|$ is the norm of this operator in $\mathscr{L}(D(A^{\alpha-1/2}), D(A^\alpha))$. Owing to (2.34)

$$\int_{-\infty}^{t} \|e^{-(t-\tau)A}\|\, d\tau \le \int_{-\infty}^{t} \left(\frac{1}{2(t-\tau)} + \lambda_1\right)^{1/2} \exp(-\lambda_1(t-\tau))\, d\tau \le \frac{\kappa_1}{\sqrt{\lambda_1}},$$

where

$$\kappa_1 = \int_0^\infty e^{-s}\left(1 + \frac{1}{2s}\right)^{1/2} ds < \infty. \tag{2.36}$$

We conclude that the integral in (2.35) exists and

$$|A^\alpha \xi(t)| \le \frac{\kappa_1}{\sqrt{\lambda_1}}|\sigma|_\infty, \qquad \forall t \in \mathbb{R}. \tag{2.37}$$

Thus this function ξ is in $L^\infty(\mathbb{R}; D(A^\alpha))$; the fact that ξ is solution of (2.28) and that $\xi \in \mathscr{C}(\mathbb{R}; D(A^\alpha))$ is then standard. □

Remark 2.1. We infer from (2.35) a useful integral expression of $\mathscr{T}\Phi(y_0) = z(0; y_0, \Phi)$ which is $\xi(0)$ in (2.35) when $\sigma = -QR_\theta(y + \Phi(y))$

$$\mathscr{T}\Phi(y_0) = -\int_{-\infty}^{0} e^{\tau A} QR_\theta(y(\tau) + \Phi(y(\tau)))\, d\tau,$$
$$\forall y_0 \in PD(A^\alpha), \quad \Phi \in \mathscr{F}_{b,l}^\alpha. \tag{2.38}$$

3. Existence of an Inertial Manifold

We are still interested in the equation of Section 2, either in its initial form (2.11) or in its prepared form (2.15). We construct an inertial manifold of this equation by constructing a fixed point of the operator $\mathscr{T}$. Section 3.1 contains the statement of the main result. Sections 3.2 and 3.3 provide the proof of this result.

3.1. The Result of Existence

All the assumptions of Sections 2.2 and 2.3 are retained. We search the inertial manifold of (2.11), (2.15) in the form of the graph of a function Φ in $\mathscr{F}_{b,l}^{\alpha}$

$$\mathscr{M} = \{\{y, \Phi(y)\}, y \in PD(A^{\alpha})\}. \tag{3.1}$$

As indicated in the Introduction to this chapter, an inertial manifold $\mathscr{M}$ for (2.11), (2.16) is a finite-dimensional manifold enjoying the following three properties:

$\mathscr{M}$ is Lipschitz (3.2)

$\mathscr{M}$ is positively invariant for the semigroup, i.e.,
$S(t)\mathscr{M} \subset \mathscr{M}, \forall t \geq 0.$ (3.3)

$\mathscr{M}$ attracts exponentially all the orbits of (2.11), (2.15). (3.4)

For the sake of simplicity in the notations we set $\lambda_N = \lambda$, $\lambda_{N+1} = \Lambda$. Therefore we can write

$$|A^{\alpha+1/2}z|^2 = (A^{\alpha+1}z, A^{\alpha}z) \geq \Lambda|A^{\alpha}z|^2, \qquad \forall z \in QD(A^{\alpha}), \tag{3.5}$$

$$|A^{\alpha+1/2}y|^2 = (A^{\alpha+1}y, A^{\alpha}y) \leq \lambda|A^{\alpha}y|^2, \qquad \forall y \in PD(A^{\alpha}). \tag{3.6}$$

The assumptions recapitulated in the statement of Theorem 3.1 are those of Sections 2.2 and 2.3. We supplement them with the following technical hypotheses appearing in the course of the proof: we are given l, $0 < l \leq \frac{1}{8}$ and we assume that

$$\Lambda > M_2^2\left\{\frac{1+l}{l} + 4\kappa_4 + 11\right\}^{1/2}, \tag{3.7}$$

$$\Lambda - \lambda > 2M_2\frac{(1+l)}{l}(\lambda^{1/2} + \Lambda^{1/2}), \tag{3.8}$$

where M_2 is defined in (2.19) while κ_4 is an absolute constant appearing below.

Remark 3.1. We recall that $P = P_N$, $\Lambda = \lambda_{N+1}$, $\lambda = \lambda_N$, for some appropriate N. In this respect, condition (3.7) means that Λ is sufficiently large whereas condition (3.8) will appear to be a *spectral gap* condition; see the examples in Section 4.

We then have

Theorem 3.1. *The hypotheses are those of* Sections 2.2 *and* 2.3, *in particular,* (2.9), (2.12), (2.13), (2.16), (2.20). *We also assume that l is given,* $0 < l \leq \frac{1}{8}$, *and that* (3.5)–(3.8) *hold.*

Then there is a $b > 0$ such that:

(i) *$\mathscr{T}$ is a strict contraction of $\mathscr{F}^{\alpha}_{b,l}$ into itself and by the Contraction Principle it possesses a unique fixed point $\Phi \in \mathscr{F}^{\alpha}_{b,l}$;*
(ii) *the graph $\mathscr{M}$ of Φ is an inertial manifold for equations* (2.11) and (2.15).[1]

Remark 3.2. The inertial manifold $\mathscr{M}$ necessarily contains the universal attractor $\mathscr{A}$. Indeed, let $u_* \in \mathscr{A}$; then for any $t > 0$ there exists $u^t \in \mathscr{A}$ such that $u_* = S(t)u^t$. By the exponential decay property (see (3.47)), there exists $\eta_1, \eta_2 > 0$ such that

$$\operatorname{dist}(u_*, \mathscr{M}) = \operatorname{dist}(S(t)u^t, \mathscr{M}) \leq \operatorname{dist}(u^t, \mathscr{M})\eta_1 \cdot \exp(-\eta_2 t).$$

Since $\mathscr{A}$ is bounded, by letting $t \to +\infty$, we find $\operatorname{dist}(u_*, \mathscr{M}) = 0$, i.e., $u_* \in \mathscr{M}$ since $\mathscr{M}$ is closed.

Remark 3.3. See at the end of Section 3 a slightly different form of Theorem 3.1 (Theorem 3.2).

Remark 3.4. As appears in the course of the proof of Theorem 3.1 (see Section 3.4), the hypotheses of Theorem 3.1 imply that Theorem 1.2 (i.e., the cone property for the difference of two solutions) applies to equation (2.15) and thus to equation (2.11), inside the absorbing set.

3.2. First Properties of $\mathscr{T}$

We start with

Lemma 3.1. *For every $\Phi \in \mathscr{F}$,*

$$\operatorname{Supp} \mathscr{T}\Phi \subset \{y \in PD(A^\alpha), |A^\alpha y| \leq 2\rho\}. \tag{3.9}$$

Proof. Let y_0 be given such that $|A^\alpha y_0| > 2\rho$. Then, setting $u_0 = y_0 + \Phi(y_0)$,

$$|A^\alpha u_0| = (|A^\alpha y_0|^2 + |A^\alpha \Phi(y_0)|^2)^{1/2} \geq |A^\alpha y_0| > 2\rho,$$

so that $\theta_\rho(|A^\alpha u_0|) = 0$. By continuity, for t small, $\theta_\rho(|A^\alpha u(t)|) = 0$, $u(t) = y(t) + \Phi(y(t))$, and then y satisfies

$$\frac{dy}{dt} + Ay = -PR_\theta(u) = 0. \tag{3.10}$$

Hence

$$\frac{1}{2}\frac{d}{dt}|A^\alpha y|^2 + |A^{\alpha+1/2} y|^2 = 0,$$

$$\frac{d}{dt}|A^\alpha y| + \lambda_1 |A^\alpha y| \leq 0,$$

[1] We have $S_\theta(t)\mathscr{M} \subset \mathscr{M}$, $\forall t \geq 0$, but the property $S(t)\mathscr{M} \subset \mathscr{M}$ is not satisfied for all $t \geq 0$. Strictly speaking condition (3.3) is not satisfied for (2.11) (i.e., $S(t)$). However, we have $S(t)(\mathscr{M} \cap \mathscr{B}_0) \subset \mathscr{M}$, $\forall t \geq 0$, since $\mathscr{B}_0$ is positively invariant.

and for $\tau < 0$, $|\tau|$ sufficiently small,

$$|A^\alpha y(\tau)| \geq |A^\alpha y_0| \exp(-\tau\lambda_1) \geq |A^\alpha y_0| > 2\rho.$$

Consequently, $|A^\alpha y(\tau)| > 2\rho$, $\forall \tau < 0$, and

$$|A^\alpha(y(\tau) + \Phi(y(\tau)))| \geq |A^\alpha y(\tau)| > 2\rho, \qquad \forall \tau < 0.$$

The equation defining z reduces for $t < 0$ to

$$\frac{dz}{dt} + Az = -QR_\theta(y + \Phi(y)) = 0;$$

the unique solution of this equation remaining bounded as $t \to -\infty$, vanishes on $(-\infty, 0]$; in particular,

$$z(0) = z(0; y_0, \Phi) = \mathscr{T}\Phi(y_0) = 0,$$

The lemma is proved. □

The following technical result will be useful:

Lemma 3.2. *For every $\sigma \in \mathbb{R}$ and $\tau < 0$ the norm of $(AQ)^\sigma e^{\tau AQ}$ in $\mathscr{L}(QH)$ is bounded by*

$$\kappa_2(\sigma)|\tau|^{-\sigma} \quad \text{if} \quad -\frac{\sigma}{\Lambda} \leq \tau < 0,$$

$$\Lambda^\sigma e^{\tau\Lambda} \quad \text{if} \quad \tau < -\sigma/\Lambda.$$

If $\sigma < 1$

$$\int_{-\infty}^{0} |(AQ)^\sigma e^{\tau AQ}|_{\mathscr{L}(QH)}\, d\tau \leq \kappa_3(\sigma)\Lambda^{\sigma-1}, \tag{3.11}$$

where $\kappa_2(\sigma)$, $\kappa_3(\sigma)$ denote appropriate constants depending on σ.

Proof. An element v of QH is written (see (2.20)) $v = \sum_{j=N+1}^{\infty} b_j w_j$. Thus

$$(AQ)^\sigma e^{\tau AQ} v = \sum_{j=N+1}^{\infty} \lambda_j^\sigma e^{\tau\lambda_j} b_j w_j,$$

$$|(AQ)^\sigma e^{\tau AQ} v|^2 = \sum_{j=N+1}^{\infty} (\lambda_j^\sigma e^{\tau\lambda_j})^2 b_j^2$$

$$\leq \operatorname*{Sup}_{r\geq\Lambda} (r^\sigma e^{\tau r})^2 \cdot |v|^2,$$

$$|(AQ)^\sigma e^{\tau AQ}|_{\mathscr{L}(QH)} \leq \operatorname*{Sup}_{r\geq\Lambda} (r^\sigma e^{\tau r}). \tag{3.12}$$

The supremum in (3.12) is determined by elementary computations. We find $\kappa_2|\tau|^{-\sigma}$, $\kappa_2 = \sigma^\sigma e^{-\sigma}$ if $\sigma \geq 0$ and $-\sigma/\Lambda \leq \tau < 0$. If $\sigma \geq 0$ and $\tau \leq -\sigma/\Lambda$ or if $\sigma < 0$, $\tau < 0$, this supremum is $\Lambda^\sigma e^{\tau\Lambda}$.

Using this estimate, we easily prove (3.11) with $\kappa_3 = 1$ if $\sigma < 0$ and, for

$0 \leq \sigma < 1$,

$$\kappa_3 = e^{-\sigma} + \frac{\kappa_2}{1-\sigma}(\sigma)^{1-\sigma}. \qquad \square \tag{3.13}$$

Lemma 3.3. *There exists a constant M_3 such that*

$$\operatorname*{Sup}_{y_0 \in PD(A^\alpha)} |A^\alpha \mathcal{T}\Phi(y_0)| \leq M_3 \Lambda^{-1/2}. \tag{3.14}$$

PROOF. Due to the expression (2.38) of $\mathcal{T}$,

$$\begin{aligned} |A^\alpha \mathcal{T}\Phi(y_0)| &= \left| \int_{-\infty}^{0} A^\alpha e^{\tau AQ} R_\theta(y + \Phi(y))\, d\tau \right| \\ &\leq \int_{-\infty}^{0} |(AQ)^\alpha e^{\tau AQ} R_\theta(y + \Phi(y))|\, d\tau \\ &\leq \text{(with (2.17) and (3.11))} \\ &\leq M_1 \int_{-\infty}^{0} |(AQ)^{1/2} e^{\tau AQ}|_{\mathcal{L}(QH)}\, d\tau \\ &\leq \kappa_4 M_1 \Lambda^{-1/2}. \end{aligned}$$

$\square$

3.3. Utilization of the Cone Property

We now use the cone property in its generalized form given in Section 1.2 for showing that $\mathcal{T}\Phi$ is Lipschitz in $D(A^\alpha)$.

Lemma 3.4. *Let there be given $\gamma > 0$ and assume that*

$$\Lambda > \left(M_2 \frac{(1+l)}{2\gamma}\right)^2, \tag{3.7'}$$

$$\Lambda - \lambda > M_2(1+l)\left(\lambda^{1/2} + \frac{1}{\gamma}\Lambda^{1/2}\right). \tag{3.8'}$$

Then for every Φ in $\mathcal{F}_{b,l}^\alpha$

$$|A^\alpha(\mathcal{T}\Phi(y_{01}) - \mathcal{T}\Phi(y_{02}))| \leq \gamma |A^\alpha(y_{01} - y_{02})|, \qquad \forall y_{01}, y_{02} \in P(D(A^\alpha)). \tag{3.15}$$

In particular, if (3.7), (3.8) *are satisfied,* (3.15) *holds with* $\gamma = l/2$.

PROOF. Let there be given Φ in $\mathcal{F}_{b,l}^\alpha$ and y_{01}, y_{02} in $PD(A^\alpha)$. We set

$$\begin{aligned} &y_i(t) = y(t; y_{0i}, \Phi), \qquad z_i(t) = z(t; y_{0i}, \Phi), \\ &u_i(t) = y_i(t) + \Phi(y_i(t)), \qquad i = 1, 2, \\ &y(t) = y_1(t) - y_2(t), \qquad z(t) = z_1(t) - z_2(t). \end{aligned}$$

Hence

$$\frac{dy_i}{dt} + Ay_i + PR_\theta(u_i) = 0, \qquad i = 1, 2, \tag{3.16}$$

$$\frac{dy}{dt} + Ay = -(PR_\theta(u_1) - PR_\theta(u_2)), \tag{3.17}$$

$$\begin{aligned}\frac{1}{2}\frac{d}{dt}|A^\alpha y|^2 + |A^{\alpha+1/2}y|^2 &= -(A^{\alpha-1/2}(PR_\theta(u_1) - PR_\theta(u_2)), A^{\alpha+1/2}y)\\ &\geq -|A^{\alpha-1/2}(PR_\theta(u_1) - PR_\theta(u_2))||A^{\alpha+1/2}y|\\ &\geq \text{(with Lemma 2.2)}\\ &\geq -M_2|A^\alpha(u_1 - u_2)||A^{\alpha+1/2}y|.\end{aligned}$$

We have

$$|A^\alpha(u_1 - u_2)| = |A^\alpha y + A^\alpha(\Phi(y_1) - \Phi(y_2))|,$$

$$|A^\alpha(u_1 - u_2)| \leq (1 + l)|A^\alpha y|. \tag{3.18}$$

Therefore

$$\frac{1}{2}\frac{d}{dt}|A^\alpha y|^2 \geq -|A^{\alpha+1/2}y|^2 - M_2(1 + l)|A^{\alpha+1/2}y||A^\alpha y|,$$

and using (3.6),

$$\frac{1}{2}\frac{d}{dt}|A^\alpha y|^2 \geq -(\lambda + M_2(1 + l)\lambda^{1/2})|A^\alpha y|^2. \tag{3.19}$$

Similarly,

$$\frac{dz_i}{dt} + Az_i + QR_\theta(u_i) = 0, \qquad i = 1, 2, \tag{3.20}$$

$$\frac{dz}{dt} + Az = -(QR_\theta(u_1) - QR_\theta(u_2)), \tag{3.21}$$

$$\begin{aligned}\frac{1}{2}\frac{d}{dt}|A^\alpha z|^2 + |A^{\alpha+1/2}z|^2 &= -(A^{\alpha-1/2}(R_\theta(u_1) - R_\theta(u_2)), A^{\alpha+1/2}z)\\ &\leq M_2|A^\alpha(u_1 - u_2)||A^{\alpha+1/2}z|,\end{aligned} \tag{3.22}$$

$$\frac{1}{2}\frac{d}{dt}|A^\alpha z|^2 \leq -|A^{\alpha+1/2}z|^2 + M_2(1 + l)|A^\alpha y||A^{\alpha+1/2}z|.$$

We want to prove that a relation similar to (1.9) holds when

$$|A^\alpha z| \geq \gamma|A^\alpha y|. \tag{3.23}$$

The function $X \to -X^2 + M_2(1 + l)|A^\alpha y|X$ is decreasing for $X \geq$

$(M_2/2)(1 + l)|A^\alpha y|$. Thanks to (3.5) and (3.7′)

$$|A^{\alpha+1/2}z| \geq \Lambda^{1/2}|A^\alpha z| \geq \gamma\Lambda^{1/2}|A^\alpha y| \geq \frac{M_2}{2}(1 + l)|A^\alpha y|,$$

and we can replace $|A^{\alpha+1/2}z|$ by $\Lambda^{1/2}|A^\alpha z|$ in the right-hand side of (3.22). This yields

$$\frac{1}{2}\frac{d}{dt}|A^\alpha z|^2 \leq -\Lambda|A^\alpha z|^2 + M_2(1 + l)\Lambda^{1/2}|A^\alpha y||A^\alpha z|. \tag{3.24}$$

The inequalities (3.19) and (3.24) are similar to (1.8) and (1.9). Hence using Theorem 1.2 we conclude that the cone property is valid for $u(t) = y(t) + z(t)$[1] provided (3.8′), the present analog of (1.4), holds; the cone is here

$$\mathscr{C}_\gamma = \{\xi \in D(A^\alpha), |A^\alpha Q\xi| \leq \gamma|A^\alpha P\xi|\}.$$

We now apply Theorem 1.2 to u. Two possibilities can occur on $(-\infty, 0)$:

(i) Either there exists $t_0 < 0$ such that $|A^\alpha z(t_0)| \leq \gamma|A^\alpha y(t_0)|$.
In this case Theorem 1.2 implies that $|A^\alpha z(t)| \leq \gamma|A^\alpha y(t)|$, $\forall t \geq t_0$, and in particular,

$$\begin{aligned}|A^\alpha z(0)| &= |A^\alpha(\mathscr{T}\Phi(y_{01}) - \mathscr{T}\Phi(y_{02}))| \\ &\leq \gamma|A^\alpha y(0)| = \gamma|A^\alpha(y_{01} - y_{02})|\end{aligned}$$

and (3.15) is proved in this case.

(ii) Or $|A^\alpha z(t)| > \gamma|A^\alpha y(t)|$, $\forall t < 0$.
In this case we apply the first inequality (1.14) on some interval $[t_0, t]$, $-\infty < t_0 < t < 0$, and find

$$|A^\alpha z(t)| \leq |A^\alpha z(t_0)| \exp(-\nu(t - t_0)), \qquad \nu = \Lambda - \frac{M_2(1 + l)}{\gamma}\Lambda^{1/2} > 0.$$

Since $|A^\alpha z(t_0)|$ is bounded, by letting $t_0 \to -\infty$ we obtain $z(t) = 0$, and $|A^\alpha z(t)| \leq (1/\gamma)|A^\alpha z(t)| = 0$.

This occurs only when $y_{01} = y_{02}$, $z_1(0) = z_2(0)$, and (3.15) is trivial in this case.

The proof is complete. □

Lemmas 3.1, 3,3, and 3.4 readily imply

[1] Note that u is not the difference of two solutions of (2.15) but rather the difference of the two solutions u_1, u_2 of

$$\frac{dv}{dt} + Av + R_\theta(Pv + \Phi(Pv)) = 0.$$

Proposition 3.1. *If* $b \geq M_3\Lambda^{-1/2}$, *and* (3.7), (3.8) *are satisfied,* $\mathscr{T}$ *maps* $\mathscr{F}_{b,l}^{\alpha}$ *into itself.*

We now prove that $\mathscr{T}$ is a Lipschitz mapping in $\mathscr{F}_{b,l}^{\alpha}$. The proof has some common points with that of Lemma 3.4 and Theorem 1.2 albeit the cone property is not explicitly used.

Proposition 3.2. *We assume that for* $\eta > 0$ *given,* (3.7) *is satisfied and*

$$\Lambda \geq \left(\frac{2M_2\kappa_4}{\eta}\right)^2, \qquad \kappa_4 = \kappa_3(\tfrac{1}{2}), \tag{3.7''}$$

$$\Lambda - \lambda > M_2(1 + l)(\lambda^{1/2} + \Lambda^{1/2}) + \frac{2M_2}{\eta}\lambda^{1/2}. \tag{3.8''}$$

Then

$$d(\mathscr{T}\Phi_1, \mathscr{T}\Phi_2) \leq \eta d(\Phi_1, \Phi_2), \qquad \forall \Phi_1, \Phi_2 \in \mathscr{F}_{b,l}^{\alpha}. \tag{3.25}$$

In particular, if (3.7), (3.8) *are satisfied,* (3.25) *is valid with* $\eta = \frac{1}{2}$.

Proof. Let Φ_1, Φ_2 be given in $\mathscr{F}$, let $y_0 \in PD(A^\alpha)$,

$$y_i(t) = y_i(t; y_0, \Phi_i), \qquad z_i(t) = z_i(t; y_0, \Phi_i),$$

$$u_i(t) = y_i(t) + \Phi_i(y_i(t)), \qquad i = 1, 2,$$

$$y = y_1 - y_2, \qquad z = z_1 - z_2.$$

Then (3.17), (3.21) are valid but instead of (3.18)

$$\begin{aligned}|A^\alpha(u_1 - u_2)| &= |A^\alpha y + A^\alpha(\Phi_1(y_1) - \Phi_2(y_2))|\\ &\leq |A^\alpha y| + |A^\alpha(\Phi_1(y_1) - \Phi_1(y_2))|\\ &\quad + |A^\alpha(\Phi_1(y_2) - \Phi_2(y_2))|,\end{aligned}$$

$$|A^\alpha(u_1 - u_2)| \leq (1 + l)|A^\alpha y| + d(\Phi_1, \Phi_2). \tag{3.26}$$

Also, using Lemma 2.2,

$$|A^{\alpha-1/2}(R_\theta(u_1) - R_\theta(u_2))| \leq M_2(1 + l)|A^\alpha y| + M_2 d(\Phi_1, \Phi_2). \tag{3.27}$$

Thus (3.19) is replaced by

$$\frac{1}{2}\frac{d}{dt}|A^\alpha y|^2 \geq -\tilde{\lambda}|A^\alpha y|^2 - M_2\lambda^{1/2}d(\Phi_1, \Phi_2)|A^\alpha y|,$$

$$\frac{d}{dt}|A^\alpha y| + \tilde{\lambda}|A^\alpha y| \geq -M_2\lambda^{1/2}d(\Phi_1, \Phi_2), \tag{3.28}$$

$$\tilde{\lambda} = \lambda + M_2(1 + l)\lambda^{1/2}.$$

Now thanks to (2.38)

$$\mathscr{T}\Phi_1(y_0) - \mathscr{T}\Phi_2(y_0) = I_1 + I_2, \tag{3.29}$$

$$I_1 = -\int_{-\infty}^{0} e^{\tau AQ}(QR_\theta(y_1 + \Phi_1(y_1)) - QR_\theta(y_2 + \Phi_1(y_2)))\, d\tau,$$

$$I_2 = -\int_{-\infty}^{0} e^{\tau AQ}(QR_\theta(y_2 + \Phi_1(y_2)) - QR_\theta(y_2 + \Phi_2(y_2)))\, d\tau,$$

Due to Lemma 2.2

$$\begin{aligned}|A^{\alpha-1/2}(R_\theta(y_2 + \Phi_1(y_2)) - R_\theta(y_2 + \Phi_2(y_2)))| &\le M_2|A^\alpha(\Phi_1(y_2) - \Phi_2(y_2))| \\ &\le M_2 d(\Phi_1, \Phi_2).\end{aligned}$$

We can therefore bound the norm of $A^\alpha I_2$ by

$$M_2 d(\Phi_1, \Phi_2)\int_{-\infty}^{0} |(AQ)^{1/2}e^{\tau AQ}|_{\mathscr{L}(QH)}\, d\tau,$$

and with Lemma 3.2

$$|A^\alpha I_2| \le M_2\kappa_4\Lambda^{-1/2}d(\Phi_1, \Phi_2), \qquad (\kappa_4 = \kappa_3(\tfrac{1}{2})). \tag{3.30}$$

In order to estimate $|A^\alpha I_1|$, we identify I_1 as $\tilde{z}(0)$, where $\tilde{z}$ is the solution bounded on $(-\infty, 0]$ (see Lemma 2.3) of

$$\frac{d\tilde{z}}{dt} + A\tilde{z} = -QR_\theta(y_1 + \Phi_1(y_1)) + QR_\theta(y_2 + \Phi_1(y_2)). \tag{3.31}$$

Exactly as for (3.21), (3.22) we find that

$$\frac{1}{2}\frac{d}{dt}|A^\alpha\tilde{z}|^2 \le -|A^{\alpha+1/2}\tilde{z}|^2 + M_2(1 + l)|A^\alpha y||A^{\alpha+1/2}\tilde{z}|.$$

Using (3.7) we then prove, exactly as we proved (3.24), that if[1]

$$|A^\alpha\tilde{z}| \ge |A^\alpha y|, \tag{3.32}$$

then

$$\frac{1}{2}\frac{d}{dt}|A^\alpha\tilde{z}|^2 \le -\Lambda|A^\alpha\tilde{z}|^2 + M_2(1 + l)\Lambda^{1/2}|A^\alpha y||A^\alpha\tilde{z}|.$$

Hence, at every time t where (3.32) holds,

$$\frac{d}{dt}|A^\alpha\tilde{z}| + \tilde{\Lambda}|A^\alpha\tilde{z}| \le 0, \qquad \tilde{\Lambda} = \Lambda - M_2(1 + l)\Lambda^{1/2}. \tag{3.33}$$

The relations (3.28), (3.33) are similar to (1.8), (1.9) and we will now use the

[1] Since $l \le 1$, (3.7) implies $\Lambda^{1/2} \ge (M_2/2)(1 + l)$.

same type of argument as in the proof of the cone property (although we do not use Theorem 1.2 itself).

We first observe that $y(0) = 0$ and that if $\tilde{z}(0) = 0$, then $I_2 = 0$. Hence we can assume that $\tilde{z}(0) \neq 0$, so that

$$|A^\alpha \tilde{z}(t)| > |A^\alpha y(t)| \tag{3.34}$$

for $t < 0$ sufficiently small in absolute value.

At this point we observe that there are two possibilities:

(i) Inequality (3.34) holds for every $t < 0$.

In this case let $t_0 < t \le 0$; (3.33) yields

$$|A^\alpha \tilde{z}(t)| \le |A^\alpha \tilde{z}(t_0)| \exp(-\tilde{\Lambda}(t - t_0)).$$

Since $\tilde{\Lambda} > 0$ because of (3.7) and $|A^\alpha z(t_0)|$ remains bounded as $t_0 \to -\infty$, we obtain, by passing to the limit $t_0 \to -\infty$,

$$|A^\alpha \tilde{z}(t)| = 0, \qquad \forall t \le 0.$$

In particular, $\tilde{z}(0) = 0$ which contradicts the hypothesis $\tilde{z}(0) \neq 0$.

(ii) Inequality (3.34) is not valid for all $t < 0$.

In this case there exists $t_0 < 0$ such that (3.34) is valid for $t_0 < t \le 0$, and $|A^\alpha z(t_0)| = |A^\alpha y(t_0)|$.

We substract (3.28) from (3.33) and we find that on $(t_0, 0)$ we have

$$\frac{d}{dt}(|A^\alpha \tilde{z}| - |A^\alpha y|) + (\tilde{\Lambda} - \tilde{\lambda})(|A^\alpha \tilde{z}| - |A^\alpha y|) \le M_2 \lambda^{1/2} d(\Phi_1, \Phi_2).$$

We use the Gronwall lemma; taking into account that $|A^\alpha z(t_0)| - |A^\alpha y(t_0)| = 0$ and $A^\alpha y(0) = 0$, we obtain

$$|A^\alpha \tilde{z}(0)| \le \frac{M_2 \lambda^{1/2}}{\tilde{\Lambda} - \tilde{\lambda}} d(\Phi_1, \Phi_2).$$

Hence in all cases

$$|A^\alpha I_1| \le \frac{M_2 \lambda^{1/2}}{\Lambda - \lambda - M_2(1 + l)(\lambda^{1/2} + \Lambda^{1/2})} d(\Phi_1, \Phi_2). \tag{3.35}$$

Finally,

$$|A^\alpha(\mathscr{T}\Phi_1(y_0) - \mathscr{T}\Phi_2(y_0))|$$
$$\le \left(M_2 \kappa_4 \Lambda^{-1/2} + \frac{M_2 \lambda^{1/?}}{\Lambda - \lambda - M_2(1 + l)(\lambda^{1/2} + \Lambda^{1/2})} \right) d(\Phi_1, \Phi_2). \tag{3.36}$$

When (3.7″) and (3.8″) are satisfied, we easily infer (3.25) from (3.36). It is also clear that (3.7″), (3.8″) hold with $\eta = \frac{1}{2}$ when (3.7), (3.8) are satisfied.

The proposition is proved. □

3.4. Proof of Theorem 3.1 (End)

Propositions 3.1 and 3.2 show that $\mathscr{T}$ is a strict contraction from the complete metric space $\mathscr{F}_{b,l}^{\alpha}$ into itself. The Contraction Principle implies that $\mathscr{T}$ has a unique fixed point Φ in $\mathscr{F}$. It is clear from the construction of Φ that the graph $\mathscr{M}$ of Φ is positively invariant for the semigroup $S_\theta(\cdot)$ of the prepared equation (and it is positively invariant for the semigroup $S(\cdot)$ of the initial equation in the absorbing set $\mathscr{B}_0$). In order to conclude the proof of Theorem 3.1 we must show that the graph $\mathscr{M}$ of Φ attracts *exponentially* all the orbits, of both the initial and prepared equations.

We start with some preliminary results and then prove the desired result.

Lipschitz Property

Let $u_1(\cdot)$, $u_2(\cdot)$ be two solutions of (2.15) and let $u(\cdot) = u_1(\cdot) - u_2(\cdot)$. Obviously,

$$\frac{du}{dt} + Au + R_\theta(u_1) - R_\theta(u_2) = 0, \tag{3.37}$$

$$\begin{aligned}\frac{d}{dt}\frac{|A^\alpha u|^2}{2} + |A^{\alpha+1/2}u|^2 &= -(A^{\alpha-1/2}(R_\theta(u_1) - R_\theta(u_2)), A^{\alpha+1/2}u)\\ &\leq \text{(with Lemma 2.2)}\\ &\leq M_2|A^\alpha u|\,|A^{\alpha+1/2}u|\\ &\leq \tfrac{1}{2}|A^{\alpha+1/2}u|^2 + \frac{M_2^2}{2}|A^\alpha u|^2.\end{aligned}$$

Hence

$$\begin{gathered}\frac{d}{dt}|A^\alpha u|^2 + |A^{\alpha+1/2}u|^2 \leq M_2^2|A^\alpha u|^2,\\ |A^\alpha u(t)|^2 \leq |A^\alpha u(0)|^2 \exp(M_2^2 t),\end{gathered} \tag{3.38}$$

and, in particular,

$$\begin{gathered}|A^\alpha u(t)|^2 \leq 2|A^\alpha u(0)|^2 \quad \text{for} \quad 0 \leq t \leq 2t_0,\\ t_0 = \frac{1}{2M_2^2}\log 2. \qquad \square\end{gathered} \tag{3.39}$$

The Cone Property

Let us show that the cone property applies to the difference $u(\cdot) = u_1(\cdot) - u_2(\cdot)$ of two solutions of (2.15) in the space $D(A^\alpha)$, the cone $\mathscr{C}_\gamma$ being

$$\mathscr{C} = \{\xi \in D(A^\alpha), |A^\alpha Q\xi| \leq \tfrac{1}{8}|A^\alpha P\xi|\}.$$

The analogs of (1.8), (1.9) are easily derived ($y = Pu$, $z = Qu$)

$$\frac{du}{dt} + Au + R_\theta(u_1) - R_\theta(u_2) = 0,$$

$$\frac{dy}{dt} + Ay = -P(R_\theta(u_1) - R_\theta(u_2)),$$

$$\frac{1}{2}\frac{d}{dt}|A^\alpha y|^2 + |A^{\alpha+1/2}y|^2 \geq \text{(with Lemma 2.2)} \tag{3.40}$$

$$\geq -M_2|A^\alpha u||A^{\alpha+1/2}y|$$

$$\geq -M_2(|A^\alpha y| + |A^\alpha z|)|A^{\alpha+1/2}y|,$$

$$\frac{1}{2}\frac{d}{dt}|A^\alpha y|^2 \geq -(\lambda + M_2\lambda^{1/2})|A^\alpha y|^2 - M_2\lambda^{1/2}|A^\alpha y||A^\alpha z|.$$

Similarly,

$$\frac{dz}{dt} + Az = -Q(R_\theta(u_1) - R_\theta(u_2)),$$

$$\frac{1}{2}\frac{d}{dt}|A^\alpha z|^2 + |A^{\alpha+1/2}z|^2 \leq M_2|A^\alpha u||A^{\alpha+1/2}z|,$$

$$\frac{1}{2}\frac{d}{dt}|A^\alpha z|^2 \leq -|A^{\alpha+1/2}z|^2 + M_2(|A^\alpha y| + |A^\alpha z|)|A^{\alpha+1/2}z|.$$

When $|A^\alpha z| \geq \frac{1}{8}|A^\alpha z|$ we can replace, in the right-hand side of the last inequality, $|A^{\alpha+1/2}z|$ by $\Lambda^{1/2}|A^\alpha z|$ provided

$$|A^{\alpha+1/2}z| \geq \Lambda^{1/2}|A^\alpha z| \geq \frac{M_2}{2}(|A^\alpha y| + |A^\alpha z|).$$

A sufficient condition for that is

$$\frac{1}{8}\left(\Lambda^{1/2} - \frac{M_2}{2}\right) \geq \frac{M_2}{2},$$

and this is indeed guaranteed by (3.7). Thus outside $\mathscr{C}$

$$\frac{1}{2}\frac{d}{dt}|A^\alpha z|^2 \leq -(\Lambda - M_2\Lambda^{1/2})|A^\alpha z|^2 + M_2\Lambda^{1/2}|A^\alpha y||A^\alpha z|. \tag{3.41}$$

Condition (1.4) reads here

$$\Lambda - \lambda \geq M_2(\tfrac{1}{8}\lambda^{1/2} + 8\Lambda^{1/2}) + M_2(\lambda^{1/2} + \Lambda^{1/2})$$

and this is guaranteed by (3.8) ($l \leq \frac{1}{8}$). Finally, we observe that the number ν

in (1.14) is

$$\nu = \Lambda - 9M_2\Lambda^{1/2} \qquad (> 0). \qquad \square \tag{3.42}$$

Now consider a solution of (2.15) starting from a point $u_0 \in D(A^\alpha)$, $u(0) = u_0$. Since $\mathscr{M}$ has finite dimension there exists a point $v_0 \in \mathscr{M}$ such that

$$|A^\alpha(u_0 - v_0)| = \operatorname{dist}(u_0, \mathscr{M}) = \operatorname*{Inf}_{w \in \mathscr{M}} |A^\alpha(u_0 - w)|. \tag{3.43}$$

We want to compare the orbit $u(\cdot)$ to the solution $v(\cdot)$ of (2.15) starting from v_0, $v(0) = v_0$; of course, the orbit $v(\cdot)$ lies in $\mathscr{M}$ and thus at every time $t \geq 0$

$$\operatorname{dist}(u(t), \mathscr{M}) \leq |A^\alpha(u(t) - v(t))|. \tag{3.44}$$

We proved that Theorem 1.2 (the cone property) applies to the difference $w = u - v$. Let then t be given, $t_0 \leq t \leq 2t_0$:

If $w(t) \notin \mathscr{C}$, then $w(\tau) \notin \mathscr{C}$ for $0 \leq \tau \leq t$, and we infer from the second inequality (1.14) that

$$|A^\alpha(u(t) - v(t)| \leq (65)^{1/2}|A^\alpha(u(0) - v(0))| \exp(-\nu t).$$

Hence with (3.44)

$$\operatorname{dist}(u(t), \mathscr{M}) \leq 9 \operatorname{dist}(u(0), \mathscr{M}) \cdot \exp(-\nu t_0)$$

and

$$\operatorname{dist}(S_\theta(t)u_0, \mathscr{M}) \leq \tfrac{1}{2} \operatorname{dist}(u_0, \mathscr{M}), \tag{3.45}$$

provided

$$\Lambda \geq 9M_2\Lambda^{1/2} + 10M_2^2, \tag{3.46}$$

which is a consequence of (3.7) and (3.8) ($\Lambda^{1/2} \geq 11M_2$ by (3.7)).

If $w(t) \in \mathscr{C}$, we write

$$\begin{aligned} \operatorname{dist}(u(t), \mathscr{M}) &\leq |A^\alpha\{u(t) - (Pu(t) + \Phi(Pu(t)))\}| \\ &\leq |A^\alpha(Qu(t) - \Phi(Pu(t)))|. \end{aligned}$$

Recall that $Qv(t) = \Phi(Pv(t))$ since $v(\cdot)$ lies on $\mathscr{M}$ and apply (3.39)

$$\begin{aligned} \operatorname{dist}(u(t), \mathscr{M}) &\leq |A^\alpha Q(u(t) - v(t))| + |A^\alpha(\Phi(Pv(t)) - \Phi(Pu(t)))| \\ &\leq (\text{since } w(t) \in \mathscr{C} \text{ and } l \leq \tfrac{1}{8}) \\ &\leq (\tfrac{1}{8} + l)|A^\alpha P(u(t) - v(t))| \\ &\leq \tfrac{1}{2}|A^\alpha(u_0 - v_0)| = \tfrac{1}{2} \operatorname{dist}(u(0), \mathscr{M}). \end{aligned}$$

We conclude that (3.45) holds in both cases for any $t \in [t_0, 2t_0]$. By iteration

$$\operatorname{dist}(S_\theta(nt_0)u_0, \mathscr{M}) \leq 2^{-n} \operatorname{dist}(u_0, \mathscr{M})$$

and for $t > 0$ arbitrary we write $t = nt_0 + t_1$, $n \in \mathbb{N}$, $t_0 \leq t_1 < 2t_0$, and find

that

$$\begin{aligned}\operatorname{dist}(S_\theta(t)u_0, \mathscr{M}) &\leq 2^{-(n+1)} \operatorname{dist}(u_0, \mathscr{M})\\ &\leq \exp(-(n+1)\log 2)\operatorname{dist}(u_0, \mathscr{M})\\ &\leq \exp\left(\left(1-\frac{t}{t_0}\right)\log 2\right)\operatorname{dist}(u_0, \mathscr{M})\end{aligned}$$

$$\operatorname{dist}(S_\theta(t)u_0, \mathscr{M}) \leq 2\exp(-2M_2^2 t)\operatorname{dist}(u_0, \mathscr{M}).$$

Hence

$$\operatorname{dist}(S_\theta(t)u_0, \mathscr{M}) \leq \eta_1 \exp(-\eta_2 t)\operatorname{dist}(u_0, \mathscr{M}). \tag{3.47}$$

and we obtain (3.4) for the prepared equation (2.15) with coefficients $\eta_1 = 2$, $\eta_2 = 2M_2^2$ which depend only on the equation.

For the initial equation (2.11), let $u_0 \in H$ and $t_0 = t_0(u_0)$, be such that $S(t)u_0 \in \mathscr{B}_0$, $\forall t \geq t_0$; then for $t \geq t_0$:

$$\begin{aligned}\operatorname{dist}(S(t)u_0, \mathscr{M}) &= \operatorname{dist}(S_\theta(t-t_0)S(t_0)u_0, \mathscr{M})\\ &\leq \eta_1 \exp(-\eta_2(t-t_0))\operatorname{dist}(S(t_0)u_0, \mathscr{M});\\ &\leq \eta_1' \exp(-\eta_2(t-t_0))\end{aligned}$$

with $\eta_1' = \eta_1\, d(\mathscr{B}_0, \mathscr{M})$. Hence (3.4) in this case with a decay rate which depends on u_0 and can be chosen so that it is uniformly valid for u_0 in a ball of H, $|u_0| \leq R$.

This concludes the proof of Theorem 3.1. □

3.5. Another Form of Theorem 3.1

We can give a slightly different form of Theorem 3.1 by replacing $D(A^{\alpha-1/2})$ by $D(A^{\alpha-\gamma})$, $0 \leq \gamma \leq \frac{1}{2}$, in (2.9):

R is Lipschitzian on the bounded sets of $D(A^\alpha)$ with values in $D(A^{\alpha-\gamma})$, for some $\alpha \in \mathbb{R}$ and some γ, $0 \leq \gamma \leq \frac{1}{2}$. (3.48)

Lemmas 2.1 and 2.2 extend easily to this case

$$\operatorname*{Sup}_{u \in D(A^\alpha)} |A^{\alpha-\gamma}R_\theta(u)| \leq M_1, \tag{3.49}$$

$$\operatorname*{Sup}_{u_1, u_2 \in D(A^\alpha)} |A^{\alpha-\gamma}(R_\theta(u_1) - R_\theta(u_2))| \leq M_2 |A^\alpha(u_1 - u_2)|. \tag{3.50}$$

Assuming (2.12), (2.13), (2.16), (2.20) as before, the proof of Theorem 3.1 will extend to this case. A perusal of this proof shows that we must replace $\lambda^{1/2}$ and $\Lambda^{1/2}$ by λ^γ and Λ^γ in (3.8), i.e., replace (3.8) by

$$\Lambda - \lambda > 8M_2\frac{(1+l)}{l}(\lambda^\gamma + \Lambda^\gamma) \tag{3.51}$$

Theorem 3.2. *The hypotheses are those of Theorem* 3.1, (3.48), (3.51) *replacing, respectively*, (2.9) *and* (3.8).

Then the conclusions of Theorem 3.1 *are valid without any modification.*

4. Examples

We briefly describe three examples to which Theorem 3.1 applies. Note also that the cone property (in the form of Theorem 1.2) applies to the difference of two solutions: the proof of this point was part of the proof of Theorem 3.1 (see Section 3.4).

4.1. Example 1: The Kuramoto–Sivashinsky Equation

The first example is the Kuramoto–Sivashinsky equation (see Section III.4.1) which will be treated with two different values of α: $\alpha = \frac{1}{4}$ first, and then $\alpha = 0$.

The notations are those of Section III.4.1, H is the subspace of $\dot{L}^2(-L/2, L/2)$ consisting of odd functions; we denote by $D(A)$ the space $\dot{H}^4_{\text{per}} \cap H$, and for $v \in D(A)$, $Av = v(d^4v/dx^4) = vD^4v$. For every $v \in D(A)$,

$$R(v) = vDv + D^2v. \tag{4.1}$$

For $N \in \mathbb{N}$, we choose $P = P_N =$ the orthogonal projector in H onto the space spanned by $w_1, \ldots, w_N$, the eigenvectors of A corresponding to the first N eigenvalues $\lambda_1 \leq \cdots \leq \lambda_N$. We recall (see Remark III.4.2) that $\lambda_N = \lambda_1 N^4$, $\lambda_1 = 4\Pi^4/L^4$.

The hypotheses (2.12), (2.13) have been proved in Section III.4.2; the proof of (2.16) necessitates only minor modifications of the arguments used for proving (2.12), (2.13). Concerning R, the main hypothesis to check is (2.9). In fact we verify (2.10) directly and then (2.9).

For $\alpha = \frac{1}{4}$, let u, $v \in D(A^{1/4})$. We observe that $D(A^{1/4}) = \dot{H}^1_{\text{per}} \cap H$, and we write

$$\begin{aligned}(R(u), v) &= \int_{-L/2}^{L/2} \{u(Du)v + (D^2u)v\}\, dx \\ &= \int_{-L/2}^{L/2} \{u(Du)v - (Du)(Dv)\}\, dx,\end{aligned}$$

$$|(R(u), v)| \leq |u|_{L^4}|Du|\,|v|_{L^4} + |Du|\,|Dv|.$$

Since $D(A^{1/4}) \subset H^1(\Omega) \subset H^{1/4}(\Omega) \subset L^4(\Omega)$ $(\Omega = (-L/2, L/2))$, there exists a constant c_1' such that

$$\begin{aligned}&|(R(u), v)| \leq c_1'|A^{1/4}u|\,|A^{1/4}v|(1 + |A^{1/4}u|),\\ &|A^{1/4}R(u)| \leq c_1'|A^{1/4}u|(1 + |A^{1/4}u|).\end{aligned} \tag{4.2}$$

Hence (2.10).

Similarly for (2.9), if $u_1, u_2, v \in D(A^{1/4})$, and $u = u_1 - u_2$,

$$(R(u_1) - R(u_2), v) = \int_{-L/2}^{L/2} \{(u_1(Du_1) - u_2(Du_2))v + (D^2u)v\}\, dx$$
$$= \frac{1}{2}\int_{-L/2}^{L/2} \{(u_2^2 - u_1^2)Dv - 2(Du)(Dv)\}\, dx,$$
$$|(R(u_1) - R(u_2), v)| \le \tfrac{1}{2}|u_1 - u_2|_{L^4}|u_1 + u_2|_{L^4}|Dv| + |Du|\,|Dv|$$
$$\le c_2'\{|A^{1/4}(u_1 + u_2)| + 1\}|A^{1/4}u|\,|A^{1/4}v|.$$

Thus

$$|A^{-1/4}(R(u_1) - R(u_2))| \le c_2'\{|A^{1/4}(u_1 + u_2)| + 1\}|A^{1/4}u| \tag{4.3}$$

and (2.9) follows.

We recall that $\lambda = \lambda_N$, $\Lambda = \Lambda_{N+1}$. Conditions (3.7), (3.8) are easily verified for large N since $\lambda_N = \lambda_1 N^4$.

In the case $\alpha = 0$, the main point is again to verify (2.9), (2.10). Let $u \in H$ and $v \in D(A^{1/2})\ (= \dot{H}^2_{\text{per}} \cap H)$. We have

$$(R(u), v) = \int_{-L/2}^{L/2} \{u(Du)v + (D^2u)v\}\, dx$$
$$= -\int_{-L/2}^{L/2} (\tfrac{1}{2}u^2 Dv - uD^2v)\, dx,$$
$$|(R(u), v)| \le \tfrac{1}{2}|u|^2\,|Dv|_{L^\infty} + |u|\,|D^2v|.$$

Since $H^1(\Omega) \subset L^\infty(\Omega)$ and $D(A^{1/2}) \subset \dot{H}^2_{\text{per}}(\Omega)$, there exists a constant c_3' such that

$$|(R(u), v)| \le c_3'(|u| + 1)|u|\,|A^{1/2}v|.$$

Hence (2.10)

$$|A^{1/2}R(u)| \le c_3'(|u| + 1)|u|. \tag{4.4}$$

Similarly, for (2.9), let $u_1, u_2 \in H$, $v \in D(A^{1/2})$ and set $u = u_1 - u_2$

$$(R(u_1) - R(u_2), v) = \int_{-L/2}^{L/2} \{(u_1(Du_1) - u_2(Du_2))v + (D^2u)v\}\, dx$$
$$= \int_{-L/2}^{L/2} \{\tfrac{1}{2}(u_2^2 - u_1^2)Dv + u(D^2v)\}\, dx,$$
$$|(R(u_1) - R(u_2), v)| \le \{\tfrac{1}{2}|u_1 + u_2| + 1\}|u|\,|Dv|_{L^\infty}$$
$$\le c_3'(|u_1 + u_2| + 1)|u|\,|A^{1/2}v|,$$
$$|A^{-1/2}(R(u_1) - R(u_2))| \le c_3'(|u_1 + u_2| + 1)|u| \tag{4.5}$$

and (2.9) is proved.

The conditions (3.7), (3.8) are the same when $\alpha = 0$ as when $\alpha = \frac{1}{4}$, and are satisfied for large N.

In conclusion

Theorem 4.1. *The cone property applies to the Kuramoto–Sivashinsky equation (difference of two solutions) in both $D(A^{1/4})$ (H^1-norm) and H (L^2-norm), when N is sufficiently large.*

The Kuramoto–Sivashinsky equation possesses an inertial manifold of the form given by Theorem 3.1, in both $D(A^{1/4})$ and H.

Remark 4.1. For a refined evaluation of the dimension of the inertial manifold (and the values of N for which the cone property is satisfied) the reader is referred to C. Foias, B. Nicolaenko, G. Sell, and R. Temam [1], [2] where Theorem 4.1 was first proved.

4.2. Example 2: Approximate Inertial Manifolds for the Navier–Stokes Equations

This example is a modified Navier–Stokes equation in space dimension n with a higher-order viscosity term

$$\frac{\partial u}{\partial t} + \varepsilon(-\Delta)^r u - \nu\Delta u + (u\cdot\nabla)u + \nabla p = f, \tag{4.6}$$

$$\nabla\cdot u = 0. \tag{4.7}$$

The functions $u = u(x, t)$ and $p = p(x, t)$ are defined on $\mathbb{R}^n \times \mathbb{R}_+$, taking, respectively, values in $\mathbb{R}^n$ and $\mathbb{R}$ (note that $u = (u_1, \ldots, u_n)$); ε and ν are strictly positive numbers, $r > 1$. Note that (4.6) reduces to the usual Navier–Stokes equations when $\varepsilon = 0$, but we do not treat the case $\varepsilon = 0$ here.

Although other boundary conditions can also be considered, we restrict ourselves to the space-periodic case. We assume that u and p are periodic in each direction $x_1, \ldots, x_n$ with period $L > 0$

$$\begin{cases} u(x + Le_i, t) = u(x, t), \qquad i = 1, \ldots, n, \\ p(x + Le_i, t) = p(x, t), \end{cases} \tag{4.8}$$

where $\{e_i, \ldots, e_n\}$ is the natural basis of $\mathbb{R}^n$. Furthermore, we assume that

$$\int_\Omega f(x)\,dx = 0, \qquad \int_\Omega u(x, t)\,dx = 0, \qquad \int_\Omega p(x, t)\,dx = 0, \tag{4.9}$$

where $\Omega = (0, L)^n$ is the n-cube.

Proceeding exactly as in Sections III.2 and III.3, we can show that (4.6)–(4.9) reduces to an evolution equation for u of the form (2.11). Furthermore, the results of Section III.3 apply and provide a proof of (2.12), (2.13), and (2.16). We omit this point; we will just briefly describe the functional setting and check the hypotheses (2.9), (2.10).

For the functional setting

$$
\begin{aligned}
H &= \{v \in \dot{L}^2(\Omega)^n, \operatorname{div} v = 0, v_{i|x_i=L} = v_{i|x_i=0}\}, \\
D(A) &= \dot{H}^{2r}_{\text{per}}(\Omega)^n \cap H, \\
Au &= \varepsilon(-\Delta)^r u, \qquad \forall u \in D(A), \\
R(u) &= -\nu\Delta u + B(u, u), \\
(B(u, v), w) &\leq \int_\Omega ((u \cdot \nabla)v)w \, dx, \qquad \forall u, v, w \in D(A).
\end{aligned}
$$

We choose $\alpha = 0$ and $r > 1 + n/2$ for the moment. Since (2.9), (2.10) are obvious for the linear operator $-\nu\Delta$ we need only check these properties for B: let $u \in H$ and let $v \in D(A^{1/2}) \subset \dot{H}^2_{\text{per}}$). We write, using III.(2.21),

$$
\begin{aligned}
(B(u, u), v) &= -(B(u, v), u) = -\int_\Omega ((u \cdot \nabla)v)u \, dx, \\
|(B(u, u), v)| &\leq |u|^2 \, |\nabla v|_{L^\infty(\Omega)}.
\end{aligned}
$$

Since $D(A^{1/2}) \subset \dot{H}^r_{\text{per}}(\Omega)$ and $H^{r-1}(\Omega) \subset L^\infty(\Omega)$ $(r - 1 > n/2)$, there exists a constant c_4' such that

$$
\begin{aligned}
|(B(u, u), v)| &\leq c_4' |u|^2 \, |A^{1/2} v|, \\
|A^{-1/2} B(u, u)| &\leq c_4' |u|^2.
\end{aligned}
\tag{4.10}
$$

Hence (2.10). Similarly for (2.9), let u_1, $u_2 \in H$, $v \in D(A^{1/2})$, $u = u_1 - u_2$, $\bar{u} = (u_1 + u_2)/2$,

$$
\begin{aligned}
B(u_1, u_1) - B(u_2, u_2) &= B(\bar{u}, u) + B(u, \bar{u}), \\
((B(u_1, u_1) - B(u_2, u_2), v) &= -(B(\bar{u}, v), u) - (B(u, v), \bar{u}), \\
|(B(u_1, u_1) - B(u_2, u_2), v)| &\leq 2|\bar{u}| \, |u| \, |\nabla v|_{L^\infty(\Omega)} \\
&\leq 2c_4' |\bar{u}| \, |u| \, |A^{1/2} v|.
\end{aligned}
$$

Hence (2.9)

$$
|A^{-1/2}(B(u_1, u_1) - B(u_2, u_2))| \leq c_4' |u_1 + u_2| \, |u_1 - u_2|. \tag{4.11}
$$

We have $\lambda = \lambda_N$, $\Lambda = \lambda_{N+1}$, and we infer from classical spectral results (see R. Courant and D. Hilbert [1]) that $\lambda_N \sim \lambda_1 N^{2r/n}$ as $N \to \infty$; (3.7) follows from Lemma 4.1 below.

In conclusion, we have

Theorem 4.2. *We assume that $r > n$, $n \geq 2$, $\alpha = 0$. The cone property applies to the modified Navier–Stokes equations* (4.6), (4.7), *when N is sufficiently large.*

The modified Navier–Stokes equations possesses in H an inertial manifold of the form given by Theorem 3.1.

Remark 4.2.

(i) With the same assumption on r, we can prove the existence of an inertial manifold in $D(A^\alpha)$, $\alpha \neq 0$; for that purpose we only need to check (2.9), (2.10). This is left as an exercise for the reader. The case $\alpha = 1$ is treated in C. Foias, G. Sell, and R. Temam [1], [2].

(ii) Theorem 4.2 does not apply when $\varepsilon = 0$ (classical Navier–Stokes equations), even in space dimension $n = 2$. Hence the inertial manifold provided by Theorem 4.2 can be considered, for small ε, as an approximate I.M. for the Navier–Stokes equations. For another construction of approximate inertial manifolds for the Navier–Stokes equations, see H.O. Kreiss [1], and C. Foias, O. Manley, and R. Temam [2], [3].

In the proof of Theorem 4.2 we used the following simple lemma:

Lemma 4.1. *If* $\lambda = \lambda_N$ *and* $\Lambda = \lambda_{N+1}$, *with* $\lambda_N \sim cN^\alpha$ *as* $N \to \infty$, *and* $\alpha > 2$, *then condition* (3.8) *is satisfied for arbitrarily large* N's.

Similarly, if $\alpha > 1/(1-\gamma)$, (3.51) *is satisfied for arbitrarily large* N's.

PROOF. We just prove the result concerning (3.8). We argue by contradiction and assume that there exists m_0 such that

$$\lambda_{m+1} - \lambda_m \leq \delta(\lambda_m^{1/2} + \lambda_{m+1}^{1/2}), \qquad \forall m \geq m_0,$$

($\delta = 8M_2(1+l)^2/l$). By summation we find

$$\lambda_{m+1} \leq \lambda_{m_0} + \delta \sum_{j=m_0}^{m} (\lambda_j^{1/2} + \lambda_{j+1}^{1/2}), \qquad \forall m > m_0.$$

Since $\lambda_m \sim cm^\alpha$, the right-hand side of this inequality is bounded by an expression $\lambda_{m_0} + c_1 m^{1+\alpha/2}$, and this is not consistent with $\lambda_m \sim cm^\alpha$ if $\alpha > 2$. □

4.3. Example 3: Reaction–Diffusion Equations

We consider now an application of Theorem 3.2: a simple reaction–diffusion equation corresponding to a locally Lipschitz nonlinearity

$$\frac{du}{dt} + Au + R(u) = 0. \tag{4.12}$$

Assume that (3.48) holds with $\alpha = \gamma = 0$;

R is Lipschitz on the bounded sets of H with values into itself. (4.13)

Then if (2.12), (2.13), (2.16), (2.20) and (3.7), (3.51) hold, Theorem 3.2 is applicable. We do not emphasize on (2.12), (2.13), (2.16), (2.20); (3.7) is satisfied for N sufficiently large, $\lambda_{N+1} = \Lambda$. For (3.51) consider the case where $D(A) = H^2_{\text{per}}(\Omega)$, $\Omega = (0, L)^n$, $Au = -\nu\Delta u$. Then the eigenvalues of A have the form

$$\nu(m_1^2 + \cdots + m_n^2),\ m_1, \ldots, m_n \in \mathbb{N}. \tag{4.14}$$

In order to apply Theorem 3.2 (i.e., to check (3.51)), we need to verify that the ordered set of numbers of the form (4.14) contains arbitrarily large gaps: in space dimension $n = 1$ this is obvious; and in space dimension $n = 2$ this follows from J. Richards [1], G.H. Hardy and E.M. Wright [1]. Theorem 3.2 is applicable in this case and provides the existence of an inertial manifold.

4.4. Example 4: The Ginzburg–Landau Equation

We conclude with the study of inertial manifolds for the Ginzburg–Landau equation, i.e., the nonlinear Schrödinger equation considered in Section IV.5. We restrict ourselves to space dimension $n = 1$.

The notations are those of Section IV.5.1, $H = \mathbb{L}^2(\Omega) = L^2(\Omega)^2$, the complexified $L^2(\Omega)$ space; A and $D(A)$ are the same as in Section IV.5.1, and we set here $R(u) = B(u; u) - \gamma u$. We denote by A_0 the symmetric part of A and by A_1 its antisymmetric part.[1]

For $N \in \mathbb{N}$, we pose $P = P_N =$ the orthogonal projector in H onto the space spanned by $w_1, \ldots, w_N$, the eigenvectors of A corresponding to the first N eigenvalues $\lambda_1 \leq \cdots \leq \lambda_N$. It is elementary (see R. Courant and D. Hilbert [1]) that $\lambda_N \sim c\lambda_1 N^2$, as $N \to \infty$ and therefore (3.48) is satisfied for arbitrarily large values of N if $\gamma < \frac{1}{2}$.

The hypotheses (2.12), (2.13) have been proved in Section IV.5.2; the proof of (2.16) necessitates only minor modifications of the arguments used for proving (2.12), (2.13). Concerning R, the main hypothesis to check is (2.9). We choose $\alpha = 1/8$, and show that R is Lipschitz from bounded sets of $D(A^{1/8})$ with values in $D(A^{-1/8})$; this proves (3.48) with $\gamma = 1/4$.

It suffices to prove the Lipschitz property for B; we recall that $D(A) \subset H^2(\Omega)$ and $D(A^{1/8}) \subset H^{1/4}(\Omega) \subset L^4(\Omega)$. Then if $u, v, w \in D(A^{1/8})$, $u = u_1 + iu_2$, $v = v_1 + iv_2$, $w = w_1 + iw_2$:

$$(B(u; u) - B(v; v), w) = \text{(see Section IV.5.1)}$$

$$= \int_\Omega |u|^2 \{(\kappa u_1 - \beta u_2)w_1 + (\beta u_1 + \kappa u_2)w_2\}\, dx$$

$$- \int_\Omega |v|^2 \{(\kappa v_1 - \beta v_2)w_1 + (\beta v_1 + \kappa v_2)w_2\}\, dx.$$

With the Hölder inequality we easily obtain

$$|(B(u; u) - B(v; v), w)| \leq c(|u|_{L^4(\Omega)} + |v|_{L^4(\Omega)})^2 |u - v|_{L^4(\Omega)} |w|_{L^4(\Omega)},$$

and since $D(A^{1/8}) \subset L^4(\Omega)$,

$$|(B(u; u) - B(v; v), w)| \leq c'(|A^{1/8}u| + |A^{1/8}v|)^2 |A^{1/8}(u - v)|\,|A^{1/8}w|,$$

$$|A^{-1/8}(B(u; u) - B(v; v))| \leq c'(|A^{1/8}u| + |A^{1/8}v|)^2 |A^{1/8}(u - v)|. \tag{4.15}$$

[1] We need here to extend slightly the hypotheses of Theorems 3.1 and 4.1 and observe that the conclusions hold with the same proof, when $A = A_0 + A_1$, A_0 satisfying the hypotheses made on A (A_0 symmetric) and A_1 antisymmetric, commuting with A_0, $A_1 \in \mathscr{L}(D(A_0), H)$. The w_j are then the eigenvectors of A_0 (F. Demengel and J.M. Ghidaglia).

Hence (2.9); actually (4.15) proves (2.9) for any value of α, $\frac{1}{8} \leq \alpha \leq \frac{3}{8}$.

In conclusion

Theorem 4.3. *The cone property applies to the Ginzburg–Landau equation (difference of two soltutions) in space dimension one, in the norm of* $D(A^\alpha)$, $\frac{1}{8} \leq \alpha \leq \frac{3}{8}$.

Under the same hypotheses, this equation possesses an inertial manifold of the form given by Theorem 3.1, *in* $D(A^\alpha)$, *for the same values of* α.

An alternative proof of Theorem 4.3 is given in P. Constantin, C. Foias, B. Nicolaenko, and R. Temam [3]; this article also contains an estimate of the dimension of the manifold in terms of the data.

Remark 4.3. Further properties on inertial manifolds can be found in the following references: S.N. Chow, K. Lu, and G. Sell [1] (further regularity properties of inertial manifolds), J. Mallet-Paret and G. Sell [1] (inertial manifolds for reaction–diffusion equations in higher dimension (see Example 3, Section 4.3); see also S.N. Chow, X.B. Lin, and K. Lu [1]. In P. Constantin, C. Foias, B. Nicolaenko, and R. Temam [1], [2], a more geometrical construction of inertial manifolds is given, based on a concept of integral manifolds. Related theories of invariant manifolds can be found in D. Henry [1] (Chaps. 6, 9), R. Mañé [3], D.A. Kamaev [3], and X. Mora [2].

The existence of inertial manifolds for specific equations is proved in B. Nicolaenko, B. Scheurer, and R. Temam [3], M. Marion [3]. A nonexistence result for an equation related to the weakly damped sine–Gordon equation appears in X. Mora and J. Solà-Morales [1].

In cases where the existence of an inertial manifold does not follow from Theorem 3.1, by lack of the spectral gap condition (3.8), other approaches are being explored:

generic result of existence of inertial manifolds (genericity with respect to certain functions appearing in the equation); see J. Hale and G. Sell [1];

obtention of *approximate inertial manifolds*; a first form of this concept, for the Navier–Stokes equations, appears in Example 2, Section 4.2; see also a completely different approach in C. Foias, O. Manley, and R. Temam [2], [3].

5. Approximation and Stability of the Inertial Manifold with Respect to Perturbations

The inertial manifold which we have constructed in Section 3 is stable with respect to certain perturbations of the evolution equation. In order to give the main ideas, without concern for technical difficulties, we will restrict ourselves in this section to giving a specific example of such stability results; namely, the stability with respect to the perturbations corresponding to a Galerkin approximation of (2.11), or better (2.15), associated with the eigenfunctions $\{w_j\}$ of A. Thus for any $M \geq 1$, the perturbed equation reads

$$\frac{du_M}{dt} + Au_M + P_M R_\theta(u_M) = 0, \tag{5.1}$$

where u_M takes its values in $P_M D(A)$, P_M denoting as before the projector in H onto the space spanned by $w_1, \ldots, w_M$.

Let us assume that the hypotheses of Theorem 3.1 are verified by (2.11), (2.15), when N is sufficiently large ($P = P_N$, $\Lambda = \lambda_{N+1}$, $\lambda = \lambda_N$). It is then easy to see that for every $M \geq N$, (5.1) satisfies the hypotheses of Theorem 3.1; hence there exists an inertial manifold $\mathscr{M}_M$ for (5.1) which is obtained as the graph of a Lipschitz function

$$\Phi_M: PD(A^\alpha) \to QP_M D(A^\alpha) \subset QD(A^\alpha).$$

Our object now is to investigate this point in more detail and study the convergence as $M \to \infty$, of Φ_M to the function Φ defining the inertial manifold $\mathscr{M}$ of the exact equation.

We will prove the following result.

Theorem 5.1. *The assumptions are those of Theorem* 3.1. *In particular, we are given* l, $0 < l \leq \frac{1}{8}$, *and an integer* N *such that* (3.7), (3.8) *hold* ($P = P_N$, $\Lambda = \lambda_{N+1}$, $\lambda = \lambda_N$). *Furthermore, we assume that the analog of* (2.12), (2.13), (2.16) *for* (5.1) *are valid for every* M.

Then for every $M > N$ *equation* (5.1) *has an inertial manifold* $\mathscr{M}_M$ *which is constructed as the graph of a Lipschitz function*

$$\Phi_M: PD(A^\alpha) \to QP_M D(A^\alpha) = (P_M - P_N)D(A^\alpha) \subset QD(A^\alpha).$$

Moreover, the Lipschitz constant l *for* Φ_M *is the same as that for the function* $\Phi: PD(A^\alpha) \to QD(A^\alpha)$ *constructed in Theorem* 3.1. *Finally, we have*

$$\operatorname{dist}(\Phi_M, \Phi) \leq \kappa_6 (\lambda_{N+1} \lambda_{M+1})^{-1/4}, \tag{5.2}$$

where

$$\operatorname{dist}(\Phi_M, \Phi) = \sup_{y_0 \in PD(A^\alpha)} |A^\alpha \Phi_M(y_0) - A^\alpha \Phi(y_0)|,$$

and κ_6 *is a constant defined below.*

Proof. It is straightforward to verify that (5.1) satisfies the analog of properties (2.9), (2.10); (2.12), (2.13), (2.16) have been explicitly assumed in the statement of the theorem while conditions (3.5)–(3.8) are just the same for (5.1) and (2.11), and thus hold true for every $M \geq N$. We conclude that the construction of Theorem 3.1 can be applied to (5.1) and we will now review this construction in more detail.

The set $\mathscr{F}^{\alpha}_{b,l}$ is replaced by the set $\mathscr{F}^{\alpha}_{b,l,M}$ of functions

$$\Phi_M: PD(A^\alpha) \to QP_M D(A^\alpha) \subset QD(A^\alpha),$$

which satisfy

$$\begin{cases} \text{Supp}\,\Phi_M \subset \{y \in PD(A^\alpha), |A^\alpha y| \le 2\rho\}, \\ |A^\alpha \Phi_M(y)| \le b, \qquad \forall y \in PD(A^\alpha), \\ |A^\alpha(\Phi_M(y_1) - \Phi_M(y_2))| \le l|A^\alpha(y_1 - y_2)|, \qquad \forall y_1, y_2 \in PD(A^\alpha). \end{cases} \tag{5.3}$$

Notice that we have $\mathscr{F}^{\alpha}_{b,l,M} \subset \mathscr{F}^{\alpha}_{b,l}$ for all $M \ge N$. The operator $\mathscr{T}$ providing the fixed point is now replaced by $\mathscr{T}_M$ (compare with (2.38))

$$\mathscr{T}_M \Phi_M(y_0) = -\int_{-\infty}^{0} e^{\tau AQP_M} QP_M R_\theta(y_M + \Phi_M(y_M))\, d\tau. \tag{5.4}$$

In (5.4), $y_M = y_M(\tau; y_0, \Phi_M)$ is the solution of (2.23). Because the P_M's and Q_M's are orthogonal commuting projections ($Q_M = I - P_M$), we have $QP_M = P_M - P$.

Let us rewrite (2.23), (2.24) as they apply to (5.1)

$$\frac{dy_M}{dt} + Ay_M + PR_\theta(y_M + \Phi_M(y_M)) = 0, \qquad y_M(0) = y_0, \tag{5.5}$$

$$\frac{dz_M}{dt} + Az_M + QP_M R_\theta(y_M + \Phi_M(y_M)) = 0. \tag{5.6}$$

We recall that (5.6) has a unique solution $z_M = z_M(\tau; y_0, \Phi_M)$ which is bounded as $\tau \to -\infty$ and that, in fact,

$$\mathscr{T}_M \Phi_M(y_0) = z_M(0; y_0, \Phi_M).$$

Now since $P_M e^{\tau AQ} = e^{\tau AQP_M} = e^{\tau AQP_M} P_M$, it follows from (2.38) and (5.4) that $\mathscr{T}_M$ is nothing other than the restriction of $P_M \mathscr{T}$ to $\mathscr{F}^{\alpha}_{b,l,M}$. Having noticed this point, it becomes easy to derive properties of $\mathscr{T}_M$ from those of $\mathscr{T}$ and to compare the fixed points Φ_M and Φ of $\mathscr{T}_M$ and $\mathscr{T}$.

Let $\Phi_M \in \mathscr{F}^{\alpha}_{b,l,M} \subset \mathscr{F}^{\alpha}_{b,l}$ and $y_0 \in PD(A^\alpha)$. Let $y_M = y_M(t; y_0, \Phi_M)$ denote the solution of (2.23) *and* (5.5), these two equations being identical on $PD(A)$ for $\Phi \in \mathscr{F}^{\alpha}_{b,l,M}$. From (2.24) and (5.6) we can see that

$$\mathscr{T}\Phi_M(y_0) - \mathscr{T}_M \Phi_M(y_0) = Q_M \mathscr{T}\Phi_M(y_0). \tag{5.7}$$

In addition for all $M > N$, all $\Phi_M \in \mathscr{F}^{\alpha}_{b,l,M}$ and all $y_0 \in PD(A^\alpha)$, we have

$$\begin{aligned} |A^\alpha \mathscr{T}\Phi_M(y_0) - A^\alpha \mathscr{T}_M \Phi_M(y_0)| &= |Q_M A^\alpha \mathscr{T}\Phi_M(y_0)| \\ &= |Q_M A^{-1/4} A^{\alpha+1/4} \mathscr{T}\Phi_M(y_0)| \\ &\le \lambda_{M+1}^{-1/4} |A^{\alpha+1/4} \mathscr{T}\Phi_M(y_0)|. \end{aligned}$$

A uniform bound of $|A^{\alpha+1/4}\mathscr{T}\Phi_M(y_0)|$ can be derived as in Lemma 3.3, using

Lemma 3.2,

$$
\begin{aligned}
|A^{\alpha+1/4}\mathcal{T}\Phi_M(y_0)| &= \left|\int_{-\infty}^{0} A^{\alpha+1/4}e^{\tau AQ}R_\theta(y+\Phi_M(y))\,d\tau\right| \\
&\le \int_{-\infty}^{0} |(AQ)^{3/4}e^{\tau AQ}A^{\alpha-1/2}R_\theta(y+\Phi_M(y))|\,d\tau \\
&\le \int_{-\infty}^{0} |(AQ)^{\alpha+1/4}e^{\tau AQ}R_\theta(y+\Phi_M(y))|\,d\tau \\
&\le \text{(with (2.17) and (3.11))} \\
&\le M_1\int_{-\infty}^{0} |(AQ)^{3/4}e^{\tau AQ}|_{\mathcal{L}(QH)}\,d\tau \\
&\le \kappa_3 M_1\Lambda^{-1/4}.
\end{aligned}
$$

Hence

$$
\operatorname*{Sup}_{y_0\in PD(A^\alpha)} |A^\alpha\mathcal{T}\Phi_M(p_0)| \le \kappa_3 M_1\Lambda^{-1/4}, \tag{5.8}
$$

$$
\operatorname{dist}(\mathcal{T}\Phi_M, \mathcal{T}_M\Phi_M) \le \kappa_3 M_1\Lambda^{-1/4}\lambda_{M+1}^{-1/4} = \kappa_3 M_1\lambda_{N+1}^{-1/4}\lambda_{M+1}^{-1/4}, \quad \forall\Phi_M\in\mathcal{F}_{b,l,M}^{\alpha}. \tag{5.9}
$$

We then compare Φ and Φ_M

$$
\Phi = \mathcal{T}\Phi, \qquad \Phi_M = \mathcal{T}_M\Phi_M
$$

and by substracting

$$
\begin{aligned}
\Phi - \Phi_M &= \mathcal{T}\Phi - \mathcal{T}_M\Phi_M, \\
\operatorname{dist}(\Phi, \Phi_M) &\le \operatorname{dist}(\mathcal{T}\Phi, \mathcal{T}\Phi_M) + \operatorname{dist}(\mathcal{T}\Phi_M, \mathcal{T}_M\Phi_M) \\
&\le l\,\operatorname{dist}(\Phi, \Phi_M) + \kappa_3 M_1(\lambda_{N+1}\lambda_{M+1})^{-1/4}, \qquad (5.10) \\
\operatorname{dist}(\Phi, \Phi_M) &\le \frac{1}{1-l}\kappa_3 M_1(\lambda_{N+1}\lambda_{M+1})^{-1/4}.
\end{aligned}
$$

This proves (5.2) and the theorem. □

Remark 5.1.

(i) As indicated above, we can consider other perturbations of (2.16) apart from (5.1); for example, we can consider operators R depending in an appropriate manner on a parameter ξ, $R = R(\xi, u)$, and study the dependence of $\mathcal{T} = \mathcal{T}(\xi)$ on ξ. Under appropriate assumptions, $\mathcal{T} = \mathcal{T}(\xi)$ and $\Phi = \Phi(\xi)$ depend continuously on ξ.

(ii) The replacement of $\mathcal{T}$ by $\mathcal{T}_M = P_M\mathcal{T}$ is a step in the finite-dimensional approximation of $\mathcal{T}$ and also a step in the approximation of the dynamics of (2.11), (2.16); other methods of approximation could be implemented at this level.

CHAPTER IX

Inertial Manifolds and Slow Manifolds. The Non-self-adjoint Case

Introduction

This chapter and the following chapter are based on a presentation of Inertial Manifolds (I.M.) which is different from that contained in Chapter VIII. As we said in the Preface to the Second Edition, Chapters IX and X can be read independently of Chapter VIII. The present chapter contains a presentation of inertial manifolds which is self-contained. The main result of this chapter is an existence result for inertial manifolds (see Theorem 2.1 and also Theorem 3.1) which generalizes Theorem 3.1 of Chapter VIII. Although the results are very similar in cases where both theorems apply, the approach in this chapter offers several advantages.

(i) The basic linear operator A in (1.1) need not be self-adjoint any more (compare to VIII.(2.11)).

(ii) The function spaces can be Banach spaces, and not necessarily Hilbert spaces.

(iii) Furthermore, this method paves the way for Chapter X, for the solution of certain interesting questions concerning the approximation of inertial manifolds. In particular, (i) allows us to relate the concept of inertial manifolds to that of Slow Manifolds appearing in meteorology (see Section 4 and the comments and references therein).

The chapter is organized as follows. Section 1 contains the description of the framework and the general hypotheses. Section 2 contains the statement and the proof of the main results in the Lipschitz case (Theorem 2.1). Section 3 contains some complements and applications. In particular, we consider the locally Lipschitz case (Theorem 3.1) and show how one can estimate the dimension of the inertial manifold in terms of physically relevant quantities.

Finally, in Section 4, we give an application of the general results to an example related to Slow Manifolds, a related concept appearing in meteorology.

In this chapter we follow A. Debussche and R. Temam [2], [3]. For other results on inertial manifolds in the non-self-adjoint case, see G. Sell and Y. You [1], and R. Rosa and R. Temam [1].

1. The Functional Setting

The general form of the abstract equation that we consider in this chapter is the same as in Chapter VIII.(2.11), the hypotheses are however very different. The only slight change in the form of the equation is, for the sake of convenience, to replace R by $-R$, so that the equation now reads

$$\frac{du}{dt} + Au = R(u). \tag{1.1}$$

In Section 1.1 we present the general hypotheses related to equation (1.1), then, in Section 1.2, we describe our approach to the construction of inertial manifolds.

1.1. Notations and Hypotheses

We are given three Banach spaces E, F, $\mathscr{E}$, such that

$$E \subset F \subset \mathscr{E}, \tag{1.2}$$

each space being dense in the following one, the injections being continuous. The norms on $\mathscr{E}$, E, and F are denoted by $|\cdot|_{\mathscr{E}}$, $|\cdot|_E$, and $|\cdot|_F$.

We assume that R is a $\mathscr{C}^1$ nonlinear function from E into F which satisfies the following boundedness and Lipschitz properties:

$$|R(u)|_F \leq M_0, \qquad \forall u \in E, \tag{1.3}$$

$$|R(u) - R(v)|_F \leq M_1 |u - v|_E, \qquad \forall u, v \in E; \tag{1.4}$$

here M_0 and M_1 are positive constants.

Concerning the linear operator A in (1.1), we assume that A is a linear densely defined operator in $\mathscr{E}$. Furthermore, we assume that the linear equation

$$\begin{cases} \dfrac{du}{dt} + Au = 0, \\ u(0) = u_0, \end{cases} \tag{1.5}$$

defines a strongly continuous linear semigroup $\{e^{-At}\}_{t \geq 0}$ on $\mathscr{E}$ such that

$$e^{-At}F \subset E, \qquad \text{for all} \quad t > 0. \tag{1.6}$$

The Eigenprojectors P_n, Q_n

We suppose that we are given a sequence $\{P_n\}_{n\in\mathbb{N}}$ of eigenprojectors of A and two sequences of numbers $\{\lambda_n\}_{n\in\mathbb{N}}$, $\{\Lambda_n\}_{n\in\mathbb{N}}$ satisfying, for some $\lambda_* > 0$[1],

$$\Lambda_n \geq \lambda_n \geq \lambda_* \geq 0, \qquad \forall n \geq 0, \tag{1.7}$$

$$\lambda_n \to \infty \qquad \text{as} \quad n \to \infty, \tag{1.8}$$

$$\frac{\Lambda_n}{\lambda_n} \quad \text{is bounded as} \quad n \to \infty. \tag{1.9}$$

We suppose that if $Q_n = I - P_n$, then

$$P_n\mathscr{E} \text{ and } Q_n\mathscr{E} \text{ are invariant under } e^{-At} \text{ for all } t \geq 0, \tag{1.10}$$

$$\{e^{-At}\}_{t\geq 0} \text{ can be extended to a group } \{e^{-At}\}_{t\in\mathbb{R}} \text{ on } P_n\mathscr{E}. \tag{1.11}$$

We also assume that these projectors define an *exponential dichotomy of* $\{e^{-At}\}_{t\geq 0}$ in the sense that there exist two positive constants k_1, k_2 and α, $0 \leq \alpha < 1$, independent of n, such that

$$\begin{cases} |e^{-At}P_n|_{\mathscr{L}(E)} \leq k_1 e^{-\lambda_n t}, \\ |e^{-At}P_n|_{\mathscr{L}(F,E)} \leq k_1 \lambda_n^\alpha e^{-\lambda_n t}, \qquad \forall t \leq 0. \end{cases} \tag{1.12}$$

$$\begin{cases} |e^{-At}Q_n|_{\mathscr{L}(F,E)} \leq k_2\left(\dfrac{1}{t^\alpha} + \Lambda_n^\alpha\right)e^{-\Lambda_n t}, \qquad \forall t > 0, \\ |A^{-1}e^{-At}Q_n|_{\mathscr{L}(F,E)} \leq k_2 \Lambda_n^{\alpha-1} e^{-\Lambda_n t}, \\ |e^{-At}Q_n|_{\mathscr{L}(E)} \leq k_2 e^{-\Lambda_n t}, \qquad \forall t \geq 0. \end{cases} \tag{1.13}$$

Of course, since P_n and Q_n are eigenprojectors of A, they commute with A,

$$P_n A = A P_n, \qquad Q_n A = A Q_n, \qquad \forall n. \tag{1.14}$$

Finally, concerning the nonlinear equation (1.1), we assume that the initial value problem

$$\begin{cases} \dfrac{du}{dt} + Au = R(u), \\ u(0) = u_0, \end{cases} \tag{1.15}$$

defines a continuous semigroup $\{S(t)\}_{t\geq 0}$ on E.

Remark 1.1. Hypotheses (1.2)–(1.15) are standing hypotheses in this chapter (and in the next one). We show in Section 3 below how the hypotheses above can be verified, especially (1.5)–(1.13); in particular, we show how to recover the Hilbert case and we present a non-self-adjoint case related to the slow manifolds.

[1] The hypothesis $\lambda_n \geq \lambda_* > 0$ is made for the sake of simplicity. Indeed, it follows from (1.8) that $\lambda_n \geq \lambda_* > 0$ for n sufficiently large and this is, in fact, sufficient.

We recall that an *Inertial Manifold* (I.M.) for equation (1.15) (or for the semigroup $\{S(t)\}_{t\geq 0}$) is a finite-dimensional Lipschitz manifold $\mathcal{M}$, which enjoys the following properties:

(i) $\mathcal{M}$ is positively invariant for the semigroup (i.e., $S(t)\mathcal{M} \subset \mathcal{M}$, $\forall t \geq 0$); and
(ii) $\mathcal{M}$ attracts all orbits of (1.1) at an exponential rate.

1.2. Construction of the Inertial Manifold

For some fixed n we apply the projectors P_n, Q_n to (1.15) and setting $P_n u = y$, $Q_n u = z$, and using (1.14), we see that equation (1.15) is equivalent to a coupled system for y and z,

$$\begin{cases} \dfrac{dy}{dt} + Ay = P_n R(y + z), \\ y(0) = y_0, \end{cases} \tag{1.16}$$

$$\begin{cases} \dfrac{dz}{dt} + Az = Q_n R(y + z), \\ z(0) = z_0, \end{cases} \tag{1.17}$$

where $y_0 = P_n u_0$ and $z_0 = Q_n u_0$.

Our construction of the inertial manifold will be based on the Lyapunov–Perron method as in Chapter VIII, and we will obtain $\mathcal{M}$ as the graph of a function $\Phi: P_n E \to Q_n E$. The function Φ will be obtained, as in Chapter VII, as the fixed point of a mapping $\mathcal{T}$ and we now explain the construction of $\mathcal{T}$.

Assuming that Φ is known and that its graph is invariant for $\{S(t)\}_{t\geq 0}$, then for an orbit $u(t)$, $t \geq 0$, lying on $\mathcal{M}$ (i.e., if $u_0 \in \mathcal{M}$), we have

$$z(t) = \Phi(y(t)), \tag{1.18}$$

and we infer from (1.16) and (1.17) that y satisfies

$$\begin{cases} \dfrac{dy}{dt} + Ay = P_n R(y + \Phi(y)), \qquad t > 0, \\ y(0) = y_0, \end{cases} \tag{1.19}$$

and the function $t \to \Phi(y(t))$ $(= z(t))$ satisfies

$$\frac{d\Phi(y)}{dt} + A\Phi(y) = Q_n R(y + \Phi(y)), \qquad t > 0. \tag{1.20}$$

By integration of (1.20) from t_0 to t, $0 < t_0 < t$, using the variation of constants formula, we see that

$$\Phi(y(t)) = e^{-A(t-t_0)}\Phi(y(t_0)) + \int_{t_0}^{t} e^{-A(t-s)} Q_n R(y(s) + \Phi(y(s)))\, ds. \tag{1.21}$$

Now (1.19) is a finite-dimensional system of ordinary differential equations, and since Φ is Lipschitz (and A is linear), it has a unique solution y defined for all $t \in \mathbb{R}$. Then we can also consider (1.20) and write (1.21) for $t_0 < t$, not necessarily positive. We will also assume that Φ is bounded (see Section 2 below); hence letting $t_0 \to -\infty$ in (1.21), we infer that

$$\Phi(y(t)) = \int_{-\infty}^{t} e^{-A(t-s)} Q_n R(y(s) + \Phi(y(s)))\, ds. \tag{1.22}$$

In particular, at $t = 0$,

$$\Phi(y_0) = \int_{-\infty}^{0} e^{As} Q_n R(y(s) + \Phi(y(s)))\, ds. \tag{1.23}$$

Hence, it appears natural to construct Φ as the fixed point of the mapping $\psi \to \mathscr{T}\psi$ where $\mathscr{T}\psi$ is defined (compare to (1.23)) by

$$\mathscr{T}\psi(y_0) = \int_{-\infty}^{0} e^{As} Q_n R(y(s) + \psi(y(s)))\, ds. \tag{1.24}$$

Here ψ is a Lipschitz bounded function from $P_n E$ into $Q_n E$ and y is the solution of

$$\begin{cases} \dfrac{dy}{dt} + Ay = P_n R(y + \psi(y)), \\ y(0) = y_0. \end{cases} \tag{1.25}$$

Remark 1.2. Even in the case where $\psi = \Phi$, the solution y of (1.25) is not the same as the solution of (1.16), unless $u_0 \in \mathscr{M}$, i.e., $z_0 = \Phi(y_0)$. In general, these two functions are different and we will denote them as $\bar{y}$ and y when they are simultaneously used.

We conclude this section with the precise definition of $\mathscr{T}$ and of the spaces to which ψ and Φ belong.

For suitable $l, b > 0$, we set

$$\mathscr{F} = \mathscr{F}_{l,b} = \left\{ \psi \colon P_n E \to Q_n E,\ \mathrm{Lip}\, \psi \le l,\ |\psi|_\infty = \operatorname*{Sup}_{y \in P_n E} |\psi(y)|_E \le b \right\}.$$

Of course, $\mathscr{F}_{l,b}$ is a complete metric space for the distance

$$d(\psi_1, \psi_2) = \operatorname*{Sup}_{y \in P_n E} |\psi_1(y) - \psi_2(y)|_E = |\psi_1 - \psi_2|_\infty. \tag{1.26}$$

It is clear that for $\psi \in \mathscr{F}_{l,b}$, (1.25) is an ordinary (finite-dimensional) system of differential equations which possesses a unique solution $y = y(t)$, defined for all $t \in \mathbb{R}$. It is easy to see (we will prove more below), that for $y_0 \in P_n E$, $\mathscr{T}\psi(y_0)$ given by (1.24) belongs to $Q_n E$. Hence $\mathscr{T}\psi$ is a mapping from $P_n E$ into $Q_n E$. In fact, under suitable hypotheses, we show in Section 2 that $\mathscr{T}\psi$ belongs to $\mathscr{F}_{l,b}$ and that $\mathscr{T}$ is a strict contraction. By the Contraction Principle, we see that $\mathscr{T}$ possesses a fixed point Φ and we show that the graph of Φ is an inertial manifold.

2. The Main Result (Lipschitz Case)

In this section we state and prove one of our main results, namely, the existence of inertial manifolds in the globally Lipschitz case, i.e., when R satisfies (1.3) and (1.4). We refer the reader to Section 3 for the locally Lipschitz case.

2.1. Existence of Inertial Manifolds

Our aim is to prove the following:

Theorem 2.1. *We assume that the general hypotheses* (1.2)–(1.15) *are satisfied. Then there exist two constants c_1, c_2 depending only on k_1, k_2, α, l, b such that*

$$\Lambda_n - \lambda_n \geq c_1(M_0 + M_1 + M_1^2)(\Lambda_n^\alpha + \lambda_n^\alpha), \tag{2.1}$$

and

$$\lambda_n^{1-\alpha} \geq c_2(M_0 + M_1), \tag{2.2}$$

then the mapping $\mathscr{T}$ defined by (1.24) *and* (1.25) *is a strict contraction in $\mathscr{F}_{l,b}$. The graph $\mathscr{M}$ of its fixed point Φ is an inertial manifold for equation* (1.15). *Furthermore, Φ is a continuously differentiable function* (*and $\mathscr{M}$ is of class $\mathscr{C}^1$*).

Theorem 2.1 is proved in Sections 2.2 and 2.3.

Remark 2.1. Condition (2.1) will be refered to as the *spectral gap condition.*

A more precise form of the constants c_1, c_2 appears in the proof of Theorem 2.1 (see (2.48) and (2.49)). Also we refer the reader to R. Rosa and R. Temam [1] for another form of Theorem 2.1 where the constants c_1 and c_2 are fully explicit.

Remark 2.2. We call the *inertial form* of equation (1.15) a finite-dimensional dynamical system which produces the same dynamics.

When Theorem 2.1 applies, equation (1.19) is an *inertial form* of (1.15).

Remark 2.3. Another interesting concept related to the inertial manifold is that of *asymptotic completeness* (see P. Constantin, C. Foias, B. Nicolaenko, and R. Temam [2], and C. Foias, G. Sell, and E. Titi [1]). The inertial manifold $\mathscr{M}$ is said to be asymptotically complete if the following holds:

$$\begin{cases} \text{For every } u_0 \in E, \text{ there exists } \bar{u}_0 \in \mathscr{M} \text{ and } \tau \in \mathbb{R}, \\ \text{such that } |S(t)u_0 - S(t+\tau)\bar{u}_0|_E \to 0, \text{ as } t \to \infty. \end{cases} \tag{2.3}$$

It can be shown that the manifold $\mathcal{M}$ given by Theorem 2.1 is asymptotically complete.

2.2. Properties of $\mathcal{T}$

For the sake of simplicity, we set

$$\Lambda = \Lambda_n, \qquad \lambda = \lambda_n, \qquad P = P_n, \qquad Q = Q_n.$$

We start the proof of Theorem 2.1 by showing that $\mathcal{T}$ maps $\mathcal{F}_{b,l}$ into itself.

Lemma 2.1. *For ψ in $\mathcal{F}_{b,l}$*

$$|\mathcal{T}\psi(y_0)|_E \leq b, \qquad \forall y_0 \in PE, \tag{2.4}$$

provided the constant c_2 in (2.2) is sufficiently large.

Proof. We write

$$\mathcal{T}\psi(y_0) = \int_{-\infty}^{0} e^{As} QR(y(s) + \psi(y(s)))\, ds,$$

$$|\mathcal{T}\psi(y_0)|_E \leq \int_{-\infty}^{0} |e^{As} QR(y(s) + \psi(y(s)))|_E\, ds$$

$$\leq \int_{-\infty}^{0} |e^{As} Q|_{\mathcal{L}(F,E)} |R(y(s) + \psi(y(s)))|_F\, ds.$$

Thus, by (1.3) and (1.13),

$$|\mathcal{T}\psi(y_0)|_E \leq M_0 k_2 \int_{-\infty}^{0} (|s|^{-\alpha} + \Lambda^\alpha) e^{\Lambda s}\, ds.$$

We observe that for α, $a > 0$,

$$\int_{-\infty}^{0} |s|^{-\alpha} e^{as}\, ds = a^{\alpha-1}\gamma_\alpha, \qquad \gamma_\alpha = \int_{0}^{\infty} s^{-\alpha} e^{-s}\, ds. \tag{25}$$

Hence

$$|\mathcal{T}\psi(y_0)|_E \leq k_2 M_0 \Lambda^{\alpha-1}(1 + \gamma_\alpha)$$

and this is bounded by b, thanks to (1.7) and (2.2), provided

$$c_2 \geq \frac{k_2(1 + \gamma_\alpha)}{b}. \tag{2.6}$$ □

Lemma 2.2. *For ψ in $\mathcal{F}_{b,l}$ and for every y_{01}, y_{02} in PE,*

$$|\mathcal{T}\psi(y_{01}) - \mathcal{T}\psi(y_{02})|_E \leq l |y_{01} - y_{02}|_E, \tag{2.7}$$

provided the constant c_1 in (2.1) is sufficiently large.

PROOF. For $i = 1, 2$, we consider the functions $y = y_i$, which are solutions of (1.25) with $y_i(0) = y_{0i}$. From the definition (1.24) of $\mathscr{T}$ we then have

$$\mathscr{T}\psi(y_{01}) - \mathscr{T}\psi(y_{02})$$
$$= \int_{-\infty}^{0} e^{As}Q[R(y_1(s) + \psi(y_1(s))) - R(y_2(s) + \psi(y_2(s)))]\, ds,$$

$$|\mathscr{T}\psi(y_{01}) - \mathscr{T}\psi(y_{02})|_E$$
$$\leq \int_{-\infty}^{0} |e^{As}Q|_{\mathscr{L}(F,E)}|R(y_1(s) + \psi(y_1(s))) - R(y_2(s) + \psi(y_2(s)))|_F\, ds$$
$$\leq \text{(by (1.4), (1.13) and } \psi \in \mathscr{F}_{l,b}),$$
$$\leq k_2 M_1 \int_{-\infty}^{0} (|s|^{-\alpha} + \Lambda^\alpha)e^{\Lambda s}(|y_1(s) - y_2(s)|_E + |\psi(y_1(s)) - \psi(y_2(s))|_E)\, ds$$
$$\leq k_2 M_1(1 + l) \int_{-\infty}^{0} e^{\Lambda s}(|s|^{-\alpha} + \Lambda^\alpha)|y(s)|_E\, ds, \tag{2.8}$$

where $y(s) = y_1(s) - y_2(s)$. An estimate of $|y(s)|_E$ is given in Lemma 2.3 hereafter; this yields (see (2.12))

$$|y(s)|_E \leq 2k_1|y_{01} - y_{02}|_E e^{-\tilde{\lambda}s},$$

where $\tilde{\lambda} = \lambda + k_1 M_1(1 + l)\lambda^\alpha$. Hence, from the last inequality (2.8) we see that

$$|\mathscr{T}\psi(y_{01}) - \mathscr{T}\psi(y_{02})|_E$$
$$\leq 2k_1 k_2 M_1(1 + l)|y_{01} - y_{02}|_E \int_{-\infty}^{0} (|s|^{-\alpha} + \Lambda^\alpha)e^{(\Lambda - \tilde{\lambda})s}\, ds.$$

Due to (2.4),

$$\int_{-\infty}^{0} (|s|^{-\alpha} + \Lambda^\alpha)e^{(\Lambda - \tilde{\lambda})s}\, ds = (\Lambda - \tilde{\lambda})^{\alpha - 1}\gamma_\alpha + \Lambda^\alpha(\Lambda - \tilde{\lambda})^{-1}$$
$$\leq \Lambda^\alpha(\Lambda - \tilde{\lambda})^{-1}(1 + \gamma_\alpha);$$

here we have used

$$0 \leq \Lambda - \tilde{\lambda} = \Lambda - \lambda - k_1 M_1(1 + l)\lambda^\alpha \leq \Lambda,$$

which follows from (2.1), provided

$$c_1 \geq k_1(1 + l). \tag{2.9}$$

Thus

$$|\mathscr{T}\psi(y_{01}) - \mathscr{T}\psi(y_{02})|_E \leq 2k_1 k_2 M_1(1 + l)(1 + \gamma_\alpha)\Lambda^\alpha(\Lambda - \tilde{\lambda})^{-1}|y_{01} - y_{02}|_E, \tag{2.10}$$

and this is bounded by $l|y_{01} - y_{02}|_E$, thanks to (2.1) and provided

$$\Lambda - \tilde{\lambda} = \Lambda - \lambda - k_1 M_1(1 + l)\lambda^\alpha \geq 2k_1 k_2 M_1 l^{-1}(1 + l)(1 + \gamma_\alpha)\Lambda^\alpha,$$

i.e., beside (2.9),

$$c_1 \geq 2k_1 k_2 M_1 l^{-1}(1 + l)(1 + \gamma_\alpha). \tag{2.11}$$

□

The proof of Lemma 2.2 will be complete after we prove Lemmas 2.3 and 2.4 hereafter.

Lemma 2.3. *If y_1 and y_2 are two solutions of (1.25) with $y_i(0) = y_{0i} \in PE$, $i = 1, 2$, and $\psi \in \mathscr{F}_{b,l}$, and if $y(t) = y_1(t) - y_2(t)$, then*

$$|y(t)|_E \leq 2k_1 |y_{01} - y_{02}|_E e^{-\tilde{\lambda} t}, \qquad \forall t \leq 0, \tag{2.12}$$

where $\tilde{\lambda} = \lambda + k_1 M_1 (1 + l)\lambda^\alpha$.

PROOF. By the variation of constants formula, we write the solutions y_1, y_2 of (1.25) in the form

$$y_i(t) = e^{-At} y_{0i} + \int_0^t e^{-A(t-s)} PR(y_i(s) + \psi(y_i(s)))\, ds. \tag{2.13}$$

Substracting these relations, we find

$$y(t) = e^{-At}(y_{01} - y_{02}) + \int_0^t e^{-A(t-s)} P[R(y_1(s) + \psi(y_1(s))) - R(y_2(s) + \psi(y_2(s)))]\, ds.$$

Hence, upon using (1.4), (1.12) and $\psi \in \mathscr{F}_{b,l}$.

$$\begin{aligned}|y(t)|_E &\leq |e^{-At}P|_{\mathscr{L}(E)} |y_{01} - y_{02}|_E \\ &\quad + \int_t^0 |e^{-A(t-s)}P|_{\mathscr{L}(F,E)} |R(y_1(s) + \psi(y_1(s))) \\ &\quad - R(y_2(s) + \psi(y_2(s)))|_F\, ds \\ &\leq k_1 |y_{01} - y_{02}|_E e^{-\lambda t} + k_1 M_1 (1 + l)\lambda^\alpha \int_t^0 e^{-\lambda(t-s)} |y(s)|\, ds,\end{aligned}$$

($t < 0$). We apply Lemma 2.4 below with $f(t) = |y(t)|_E$, and we conclude that

$$|y(t)|_E \leq 2k_1 |y_{01} - y_{02}|_E e^{-(\lambda + k_1 M_1 (1+l)\lambda^\alpha)t} = 2k_1 |y_{01} - y_{02}|_E e^{-\tilde{\lambda} t}, \qquad t < 0. \ \square$$

Lemma 2.4. *If a function $f \geq 0$ satisfies for $t \leq 0$,*

$$f(t) \leq a e^{-\gamma' t} + b \int_t^0 e^{-\gamma(t-s)} f(s)\, ds, \tag{2.14}$$

with $a, b, \gamma, \gamma' > 0$, $\gamma + b > \gamma'$, then, for $t < 0$,

$$\int_t^0 e^{\gamma s} f(s)\, ds \leq \frac{a}{\gamma - \gamma' + b} e^{-bt}, \tag{2.15}$$

$$f(t) \le a \frac{\gamma - \gamma' + 2b}{\gamma - \gamma' + b} e^{-(b+\gamma)t}. \tag{2.16}$$

PROOF. We set

$$r(t) = \int_t^0 e^{\gamma s} f(s)\, ds,$$

and infer from (2.16) that

$$-r' \le a e^{(\gamma - \gamma')t} + br.$$

By Gronwall's lemma and $r(0) = 0$, we find

$$r(t) \le \frac{a}{\gamma - \gamma' + b}(e^{-bt} - e^{(\gamma - \gamma')t}),$$

$$r(t) \le \frac{a}{\gamma - \gamma' + b} e^{-bt}.$$

Hence

$$f(t) \le a e^{-\gamma' t} + \frac{ab}{\gamma - \gamma' + b} e^{-(b+\gamma)t}$$

$$\le a \frac{\gamma - \gamma' + 2b}{\gamma - \gamma' + b} e^{-(b+\gamma)t}. \qquad \square$$

Lemmas 2.1 and 2.2 show that $\mathscr{T}$ maps $\mathscr{F}_{b,l}$ into itself. We now prove that $\mathscr{T}$ is a strict contraction in $\mathscr{F}_{b,l}$.

Lemma 2.5. *If ψ_1, ψ_2 are in $\mathscr{F}_{b,l}$ and $\delta > 0$, then*

$$|\mathscr{T}\psi_1 - \mathscr{T}\psi_2|_\infty \le \delta |\psi_1 - \psi_2|_\infty, \tag{2.17}$$

provided the constants c_1 and c_2 in (2.1) and (2.2) are sufficiently large.

PROOF. For y_0 in PE, we denote by y_1 and y_2 the solutions of (1.25) corresponding to $\psi = \psi_1$ and $\psi = \psi_2$. Then, as in (2.13), the variation of constants formula gives

$$y_i(t) = e^{-At} y_0 + \int_0^t e^{-A(t-s)} PR(y_i(s) + \psi_i(y_i(s)))\, ds, \qquad i = 1, 2. \tag{2.18}$$

Then

$$\mathscr{T}\psi_i(y_0) = \int_{-\infty}^0 e^{As} QR(y_i(s) + \psi_i(y_i(s)))\, ds,$$

and setting $y = y_1 - y_2$, we find

$$\mathscr{T}\psi_1(y_0) - \mathscr{T}\psi_2(y_0)$$
$$= \int_{-\infty}^{0} e^{As}Q[R(y_1(s) + \psi_1(y_1(s))) - R(y_2(s) + \psi_2(y_2(s)))]\, ds,$$

$$|\mathscr{T}\psi_1(y_0) - \mathscr{T}\psi_2(y_0)|_E$$
$$\leq \int_{-\infty}^{0} |e^{As}Q|_{\mathscr{L}(F,E)} |R(y_1(s) + \psi_1(y_1(s))) - R(y_2(s) + \psi_2(y_2(s)))|_F\, ds$$
$$\leq \text{(with (1.13))}$$
$$\leq k_2 \int_{-\infty}^{0} (|s|^{-\alpha} + \Lambda^\alpha) e^{\Lambda s} r(s)\, ds;$$

here

$$r(s) = |R(y_1(s) + \psi_1(y_1(s))) - R(y_2(s) + \psi_2(y_2(s)))|_F$$
$$\leq \text{(with (1.5) and } \psi_1 \in \mathscr{F}_{b,l})$$
$$\leq M_1(|y_1(s) - y_2(s)|_E + |\psi_2(y_1(s)) - \psi_2(y_2(s))|_E$$
$$+ |\psi_1(y_1(s)) - \psi_2(y_1(s))|_E)$$
$$\leq M_1[|\psi_1 - \psi_2|_\infty + (1 + l)|y(s)|_E], \tag{2.19}$$

with $y(s) = y_1(s) - y_2(s)$. Hence

$$|\mathscr{T}\psi_1(y_0) - \mathscr{T}\psi_2(y_0)|_E$$
$$\leq M_1 k_2 \int_{-\infty}^{0} (|s|^{-\alpha} + \Lambda^\alpha) e^{\Lambda s}[|\psi_1 - \psi_2|_\infty + (1 + l)|y(s)|_E]\, ds$$
$$\leq \text{(with (2.4))}$$
$$\leq M_1 k_2 (1 + \gamma_\alpha) \Lambda^{\alpha - 1} |\psi_1 - \psi_2|_\infty$$
$$+ M_1 k_2 (1 + l) \int_{-\infty}^{0} (|s|^{-\alpha} + \Lambda^\alpha) e^{\Lambda s} |y(s)|_E\, ds. \tag{2.20}$$

Now we ought to estimate $|y(s)|_E$. We infer from (2.18), (1.4), and (1.12) that

$$y(t) = \int_0^t e^{-A(t-s)} P[R(y_1(s) + \psi_1(y_1(s))) - R(y_2(s) + \psi_2(y_2(s)))]\, ds,$$

$$|y(t)|_E \leq k_1 M_1 \lambda^\alpha \int_t^0 e^{-\lambda(t-s)} r(s)\, ds,$$

$r(s)$ as in (2.19); therefore

$$|y(t)|_E \leq k_1 M_1^2 \lambda^\alpha \int_t^0 e^{-\lambda(t-s)} [|\psi_1 - \psi_2|_\infty + (1 + l)|y(s)|_E]\, ds,$$

$$|y(t)|_E \le k_1 M_1^2 \lambda^{\alpha-1} |\psi_1 - \psi_2|_\infty e^{-\lambda t} + k_1 M_1^2 (1 + l) \lambda^\alpha \int_t^0 e^{-\lambda(t-s)} |y(s)|_E \, ds.$$

We can apply Lemma 2.4 with $f(t) = |y(t)|_E$, and we find

$$|y(t)|_E \le 2k_1 M_1^2 \lambda^{\alpha-1} |\psi_1 - \psi_2|_\infty e^{-\lambda' t},$$

where $\lambda' = \lambda + k_1 M_1^2 (1 + l) \lambda^\alpha$. Finally, using (2.2) and assuming $c_2 \ge 1$,

$$|y(t)|_E \le 2k_1 M_1 |\psi_1 - \psi_2|_\infty e^{-\lambda' t}. \tag{2.21}$$

Now from (2.20) and (2.21),

$$\begin{aligned}
&|\mathscr{T}\psi_1(y_0) - \mathscr{T}\psi_2(y_0)|_E \\
&\quad \le M_1 k_2 (1 + \gamma_\alpha) \Lambda^{\alpha-1} |\psi_1 - \psi_2|_\infty \\
&\qquad + 2k_1 k_2 M^2 (1 + l) |\psi_1 - \psi_2|_\infty \int_{-\infty}^0 (|s|^{-\alpha} + \Lambda^\alpha) e^{(\Lambda - \lambda')s} \, ds \\
&\quad \le (\text{with (2.4) and } 0 < \Lambda - \lambda' \le \Lambda) \\
&\quad \le M_1 k_2 (1 + \gamma_\alpha) \Lambda^{\alpha-1} |\psi_1 - \psi_2|_\infty \\
&\qquad + 2k_1 k_2 M_1^2 (1 + l) |\psi_1 - \psi_2|_\infty (1 + \gamma_\alpha) \Lambda^\alpha (\Lambda - \lambda')^{-1}.
\end{aligned} \tag{2.22}$$

With (1.7), (2.1), and (2.2), we obtain

$$|\mathscr{T}\psi_1(y_0) - \mathscr{T}\psi_2(y_0)|_E \le \delta |\psi_1 - \psi_2|_\infty, \qquad \forall y_0 \in PE, \tag{2.23}$$

provided

$$\begin{aligned}
c_1 &\ge \frac{4}{\delta} k_1 k_2 (1 + l)(1 + \gamma_\alpha), \\
c_2 &\ge \frac{2}{\delta} k_2 (1 + \gamma_\alpha).
\end{aligned} \tag{2.24}$$

Lemma 2.5 is proved; we only need $\delta < 1$ and we take, e.g., $\delta = \frac{1}{2}$. □

We have shown that, under hypotheses (2.1) and (2.2), $\mathscr{T}$ maps $\mathscr{F}_{b,l}$ into itself and is Lipschitz with constant $\frac{1}{2}$. By the contraction principle $\mathscr{T}$ possesses a unique fixed point Φ in $\mathscr{F}_{b,l}$,

$$\mathscr{T}\Phi = \Phi. \tag{2.25}$$

We will study the properties of Φ and conclude the proof of Theorem 2.1 by showing that its graph $\mathscr{M}$ is an inertial manifold. First we conclude this section by showing that the graph $\mathscr{M}$ of Φ is positively invariant by the semigroup $\{S(t)\}_{t \ge 0}$, which was the idea behind this construction.

Let us assume that the initial data u_0 in (1.15) belongs to $\mathscr{M}$, i.e.,

$$u_0 = y_0 + z_0, \qquad y_0 \in PE, \qquad z_0 = \Phi(y_0) \in QE. \tag{2.26}$$

Then the function y given by (1.25) satisfies, for all $t \in \mathbb{R}$,

$$\begin{cases} \dfrac{dy}{dt} + Ay = PR(y + \Phi(y)), \\ y(0) = y_0. \end{cases} \tag{2.27}$$

It is clear that, for any $t_0 > 0$, the solution $\tilde{y}$ of

$$\begin{cases} \dfrac{d\tilde{y}}{dt} + A\tilde{y} = PR(\tilde{y} + \Phi(\tilde{y})), \\ \tilde{y}(0) = y(t_0), \end{cases} \tag{2.28}$$

is given by

$$\tilde{y}(t) = y(t + t_0).$$

Therefore

$$\begin{aligned} \Phi(y(t_0)) = \mathcal{T}\Phi(y(t_0)) &= \int_{-\infty}^{0} e^{As}QR(\tilde{y}(s) + \Phi(\tilde{y}(s)))\, ds \\ &= \int_{-\infty}^{0} e^{As}QR(y(s + t_0) + \Phi(y(s + t_0)))\, ds \\ &= \int_{-\infty}^{t_0} e^{-A(t_0 - s)}QR(y(s) + \Phi(y(s)))\, ds. \end{aligned}$$

This shows that $z(t_0) = \Phi(y(t_0))$ satisfies

$$\frac{dz}{dt} + Az = QR(y + \Phi(y)), \tag{2.29}$$

for all $t_0 > 0$; furthermore,

$$z(0) = \Phi(y(0)) = z_0. \tag{2.30}$$

Combining (2.27), (2.29), and (2.30), we conclude that $y(t) + z(t) = y(t) + \Phi(y(t))$ is solution of (1.15); by the uniqueness of solution of (1.15),

$$u(t) = y(t) + z(t), \qquad z(t) = \Phi(y(t)), \tag{2.31}$$

and $u(t) \in \mathcal{M}$, $\forall t \geq 0$.

2.3. Smoothness Property of Φ (Φ is $\mathscr{C}^1$)

In this section we show that Φ is a $\mathscr{C}^1$ function from PE into QE, not simply a Lipschitz function. The proof is based on the utilization of the fiber contraction theorem (see M.W. Hirsch and L.C. Pugh [1], and for its utilization for the regularity of inertial manifolds, see F. Demengel and J.M. Ghidaglia [1], and A. Debussche and R. Temam [2], [3]).

We briefly recall the theorem and then describe its utilization.

Fiber Contraction Theorem (M.W. Hirsch and L.C. Pugh [1]). *Let X and Y be two metric spaces and let f be a continuous mapping from X into itself which possesses an attractive fixed point p* (i.e., $f(p) = p$ and $\forall x \in X$, $f^n(x) \to p$ *as* $n \to \infty$). *We are also given a continuous mapping g from $X \times Y$ into Y.*

Assume that $q \in Y$ is a fixed point of g_p, where $g_x(\cdot) = g(x, \cdot)$, such that

$$\limsup_{n\to\infty} \operatorname{Lip}(g_{f^n(x)}) < 1, \qquad \forall x \in X. \tag{2.32}$$

Then $(p, q) \in X \times Y$ is an attractive fixed point of the mapping F defined by

$$F\colon X \times Y \to X \times Y, \qquad F(x, y) = (f(x), g_x(y)).$$

We will apply this theorem in the case where X, Y are complete metric spaces and

$$f \text{ is a strict contraction in } X, \tag{2.33}$$

$$\begin{gathered} g \text{ is a continuous mapping from } X \times Y \text{ into } Y \\ \text{and } g_x \text{ is a strict contraction, } \forall x \in X. \end{gathered} \tag{2.34}$$

Hence all the hypotheses of the theorem are satisfied, f possesses a unique attractive fixed point p, g_p possesses a unique attractive fixed point q, and (p, q) is an attractive fixed point of F, $F(x, y) = (f(x), g_x(y))$.

Consider

$$\mathscr{G}_l = \left\{\Delta\colon PE \to \mathscr{L}(PE, QE),\ \operatorname*{Sup}_{y \in PE} |\Delta(y)|_{\mathscr{L}(PE,QE)} \le l\right\},$$

which is a complete metric space for the distance

$$d(\Delta, \Delta') = \operatorname*{Sup}_{y \in PE} |\Delta(y) - \Delta'(y)|_{\mathscr{L}(PE,QE)}.$$

For every $\psi \in \mathscr{F}_{l,b}$, we define the mapping $\mathscr{T}_\psi$ on $\mathscr{G}_l$ by

$$\begin{aligned} T_\psi(\Delta)(y_0)\eta_0 = \int_{-\infty}^{0} & e^{As} QDR(y(s) + \psi(y(s))) \\ & \cdot (\eta(s) + \Delta(y(s))\eta(s))\, ds, \qquad \forall y_0, \eta_0 \in PE. \end{aligned} \tag{2.35}$$

In (2.35), DR is the Fréchet differential of R, y is the solution of (1.25), and η is solution of the following linear differential equation (namely, the linearized form of (1.25)):

$$\begin{cases} \dfrac{d\eta}{dt} + A\eta = PDR(y + \psi(y))\cdot(\eta + \Delta(y)\eta), \\ \eta(0) = \eta_0. \end{cases} \tag{2.36}$$

Computations similar to those in Section 2.2 show that if (2.1) holds with

$$c_1 \ge 2k_1(1 + l) + \frac{8}{\delta} K_1 k_2 (1 + \gamma_\alpha), \tag{2.37}$$

then T_ψ is a contraction with constant δ (we choose $\delta = \frac{1}{2}$).

Hence (2.33) and (2.34) are satisfied by $f = \mathscr{T}$ and $g_\psi = T_\psi$; denoting by Δ_Φ the fixed point of T_Φ (and Φ being the fixed point of $\mathscr{T}$), we conclude that (Φ, Δ_Φ) is an attractive fixed point of

$$T: \mathscr{F}_{l,b} \times \mathscr{G}_l \to \mathscr{F}_{l,b} \times \mathscr{G}_l,$$
$$(\psi, \Delta) \to (\mathscr{T}\psi, T_\psi(\Delta)).$$

If φ is any $\mathscr{C}^1$ function in $\mathscr{F}_{l,b}$, then it can be checked by elementary calculations that[1] $\mathscr{T}\varphi$ belongs also to $\mathscr{C}^1 \cap \mathscr{F}_{l,b}$ and

$$T(\varphi, D\varphi) = (\mathscr{T}\varphi, D\mathscr{T}\varphi).$$

Hence, we obtain recursively,

$$T^n(\varphi, D\varphi) = (T^n\varphi, D\mathscr{T}^n\varphi).$$

As $n \to \infty$, $\mathscr{T}^n\varphi$ converges to Φ in $\mathscr{F}_{l,b}$, $D\mathscr{T}^n\varphi$ converges to Δ_Φ in $\mathscr{G}_l$; then, by elementary properties of the convergence of differentiable functions, Φ is Fréchet differentiable and

$$\mathrm{D}\Phi = \Delta_\Phi. \tag{2.38}$$

2.4. Proof of Theorem 2.1

We now conclude the proof of Theorem 2.1. We have shown that the graph $\mathscr{M}$ of Φ is positively invariant and that $\mathscr{M}$ and Φ are of class $\mathscr{C}^1$; there remains to show that $\mathscr{M}$ is exponentially attracting.

First, we deduce from equation (2.29) concerning an orbit lying on $\mathscr{M}$ $(z(t) = \Phi(y(t)))$, that

$$\frac{d}{dt}\Phi(y) + A\Phi(y) = QR(y + \Phi(y)), \qquad \forall t > 0.$$

Since Φ is Fréchet differentiable, this yields

$$D\Phi(y)\frac{dy}{dt} + A\Phi(y) = QR(y + \Phi(y)), \tag{2.39}$$

and, with (1.25),

$$D\Phi(y)(-Ay + PR(y + \Phi(y))) + A\Phi(y) = QR(y + \Phi(y)). \tag{2.40}$$

Equation (2.40) is valid for $y = y(t)$, $\forall t \geq 0$. In particular, for $t = 0$,

$$D\Phi(y_0)(-Ay_0 + PR(y_0 + \Phi(y_0))) + A\Phi(y_0) = QR(y_0 + \Phi(y_0)). \tag{2.41}$$

Since $y_0 \in PE$ is arbitrary, (2.41) is in fact a semilinear hyperbolic equation (in infinite dimension, with n variables), satisfied by Φ.[2]

[1] The definition of T_ψ has been arranged for that purpose.

[2] This equation, called Sacker's equation, has also been used to construct inertial manifolds. Here we use it only to show that $\mathscr{M}$ is exponentially attracting.

Now let u be any solution of (1.25) and let $y = Pu$, $z = Qu$; u_0 does not belong necessarily to $\mathcal{M}$. We show that this orbit is attracted by $\mathcal{M}$ by comparing it to the companion curve $t \to y(t) + \Phi(y(t))$ lying on $\mathcal{M}$ (see Section 2.2). We observe that y and z satisfy (1.16) and (1.17) and we write

$$\begin{aligned}\frac{d}{dt}(z - \Phi(y)) &= -Az + QR(y+z) - D\Phi(y)(-Ay + PR(y+z)) \\ &= \text{(with (2.40))} \\ &= -A(z - \Phi(y)) + QR(y+z) - QR(y + \Phi(y)) \\ &\quad - D\Phi(y)(PR(y+z) - PR(y+\Phi(y))). \end{aligned} \tag{2.42}$$

By the variation of constants formula,

$$\begin{aligned} z(t) - \Phi(y(t)) &= e^{-At}(z_0 - \Phi(y_0)) \\ &\quad + \int_0^t e^{-A(t-s)} Q[R(y(s) + z(s)) - R(y(s) + \Phi(y(s))) \\ &\quad - D\Phi(y(s))(PR(y(s)+z(s)) - PR(y(s) + \Phi(y(s))))]\, ds. \end{aligned}$$

We use $D\Phi \in \mathcal{G}_l$, (1.12) (at $t = 0$), and (1.4)

$$\begin{aligned} &|D\Phi(y(s))(PR(y(s) + z(s)) - PR(y(s) + \Phi(s))))|_E \\ &\leq l|PR(y(s) + z(s)) - PR(y(s) + \Phi(y(s)))|_E \\ &\leq k_1 l \lambda^\alpha |R(y(s) + z(s)) - R(y(s) + \Phi(y(s)))|_F \\ &\leq k_1 M_1 l \lambda^\alpha |z(s) - \Phi(y(s))|_E \\ &\leq \text{(since } \Lambda \geq \lambda \text{ and } t \geq s) \\ &\leq k_1 M_1 l((t-s)^{-\alpha} + \Lambda^\alpha)|z(s) - \Phi(y(s))|_E. \end{aligned}$$

Hence using again (1.4) and (1.13)

$$\begin{aligned} &|z(t) - \Phi(y(t))|_E \\ &\quad \leq k_2 e^{-\Lambda t}|z_0 - \Phi(y_0)|_E \\ &\quad + (k_1 l + k_2) M_1 \int_0^t ((t-s)^{-\alpha} + \Lambda^\alpha) e^{-\Lambda(t-s)} |z(s) - \Phi(y(s))|_E\, ds, \\ &e^{(\Lambda t/2)}|z(t) - \Phi(y(t))|_E \leq k_2 e^{-(\Lambda t/2)}|z_0 - \Phi(y_0)|_E \\ &\quad + (k_1 l + k_2) M_1 \int_0^t ((t-s)^{-\alpha} + \Lambda^\alpha) e^{-\Lambda(t-s)/2} e^{\Lambda t/2} |z(s) - \Phi(y(s))|_E\, ds. \end{aligned} \tag{2.43}$$

We set $g(s) = e^{\Lambda s/2}|z(s) - \Phi(y(s))|_E$ and

$$G(t) = \operatorname*{Sup}_{s \in [0,t]} g(s).$$

We infer from (2.43) that for every $t > 0$,

$$\begin{aligned} g(t) &\le k_2|z_0 - \Phi(y_0)|_E \\ &\quad + (k_1 l + k_2)M_1 G(t) \int_0^t ((t-s)^{-\alpha} + \Lambda^\alpha)e^{-\Lambda(t-s)/2}\, ds \\ &\le k_2|z_0 - \Phi(y_0)|_E \\ &\quad + (k_1 l + k_2)M_1 G(t) \int_{-\infty}^0 (|s|^{-\alpha} + \Lambda^\alpha)e^{\Lambda s/2}\, ds \\ &\le \text{(with (2.4))} \\ &\le k_2|z_0 - \Phi(y_0)|_E \\ &\quad + (k_1 l + k_2)M_1((\tfrac{1}{2})^{\alpha-1}\gamma_\alpha + 2)\Lambda^{\alpha-1}G(t). \end{aligned}$$

We deduce that if (2.2) is satisfied with

$$c_2 \ge 2(k_1 l + k_2)((\tfrac{1}{2})^{\alpha-1}\gamma_\alpha + 2), \tag{2.44}$$

then

$$g(t) \le G(t) \le 2k_2|z_0 - \Phi(y_0)|_E, \qquad \forall t > 0,$$

i.e.,

$$|z(t) - \Phi(y(t))|_E \le 2k_2|z_0 - \Phi(y_0)|_E e^{-\Lambda t/2}. \tag{2.45}$$

This shows that the distance in E of $u(t)$ to $\mathscr{M}$ is majorized by $|z(t) - \Phi(y(t))|_E$, i.e.,

$$\operatorname{dist}(S(t)u_0, \mathscr{M}) \le 2k_2|z_0 - \Phi(y_0)|_E e^{-\Lambda t/2}. \tag{2.46}$$

Note that this distance converges exponentially to 0 *at a rate* $\Lambda/2$ *independent of* u_0 *and with a factor* $2k_2|z_0 - \Phi(y_0)|_E$ *depending boundedly on* u_0

$$\begin{aligned} 2k_2|z_0 - \Phi(y_0)|_E &\le 2k_2(|z_0|_E + b) \\ &\le \text{(by (1.13))} \\ &\le 2k_2(k_2|u_0|_E + b). \end{aligned} \tag{2.47}$$

Remark 2.4. Gathering all the conditions on c_1 and c_2 (namely, (2.6), (2.9), (2.10), (2.11), (2.21), (2.24), (2.37), and (2.44)), we see that we can take for (2.1) and (2.2)

$$c_1 = 16k_1k_2(1 + \gamma_\alpha) + 2k_1(1 + l)(1 + k_2M_1l^{-1}(1 + \gamma_\alpha) + 8k_2(1 + \gamma_\alpha)), \tag{2.48}$$

$$c_2 = \left(4 + \frac{1}{b}\right)k_2(1 + \gamma_\alpha) + 1. \tag{2.49}$$

3. Complements and Applications

In this section we give some complements and applications. In Section 3.1 we consider the case where R does not satisfy (1.3) and (1.4) but is only Lipschitz and bounded on bounded sets of E and give, in this case, an existence result for the inertial manifold (Theorem 3.1). In Section 3.2 we show how the hypotheses of Theorems 2.1 and 3.1 are verified in cases of interest, and we give some estimates on the dimension n of the inertial manifold.

3.1. The Locally Lipschitz Case

We consider here the case where (1.3) and (1.4) are not satisfied and replaced by the more usual hypotheses

$$|R(u)|_F \leq d_0(r), \tag{3.1}$$

$$|R(u) - R(v)|_F = d_1(r), \tag{3.2}$$

for all u, v in E, such that

$$|u|_E \leq r, \qquad |v|_E \leq r.$$

In this case, we also assume that equation (1.15) posseses an absorbing set in E included in the ball of E centered at 0 of radius ρ.

All other hypotheses being unchanged, we can prove in this case the existence of an inertial manifold for the *prepared equation*. The preparation of equation (1.15) consists in truncating the nonlinear term R outside the absorbing ball, which essentially does not affect the dynamics; namely, we replace R by R_θ

$$R_\theta(u) = \theta\left(\frac{|u|_E^2}{\rho^2}\right) R(u), \tag{3.3}$$

where θ is a $\mathscr{C}^1$ function from $\mathbb{R}_+$ into [0, 1], equal to 1 on [0, 1] and to 0 on $[4, \infty]$.

The prepared form of equation (1.15) reads

$$\begin{cases} \dfrac{du}{dt} + Au = R_\theta(u), \\ u(0) = u_0. \end{cases} \tag{3.4}$$

We assume that the initial-value problem (3.4) is well posed in E and we denote by $\{S_\theta(t)\}_{t \geq 0}$ the corresponding semigroup. We also assume that the ball of E centered at 0 of radius ρ is an absorbing set for (3.4) as well, so that (3.4) and (1.15) have the same long-term dynamics and the same attractors.

We discuss now the existence of an inertial manifold for the prepared equation and indicate below the relations between an inertial manifold for (3.4) and an inertial manifold for (1.15).

We observe that

Lemma 3.1. *For every $u, v \in E$,*

$$|R_\theta(u)|_F \leq M_0, \tag{3.5}$$

$$|R_\theta(u) - R_\theta(v)|_F \leq M_1, \tag{3.6}$$

with

$$M_0 = d_0(2\rho), \tag{3.7}$$

$$M_1 = 2\frac{L_\theta}{\rho} d_0(2\rho) + d_1(2\rho), \tag{3.8}$$

where L_θ is the Lipschitz constant of θ.

PROOF. Relation (3.5) is obvious since $R_\theta = 0$ outside the ball of E centered at 0 of radius 2ρ; (3.6) is also obvious if $|u|_E \leq 2\rho$ and $|v|_E \leq 2\rho$, or if $|u|_E > 2\rho$ and $|v|_E > 2\rho$. Now if, say, $|u|_E \leq 2\rho$ and $|v|_E > 2\rho$, we denote by u_* the intersection of the segment $[u, v]$ with the sphere of E centered at 0 of radius 2ρ and we write

$$\begin{aligned}
|R_\theta(u) - R_\theta(v)|_F &= |R_\theta(u) - R_\theta(u_*)|_F \\
&\leq \theta\left(\frac{|u|_E^2}{\rho^2}\right)|R(u) - R(u_*)|_F + \left|\theta\left(\frac{|u|_E^2}{\rho^2}\right) - \theta\left(\frac{|u_*|^2}{\rho^2}\right)\right| |R(u_*)|_F \\
&\leq d_1(2\rho)|u - u_*|_E + \frac{L_\theta}{\rho^2}(|u|_E + |u_*|_E)(|u|_E + |u_*|_E) d_0(2\rho) \\
&\leq \left(d_1(2\rho) + \frac{2L_\theta}{\rho} d_0(2\rho)\right)|u - u_*|_E \\
&\leq M_1 |u - v|_E.
\end{aligned}$$

□

From Theorem 2.1 we promptly infer the

Theorem 3.1. *We assume that the hypotheses* (1.2), (1.5)–(1.14), (3.1), *and* (3.2) *are satisfied and that* (2.1) *and* (2.2) *hold with M_0, M_1 given by* (3.4).

Then the prepared equation (3.4) *possesses an inertial manifold $\mathscr{M} = \mathscr{M}_\theta$ of class $\mathscr{C}^1$, which is the graph of the function $\Phi = \Phi_\theta \in \mathscr{F}_{b,l}$, given by Theorem* 2.1.

Remark 3.1. We discuss the relation of $\mathscr{M}_\theta$ with an inertial manifold $\mathscr{M}$ for (1.15); $\mathscr{M}_\theta$ has all the properties required of an inertial manifold for (1.15), except that it is not positively invariant for the semigroup $\{S(t)\}_{t\geq 0}$; it is positively invariant for the semigroup $\{S_\theta(t)\}_{t\geq 0}$.

For (1.15) we need a generalization of the inertial manifold, defined on some subset of PE.

Definition 3.1. *In the non-Lipschitz case, $\mathcal{M}$ is an inertial manifold for* (1.15) (*or* $\{S(t)\}_{t\geq 0}$) *if*:

(i) *$\mathcal{M}$ is the graph of Φ, where Φ is a Lipschitz mapping on an open set $\mathcal{O}$ of PE into QE;*
(ii) *$S(t)\mathcal{M} \subset \mathcal{M}$, $\forall t \geq 0$; and*
(iii) *there exist two positive constants K, K', K depending boundedly on $|u_0|_E$, K' independent of u_0, such that*

$$\operatorname{dist}(S(t)u_0, \mathcal{M}) \leq K \exp(-K'(t)), \qquad \forall t \geq 0.$$

If the open ball $\mathcal{B}$ of E centered at 0 of radius ρ is an absorbing set and a positively invariant set for both (1.15) and (3.1), then it can easily be verified that for $\mathcal{O} = P\mathcal{B}$, the part $\mathcal{M}$ of $\mathcal{M}_\theta$ above $\mathcal{O}$ is an inertial manifold for (1.15). If the absorbing set is closed or is not positively invariant, then $\mathcal{M} \subset \mathcal{M}_\theta$ can be obtained, for some suitable set $\mathcal{O}$; for the details of the construction of $\mathcal{O}$ in this case the reader is referred to C. Foias, B. Nicolaenko, G. Sell, and R. Temam [2].

3.2. Dimension of the Inertial Manifold

We give some estimate on the dimension of the inertial manifold given by Theorems 2.1 or 3.1 (Section 3.2.3). Before, we show in Sections 3.2.1 and 3.2.2 how the hypotheses of Theorems 2.1 and 3.1 are verified in cases of interest.

3.2.1. The Function Spaces and the Operator A

We restrict ourselves to the case where the spaces E, F, $\mathcal{E}$ are Hilbert spaces.

In the self-adjoint case, we set $\mathcal{E} = H$, and then A is a self-adjoint unbounded operator in H as in Chapter II, and in most of the equations considered in this book. In this case, we set $F = D(A^\gamma)$ and $E = D(A^{\gamma+\alpha})$, for some $\gamma \geq 0$ and α as before, $0 \leq \alpha < 1$, $\lambda_n = \mu_n$, $\Lambda_n = \mu_{n+1}$, where the μ_n are the eigenvalues of A. Hypotheses (1.2)–(1.14) are easily verified. For the exponential dichotomy we recall that if w_m, $m \geq 0$, are the eigenvectors of A ($Aw_m = \mu_m w_m$), and if $u_0 = \sum_{m=1}^{\infty} u_{0m} w_m$, then

$$e^{-At}u_0 = \sum_{m=1}^{\infty} e^{-\mu_m t} u_{0m} w_m, \tag{3.9}$$

and (1.12) and (1.13) hold with k_1, k_2 replaced by 1 in (1.12) and in the second and third inequality (1.13). Of course P_n, the eigenprojector associated with $\mu_1, \ldots, \mu_n$ is orthogonal.

With a similar choice of spaces we can consider the case where $\mathcal{E} = H$, $A = A_0 + A_1$, A_0 is a self-adjoint positive operator like the operator A before (eigenvectors w_m, eigenvalues μ_m, $A_0 w_m = \mu_m w_m$), and A_1 is a skew-symmetric unbounded operator with domain $D(A_0)$ in H and such that

$$A_0 A_1 = A_1 A_0. \tag{3.10}$$

Then we set $F = D(A^\gamma)$, $E = D(A^{\gamma+\alpha})$ as before and P_n is the eigenprojector of A_0 associated with the eigenvalues $\mu_1, \ldots, \mu_n$. This is an orthogonal projector and, thanks to (3.10), it is also an eigenprojector of A.

Another case of interest (see Section 4) is when $\mathscr{E} = H$ as before, $A = A_0 + b$, A_0 is again self-adjoint positive unbounded in H, and b is a linear unbounded operator in H dominated by A_0 in the sense that $D(A_0^\alpha) \subset D(b)$ and $bA^{-\alpha}$ is bounded. Then we can take $F = H$, $E = D(A_0^\alpha)$ and if μ_n and w_n are the eigenvalues and eigenvectors of A_0,

$$\lambda_n = \mu_n + c'\mu_n^\alpha,$$

$$\Lambda_n = \mu_{n+1} - c'\mu_{n+1}^\alpha,$$

for some $c' > 0$; see Section 4 for more details.

3.2.2. The Hypotheses (2.1) *and* (2.2)

In all the cases in Section 3.2.1, $\lambda_n = \mu_n$, or $\lambda_n \sim \mu_n$ as $n \to \infty$. Now, for large classes of operators A_0 defined by an elliptic boundary-value problem, the behavior of μ_n for large n's is known

$$\mu_n \sim c\mu_0 n^p \qquad \text{as} \quad n \to \infty, \tag{3.11}$$

for some c, p, $\mu_0 > 0$. In fact (see S. Agmon [1], R. Courant and D. Hilbert [1], or G. Métivier [1] for the Stokes operator), we know for many elliptic operators that

$$p = 2r/m \tag{3.12}$$

where $2r$ is the order of the operator and m the space dimension. For some particular geometrical domains, more precise information on the eigenvalues is available.

If $\lambda_n \sim cn^p$ as $n \to \infty$ and $p(1-\alpha) > 1$ (see (3.13)), then it is elementary to show that (2.1) and (2.2) are satisfied for n sufficiently large (see Lemma VIII.4.1). However, the asymptotic behavior of λ_n, $\lambda_n \sim cn^p$ as $n \to \infty$, is not sufficient to actually determine the values of n for which conditions (2.1) and (2.2) are satisfied. Some indications can be obtained if more information on the λ_n is available, for instance:

Lemma 3.2. *If $\lambda_n = cn^p$ for all n and if*

$$p(1-\alpha) > 1, \tag{3.13}$$

then, for any r_1, r_2,

$$\lambda_{n+1} - \lambda_n \geq r_1(\lambda_{n+1}^\alpha + \lambda_n^\alpha), \tag{3.14}$$

$$\lambda_n^{1-\alpha} \geq r_2, \tag{3.15}$$

for some n such that

$$n \geq c_3\{1 + \max(r_1^{1/(p-1-p\alpha)}, r_2^{1/p(\alpha-1)})\}, \tag{3.16}$$

where c_3 is an appropriate constant.

PROOF. Choose n_0 such that

$$c^{-1/p} r_2^{1/p(1-\alpha)} \leq n_0 < 1 + c^{-1/p} r_2^{1/p(1-\alpha)},$$

so that (3.15) is satisfied for all $n \geq n_0$. The lemma is proved if

$$\lambda_{n_0+1} - \lambda_{n_0} \geq r_1(\lambda_{n_0+1}^\alpha + \lambda_{n_0}^\alpha).$$

If not, we observe that (3.14) and (3.15) are satisfied by $n \geq N + 1$,

$$N = \max\{n \geq n_0, \lambda_{j+1} - \lambda_j < r_1(\lambda_{j+1}^\alpha + \lambda_j^\alpha) \text{ for } n_0 \leq j \leq n\},$$

if $N < \infty$. To show that $N < \infty$ and estimate N we add the inequalities

$$\lambda_{j+1} - \lambda_j \leq r_1(\lambda_{j+1}^\alpha + \lambda_j^\alpha)$$

for $j = n_0, \ldots, N$; this yields

$$\begin{aligned} \lambda_{N+1} - \lambda_{n_0} &\leq r_1 \sum_{j=n_0}^{N} (\lambda_{j+1}^\alpha + \lambda_j^\alpha) \\ &\leq 2r_1 \sum_{j=n_0}^{N+1} \lambda_j^\alpha; \end{aligned}$$

hence

$$\begin{aligned} c(N+1)^p &\leq cn_0^p + 2r_1 \sum_{j=n_0}^{N+1} c^\alpha j^{p\alpha}, \\ &\leq cn_0^p + \frac{2r_1 c^\alpha (N+2)^{1+p\alpha}}{1+p\alpha}. \end{aligned} \tag{3.17}$$

If $N + 1 \leq 2^{1/p} n_0$, (3.16) is proved. If not, we infer from (3.17) that

$$(N+1)^p \leq \frac{2r_1 c^{\alpha-1}}{1+p\alpha}(N+1)^{1+p\alpha},$$

$$N + 1 \leq \left(\frac{4r_1 c^{\alpha-1}}{1+p\alpha}\right)^{1/(p-1-p\alpha)},$$

and (3.16) follows as well.

Remark 3.2. *Lemma* 3.1 can be extended to the case where enough information is available on the second term of the asymptotic expansion of λ_n,

$$\lambda_n = cn^p(1 + \varepsilon_n) \quad \text{as} \quad n \to \infty.$$

The dimension of the inertial manifold n is then larger than or equal to n_0, such that $|\varepsilon_n| \leq \frac{1}{2}$ for $n \geq n_0$.

Remark 3.3. The following remark (A. Debussche), related to Remark 2.2 and Lemma 3.1, is useful.

Assume, as is often the case for boundary-value problems, that (1.1) is of the form

$$\frac{du}{dt} + \nu A u = R(u), \tag{3.18}$$

with $\nu > 0$, and A self-adjoint as before. Let μ_n, $n \geq 1$, denote the eigenvalues of A with (3.11), and let $\lambda_n = \nu\mu_n$.

Then, as in Lemma 3.1, if (3.13) holds, for any $r_1, r_2 > 0$ given, we can find and estimate n's such that (3.14) and (3.15) hold. Indeed, we first find n_* depending on A, but not on ν, such that

$$\frac{c}{2}\mu_0 n^p \leq \mu_n \leq \frac{3c}{2}\mu_0 n^p, \qquad \text{for} \quad n \geq n_*.$$

Then we determine $n_0 = n_0(\nu) \geq n_*$ such that

$$\lambda_n^{1-\alpha} = \nu^{1-\alpha}\mu_n^{1-\alpha} \geq \nu^{1-\alpha}\left(\frac{c\mu_0 n^p}{2}\right)^{1-\alpha} \geq r_2,$$

namely,

$$n_0 = \max\left\{n_*, \left(\frac{2}{\nu c \mu_0}\right)^{1/p} r_2^{1/p(1-\alpha)}\right\}.$$

Then we proceed essentially as in Lemma 3.1. The estimate on N that we obtain depends explicitly on ν and it depends in a nonexplicit way on A through n_* as explained in Remark 3.2.

3.2.3. Dimension of the Inertial Manifolds

Estimates on the dimension of the inertial manifolds can be obtained when enough information is available on the λ_n, so that one can estimate the values of n for which (2.1) and (2.2) are satisfied. This may follow from simple arguments as in Lemma 3.1 (and Remark 3.2) or by using specific information on the eigenvalues; the reader is referred for details and examples to R. Temam and S. Wang [1] where the λ_n are the eigenvalues of the Laplace–Beltrami operator on the sphere or to R. Temam and X. Wang [1] which concerns the Kuramoto–Sivashinsky equation.

Typically, the dimension n of the inertial manifold is estimated in terms of M_0 and M_1 (in (1.3), (1.4) or in (3.1), (3.2)). Usually these numbers are related to physically relevant quantities, like a Grashoff or a Reynolds number, in particular, in the case of (3.1), (3.2) where these numbers are related to the radius of the absorbing ball of the semigroup (see (3.7), (3.8) and the estimates on the size of the absorbing balls in Chapter III).

4. Inertial Manifolds and Slow Manifolds

Our aim in this section is to present an application of Theorem 2.1 to a case where the basic linear operator A is not self-adjoint. This example is related to the concept of a slow manifold which appears in meteorology (see, e.g., R. Daley [1], A. Kasahara [1], C.E. Leith [1], B.A. Machenhauer [1], J.J. Tribbia [1]–[3], R. Vautard and B. Legras [1], A. Debussche and R. Temam [1], and R. Temam [8]).

4.1. The Motivation

The following situation occurs in meteorology and oceanography. We consider in a Hilbert space H, an abstract equation like (1.1),

$$\frac{du}{dt} + A_0 u = R_0 u, \tag{4.1}$$

$$u(0) = u_0. \tag{4.2}$$

Here A_0 is a linear closed self-adjoint positive unbounded operator in H of the type encountered in most equations in this book (as in Chapter II); R_0 is a $\mathscr{C}^1$ mapping from $D(A^\alpha)$ into H, for some α, $0 \leq \alpha < 1$. We assume that the initial-value problem (4.1), (4.2) is well-posed in $D(A_0^\alpha)$, so that we can define a semigroup of operators $\{S_0(t)\}_{t \geq 0}$ in $D(A_0^\alpha)$, $S_0(t)$: $u(0) \to u(t)$. In meteorology or oceanography, equations like (4.1) occur in which u consists of the vector fields for horizontal velocity, temperature, and humidity (for the air) or salinity (for the sea).

Let $\bar{u}$ be a stationary solution of (4.1)

$$A_0 \bar{u} = R_0(\bar{u}), \tag{4.3}$$

and let $v = u - \bar{u}$, which satisfies

$$\frac{dv}{dt} + A_0 v = R_0(\bar{u} + v) - R_0(\bar{u}),$$

or

$$\frac{dv}{dt} + Av = R(v), \tag{4.4}$$

$$v(0) = v_0; \tag{4.5}$$

here $v_0 = u_0 - \bar{u}$ and

$$Av = A_0 v - DR_0(\bar{u}) \cdot v, \tag{4.6}$$

$$R(v) = R_0(\bar{u} + v) - R_0(\bar{u}) - DR_0(\bar{u}) \cdot v, \tag{4.7}$$

$DR_0(u)$ being the Fréchet differential of R_0 at $\bar{u}$. We have

$$R(0) = 0, \qquad DR(0) = 0, \tag{4.8}$$

$DR(0)$ denoting the Fréchet differential of R at 0.

We want to show how one can apply Theorem 2.1 to equations like (4.4) and (4.5); the function spaces will be Hilbert spaces, $\mathscr{E} = F = H$ and $E = D(A_0^\alpha)$. However, the operator A is, in general, non-self-adjoint, its eigenvalues are complex, and the construction of the inertial manifold will be based on the *generalized eigenvectors* of A.

4.2. The Abstract Equation

We consider a Hilbert space H and an abstract equation in H, of the form (4.4). The operator A in (4.4) is of the form

$$A = A_0 + b, \tag{4.9}$$

where A_0 is as before a self-adjoint operator in H; we denote by μ_j and w_j its eigenvalues and eigenvectors,

$$\begin{cases} A_0 w_j = \mu_j w_j, & j \geq 1, \\ 0 < \mu_1 \leq \mu_2 \leq \ldots, \mu_j \to \infty & \text{as} \quad j \to \infty. \end{cases} \tag{4.10}$$

We assume also that

$$\mu_j \sim c_0 j^p \qquad \text{as} \quad j \to \infty, \tag{4.11}$$

and

$$p(1 - \alpha) > 1. \tag{4.12}$$

As we shall see in the examples, hypothesis (4.12) related to the spectral gap condition will be the most restrictive one.

The operator b is a linear unbounded operator in H with domain $D(b)$; it is dominated by A_0 in the following sense:

$$D(A_0^\alpha) \subset D(b), \qquad \text{for some } \alpha, \qquad 0 \leq \alpha < 1, \tag{4.13}$$

$$bA_0^{-\alpha} \text{ is bounded in } H. \tag{4.14}$$

Of course, because of (4.13), $D(A) = D(A_0)$.

Concerning the eigenvectors and eigenvalues of A, we have, thanks to (4.13) and (4.14), the following important properties which we state without proof.

(i) The spectrum of $A = A_0 + b$ consists of eigenvalues of finite multiplicity. Moreover, if we denote by $\nu_j, j \geq 1$, these eigenvalues are arranged so that the sequence $\{\text{Re } \nu_j\}_{j \in \mathbb{N}^*}$ is nondecreasing, then

$$\text{Re}(\nu_j) \sim \mu_j \qquad \text{as} \quad j \to \infty. \tag{4.15}$$

(ii) There exists a constant $c_0 > 0$ such that

$$\sigma(A) \subset \bigcup_{j \in \mathbb{N}^*} B(\mu_j, c_0 \mu_j^\alpha), \tag{4.16}$$

where $\sigma(A)$ is the spectrum of A, and $B(\mu_j, c_0\mu_j^\alpha)$ is the ball of $\mathbb{C}$ centered at μ_j of radius $c_0\mu_j^\alpha$. Moreover, if $\lambda \notin \bigcup_{j \in \mathbb{N}} B(\mu_j, c_0\mu_j^\alpha)$,

$$\|(A - \lambda)^{-1}\|_{\mathscr{L}(H)} \leq 2\|(A_0 - \lambda)^{-1}\|_{\mathscr{L}(H)}, \tag{4.17}$$

$$\|(A - \lambda)^{-1} - (A_0 - \lambda)^{-1}\|_{\mathscr{L}(H)} \leq c_0 \sup_{i>0} \frac{\mu_i^\alpha}{|\lambda - \mu_i|} \sup_{i>0} \frac{1}{|\lambda - \mu_i|}. \tag{4.18}$$

Let $\{v_j\}_{j \in \mathbb{N}}$ denote a system of generalized eigenvectors (root vectors) of A associated with the v_j's:

$$\forall j \in \mathbb{N}^*, \quad \exists k_j \geq 1, \qquad (A - v_j)^{k_j} v_j = 0. \tag{4.19}$$

(iii) We have

$$\text{The system } \{v_j\}_{j \in \mathbb{N}^*} \text{ is total in } H. \tag{4.20}$$

The proof of (4.15) follows from results of I.C. Gohberg and M.G. Krein [1]; see A. Debussche and R. Temam [2] for more details and for the proofs of (4.17)–(4.20) which are not related to the tools and methods used in this book.

Finally, thanks to (4.12), one can prove exactly, as in Lemma VIII.4.1, that

$$\text{There exists a sequence } \{\mu_{n_k}\}_{k \in \mathbb{N}} \text{ such that}$$

$$\frac{\mu_{n_k+1} - \mu_{n_k}}{\mu_{n_k+1}^\alpha + \mu_{n_k}^\alpha} \to \infty \qquad \text{as} \quad k \to \infty. \tag{4.21}$$

We set $E = D(A^\alpha)$, $F = \mathscr{E} = H$, and we assume that R is a $\mathscr{C}^1$ mapping from $D(A^\alpha)$ into H which satisfies (1.3) and (1.4) (the same α as in (4.13) and (4.14)).[1] Then all hypotheses of Section 1.1 are satisfied; (1.2)–(1.6) and (1.15) are now obvious (or have been assumed), and there remains to consider the hypotheses of exponential dichotomy.

We define the operators P_n and Q_n as follows:

P_n is the projector in H (or $D(A)$) onto the space spanned by $\{v_1, \ldots, v_n\}$, parallel to the space spanned by $\{v_{n+1}, \ldots\}$; and

$Q_n = I - P_n$ is the projector in H (or $D(A)$) onto the space spanned by $\{v_{n+1}, \ldots\}$, parallel to the space spanned by $\{v_1, \ldots, v_n\}$.

Of course, P_n and Q_n are eigenprojectors of A and they satisfy (1.14). However, in this truly non-self-adjoint case, they are not orthogonal projectors.

Now, for the sequence n_k given by (4.21), we can prove the following

[1] Hence if equation (4.4) is derived from (4.1) as indicated in Section 4.1, (1.4) will not be satisfied in general, and it will be necessary first to consider a prepared form of (4.4).

properties, essentially equivalent to (1.12) and (1.13) (see again A. Debussche and R. Temam [2] for the details).

Lemma 4.1. *Assume that the hypotheses above hold and, in particular,* (4.12). *Then for every* $\beta > 0$, *there exists a constant* $c' > 0$ *depending on* β *but not on* n_k *or* t *such that*

$$\|A^\beta e^{-At} P_{n_k}\|_{\mathscr{L}(H)} \leq c'(\mu_{n_k} + 2c_0\mu_{n_k}^\alpha)^\beta e^{-(\mu_{n_k}+2c_0\mu_{n_k}^\alpha)t}, \qquad \forall t \leq 0, \quad \forall n_k, \tag{4.22}$$

$$\|e^{-At}(I - P_{n_k})\|_{\mathscr{L}(H)} \leq c' e^{-(\mu_{n_k+1}-2c_0\mu_{n_k+1}^\alpha)t}, \qquad \forall t \geq 0, \quad \forall n_k, \tag{4.23}$$

$$\|A^\beta e^{-At}(I - P_{n_k})\|_{\mathscr{L}(H)} \leq c'\left(\frac{1}{t^\beta} + \mu_{n_{k+1}}^\beta\right) e^{-(\mu_{n_k}-2c_0\mu_{n_k+1}^\alpha)t}, \qquad \forall t > 0, \quad \forall n_k, \tag{4.24}$$

Also

$$|A^\beta z| \geq c'(\mu_{n_k+1} - 2c_0\mu_{n_k+1}^\alpha)|z|, \tag{4.25}$$

for all z in $(I - P_{n_k})D(A^\beta)$.

Then it is easy to see that (1.12) and (1.13) are satisfied with $n = k$ and

$$\begin{aligned} \Lambda_k &= \mu_{n_k+1} - 2c_0\mu_{n_k+1}^\alpha, \\ \lambda_k &= \mu_{n_k} + 2c_0\mu_{n_k}^\alpha, \end{aligned} \tag{4.26}$$

and $k_1 = k_2 = c'$.

All the other hypotheses (1.7)–(1.11) follow promptly; (1.7) is satisfied for n sufficiently large, $n \geq n_1$, which is not restrictive at all.

Then Theorem 2.1 applies without any restriction (except to require that $n \geq n_1$). Furthermore, in view of (4.8), we show that

$$0 \in \mathscr{M} \text{ and } \mathscr{M} \text{ is tangent at 0 to the space } PE, \tag{4.27}$$

a property which appears naturally in the context of slow manifolds.

Remark 4.1. For more details on the relations with slow manifolds, the reader is referred to A. Debussche and R. Temam [1], [2] and and R. Temam [8].

4.3. An Equation of Navier–Stokes Type

In this section we give an application of Theorem 2.1 in both the self-adjoint and non-self-adjoint cases. We present the equations in the nonprepared case; it is understood that the equations must be prepared.

We consider, in space dimension $l \geq 2$, the Navier–Stokes equations with a higher-order viscosity term

$$\frac{\partial u}{\partial t} + \varepsilon(-\Delta)^r u - \nu\Delta u + (u \cdot \nabla)u + \nabla p = f, \tag{4.28}$$

$$\nabla \cdot u = 0, \tag{4.29}$$

The equation is the same as (4.6) and (4.7) in Chapter VIII, but the treatment will be slightly different. The functions $u = u(x, t)$ and $p = p(z, t)$ are defined on $\mathbb{R}^l \times \mathbb{R}_+$, taking, respectively, values in $\mathbb{R}^l$ and $\mathbb{R}$, $u = (u_1, \dots, u_l)$; ε and v are strictly positive numbers, $r > 1$. Of course, (4.28) reduces to the usual Navier–Stokes equations when $\varepsilon = 0$, but here we assume $\varepsilon > 0$.

We restrict ourselves to the space-periodic case although other boundary conditions can be considered. We assume that u and p are periodic in each direction $x_1, \dots, x_l$ with period $L > 0$,

$$\begin{cases} u(x + Le_i, t) = u(x, t), \qquad i = 1, \dots, l, \\ p(x + Le_i, t) = p(x, t), \end{cases} \tag{4.30}$$

where $\{e_1, \dots, e_l\}$ is the natural basis of $\mathbb{R}^l$. Furthermore, we assume that

$$\int_\Omega f(x)\,dx = 0, \qquad \int_\Omega u(x, t)\,dx = 0, \qquad \int_\Omega p(x, t)\,dx = 0, \tag{4.31}$$

where $\Omega = (0, L)^l$ is the l-cube.

By classical results, (4.28)–(4.31) reduce to an evolution equation for u of the form (1.1). Let $H^m_{\text{per}}(\Omega)$ denote the restriction to Ω of the Ω-periodic functions v from $\mathbb{R}^l$ into $\mathbb{R}$ which are locally in $H^m(\mathbb{R}^l)$. Let $\dot{H}^m_{\text{per}}(\Omega)$ denote the subspace of $H^m_{\text{per}}(\Omega)$ consisting of the functions v which satisfy $\int_\Omega v(x)\,dx = 0$. Then the spaces $H^m_{\text{per}}(\Omega)$ and $\dot{H}^m_{\text{per}}(\Omega)$ are both Hilbert subspaces of $H^m(\Omega)$.

For the application of Theorem 2.1, we define $F = \mathscr{E} = H$ to be the subspace of $L^2(\Omega)^l$ consisting of the restrictions to Ω of the locally L^2 vector functions with free divergence and vanishing average on Ω. We set $D(A_0) = \dot{H}^{4l}_{\text{per}}(\Omega) \cap H$ and $D(A_0^{1/2}) = \dot{H}^{2l}_{\text{per}}(\Omega) \cap H$. Then

$$A_0 u = \varepsilon(-\Delta)^r u, \qquad \text{for all} \quad u \in D(A_0),$$

and $R_0(u)$ is defined by

$$\bullet \quad R_0(u), v) = -v \int_\Omega \Delta u \cdot v\,dx + \int_\Omega ((u \cdot \nabla)u)v\,dx - f, \qquad \forall u, v \in D(A),$$

where f is given in H.

We choose $E = D(A_0^\alpha)$ and

$$\alpha = \frac{1}{4} \quad \text{for } l = 2, 3, \qquad \alpha = \frac{1}{16} + \frac{1}{4l} \quad \text{for } l \geq 4.$$

It is easy to check, as in Section VIII.4.2, that for $r = 2l$, R is a $\mathscr{C}^1$ mapping from $D(A_0^\alpha)$ into H.

Then equation

$$\frac{du}{dt} + A_0 u = R_0(u), \tag{4.32}$$

$$u(0) = u_0, \tag{4.33}$$

satisfies all the hypotheses of Theorem 2.1 except (1.3) and (1.4) which are replaced by (3.1) and (3.2). Hence Theorem 2.1 applies to the prepared form of this equation, as in Theorem 3.1. The existence result of the inertial manifold is not much different than that resulting from Theorem VIII.4.2: here $r = 2l$ and we obtain the inertial manifolds in $D(A_0^\alpha)$, while in Chapter VIII, $r > l$ and the inertial manifold is obtained in H. It is noteworthy that several other choices are possible here for r and for E and F; $E = D(A_0^{\alpha+\gamma})$, $F = \mathscr{E} = D(A_0^\gamma)$, with another value of α and suitable values of γ. Theorems 2.1 and 3.1 then give inertial manifolds in $D(A_0^{\alpha+\gamma})$; this is left as an exercise to the reader.

Now, let us discuss the non-self-adjoint case corresponding to the slow manifold motivation as in Section 4.1.

We consider a stationary solution $\bar{u}$, $\bar{p}$ of (4.28) and (4.29), i.e.,

$$\varepsilon(-\Delta)^r\bar{u} - \nu\Delta\bar{u} + (\bar{u}\cdot\nabla)\bar{u} + \nabla\bar{p} = f, \tag{4.34}$$

$$\nabla\cdot\bar{u} = 0, \tag{4.35}$$

(and (4.31) is satisfied). The existence of $\bar{u}$, $\bar{p}$, follows, e.g., from J.L. Lions [2] and R. Temam [3]. Let $\bar{u}$ be any such solution and set $v = u - \bar{u}$ and $q = p - \bar{p}$. We have

$$\frac{\partial v}{\partial t} + \varepsilon(-\Delta)^r v - \nu\Delta v + (\bar{u}\cdot\nabla)v + (v\cdot\nabla)\bar{u} + (v\cdot\nabla)v + \nabla q = 0, \tag{4.36}$$

$$\nabla\cdot v = 0, \tag{4.37}$$

$$v(x + Le_i, t) = v(x, t), \qquad q(x, Le_i, t) = q(x, t), \qquad i = 1, \dots, l. \tag{4.38}$$

Equations (4.36)–(4.38) amount to an equation for v of type (4.4). Here we have $D(A) = D(A_0)$ and

$$Av = \varepsilon(-\Delta)^r v + \Pi((\bar{u}\cdot\nabla)v + (v\cdot\nabla)\bar{u}), \qquad \forall v \in D(A),$$

where Π is the projector in $L^2(\Omega)^l$ onto the space H.

Theorem 2.1 applies to the prepared form of this equation (i.e., in the form of Theorem 3.1). The inertial manifold is obtained as a graph above a root space of A, namely, the space spanned by the root vectors $v_1, \dots, v_n$ of A, with n sufficiently large. Also, as indicated before, and in view of (4.8), the inertial manifold $\mathscr{M}$ contains 0 in $D(A^\alpha)$ and is tangent at 0 to PE.

As indicated before, a similar concept appears, in the meteorology literature, for the equations of meteorology, under the name of *slow manifold*.

CHAPTER X

Approximation of Attractors and Inertial Manifolds. Convergent Families of Approximate Inertial Manifolds

Introduction

Our aim in this chapter is to study the approximation of attractors and inertial manifolds by smooth finite-dimensional manifolds, called approximate inertial manifolds. There is a vast and growing literature on approximation of attractors and inertial manifolds, approximate dynamics, and related algorithms. These important topics could justify, by themselves, the writing of one or several books; we will not try to give an exhaustive presentation or even an exhaustive bibliography on this question (see, however, a few references in Remark 3.2 at the end of this chapter).

Although the concept of approximate inertial manifolds (AIM) has considerably developed, and numerous methods of the construction of AIMs are now available, the following presentation is different from previous ones: *inertial manifolds are constructed by a discretized version of the fixed-point technique used in Chapter IX*. The specific interest of the construction presented in this chapter is twofold.

(i) Our construction allows us to obtain *approximations of the attractor of exponential order*, i.e., the attractor lies in a neighborhood of the AIM which is exponentially small with respect to the dimension of the AIM. This result is sharp, see the comments in Remark 2.2.

These high-order AIMs are constructed with very general hypotheses: we assume little regularity on the data (see Remark 2.2) and, also, no hypothesis of the spectral gap type is needed.

(ii) When the spectral gap condition is satisfied and an (exact) inertial manifold (IM) exists, our approach allows us to construct a *whole sequence of approximate inertial manifolds which converge, in an appropriate topology, to the exact inertial manifold.*

This chapter is organized as follows. In Section 1 we introduce the notations and hypotheses and describe in an informal way the construction of the manifolds. This construction is based on a finite-difference approximation of the evolution equation using an implicit Euler scheme. In Section 2 we state and prove the result on the exponential approximation of the attractor. In Section 3 we state and prove the result on convergent families of approximate inertial manifolds. The presentation is based on A. Debussche and R. Temam [4]; see also R. Rosa [2].

1. Construction of the Manifolds

In this chapter, as in Chapter IX, we are interested in the long-time behavior of the solutions of an evolution equation of the first order in time written as IX.(1.1):

$$\frac{du}{dt} + Au = R(u), \tag{1.1}$$

$$u(0) = u_0. \tag{1.2}$$

We retain all the notations and hypotheses of Section 1.1 of Chapter IX and we write $M = \max(M_0, M_1)$. Furthermore (this was not explicitly needed nor stated in Chapter IX), we assume that equation (1.1) possesses a global attractor $\mathscr{A}$.

In this chapter, we will construct some smooth finite-dimensional manifolds by considering a discretized version $\mathscr{T}_\tau$ of the operator $\mathscr{T}$ which appeared in the previous chapter. There is much flexibility for the construction of $\mathscr{T}_\tau$. Our construction of $\mathscr{T}_\tau$ is based on the approximation of the evolution equation by an implicit Euler scheme. The framework and notations are made explicit in Section 1.1; then we describe the construction of the manifolds $\mathscr{M}_n$ in Section 1.2.

1.1. Approximation of the Differential Equation

We use the notations of Section IX.1.1 and, omiting the indices n, we write

$$P = P_n, \qquad Q = Q_n.$$

Setting $\bar{y} = Pu$, $\bar{z} = Qu$, we recall that, by application of the operators P and Q to (1.1) and (1.2), this equation is equivalent to the coupled system

$$\frac{d\bar{y}}{dt} + A\bar{y} = PR(\bar{y} + \bar{z}), \tag{1.3}$$

$$\bar{y}(0) = \bar{y}_0, \tag{1.4}$$

$$\frac{d\bar{z}}{dt} + A\bar{z} = QR(\bar{y} + \bar{z}), \tag{1.5}$$

$$\bar{z}(0) = \bar{z}_0. \tag{1.6}$$

Here $\bar{y}_0 = Pu_0$ and $\bar{z}_0 = Qu_0$.

In Chapter IX we considered for $y_0 \in PE$ and $\psi \in \mathscr{F}_{l,b}$,

$$T\psi(y_0) = \int_{-\infty}^{0} e^{As} QR(y(s) + \psi(y(s)))\, ds, \tag{1.7}$$

where y is the solution defined for $t \in \mathbb{R}$, of the differential equation

$$\frac{dy}{dt} + Ay = PR(y + \psi(y)), \tag{1.8}$$

$$y(0) = y_0. \tag{1.9}$$

Let $\tau > 0$ be fixed, corresponding to a time-discretization mesh. For y_0 and ψ given, $y_0 \in PE$, $\psi \in \mathscr{F}_{l,b}$, we consider a family of y_k's, $k \in \mathbb{N}$, where y_k approximates $y(-k\tau)$. The y_k are recursively defined by

$$\frac{y_{k+1} - y_k}{-\tau} + Ay_k = PR(y_k + \psi(y_k)), \qquad k \geq 1, \tag{1.10}$$

i.e.,

$$y_{k+1} = \mathscr{R}_\tau y_k + \mathscr{S}_\tau PR_k, \tag{1.11}$$

where

$$R_k = R(y_k + \psi(y_k)), \tag{1.12}$$

and

$$\mathscr{R}_\tau = (I + \tau A), \qquad \mathscr{S}_\tau = -\tau I. \tag{1.13}$$

For further generality, we will consider operators $\mathscr{R}_\tau$ and $\mathscr{S}_\tau$ more general than (1.13). More precisely we consider a sequence of y_k's defined by (1.10), (1.11), and (1.12) (y_0 as in (1.9)), where $\mathscr{R}_\tau$ and $\mathscr{S}_\tau$ are linear unbounded operators in $\mathscr{E}$ which satisfy

$$|\mathscr{R}_\tau P_n|_{\mathscr{L}(E)} \leq e^{\tau\lambda_n}, \tag{1.14}$$

$$|\mathscr{S}_\tau P_n|_{\mathscr{L}(F,E)} \leq k_3 \lambda_n^{\alpha-1}(e^{\tau\lambda_n} - 1), \tag{1.15}$$

for some $k_3 > 0$.

We now define the step function, $y_\tau = y_\tau^N$, aimed at approximating y, by setting for some $N \in \mathbb{N}$,

$$y_\tau(s) = y_k \quad \text{for} \quad -(k+1)\tau < s \leq -k\tau, \quad k = 0, \ldots, N-1, \tag{1.16}$$

$$y_\tau(s) = y_N \qquad \text{for} \quad s \leq -N\tau. \tag{1.17}$$

The approximation $\mathcal{T}_N^\tau$ of $\mathcal{T}$ is defined by replacing y by y_τ in the right-hand side of (1.7); namely, we set

$$\mathcal{T}_N^\tau\psi(y_0) = \int_{-\infty}^{0} e^{As}QR(y_\tau(s) + \psi(y_\tau(s)))\,ds, \tag{1.18}$$

i.e.,

$$\mathcal{T}_N^\tau\psi(y_0) = A^{-1}(I - e^{-Ar})\sum_{k=0}^{N-1} e^{-kA\tau}QR(y_k + \psi(y_k)) + A^{-1}e^{-NA\tau}QR(y_n + \psi(y_N)). \tag{1.19}$$

Remark 1.1. The hypotheses IX.(1.2) to IX.(1.14), and the hypotheses above especially (1.14) and (1.15) will be standing hypotheses in this chapter. Some specific hypotheses will be added as needed, in particular, we will consider one of the two groups of hypotheses hereafter.

Now we consider two alternate groups of hypotheses which state that (1.11) is consistent with (1.8).

For any ψ in $\mathcal{F}_{l,b}$ and y_0 in P_nE, let y be the solution of (1.8) and (1.9) and set $\tilde{y}_k = y(-k\tau)$; then we assume

$$\begin{cases} \tilde{y}_{k+1} = \mathcal{R}_\tau\tilde{y}_k + \mathcal{S}_\tau P_n R(\tilde{y}_k + \psi(\tilde{y}_k)) + \varepsilon_k \\ \text{with} \\ |\varepsilon_k|_E \leq \alpha_1(\lambda_n)\tau^2 e^{\tau(k+1)(\lambda_n + k_4\lambda_n^\alpha)}, \end{cases} \tag{1.20}$$

for $k = 0, \ldots, N-1$. We also assume that the time derivative of y does not grow too fast; namely, for $t \in (-\infty, 0]$

$$\left|\frac{dy}{dt}\right|_E \leq \alpha_2(\lambda_n)e^{-(\lambda_n + k_5\lambda_n^\alpha)t}. \tag{1.21}$$

Here k_4, k_5 do not depend on N, τ, or n whereas $\alpha_1(\lambda_n)$, $\alpha_2(\lambda_n)$ depend on λ_n but not on N or τ.

The second group of hypotheses is related to the case where $u = u(t)$ is a complete trajectory lying on the attractor $\mathcal{A}$. In this case, we set as before $\bar{y} = P_nu$, and we write also $\bar{y}_k = \bar{y}(-k\tau) = P_nu(-k\tau)$, for $k \in \mathbb{N}$. In this case, we assume the following:

$$\begin{cases} \bar{y}_{k+1} = \mathcal{R}_\tau\bar{y}_k + \mathcal{S}_\tau P_n R(u(-k\tau)) + \varepsilon_k, \\ \text{with} \\ |\varepsilon_k|_E \leq \tau^2\beta_1, \qquad \text{for} \quad k = 0, \ldots, N-1, \end{cases} \tag{1.22}$$

$$\left|\frac{d\bar{y}}{dt}\right|_E \leq \beta_2, \qquad \text{for all } t \leq 0. \tag{1.23}$$

Here β_1, β_2 do not depend on N, τ, or n, nor on the orbit $u = u(t)$ lying on $\mathcal{A}$.

1.2. The Approximate Manifolds

In Chapter IX the inertial manifold $\mathcal{M}$ was obtained as the graph of a function Φ from P_nE into Q_nE and Φ was obtained as the fixed point of $\mathcal{T}$; by the contraction principle Φ the limit of any sequence $\Phi_N \in \mathcal{F}_{l,b}$ defined by

$$\Phi_{N+1} = \mathcal{T}\Phi_N,$$

with $\Phi_0 \in \mathcal{F}_{l,b}$ arbitrary.

The approximate manifolds $\mathcal{M}_N$ be defined in a similar manner. Namely, given a sequence of positive numbers τ_N, $\mathcal{M}_N$ will be the graph of $\Phi_N \in \mathcal{F}_{l,b}$ where the Φ_N are recursively defined by

$$\begin{cases} \Phi_0 = 0, \\ \Phi_{N+1} = \mathcal{T}_N^{\tau_N}(\Phi_N), \qquad N \geq 0. \end{cases} \tag{1.24}$$

Of course, for this to make sense, we must verify that $\mathcal{T}_N^{\tau_N}$ maps $\mathcal{F}_{l,b}$ into itself, which we will do in Sections 2 and 3 under suitable hypotheses.

Remark 1.2. It is interesting to observe that

$$\mathcal{T}_0^{\tau}\Phi(y_0) = A^{-1}Q_nR(y_0 + \Phi(y_0)). \tag{1.25}$$

Hence for any τ_1,

$$\Phi_1(y_0) = A^{-1}Q_nR(y_0), \tag{1.26}$$

and the graph $\mathcal{M}_1$ of Φ_1 is the approximate inertial manifold defined in C. Foias, O. Manley, and R. Temam [1], [2] in the context of the Navier–Stokes equations and subsequently used in many other articles, both theoretical and numerical (see, e.g., R. Temam [9]).

We define $\mathcal{M}_N$ to be the graph of Φ_N. In defining $\mathcal{T}_N^{\tau}$, τ corresponds to the time step of the numerical scheme (1.11), and $-N\tau$ is the time from which we consider the function y_τ as a constant. Therefore it seems reasonable to hope that if the τ_N converge to zero and $N\tau_N$ to infinity when N converges to infinity, the manifold $\mathcal{M}_N$ will become increasingly accurate. We will see in Section 3 that if the spectral condition holds, the manifolds $\mathcal{M}_N$ converge to the exact inertial manifold given by Theorem IX.2.1, if $\tau_N \to 0$ and $N\tau_N \to \infty$. The construction will still be of interest when the ratio

$$\frac{\Lambda_n - \lambda_n}{\Lambda_n^\alpha + \lambda_n^\alpha}$$

is not sufficiently large, either because n is too small or because the spectral gap condition does not hold; in this case, we will show in Section 2 how the $\mathcal{M}_N$ can be used to approximate the attractor $\mathcal{A}$ itself.

We conclude this section with two simple lemmas which will be repeatedly used.

Lemma 1.1. *Assume that* (1.20) *holds.*

(i) *Let ψ be in $\mathscr{F}_{l,b}$ and let y_0 be in P_nE, and define y_k, $k = 0, \dots, N$, by* (1.11). *Then we have*

$$1 + |y_k|_E \le e^{k\tau(\lambda_n + k_3 M \lambda_n^\alpha)}(1 + |y_0|_E), \tag{1.27}$$

for all $k \in \mathbb{N}$, where $M = \max(M_0, M_1)$.

(ii) *Let ψ_i be in $\mathscr{F}_{l,b}$, and let y_{0i} be in P_nE, and define $y_{k,i}$, $k = 0, \dots, N$,* by (1.11), *with $y_0 = y_{0,i}$, $i = 1, 2$. Then we have*

$$\begin{aligned} |y_{k,1} - y_{k,2}|_E \le e^{k\tau(\lambda_n + Mk_3(1+l)\lambda_n^\alpha)}|y_{0,1} - y_{0,2}|_E \\ + k\tau M k_3 \lambda_n^\alpha |\psi_1 - \psi_2|_\infty e^{k\tau(\lambda_n + Mk_3(1+l)\lambda_n^\alpha)}. \end{aligned} \tag{1.28}$$

Proof. By (1.11), (1.14), (1.15), and $e^s \le 1 + se^s$ ($s \ge 0$), we have

$$\begin{aligned} |y_{k+1}|_E &\le e^{\tau\lambda_n}|y_k|_E + M_0 k_3 \lambda_n^{\alpha-1}(e^{\tau\lambda_n} - 1) \\ &\le e^{\tau\lambda_n}|y_k|_E + M_0 k_3 \lambda_n^\alpha \tau e^{\tau\lambda_n}, \end{aligned} \tag{1.29}$$

$$1 + |y_{k+1}|_E \le (1 + |y_k|_E)e^{\tau\lambda_n}(1 + M_0 k_3 \lambda_n^\alpha \tau),$$

$$1 + |y_{k+1}|_E \le (1 + |y_k|_E)e^{\tau(\lambda_n + M_0 k_3 \lambda_n^\alpha)}. \tag{1.30}$$

Hence (1.27).

For (1.28) we write $y_k = y_{k,1} - y_{k,2}$ and

$$\begin{aligned} y_{k+1} &= \mathscr{R}_\tau y_k + \mathscr{S}_\tau P_n(R(y_{k,1} + \psi_1(y_{k,1})) - R(y_{k,2} + \psi_2(y_{k,2}))), \\ |y_{k+1}|_E &\le e^{\tau\lambda_n}|y_k|_E + M_1 k_3 \lambda_n^{\alpha-1}(e^{\lambda_n\tau} - 1)((1 + l)|y_k|_E + |\psi_1 - \psi_2|_\infty) \\ &\le |y_k|_E e^{\lambda_n\tau}(1 + M_1 k_3(1 + l)\lambda_n^\alpha \tau) + M_1 k_3 \lambda_n^\alpha \tau |\psi_1 - \psi_2|_\infty e^{\lambda_n\tau} \\ &\le |y_k|_E e^{\lambda_n\tau + M_1 k_3(1+l)\lambda_n^\alpha \tau} + M_1 k_3 \lambda_n^\alpha \tau |\psi_1 - \psi_2|_\infty e^{\lambda_n\tau}. \end{aligned}$$

From this we easily derive (1.28) by induction. □

Lemma 1.2. *Assume that* (1.20) *holds; let ψ be in $\mathscr{F}_{l,b}$, y_0 in P_nE, and let e_k, $k = 0, \dots, N$, be a sequence in P_nE. We define y_k^*, $k = 0, \dots, N$, by*

$$y_{k+1}^* = \mathscr{R}_\tau y_k^* + \mathscr{S}_\tau P_n R(y_k^* + \psi(y_k^*)) + e_k,$$

$$y_0^* = y_0.$$

Then

$$|y_k^* - y_k|_E \le \sum_{j=0}^{k-1} e^{(k-1-j)\tau(\lambda_n + Mk_3(1+l)\lambda_n^\alpha)}|e_j|_E,$$

for $k = 1, \dots, N$.

The proof is left to the reader.

2. Approximation of Attractors

In this section we study under what conditions the manifolds $\mathcal{M}_N$, graphs of Φ_N, can approximate the global attractor $\mathcal{A}$, in particular in low dimension. Hence, we do not assume that $(\Lambda_n - \lambda_n)/(\Lambda_n^\alpha + \lambda_n^\alpha)$ is large, either because n is not sufficiently large or because the spectral gap condition IX.(2.1) is not satisfied, i.e.,

$$\limsup_{n\to\infty} \frac{\Lambda_n - \lambda_n}{\Lambda_n^\alpha + \lambda_n^\alpha} < +\infty.$$

We will prove that with a suitable choice of the τ_N, the attractor is contained in a thin neighborhood of $\mathcal{M}_N$; the thickness (or at least a bound on the thickness) decreases as N grows and converges to a small number. However, in Section 3 the point of view will be different; assuming IX.(2.1) and taking a different choice of the τ_N we will obtain a similar result, except that the thickness converges then to zero and the family $\mathcal{M}_N$, $N \in \mathbb{N}$, converges to the exact inertial manifold $\mathcal{M}$. In Section 3 we will assume naturally that $\tau_N \to 0$ and $N\tau_N \to \infty$, but in this section it is preferable to let τ_N go to zero but to take $N\tau_N$ constant (see Remark 2.3).

In Section 2.1 we show that $\mathcal{T}_N^\tau$ maps $\mathcal{F}_{l,b}$ into itself. In Section 2.2 we estimate the distance of the manifold to the attractor and in Section 2.3 we give the final result of this section.

2.1. Properties of $\mathcal{T}_N^\tau$

We give sufficient conditions which ensure that $\mathcal{T}_N^\tau$ maps $\mathcal{F}_{l,b}$ into itself.

Lemma 2.1. *Assume that* (1.14), (1.15), *and* IX.(1.13) *hold. Then there exist two constants c_1, c_2 such that if*

$$N\tau \le \frac{c_1}{\lambda_n^\alpha} \qquad \text{and} \qquad \lambda_n \ge c_2, \tag{2.1}$$

then there exists l and b_0 such that $\mathcal{T}_N^\tau$ maps $\mathcal{F}_{l,b}$ into itself for all $b \ge b_0$.

Proof. Let there be given ψ in $\mathcal{F}_{l,b}$ and y_0 in P_nE. We want to show first that $|\mathcal{T}_N^\tau\psi(y_0)|_E \le b$. For the sake of simplicity, we write P, Q, λ, Λ instead of P_n, Q_n, λ_n, Λ_n.

Consider the function y_τ defined by (1.16) and (1.17). From the expression (1.18) of $\mathcal{T}_N^\tau$ and from IX.(1.3), IX.(1.13) (and $M_0 \le M$) we infer

$$|\mathcal{T}_N^\tau\psi(y_0)|_E \le Mk_2 \int_{-\infty}^0 (|s|^{-\alpha} + \Lambda^\alpha)e^{\Lambda s}\, ds$$

$$\le \text{(with IX.(2.4))}$$

$$\leq Mk_2(1 + \gamma_\alpha)\Lambda^{\alpha-1}$$
$$\leq Mk_2(1 + \gamma_\alpha)c_2^{\alpha-1}, \tag{2.2}$$

and this is less than b provided

$$b \geq b_0 = Mk_1(1 + \gamma_\alpha)c_2^{\alpha-1}. \tag{2.3}$$

To show that $\mathscr{T}_N^\tau$ is Lipschitz with constant l, consider $\psi \in \mathscr{F}_{l,b}$ and $y_{0,1}$, $y_{0,2}$ in PE. We denote by $y_{k,i}$, $k = 0, \ldots, N$, and $y_{\tau,i}$, the vectors y_k and the functions y_τ given by (1.11) and (1.16), (1.17) for $y_0 = y_{0,i}$, $i = 1, 2$. Then by (1.18),

$$\mathscr{T}_N^\tau\psi(y_{0,1}) - \mathscr{T}_N^\tau\psi(y_{0,2})$$
$$= \int_{-N\tau}^0 e^{As}Q[R(y_{\tau,1}(s) + \psi(y_{\tau,1}(s))) - R(y_{\tau,2}(s) + \psi(y_{\tau,2}(s)))]\,ds$$
$$+ A^{-1}e^{-AN\tau}Q[R(y_{N,1} + \psi(y_{N,1})) - R(y_{N,2} + \psi(y_{N,2}))].$$

Thanks to IX.(1.4) and IX.(1.13) and by definition of $\mathscr{F}_{l,b}$

$$|\mathscr{T}_N^\tau\psi(y_{0,1}) - \mathscr{T}_N^\tau\psi(y_{0,2})|_E$$
$$\leq Mk_2(1 + l)\int_{-N\tau}^0 (|s|^{-\alpha} + \Lambda^\alpha)e^{\Lambda s}|y_{\tau,1}(s) - y_{\tau,2}(s)|_E\,ds$$
$$+ Mk_2(1 + l)\Lambda^{\alpha-1}e^{-\Lambda N\tau}|y_{N,1} - y_{N,2}|_E.$$

Lemma 1.1(ii) (with $\psi_1 = \psi_2 = \psi$) gives

$$|y_{k,1} - y_{k,2}|_E \leq e^{k\tau(\lambda + Mk_3(1+l)\lambda^\alpha)}|y_{0,1} - y_{0,2}|_E,$$

which implies for all $s \leq 0$,

$$|y_{\tau,1}(s) - y_{\tau,2}(s)|_E \leq e^{-s(\lambda + Mk_3(1+l)\lambda^\alpha)}|y_{0,1} - y_{0,2}|_E. \tag{2.4}$$

Therefore

$$|\mathscr{T}_N^\tau\psi(y_{0,1}) - \mathscr{T}_N^\tau\psi(y_{0,2})|_E$$
$$\leq Mk_2(1 + l)|y_{0,1} - y_{0,2}|_E\int_{-N\tau}^0 (|s|^{-\alpha} + \Lambda^\alpha)e^{(\Lambda - \lambda - Mk_3(1+l)\lambda^\alpha)s}\,ds$$
$$+ Mk_2(1 + l)|y_{0,1} - y_{0,2}|_E\Lambda^{\alpha-1}e^{-(\Lambda - \lambda - Mk_3(1+l)\lambda^\alpha)N\tau}$$
$$\leq (\Delta_1 + \Delta_2)|y_{0,1} - y_{0,2}|_E, \tag{2.5}$$

with

$$\Delta_1 = MK_2(1 + l)\int_{-N\tau}^0 (|s|^{-\alpha} + \Lambda^\alpha)e^{-Mk_3(1+l)\lambda^\alpha s}\,ds, \tag{2.6}$$
$$\Delta_2 = Mk_2(1 + l)\Lambda^{\alpha-1}e^{Mk_3(1+l)\lambda^\alpha N\tau}. \tag{2.7}$$

Thanks to IX.(1.7) and $N\tau \leq c_1 \lambda^{-\alpha}$, we write $\Delta_1 + \Delta_2 \leq \Delta_3 + \Delta_4 + \Delta_5$,

$$\Delta_3 = Mk_2(1 + l)e^{Mk_3(1+l)c_1} \int_{-N\tau}^{0} |s|^{-\alpha}\, ds,$$

$$\Delta_4 = \Lambda^\alpha \int_{-N\tau}^{0} e^{Mk_3(1+l)\lambda^\alpha s}\, ds,$$

$$\Delta_5 = Mk_2(1 + l)\Lambda^{\alpha-1} e^{Mk_3(1+l)c_1},$$

and $\Delta_3 + \Delta_4 + \Delta_5 \leq \Delta$, with

$$\Delta = \left(Mk_2(1 + l)\left(\frac{\lambda^{\alpha(\alpha-1)} c_1^{1-\alpha}}{1-\alpha} + \Lambda^{\alpha-1} \right) + \frac{k_2}{k_3}\left(\frac{\Lambda}{\lambda}\right)^\alpha \right) e^{Mk_3(1+l)c_1}. \tag{2.8}$$

Hence

$$|\mathcal{T}_N^\tau \psi(y_{0,1}) - \mathcal{T}_N^\tau \psi(y_{0,2})|_E \leq \Delta |y_{0,1} - y_{0,2}|_E \leq l |y_{0,1} - y_{0,2}|_E,$$

provided $\Delta \leq l$.

We now choose c_1 and c_2 to ensure that $\Delta \leq l$. Choose first $\delta_0 > 0$ arbitrary; then for $\Lambda \geq \lambda \geq c_2$ and $c_1 \leq \delta_0$,

$$Mk_2\left(\frac{\lambda^{\alpha(\alpha-1)} c_1^{1-\alpha}}{1-\alpha} + \Lambda^{\alpha-1} \right) e^{Mk_3 c_1} \leq Mk_2\left(\frac{c_2^{\alpha(\alpha-1)} \delta_0^{1-\alpha}}{1-\alpha} + c_2^{\alpha-1} \right) e^{Mk_3 \delta_0},$$

and we now choose c_2 sufficiently large so that this quantity is $\leq \frac{1}{2}$.

Therefore, with this choice of δ_0, c_2 and for $c_1 \leq \delta$,

$$\Delta \leq \left(\frac{l}{2} + c_1' \right) e^{Mk_3 c_1 l}$$

with

$$c_1' = \frac{1}{2} + \frac{k_2}{k_3} \sup_k \left(\frac{\Lambda_k}{\lambda_k} \right)^\alpha.$$

Finally, we achieve

$$\left(\frac{l}{2} + c_1' \right) e^{Mk_3 c_1 l} \leq l$$

by choosing $l = 6c_1'$ and $c_1 = \min(\delta_0, \log(3/2)/6Mk_3 c_1')$.

Lemma 2.1 is proved. □

2.2. Distance to the Attractor

For the rest of Section 2 we retain the values of l and b given by Lemma 2.1,

$$\lambda_n \geq c_2 \tag{2.9}$$

and we choose a sequence of $\tau_N > 0$,

$$\tau_N \le \frac{c_1}{N\lambda_n^\alpha}. \tag{2.10}$$

Hence, by Lemma 2.1, we have constructed the sequence of functions Φ_N; we call $\mathscr{M}_N$ the graph of Φ_N.

Ultimately, we wish to give an estimate of the thickness required so that a neighborhood of $\mathscr{M}_N$ contains the attractor. This can be measured by

$$\max_{u_0 = y_0 + z_0 \in \mathscr{A}} |\Phi_N(y_0) - z_0|_E$$

The following lemma shows how the mapping $\mathscr{T}_N^\tau$ decreases the quantity

$$\max_{u_0 = y_0 + z_0 \in \mathscr{A}} |\psi(y_0) - z_0|_E$$

for any given ψ in $\mathscr{F}_{l,b}$.

Lemma 2.2. *Under the same hypotheses as in Lemma* 2.1, *and* (1.22) *and* (1.23), *there exist three constants c_3, c_4, c_5 such that for any ψ in $\mathscr{F}_{l,b}$*

$$\begin{aligned}\operatorname*{Sup}_{u_0 = y_0 + z_0 \in \mathscr{A}} |\mathscr{T}_N^\tau \psi(y_0) - z_0|_E &\le c_3 \lambda_n^{\alpha - 1} \operatorname*{Sup}_{u_0 = y_0 + z_0 \in \mathscr{A}} |\psi(y_0) - z_0|_E \\ &\quad + c_4(\Lambda_n^{\alpha-1}\beta_2 + \lambda_n^{-1}\beta_1)\tau + c_5\left(\frac{(N\tau)^{-\alpha}}{\Lambda_n} + \Lambda_n^{\alpha-1}\right)e^{-\Lambda_n N\tau}.\end{aligned}$$

Proof. Let ψ be in $\mathscr{F}_{l,b}$ and let $u_0 = y_0 + z_0$ be a point on the attractor $\mathscr{A}$, $y_0 = Pu_0$, $z_0 = Qu_0$. We call $u = u(t)$, the orbit lying on $\mathscr{A}$ such that $u(0) = u_0$,[1] and we write $y = Pu$, $z = Qu$ and, as in (1.22), $\bar{y}_k = y(-k\tau)$. We also denote by y_k, $k = 0, \ldots, N$, the sequence constructed by (1.10) starting with this y_0, and we denote by y_τ the corresponding function defined by (1.16) and (1.17). We recall that

$$\mathscr{T}_N^\tau \psi(y_0) = \int_{-\infty}^0 e^{As} QR(y_\tau(s) + \psi(y_\tau(s)))\, ds,$$

$$z_0 = \int_{-\infty}^0 e^{As} QR(y(s) + z(s))\, ds.$$

Therefore

$$\begin{aligned}&|\mathscr{T}_N^\tau \psi(y_0) - z_0|_E \\ &\quad \le \int_{-\infty}^0 |e^{As} Q(R(y_\tau(s) + \psi(y_\tau(s))) - R(y(s) + z(s))|_E\, ds \\ &\quad \le \text{(with IX.(1.3), IX.(1.4), IX.(1.13))}\end{aligned}$$

[1] $u(t) = S(t)u_0$, $u_0 = S(t)u(-t)$ for $t > 0$.

$$\leq k_2 \int_{-\infty}^{0} (|s|^{-\alpha} + \Lambda^\alpha) e^{\Lambda s} |R(y_\tau(s) + \psi(y_\tau(s))) - R(y(s) + z(s))|_F \, ds$$

$$\leq k_2 M \int_{-N\tau}^{0} (|s|^{-\alpha} + \Lambda^\alpha) e^{\Lambda s} ((1 + l)|y(s) - y_\tau(s)|_E + |\psi(y(s)) - z(s)|_E) \, ds$$

$$+ 2k_2 M_0 \int_{-\infty}^{-N\tau} (|s|^{-\alpha} + \Lambda^\alpha) e^{\Lambda s} \, ds. \tag{2.11}$$

Thanks to (1.22), $\bar{y}_k$ satisfies

$$\bar{y}_{k+1} = \mathscr{R}_\tau \bar{y}_k + \mathscr{S}_\tau PR(\bar{y}_k + z(-k\tau)) + \varepsilon_k.$$

We apply Lemma 1.2 with $y_k^* = \bar{y}_k$ and

$$e_k = \varepsilon_k + \mathscr{S}_\tau P(R(\bar{y}_k + z(-k\tau)) - R(\bar{y}_k + \psi(\bar{y}_k)));$$

this yields

$$|y_k - \bar{y}_k|_E$$

$$\leq \sum_{j=0}^{k-1} e^{(k-1-j)\tau(\lambda + Mk_3(1+l)\lambda^\alpha)} (|\varepsilon_j|_E + |\mathscr{S}_\tau P(R(\bar{y}_j + z(-j\tau)) - R(\bar{y}_j + \psi(\bar{y}_j)))|_E$$

$\leq$ (with (1.15) and (1.22))

$$\leq \sum_{j=0}^{k-1} (\tau^2 \beta_1 + M_1 k_3 \lambda^{\alpha-1}(e^{\tau\lambda} - 1)|\psi(\bar{y}_j) - z(-j\tau)|_E) e^{(k-1-j)\tau(\lambda + Mk_3(1+l)\lambda^\alpha)}$$

$$\leq (\tau^2 \beta_1 + M_1 k_3 \lambda^{\alpha-1}(e^{\tau\lambda} - 1) \operatorname*{Sup}_{y_* + z_* \in \mathscr{A}} |\psi(y_*) - z_*|_E) \frac{e^{k\tau(\lambda + Mk_3(1+l)\lambda^\alpha)}}{e^{\tau(\lambda + Mk_3(1+l)\lambda^\alpha)} - 1}$$

$\leq$ (from $1 + s \leq e^s \leq 1 + se^s$, $s > 0$)

$$\leq \left(\frac{\tau\beta_1}{\lambda(1 + Mk_3\lambda^{\alpha-1})} + Mk_3\lambda^{\alpha-1}\sigma \right) e^{k\tau(\lambda + Mk_3(1+l)\lambda^\alpha)},$$

where $\sigma = \operatorname{Sup}_{u_* = y_* + z_* \in \mathscr{A}} |\psi(y_*) - z_*|_E$. For any s in $(-(k+1)\tau, -k\tau)$, $0 \leq k \leq N - 1$, this gives using (1.23),

$$|y(s) - y_\tau(s)|_E \leq |y(s) - y(-k\tau)|_E + |y_k - \bar{y}_k|_E$$

$$\leq \tau\beta_2 + \left(\frac{\tau\beta_1}{\lambda + Mk_3\lambda^\alpha} + Mk_3\lambda^{\alpha-1}\sigma \right) e^{-s(\lambda + Mk_3(1+l)\lambda^\alpha)}. \tag{2.12}$$

Inserting this in (2.11) we obtain thanks to IX.(2.4)

$$|\mathscr{T}_N^\tau \psi(y_0) - z_0|_E \leq k_2 M \tau \beta_2 (1 + l) \int_{-N\tau}^{0} (|s|^{-\alpha} + \Lambda^\alpha) e^{\Lambda s} \, ds$$

$$+ k_2 M(1 + l) \left(\frac{\tau\beta_1}{\lambda + Mk_3\lambda^{\alpha-1}} + Mk_3\lambda^{\alpha-1}\sigma \right) \int_{-N\tau}^{0} (|s|^{-\alpha} + \Lambda^\alpha) e^{-sMk_3(1+l)\lambda^\alpha} \, ds$$

$$+ k_2 M(1 + \gamma_\alpha)\Lambda^{\alpha+1}\sigma + 2k_2 M_0((N\tau)^{-\tau} + \Lambda^\alpha) \frac{e^{-\Lambda N\tau}}{\Lambda}.$$

From the proof of Lemma 2.1, we know that

$$\Delta_1 = k_2 M(1 + l) \int_{-N\tau}^{0} (|s|^{-\alpha} + \Lambda^{\alpha}) e^{-MK_3(1+l)\lambda^{\alpha} s\, ds} \leq l;$$

therefore

$$\begin{aligned} &|\mathscr{T}_N^{\tau}\psi(y_0) - z_0|_E \\ &\quad \leq k_2 M\tau\beta_2(1 + l)(1 + \gamma_\alpha)\Lambda^{\alpha-1} + k_2 M(1 + l)l\lambda^{-1}\tau\beta_1 + k_3 M\lambda^{\alpha-1}\sigma l \\ &\quad\quad + k_2 M(1 + \gamma_\alpha)\Lambda^{\alpha-1}\sigma + 2k_2 M \frac{(N\tau)^{-\alpha} + \Lambda^{\alpha}}{\Lambda} e^{-\Lambda N\tau}, \end{aligned}$$

and the lemma is proved. □

Remark 2.1. We deduce from Lemma 2.2 that, under the hypotheses (2.9) and (2.10), the Φ_N satisfy

$$\begin{aligned} \operatorname*{Sup}_{u_0 = y_0 + z_0 \in \mathscr{A}} |\Phi_{N+1}(y_0) - z_0|_E &\leq c_3 \lambda_n^{\alpha-1} \operatorname*{Sup}_{u_0 = y_0 + z_0 \in \mathscr{A}} |\Phi_N(y_0) - z_0|_E \\ &\quad + c_4(\Lambda_n^{\alpha-1}\beta_2 + \lambda_n^{-1}\beta_1)^{\tau_N} + c_5\left(\frac{(N\tau_N)^{-\alpha}}{\Lambda_n} + \Lambda_n^{\alpha-1}\right) e^{-\Lambda_n N\tau_N}. \end{aligned} \tag{2.13}$$

2.3. The Main Result

We want to derive some information on the semidistance in E of $\mathscr{A}$ to $\mathscr{M}_N$ (see I.(1.10)):

$$d_E(\mathscr{A}, \mathscr{M}_N) = \operatorname*{Sup}_{v \in \mathscr{A}} \operatorname*{Inf}_{w \in \mathscr{M}_N} |v - w|_E.$$

Our estimate given in Theorem 2.1 is based on (2.13). As explained before, in Section 3, we will let $\tau_N \to 0$ and $N\tau_N \to \infty$ as $N \to \infty$. However, in the present case, we let only $\tau_N \to 0$, and keep $N\tau_N$ bounded (from above and from below); some heuristic explanation for this is given in Remark 2.3. Our result is the following theorem.

Theorem 2.1. *Beside the standing hypotheses of Sections* 1 *and* IX.1, *in particular,* IX.(1.12), IX.(1.13), (1.14) and (1.15), *we assume* (1.22) *and* (1.23) *and we assume that*

$$\lambda_n \geq \max(c_1^{-1}, c_2, (2c_3)^{1/(1-\alpha)}), \tag{2.14}$$

and that the sequence τ_N, $N \in \mathbb{N}$ *satisfies*

$$c_6 \leq \tau_N N \lambda_n^{\alpha} \leq c_1, \qquad \text{for all } N, \tag{2.15}$$

where c_6 *is any fixed constant* $< c_1$, *and* $c_1 - c_5$ *are as in Lemmas* 2.1 *and* 2.2.

Then the manifolds $\mathscr{M}_N$, $N \in \mathbb{N}$, *defined by* (1.24) (*and* $\mathscr{M}_N = \text{graph } \Phi_N$), *satisfy*

$$
\begin{aligned}
d_E(\mathscr{A}, \mathscr{M}_N) &= \operatorname*{Sup}_{u \in \mathscr{A}} \operatorname*{Inf}_{w \in \mathscr{M}} |v - w|_E \\
&\leq (c_3 \lambda_n^{\alpha-1})^N \operatorname*{Sup}_{u_0 \in \mathscr{A}} |Q_n u_0|_E \\
&\quad + c_4(\Lambda_n^{\alpha-1}\beta_2 + \lambda_n^{-1}\beta_1) \sum_{j=0}^{N-1} (c_3 \Lambda_n^{\alpha-1})^j \tau_{N-1-j} \\
&\quad + 4c_5 \Lambda_n^{1-\alpha} e^{-c_6 \Lambda_n^{\alpha-1}}.
\end{aligned}
\tag{2.16}
$$

In particular, for N sufficiently large, n being fixed according to (2.14),

$$
d_E(\mathscr{A}, \mathscr{M}_N) \leq 8c_5 \Lambda_n^{\alpha-1} e^{-c_6 \Lambda_n^{\alpha-1}}. \tag{2.17}
$$

Proof. Using (2.14), (2.15), and setting

$$
r_N = \operatorname*{Sup}_{u_0 = y_0 + z_0 \in \mathscr{A}} |\Phi_N(y_0) - z_0|_E, \tag{2.18}
$$

we infer from (2.13) and I.(1.7) that

$$
\begin{gathered}
r_{N+1} \leq \delta \tau_N + \sigma_N, \\
\sigma_N = c_4(\Lambda_n^{\alpha-1}\beta_2 + \lambda_n^{-1}\beta_1)\tau_N + 2c_5 \Lambda_n^{\alpha-1} e^{-c_6 \Lambda_n^{1-\alpha}},
\end{gathered}
\tag{2.19}
$$

with $\delta = c_3 \lambda_n^{\alpha-1} < \frac{1}{2}$. Hence, by reiteration and using (2.14),

$$
r_N \leq \delta^N \tau_0 + \sum_{j=0}^{N-1} \sigma_{N-j-1} \delta^j, \tag{2.20}
$$

$r_0 = \operatorname{Sup}_{u_0 \in \mathscr{A}} |Q_n u_0|$. For any $k < N$, the sum in the right-hand side of (2.20) is majorized by

$$
\begin{aligned}
&\frac{2c_5}{1 - c_3 \lambda_n^{\alpha-1}} \Lambda_n^{\alpha-1} e^{-c_6 \Lambda_n^{\alpha-1}} + \frac{c_4}{1-\delta}(\Lambda_n^{\alpha-1}\beta_2 + \lambda_n^{-1}\beta_1) \cdot \operatorname*{Sup}_{N-1-k \leq j \leq N-1} \tau_j \\
&\qquad + \sum_{j=0}^{k-1} \left(\operatorname*{Sup}_{l \geq 0} \tau_l \right) \cdot \frac{\delta^{k+1}}{1-\delta} c_4(\Lambda_n^{\alpha-1}\beta_2 + \lambda_n^{-1}\beta).
\end{aligned}
$$

Hence, as $N \to \infty$ (n fixed), the right-hand side of (2.20) converges to

$$
\frac{2c_5}{1 - c_3 \lambda_n^{\alpha-1}} \Lambda_n^{\alpha-1} e^{-c_6 \Lambda_n^{1-\alpha}} \leq 4c_5 \Lambda_n^{\alpha-1} e^{-c_6 \Lambda_n^{1-\alpha}};
$$

for N sufficiently large, the right-hand side of (2.20) is indeed bounded by $8c_5 \Lambda_n^{\alpha-1} e^{-c_6 \Lambda_n^{1-\alpha}}$, and the theorem is proved. □

Remark 2.2. Theorem 2.1 provides a finite-dimensional smooth manifold $\mathscr{M}_N$ which, for suitable N, approximates the attractor $\mathscr{A}$ at an exponential order, i.e.,

$$
d_E(\mathscr{A}, \mathscr{M}_N) \leq c_1' e^{-c_2' n^q}, \tag{2.21}
$$

$c_1', c_2' > 0$ constants, n being the dimension of $\mathcal{M}_N$.[1] Most approximating manifolds usually approximate the attractor at a polynomial order, the error being of the form $c\lambda_n^{-p}$ (or $c'n^{-q}$); see, e.g., among many references, C. Foias, O. Manley, and R. Temam [2], [3]; C. Foias, M.S. Jolly, I.G. Kevrekidis, G.R. Sell, and E.S. Titi [1], and M. Marion [4].

The existence of an approximate manifold of exponential order was proved in C. Foias, O. Manley, and R. Temam [3], but the proof there was not constructive. Manifolds of exponential order which are not graphs are obtained in C. Foias and R. Temam [7], [9]. Finally, using the Gevrey class regularity of solutions (see C. Foias and R. Temam [8]), one can obtain for specific equations manifolds which approximate the attractor at an exponential rate but, in the actual applications to evolution partial differential equations, more regularity is required on the data (Gevrey-type regularity instead of L^2-regularity); see, e.g., D. Jones and E.S. Titi [1], and for Gevrey class regularity, see C. Foias and R. Temam [8].

Remark 2.3. It would be natural to let $N\tau_N$ go to ∞ as $N \to \infty$, since this means that (1.10) indeed approximates (1.9) on the whole interval $(-\infty, 0)$. We will do so in Section 4 but this is not suitable in the present situation. Indeed, when $(\Lambda_n - \lambda_n)/(\Lambda_n^\alpha + \lambda_n^\alpha)$ is not sufficiently large, either because the spectral gap condition is not satisfied or because n is not sufficiently large, it is preferable to let τ_N go to zero and to take $N\tau_N$ constant. Since the scheme (1.10) expends the errors exponentially at the rate $\lambda_n + c_1'\lambda_n^\alpha$ for some c_1', while formula (1.19) reduces them at the rate $\Lambda_n - c_2'\Lambda_n^\alpha$ for some c_2', if $\lambda_n c_2'\lambda_n^\alpha$ is larger than $\Lambda_n - c_2'\Lambda_n^\alpha$, taking $N\tau_N$ too large will only increase the discretization error.

Remark 2.4. A procedure to accelerate the convergence as $N \to \infty$ is indicated in A. Debussche and R. Temam [4] where further details are given.

3. Convergent Families of Approximate Inertial Manifolds

Whereas in Section 2 we tried to construct low-order smooth approximations of the global attractor, the point of view here is different: we assume that the spectral gap condition IX.(2.1) is satisfied, that an exact inertial manifold exists, and we construct a sequence of approximate inertial manifolds which converge to the exact inertial manifold. Sequences of approximate inertial manifolds were constructed elsewhere, but it is not known whether they converge; see, e.g., A. Debussche and M. Marion [1], K. Promislow and R.

[1] $\Lambda_n \sim cn^p$ for n large is accounted for in (2.21).

Temam [1], R. Temam [7]; see also F. Demengel and J.M. Ghidaglia [1] in a different context.

The manifolds are constructed as in Section 1. The hypotheses are the standing hypotheses of Section IX.1 and Section 1 in this chapter, we retain (1.20), (1.21) and not (1.22), (1.23); we also assume IX.(2.1) and IX.(2.2) which guarantee, according to Theorem IX.2.1, the existence of an exact inertial manifold.

In Section 3.1 we derive properties of $\mathcal{T}_N^\tau$ which are better, and easier to derive than in Section 2.1 because the spectral gap condition is available. Section 3.3 gives the convergence result.

3.1. Properties of $\mathcal{T}_N^\tau$

We consider the mapping $\mathcal{T}_N^\tau$ defined by (1.18) and (1.19) and, in this section we show that $\mathcal{T}_N^\tau$ maps $\mathcal{F}_{l,b}$ into itself; in particular (see below), this will allow us to construct the sequence Φ_N in (1.24).

Lemma3.1. *Besides the standing hypotheses and* (1.22) *and* (1.23), *we assume that*

$$\Lambda_n - \lambda_n \geq c_7(\Lambda_n^\alpha + \mu_n^\alpha), \tag{3.1}$$

for some appropriate constant c_7.

Then, for any N *and* $\tau > 0$, $\mathcal{T}_N^\tau$ *maps* $\mathcal{F}_{l,b}$ *into itself and is a strict contraction in* $\mathcal{F}_{l,b}$ *with a constant less than* $\frac{1}{2}$.

Proof. Let $\psi \in \mathcal{F}_{l,b}$ and $y_0 \in PE$ (indices n are temporarily omitted); first we want to show that $|\mathcal{T}_N^\tau\psi(y_0)|_E \leq b$. We start from the expression (1.18) of $\mathcal{T}_N^\tau$ and use IX.(1.3) and IX.(1.13)

$$\begin{aligned} |\mathcal{T}_N^\tau\psi(y_0)|_E &\leq k_2 M_0 \int_{-\infty}^0 (|s|^{-\alpha} + \Lambda^\alpha)e^{\Lambda s}\, ds \\ &\leq (\text{with IX.(2.4)}) \\ &\leq k_2 M_0(1 + \gamma_\alpha)\Lambda_n^{\alpha-1}. \end{aligned}$$

This is less than b if

$$\Lambda^{1-\alpha} \geq k_2 M_0(1 + \gamma_\alpha)b^{-1}. \tag{3.2}$$

Note that (3.1) implies

$$\Lambda^{1-\alpha} \geq c_7, \tag{3.3}$$

and (3.2) follows, provided $c_7 \geq k_2 M_0(1 + \gamma_\alpha)$.

Then we want to show that, for given $\psi \in \mathcal{F}_{l,b}$, $\mathcal{T}_N^\tau\psi$ is Lipschitz with constant l. Hence, we consider $y_{0,1}, y_{0,2} \in PE$ and the corresponding y_k, y_τ as in (1.10) and (1.16), which we denote by $y_{k,i}, y_{\tau,i}$, $i = 1, 2$. We infer from (1.28) (with $\psi_1 = \psi_2 = \psi$), that

$$|y_{\tau,1}(s) - y_{\tau,2}(s)|_E \le e^{-s(\lambda+Mk_3(1+l)\lambda^\alpha)}|y_{0,1} - y_{0,2}|_E, \tag{3.4}$$

and using (1.18) we write

$$\begin{aligned}&\mathscr{T}_N^\tau\psi(y_{0,1}) - \mathscr{T}_N^\tau\psi(y_{0,2})\\&\quad= \int_{-\infty}^0 e^{As}Q(R(y_{\tau,1}(s) + \psi(y_{\tau,1}(s))) - R(y_{\tau,2}(s) + \psi(y_{\tau,2}(s))))\,ds.\end{aligned}$$

Hence with IX.(1.4), IX.(1.13))

$$\begin{aligned}&|\mathscr{T}_N^\tau\psi(y_{0,1}) - \mathscr{T}_N^\tau\psi(y_{0,2})|_E\\&\quad\le k_2M_1(1+l)\int_{-\infty}^0 (|s|^{-\alpha} + \Lambda^\alpha)e^{\Lambda s}|y_{\tau,1}(s) - y_{\tau,2}(s)|_E\,ds\\&\quad\le \Delta|y_{0,1} - y_{0,2}|_E,\end{aligned}$$

with

$$\Delta = k_2M_1(1+l)\int_{-\infty}^0 (|s|^{-\alpha} + \Lambda^\alpha)e^{\tilde\Lambda s}\,ds,$$

$$\tilde\Lambda = \Lambda - \lambda - Mk_3(1+l)\lambda^\alpha;$$

$\tilde\Lambda > 0$ provided $c_7 \ge Mk_3(1+l)$, in which case,

$$\begin{aligned}\Delta &= k_2M(1+l)(\gamma_\alpha\tilde\Lambda^{\alpha-1} + \Lambda^\alpha\tilde\Lambda^{-1})\\&\le k_2M(1+l)\Lambda^\alpha\tilde\Lambda^{-1},\end{aligned}$$

and $\Delta \le l$ if

$$\tilde\Lambda = \Lambda - \lambda - Mk_3(1+l)\lambda^\alpha \ge k_2M\frac{(1+l)}{l}\Lambda^\alpha(1+\gamma_\alpha),$$

i.e., if $c_7 \ge \max(Mk_3(1+l), k_2M[(1+l)/l](1+\gamma_\alpha))$.

With the conditions above we know that $\mathscr{T}_N^\tau$ maps $\mathscr{F}_{l,b}$ into itself. Finally, we want to show that $\mathscr{T}_N^\tau$ is a strict contraction with constant $\le \frac{1}{2}$. For that purpose we consider $\psi_1, \psi_2 \in \mathscr{F}_{l,b}$, $y_0 \in PE$, and we consider the corresponding y_k, y_τ as in (1.10), (1.16), which we denote by $y_{k,i}$, $y_{\tau,i}$, $i = 1, 2$. Again from (1.28) (with $y_{0,1} = y_{0,2}$), we infer that

$$|y_{\tau,1}(s) - y_{\tau,2}(s)|_E \le sMk_3\lambda^\alpha|\psi_1 - \psi_2|_\infty e^{-s(\lambda+Mk_3(1+l)\lambda^\alpha)}. \tag{3.5}$$

Hence with (1.18), IX.(1.3), IX.(1.4), and IX.(1.13),

$$\begin{aligned}&\mathscr{T}_N^\tau\psi_1(y_0) - \mathscr{T}_N^\tau\psi_2(y_0)\\&\quad= \int_{-\infty}^0 e^{As}Q(R(y_{\tau,1}(s) + \psi_1(y_{\tau,1}(s))) - R(y_{\tau,2}(s) + \psi_2(y_{\tau,2}(s))))\,ds,\end{aligned}$$

$$\begin{aligned}
&|R(y_{\tau,1} + \psi_1(y_{\tau,1})) - R(y_{\tau,2} + \psi_2(y_{\tau,2}))|_F \\
&\qquad \le M|\psi_1(y_{\tau,1}) - \psi_2(y_{\tau,1})|_F + M(1+l)|y_{\tau,1} - y_{\tau,2}|_E \\
&\qquad \le M|\psi_1 - \psi_2|_\infty + M(1+l)|y_{\tau,1} - y_{\tau,2}|_E, \\
&|\mathscr{T}_N^\tau \psi_1(y_0) - \mathscr{T}_N^\tau \psi_2(y_0)|_E \le (\Delta_1 + \Delta_2)|\psi_1 - \psi_2|_\infty,
\end{aligned} \tag{3.6}$$

$$\Delta_1 = k_2 M \int_{-\infty}^0 (|s|^{-\alpha} + \Lambda^\alpha) e^{\Lambda s}\, ds,$$

$$\Delta_2 = k_2 k_3 M^2 \lambda^\alpha (1+l) \int_{-\infty}^0 (|s|^{1-\alpha} + \Lambda^\alpha |s|) e^{\tilde{\Lambda} s}\, ds,$$

$\tilde{\Lambda}$ as above. Thanks to IX.(2.4) and $\gamma_{-1} = 1$,

$$\Delta_1 = k_2 M (1 + \gamma_\alpha) \Lambda^{\alpha - 1},$$

$$\Delta_2 = k_2 k_3 M^2 \lambda^\alpha (1+l)(\gamma_{\alpha-1} \tilde{\Lambda}^{\alpha-2} + \Lambda^\alpha \tilde{\Lambda}^{-2}).$$

According to (3.3), we have $\Delta_1 \le \frac{1}{4}$ if $c_7 \ge 4k_2 M(1 + \gamma_\alpha)$; for $\Delta_2 \le \frac{1}{4}$ it is then sufficient to have

$$k_2 k_3 M^2 (1+l)(1 + \gamma_{\alpha-1}) \Lambda^{2\alpha} \tilde{\Lambda}^{-2} \le \tfrac{1}{4},$$

which follows promptly from (3.1), if c_7 is sufficiently large.

This concludes the proof of Lemma 3.1. □

3.2. Distance to the Exact Inertial Manifold

Lemma 3.1 implies that the sequence (1.24) of functions Φ_N is well defined. Our aim is now to compare Φ_N to the function Φ given by Theorem IX.2.1; for that purpose, we will need the following technical result.

Lemma 3.2. *Besides the standing hypotheses, we assume* (1.22), (1.23), IX.(2.1), IX.(2.2), *and* (3.1).

Then there exist two constants $c_8, c_9, c_8 \ge c_7$, *such that if*

$$\Lambda_n - \lambda_n \ge c_8(\Lambda_n^\alpha + \lambda_n^\alpha), \tag{3.7}$$

then for all ψ *in* $\mathscr{F}_{l,b}$ *and all* N, τ,

$$|\mathscr{T}_N^\tau \psi - \Phi|_\infty \le \tfrac{1}{2}|\psi - \Phi|_\infty + \varepsilon(N, \tau), \tag{3.8}$$

with

$$\varepsilon(N,\tau) = c_9 \left(\alpha_2(\lambda_n)\tau + \frac{\alpha_1(\lambda_n)}{\Lambda_n^\alpha}\tau + \frac{\alpha_2(\lambda_n)}{\Lambda_n^\alpha} e^{-\Lambda_n^\alpha N \tau} \right).$$

PROOF. Due to Lemma 3.1,

$$\begin{aligned}|\mathcal{T}_N^\tau\psi - \Phi|_\infty &\le |\mathcal{T}_N^\tau\psi - \mathcal{T}_N^\tau\Phi|_\infty + |\mathcal{T}_N^\tau\Phi - \Phi|_\infty\\ &\le \tfrac{1}{2}|\psi - \Phi|_\infty + |\mathcal{T}_N^\tau\Phi - \Phi|_\infty,\end{aligned}$$

and it suffices to estimate $|\mathcal{T}_N^\tau\Phi - \Phi|_\infty$.

Recall that $\mathcal{T}\Phi = \Phi$, so that using the definitions of $\mathcal{T}_N^\tau$ and $\mathcal{T}$ (see IX.(1.24) and IX.(1.18)), we have, for any $y_0 \in PE$,

$$\mathcal{T}_N^\tau\Phi(y_0) - \Phi(y_0) = \int_{-\infty}^0 e^{As}Q(R(y_\tau(s) + \Phi(y_\tau(s))) - R(y(s) + \Phi(y(s))))\,ds. \tag{3.9}$$

In this lemma, $y = y(s)$ is given by IX.(1.19) and the function y_τ and the sequence y_k is constructed with (1.16) and (1.10) in which ψ is replaced by Φ.

We set $\tilde{y}_k = y(-k\tau)$; according to (1.10) and (1.20)

$$\begin{aligned}y_{k+1} &= \mathcal{R}_\tau y_k + \mathcal{S}_\tau PR(y_k + \Phi(y_k)),\\ \tilde{y}_{k+1} &= \mathcal{R}_\tau\tilde{y}_k + \mathcal{S}_\tau PR(\tilde{y}_k + \Phi(\tilde{y}_k)) + \varepsilon_k.\end{aligned}$$

We apply Lemma 1.2 with ψ, y_k, y_k^* replaced by Φ, y_k, $\tilde{y}_k$; this yields

$$\begin{aligned}|y_k - \tilde{y}_k|_E &\le \sum_{j=0}^{k-1} e^{(k-1-j)\tau(\lambda+Mk_3(1+l)\lambda^\alpha)|}|\varepsilon_j|_E\\ &\le \text{(with (1.20))}\\ &\le \alpha_1(\lambda)\tau^2ke^{k\tau(\lambda+c_1'\lambda^\alpha)},\end{aligned}$$

with $c_1' = \max(k_4, Mk_3(1 + l))$. From this and (1.21) we conclude that for $-N\tau \le s \le 0$

$$|y(s) - y_\tau(s)|_E \le \tau\alpha_2(\lambda)e^{-s(\lambda+k_5\lambda^\alpha)} + \tau\alpha_1(\lambda)|s|e^{-s(\lambda+c_1'\lambda^\alpha)}.$$

Similarly, if $s < -N\tau$,

$$|y(s) - y_\tau(s)|_E \le |s + N\tau|\alpha_2(\lambda)e^{-s(\lambda+k_5\lambda^\alpha)} + \tau^2\alpha_1(\lambda)|s|e^{-s(\lambda+c_1'\lambda^\alpha)}.$$

Now we return to (3.9) and use IX.(1.4) and IX.(1.13)

$$\begin{aligned}|\mathcal{T}_N^\tau\Phi(y_0) - \Phi(y_0)|_E &\le k_2M(1 + l)\int_{-\infty}^0 (|s|^{-\alpha} + \Lambda^\alpha)|y_\tau(s) - y(s)|_E\,ds\\ &\le k_2M(1 + l)\left[\tau\alpha_2(\lambda)\int_{-N\tau}^0 (|s|^{-\alpha} + \Lambda^\alpha)e^{(\Lambda-\lambda-k_5\lambda^\alpha)s}\,ds\right.\\ &\quad + \tau\alpha_1(\lambda)\int_{-\infty}^0 (|s|^{1-\alpha} + \Lambda^\alpha|s|)e^{(\Lambda-\lambda-c_1'\lambda^\alpha)s}\,ds\\ &\quad \left.+ \tau\alpha_2(\lambda)\int_{-\infty}^{-N\tau} |s + N\tau|(|s|^{-\alpha} + \Lambda^\alpha)e^{(\Lambda-\lambda-k_5\lambda^\alpha)s}\,ds\right].\end{aligned}$$

Provided $c_8 \geq \max(k_5, 1, c_7, c_1')$, the right-hand side of the last inequality is bounded by

$$c_9\left(\tau\alpha_2(\lambda)\frac{\Lambda^\alpha(1+\gamma_\alpha)}{\Lambda-\lambda-k_5\lambda^\alpha}+\tau\alpha_1(\lambda)\frac{\Lambda^\alpha}{(\Lambda-\lambda-c_1'\lambda^\alpha)^2}\right.$$
$$\left.+\alpha_2(\lambda)\frac{\Lambda^\alpha(1+\gamma_{\alpha-1})}{(\Lambda-\lambda-k_5\lambda^\alpha)^2}e^{-(\Lambda-\lambda-k_5\lambda^\alpha)N\tau}\right)$$
$$\leq c_9\left(\alpha_2(\lambda)\tau+\frac{\alpha_1(\lambda)}{\Lambda^\alpha}\tau+\frac{\alpha_2(\lambda)}{\Lambda^\alpha}e^{-\Lambda^\alpha N\tau}\right),$$

where the constant c_9 can be easily determined. □

3.3. Convergence to the Exact Inertial Manifold

It is now easy to conclude.

Theorem 3.1. *The hypotheses are the standing hypotheses (see Remark* 1.1), (1.20), (1.21), IX.(2.1), IX.(2.2), *and* (3.7).

Then the family Φ_N converges to Φ as $N \to \infty$, in the topology defined by $|\cdot|_\infty$ provided $\tau_N \to 0$ and $N\tau_N \to \infty$ as $N \to \infty$.

Proof. We infer from (3.8) that

$$|\Phi_N-\Phi|_\infty = |\mathcal{T}_N^\tau\Phi_{N-1}-\Phi|_\infty \leq \tfrac{1}{2}|\Phi_{N-1}-\Phi|_\infty+\varepsilon(N,\tau_N).$$

Hence, recursively,

$$|\Phi_N-\Phi|_\infty \leq (\tfrac{1}{2})^N|\Phi|_\infty+\sum_{j=1}^N \varepsilon(j,\tau_j)(\tfrac{1}{2})^{N-j}.$$

Since $\varepsilon(j,\tau_j)\to 0$ as $j\to\infty$, if $\tau_j\to 0$ and $j\tau_j\to\infty$, we conclude as in the proof of Theorem 2.1, that

$$|\Phi_N-\Phi|_\infty \to 0 \quad \text{as} \quad N\to\infty.$$

The proof is complete. □

Remark 3.1. (i) It is shown in A. Debussche and R. Temam [4] that the sequence Φ_N converges to Φ in the $\mathscr{C}^1$ topology.

(ii) The rate of convergence of Φ_N to Φ (in the norm $|\ |_\infty$) can be increased by using more points and by changing the integration formula (1.10) (Euler's scheme); see also A. Debussche and R. Temam [4].

(iii) We refer the reader to the same reference for the verification of the main hypotheses (1.14), (1.15), (1.20)–(1.23), in the case where $\mathscr{R}_\tau$ and $\mathscr{S}_\tau$ correspond to the Euler scheme, i.e., when they are given by (1.13).

We refrain here from developing these technical points.

Remark 3.2. As mentioned in the Introduction to this chapter we do not try to give an exhaustive bibliography on the topics of approximation of attractors and inertial manifolds, approximate dynamics, and related algorithms. Let us at least mention a few references: C. Foias, M.S. Jolly, J.G. Kevrekidis, G.R. Sell, and E. Titi [1], G.R. Sell [1], R. Temam [7], [8], [10]; for numerical and computational issues, see M. Marion and R. Temam [1], [2], and the articles by D. Jones, L. Margolin, and E. Titi, by M.S. Jolly, and by A. Debussche, T. Dubois, and R. Temam in R. Temam [9]; see also the many references quoted in all these articles.

It is noteworthy also to mention the new concept of inertial manifold with delay introduced by A. Debussche and R. Temam in [5]; whereas inertial manifolds correspond to the concepts of slow manifolds and *balanced models* in meteorology, inertial manifolds with delay corresponds to *unbalanced models* in meteorology; see A. Debussche, and R. Temam [5]. □

APPENDIX

Collective Sobolev Inequalities

Introduction

The Lieb–Thirring inequalities (see E. Lieb and W. Thirring [1], and also B. Simon [1], M. Cwikel [1]) are improvements of the classical Sobolev–Gagliardo–Nirenberg inequalities which have been introduced for the estimate of the number of bound states and their energy; and they provide in theoretical physics a proof of the stability of matter with a constant with the right order of magnitude. From the mathematical point of view they give improved estimates for the trace of some linear Schrödinger-type operators; and, in our case, they have a natural role to play in the estimate of the traces of the linear operators which appear in the linearization of the nonlinear evolution equations. They are constantly used in Chapter VI.

In this Appendix we give a self-contained presentation of these estimates and of several generalizations that are needed for our applications. We generalize the method of E. Lieb and W. Thirring [1] with some simplifications due to the fact that we restrict ourselves to differential operators defined in a bounded domain. The generalizations that we prove are of two types. First, we allow for more general operators and boundary conditions than in the references quoted above. In the first sections (1, 2, 3) elliptic operators of order $2m$, $m \geq 1$, are considered and the boundary conditions can also be of the Dirichlet type or of some other types. Then, in Section 4, we derive Lieb–Thirring-type inequalities in the spaces $H^m(\Omega)^k$, $m \geq 1$, without reference to any boundary condition. Of course, since the Lieb–Thirring inequality is not valid for instance in $H^1(\Omega)$, the inequalities that we prove at that point are slightly different. However, our inequalities contain those of E. Lieb and W. Thirring [1] as a particular case; also, as far as the applications that we have in view are concerned, they provide estimates of the dimension of the attractors as sharp and of the same order as in the case of Dirichlet boundary conditions (see Chapter VI).

We set the notations and hypotheses in Section 1. Section 2 contains the proof of spectral estimates for Schrödinger-type operators, including the so-called Birman–Schwinger inequality. Section 3 contains the statement and the proof of a first generalization of the Lieb–Thirring inequality. Then Section 4 gives another, more interesting generalization. Some examples are briefly described in Section 5 but, of course, many other examples appear in Chapter VI.

1. Notations and Hypotheses

1.1. The Operator $\mathfrak{A}$

Let Ω denote an open bounded set of $\mathbb{R}^n$ with a $\mathscr{C}^\infty$ boundary Γ (see II.(1.2)).

We are given a self-adjoint differential operator $\mathfrak{A}$ on Ω with values in $\mathbb{R}^k$, homogeneous of order $2m$,

$$\mathfrak{A}u = \sum_{\substack{\alpha,\beta\in\mathbb{N}^n\\ [\alpha]=[\beta]=m}} (-1)^m D^\beta(a_{\alpha\beta}(x)D^\alpha u), \qquad u \in \mathscr{C}_0^\infty(\Omega;\mathbb{R}^k), \tag{1.1}$$

where $\alpha = (\alpha_1, \dots, \alpha_n) \in \mathbb{N}^n$, $[\alpha] = \alpha_1 + \cdots + \alpha_n$, and the coefficients $a_{\alpha\beta}$ of the operator are matrix functions which are $\mathscr{C}^{2m}$ from $\bar{\Omega}$ into the space of $k \times k$ symmetric matrices and $a_{\alpha\beta} = a_{\beta\alpha}$. To this operator we associate the bilinear symmetric form

$$a(u, v) = \sum_{\substack{\alpha,\beta\in\mathbb{N}^n\\ [\alpha]=[\beta]=m}} \int_\Omega a_{\alpha\beta} D^\alpha u D^\beta v \, dx, \tag{1.2}$$

which is defined and continuous on $H^m(\Omega)^k \times H^m(\Omega)^k$. It is assumed that a is semicoercive in the following sense: there exists $\delta > 0$ such that

$$a(u, u) \geq \delta \sum_{\substack{\alpha\in\mathbb{N}^n\\ [\alpha]=m}} |D^\alpha u|^2, \qquad \forall u \in H^m(\Omega)^k, \tag{1.3}$$

where $|\cdot|$ denotes the norm in the space $L^2(\Omega)^k$, and as before $(\cdot, \cdot)$ denotes the scalar product in this space; $|\Omega|$ is the measure of Ω.

For every $u \in H^m(\Omega)^k$, we set

$$[\![u]\!]^2 = \sum_{[\alpha]=m} |D^\alpha u|^2, \qquad \|u\|^2 = [\![u]\!]^2 + \frac{1}{|\Omega|^{2m/n}} |u|^2.$$

It follows readily from the Poincaré inequality II.(1.39) that $[\![\cdot]\!]$ is a norm on $H^m(\Omega)^k$ equivalent to the usual one. On the other hand, (1.3) implies

$$a(u, u) + \lambda|u|^2 \geq \delta\|u\|^2, \qquad \forall u \in H^m(\Omega)^k, \tag{1.4}$$

$$\lambda = \frac{\delta}{|\Omega|^{2m/n}}.$$

We will consider the restriction of a to a closed subspace V of $H^m(\Omega)^k$; a is semicoercive on V in the sense of (1.4). We denote by H the closure of V in $L^2(\Omega)^k$.

As usual the bilinear form a (restricted to V) defines a linear unbounded operator A in H with domain $D(A) \subset H$. This operator is self-adjoint; $A + \lambda$ is self-adjoint, positive, invertible. Since by Rellich's theorem the embedding of $H^1(\Omega)$ into $L^2(\Omega)$ is compact, the inverse $(A + \lambda)^{-1}$ is compact in H. We know that we can define the powers A^s of s, $s \in \mathbb{R}$, and V is nothing other than $D(A^{1/2})$.

Since $(A + \lambda)^{-1}$ is self-adjoint, compact > 0, there exists an orthonormal family $w_j, j \in \mathbb{N}$, of H and a sequence $\lambda_j, j \in \mathbb{N}$, such that

$$\begin{cases} -\lambda < \lambda_1 \leq \lambda_2 \leq \ldots, & \lambda_j \to \infty \quad \text{as} \quad j \to \infty, \\ A w_j = \lambda_j w_j, & \forall j. \end{cases} \tag{1.5}$$

We make on λ_j and w_j the following assumptions (see Sections 4 and 5 for examples):

There exists a constant c_1 (depending on Ω, a, and V) such that

$$\lambda_j \geq c_1 \left(\frac{j}{|\Omega|}\right)^{2m/n} - \lambda, \qquad \forall j \in \mathbb{N}, \quad |\Omega| = \text{volume of } \Omega. \tag{1.6}$$

There exists a constant c_2 (depending on Ω, a, and V) such that

$$\operatorname*{Sup}_{x \in \Omega} |w_j(x)|^2 \leq c_2 |\Omega|^{-1}, \qquad \forall j \in \mathbb{N}.^1 \tag{1.7}$$

It is also assumed that for $r > \lambda$ the operator $(A + r)^{-1}$ corresponds to the solution of an elliptic boundary-value problem; hence it can be extended to all the spaces $L^s(\Omega)^k$, $1 < s < \infty$, and we assume that it possesses the maximal $2m$-regularity property in these spaces (i.e., the usual regularity for elliptic systems of the S. Agmon, A. Douglis, and L. Nirenberg [1] type). More precisely,

For every $r > \lambda$ the operator $(A + r)^{-1} \in \mathcal{L}(V', V)$ can be extended as a linear continuous operator (still denoted by $(A + r)^{-1}$) from $L^s(\Omega)^k$ into $V \cap W^{2m,s}(\Omega)^k$, $\forall s$, $1 < s < \infty$.[2] Furthermore, this operator considered as an operator in $L^2(\Omega)^k$ is positive. (1.8)

[1] In physical terms note that, by the orthonormality condition, $|w_j(x)|^2$ has dimension L^{-n}, L = a length = (say) $|\Omega|^{1/n}$, while λ_j, like λ, has dimension L^{-2m}.

[2] In the abstract setting this extension is not unique. This is already obvious in the Hilbertian case when $s = 2$. However, the extension of the operator $(A + r)^{-1}$ will be, in general, well defined by the corresponding elliptic boundary-value problem.

1.2. The Schrödinger-Type Operators

We now consider a scalar function f which belongs to $L^p(\Omega)$ for some appropriate p such that

$$\text{The multiplication by } |f|^{1/2} \text{ defines a linear continuous operator from } H^{m_0}(\Omega) \text{ into } L^2(\Omega) \text{ for some } m_0 < m. \tag{1.9}$$

A condition on p which guarantees (1.9) is

$$f \in L^p(\Omega) \text{ with } p > n/2m_0 > n/2m \text{ when } n/2m \geq 1 \text{ and } p \text{ arbitrary} > 1 \text{ (and } m > m_0 > n/2) \text{ when } n/2m < 1. \tag{1.10}$$

Indeed, by the Holder inequality,

$$\int_\Omega |f|\varphi^2\, dx \leq \left(\int_\Omega |f|^p\, dx\right)^{1/p} \left(\int_\Omega |\varphi|^{2p'}\, dx\right)^{1/p'},$$

where $p' = p/(p-1)$ is the conjugate exponent of p. If $m > n/2$, then we can find $m_0, m > m_0 > n/2$, and since, by the Sobolev embedding theorem, $H^{m_0}(\Omega)$ is included in $\mathscr{C}(\bar{\Omega})$ with a continuous injection

$$\left(\int_\Omega |\varphi|^{2p'}\, dx\right)^{1/p'} \leq |\Omega|^{1/p'} |\varphi|^2_{\mathscr{C}(\bar{\Omega})} \leq c_1' \|\varphi\|^2_{H^{m_0}(\Omega)}$$

and (1.9) follows in this case. If $m \leq n/2$ and $p > n/2m$, then we can find m_0 satisfying $p > n/2m_0 > n/2m$. Due to the Sobolev embeddings, we then have $H^{m_0}(\Omega) \subset L^q(\Omega)$, $1/q = \frac{1}{2} - m_0/n < 1/2p' = \frac{1}{2} - 1/2p$; therefore, since Ω is bounded,

$$|\varphi|_{L^{2p'}(\Omega)} \leq c_3' |\varphi|_{L^q(\Omega)} \leq c_4' |\varphi|_{H^{m_0}(\Omega)};$$

hence (1.9). □

Thanks to (1.9), (1.10) the expression

$$a(u, v) = \int_\Omega fuv\, dx \tag{1.11}$$

makes sense for every $u, v \in H^m(\Omega)^k$ and (1.11) defines a bilinear continuous form on $H^m(\Omega)^k$ (or V). We also have with (1.3)

$$\begin{aligned} a(u, u) + \int_\Omega fu^2\, dx &\geq \delta \sum_{[\alpha]=m} |D^\alpha u|^2 - c_3 \|u\|^2_{H^{m_0}(\Omega)^k} \\ &\geq \delta \|u\|^2 - \lambda |u|^2 - c_3 \|u\|^2_{H^{m_0}(\Omega)^k}, \qquad \forall u \in V, \end{aligned} \tag{1.12}$$

where c_3 is the norm of f in $\mathscr{L}(H^m(\Omega), L^2(\Omega))$. According to the classical Lemma 1.1 below, for every $\varepsilon > 0$, there exists a constant c_ε such that for every u in $H^m(\Omega)^k$

$$|u|_{H^{m_0}(\Omega)^k} \leq \varepsilon \|u\|_{H^m(\Omega)^k} + c_\varepsilon |u|.$$

In particular, there exists a constant c_5' such that for every u in V,

$$\|u\|^2_{H^{m_0}(\Omega)^k} \leq \frac{\delta}{2c_3}\|u\|^2 + c_5'|u|^2, \tag{1.13}$$

and we infer from (1.12) that

$$a(u, u) + \int_\Omega fu^2\, dx \geq \frac{\delta}{2}\|u\|^2 - c_4|u|^2, \qquad \forall u \in V, \tag{1.14}$$

$c_4 = c_3 c_5' + \lambda$. Consequently, the form

$$a(u, v) + \int_\Omega (f + r)uv\, dx \tag{1.15}$$

is bilinear continuous coercive on V for $r \geq c_4$, and the operator $(A + f + r)^{-1}$ is self-adjoint positive compact in H for such values of r.

We consider the quadratic form associated with (1.11)

$$Q_f(u) = a(u, u) + \int_\Omega (\lambda + f)u^2\, dx \tag{1.16}$$

and the numbers $\mu_j = \mu_j(f)$ defined by the min–max method

$$\mu_j(f) = \underset{\psi_1, \ldots, \psi_{j-1} \in V}{\text{Min}} \quad \underset{\substack{\varphi \in V \\ |\varphi| = 1 \\ (\varphi, \psi_i) = 0 \\ i = 1, \ldots, j-1}}{\text{Max}} \; Q_f(\varphi), \tag{1.17}$$

where $(\cdot, \cdot)$ is the usual scalar product in $L^2(\Omega)^k$. The numbers $\mu_j(f)$ are the eigenvalues of $A + f$, they form an increasing sequence bounded from below by $-c_4$, and converging to ∞ as $j \to \infty$ (see R. Courant and D. Hilbert [1], M. Reed and B. Simon [1])

$$\begin{cases} \mu_1(f) \leq \mu_2(f) \leq \mu_3(f), \ldots, \\ \mu_j(f) \to \infty \quad \text{as} \quad j \to \infty. \end{cases} \tag{1.18}$$

If f' is another function satisfying the same assumptions as f and $f \leq f'$ almost everywhere in Ω, it is then clear from (1.16), (1.17) that

$$\mu_j(f) \leq \mu_j(f'), \qquad \forall j \in \mathbb{N}. \tag{1.19}$$

Considering, in particular, the positive and negative parts of f, $f_+ = \max(f, 0)$, $f_- = \max(-f, 0)$, we see that $f = f_+ - f_- \geq -f_-$ and

$$\mu_j(-f_-) \leq \mu_j(f), \qquad \forall j \in \mathbb{N}. \tag{1.20}$$

This implies that for certain purposes we can restrict ourselves to negative functions f ($f \leq 0$ a.e. in Ω), which we will do in *most* of Section 2. The aim of Section 2 is to estimate in terms of f certain quantities related to the $\mu_j(f)$, but before that we conclude this section by stating Lemma 1.1 which was used above.

Lemma 1.1. *Let Ω be an open set of $\mathbb{R}^n$ and assume that $0 < m_0 < m$. Then for every $\varepsilon > 0$ there exists a constant c_ε depending on ε, m_0, m, and Ω such that*

$$\|u\|_{H^{m_0}(\Omega)} \leq \varepsilon \|u\|_{H^m(\Omega)} + c(\varepsilon)|u|, \qquad \forall u \in H^m(\Omega) \tag{1.21}$$

PROOF. The space $H^{m_0}(\Omega)$ is an interpolation space between $H^m(\Omega)$ and $L^2(\Omega)$

$$H^{m_0}(\Omega) = [H^m(\Omega), L^2(\Omega)]_\theta, \qquad \theta = \frac{m_0}{m}.$$

Hence (see Section II.1.3) there exists a constant c_1' depending on m_0, m, Ω such that

$$\|u\|_{H^{m_0}(\Omega)} \leq c_1' \|u\|_{H^m(\Omega)}^\theta |u|^{1-\theta}, \qquad \forall u \in H^m(\Omega).$$

Thanks to the Young inequality we then obtain, for $\alpha > 0$ arbitrary,

$$\|u\|_{H^{m_0}(\Omega)} \leq \theta\alpha^{1/\theta} \|u\|_{H^m(\Omega)} + (1-\theta)\left(\frac{c_1'}{\alpha}\right)^{1/(1-\theta)} |u|.$$

Setting $\varepsilon = \theta\alpha^{1/\theta}$, i.e., $\alpha = (\varepsilon/\theta)^\theta$, we find (1.21) with

$$c_\varepsilon = (1-\theta)(c_1')^{1-\theta}(\theta/\varepsilon)^{1/(1-\theta)}.$$ □

Of course, (1.21) induces a similar result on the product spaces $H^{m_0}(\Omega)^k$, $H^m(\Omega)^k$.

2. Spectral Estimates for Schrödinger-Type Operators

We continue the study of the Schrödinger-type operators $A + f$ introduced in Section 1. The notations are the same as in Section 1 and our aim here is to estimate the sums

$$\sum_{j=1}^{\infty} |\mu_j(f)|^\gamma, \qquad \gamma > 0, \tag{2.1}$$

in terms of f (and γ). This estimate is proved in Proposition 2.2.

2.1. The Birman–Schwinger Inequality

For every $r \in \mathbb{R}$ we denote by $N_r(f)$ *the number of eigenvalues* $\mu_j(f) \leq r$, counted with their multiplicity. It is easy to check that

$$\sum_{j=1}^{\infty} |\mu_j(f)|^\gamma = \gamma \int_0^\infty r^{\gamma-1} N_{-r}(f)\, dr, \tag{2.2}$$

and the sum (2.1) is thus naturally connected to the function $r \to N_r(f)$.

We set $A_\lambda = A + \lambda$. Our first observation is the following one:

Lemma 2.1. *The assumptions are those of Section* 1 *and we also assume that* $f \leq 0$, f *not identically equal to* 0, *and* $r < 0$ *are given. Then the number* $N_r(f)$ *of eigenvalues of* $A_\lambda + f$ *less than or equal to* r *is equal to the number of* κ's *in* $]0, 1]$ *such that* r *is an eigenvalue of* $A_\lambda + \kappa f$ *(counting multiplicities).*

PROOF. We set $N_r(f) = N$ and we write momentarily $\mu_j(\kappa) = \mu_j(\kappa f)$ so that $\mu_j(1) = \mu_j(f)$ and $\mu_j(0) = \lambda_j$. We have

$$\mu_1(1) \leq \mu_2(1) \leq \cdots \leq \mu_N(1) \leq r < \mu_{N+1}(1) < \cdots.$$

By classical results on eigenvalues (see, for instance, T. Kato [1]) the numbers $\mu_j(\kappa)$ depend continuously on κ and decrease strictly as κ increases; note that the decay property follows easily from (1.18) ($f \leq 0$) while the strict decay property is more delicate and demands that f does not vanish identically. As κ converges to 0, $\mu_1(\kappa)$ converges to $\lambda_1 + \lambda$ (> 0). Thus as κ decreases from 1 to 0, for each j, $1 \leq j \leq N$, $\mu_j(\kappa)$ increases from $\mu_j(1)$ ($\leq r \leq 0$) to $\lambda_j + \lambda$ ($\geq \lambda_1 + \lambda > 0$) and $\mu_j(\kappa)$ takes once and only once the value r. □

When $f \leq 0$ Lemma 2.1 reduces the determination of $N_r(f)$ to the determination of the number of $\kappa \in]0, 1]$ such that r is an eigenvalue of $A_\lambda + \kappa f$. This last problem in its turn will be reduced to the determination of the eigenvalues of an auxiliary operator that we now define.

For f given, satisfying the assumptions of Section 1 and in particular (1.9), (1.10), we consider the linear operator in $L^2(\Omega)^k$

$$G_r = |f|^{1/2}(A_\lambda - r)^{-1}|f|^{1/2}, \qquad r \leq 0. \tag{2.3}$$

In (2.3), $(A_\lambda - r)^{-1}$ is understood in the sense of (1.8). We have

Lemma 2.2. *Under the assumptions of Section* 1, *in particular* (1.8) *and* (1.10), G_r *is, for* $r < 0$, *defined, linear compact, and self-adjoint in* $L^2(\Omega)^k$.

PROOF. By (1.10), $|f|^{1/2}$ belongs to $L^{2p}(\Omega)^k$, $p > 1$, and if φ belongs to a bounded set of $L^2(\Omega)^k$, then $|f|^{1/2}\varphi$ belongs to a bounded set of $L^s(\Omega)^k$, $s = 2p/(1 + p) > 1$. By application of (1.8), $\xi = (A_\lambda - r)^{-1}(|f|^{1/2}\varphi)$ belongs to a bounded set of $W^{2m,s}(\Omega)^k$. First we consider the case $n < 2m$. Then by (1.10) $p > 1$ is arbitrary and by the Sobolev embeddings and the compactness theorems for Sobolev spaces, $W^{2m,s}(\Omega)$ is compactly embedded into $\mathscr{C}(\bar{\Omega})$ since $1/s - 2m/n < 1 - 2m/n < 0$; thus ξ belongs to a compact set of $\mathscr{C}(\bar{\Omega})^k$ and $|f|^{1/2}\xi$ belongs to a compact set of $L^{2p}(\Omega)^k$ and therefore of $L^2(\Omega)^k$. The compactness of G_r in $L^2(\Omega)^k$ follows in this case.

When $n \geq 2m$, we have

$$p > \frac{n}{2m_0} > \frac{n}{2m}, \qquad \frac{1}{s} - \frac{2m}{n} = \frac{1+p}{2p} - \frac{2m}{n} < \frac{1}{2} - \frac{m_0}{n}$$

and by the Sobolev embeddings and the compactness theorems for Sobolev spaces, ξ belongs to a compact set of $W^{2m_0,s}(\Omega)^k$ and thus to a compact set of $L^{q_0}(\Omega)^k$ or even $L^q(\Omega)^k$ $(1/q_0 = \frac{1}{2} - m_0/n,\ 1/q = \frac{1}{2} - m/n)$. The product $|f|^{1/2}\xi$ then belongs to a compact set of $L^\sigma(\Omega)^k$, $1/\sigma = \frac{1}{2} - m/n + 1/2p < \frac{1}{2}$ (by (1.10)). Therefore $|f|^{1/2}\xi$ belongs to a compact set of $L^2(\Omega)^k$ and the compactness of G_r also follows in this case.

It is obvious that G_r is self-adjoint. The positivity follows from the last assumption in (1.8). The lemma is proved. □

We now return to the question of estimating $N_r(f)$ with

Lemma 2.3. *The assumptions are those of Lemma 2.1. If $\psi \in V$ and $0 < \kappa \le 1$ satisfy*

$$(A_\lambda + \kappa f)\psi = r\psi, \tag{2.4}$$

then $\varphi = -|f|^{1/2}\psi \in L^2(\Omega)^k$ and satisfies

$$G_r\varphi = \frac{1}{\kappa}\varphi. \tag{2.5}$$

Conversely, if $\varphi \in L^2(\Omega)^k$ and $0 < \kappa \le 1$ satisfies (2.5), *then $\psi = -|f|^{1/2}\varphi$ is in V and satisfies* (2.4).

Proof. If $\psi \in V$ satisfies (2.4) we set $\varphi = -|f|^{1/2}\psi$. If $n < 2m$, the Sobolev embedding $H^m(\Omega) \subset \mathscr{C}(\bar{\Omega})$ shows that $\psi \in \mathscr{C}(\bar{\Omega})^n$ and $\varphi \in L^{2p}(\Omega)^k \subset L^2(\Omega)^k$; if $n \ge 2m$, $H^m(\Omega) \subset L^q(\Omega)$, $1/q = \frac{1}{2} - m/n$ for $n > 2m$, q is finite arbitrary for $n = 2m$, then φ is in $L^\sigma(\Omega)^k \subset L^2(\Omega)^k$ since

$$\frac{1}{\sigma} = \frac{1}{2p} + \frac{1}{q} = \frac{1}{2p} + \frac{1}{2} - \frac{m}{n} < \frac{1}{2}.$$

In all cases we have $\varphi \in L^2(\Omega)^k$ and (2.5) then follows by an easy algebraic computation:

$$\begin{aligned} (A_\lambda + \kappa f)\psi &= r\psi, \\ (A_\lambda - r)\psi &= -\kappa f\psi = \kappa|f|\psi = -\kappa|f|^{1/2}\varphi, \\ |f|^{1/2}(A_\lambda - r)^{-1}|f|^{1/2}\varphi &= -\frac{1}{\kappa}|f|^{1/2}\psi = \frac{1}{\kappa}\varphi. \end{aligned}$$

Conversely, if $\varphi \in L^2(\Omega)^k$ satisfies (2.5), we set

$$\psi = -\kappa\xi = -\kappa(A_\lambda - r)^{-1}(|f|^{1/2}\varphi) \in W^{2m,s}(\Omega)^k,$$

where ξ, s are as in Lemma 2.2. The derivatives of ψ of order m are in $W^{m,s}(\Omega)^k$. Since in all cases

$$\frac{1}{s} - \frac{m}{n} = \frac{1}{2p} + \frac{1}{2} - \frac{m}{n} < \frac{1}{2}.$$

$W^{m,s}(\Omega) \subset L^2(\Omega)$ and ψ is in $H^m(\Omega)^k$. It is clear that ψ is in V and (2.4) follows as above. □

Our aim is now to prove the following inequality called the Birman–Schwinger inequality (M.S. Birman [1], J. Schwinger [1]).

Proposition 2.1. *The assumptions are those of Section* 1, *and* $r < 0$ *is given.*

For any $t \in [0, 1]$ *and any* $k \geq 1$ *we have*

$$N_r(f) \leq \operatorname{Tr}((f - (1-t)r)_-^{1/2}(A_\lambda - tr)^{-1}(f - (1-t)r)_-^{1/2})^k. \tag{2.6}$$

Proof. By the previous results $N_r(f)$ is equal to the number of $\kappa \geq 1$ which are eigenvalues of G_r. Therefore, denoting by σ_j the eigenvalues of G_r, we have

$$\begin{aligned} N_r(f) = \sum_{\sigma_j \geq 1} 1 \leq \sum_{\sigma_j \geq 1} \sigma_j^k \\ \leq \sum_{j=1}^{\infty} |\sigma_j|^k \\ \leq \operatorname{Tr} G_r^k, \end{aligned}$$

and (2.6) follows for $t = 1$.

For $t \in [0, 1[$ we observe that

$$(A_\lambda + f)\psi = \kappa\psi \quad \text{and} \quad \kappa \leq r$$

is equivalent to

$$(A_\lambda + (f - (1-t)r))\psi = \rho\psi, \quad \text{with} \quad \rho = \kappa - (1-t)r \leq tr.$$

Hence

$$\begin{aligned} N_r(f) &= N_{tr}(f - (1-t)r) \\ &\leq (\text{by } (1.19)) \\ &\leq N_r(-(f - (1-t)r)_-), \end{aligned}$$

and from the previous result ((2.6) for $t = 1$) this is bounded by

$$\operatorname{Tr}\{(f - (1-t)r)_-^{1/2}(A_\lambda - tr)^{-1}(f - (1-t)r)_-^{1/2}\}^k.$$

The proof is complete. □

2.2. The Spectral Estimate

We will use the Birman–Schwinger inequality with $t = \frac{1}{2}$ and r replaced by $-r$

$$N_{-r}(f) \leq \operatorname{Tr}\left\{\left(f + \frac{r}{2}\right)_-^{1/2}\left(A_\lambda + \frac{r}{2}\right)^{-1}\left(f + \frac{r}{2}\right)_-^{1/2}\right\}^k, \tag{2.7}$$

$r > 0$, $k \geq 1$. According to a convexity inequality proved in E. Lieb and

W. Thirring [1] App. B),[1] the right-hand side of (2.7) is less than or equal to

$$\operatorname{Tr}\left(f+\frac{r}{2}\right)_-^{k/2}\left(A_\lambda+\frac{r}{2}\right)^{-k}\left(f+\frac{r}{2}\right)_-^{k}. \tag{2.8}$$

Using the eigenfunctions w_j of A we write

$$\left(A_\lambda+\frac{r}{2}\right)^{-1}u=\sum_{j=1}^{\infty}\left(\lambda_j+\lambda+\frac{r}{2}\right)^{-1}(u, w_j)w_j, \qquad \forall u\in V,$$

$$\left(A_\lambda+\frac{r}{2}\right)^{-k}u=\sum_{j=1}^{\infty}\left(\lambda_j+\lambda+\frac{r}{2}\right)^{-k}(u, w_j)w_j,$$

and setting $g=(f+r/2)_-$,

$$g^{k/2}\left(A_\lambda+\frac{r}{2}\right)^{-k}g^{k/2}u=\sum_{j=1}^{\infty}\left(\lambda_j+\lambda+\frac{r}{2}\right)^{-k}(g^{k/2}u, w_j)g^{k/2}w_j.$$

Hence the kernel of $g^{k/2}(A_\lambda+r/2)^{-k}g^{k/2}$ is

$$G(x, y)=\sum_{j=1}^{\infty}\left(\lambda_j+\lambda+\frac{r}{2}\right)^{-k}g^{k/2}(y)w_j(y)g^{k/2}(x)w_j(x),$$

and its trace is

$$\int_\Omega G(x, x)\,dx=\int_\Omega\sum_{j=1}^{\infty}\left(\lambda_j+\lambda+\frac{r}{2}\right)^{-k}g^k(x)|w_j(x)|^2\,dx. \tag{2.9}$$

We majorize this expression by using assumption (1.7) and we obtain

$$N_{-r}(f)\le c_2|\Omega|^{-1}\int_\Omega\sum_{j=1}^{\infty}\left(\lambda_j+\lambda+\frac{r}{2}\right)^{-k}g^k(x)\,dx. \tag{2.10}$$

Then using assumption (1.6) we write

$$\sum_{j=1}^{\infty}\left(\lambda_j+\lambda+\frac{r}{2}\right)^{-k}\le\sum_{j=1}^{\infty}\left(c_1\left(\frac{j}{|\Omega|}\right)^{2m/n}+\frac{r}{2}\right)^{-k}.$$

Since the function $s\to(c_1(s/|\Omega|)^{2m/n}+r/2)$ is decreasing, the last expression is less than or equal to

$$\int_0^\infty\frac{ds}{\left(c_1\left(\frac{s}{|\Omega|}\right)^{2m/n}+\frac{r}{2}\right)^k}=\frac{2^k}{r^k}\int_0^\infty\frac{ds}{\left(\frac{2c_1}{r}\left(\frac{s}{|\Omega|}\right)^{2m/n}+1\right)^k}$$

$$=|\Omega|\frac{2^k}{r^k}\left(\frac{r}{2c_1}\right)^{n/2m}\int_0^\infty\frac{ds}{(s^{2m/n}+1)^k}.$$

[1] If B and C are linear positive operators in a Hilbert space, then

$$\operatorname{Tr}(B^{1/2}CB^{1/2})^k\le\operatorname{Tr}B^{k/2}C^kB^{k/2}.$$

The last integral is finite and denoted by $\kappa_1'(m, n, k)$, provided

$$k > \frac{n}{2m}, \tag{2.11}$$

in which case we can write

$$\begin{aligned} N_{-r}(f) &\leq \kappa_2' r^{(n/2m)-k} \int_\Omega \left(f + \frac{r}{2}\right)_-^k dx, \\ \kappa_2' &= c_2 c_1^{-(n/2m)} 2^{k-(n/2m)} \kappa_1'(m, n, k). \end{aligned} \tag{2.12}$$

We can now prove the estimate announced at the beginning of this section.

Proposition 2.2. *The assumptions are those of Section* 1 *and we are given* $k \geq 1$ *and* $\gamma > 0$ *satisfying (compare with* (2.11))

$$\frac{n}{2m} < k < \gamma + \frac{n}{2m}. \tag{2.13}$$

Then there exists a constant κ_3' *such that*

$$\sum_{j=1}^{\infty} |\mu_j(f)|^\gamma \leq \kappa_3' \int_\Omega (f_-(x))^{\gamma+(n/2m)}\, dx. \tag{2.14}$$

Proof. According to (2.2)

$$\sum_{j=1}^{\infty} |\mu_j(f)|^\gamma = \gamma \int_0^\infty r^{\gamma-1} N_{-r}(f)\, dr.$$

With (2.12) we majorize this integral by

$$\begin{aligned} &\kappa_2'\gamma \int_0^\infty r^{\gamma-1+(n/2m)-k} \int_\Omega \left(f(x) + \frac{r}{2}\right)_-^k dx\, dr \\ &= \kappa_2'\gamma \int_\Omega \int_0^\infty r^{\gamma-1+(n/2m)-k} \left(f(x) + \frac{r}{2}\right)_-^k dr\, dx. \end{aligned}$$

The integration is limited to the points x, r such that $f(x) + r/2 < 0$, $f(x) < -r/2 < 0$. For x fixed such that $f(x) < 0$, we set $r = 2f_-(x)\rho$ and write

$$\left(f(x) + \frac{r}{2}\right)_-^k = (\rho - 1)_-^k f_-(x)^k,$$

$$\begin{aligned} \int_0^\infty r^{\gamma-1+(n/2m)-k} \left(f(x) + \frac{r}{2}\right)_-^k dr = \int_0^\infty & 2^{\gamma+(n/2m)-k} (f_-(x))^{\gamma+(n/2m)} \\ & \times \rho^{\gamma-1+(n/2m)-k} (\rho - 1)_-^k\, d\rho. \end{aligned}$$

This integral is finite and equal to the product of $f_-(x)^{\gamma+(n/2m)}$ by a constant κ_4', thanks to (2.13).

The result is proved with $\kappa_3' = \gamma\kappa_2'\kappa_4'$. □

Remark 2.1. For $0 < \gamma \leq 1$, an estimate of

$$\sum_{\mu_j<0} |\mu_j(f)|^\gamma$$

can be derived as follows: since

$$\mu_1 \leq \mu_2 \leq \cdots \leq \mu_N \leq 0 < \mu_{N+1} \cdots,$$

we have

$$\sum_{\mu_j<0} |\mu_j|^\gamma \geq |\mu_1|^{1-\gamma} \sum_{\mu_j<0} |\mu_j|,$$

$$\sum_{\mu_j<0} |\mu_j| \leq |\mu_1|^{1-\gamma} \cdot \left(\sum_{\mu_j<0} |\mu_j|^\gamma \right)$$

$$\leq \left(\sum_{\mu_j<0} |\mu_j|^\gamma \right)^{(1-\gamma)/\gamma} \cdot \left(\sum_{\mu_j<0} |\mu_j|^\gamma \right),$$

$$\sum_{\mu_j<0} |\mu_j| \leq \left(\sum_{\mu_j<0} |\mu_j|^\gamma \right)^{1/\gamma}.$$

We then infer from (2.14) that

$$\sum_{\mu_j<0} |\mu_j| \leq (\kappa_3')^{1/\gamma} \left(\int_\Omega (f_-(x))^{\gamma+(n/2m)}\, dx \right)^{1/\gamma}, \qquad 0 < \gamma \leq 1. \tag{2.15}$$

3. Generalization of the Sobolev–Lieb–Thirring Inequality (I)

We now prove (Theorem 3.1) one of the main results of this Appendix, i.e., the first generalization of the Sobolev–Lieb–Thirring inequalities. The reader is referred to Remarks 3.1–3.3 for several comments concerning Theorem 3.1. See also, in Section 4, another generalization of these inequalities and some consequences.

Theorem 3.1. *Let there be given a, V, H, satisfying the assumptions of Section* 1 *and, in particular,* (1.2)–(1.4) *and* (1.6)–(1.8).

Let φ_j, $1 \leq j \leq N$, *be a finite family of V which is orthonormal in H and set, for almost every* $x \in \Omega$,

$$\rho(x) = \sum_{j=1}^N |\varphi_j(x)|^2. \tag{3.1}$$

Then for every p satisfying

$$\max\left(1, \frac{n}{2m}\right) < p \leq 1 + \frac{n}{2m}, \tag{3.2}$$

there exists a constant κ_1 *such that*

$$\left(\int_\Omega \rho(x)^{p/(p-1)}\,dx\right)^{2m(p-1)/n} \leq \kappa_1\left\{\sum_{j=1}^N a(\varphi_j, \varphi_j) + \int_\Omega \rho(x)\,dx\right\}. \tag{3.3}$$

The constant κ_1 *depends on* m, n, p, Ω, *and* a, *but is independent of the family* φ_j *and of* N.[1,2]

PROOF. (i) We consider, as in Section 1.2, the Schrödinger-type operator $A + f$ with

$$f = -\alpha\rho^\beta,$$

$\alpha > 0$ chosen later and

$$\beta = \frac{1}{p-1}. \tag{3.4}$$

We see that $f \in L^p(\Omega)$, the conditions on p given in (1.10) being satisfied: this is obvious if $n \leq 2m$ since by the Sobolev embeddings $H^m(\Omega) \subset L^\infty(\Omega)$ (for $n < 2m$) or $H^m(\Omega) \subset L^q(\Omega)$, $\forall q$, $1 < q < \infty$ (for $n = 2m$); for $n > 2m$, $H^m(\Omega) \subset L^q(\Omega)$, $q = 2n/(n - 2m)$, and (3.2), (3.4) imply that $f \in L^p(\Omega)$ (i.e., $2\beta p \leq q$) and $p > n/2m$.

Since (1.9), (1.10) are satisfied we infer from (1.11) that $B = A + f$ defines a linear continuous operator from $D(A)$ into H.

(ii) We consider, as in Section V.1, the exterior product space $\bigwedge^N H$ which is endowed with its natural scalar product defined as follows:

$$(u_1 \wedge \cdots \wedge u_N; v_1 \wedge \cdots \wedge v_N) = \det\{(u_i, v_j)\}_{1 \leq i,j \leq N} \tag{3.5}$$

for every family of elements u_i, v_j in H.

As in Chapter V we associate with the linear unbounded operator A of H, the linear unbounded operator A_N of $\bigwedge^N H$:[3]

$$A_N(u_1 \wedge \cdots \wedge u_N) = (Au_1 \wedge u_2 \wedge \cdots \wedge u_N + \cdots + u_1 \wedge \cdots \wedge u_{N-1} \wedge Au_N),$$
$$\forall u_1, \ldots, u_N \in D(A). \tag{3.6}$$

It was shown in Section V.1 that the right-hand side on (3.6) depends only on $u_1 \wedge \cdots \wedge u_N$ and the notation $A_N(u_1 \wedge \cdots \wedge u_N)$ is then justified. The domain of A_N in $\bigwedge^N H$ contains (and is in fact equal to) $\bigwedge^N D(A)$.

If $u_1, \ldots, u_N, v_1, \ldots, v_N$ belong to $D(A)$ we consider

$$a_N(u_1 \wedge \cdots \wedge u_N; v_1 \wedge \cdots \wedge v_N) = (A_N(u_1 \wedge \cdots \wedge u_N); v_1 \wedge \cdots \wedge v_N). \tag{3.7}$$

[1] Of course, $\int_\Omega \rho(x)\,dx = \sum_{j=1}^N \int_\Omega |\varphi_j|^2\,dx = N$.

[2] The best constant κ_1 in (3.3) depends on the shape of Ω but not on its size; see Remark 3.3.

[3] This operator A_N acting on $\bigwedge^N H$ should not be confused with the operator A_λ considered in Section 2 ($A_\lambda = A + \lambda$).

Using (3.5), (3.6) we see that the expression in (3.7) is the sum of the determinants of the matrices

$$\left\{\begin{array}{c}\begin{pmatrix} a(u_1, v_1) & \cdots & a(u_1, v_N) \\ (u_2, v_1) & \cdots & (u_2, u_N) \\ & & \\ & & \\ (u_N, v_1) & \cdots & (u_N, v_N) \end{pmatrix} \\ \cdots\cdots\cdots\cdots\cdots\cdots \\ \begin{pmatrix} (u_1, v_1) & \cdots & (u_1, v_N) \\ & & \\ & & \\ (u_{N-1}, v_1) & \cdots & (u_{N-1}, v_N) \\ a(u_N, v_1) & \cdots & a(u_N, v_N) \end{pmatrix}\end{array}\right. . \tag{3.8}$$

Hence a_N can be extended as a symmetric bilinear continuous form on $\bigwedge^N D(A^{1/2}) = \bigwedge^N V$. Also, if $v_j = u_j$ and the u_j are orthonormals in H, then the determinants above reduce to

$$a(u_1, u_1) + \cdots + a(u_N, u_N). \tag{3.9}$$

This shows that $a_N + N\lambda$ is coercive on $\bigwedge^N V$ and $A_N + N\lambda$ is positive on H. We conclude that $(A_N + N\lambda)^{-1}$ exists, is self-adjoint compact in $\bigwedge^N H$, and possesses a sequence of eigenvectors which is orthonormal and complete in $\bigwedge^N H$. With w_j, λ_j denoting, as in Section 1, the eigenvectors and eigenvalues of A, it is easy to see that

$$A_N(w_{i_1} \wedge \cdots \wedge w_{i_N}) = (\lambda_{i_1} + \cdots + \lambda_{i_N})(w_{i_1} \wedge \cdots \wedge w_{i_N})$$

for any sequence $i_1, \ldots, i_N \in \mathbb{N}^N$ and thus the $w_{i_1} \wedge \cdots \wedge w_{i_N}$ (which are different from 0) provide a sequence of eigenvectors of A_N which is orthonormal and complete in $\bigwedge^N H$. Hence these vectors span the eigenspaces of A_N and the sums

$$\lambda_{i_1} + \cdots + \lambda_{i_N}{}^{1}$$

produce all the eigenvalues of A_N. In particular,

$$\text{The smallest eigenvalue of } A_N \text{ is } \lambda_1 + \cdots + \lambda_N. \tag{3.10}$$

(iii) Then we consider the linear unbounded operator $B = A + \lambda + f$ of H and, exactly as in (3.6), (3.7), we associate with B the linear unbounded operator B_N and the bilinear form b_N. Due to (1.9), (1.10) the operator $u \to fu$ can be extended as a compact operator in $D(A)$ and thus B is linear continuous from $D(A)$ in H, and B_N is linear continuous from $\bigwedge^N D(A)$ into $\bigwedge^N H$. The

[1] $\lambda_{i_j} = \lambda_{i_N}$ is allowed for $j \neq k$ if and only if $w_{i_j} \neq w_{i_k}$.

expression of

$$b_N(u_1 \wedge \cdots \wedge u_N; v_1 \wedge \cdots \wedge v_N) = (B_N(u_1 \wedge \cdots \wedge u_N); v_1 \wedge \cdots \wedge v_N)$$

is similar to that of $a_N(u_1 \wedge \cdots \wedge u_N; v_1 \wedge \cdots \wedge v_N)$ (i.e., (3.8) with a replaced by b).

Because of (1.14) the operator $B_N + rI_N$ is positive for $r \geq c_4 - \lambda$; hence $(B_N + rI_N)^{-1}$ exists and is self-adjoint compact for such values of r. We conclude that the quadratic form $b_N(\psi, \psi)$ is bounded from below on the set

$$\{\psi \in \textstyle\bigwedge^N V, \|\psi\|_{\bigwedge^N H} = 1\}. \tag{3.11}$$

The minimum of $b_N(\psi, \psi)$ on this set, denoted e, is the smallest eigenvalue of B_N. But, exactly as we did for (3.10), we can prove that the eigenvalues of B_N are the numbers

$$\mu_{i_1}(f) + \cdots + \mu_{i_N}(f),$$

the $\mu_j(f)$ denoting, as in Sections 1 and 2, the eigenvalues of $B = A + \lambda + f$. In particular, this is the case for

$$e = \mu_1(f) + \cdots + \mu_N(f). \tag{3.12}$$

It is easy to notice that

$$\sum_{\mu_j(f)<0} \mu_j(f) \leq e. \tag{3.13}$$

Considering now the φ_i in the statement of the theorem, we observe that $\psi = \varphi_1 \wedge \cdots \wedge \varphi_N$ is of norm 1 in $\bigwedge^N H$ since the family φ_i is orthonormal in H. Therefore, by the definition of e and (3.13),

$$b_N(\varphi_1 \wedge \cdots \wedge \varphi_N; \varphi_1 \wedge \cdots \wedge \varphi_N) \geq e \geq \sum_{\mu_j(f)<0} \mu_j(f). \tag{3.14}$$

We have an expression of $b_N(\varphi_1 \wedge \cdots \wedge \varphi_N; \varphi_1 \wedge \cdots \wedge \varphi_N)$ similar to (3.9)

$$\begin{aligned} b_N(\varphi_1 \wedge \cdots \wedge \varphi_N; \varphi_1 \wedge \cdots \wedge \varphi_N) &= \sum_{j=1}^N b(\varphi_j, \varphi_j) \\ &= \sum_{j=1}^N a(\varphi_j, \varphi_j) + \sum_{j=1}^N \int_\Omega f(x)\varphi_j^2(x)\,dx. \end{aligned}$$

Since $f = -\alpha\rho^\beta$ we find with (3.1) and (3.4) that

$$\begin{aligned} b_N(\varphi_1 \wedge \cdots \wedge \varphi_N; \varphi_1 \wedge \cdots \wedge \varphi_N) = \sum_{j=1}^N a(\varphi_j, \varphi_j) + \lambda \sum_{j=1}^N |\varphi_j|^2 \\ - \int_\Omega \alpha\rho^{p/(p-1)}\,dx. \end{aligned}$$

The right-hand side of (3.14) is bounded from below using (2.15) that we apply with $\gamma = p - n/2m$, $f(x) = -\alpha\rho(x)^\beta$ (note that $0 < \gamma \leq 1$, thanks to (3.2)).

We obtain

$$\sum_{\mu_j(f)<0} \mu_j(f) = - \sum_{\mu_j(f)<0} |\mu_j(f)|$$
$$\leq -(\kappa_3')^{1/\gamma}\left(\int_\Omega (\alpha\rho(x)^\beta)^p\,dx\right)^{1/\gamma}$$
$$\leq -(\kappa_3')^{1/\gamma}\alpha^{p/\gamma}\left(\int_\Omega \rho(x)^{p/(p-1)}\,dx\right)^{1/\gamma}.$$

Finally,

$$-\left(\kappa_3'\alpha^p\int_\Omega \rho^{p/(p-1)}\,dx\right)^{1/\gamma} \leq \sum_{j=1}^N a(\varphi_j,\varphi_j) + \lambda\sum_{j=1}^N |\varphi_j|^2 - \alpha\int_\Omega \rho^{p/(p-1)}\,dx. \tag{3.15}$$

At this point we choose α such that

$$\alpha\left(\int_\Omega \rho^{p/(p-1)}\,dx\right) = 2\left(\kappa_3'\alpha^p\int_\Omega \rho^{p/(p-1)}\,dx\right)^{1/\gamma},$$

i.e.,

$$\alpha^\gamma\left(\int_\Omega \rho^{p/(p-1)}\,dx\right)^\gamma = 2^\gamma\kappa_3'\alpha^p\int_\Omega \rho^{p/(p-1)}\,dx,$$

$$\alpha = 2^{\gamma/(\gamma-p)}(\kappa_3')^{1/(\gamma-p)}\left(\int_\Omega \rho(x)^{p/(p-1)}\,dx\right)^{(1-\gamma)/(\gamma-p)}. \tag{3.16}$$

Therefore

$$\sum_{j=1}^N a(\varphi_j,\varphi_j) + \lambda\int_\Omega \rho\,dx \geq \tfrac{1}{2}\alpha\int_\Omega \rho^{p/(p-1)}\,dx$$
$$\geq 2^{p/(p-\gamma)}\kappa_3^{1/(\gamma-p)}\left(\int_\Omega \rho^{p/(p-1)}\,dx\right)^{(p-1)/(p-\gamma)},$$

and this is just (3.3) provided we set

$$\kappa_1 = 2^{p/(\gamma-p)}(\kappa_3')^{1/(\gamma-p)} = 2^{-(np)/(2m)}(\kappa_3')^{(2m/n)}.$$

The proof is complete. □

Remark 3.1.

(i) For the applications of Theorem 3.1, the hypotheses (1.6), (1.7) are not easy to verify and (1.7) may not be true for $m > 1$. We give in Section 4 an improved form of Theorem 3.1 which does not necessitate these hypotheses.

(ii) If $V = H_0^m(\Omega)^k$, then thanks to the Poincaré inequality (see II.(1.39)), there exists a constant $c = c(\Omega)$ depending on Ω such that

$$|\varphi|^2 \leq c(\Omega)|\Omega|^{2m/n}[\![\varphi]\!]^2, \qquad \forall\varphi\in H_0^m(\Omega)^k. \tag{3.17}$$

In this case (3.3) and (1.3) yield

$$\int_\Omega \rho(x)\,dx = \sum_{j=1}^N |\varphi_j|^2 \le \frac{c(\Omega)|\Omega|^{2m/n}}{\delta} \sum_{j=1}^N a(\varphi_j, \varphi_j),$$

$$\left(\int_\Omega \rho(x)^{p/(p-1)}\,dx\right)^{2m(p-1)/n} \le \kappa_1\left(1 + \frac{c(\Omega)|\Omega|^{2m/n}}{\delta}\right)\sum_{j=1}^N a(\varphi_j, \varphi_j). \quad (3.18)$$

The initial version of these inequalities given in E. Lieb and W. Thirring [1] is of the form (3.18) with $\Omega = \mathbb{R}^n$, $m = 1$.

(iii) When p assumes its largest value, $p = 1 + n/2m$ (see (3.2)), (3.3), (3.18) become

$$\int_\Omega \rho(x)^{1+(2m/n)}\,dx \le \kappa_1\left\{\sum_{j=1}^N a(\varphi_j, \varphi_j) + \int_\Omega \rho(x)\,dx\right\}, \quad (3.19)$$

$$\int_\Omega \rho(x)^{1+(2m/n)}\,dx \le \kappa_1\left(1 + \frac{c(\Omega)|\Omega|^{2m/n}}{\delta}\right)\sum_{j=1}^N a(\varphi_j, \varphi_j). \quad (3.20)$$

Remark 3.2. For $N = 1$, and say $k = 1$, $V = H_0^m(\Omega)$, (3.3) is a particular case of the Gagliardo–Nirenberg inequalities which imply (when they are valid) the Sobolev inequalities (see E. Gagliardo [1], L. Nirenberg [1])

$$\begin{aligned}\left(\int_\Omega |\varphi|^{2p/(p-1)}\,dx\right)^{2m(p-1)/n} &\le c_1'\left(\int_\Omega |\varphi|^2\,dx\right)^s\left(\int_\Omega \sum_{[\alpha]=m} |D^\alpha\varphi|^2\,dx\right)\\ &\le \text{(with (1.3))}\\ &\le c_2'\left(\int_\Omega |\varphi|^2\,dx\right)^s a(\varphi, \varphi)\\ &\le c_2' a(\varphi, \varphi),\end{aligned}$$

where $\varphi = \varphi_1 \in H_0^m(\Omega)$, $s = (\alpha m/n)p - 1$ and, as in (1.9), $p > \max(1, n/2m)$. For instance, for $m = 1$, $k = 1$, $p = 2$, $n = 2$ or 3, $V = H_0^1(\Omega)$, $a(\varphi, \varphi) = \int_\Omega (\operatorname{grad} \varphi)^2\,dx$, we recover

$$\int_\Omega |\varphi|^4\,dx \le c_3' \int_\Omega \varphi^2\,dx \cdot \int_\Omega |\operatorname{grad} \varphi|^2\,dx, \qquad \forall \varphi \in H_0^1(\Omega) \quad \text{if } n = 2,$$

$$\left(\int_\Omega \varphi^4\,dx\right)^{2/3} \le c_4'\left(\int_\Omega \varphi^2\,dx\right)^{1/3} \cdot \int_\Omega |\operatorname{grad} \varphi|^2\,dx, \qquad \forall \varphi \in H_0^1(\Omega) \quad \text{if } n = 3.$$

If we start from the Gagliardo–Nirenberg inequality above and apply it to each of the functions φ_j given in the statement of Theorem 3.1 we find

$$\left(\int_\Omega |\varphi_j|^{2p/(p-1)}\,dx\right)^{2m(p-1)/n} \le c_5 a(\varphi_j, \varphi_j), \qquad j = 1, \ldots, N.$$

By adding these inequalities for $j = 1, \ldots, N$ we obtain

$$\sum_{j=1}^{N} \left(\int_{\Omega} |\varphi_j|^{2p/(p-1)}\, dx \right)^{2m(p-1)/n} \leq c_5' \sum_{j=1}^{N} a(\varphi_j, \varphi_j).$$

It is elementary that for some constant $c = c(N) > 0$

$$\sum_{j=1}^{N} \left(\int_{\Omega} |\varphi_j|^{2p/(p-1)}\, dx \right)^{2m(p-1)/n} \geq c(N) \left(\int_{\Omega} \left(\sum_{j=1}^{N} |\varphi_j|^2 \right)^{p/(p-1)} dx \right)^{2m(p-1)/n}$$

$$\geq c(N) \left(\int_{\Omega} \rho^{p/(p-1)}\, dx \right)^{2m(p-1)/n}.$$

In this manner we recover an analog of (3.3) with κ replaced by $c_5' c(N)^{-1}$. The important advantage of (3.3) is that κ is independent of N while $c(N) \to 0$ as $N \to \infty$ and $c_5' c(N)^{-1} \to +\infty$ as $N \to \infty$. *This improvement plays an essential role in the obtention of the sharp physical estimates in Chapter* VI.

Remark 3.3. The best constant κ in (3.3) may depend on the shape of Ω but not on its size. Assume for simplicity that $0 \in \Omega$ and let $\kappa_1(r)$ be the best constant in (3.3) for $r\Omega$, $r > 0$ (so that $\kappa_1(1) = \kappa_1$). If φ_j is an orthonormal family for the space H corresponding to Ω (denoted H_r), then $\psi_j(y) = r^{n/2} \varphi_j(ry)$ belongs to V_r (the space V for $r\Omega$) and is orthonormal in H_r. We have

$$a_r(\psi_j, \psi_j) = r^{2m} a(\varphi_j, \varphi_j),$$

$$\left(\int_{\Omega} \left(\sum_{j=1}^{N} \varphi_j(x)^2 \right)^{p/(p-1)} dx \right)^{2m(p-1)/n} = r^{2m} \left(\int_{\lambda\Omega} \left(\sum_{j=1}^{N} \psi_j(y)^2 \right)^{p/(p-1)} dy \right)^{2m(p-1)/n}$$

Thus

$$\left\{ \int_{\lambda\Omega} \left(\sum_{j=1}^{N} \psi_j(y)^2 \right)^{p/(p-1)} dy \right\}^{2m(p-1)/n} \leq \kappa_1 \left\{ \sum_{j=1}^{N} a_r(\varphi_j, \varphi_j) + \sum_{j=1}^{N} |\psi_j|^2 \right\}$$

and $\kappa_1(r) \leq \kappa_1(1) = \kappa_1$. We could prove in a similar manner that $\kappa_1 \leq \kappa_1(r)$, $\forall r > 0$, and finally $\kappa_1(r) = \kappa_1$, $\forall r > 0$.

4. Generalization of the Sobolev–Lieb–Thirring Inequality (II)

As indicated in Remark 3.1(i), Theorem 3.1 is not convenient for the applications since the hypotheses (1.6), (1.7) are not easy to verify when $m > 1$ and (1.7) may not be true. Theorem 4.1 proved hereafter is an improved form of Theorem 3.1 which does not necessitate (1.6), (1.7). We achieve this improvement in two steps: first, we consider the case where $V = H^m_{\text{per}}(\Omega)^k$, $\Omega = (]0, L[)^n$, and the coefficients of $\mathfrak{A}$ do not depend on x; the eigenfunctions of A are easily expressed in terms of the trigonometric polynomials, (1.6), (1.7) are easy to verify, and we apply Theorem 3.1. In particular, we obtain (3.3) and (3.8) for functions $\varphi_j \in H_0^m(\Omega)^k \subset H_{\text{per}}^m(\Omega)^k$. The second step of the proof is based on an extension argument which allows us to reduce the general case to this particular case.

The proof below follows J.M. Ghidaglia, M. Marion, and R. Temam [1].

There is no need here to consider a general operator $\mathfrak{A}$ and we will simply set

$$a(u, v) = \sum_{\substack{\alpha \in \mathbb{N}^n \\ [\alpha]=m}} \int_\Omega D^\alpha u D^\alpha v \, dx, \qquad \forall u, v \in H^m(\Omega)^k, \tag{4.1}$$

$$\mathfrak{A}u = (-1)^m \sum_{\substack{\alpha \in \mathbb{N}^n \\ [\alpha]=m}} D^{2\alpha} u = (-1)^m \Delta^m u. \tag{4.2}$$

Hence (1.3) holds with $\delta = 1$.

For simplicity of notation we denote by D^m the family of derivatives D^α of order m

$$D^\alpha, \qquad \alpha \in \mathbb{N}^n, \qquad [\alpha] = \alpha_1 + \cdots + \alpha_n = m. \tag{4.3}$$

In particular,

$$|D^m u|^2 = \sum_{\substack{\alpha \in \mathbb{N}^n \\ [\alpha]=m}} |D^\alpha u|^2 = a(u, u). \tag{4.4}$$

4.1. The Space-Periodic Case

Our aim is to show

Proposition 4.1. *The hypotheses* (*and conclusions*) *of Theorem* 3.1 *hold true if a is given by* (4.1), *and* $V = H_{\text{per}}^m(\Omega)^k$, $\Omega = x_0 + (]0, L[)^n$.

Proof. The main hypotheses to verify are (1.6), (1.7), (1.8); (1.8) follows immediately from the regularity theory of elliptic equations: the matrix of coefficients of $\mathfrak{A}$ is diagonal and for each component of u the boundary condition is the space periodicity (see S. Agmon, A. Douglis, and L. Nirenberg [1]).

We observed in Remark 3.3 that (3.3) is invariant by translation and homothety of Ω. Hence we can assume that $\Omega = (]0, 2\pi[)^n$. The eigenfunctions w_j of A are easy to determine: they are the functions

$$\varphi_{i,l}, \psi_{i,l}, \qquad 1 \le i \le k, \quad l \in \mathbb{N}^n,$$

$$\varphi_{i,l} = \frac{1}{(2\pi)^{n/2}} \cos(l \cdot x) e_i, \qquad \psi_{i,l} = \frac{1}{(2\pi)^{n/2}} \sin(l \cdot x) e_i,$$

where $(e_1, \dots, e_k)$ is the canonical basis of $\mathbb{R}^k$. The corresponding eigenvalues are the numbers

$$\Lambda_{i,l} = \sum_{\substack{\alpha \in \mathbb{N}^n \\ [\alpha]=m}} l_1^{2\alpha_1} \dots l_n^{2\alpha_n}.$$

If we order these numbers counting multiplicities

$$0 = \lambda_1 \leq \lambda_2 \leq \dots, \lambda_j \to \infty \quad \text{as} \quad j \to \infty,$$

and denote by w_j, $j \in \mathbb{N}$, the corresponding sequence of eigenfunctions, (1.7) is then clearly satisfied and it remains to verify (1.6).

For the proof of (1.6), we define for every $p \in \mathbb{N}$,

$$\mathscr{E}_p = \left\{ l \in \mathbb{N}^n, \sum_{\substack{\alpha \in \mathbb{N}^n \\ [\alpha]=m}} l_1^{2\alpha_1} + \dots + l_n^{2\alpha_n} \leq p \right\},$$

and we denote by N_p the cardinal of $\mathscr{E}_p$.

If $l \in \mathscr{E}_p$, then $l_i^{2m} \leq p$ and $l_i \leq p^{1/2m}$, $\forall i = 1, \dots, n$, so that

$$\mathscr{E}_p \subset [0, p^{1/2m}]^n$$

and

$$N_p \leq 2^n(1 + p^{1/2m})^n. \tag{4.5}$$

On the other hand, by definition $\mathscr{E}_p$ consists of $\lambda_1, \dots, \lambda_{N_p}$; thus

$$\lambda_{N_p} \leq p, \qquad \lambda_{N_p+1} \geq p + 1. \tag{4.6}$$

Since the function $q \to \lambda_q$ is increasing,

$$\lambda_{1+2^n(1+p^{1/2m})^n} \geq \lambda_{1+N_p} \geq p + 1.$$

Now for every $j \in \mathbb{N}$, $j \geq 1$, we denote by $q = q(j)$ the largest integer q such that $1 + 2^n(1 + q^{1/2m})^n \leq j$; therefore

$$j < 1 + 2^n(1 + (1 + q)^{1/2m})^n,$$

$$\lambda_j \geq \lambda_{1+2^n(1+q^{1/2m})^n} \geq q + 1 \geq \left(\frac{(j-1)^{1/n}}{2} - 1 \right)^{2m};$$

(1.6) follows. □

Remark 4.1. Since $H_0^m(\Omega) \subset H_{\text{per}}^m(\Omega)$, Ω as in Proposition 4.1, we infer from Remark 3.1(ii) and Proposition 4.1 that

If φ_j, $j = 1, \dots, N$, is a family of elements of $H_0^m(\Omega)^k$, which is orthonormal in $L^2(\Omega)^k$, $\Omega = x_0 + (]0, L[)^n$, then

$$\left(\int_\Omega \rho(x)^{p/(p-1)} \, dx \right)^{2m(p-1)/n} \leq \kappa_1'' \sum_{j=1}^{N} a(\varphi_j, \varphi_j), \tag{4.7}$$

where ρ, m, n, p are as in Theorem 3.3, and κ_1'' is a constant depending only on m, n, p, and k.

4.2. The General Case

We now consider the general case. We are given an open bounded set Ω of $\mathbb{R}^n$ with boundary Γ, and we assume that Ω is of class $\mathscr{C}^m$ or more simply that it enjoys the m-prolongation property stated in II.(1.9). The form a is that given in (4.1).

Theorem 4.1. *The hypotheses are the same as in Theorem 3.1. We assume, furthermore, that Ω is of class $\mathscr{C}^{2m}$ or that it enjoys the prolongation property* II.(1.9), *and we denote by $|\Omega|$ the volume of Ω.*

Let φ_j, $1 \le j \le N$, be a finite family of $H^m(\Omega)^k$ which is orthonormal in $L^2(\Omega)^k$ and set, for almost every $x \in \Omega$,

$$\rho(x) = \sum_{j=1}^{N} |\varphi_j(x)|^2. \tag{4.8}$$

Then for every p satisfying

$$\max\left(1, \frac{n}{2m}\right) < p \le 1 + \frac{n}{2m}, \tag{4.9}$$

there exists a constant κ_2 such that

$$\left(\int_\Omega \rho(x)^{p/(p-1)}\, dx\right)^{2m(p-1)/n} \le \frac{\kappa_2}{|\Omega|^{2m/n}} \int_\Omega \rho(x)\, dx + \kappa_2 \sum_{j=1}^{N} \int_\Omega |D^m \varphi_j|^2\, dx.^1 \tag{4.10}$$

The constant κ_2 depends on m, n, p; it also depends on the shape of Ω but not on its size. It is independent of the family φ_j and of N.

The proof of Theorem 4.1 is given below in Section 4.3; before that we give a corollary and some remarks.

Remark 4.2. If p assumes its largest value, $p = 1 + n/2m$ (see (4.9)), (4.10) becomes

$$\int_\Omega \rho(x)^{1+(2m/n)}\, dx \le \kappa_2 \sum_{j=1}^{N} \int_\Omega |D^m \varphi_j|^2\, dx. \tag{4.11}$$

Remark 4.3. As in Remark 3.1(ii), using the Poincaré inequality we see that if the functions φ_j belong to $H_0^m(\Omega)^k$ (4.10) yields

$$\left(\int_\Omega \rho(x)^{p/(p-1)}\, dx\right)^{2m(p-1)/n} \le \kappa_3 \sum_{j=1}^{N} \int_\Omega |D^m \varphi_j|^2\, dx, \tag{4.12}$$

$\kappa_3 = \kappa_1(1 + c(\Omega)|\Omega|^{2m/n})$, $c(\Omega)$ as in (3.17); for $p = n/2m + 1$, (4.12) becomes

$$\int_\Omega \rho(x)^{1+(2m/n)}\, dx \le \kappa_3 \sum_{j=1}^{N} \int_\Omega |D^m \varphi_j|^2\, dx. \tag{4.13}$$

[1] Of course, $\int_\Omega \rho(x)\, dx = N$.

Remark 4.4. The inequality proved in E. Lieb and W. Thirring [1] is (4.12) with $m = 1$ and $\Omega = \mathbb{R}^n$.

An extension of Theorem 4.1 to the unbounded case, not needed in this volume, can be found in J.M. Ghidaglia, M. Marion, and R. Temam [1]. See also A.A. Ilyin [1] for Lieb-Thirring inequalities on the sphere and other results. □

Theorem 4.1 implies the following useful corollary:

Corollary 4.1. *We assume, as in Theorem* 4.1, *that* Ω *is an open bounded set of* $\mathbb{R}^n$ *which is of class* $\mathscr{C}^m$ *or which enjoys the prolongation property* II.(1.9), *and we denote by* $|\Omega|$ *the volume of* Ω.

Then if φ_j, $1 \le j \le N$, *is a finite family of* $H^m(\Omega)^k$ *which is orthonormal in* $L^2(\Omega)^k$, *we have*

$$|\Omega|^{2m/n} \sum_{j=1}^{N} \int_\Omega |D^m \varphi_j|^2 \, dx \ge \kappa_4 N^{1+2m/n} - \kappa_5 N. \tag{4.14}$$

If the functions φ_j *belong to* $H_0^m(\Omega)^k$ *then* (4.14) *holds with* $\kappa_5 = 0$.

The constants κ_4, κ_5 *depend on* m, n *and on the shape of* Ω. *They are independent of the size of* Ω, *of* N, *and of the* φ_j*'s.*

Proof. The notations are those of Theorem 4.1. By the Hölder inequality

$$\begin{aligned} N = \int_\Omega \rho \, dx &\le |\Omega|^{2m/(n+2m)} \left(\int_\Omega \rho^{1+(2m/n)} \right)^{n/(n+2m)} \\ &\le \text{(by (4.10), (4.11))} \\ &\le |\Omega|^{2m/(n+2m)} \left(\frac{\kappa_2 N}{|\Omega|^{2m/n}} + \kappa_2 \sum_{j=1}^{N} \int_\Omega |D^m \varphi_j|^2 \, dx \right)^{n/(n+2m)}. \end{aligned}$$

Hence

$$\sum_{j=1}^{N} \int_\Omega |D^m \varphi_j|^2 \, dx \ge \frac{1}{\kappa_2} \frac{N^{1+2m/n}}{|\Omega|^{2m/n}} - \frac{N}{|\Omega|^{2m/n}},$$

and we obtain (4.14) with $\kappa_4 = 1/\kappa_2$, $\kappa_5 = 1$. In order to obtain (4.14) with $\kappa_5 = 0$ when the φ_j belong to $H_0^m(\Omega)^k$, it suffices to use (4.12) instead of (4.11). □

Remark 4.5. Inequalities similar to (4.14) have been derived in Chapter VI (see Lemma 2.1) in a different more algebraic manner which relies more directly on the results and methods of Chapter V.

We now give the proof of Theorem 4.1.

4.3. Proof of Theorem 4.1

(i) The closure of Ω is included in a hypercube Q_1 of edge $2L$, of the form $x_0 + (]-L, L[)^n$. In its turns this hypercube Q_1 is included in a larger hypercube $Q = x_0 + (]-2L, 2L[)^n$.

The hypotheses ensure the existence of linear continuous operators R_1, R mapping, respectively, $H^m(\Omega)^k$ into $H_0^m(Q_1)^k$ and into $H_0^m(Q)^k$ with

$$(R\varphi)(x) = (R_1\varphi)(x) = \varphi(x), \qquad \forall\varphi \in H^m(\Omega)^k, \quad \text{a.e.} \quad x \in \Omega, \tag{4.15}$$

$$(R\varphi)(x) = 0, \qquad \forall\varphi \in H^m(\Omega)^k, \quad \text{a.e.} \quad x \in Q\backslash Q_1. \tag{4.16}$$

Indeed II.(1.9) (or Ω of class $\mathscr{C}^m$) guarantees the existence of a prolongation operator $\Pi \in \mathscr{L}(H^m(\Omega), H^m(\mathbb{R}^n))$ and $R_1\varphi$ is just the product of $\Pi\varphi$ with a $\mathscr{C}^\infty$ cut-off function $\theta = 1$ in a neighborhood of Ω and whose support is compactly embedded in Q_1. Then $R\varphi$ is the function equal to $R_1\varphi$ on Q_1 and to 0 and $Q\backslash Q_1$.

We can also choose the operator Π above so that, furthermore, $\Pi \in \mathscr{L}(L^2(\Omega)^k, L^2(\mathbb{R}^n)^k)$ and then consequently $R \in \mathscr{L}(L^2(\Omega)^k, L^2(Q)^k)$. We deduce from this the existence of two numbers $r_0, r_1 > 0$, such that

$$\int_Q |Ru|^2\,dx \le r_0^2 \int_\Omega |u|^2\,dx, \tag{4.17}$$

$$\int_Q |D^mRu|^2\,dx \le r_1^2 \int_\Omega \left(|D^mu|^2 + \frac{1}{|\Omega|^{2m/n}}|u|^2\right)dx. \tag{4.18}$$

It is easy to check that r_0, r_1 are invariant by homothety, i.e., they are the same for Ω, Q and $\lambda\Omega$, λQ, $\forall\lambda > 0$; they depend therefore on the shape of Ω but not on its size.

(ii) We then consider a ball B_η, included in $Q\backslash\bar{Q}_1$, of radius ηL, $\eta > 0$, and we denote by w_j, $j = 1, \ldots, m$, the sequence of eigenfunctions of the operator

$$\mathfrak{A}u = (-1)^m \sum_{\substack{\alpha\in\mathbb{N}^n \\ [\alpha]=m}} D^{2\alpha}u = (-1)^m\Delta^m \tag{4.19}$$

associated with the Dirichlet boundary conditions. Denoting λ_j the eigenvalue associated with w_j, we thus have

$$w_j \in H_0^m(B_\eta)^k, \qquad \mathfrak{A}w_j = \lambda_j w_j, \qquad j = 1, \ldots. \tag{4.20}$$

We assume, this is licit, that the sequence w_j is orthonormal in $L^2(B_\eta)^k$, and we again denote by w_j the function equal to w_j on B_η and to 0 on $Q\backslash B_\eta$; we observe that the prolongated function w_j belongs to $H_0^m(\Omega)^k$.

(iii) The proof now consists of applying Theorem 3.1 to a well-chosen family of elements $\psi_1, \ldots, \psi_N$ of $H_0^m(\Omega)^k$. These elements are searched of the form

$$\begin{cases} \psi_1 = R\varphi_1 + \alpha_{11} w_1, \\ \psi_2 = R\varphi_2 + \alpha_{21} w_1 + \alpha_{22} w_2, \\ \vdots \\ \psi_N = R\varphi_N + \alpha_{N1} w_1 + \cdots + \alpha_{NN} w_N. \end{cases} \tag{4.21}$$

We will also request that $\alpha_{ii} > 0$, $\forall i$, and

$$(\psi_i, \psi_j)_{L^2(Q)^k} = 2r_0^2 \delta_{ij}. \tag{4.22}$$

We recall that the supports of the functions φ_i and w_j do not intersect and the families $\{\varphi_i\}_{1 \le i \le N}$, $\{w_j\}_{1 \le j \le N}$ are, respectively, orthonormal. It is then easy to express (4.22) in terms of the φ_j and α_{ij}; we find

$$(\psi_i, \psi_j)_{L^2(Q)^k} = (R\psi_i, R\psi_j)_{L^2(Q)} + \sum_{s=1}^{N} \alpha_{is}\alpha_{js} = 2r_0^2 \delta_{ij}. \tag{4.23}$$

Considering the matrices (b_{ij}), (α_{ij})

$$b_{ij} = 2r_0^2 \delta_{ij} - (R\psi_i, R\psi_j)_{L^2(Q)^k}, \tag{4.24}$$

$\alpha_{ij} = 0$ for $j > i$, and defined above for $j \le i$, we see that (4.23) is equivalent to $b = \alpha \cdot {}^t\alpha$, and b being given, we ought to find the lower triangular matrix α such that $b = \alpha \cdot {}^t\alpha$. This is easy; α exists if b is symmetric positive definite. If we request, furthermore, that $\alpha_{ii} > 0$, then α is unique. It is clear that b is symmetric, and in order to check that b is positive we consider $\xi \in \mathbb{R}^N$ and form

$$q(\xi) = \sum_{i,j=1}^{N} b_{ij}\xi_i\xi_j = 2r_0^2 \sum_{i=1}^{N} \xi_i^2 - \sum_{i,j=1}^{N} \xi_i\xi_j(R\psi_i, R\psi_j)_{L^2(Q)}.$$

But

$$\begin{aligned} \sum_{i,j=1}^{N} \xi_i\xi_j(R\psi_i, R\psi_j)_{L^2(Q)} &= \left| R\left(\sum_{i=1}^{N} \xi_i\varphi_i\right)\right|_{L^2(Q)^k}^2 \\ &\le \text{(by (4.17))} \\ &\le r_0^2 \left|\sum_{i=1}^{N} \xi_i\varphi_i\right|_{L^2(Q)^k}^2 \\ &\le \text{(since the } \varphi_i \text{ are orthonormal in } L^2(Q)^k) \\ &\le r_0^2 \sum_{i=1}^{N} \xi_i^2. \end{aligned}$$

We conclude that $q(\xi) \ge r_0^2 \sum_{i=1}^{N} \xi_i^2$ and the matrix b is positive denfinite.

(iv) Now we observe that the functions $\psi_j/\sqrt{2}r_0$, $j = 1, \ldots, N$, are orthonormal in $L^2(Q)^k$ and belong to $H_0^m(\Omega)^k$. Theorem 3.1 is then applicable with

$H = L^2(Q)^k$, $V = H_0^m(\Omega)^k$, $a(u, v)$ equal to the bilinear form (4.1), m, n, p, N unchanged. This gives

$$\left(\int_Q \left(\sum_{j=1}^N |\psi_i(x)|^2\right)^{p/(p-1)} dx\right)^{2m(p-1)/n} \leq \kappa_1 (2r_0^2)^{(2mp/n)-1} \sum_{j=1}^n \int_Q |D^m \psi_j(x)|^2 \, dx. \tag{4.25}$$

Since $\psi_j = \varphi_j$ on Ω, the left-hand side of (4.25) is larger than

$$\left(\int_\Omega \left(\sum_{j=1}^N |\varphi_j(x)|^2\right)^{p/(p-1)} dx\right)^{2m(p-1)/n}$$

In order to majorize the right-hand side of (4.25) we recall that the supports of the functions φ_j and w_i do not intersect and that

$$\sum_{[\alpha]=m} \int_Q D^\alpha w_i D^\alpha w_j \, dx = (-1)^m \int_Q \mathscr{A} w_i w_j \, dx = \lambda_i \delta_{ij}.$$

Using relations (4.21) we then write

$$\begin{aligned}\int_Q |D^m \psi_j|^2 \, dx &= \int_Q \left| D^m R\varphi_j + \sum_{i=1}^j \alpha_{ji} D^m w_i \right|^2 dx \\ &= \int_Q \left(|D^m R\varphi_j|^2 + \sum_{i=1}^j \alpha_{ji}^2 |D^m w_i|^2 \right) dx.\end{aligned}$$

But

$$\begin{aligned}\alpha_{ji} &= (\psi_j, w_i)_{L^2(Q)}, \\ \alpha_{ji}^2 &\leq |\psi_j|_{L^2(Q)}^2 |w_i|_{L^2(Q)}^2 = 2r_0^2.\end{aligned}$$

Finally, using (4.18),

$$\left(\int_\Omega \left(\sum_{j=1}^N |\varphi_j(x)|^2\right)^{p/(p-1)} dx\right)^{2m(p-1)/n}$$
$$\leq \kappa_1 (2r_0^2)^{(2mp/n)-1} \cdot \sum_{j=1}^N \left(r_1^2 \int_\Omega |D^m \varphi_j|^2 \, dx + \frac{r_1^2}{|\Omega|^{2m/n}} \int_\Omega |\varphi_j|^2 \, dx + 2r_0^2 \sum_{i=1}^j \lambda_i \right). \tag{4.26}$$

In (4.26) η does not appear explicitly except in λ_i which we now write as $\lambda_i(B_\eta)$ to emphasize the dependence on η. But we know that $\lambda_i(B_\eta) = \eta^{-2m}\lambda_i(B_1)$. Hence if we let $\eta \to \infty$ in (4.26) we obtain exactly the same inequality with the λ_i's removed We have thus proved (4.10) with

$$\kappa_2 = \kappa_1 r_1^2 (2r_0^2)^{(2mp/n)-1}.$$

Like r_0, r_1 and κ_1, κ_2 depends only on m, n, p and on the shape of Ω. It does not depend on the size of Ω, or on the family φ_j, or on N. □

5. Examples

In this section we describe some examples of the application of Theorem 4.1. For each example we describe the differential operator $\mathfrak{A}$ and the form a, and write explicitly the inequalities following from Theorem 3.1 or Theorem 4.1 and Remark 4.2 (the Dirichlet case). Many more examples appear in Chapter VI.

EXAMPLE 5.1. In this example $\mathfrak{A} = -\Delta$, Ω is a bounded domain of $\mathbb{R}^n$, the space V is $H_0^1(\Omega)$ ($k = 1, m = 1$). We set

$$a(u, v) = \int_\Omega \operatorname{grad} u \cdot \operatorname{grad} v \, dx. \tag{5.1}$$

By application of (4.12) we find

$$\left(\int_\Omega \left(\sum_{j=1}^N |\varphi_j(x)|^2\right)^{p/(p-1)} dx\right)^{2(p-1)/n} \leq \kappa \sum_{j=1}^N \int_\Omega |\operatorname{grad} \varphi_j|^2 \, dx, \tag{5.2}$$

where $\varphi_1, \ldots, \varphi_N$ belong to $H_0^1(\Omega)$ and are orthonormal in $L^2(\Omega)$, and

$$\frac{n}{2} < p \leq 1 + \frac{n}{2}.$$

For $p = 1 + n/2$

$$\int_\Omega \left(\sum_{j=1}^N |\varphi_j(x)|^2\right)^{1+(2/n)} dx \leq \kappa \sum_{j=1}^N \int_\Omega |\operatorname{grad} \varphi_j|^2 \, dx. \tag{5.3}$$

EXAMPLE 5.2.

(i) The following example is useful for the Navier–Stokes equations: Ω is an arbitrary bounded domain of $\mathbb{R}^n$, $k = n$, $\mathfrak{A} = -\Delta$,

$$\begin{aligned} a(u, v) &= \sum_{i=1}^n \int_\Omega D_i u \cdot D_i v \, dx, \\ V &= \{v \in H_0^1(\Omega)^n, \operatorname{div} v = 0\}, \\ H &= \text{the closure of } V \text{ in } L^2(\Omega)^k \\ &= \{v \in L^2(\Omega)^n, \operatorname{div} v = 0, v \cdot \mathbf{v} = 0 \text{ on } \partial\Omega\}, \end{aligned}$$

$\mathbf{v}$ being the unit outward normal on $\partial\Omega$ (see (2.8), Chapter III).

Given a family $\{\varphi_j\}$ of elements of V which is orthonormal in H, we just consider the φ_j as a family of elements of $H_0^1(\Omega)^k$ which is orthonormal in $L^2(\Omega)^k$ and we obtain the inequalities (5.2), (5.3) by the arguments of Example 5.1 which extend obviously to systems.

(ii) We extend Example 5.1 to more general differential operators. Hence

Ω is an arbitrary bounded domain of $\mathbb{R}^n$, $k = m = 1$,

$$\mathfrak{A}u = -\sum_{i,j=1}^{n} D_i(a_{ij}(x)D_j u),$$

$$a(u, v) = \sum_{i,j=1}^{n} \int_\Omega a_{ij}(x)D_i u(x)D_j v(x)\,dx.$$

The coefficients a_{ij} are assumed to be smooth functions in Ω and we assume that there exists $\delta > 0$ such that

$$\sum_{i,j=1}^{n} a_{ij}(x)\xi_i\xi_j \geq \delta \sum_{i=1}^{n} \xi_i^2, \qquad \forall \xi = (\xi_1, \ldots, \xi_n) \in \mathbb{R}^n.$$

This guarantees (1.3), i.e.,

$$\delta \int_\Omega |\text{grad}\, u|^2\,dx \leq a(u, u). \tag{5.4}$$

If φ_j is a family of elements of $H_0^1(\Omega)$ which is orthonormal in $L^2(\Omega)$, then (5.2) and (5.3) are valid and the combination of these inequalities with (5.4) provides the analogs of (3.3) and (3.17).

Example 5.3. Here $\mathfrak{A} = -\Delta$, Ω is the cube $(]-L, L[)^n$, $k = 1$, $m = 1$. The space V is of the space of periodic functions in Ω with 0 average

$$\int_\Omega u(x)\,dx = 0. \tag{5.5}$$

The periodicity condition means that u has the same trace on corresponding points of $\partial\Omega$. We set

$$a(u, v) = \int_\Omega \text{grad}\, u \cdot \text{grad}\, v\,dx$$

and (5.5) ensures that a is coercive on V (see II.(1.32)).

Hence if $\varphi_1, \ldots, \varphi_N$ belong to $H^1(\Omega)$, have 0 average on Ω, are Ω-periodic, and are orthonormal in $L^2(\Omega)$, then (4.10) and (4.11) are valid. By utilization of the Poincaré inequality as in Remark 4.2 we then obtain (5.2) and (5.3).

Example 5.4. In this example $n = 1$, $\Omega =]-L, L[$, $\mathscr{A} = D^4$, $k = 1$, $m = 2$,

$$a(u, v) = \int_\Omega D^2 u D^2 v\,dx. \tag{5.6}$$

The space V is the space of even functions u in $H^2(\Omega)$ such that

$$\frac{du}{dx}(L) = \frac{du}{dx}(-L) \qquad \left(D = \frac{d}{dx}\right), \tag{5.7}$$

and

$$\int_{-L}^{L} u(x)\,dx = 0. \tag{5.8}$$

Note that since the functions in V are even, we also have

$$u(L) = u(-L) \tag{5.9}$$

and

$$\int_{-L}^{L} xu(x)\,dx = 0. \tag{5.10}$$

Without using (5.7)–(5.10), we can apply Theorem 4.1 to a family of functions φ_j in $H^2(\Omega)$ which is orthonormal in $L^2(\Omega)$. We obtain

$$\left(\int_{-L}^{L}\left(\sum_{j=1}^{N}|\varphi_j(x)|^2\right)^{p/(p-1)} dx\right)^{4(p-1)} \le \kappa_2 \sum_{j=1}^{N}\int_{-L}^{L}\left|\frac{d^2\varphi_j}{dx^2}\right|^2 dx + \kappa_2 \sum_{j=1}^{N}\int_{-L}^{L}|\varphi_j|^2\,dx.$$

If the functions φ_j belong to V we can take advantage of (5.7)–(5.10), using the Poincaré inequality as in Remark 4.2. We find

> If $\varphi_1, \ldots, \varphi_N$ belong to $H^2(\Omega)$, are even, and satisfy (5.7)–(5.10), then

$$\left(\int_{-L}^{L}\left(\sum_{j=1}^{N}|\varphi_j(x)|^2\right)^{p/(p-1)} dx\right)^{4(p-1)} \le \kappa \sum_{j=1}^{N}\int_{-L}^{L}\left|\frac{d^2\varphi_j}{dx^2}\right|^2 dx, \tag{5.11}$$

for $1 < p \le \frac{5}{4}$, the constant κ depending only on p. For $p = \frac{5}{4}$ we obtain

$$\left(\int_{-L}^{L}\left(\sum_{j=1}^{n}|\varphi_j(x)|^2\right)^{5} dx\right) \le \kappa \sum_{j=1}^{n}\int_{-L}^{L}\left|\frac{d^2\varphi_j}{dx^2}\right|^2 dx, \tag{5.12}$$

where κ is now an absolute constant.

Bibliography

For this second edition we did not attempt to produce an exhaustive bibliography. We only updated the references from the first edition and included new references directly related to the additional material.

The interested reader can consult the lists of references on nonlinear phenomena which are available in the form of preprints or by electronic mail; in particular the list compiled by the AHPCRC at University of Minnesota and edited by G. Sell [1], or the list compiled by UK Nonlinear News and available on e-mail at ucesnl@ucl. ac.uk.

F. Abergel

[1] Attractor for a Navier–Stokes flow in an unbounded domain, *Math. Model. Numer. Anal.* (M2AN), **23**, 3 (1989), 359–370.

[2] Existence and finite dimensionality of the global attractor for evolution equations on unbounded domains, *J. Differential Equations*, **83**, 1 (1990), 85–108.

J. Ablowitz and H. Segur

[1] *Solitons and the Inverse Scattering Transform*, SIAM Studies in Applied Mathematics, Vol. 4, SIAM, Philadelphia, 1981.

M. Abounouh

[1] Asymptotic behavior of a weakly damped Schrödinger equation in dimension two, *Appl. Math. Lett.*, **6**, 6 (1993), 29–32.

R. Abraham and J.E. Marsden

[1] *Foundations of Mechanics*, 2nd edn., Benjamin/Cummings, London, 1978.

R.S. Adams

[1] *Sobolev Spaces*, Academic Press, New York, 1975.

S. Agmon

[1] *Lectures on Elliptic Boundary Value Problems*, Mathematical Studies, Van Nostrand, New York, 1965.

S. Agmon, A. Douglis, and L. Nirenberg

[1] Estimates near the boundary for solutions of elliptic partial differential equations satisfying general boundary conditions, I, II, *Comm. Pure Appl. Math.*, **12** (1959), 623–727; **17** (1964), 35–92.

S. Agmon and L. Nirenberg
[1] Lower bounds and uniqueness theorems for solutions of differential equations in a Hilbert space, *Comm. Pure Appl. Math.*, **20** (1967), 207–229.

M.S. Agranovitch
[1] Series in the root vectors of operators that are very close to being self-adjoint, *Funct. Anal. Appl.*, **4** (1977), 296–299, Translated from *Funktsional'nyi Analizi Eg o Prilozheniga*, **11** (1977), 65–67.

G.P. Agrawal
[1] *Nonlinear Fiber Optics*, Academic Press, San Diego, CA, 1989.

G.P. Agrawal and W. Boyd
[1] *Contemporary Nonlinear Optics*, Academic Press, New York, 1992.

S.A. Akmanov and M.A. Vorontsov
[1] Bistabilities, instabilities and chaos in passive nonlinear optical systems, *Nonlinear Waves*, **1**, 92–102, Springer-Verlag, Berlin, 1989.

S. Arimoto
See J. Nagumo, S. Arimoto, and S. Yoshizawa.

V.I. Arnold
[1] *Equations Différentielles Ordinaires*, Editions Mir, Moscow, 1974.

V.I. Arnold and A. Avez
[1] *Ergodic Problems of Classical Mechanics*, Benjamin, New York, 1968.

J. Arrieta, A. Carvalho, and J.K. Hale
[1] A damped hyperbolic equation with critical exponent, *Comm. Partial Differential Equations*, **17** (5–6) (1992), 841–866.

P. Atten, J.G. Caputo, B. Malraison, and Y. Gagne
[1] Détermination de dimension d'attracteurs pour différents écoulements, *J. Méc Thér. Appl.*, numéro spécial (1984), 133–156.

T. Aubin
[1] *Nonlinear Analysis on Manifolds, Monge Ampere Equations*, Springer-Verlag, New York, 1982.

A. Avez
See V.I. Arnold and A. Avez.

A. Avez and Y. Bamberger
[1] Mouvements sphériques des fluides visqueux incompressibles, *J. Méc.* **17** (1978), 107–145.

A.V. Babin
[1] Asymptotic expansion at infinity of a strongly perturbated Poiseuilbe flow, *Advances in Soviet Math*, **10**, 1992, 1–83.
[2] The asymptotic behavior as $|x| \to \infty$ of strongly perturbed Poiseuilbe flows, *Soviet Math Dokl.*, **43**, 1991, 171–175.

A.V. Babin and M.I. Vishik
[1] Attractors of partial differential equations and estimate of their dimension, *Uspekhi Mat. Nauk*, **38** (1983), 133–187 (in Russian). *Russian Math. Surveys*, **38** (1983), 151–213 (in English).
[2] Regular attractors of semigroups and evolution equations, *J. Math. Pures Appl.*, **62** (1983), 441–491.
[3] Maximal attractors of semigroups corresponding to evolution differential equations, *Mat. Sbornik*, **126** (168), 1985 (in Russian). *Math. USSR-Sbornik*, **54** (1986), 387–408 (in English).

[4] Attractors of partial differential equations in an unbounded domain, *Proc. Roy. Soc. Edinburgh*, **116A** (1990), 221–243.

[5] *Attractors of Evolution Equations*, North-Holland, Amsterdam, 1992.

R.T. Bagley, G. Mayer-Kress, and J.D. Farmer

[1] Mode locking, the Belousov-Zhabotinsky reaction and one-dimensional mappings, Los Alamos National Laboratory, 1985.

J. Ball

[1] A proof of the existence of global attractors for damped semilinear wave equations, to appear.

Y. Bamberger

See A. Avez and Y. Bamberger.

C. Bardos and L. Tartar

[1] Sur l'unicité rétrograde des équations paraboliques et quelques équations voisines, *Arch. Rational Mech. Anal.*, **50** (1973), 10–25.

G.I. Barenblatt, G. Iooss, and D.D. Joseph (Eds.)

[1] *Nonlinear Dynamics and Turbulence*, Pitman, London, 1983.

M.F. Barnsley

[1] Fractal functions and interpolation, Georgia Institute of Technology, February 1985.

[2] Conference at INRIA, Paris, January 1986.

M.F. Barnsley and S. Demko

[1] Iterated function systems and the global construction of fractals, Georgia Institute of Technology, July 1984.

G.K. Batchelor

[1] Computation of the energy spectrum in homogeneous, two-dimensional turbulence, *Phys. Fluids*, **12**, Suppl. 2 (1969), 233–239.

[2] *The Theory of Homogeneous Turbulence*, Cambridge University Press, Cambridge, 1970.

A. Bensoussan and R. Temam

[1] Equations aux derivées partielles stochastiques, *Israel J. Math.*, **11** (1972), 95–129.

[2] Equations stochastiques du type Navier–Stokes, *J. Funct. Anal.*, **13** (1973), 195–222.

H. Berestycki and B. Larrouturou

[1] Quelques aspects mathématiques de la propagation des flammes prémélangées, *Nonlinear Partial Differential Equations and Their Applications*, Collège France Seminar, Vol. X (Paris, 1987–1988), pp. 65–129, Pitman Res. Notes Math. Ser., Vol. 220, Longman Sci. Tech., Harlow, 1991.

[2] Multi-dimensional travelling-wave solutions of a flame propagation model, *Arch. Rational Mech. Anal*, **111** (1990), 33–49.

[3] A semi-linear elliptic equation in a strip arising in a two-dimensional flame propagation model, *J. Reine Angew. Math.*, **396** (1989), 14–40.

P. Bergé, Y. Pomeau, and C. Vidal

[1] *L'Ordre dans le Chaos*, Herman, Paris, 1984.

B.S. Berger and M. Rokni

[1] Lyapunov exponents and continuum kinematics, Technical Report no. 86-1, Department of Mechanical Engineering, University of Maryland, 1986.

Z.B. Berkaliev

[1] The attractor of a certain quasi-linear system of differential equations with viscoelastic terms, *Russian Math. Surveys*, **40**, no. 1 (1985), 209–210.

J.E. Billoti and J.P. La Salle
[1] Dissipative periodic processes, *Bull. Amer. Math. Soc.*, **77**, no. 6 (1971), 1082–1088.

M.S. Birman
[1] *Mat. Sbornik* **55** (97) (1961), 125; *Trans. Amer. Math. Soc., Ser. 2*, **53** (1966), 23.

A.R. Bishop, D.K. Campbell, and B. Nicolaenko (Eds.)
[1] *Nonlinear Problems: Present and Future*, North-Holland, Amsterdam, 1982.

A.R. Bishop, K. Fesser, P.S. Lomdahl, and S.E. Trullinger
[1] Influence of solitons in the initial state on chaos in the driven damped sine–Gordon system, *Physica*, **7D** (1983), 259–279.

P.J. Blennerhassett
[1] On the generation of waves by wind, *Philos. Trans. Roy. Soc. London, Ser. A*, **298** (1980), 451–494.

K.J. Blow and N.J. Doran
[1] Global and local chaos in the pumped nonlinear Schrödinger equation, *Phys. Rev. Lett.*, **52** (1984), 526–529.

J.L. Bona and R. Smith
[1] The initial-value problem for the Korteweg–de Vries equation, *Philos. Trans. Roy. Soc. London*, Ser. A, **278** (1975), 555–604.

N. Bourbaki
[1] *Algèbre*, Chap. 9, Masson, Paris, 1981.
[2] *Espaces Vectoriels Topologiques*, Chap. 1–5, Masson, Paris, 1981.

W. Boyd
See G.P. Agrawal and W. Boyd.

B. Bréfort, J.M. Ghidaglia, and R. Temam
[1] Attractors for the penalized Navier–Stokes equations, *SIAM J. Math. Anal.*, **19** (1988), 1–21.

H. Brézis
[1] *Opérateurs Maximaux Monotones*, North-Holland, Amsterdam, 1973.

H. Brézis and T. Gallouet
[1] Nonlinear Schrödinger evolution equations, *Nonlinear Anal., TMA*, **4** (1980), 677–681.

K.J. Brown, P.C. Dunne, and R.A. Gardner
[1] A semilinear parabolic system arising in the theory of superconductivity, *J. Differential Equations*, **40** (1981), 232–252.

L. Brüll
[1] On solitary waves for nonlinear Schrödinger equations in higher dimensions, *Applicable Anal.*, **22** (1986), 213–225.

L. Caffarelli, R. Kohn, and L. Nirenberg
[1] Partial regularity of suitable weak solutions of the Navier–Stokes equations, *Comm. Pure Appl. Math.*, **35** (1982), 771–831.

J.W. Cahn and J.E. Hilliard
[1] *J. Chem. Phys.*, **28** (1958), 258.

D.K. Campbell
See A.R. Bishop, D.K. Campbell, and N. Nicolaenko.

J.G. Caputo
See P. Atten, I.G., Caputo, B. Malraison, and Y. Gagne.

J. Carr
[1] *Applications of Centre Manifold Theory*, Applied Mathematics Series, Vol. 35, Springer-Verlag, New York, 1981.

A. Carvalho
See J. Arrieta, A. Carvalho, and J.K. Hale.

N. Chaffee and E.F. Infante
[1] A bifurcation problem for a nonlinear partial differential equation of parabolic type, *Applicable Anal.*, **4** (1974) 17–37.

J. Chandra
[1] *Chaos in Nonlinear Dynamical Systems*, SIAM, Philadelphia, 1984.

B. Charlet
[1] Stability and robustness for nonlinear systems feedback linearization, Ecole des Mines de Paris, Centre de Fontainebleau, 1986.

H.H. Chen
See Y.C. Lee and H.H. Chen.

G. Choquet and C. Foias
[1] Solutions d'un problème sur les itérés d'un opérateur positif sur $\mathscr{C}(K)$ et propriétés des moyennes associées, *Ann. Inst. Fourier*, **25**, fasc. 3–4 (1975), 109–129.

Y. Choquet Bruhat
[1] *Géométrie Différentielle et Systèmes Extérieurs*, Dunod, Paris, 1968.

A. Chorin, J. Marsden, and S. Smale (Eds.)
[1] *Turbulence Seminar*, Lecture Notes in Mathematics, Vol. 615, Springer-Verlag, New York, 1977.

S.N. Chow, X.B. Lin, and K. Lu
[1] Smooth invariant foliations in infinite-dimensional spaces. *J. Differential Equations*, **94**, no. 2, (1991), 266–291.

S.N. Chow, K. Lu, and G. Sell
[1] Smoothness of inertial manifolds, *J. Math. Anal. Appl.*, **169** (1992), 283–312.

S.N. Chow and K. Lu
[1] Invariant manifolds for flows in Banach spaces, *J. Differential Equations*, **74** (1988), 285–317.

K.N. Chueh, C.C. Conley, and J.A. Smoller
[1] Positively invariant regions for systems of nonlinear diffusion equations, *Indian Univ. Math. J.*, **26** (1977), 373–391.

E. Coddington and N. Levinson
[1] *Theory of Ordinary Differential Equations*, McGraw-Hill, New York, 1955.

P. Collet and J.P. Eckman
[1] *Iterated Maps on the Interval as Dynamical Systems*, Birkhäuser, Boston, 1980.

C.C. Conley
See K.N. Chueh, C.C. Conley, and J.A. Smoller.

C. Conley and J. Smoller
[1] Bifurcation and stability of stationary solutions of the Fitz-Hugh–Nagumo equations, *J. Differential Equations*, **63** (1986), 389–405.

P. Constantin
[1] Collective L^∞ estimates for families of functions with orthonormal derivatives, *Indiana Univ. Math. J.*, **36** (1987), 603–616.

P. Constantin and C. Foias
[1] Global Lyapunov exponents, Kaplan–Yorke formulas and the dimension of the attractors for two-dimensional Navier–Stokes equations, *Comm. Pure Appl. Math.*, **38** (1985), 1–27.
[2] Sur le transport des variétés de dimension finie par les solutions des équations de Navier–Stokes, *C. R. Acad. Sci. Paris*, Série I, **296**, 1 (1983), 23–26.

P. Constantin, C. Foias, O. Manley, and R. Temam
[1] Connexion entre la théorie mathématique des équations de Navier–Stokes et la théorie conventionnelle de la turbulence, *C. R. Acad. Sci. Paris, Série I*, **297** (1983), 599–602.
[2] Determining modes and fractal dimension of turbulent flows, *J. Fluid Mech.*, **150** (1985), 427–440.

P. Constantin, C. Foias, B. Nicolaenko, and R. Temam
[1] Nouveaux résultats sur les variétés inertielles pour les équations différentielles dissipatives, *C. R. Acad. Sci. Paris, Série I*, **302** (1986), 375–378.
[2] *Integral and Inertial Manifolds for Dissipative Partial Differential Equations*, Applied Mathematical Sciences, Vol. 70. Springer-Verlag, New York, 1988.
[3] Spectral barriers and inertial manifolds for dissipative partial differential equations, *J. Dynamics and Differential Equations*, **1** (1989), 45–73.

P. Constantin, C. Foias, and R. Temam
[1] *Attractors Representing Turbulent Flows*, Memoirs of AMS, Vol. 53, no. 314, 1985.
[2] On the dimension of the attractors in two-dimensional turbulence, *Physica D*, **30** (1988), 284–296.
[3] On the large-time Galerkin approximation of the Navier–Stokes equations, *SIAM J. Numer. Anal.*, **21** (1984), 615–634.

E. Conway, D. Hoff, and J. Smoller
[1] Large-time behavior of solutions of systems of nonlinear reaction–diffusion equations, *SIAM J. Appl. Math.*, **35** (1978), 1–16.

R. Courant and D. Hilbert
[1] *Methods of Mathematical Physics*, Intersciences Publishers, New York, 1953.

T.G. Cowling
[1] *Magnetohydrodynamics*, Interscience Tracts on Physics and Astronomy, New York, 1957.

M. McCracken
See J.E. Marsden and M. McCracken.

M. Cwikel
[1] Weak type estimates for singular values and the number of bound states of Schrödinger operators, *Ann. of Math.*, **106** (1977), 93–100.

R. Daley
[1] Normal mode initialization, *Rev. Geoph. Space Phys.*, **19** (1981), 450–468.
[2] The normal mode approach to the initialization problem, in *Dynamic Meteorology, Data Assimilation Methods*, L. Bengtsson, M. Ghil, and E. Kallen (Eds.), Applied Mathematical Sciences, Vol. 36, Springer-Verlag, New York, 1983.

R. Dautray and J.L. Lions
[1] *Analyse Mathematique et Calcul Numérique pour les Sciences et les Techniques*, Masson, Paris, 1985.

A. Debussche
[1] Inertial manifolds and Sacker's equation, *Differential Integral Equations*, **3** (1990), 467–486.

A. Debussche and M. Marion

[1] On the construction of families of approximate inertial manifolds, *J. Differential Equations*, to appear.

A. Debussche and R. Temam

[1] Inertial manifolds and slow manifolds, *Appl. Math. Lett.*, **4**, 4 (1991), 73–76.

[2] Inertial manifolds and the slow manifolds in meteorology, *Differential Integral Equations*, **4**, 5 (1991), 897–931.

[3] Inertial manifolds and their dimension, in *Dynamical Systems, Theory and Applications*, S.I. Andersson, A.E. Andersson, and O. Ottoson (Eds.), World Scientific, Singapore, 1993.

[4] Convergent families of approximate inertial manifolds, *J. Math. Pures Appl.* **73** (1994), 485–522.

[5] Some new generalizations of inertial manifolds, *Discrete and Continuous Dynamical Systems*, to appear.

G.F. Dell'Antonio and B. D'Onofrio

[1] Construction of a center-unstable manifold for $\mathscr{C}^1$-flows and an application to the Navier–Stokes equations, *Arch. Rational Mech. Anal.*, **93** (1986), 185–201.

Y. Demay and G. Iooss

[1] Calcul des solutions bifurquées pour le problème de Couette-Taylor avec les deux cylindres en rotation. *J. Méc. Théor. Appl.*, numéro spécial 1984, 193–216.

F. Demengel and J.M. Ghidaglia

[1] Some remarks on the smoothness of inertial manifolds, *J. Math. Anal. Appl.*, **155** (1991), 177–225.

S. Demko

See M.F. Barnsley and S. Demko.

J. Deny and J.L. Lions

[1] Les espaces du type de Beppo–Levi, *Ann. Inst. Fourier* (*Grenoble*), **5** (1954), 305–370.

G. de Vries

See D.J. Korteweg and G. de Vries.

R.C. Di Prima

See J.T. Stuart and R.C. Di Prima.

C.R. Doering, J.D. Gibbon, D. Holm, and B. Nicolaenko

[1] Low-dimensional behaviour in the complex Ginzburg–Landau equation, *Nonlinearity*, **1**, no. 2 (1988), 279–309. Alamos preprint LA-VR871546, June 1986.

C.R. Doering, J. Gibbon, and C.D. Levermore

[1] Weak and strong solutions of the complex Ginzburg–Landau equation, *Physica D*, 71, (1984), 285–318.

C.R. Doering and X. Wang

[1] article in preparation.

B. D'Onofrio

See G.F. Dell'Antonio and B. D'Onofrio.

N.J. Doran

See K.J. Blow and N.J. Doran.

A. Douady and J. Oesterlé

[1] Dimension de Hausdorff des attracteurs, *C. R. Acad. Sci. Paris, Sér A*, **290** (1980), 1135–1138.

B. Doubrovine, S. Novikov, and A. Fomenko

[1] *Géométrie Contemporaine. Méthodes et Applications*, Editions Mir, Moscow, 1982.

A. Douglis
See S. Agmon, A. Douglis, and L. Nirenberg.

N. Dunford and J.T. Schwartz
[1] *Linear Operators*, Interscience, New York, 1958.

P.C. Dunne
See K.J. Brown, P.C. Dunne, and R.A. Gardner.

D. Ebin and J. Marsden
[1] Groups of diffeomorphisms and the motion of an incompressible fluid, *Ann. of Math.*, **92** (1970), 102–163.

J.P. Eckman
[1] Roads to turbulence in dissipative dynamical systems, *Rev. Mod. Phys.*, **53** (1981), 643.
See also P. Collet and J.P. Eckman.

J.P. Eckman and D. Ruelle
[1] Ergodic theory of chaos and strange attractors, *Rev. Mod. Phys.*, **57**, 3, Part I (1985), 617–656.

D.G.B. Edelen
[1] *Applied Exterior Calculus*, Wiley, New York, 1985.

A. Eden
[1] *An abstract theory of L-exponents with applications to dimension analysis*, Ph.D. Thesis, Indiana University, Bloomington, 1988.

A. Eden, C. Foias, B. Nicolaenko, and R. Temam
[1] *Exponential Attractors for Dissipative Evolution Equations*, Masson, Paris, and Wiley, New York, 1994.

A. Eden, C. Foias, and R. Temam
[1] Local and global Lyapunov exponents, *J. Dynamics Differential Equations*, **3** (1991), 133–177.

K.J. Falconer
[1] *The Geometry of Fractal Sets*, Cambridge University Press, Cambridge, 1985.

D. Farmer
[1] Chaotic attractors of an infinite-dimensional system, *Physica* **4D** (1982), 366–393.
[2] Sensitive dependence on parameters in nonlinear dynamics, *Phys. Rev. Lett.*, **55**, 4 (1985), 351–354.
See also R.J. Bagley, G. Mayer-Kress and I.D. Farmer; D.K. Umberger and J.D. Farmer; D.K. Umberger, J.D. Farmer, and I.I. Satija.

J.D. Farmer, E. Ott, and J.A. Yorke
[1] The dimension of chaotic attractors, *Physica* **7D** (1983), 153–180.

J.D. Farmer and N.H. Packard
[1] Chaotic attractors of infinite dimensional systems. I: Delay differential equations, Preprint.

J.D. Farmer, N.H. Packard, and A.S. Perelson
[1] The immune system, adaptation and machine learning, Los Alamos National Laboratory, 1985.

P. Faurre and M. Robin
[1] *Elements d'Automatique*, Dunod, Paris, 1984.

H. Federer
[1] *Geometric Measure Theory*, Springer-Verlag, New York, 1969.

M. Feigenbaum
[1] Qualitative universality for a class of nonlinear transformations, *J. Statist. Phys.*, **19** (1978), 25–52 and **21** (1979), 669–706.

E. Feireisl
[1] Attractors for wave equations with nonlinear dissipation and critical exponent, *C. R. Acad. Sci. Paris*, Série I, **315** (1992), 551–555.
[2] Attractors for semilinear damped wave equations on $\mathbb{R}^3$, *Nonlinear Analysis TMA*, **23**, no. 2, (1994), 187–195.

E. Feireisl, P. Laurençot, F. Simondon, and H. Touré
[1] Compact attractors for reaction–diffusion equations in $\mathbb{R}^N$, *C. R. Acad. Sci. Paris*, Série I, **319** (1994), 147–151.

E. Feireisl and E. Zuazua
[1] Global attractors for semilinear wave equations with locally distributed non-linear damping and critical exponent, *Comm. Partial Differential Equations*, **18** (9–10) (1993), 1539–1555.

K. Fesser
See A.R. Bishop. K. Fesser, P.S. Lomdahl, and S.E. Trullinger.

P.C. Fife
[1] Propagator–controller systems and chemical patterns, in *Non-Equilibrium Dynamics in Chemical Systems*, C. Vidal and A. Pacault (Eds.), Springer-Verlag, New York, 1984.
[2] Understanding the patterns in the *BZ* reagent, *J. Statist. Phys.*, **39** (1985), 687–703.

R. Fitz-Hugh
[1] Mathematical models of threshold phenomena in the nerve membrane, *Bull. Math. Biophys.*, **17** (1955), 1955.

C. Flytzanis
[1] Bistability, instability and chaos in passive nonlinear optical systems, Preprint, 1985.

C. Foias
See G. Choquet and C. Foias; P. Constantin and C. Foias; P. Constantin, C. Foias, O. Manley and R. Temam; P. Constantin, C. Foias and R. Temam; A. Eden, C. Foias, and R. Temam; A. Eden, C. Foias, B. Nicolaenko, and R. Temam.

C. Foias, M.S. Jolly, I.G. Kevrekidis, G.R. Sell, and E.S. Titi
[1] On the computation of Inertial Manifolds, *Phys. Lett. A*, **131** (1988), 433–436.

C. Foias, O. Manley, and R. Temam
[1] Attractors for the Bénard problem. Existence and physical bounds on their fractal dimension, *Nonlinear Anal., TMA* **11**, (1987), 939–967.
[2] Sur l'interaction des petits et grands tourbillons dans les écoulements turbulents, *C. R. Acad. Sci. Paris*, Série I, **305** (1987), 497–500.
[3] On the interaction of small and large eddies in two-dimensional turbulent flows, *Math. Model and Numer. Anal.* M2AN **22** (1988), 93–114.

C. Foias, O. Manley, R. Temam, and Y. Treve
[1] Asymptotic analysis of the Navier–Stokes equations, *Physica* **9D** (1983), 157–188.

C. Foias, B. Nicolaenko, G.R. Sell, and R. Temam
[1] Variétés inertielles pour l'équation de Kuramoto–Sivashinski, *C. R. Acad. Sci. Paris*, Série I, **301** (1985), 285–288.

[2] Inertial manifolds for the Kuramoto–Sivashinsky equations and an estimate of their lowest dimension, *J. Math. Pures Appl.*, **67** (1988), 197–226.

C. Foias and G. Prodi
[1] Sur le comportement global des solutions non stationnaires des équations de Navier–Stokes en dimension 2, *Rend. Sem. Mat. Univ. Padova*, **39** (1967), 1–34.

C. Foias and J.C. Saut
[1] Linearization and normal form of the Navier–Stokes equations with potential forces, *Ann. Inst. H. Poincaré, Analyse Non Linéaire*, **4** (1987), 1–47.
[2] Asymptotic integration of Navier–Stokes equations with potential forces, I, *Indiana Univ. Math. J.*, **40** (1991), 305–320.

C. Foias, G.R. Sell, and R. Temam
[1] Variétés inertielles des équations différentielles dissipatives, *C. R. Acad. Sci. Paris, Sér. I*, **301** (1985), 139–142.
[2] Inertial manifolds for nonlinear evolutionary equations, *J. Differential Equations*, **73** (1988), 309–353.

C. Foias, G.R. Sell, and E. Titi
[1] Exponential tracking and approximation of inertial manifolds for dissipative equations, *J. Dynamics and Differential Equations*, **1** (1989), 199–244.

C. Foias and R. Temam
[1] Structure of the set of stationary solutions of the Navier–Stokes equations, *Comm. Pure Appl. Math.*, **30** (1977), 149–164.
[2] Some analytic and geometric properties of the solutions of the Navier–Stokes equations, *J. Math. Pures Appl.*, **58** (1979), 339–368.
[3] On the Hausdorff dimension of an attractor for the two-dimensional Navier–Stokes equations, *Phys. Lett.*, **93A**, no. 9 (1983), 451–454.
[4] Finite parameters approximative structure of actual flows, in A.R. Bishop, D.K. Campbell, and L. Nicolaaenko [1], pp. 317–327.
[5] Asymptotic numerical analysis of the Navier–Stokes equations, in G.I. Barenblatt, G. Iooss, and D.D. Joseph [1], pp. 139–155.
[6] Determination of the solutions of the Navier–Stokes equations by a set of nodal values, *Math. Comput.*, **43**, no. 167 (1984), 117–133.
[7] The algebraic approximation of attractors; the finite-dimensional case, *Physica D*, **32** (1988), 163–182.
[8] Gevrey class regularity for the solutions of the Navier–Stokes equations, *J. Funct. Anal.*, **87** (1989), 359–369.
[9] Approximation of attractors by algebraic or analytic sets, *SIAM J. Math. Anal.*, **25**, 5 (1994), 1269–1302.

A. Fomenko
See B. Doubrovine, S. Novikov, and A. Fomenko.

L. Friedlander
[1] An invariant measure for the equation $u_{tt} - u_{xx} + u^3 = 0$, *Comm. Math. Phys.*, **98** (1985), 1–16.

U. Frisch (Ed.)
[1] *Chaotic Behaviour of Deterministic Systems, Les Houches, 1983*, North-Holland, Amsterdam, 1984.

E. Gagliardo
[1] Ulteriori proprietà di alcune classi di funzioni in piu variabili, *Ricerche Mat.*, **8** (1959), 24–51.

Y. Gagne
See P. Atten, J.G. Caputo, B. Malraison, and Y. Gagne.

T. Gallouet
See H. Brézis and T. Gallouet.

C.S. Gardner
See R.M. Miura, C.S. Gardner, and M.D. Kruskal; M.D. Kruskal, R.M. Miura, C.S. Gardner, and N.J. Zabusky.

R.A. Gardner
See K.J. Brown, P.C. Dunne, and R.A. Gardner.

P. Germain
[1] *Mécanique des Milieux Continus*, Masson, Paris, 1973.

J.M. Ghidaglia
[1] *Etude d'écoulements de fluides visqueux incompressibles: comportement pour les grands temps et applications aux attracteurs*, Thèse de 3e Cycle, Université Paris Sud, Orsay, 1984.
[2] On the fractal dimension of attractors for viscous incompressible fluid flows, *SIAM J. Math. Anal.*, **17** no. 5 (1986), 1139–1157.
[3] Some backward uniqueness results, *Nonlinear Anal., TMA*, **10**, 2 (1986), 777–790.
See B. Bréfort, J.M. Ghidaglia, and R. Temam.
[4] Weakly damped forced Korteweg–de Vries equations behave as a finite-dimensional dynamical system in the long time, *J. Differential Equations* **74** (1988), 369–390.
[5] Finite-dimensional behavior for weakly damped driven Schrödinger equations, *Ann. Institut H. Poincaré, Analyse Non Linéaire*, **5** (1988), 365–405.
[6] A note on the strong convergence towards attractors of damped forced KdV equations, *J. Differential Equations* **110** (1994), 356–359.
See also F. Demengel and J.M. Ghidaglia.

J.M. Ghidaglia and B. Héron
[1] Dimension of the attractor associated to the Ginzburg–Landau equation, *Physica* **28D** (1987), 282–304.

J.M. Ghidaglia, M. Marion, and R. Temam
[1] Sur quelques inégalités fonctionnelles, *C. R. Acad. Sci. Paris*, Série I, **304** (1987), 287–290.
[2] Generalization of the Sobolev–Lieb–Thirring inequalities and applications to the dimension of attractors, *Differential Integral Equations*, **1** (1988), 1–21.

J.M. Ghidaglia and J.C. Saut (Eds.)
[1] *Equations aux dérivées partielles non linéaires dissipatives et systemes dynamiques*, Hermann, Paris, 1988.

J.M. Ghidaglia and R. Temam
[1] Attractors for damped nonlinear hyperbolic equations, *J. Math. Pures Appl.*, **66** (1987), 273–319.
[2] Dimension of the universal attractor describing the periodically driven sine–Gordon equations, *Transport Theory Statist. Phys.*, **16**, 2 and 3 (1987), 253–265.
[3] Periodic dynamical systems with application to sine–Gordon equations: estimates of the fractal dimension of the universal attractor, in *Contemporary Mathematics, AMS*, **99** (1989), 143–164.
[4] Remarks on the regularity of the solution of evolution equations and their attractors, *Ann. Scuola Norm. Sup. Pisa*, Serie IV, **14**, 3 (1987), 485–511.
[5] Long-time behavior for partly dissipative equations: the slightly compressible 2D Navier–Stokes equations, *Asymptotic Anal.*, **1** (1988), 23–49.
[6] Lower bound on the dimension of the attractor for the Navier-Stokes equations in space dimension 3, in Mechanics, Analysis and Geometry: 300 years after Lagrange, M. Francariglia ed., Elsevier, Amsterdam, 1991.

J.D. Gibbon
[1] Derivation of 3d Navier–Stokes length scales from a result of foias, Guillope' and Temam, Nonlinearity 7 (1994), 245–252.
See also C.R. Doering, J.D. Gibbon, and C.D. Levermore; C.R. Doering, J.D. Gibbon, D. Holm, and B. Nicolaenko.

T.L. Gill and W.W. Zachary
[1] Existence and finite dimensionality of attractors for the Landau–Lifschitz equations, in *Differential Equations in Mathematical Physics*, I. Knowles and Y. Saito (Eds.), Lecture Note in Mathematics, Springer-Verlag, New York, 1985.

J. Ginibre and H. Vélo
[1] Time decay of finite energy solutions of the nonlinear Klein–Gordon and Schrödinger equations, *Ann. Inst. H. Poincaré Phys. Théor*, **43**, 4 (1985), 399–442.

J. Gleick
[1] *Chaos*, Viking, New York, 1987.

I.C. Gohberg and M.G. Krein
[1] *Introduction to the Theory of Linear Non-self-adjoint Operators*, Translations of Mathematical Monographs, vol. 18, Amer. Math. Soc., Providence, RI, 1969.

J.P. Gollub
See H. Swinney and J.P. Gollub.

J.P. Gollub, E.J. Romer, and J.E. Sacolar
[1] Trajectory divergence for coupled relaxation oscillators: Measurements and models, *J. Statist. Phys.*, **23** (1980), 321–333.

O. Goubet
[1] Regularity of the attractor for the weakly damped nonlinear Schrödinger equations, *Applicable Anal.*, to appear.

R. Grappin, J. Léorat, and A. Pouquet
[1] Computation of the dimension of a model of fully developed turbulence, *J. Physique*, **47** (1986), 1127–1136.

P. Grassberger
[1] On the Hausdorff dimension of fractal attractors, *J. Stat. Phys.*, **26** (1981), 173.
See also I. Procaccia, P. Grassberger, and H.G.E. Hentschel.

P. Grassberger and I. Procaccia
[1] Measuring the strangeness of strange attractors, *Physica* **9D** (1983), 189.

J. Guckenheimer and P. Holmes
[1] *Nonlinear Oscillations, Dynamical Systems and Bifurcations of Vector Fields*, Springer-Verlag, New York, 1983.

V. Guillemin and A. Pollack
[1] *Differential Topology*, Prentice-Hall, Englewood Cliffs, NJ, 1974.

H. Haken
[1] At least one Lyapunov exponent vanishes if the trajectory of an attractor does not contain a fixed point, *Phys. Lett.*, **94A** (1983), 71–72.

J.K. Hale
[1] *Functional Differential Equations*, Applied Mathematical Sciences, Vol. 3, Springer-Verlag, New York, 1977.
[2] Asymptotic behavior and dynamics in infinite dimensions, in *Nonlinear Differential Equations*, J.K. Hale and P. Martinez-Amores (Eds.), Pitman, London, 1986, pp. 1–42.
[3] *Asymptotic Behavior of Dissipative Systems.* Mathematical Surveys and Monographs, vol. 25, AMS, Providence, 1988.

[4] Asymptotic behavior and dynamics in infinite dimensions, in J.K. Hale and P. Martinez-Amores (Eds.), *Research Notes in Mathematics*, Vol. 132, pp. 1–42, Pitman, New York, 1985.
See also J. Arrieta, A. Cavarlho, and J.K. Hale.

J.K. Hale, J.P. La Salle, and M. Slemrod
[1] Theory of a general class of dissipative processes, *J. Math. Anal. Appl.* **39** (1972), 177–191.

J.K. Hale and G. Raugel
[1] Upper semicontinuity of the attractor for a singular perturbated hyperbolic equation, *J. Differential Equations*, **73** (1988), 197–214.
[2] Lower semicontinuity of attractors of gradient systems and applications, *Ann. Mat. Pura Appl.*, **4**, 154 (1989), 281–326.
[3] Lower semicontinuity of attractors for a singularly perturbated hyperbolic equation, *J. Dynamics and Differential Equations*, **2** (1990), 19–67.
[4] A damped hyperbolic equation on thin domains, *Trans. Amer. Math. Soc.*, **329** (1992), 185–219.
[5] Partial differential equations on thin domains, *Differential Equations and Mathematical Physics* (Birmingham, AL, 1990), pp. 63–97. Math. Sci. Engrg., Vol. 186, Academic Press, Boston, MA, 1992.
[6] Reaction–diffusion equations on thin domains, *J. Math. Pures Appl.*, **71** (1992), 33–95.
[7] Convergence in gradient-like systems with applications to PDE, *Z. Angew. Math. Phys.*, **43** (1992), 63–124.
[8] Attractors for dissipative evolutionary equations, Vol. 1, 2 (Barcelona, 1991), pp. 3–22, World Scientific, River Edge, NJ, 1993.
[9] Attractors and convergence of PDE on thin L-shaped domains, *Progress in Partial Differential Equations: The Metz Surveys*, **2** (1992), 149–171. Pitman Res. Notes Math. Ser., Vol. 296, Longman. Sci. Tech., Harlow, 1993.
[10] Limits of semigroups depending on parameters, *Resenhas*, **1** (1993), 1, 1–45.
[11] Addendum: Limits of semigroups depending on parameters, *Fifth Latin American Congress of Probability and Mathematical Statistics* (Portuguese) (Sao Paulo, 1993), *Resenhas*, **1** (1994), 2–3, 361.
[12] A reaction–diffusion equation on a thin L-shaped domain, *Proc. Roy. Soc. Edinburgh*, Section A, **125** (1995), 283–327.

J.K. Hale, Xia-Biao Lin, and G. Raugel
[1] Upper semicontinuity of attractors for approximations of semigroups and partial differential equations, *Math. Comp.*, **50**, 181 (1988), 89–123.

A. Haraux
[1] Two remarks on dissipative hyperbolic problems, in *Nonlinear Partial Differential Equations and Their Applications*, Collège de France Seminar, Vol. VII, H. Brézis, J.L. Lions (Eds.), Pitman, London. 1985.

G.H. Hardy and E.M. Wright
[1] *An Introduction to the Theory of Numbers*, Oxford University Press, Oxford, 1962.

S. Hastings and J. Murray
[1] The existence of oscillatory solutions in the Field–Noyes model for the Belousov–Zhabotinski reaction, *SIAM J. Appl. Math.*, **28** (1975), 678–688.

R. Helleman
See G. Iooss, R. Helleman, and R. Stora.

M.A. Hénon
[1] A two-dimensional mapping with a strange attractor, *Comm. Math. Phys.*, **50** (1976), 69–77.

D. Henry
[1] *Geometric Theory of Semilinear Parabolic Equations*, Lecture Notes in Mathematics, Vol. 840, Springer-Verlag, New York, 1981.

H.G.E. Hentschel
See I. Procaccia, P. Grassberger, and H.G.E. Hentschel.

H.G.E. Hentschel and I. Procaccia
[1] The infinite number of generalized dimensions of fractals and strange attractors, *Physica* **8D** (1983), 435–444.

B. Héron
See J.M. Ghidaglia and B. Héron.

D. Hilbert
See R. Courant and D. Hilbert.

J.E. Hilliard
See J.W. Cahn and J.E. Hilliard.

M. Hirsch
[1] The dynamical system approach to differential equations, *Bull. Amer. Math. Soc.*, **11** (1984), 1–64.

M.W. Hirsch and L.C. Pugh
[1] Stable manifolds and hyperbolic sets, *Proc. Symp. Pure Math.*, American Mathematical Society, Providence (RI), vol. 14, **14** (1970), 133–163.

M.W. Hirsch and S. Smale
[1] *Differential Equations, Dynamical Systems and Linear Algebra*, Academic Press, New York, 1974.

A.L. Hodgkin and A.F. Huxley
[1] A qualitative description of membrane current and its application to conduction and excitation in nerve, *J. Physiol.*, **117** (1952), 500–544.

D. Hoff
See E. Conway, D. Hoff, and J. Smoller.

D.D. Holm
See Ch. Doering, J.D. Gibbon, D. Holm, and B. Nicolaenko; B. Nicolaenko, D.D. Holm, and J.M. Hyman.

P. Holmes
See J. Guckenheimer and P. Holmes.

E. Hopf
[1] On linear partial differential equations, *Lecture Series of the Symposium on Partial Differential Equations*, Berkeley 1955, Ed. University of Kansas, 1957, pp. 1–29.

L. Howard and N. Kopell
[1] Plane wave solutions to reaction–diffusion equations, *Stud. Appl. Math.*, **52** (1973), 291–328.

Huang, Yu
[1] Global attractors for semilinear wave equations with nonlinear damping and critical exponent, *Appl. Anal.*, **56** (1995), 165–174.

P. Huerre
See H.T. Moon, P. Huerre, and L.G. Redekopp.

A.F. Huxley
See A.L. Hodgkin and A.F. Huxley.

J.M. Hyman
See N. Nicolaenko, D.D. Holm, and J.M. Hyman.

A.A. Ilyin
[1] Lieb–Thirring inequalities on the N-sphere and in the plane and some applications, *Proc. Lond. Math. Soc.* 67 (1993), 159–182.
[2] Partly dissipatives semigroups generated by the Navier–Stokes system on two dimensional manifolds and their attractors, *Mat. Sb.* 184 (1993), 55–88 (in Russian), *Russ. Acad. Sci. Sb. Math.* 78 (1994), 47–76.
[3] Attractors for Navier–Stokes equations in domains with finite measure, *Nonlinear Analysis TMA*, 27 (1996), 605–616.
[4] Best constants in multiplicative inequalities for sup norms, *J. London Math. Soc.*, to appear.
[5] Best constant in Sobolev inequalities on the sphere and in Euclidean space, to appear.

E.F. Infante
See N. Chaffee and E.F. Infante.

G. Iooss and D.D. Joseph
[1] *Bifurcation of Maps and Applications*, Springer-Verlag, Berlin 1966.
See also G.I. Barenblatt, G. Iooss, and D.D. Joseph; Y. Demay and G. Iooss.

G. Iooss, R. Helleman, and R. Stora (Eds.)
[1] *Chaotic Behavior of Deterministic Systems*, North-Holland, Amsterdam, 1983.

M.C. Irwin
[1] On the stable manifold theorem, *Bull. London Math. Soc.*, **2** (1970), 196–198.

M.S. Jolly
See C. Foias, M.S. Jolly, I.G. Kevrekidis, G.R. Sell, and E.S. Titi.

M.S. Jolly, I.G. Kevrekidis, and E.S. Titi
[1] Approximate inertial manifolds for the Kuramoto–Sivashinsky equation: analysis and computations, *Physica D*, **44** (1990), 38–60.

D. Jones
[1] A remark on quasi-stationary approximate inertial manifolds for the Navier–Stokes equations, *SIAM J. Math. Anal.*, to appear.

D.S. Jones and B.D. Sleeman
[1] *Differential Equations and Mathematical Biology*, George Allen and Unwin, London, 1983.

D.A. Jones and E.S. Titi
[1] Determining finite volume elements for the 2D Navier–Stokes equations, *Physica D*, **60**, 1992, 165–174.

K. Jörgens
[1] Das Aufangswertproblem in Grossen für eine Klasse nichtlinearer Wallenglerchungen, *Math. Z.*, **77** (1961), 295–308.

D.D. Joseph
[1] *Stability of Fluid Motions*, Springer Tracts in Natural Philosophy, Vol. 27, Springer-Verlag, New York, 1976.
See also G.I. Barenblatt, G. Iooss, and D.D. Joseph

V.K. Kalantarov and O.P. Ladyzhenskaya
[1] Stabilization of solutions of a class of quasilinear parabolic equations as $t \to \infty$, *Sibirsk. Mat. Zh.*, **19** (1978), 1043–1052.

D.A. Kamaev
[1] Hyperbolic limit sets of evolutionary equations and the Galerkin method, *Russian Math. Surveys*, **35**: 3 (1980), 239–243.
[2] Hyperbolic limit sets of a class of parabolic equations and Galerkin's method, *Soviet Math. Dokl.*, **22** (1980), 344–348.

[3] Hopf's conjecture for a class of chemical kinetic equations, *J. Soviet Math.*, **25** (1984), 836–849.

J.L. Kaplan, J. Mallet-Paret, and J.A. Yorke

[1] The Lyapunov dimension of a nowhere differentiable attracting torus, *Ergodic Theory Dynamical Systems*, **4** (1984), 261–281.

J. Kaplan and J. Yorke

[1] Chaotic behavior of multidimensional difference equations, in *Functional Differential Equations and Approximation of Fixed Points*, H.O. Peitgen and H.O. Walther (Eds.), Lecture Notes in Mathematics, Vol. 730, Springer-Verlag, New York, 1979, p. 219.

A. Kasahara

[1] Nonlinear normal mode initialization and the bounded derivative method, *Rev. Geophysics Space Phys.*, **20** (1982), 385–397.

T. Kato

[1] *Perturbation Theory for Linear Operators*, Springer-Verlag, Berlin, 1966.

V.E. Katznelson

[1] Conditions under which systems of eigenvectors of some classes of operators form a basis, *Funct. Anal. Appl.*, **1** (1967), 122–132, Translated from *Funktsional'nyi Analiz i Ego Prilozheniga*, **1** (1967), 39–51.

L. Keefe

[1] Integrability and structural stability of solutions to the Ginzburg–Landau equation, *Phys. Fluids*, **29** (1986), 3135–3141.

[2] Dynamics of perturbed wavetrain solutions to the Ginzburg–Landau equation, *Stud. Appl. Math*, **73** (1995), 91–153.

A. Kelley

[1] The stable, center-stable, center, center-unstable, unstable manifolds, *J. Differential Equations*, **3** (1967), 546–570.

C.E. Kenig, G. Ponce, and L. Vega

[1] Well-posedness of the initial value problem for the Korteweg–de Vries equation, *J. Amer. Math. Soc.*, **4** 2 (1991), 323–347.

I.G. Kevrekidis

See C. Foias, M.S. Jolly, I.G. Kevrekidis, G.R. Sell, and E.S. Titi; M.S. Jolly, I.G. Kevrekidis, and E.S. Titi.

P.E. Kloeden

[1] Asymptotically stable attracting sets in the Navier–Stokes equations, *Bull. Austral. Math. Soc.*, **34** (1986), 37–52.

R. Kohn

See L. Caffarelli, R. Kohn, and L. Nirenberg.

A.N. Kolmogorov

[1] The local structure of turbulence in incompressible viscous fluid for very large Reynolds number, *C. R. Acad. Sci. URSS*, **30** (1941), 301.

[2] On degeneration of isotropic turbulence in an incompressible viscous liquid, *C. R. Acad. Sci. URSS*, **31** (1941), 538.

[3] Dissipation of energy in locally isotropic turbulence, *C. R. Acad. Sci. URSS*, **32** (1941), 16.

N. Kopell

See L. Howard and N. Kopell.

N. Kopell and D. Ruelle

[1] Bounds on complexity in reaction–diffusion systems, *SIAM J. Appl. Math.*, **46** (1986), 68–80.

D.J. Korteweg and G. de Vries

[1] On the change of form of long waves advancing in a rectangular channel and on a new type of long stationary wave, *Phil. Mag.* **39** (1895), 422–443.

R.H. Kraichnan

[1] Inertial ranges in two-dimensional turbulence, *Phys. Fluids*, **10** (1967), 1417–1423.

M.G. Krein

See I. C. Gohberg and M. G . Krein.

G. Kreiss and H.O. Kreiss

[1] Convergence to the steady state of solutions of Burgers' equation, *Appl. Numer. Math.*, **2** (1986), 161–179.

M.D. Kruskal, R.M. Miura, C.S. Gardner, and N.J. Zabusky

[1] Korteweg–de Vries equation and generalizations. V. Uniqueness and nonexistence of polynomial conservation laws, *J. Math. Phys.*, **11**, 3 (1970), 952–960.

M.D. Kruskal

See also R.M. Miura, C.S. Gardner, and M.D. Kruskal.

Y. Kuramoto

[1] Diffusion induced chaos in reactions systems, *Progr. Theoret. Phys. Suppl.*, **64** (1978), 346–367.

Y. Kuramoto and T. Tsuzuki

[1] On the formation of dissipative structures in reaction–diffusion systems, *Progr. Theoret. Phys. Suppl.*, **54** (1975), 687–699.

[2] Persistent propagation of concentration waves in dissipative media far from thermal equilibrium, *Progr. Theoret. Phys. Suppl.*, **55** (1976), 356–369.

[3] Reductive perturbation approach to chemical dynamics, *Progr. Theoret. Phys. Suppl.*, **52** (1974), 1399.

O.A. Ladyzhenskaya

[1] The *Mathematical Theory of Viscous Incompressible Flow*, 2nd ed., Gordon and Breach, New York, 1969.

[2] A dynamical system generated by the Navier–Stokes equations, *J. Soviet. Math.*, **3**, no. 4 (1975), 458–479.

[3] Dynamical system generated by the Navier–Stokes equations, *Soviet Phys. Dokl.*, **17** (1973), 647–649.

[4] On the finiteness of the dimension of bounded invariant sets for the Navier–Stokes equations and other related dissipative systems, in *The Boundary Value Problems of Mathematical Physics and Related Questions in Functional Analysis*, Seminar of the Steklov Institute, Vol. 14, Leningrad, 1982; see also *J. Soviet. Math.*, **28**, no. 5 (1985), 714–725.

[5] On the attractors of nonlinear evolution problems, in *The Boundary Value Problems of Mathematical Physics and Related Questions in Functional Analysis*, Seminar of the Steklov Institute, Vol. 18, Leningrad, 1987.

[6] On the determination of minimal global attractors for the Navier–Stokes equations and other partial differential equations, *Uspekhi Mat. Nauk*, **42**, 6 (1987), 25–60; see also *Russian Math. Surveys*, **42**, 6 (1987), 27–73.

[7] On estimates of the fractal dimension and the number of determining modes for invariants sets of dynamical systems, *Zapiskii Nauchnich Seminarovs LOMI*, **163** (1987), 105–129.

[8] *Attractors for Semigroups and Evolution Equations*, Cambridge University Press, Cambridge, 1991.

[9] First boundary value problem for the Navier–Stokes equations in domains with nonsmooth boundaries, *C. R. Acad. Sci. Paris*, Série I, **314** (1992), 253–258.

[10] Some globally stable approximations for the Navier–Stokes equations and some other equations for viscous incompressible fluids, *C. R. Acad. Sci. Paris*, Série I, **315** (1992), 387–392.
See also V.K. Kalantarov and O.A. Ladyzhenskaya.

L. Landau and E. Lifschitz
[1] *Fluid Mechanics*, Addison-Wesley, New York, 1953.
[2] *Electrodynamique des Milieux Continus, Physique Théorique*, tome VIII, MIR, Moscow, 1969.

O. Lanford
[1] Bifurcation of periodic solutions into invariant tori, in *Nonlinear Problems in the Physical Sciences*, Lecture Notes in Mathematics, Vol. 322, Springer-Verlag, New York, 1973.
[2] A computer-assisted proof of the Feigenbaum conjecture, *Bull. Amer. Math. Soc.*, **6** (1982), 427–434.

J.S. Langer
[1] *Ann. Physics*, **65** (1971), 53.

B. Larrouturou
[1] Etude mathématique et modélisation numérique de phénomènes de combustion. Thesis, University of Paris, 1986.
See also H. Berestycki and B. Larrouturou.

J.P. La Salle
[1] *The Stability of Dynamical Systems*, Regional Conference Series in Applied Mathematics, Vol. 25, SIAM, Philadelphia, 1976.
See also J.E. Billoti and J.P. La Salle; J.K. Hale, J.P. La Salle, and M. Slemrod.

J.P. La Salle and S. Lefschetz
[1] *Stability by Lyapunov's Direct Method*, Academic Press, New York, 1961.

P. Laurençot
[1] Long-time behavior for weakly damped driven nonlinear Schrödinger equations in $\mathbb{R}^N$, $N \leq 3$, *NoDEA*, **2** (1995), 357–369.
See also E. Feireisl, P. Laurençot, F. Simondon, and H. Touré.

F. Ledrappier
[1] Some relations between dimension and Lyapunov exponents, *Comm. Math. Phys.*, **81** (1981), 229–238.

Y.C. Lee and H.H. Chen
[1] Nonlinear dynamical models of plasma turbulence, *Phys. Scripta*, **T2/1** (1982), 41–47.

S. Lefschetz
See J.P. La Salle and S. Lefschetz.

B. Legras
See R. Vautard and B. Legras.

C.E. Leith
[1] Nonlinear normal mode initialization and quasi-geostrophic theory, *J. Atmos. Sci.*, **37** (1980), 958–968.

J. Léorat
See R. Grappin, J. Léorat, and A. Rouquet.

J. Leray
[1] Etude de diverses équations intégrales non linéaires et de quelques problèmes que pose l'hydrodynamique, *J. Math. Pures Appl.*, **12** (1933), 1–82.
[2] Essai sur les mouvements plans d'un liquide visqueux que limitent des parois, *J. Math. Pures Appl.*, **13** (1934), 331–418.

[3] Essai sur le mouvement d'un liquide visqueux emplissant l'espace, *Acta Math.*, **63** (1934), 193–248.

J. Leray and J.L. Lions

[1] Quelques résultats de Visik sur les problèmes elliptiques non linéaires par les méthodes de Minty–Browder, *Bull. Soc. Math. France*, **93** (1965), 97–107.

C.D. Levermore

See C.R. Doering, J. Gibbon, and C.D. Levermore.

M. Levi

[1] Beating modes in the Josephson junction, in *Chaos in Nonlinear Dynamical Systems*, J. Chandra (Ed.), SIAM, Philadelphia, 1984.

N. Levinson

[1] Transformation theory of non-differential equations of the second-order, *Ann. of Math.*, **45** (1944), 723–737.

See also E. Coddington and N. Levinson.

E. Lieb

[1] An L^p bounded for the Riesz and Bessel potentials of orthonormal functions, *J. Funct. Anal.*, **51**, no. 2 (1983), 159–165.

[2] On characteristic exponents in turbulence, *Comm. Math. Phys.*, **92** (1984), 473–480.

E. Lieb and W. Thirring

[1] Inequalities for the moments of the eigenvalues of the Schrödinger equations and their relation to Sobolev inequalities, in *Studies in Mathematical Physics: Essays in Honor of Valentine Bargmann*, E. Lieb, B. Simon, A.S. Wightman (Eds.), Princeton University Press, Princeton, NJ, 1976, pp. 269–303.

[2] Bound for the kinetic energy of fermions which proves the stability of matter, *Phys. Rev. Lett.*, **35**, no. 11 (1975), 687–689, Erratum, **35**, no. 16 (1975), 1116.

E. Lifschitz

See L. Landau and E. Lifschitz.

X.B. Lin

See S.N. Chow, X.B. Lin, and K. Lu; J.K. Hale, X.B. Lin, and G. Raugel.

J.L. Lions

[1] *Problèmes aux Limites dans les Équations aux Dérivées Partielles*, Séminaire de Mathématiques Supérieures, Université de Montréal, 1962.

[2] *Quelques Méthodes de Résolution des Problèmes aux Limites Non Linéaires*, Dunod, Paris, 1969.

[3] *Controle Optimal des Systémes Gouvernes par des Équations aux Dérivées Partielles*, Dunod, Paris, 1968.

See also R. Dautray and J.L. Lions and below.

J.L. Lions and E. Magenes

[1] *Nonhomogeneous Boundary Value Problems and Applications*, Springer-Verlag, New York, 1972.

J.L. Lions and B. Malgrange

[1] Sur l'unicité rétrograde dans les problèmes mixtes paraboliques, *Math. Scand.*, **8** (1960), 277–286.

J.L. Lions and J. Peetre

[1] Sur une classe d'espaces d'interpolation, *Publ. Inst. Hautes Etudes Scient.*, **19** (1964), 5–68.

J.L. Lions, R. Temam and S. Wang

[1] Models of the coupled atmosphere and ocean, *Computational Mechanics Advances*, J.T. Oden ed., 1 (1993), 1–119.

[2] Mathematical study of the coupled models of atmosphere and ocean, *J. Math. Pures Appl.*, 74 (1995), 105–163.

Vincent Xiaosong Liu

[1] Remarks on the Navier–Stokes equations on the two and three dimensional forces, *Comm. Partial Differ. Equ.* 19 (1994), 873–900.

P.S. Lomdahl

See A.R. Bishop, K. Fesser, P.S. Lomdahl, and S.E. Trullinger.

E.N. Lorenz

[1] Deterministic nonperiodic flow, *J. Atmospheric Sci.*, **20** (1963), 130–141.

[2] Attractors sets and quasi-geostrophic equilibrium, *J. Atmos. Sci.*, **37** (1980), 1685–1699.

[3] *The Nature and Theory of the General Circulation of the Atmosphere*, World Meteorological Organization, 1967.

K. Lu

See S.N. Chow, X.B. Lin, and K. Lu; S.N. Chow, K. Lu. and G. Sell; S.N. Chow and K. Lu.

B.A. Machenhauer

[1] On the dynamics of gravity oscillations in a shallow water model with applications to normal mode initialization, *Beitr. Phys. Atmos.*, **10** (1977), 253–271.

D. MacLaughlin

[1] Lecture at the Summer Research Conference on the connection between infinite dimensional and finite dimensional dynamical systems, Boulder, July 1987.

E. Magenes

See J.L. Lions and E. Magenes.

B. Malgrange

See J.L. Lions and B. Malgrange.

J. Mallet-Paret

[1] Negatively invariant sets of compact maps and an extension of a theorem of Cartwright. *J. Differential Equations*, **22** (1976), 331.

See also J.L. Kaplan, J. Mallet-Paret, and J.A. Yorke; J.A. Yorke, E.D. Yorke, and J. Mallet-Paret.

J. Mallet-Paret and G. Sell

[1] Inertial manifolds for reaction–diffusion equations in higher space dimensions, *J. Amer. Math. Soc.*, **1** (1988), 805–866.

P. Malliavin

[1] *Géometrie differentielle stochastique*, Séminaire de Mathématiques Supérieures, Presses de l'Université de Montreal, 1978.

B. Malraison

See P. Atten, J.G. Caputo, B. Malraison, and Y. Gagne.

B. Mandelbrot

[1] *Fractals: Form, Chance and Dimension*, Freeman, San Francisco, 1977.

[2] *The Fractal Geometry of Nature*, Freeman, San Francisco, 1982.

B. Mañé

[1] *On the Dimension of the Compact Invariant Sets of Certain Nonlinear Maps*, Lecture Notes in Mathematics, Vol. 898, Springer-Verlag, New York, 1981, pp. 230–242.

[2] Lyapunov exponents and stable manifolds for compact transformations, in *Geometrical Dynamics*, Lecture Notes in Mathematics, Vol. 1007, Springer-Verlag, New York, 1983.

[3] Reduction of semilinear parabolic equations to finite dimensional $\mathscr{C}^1$ flows, in *Geometry and Topology*, Lectures Notes in Mathematics, Vol. 597, Springer-Verlag, New York, pp. 361–378.

O. Manley
See P. Constantin, C. Foias, O. Manley, and R. Temam; C. Foias, O. Manley, and R. Temam.

P. Manneville
[1] Lyapunov exponents for the Kuramoto–Sivashinsky model.
See also Y. Pomeau and P. Manneville; A. Pumir, P. Manneville, and Y. Pomeau.

M. Marion
[1] Attractors for reaction–diffusion equations; Existence and estimate of their dimension, *Appl. Anal.*, **25** (1987), 101–147.
[2] Finite-dimensional attractors associated with partly dissipative reaction–diffusion systems, *SIAM J. Math. Anal.*, **20** (1980), 816–844.
[3] Inertial manifolds associated to partly dissipative reaction–diffusion systems, *J. Math. Anal. Appl.*, **143** (1989), 295–326.
[4] Approximate inertial manifolds for reaction–diffusion equations in high space dimension, *J. Dynamics Differential Equations*, **1** (1989), 245–267.
[5] Approximate inertial manifolds for the pattern formation Cahn–Hilliard equation, *Math. Model. Numer. Anal.* (M2AN), **23** (1989), 463–480.
See A. Debussche and M. Marion.

M. Marion and R. Temam
[1] Nonlinear Galerkin methods, *SIAM J. Numer. Anal.*, **26** (1989), 1139–1157.
[2] Nonlinear Galerkin methods: The finite elements case, *Numer. Math.*, **57** (1990), 205–226.
[3] Navier–Stokes Equations—Theory and Approximation, to appear in *Handbook of Numerical Analysis*, P.G. Ciarlet and J.L. Lions (Eds.), North-Holland, Amsterdam, 1997.

A.S. Markus
[1] Expansions in root vectors of a weakly perturbed self-adjoint operator, *Dokl. Akad. Nauk SSSR*, **142** (1962), 538–541.

J.E. Marsden
See R. Abraham and J.E. Marsden; A. Chorin, J. Marsden, and S. Smale; D. Ebin and J. Marsden.

J.E. Marsden and M. McCracken
[1] *The Hopf Bifurcation and Its Applications*, Applied Mathematics Series Vol. 19, Springer-Verlag, New York, 1976.

P. Massat
[1] Stability and fixed points of point-dissipative systems, *J. Differential Equations*, **40** (1981), 217–231.
[2] Limiting behavior for strongly damped nonlinear wave equations, *J. Differential Equations*, **48** (1983), 334–349.
[3] Attractivity properties of α-contractions, *J. Differential Equations*, **48** (1983), 326–333.

H. Matano
[1] Convergence of solutions of one-dimensional semilinear parabolic equations, *J. Math. Kyoto Univ.*, **18** (1972), 221–227.

J.C. Maxwell
[1] *Matter and Motion*, Dover, New York.

G. Mayer-Kress
See R.J. Bagley, G. Mayer-Kress, and J.D. Farmer.

V.G. Maz'ja
[1] *Sobolev Spaces*, Springer-Verlag, New York, 1985.

G. Métivier
[1] Valeurs propres d'opérateurs définis sur la restriction de systémes variationnels à des sous-espaces, *J. Math. Pures Appl.*, **57** (1978), 133–156.

D.M. Michelson
See G. Sivashinsky and D.M. Michelson.

J.W. Miles
[1] The Korteweg–de Vries equation. A historical essay, *J. Fluid Mech.*, **106** (1981), 131–147.

J. Milnor
[1] On the concept of attractor, *Comm. Math. Phys.*, **99** (1985), 177–195.

Gh. Minea
[1] Remarques sur l'unicité de la solution stationnaire d'une équation de type Navier–Stokes, *Rev. Roumaine Math. Pures Appl.*, **21**, no. 8 (1976), 1071–1075.

A. Miranville and X. Wang
[1] Upper bound on the dimension of the attractor for nonhomogeneous Navier–Stokes equations, *Discrete and Continuous Dynamical Systems*, to appear.

A. Miranville and M. Ziane
[1] On the dimension of the attractor for the Bénard problem with free surfaces, Russian J. of Math. Phys., to appear.

R. Miura, C. Gardner, and M. Kruskal
[1] Korteweg–de Vries equation and generalizations. II. Existence of conservation laws and constants of motion, *J. Math. Phys.* **9**, 8 (1968), 1204–1209.

I. Moise and R. Rosa
[1] On the regularity of the global attractor for a weakly damped, forced Korteweg-de Vries equation, Advances in Differential Equations, to appear.

H.T. Moon, P. Huerre, and L.G. Redekopp
[1] Three frequency motion and chaos in the Ginzburg–Landau equations, *Phys. Rev. Lett.*, **49** (1982), 485–460.
[2] Transitions to chaos in the Ginzburg–Landau equation, *Physica* **7D** (1983), 135–150.

X. Mora
[1] Finite-dimensional attracting manifolds in reaction–diffusion equations, *Contemp. Math.*, **17** (1983), 353–360.
[2] Finite-dimensional attracting manifolds for damped semilinear wave equations, in *Contributions to Nonlinear Partial Differential Equations II*, to appear.

X. Mora and J. Solà-Morales
[1] Existence and non-existence of finite-dimensional globally attracting invariant manifolds in semilinear damped wave equations, *Dynamics of Infinite-Dimensional Systems* (Lisbon, 1986). NATO Adv. Sci. Inst. F. Comput. Systems Sci., Vol. 37, Springer-Verlag, Berlin, 1987.

T. Mukasa
See M. Tsutsumi and T. Mukasa.

J. Murray
See S. Hastings and J. Murray.

T. Nagashima
See T. Shimada and T. Nagashima.

J. Nagumo, S. Arimoto, and S. Yoshizawa
[1] An active pulse transmission line simulating nerve axon, *Proc. IRE* **50** (1962), 2061–2070; Bistable transmission lines, *IEEE Trans. Circuit Theory*, **CT-12** (1965), 400–412.

B. Sz. Nagy
See F. Riesz and B. Sz. Nagy.

A.C. Newell
[1] *Solitons in Mathematics and Physics*, CBMS–NSF Regional Conference Series in Applied Mathematics, Vol. 48, SIAM, Philadelphia, 1985.

A.C. Newell and J.A. Whitehead
[1] Finite bandwidth, finite amplitude convection, *J. Fluid Mech.*, **38** (1969), 279–304.

B. Nicolaenko
See A.R. Bishop, D.K. Campbell, and B. Nicolaenko; P. Constantin, C. Foias, B. Nicolaenko, and R. Temam; Ch. Doering, J.D. Gibbon, D. Holm, and B. Nicolaenko; A. Eden, C. Foias, B. Nicolaenko, and R. Temam; C. Foias, B. Nicolaenko, G. Sell, and R. Temam.

B. Nicolaenko, D.D. Holm, and J.M. Hyman (Eds.)
[1] *Nonlinear Systems of Partial Differential Equations in Applied Mathematics*, Lectures in Applied Mathematics, Vol. 23, AMS, Providence, RI, 1986.

B. Nicolaenko and B. Scheurer
[1] Low-dimensional behavior of the pattern formation Cahn–Hilliard equation, in *Trends in the Theory and Practice of Nonlinear Analysis.*

B. Nicolaenko, B. Scheurer, and R. Temam
[1] Some global dynamical properties of the Kuramoto–Sivashinsky equations: Nonlinear stability and attractors, *Physica*, **16D** (1985), 155–183.
[2] Attractors for the Kuramoto–Sivashinsky equation, in B. Nicolaenko, D.D. Holm, and J.M. Hyman [1], pp. 149–170.
[3] Some global dynamical properties of a class of pattern formation equations, *Commun. Partial Differential Equations*, **14** (1989), 245–297.

G. Nicolis and I. Prigogine
[1] *Self-Organization in Non-Equilibrium Systems: From Dissipative Structures to Order Through Fluctuations*, Wiley, New York, 1977.

N.V. Nikolenko
[1] Invariant asymptotically stable tori of the perturbed Korteweg–de Vries equation, *Russian Math. Surveys*, **35** (1980), 139–207.

L. Nirenberg
[1] On elliptic partial differential equations, *Ann. Scuola Norm. Sup. Pisa*, **13** (1959), 116–162.
See S. Agmon, A. Douglis, and L. Nirenberg; S. Agmon and L. Nirenberg; L. Caffarelli, R. Kohn, and L. Nirenberg.

A. Novick-Cohen and L.A. Segel
[1] Nonlinear aspects of the Cahn–Hilliard equation, *Physica*, **10D** (1984), 277–298.

S. Novikov
See B. Doubrovine, S. Novikov, and A. Fomenko.

J. Oesterlé
See A. Douady and J. Oesterlé.

V.I. Oseledec
[1] A multiplicative ergodic theorem, Lyapunov characteristic numbers for dynamical systems, *Trudy Moskov. Mat. Obshch.*, **19** (1968), 179 (in Russian); *Moscow Math. Soc.*, **19** (1968), 197.

E. Ott
See J.D. Farmer, E. Ott, and J.A. Yorke.

E. Ott and R.N. Sudan
[1] Damping of solitary waves, *Phys. Fluids*, **13** (1970), 1432–1434.

A. Pacault
See C. Vidal and A. Pacault

N.H. Packard
See J.D. Farmer and N.H. Packard; J.D. Farmer, N.H. Packard, and A.S. Perelson.

J. Peetre
See J.L. Lions and J. Peetre.

A.S. Perelson
See J.D. Farmer. N.H. Packard. and A.S. Perelson.

N.A. Philips
[1] Variational analysis and the slow manifold, *Mon. Wea. Rev.*, **109** (1981), 2415–2426.

A. Pollack
See V. Guillemin and A. Pollack.

Y. Pomeau
See P. Bergé, Y. Pomeau, and C. Vidal; A. Pumir, P. Manneville, and Y. Pomeau.

Y. Pomeau and P. Manneville
[1] Stability and fluctuations of a spatially periodic convective flow, *J. Physique Lett.*, **40** (1979), L609–L612.

G. Ponce
See C.E. Kenig, G. Ponce, and L. Vega.

A. Pouquet
See R. Grappin, J. Léorat, and A. Pouquet.

I. Prigogine
[1] *Introduction to Thermodynamics of Irreversible Processes*, Interscience, New York, 1967.
[2] *Non-Equilibrium Statistical Mechanics*, Interscience, New York, 1962.
See also G. Nicolis and I. Prigogine.

I. Procaccia
See P. Grassberger and I. Procaccia; H.G.E. Hentschel and I. Procaccia.

I. Procaccia, P. Grassberger, and H.G.E. Hentschel
[1] Article in *Dynamical Systems and Chaos*, L. Garrido (Ed.), Lecture Notes in Physics, Vol. 179, Springer-Verlag, New York, 1982, p. 212.

G. Prodi
See C. Foias and G. Prodi.

K. Promislow and R. Temam
[1] Localization and approximation of attractors for the Ginzburg–Landau equation, *J. Dynamics Differential Equations*, **3** (1991), 491–514.

A. Pumir, P. Manneville, and Y. Pomeau
[1] On solitary waves running down an inclined plane, *J. Fluid Mech.*, **135** (1983), 27–50.

P.H. Rabinowitz
[1] *Comm. Pure Appl. Math.*, **20** (1967), 145–205.

D. Rand (ed.)
[1] *Dynamical Systems and Turbulence, Warwick* 1980, Lecture Notes in Mathematics, Vol. 898, Springer-Verlag, New York, 1981.

G. Raugel
See J. Hale and G. Raugel; J.K. Hale, Xiao-Biao Lin, and G. Raugel.

G. Raugel
[1] Continuity of attractors, in Attractors, Inertial Manifolds and Their Approximation, (Marseille–Luminy, 1987), *Math. Model. Numer. Anal.*, **23** (1989), 519–533.

G. Raugel and G. Sell
[1] Equations de Navier–Stokes dans des domaines minces en dimension trois: régularité globale, *C. R. Acad. Sci. Paris*, Série I Math., **309** (1989), 299–303.
[2] Navier–Stokes equations on thin 3D domains I. Global attractors and global regularity of solutions, *J. Amer. Math. Soc.*, **6** (1993), 503–568.
[3] Navier–Stokes equations on thin 3D domains II: Global regularity of spatially periodic conditions, *Collège de France Proceedings*, Pitman Research Notes Math. Series, Pitman, New York, 1992.

L.G. Redekopp
See H.T. Moon, P. Huerre, and L.G. Redekopp.

M. Reed and B. Simon
[1] *Methods of Modern Mathematical Physics*, Academic Press, New York, 1978.

J. Richards
[1] On the gap between numbers which are the sum of two squares, *Adv. in Math.*, **46** (1982), 1–12.

F. Riesz and B.Sz. Nagy
[1] *Leçons d'Analyse Fonctionnelle*, Akadémiai Kado. Budapest and Gauthier-Villars, Paris, 1982.

M. Robin
See P. Faurre and M. Robin.

J.C. Robinson
[1] Inertial manifolds and the strong squeezing property, to appear in *Proc. NEEDS'94*.
[2] Finite-dimensional behavior in dissipative partial differential equations, *Chaos*, **5**(1), to appear.
[3] A concise proof of the "geometric" construction of inertial manifolds, *Phys. Lett. A*, to appear.
[4] Some closure results for inertial manifolds, submitted to *J. Dynamics Differential Equations*.

M. Rokni
See B.S. Berger and M. Rokni.

E.J. Romer
See J.P. Gollub, E.J. Romer, and J.E. Sacolar.

R. Rosa
[1] Weak and strong attractors for the weakly damped Korteweg–de Vries equation, Institute for Scientific Computing and Applied Mathematics, Indiana University, Bloomington, Preprint 95, 1995.
[2] Approximate inertial manifolds of exponential order, *Discrete and Continuous Dynamical Systems*, to appear.
[3] The global attractor for the 2D Navier–Stokes flow on some unbounded domains, *Nonlinear Anal., TMA*, to appear.
See also I. Moise and R. Rosa.

R. Rosa and R. Temam
[1] Inertial manifolds and normal hyperbolicity, *Acta Appl. Math.*, to appear.

D. Ruelle
[1] Strange attractors, *Math. Intelligencer*, **2** (1979–80), 126–137.
[2] Small random perturbations of dynamical systems and the definition of attractors, *Comm. Math. Phys.*, **82** (1981), 137–151.

[3] Large volume limit of distribution of characteristic exponents in turbulence, *Comm. Math. Phys.*, **87** (1982), 287–302.
[4] Characteristics exponents for a viscous fluid subjected to time-dependent forces, *Comm. Math. Phys.*, **92** (1984), 285–300.
See also J.P. Eckman and D. Ruelle; N. Koppel and D. Ruelle.

D. Ruelle and F. Takens
[1] On the nature of turbulence, *Comm. Math. Phys.*, **30** (1971), 167; **21** (1971), 21.

J.E. Sacolar
See J.P. Gollub, E.J. Romer, and J.E. Sacolar.

I.I. Satija
See D.K. Umberger, J.D. Farmer, and I.I. Satija.

J.C. Saut
See C. Foias and J.C. Saut.

J.C. Saut and R. Temam
[1] Generic properties of Navier–Stokes equations: Genericity with respect to the boundary values, *Indiana Univ. Math. J.*, **29** (1980), 427–446.
[2] Remarks on the Korteweg–de Vries equation, *Israel J. Math.*, **24**, 1 (1976), 78–87.

V. Scheffer
[1] Hausdorff measure and the Navier–Stokes equations, *Comm. Math. Phys.*, **55** (1977), 97–112.

B. Scheurer
See B. Nicolaenko and B. Scheurer; B. Nicolaenko, B. Scheurer, and R. Temam.

L.I. Schiff
[1] Nonlinear meson theory of nuclear forces, I, *Phys. Rev.*, **84** (1951), 1–9.

J.T. Schwartz
See N. Dunford and J.T. Schwartz.

L. Schwartz
[1] *Théorie des Distributions, I, II, Hermann*, Paris, 1950–1951 (2nd edition, 1957).
[2] Distributions à valeurs vectorielles, I, II, *Ann. Inst. Fourier* (*Grenoble*), **7** (1957), 1–141 and **8** (1958), 1–209.
[3] *Les Tenseurs*, Hermann, Paris, 1975.

J. Schwinger
[1] On the bound states of a given potential, *Proc. Nat. Acad. Sci.*, **47** (1961), 122–129.

I.E. Segal
[1] The global Cauchy problem for a relativistic scalar field with power interaction, *Bull. Soc. Math. France*, **91** (1963), 129–135.
[2] Nonlinear relativistic partial differential equations, *Proc. Int. Congress Math. Moscow*, 1966, pp. 681–690.

L.A. Segel
See A. Novick–Cohen and L.A. Segel.

H. Segur
See J. Ablowitz and H. Segur.

G. Sell
See S.N. Chow, K. Lu, and G. Sell; C. Foias, G. Sell, and R. Temam; C. Foias, B. Nicolaenko, G. Sell, and R. Temam; J. Hale and G. Sell, J. Mallet-Paret and G. Sell; G. Raugel and G. Sell.

G.R. Sell
[1] *Approximation Dynamics with Applications to Numerical Analysis*, CBMS–NSF Regional Conference Series, SIAM, Philadelphia, in preparation.
See C. Foias, M.S. Jolly, I.G. Kevrekidis, C. Foias, G.R. Sell, and E.S. Titi.
[2] (Editor) References on Dynamical Systems, regularly published and updated by AHPCRC, University of Minnesota, Minneapolis.

G.R. Sell and Y. You
[1] Inertial manifolds: the non-self-adjoint case, *J. Differential Equations*, **96** (1992), 203–255.
[2] Dynamical systems and global attractors, AHPCRC Preprint 94-030, University of Minnesota, 1994.
[3] To appear.

M. Sermange and R. Temam
[1] Some mathematical questions related to the MHD equations, *Comm. Pure Appl. Math.*, **36** (1983), 635–664.

R. Shaw
[1] Strange attractors, chaotic behavior and information flow, *Z. Naturforsch.*, **A36** (1981), 80–112.

I. Shimada and T. Nagashima
[1] A numerical approach to the ergodic problem of dissipative dynamical systems, *Progr. Theoret. Phys.*, **61** (1979), 1605–1616.

B. Simon
[1] On the number of bound states of two-body Schrödinger operators—A review, in *Studies in Mathematical Physics: Essays in Honor of Valentine Bargmann*, E. Lieb, B. Simon, A.S. Wightman (Eds.), Princeton University Press, Princeton, NJ, 1976, pp. 305–326.
See also M. Reed and B. Simon.

J. Simon
[1] Existence et unicité de solutions d'équations d'évolutions sur $]-\infty, +\infty[$. *C. R. Acad. Sci. Paris, Sér. I*, **274** (1972), 1045–1047.
[2] Une majoration et quelques résultats sur le comportement à l'infini de solutions d'équations fortement non linéaires, *C. R. Acad. Sci. Paris, Sér. A*, **279** (1974), 421–424.
[3] Quelques propriétés de solutions d'équations et d'inéquations d'évolution paraboliques non linéaires, *Ann. Scuola Norm. Sup. Pisa, Série IV*, **II**, no. 4 (1975), 585–609.

H. Simondon
See E. Feireisl, P. Laurençot, F. Simondon, and H. Touré.

G. Sivashinsky
[1] Nonlinear analysis of hydrodynamic instability in laminar flames, Part I. Derivation of basic equations, *Acta Astronaut.*, **4** (1977), 1177–1206.
[2] On flame propagation under conditions of stoichiometry, *SIAM J. Appl. Math.*, **39** (1980), 67–82.

G. Sivashinsky and D.M. Michelson
[1] On irregular wavy flow of a liquid down a vertical plane, *Progr. Theoret. Phys.*, **63** (1980), 2112–2114.

B.D. Sleeman
See D.S. Jones and B.D. Sleeman.

M. Slemrod
See J.K. Hale, J.P. La Salle, and M. Slemrod.

S. Smale
[1] An infinite-dimensional version of Sard's theorem, *Amer. J. Math.*, **87** (1965), 861–866.
[2] Differential dynamical systems, *Bull. Amer. Math. Soc.*, **73**, (1967), 747–817.
[3] Dynamical systems and turbulence, in A. Chorin, J. Marsden, and S. Smale [1].
See also A. Chorin. J. Marsden, and S. Smale; M.W. Hirsch and S. Smale.

J. Smoller
[1] *Schock Waves and Reaction–Diffusion Equations*, Springer-Verlag, New York, 1983.
See K.N. Chueh, C.C. Conley, and J.A. Smoller; C. Conley and J. Smoller; E. Conway, D. Hoff, and J. Smoller.

J. Solà-Morales
See X. Mora and J. Solà-Morales.

G. Stampacchia
[1] *Equations Elliptiques du Second Ordre à Coefficients Discontinus*, Presses de l'Universitè de Montréal, 1966.

V.S. Stepanov
[1] The attractor of the equation of oscillations of a thin elastic rod, *Russian Math. Surveys*, **40**, no. 3 (1985), 245–246.

J. Stoker
[1] *Nonlinear Vibrations*, Interscience, New York, 1950.

R. Stora
See G. Iooss, R. Helleman, and R. Stora.

W.A. Strauss
[1] On the continuity of functions with values in various Banach spaces, *Pacific J. Math.*, **19**, 3 (1966), 543–555.

J. T. Stuart
[1] Taylor-vortex flow: A dynamical system, *SIAM Rev.*, **28** (1986), 315–342.

J.T. Stuart and R.C. Di Prima
[1] The Eckhaus and Benjamin–Feir resonance mechanisms, *Proc. Roy. Soc. London, Ser. A*, **362** (1978), 27–41.

R.N. Sudan
See E. Ott and R.N. Sudan.

J.B. Swift
See A. Wolf, J.B. Swift, H.L. Swinney, and J.A. Vastana.

H.L. Swinney
See A. Wolf, J.B. Swift, H.L. Swinney, and J.A. Vastana.

H. Swinney and J.P. Gollub (Eds.)
[1] *Topics in Applied Physics*, Vol. 45, Springer-Verlag, New York, 1981.

F. Takens
See D. Ruelle and F. Takens.

L. Tartar
See C. Bardos and L. Tartar.

R. Temam
[1] Behaviour at time $t = 0$ of the solutions of semilinear evolution equations, *J. Differential Equations*, **43** (1982), 73–92.
[2] *Navier–Stokes Equations, Theory and Numerical Analysis*, 3rd rev. ed., North-Holland, Amsterdam, 1984.

[3] *Navier–Stokes Equations and Nonlinear Functional Analysis*, CBMS-NSF Regional Conference Series in Applied Mathematics, SIAM, Philadelphia, 1983 second augmented edition, 1995.
[4] Infinite dimensional dynamical systems in fluid mechanics, in *Nonlinear Functional Analysis and Its Applications*, F. Browder (Ed.), AMS, Proceedings of Symposia in Pure Mathematics, Vol. 45, 1986.
[5] Attractors for Navier–Stokes equations, in *Nonlinear Partial Differential Equations and Their Applications*, Collège de France Seminar, Vol. VII, H. Brézis, J.L. Lions (Eds.), Pitman, London, 1985.
[6] Sur un problème non linéaire, *J. Math. Pures Appl.* **48** (1969), 159–172.
[7] Attractors for the Navier–Stokes equations, localization and approximation, *J. Fac. Sci. Tokyo, Sec IA*, **36** (1989), 629–647.
[8] Inertial manifolds, *The Mathematical Intelligencer*, **12**, 4 (1990), 68–74.
[9] (Editor) Inertial manifolds and their application to the simulation of turbulence. Special issue of *Theoretical and Computational Fluid Dynamics*, vol. 7, 1995.
[10] Applications of inertial manifolds to scientific computing: a new insight in multilevel methods, *Trends and Perspectives in Applied Mathematics*, Volume in honor of Fritz John, J. Marsden, and L. Sirovich (Eds.), Applied Mathematics Series, Vol. 100, pp. 315–358, Springer-Verlag, New York, 1994.
See also B. Bréfort, J.M., Ghidaglia, and R. Temam; P. Constantin, C. Foias, O. Manley, and R. Temam; P. Constantin, C. Foias, and R. Temam; A. Eden, C. Foias, and R. Temam; C. Foias, O. Manley, and R. Temam; C. Foias, O. Manley, R. Temam, and Y. Treve; C. Foias and R. Temam; J.M. Ghidaglia, M. Marion and R. Temam; J.M. Ghidaglia and R. Temam; J.L. Lions, R. Temam, and S. Wang; B. Nicolaenko, B. Scheurer, and R. Temam; J.C. Saut and R. Temam; M. Sermange and R. Temam; A. Debussche and R. Temam; M. Marion and R. Temam; K. Promislow and R. Temam; R. Rosa and R. Temam; A. Eden, C. Foias, and R. Temam; A. Eden, C. Foias, B. Nicolaenko, and R. Temam.

R. Temam and S. Wang
[1] Inertial forms of Navier–Stokes equations on the sphere, *J. Funct. Anal.*, **117** (1993), 215–242.

R. Temam and X. Wang
[1] Estimates on the lowest dimension of inertial manifolds for the Kuramoto–Sivashinsky equation in the general case, *Differential Integral Equations*, **7**, 4 (1995), 1095–1108.

R. Temam and M. Ziane
[1] Navier–Stokes equations in three-dimensional thin domains with various boundary conditions, *Advances in Differential Equations*, **1**, 4 (1996), 499–546.
[2] Asymptotic analysis for the Navier–Stokes equations in thin domains, in preparation.
[3] Navier–Stokes equations in thin spherical shells and applications, in preparation.

W. Thirring
See E. Lieb and W. Thirring.

E.S. Titi
See also C. Foias, M.S. Jolly, and I.G. Kevrekidis; C. Foias, G.R. Sell, and E.S. Titi; D. Jones and E.S. Titi; M.S. Jolly, I.G. Kevrekidis, and E.S. Titi.

H. Touré
See E. Feireisl, P. Laurençot, F. Simondon, and H. Touré.

Y. Treve
See C. Foias, O. Manley, R. Temam, and Y. Treve.

J.J. Tribbia

[1] Nonlinear initialization on an equatorial Bet-plane, *Mon. Wea. Rev.*, **107** (1979), 704–713.

[2] A simple scheme for high-order normal mode initialization, *Mon. Wea. Rev.*, **112**, (1984), 278–284.

[3] On variational normal mode initialization, *Mon. Wea. Rev.*, **110** (1982), 455–470.

S.E. Trullinger

See A.R. Bishop, K. Fesser, P.S. Lomdahl, and S.E. Trullinger.

M. Tsutsumi and T. Mukasa

[1] Parabolic regularizations for the generalized Korteweg–de Vries equation, *Funkcialaj Ekvacioj*, **14** (1971), 89–110.

S. Ulam and J. Von Neumann

[1] On combination of stochastic and deterministic processes, *Bull. Amer. Math. Soc.*, **53** (1947), 1120.

D.K. Umberger and J.D. Farmer

[1] Fat fractals on energy surface, *Phys. Rev. Lett.*, **55**, 7 (1985), 661–664.

D.K. Umberger, J.D. Farmer, and I.I. Satija

[1] A universal strange attractor underlying the quasi-periodic transition to chaos, Los Alamos National Laboratory, no. 85-0817, 1985.

J.A. Vastana

See A. Wolf, J.B. Swift, H.L. Swinney, and J.A. Vastana.

R. Vautard and B. Legras

[1] Invariant manifolds, quasi-geostrophy and initialization, *J. Atmospheric Sci.*, **43** (1986), 565–583.

L. Vega

See C.E. Kenig, G. Ponce, and L. Vega.

G. Vélo

See J. Ginibre and G. Vélo.

C. Vidal

See P. Bergé, Y. Pomeau, and C. Vidal.

C. Vidal and A. Pacault (Eds.)

[1] Nonlinear phenomena in chemical dynamics, *Proceedings of an International Conference, Bordeaux, 1981*, Springer-Verlag, New York, 1981.

M.I. Vishik

See A.V. Babin and M.I. Vishik.

J. Von Neumann

See S. Ulam and J. Von Neumann.

M.A. Vorontsov

See S.A. Akmanov and M.A. Vorontsov.

S. Wang

See J.L. Lions, R. Temam and S. Wang.

X. Wang

[1] An energy equation for the weakly damped driven nonlinear Schrödinger equations and its application to their attractors, *Discrete and Continuous Dynamical Systems*, to appear.

See C.R. Doering and X. Wang; A. Miranville and X. Wang; R. Temam and X. Wang.

G. Webb

[1] Compactness of bounded trajectories of dynamicals systems in infinite dimensional space, *Proc. Roy. Soc. Edinburgh, Sect. A*, **84** (1979), 19–33.

[2] Existence and asymptotic behavior for a strongly damped nonlinear wave equation, *Canad. J. Math.*, **32** (1980), 631–643.

J.C. Wells
[1] Invariant manifolds of nonlinear operators, *Pacific J. Math.*, **62** (1976), 285–293.

R.O. Wells
[1] *Differential Analysis on Complex Manifolds*, Springer-Verlag, New York, 1973.

J.A. Whitehead
See A.C. Newell and J.A. Whitehead.

S. Woinowsky-Krieger
[1] The effect of axial force on the vibration of hinged bars, *J. Appl. Mech.*, **17** (1960), 35–36.

A. Wolf, J.B. Swift, H.L. Swinney, and J.A. Vastana
[1] Determining Lyapunov exponents from a time series, *Physica*, **16D** (1985), 285–317.

E.M. Wright
See G.H. Hardy and E.M. Wright.

Xia-Biao Lin
See J.K. Hale, Xia-Biao Lin, and G. Raugel.

J.A. Yorke
See J.D. Farmer, E. Ott, and J.A. Yorke; J.L. Kaplan, J. Mallet-Paret, and J.A. Yorke; J. Kaplan and J. Yorke.

J.A. Yorke, E.D. Yorke, and J. Mallet-Paret
[1] Lorenz-like chaos in a partial differential equation for a heated fluid loop, 1985, Preprint.

N. Yoshida
[1] Forced oscillations of extensible beams, *SIAM J. Math. Anal.*, **16** (1985), 211–220.

S. Yoshizawa
See J. Nagumo, S. Arimoto, and S. Yoshizawa.

K. Yosida
[1] *Functional Analysis*, Springer-Verlag, New York, 1965.

Y. You
See G. Sell and Y. You.

W.W. Zachary
See T.L. Gill and W.W. Zachary.

M. Ziane
[1] Optimal bounds on the dimension of the attractors of Navier–Stokes equations, *Physica* D, to appear.
See also A. Miranville and M. Ziane.
See R. Temam and M. Ziane.

E. Zuazua
See E. Feireisl and E. Zuazua.

Index

When a topic appears in several consecutive pages, only the first page is indicated.
When a topic appears repeatedly in the book we only indicate the place where its definition is given, or the first place where it appears (with boldface numbers).

Applied Mathematical Sciences

(continued from page ii)

61. *Sattinger/Weaver:* Lie Groups and Algebras with Applications to Physics, Geometry, and Mechanics.
62. *LaSalle:* The Stability and Control of Discrete Processes.
63. *Grasman:* Asymptotic Methods of Relaxation Oscillations and Applications.
64. *Hsu:* Cell-to-Cell Mapping: A Method of Global Analysis for Nonlinear Systems.
65. *Rand/Armbruster:* Perturbation Methods, Bifurcation Theory and Computer Algebra.
66. *Hlavácek/Haslinger/Necasl/Lovísek:* Solution of Variational Inequalities in Mechanics.
67. *Cercignani:* The Boltzmann Equation and Its Applications.
68. *Temam:* Infinite-Dimensional Dynamical Systems in Mechanics and Physics, 2nd ed.
69. *Golubitsky/Stewart/Schaeffer:* Singularities and Groups in Bifurcation Theory, Vol. II.
70. *Constantin/Foias/Nicolaenko/Temam:* Integral Manifolds and Inertial Manifolds for Dissipative Partial Differential Equations.
71. *Catlin:* Estimation, Control, and the Discrete Kalman Filter.
72. *Lochak/Meunier:* Multiphase Averaging for Classical Systems.
73. *Wiggins:* Global Bifurcations and Chaos.
74. *Mawhin/Willem:* Critical Point Theory and Hamiltonian Systems.
75. *Abraham/Marsden/Ratiu:* Manifolds, Tensor Analysis, and Applications, 2nd ed.
76. *Lagerstrom:* Matched Asymptotic Expansions: Ideas and Techniques.
77. *Aldous:* Probability Approximations via the Poisson Clumping Heuristic.
78. *Dacorogna:* Direct Methods in the Calculus of Variations.
79. *Hernández-Lerma:* Adaptive Markov Processes.
80. *Lawden:* Elliptic Functions and Applications.
81. *Bluman/Kumei:* Symmetries and Differential Equations.
82. *Kress:* Linear Integral Equations.
83. *Bebernes/Eberly:* Mathematical Problems from Combustion Theory.
84. *Joseph:* Fluid Dynamics of Viscoelastic Fluids.
85. *Yang:* Wave Packets and Their Bifurcations in Geophysical Fluid Dynamics.
86. *Dendrinos/Sonis:* Chaos and Socio-Spatial Dynamics.
87. *Weder:* Spectral and Scattering Theory for Wave Propagation in Perturbed Stratified Media.
88. *Bogaevski/Povzner:* Algebraic Methods in Nonlinear Perturbation Theory.
89. *O'Malley:* Singular Perturbation Methods for Ordinary Differential Equations.
90. *Meyer/Hall:* Introduction to Hamiltonian Dynamical Systems and the N-body Problem.
91. *Straughan:* The Energy Method, Stability, and Nonlinear Convection.
92. *Naber:* The Geometry of Minkowski Spacetime.
93. *Colton/Kress:* Inverse Acoustic and Electromagnetic Scattering Theory.
94. *Hoppensteadt:* Analysis and Simulation of Chaotic Systems.
95. *Hackbusch:* Iterative Solution of Large Sparse Systems of Equations.
96. *Marchioro/Pulvirenti:* Mathematical Theory of Incompressible Nonviscous Fluids.
97. *Lasota/Mackey:* Chaos, Fractals, and Noise: Stochastic Aspects of Dynamics, 2nd ed.
98. *de Boor/Höllig/Riemenschneider:* Box Splines.
99. *Hale/Lunel:* Introduction to Functional Differential Equations.
100. *Sirovich (ed):* Trends and Perspectives in Applied Mathematics.
101. *Nusse/Yorke:* Dynamics: Numerical Explorations.
102. *Chossat/Iooss:* The Couette-Taylor Problem.
103. *Chorin:* Vorticity and Turbulence.
104. *Farkas:* Periodic Motions.
105. *Wiggins:* Normally Hyperbolic Invariant Manifolds in Dynamical Systems.
106. *Cercignani/Illner/Pulvirenti:* The Mathematical Theory of Dilute Gases.
107. *Antman:* Nonlinear Problems of Elasticity.
108. *Zeidler:* Applied Functional Analysis: Applications to Mathematical Physics.
109. *Zeidler:* Applied Functional Analysis: Main Principles and Their Applications.
110. *Diekmann/van Gils/Verduyn Lunel/Walther:* Delay Equations: Functional-, Complex-, and Nonlinear Analysis.
111. *Visintin:* Differential Models of Hysteresis.
112. *Kuznetsov:* Elements of Applied Bifurcation Theory.
113. *Hislop/Sigal:* Introduction to Spectral Theory: With Applications to Schrödinger Operators.
114. *Kevorkian/Cole:* Multiple Scale and Singular Perturbation Methods.
115. *Taylor:* Partial Differential Equations I, Basic Theory.
116. *Taylor:* Partial Differential Equations II, Qualitative Studies of Linear Equations.
117. *Taylor:* Partial Differential Equations III, Nonlinear Equations.

(continued on next page)

Applied Mathematical Sciences

(continued from previous page)

118. *Godlewski/Raviart:* Numerical Approximation of Hyperbolic Systems of Conservation Laws.
119. *Wu:* Theory and Applications of Partial Functional Differential Equations.
120. *Kirsch:* An Introduction to the Mathematical Theory of Inverse Problems.
121. *Brokate/Sprekels:* Hysteresis and Phase Transitions.
122. *Gliklikh:* Global Analysis in Mathematical Physics: Geometric and Stochastic Methods.
123. *Le/Schmitt:* Global Bifurcation in Variational Inequalities: Applications to Obstacle and Unilateral Problems.